Chemical Separations

Chemical Separations

Principles, Techniques, and Experiments

Clifton E. Meloan

Kansas State University
Manhattan, Kansas

A Wiley-Interscience Publication
JOHN WILEY & SONS, INC.
New York / Chichester / Weinheim / Brisbane / Singapore / Toronto

DISCLAIMER

SAFETY

The laboratory procedures described in this text are designed to be carried out in a suitably equipped laboratory. In common with many such procedures, they may involve hazardous materials. For the correct and safe execution of these procedures, it is essential that laboratory personnel follow standard safety precautions.

Although the greatest care has been exercised in the preparation of this information, the author, speaking for himself, and for the classroom and laboratory instructors, and the publisher, expressly disclaim any liability to users of these procedures for consequential damages of any kind arising out of or connected with their use.

The analytical procedures detailed herein, unless indicated as such, are also not to be regarded as official, but are procedures that have been found to be accurate and reproducible in a variety of laboratories. All use is at the sole risk of the reader.

APPARATUS

The items of apparatus described in this manual are intended to illustrate proper techniques to obtain a quality analysis and are not to be considered as official and/or required. Any equivalent apparatus obtained from other manufacturers may be substituted.

This book is printed on acid-free paper. ∞

Copyright © 1999 by John Wiley & Sons, Inc. All rights reserved.

Published simultaneously in Canada.

No part of this publication may be reproduced, stored in a retrieval system or transmitted in any form or by any means, electronic, mechanical, photocopying, recording, scanning or otherwise, except as permitted under Section 107 or 108 of the 1976 United States Copyright Act, without either the prior written permission of the Publisher, or authorization through payment of the appropriate per-copy fee to the Copyright Clearance Center, 222 Rosewood Drive, Danvers, MA 01923, (978) 750-8400, fax (978) 750-4744. Requests to the Publisher for permission should be addressed to the Permissions Department, John Wiley & Sons, Inc., 605 Third Avenue, New York, NY 10158-0012, (212) 850-6011, fax (212) 850-6008, E-Mail: PERMREQ @ WILEY.COM.

For ordering and customer service, call 1-800-CALL-WILEY.

Library of Congress Cataloging-in-Publication Data

Meloan, Clifton E.
   Chemical separations : principles, techniques, and experiments : a combined text, laboratory manual, and reference / Clifton E. Meloan.
     p. cm.
   "A Wiley-Interscience Publication."
   "1999."
   Includes bibliographical references and index.
   ISBN 0-471-35197-0 (alk. paper)
   1. Separation (Technology)   2. Separation (Technology) Laboratory manuals.   I. Title.
   QD63.S4M44   1999
   543'.078—dc21   99-36208
                                                 CIP

Printed in the United States of America
10 9 8 7 6 5 4 3 2 1

Dedicated To

Mr. Paul Bacher
My High School Chemistry Teacher
Burlington Senior High School
Burlington, Iowa

The one who started it all

ABOUT THE AUTHOR

Clifton E. Meloan is professor of analytical chemistry at Kansas State University, Manhattan, KS 66506, and was Science Advisor to the Food and Drug Administration for 28 years, mostly associated with the Total Diet Center in Lenexa, KS. He was also consultant for several industries. He received a B.S. degree in chemical technology from Iowa State University in 1953 and a Ph.D. in analytical chemistry from Purdue University in 1959. He has directed the research of 50 Ph.D. and 26 M.S. students. The areas of research included extraction mechanisms, reactive column gas–liquid chromatography, polymers to remove selected ions from polluted water, solid reagents to selectively detect toxic compounds, criminalistics, insect detection of chemicals, and making undergraduate chemistry laboratories real world. This research has resulted in over 160 papers in refereed journals and 1 patent. He has prepared 11 video tapes on laboratory technique for the American Chemical Society and 26 hours of video on basic chemistry for the USDA Food Safety and Inspection Service. He is the sole author of five books and the co-author of seven books, the two most recent being *Food Analysis: Theory and Practice* by Y. Pomeranz and C. E. Meloan, Chapman-Hall (1994) and the *Central American Pesticides Laboratory Training Manual* published by the A.O.A.C. International (1996). In addition he has contributed 22 chapters to 3 other books. He was selected as Distinguished Research and Teaching Professor in 1970 sponsored by the Standard Oil Co. He has twice been selected as Outstanding Educator in America and in 1995 was awarded the first Kansas State University Distinguished Teaching Chair.

PREFACE

Pick and choose.
Cover what you can in lecture,
do the experiments you have the time and apparatus for,
and leave the rest for reference.

The purposes of this book are to present the principles of operation of the most commonly used chemical separations, to describe the apparatus that is used, and to provide an experiment to illustrate the techniques required to perform each separation correctly. The samples used are commercial products and naturally occurring materials. The methods are, for the most part, official methods from the *Association of Official Analytical Chemists, International* (AOAC, Intl.); the *United States Pharmacopeia (USP);* the *American Society for Testing and Materials (ASTM);* and from university and industrial laboratories.

Most students do not know "what is available" to solve a separation problem and usually do not know how to do it. This text is intended to be "street smart" by incorporating all the manipulative techniques practicing chemists use to make the separation work well. Far more separations are described here than can be covered in a one-semester course, but this book is intended to serve as a reference for the remaining separations. The mathematical treatment is kept to a minimum, presenting only what is necessary to illustrate the principles and to show how quantitation and recoveries are determined. Individual instructors will surely supplement the theory based on their interests.

Students may select 14 experiments from about 30 available, or about one per week. The experiments are set up 5–6 at a time for 2–2.5 weeks. It is the responsibility of the student to do the experiments within that time. The directions are tape recorded so students can work at any time. Laboratory assistants are available at certain times, and although the students are encouraged to take advantage of those times, it is not mandatory. Safety in an open laboratory is always a concern. Most of these experiments have been done for several years and the lab is believed to be very safe. The open laboratory is necessary with real-world samples because it is not always possible to finish a lab in 3 hours. The author believes that learning how to solve a problem is more important than being restricted to 3 hours.

In 1966, the author became a science advisor for the FDA. It took about 20 minutes in one of the field laboratories to convince him that although he probably knew more theory than anyone there, he had been inadequately trained in the practical aspects of handling real-world samples. Subsequent visits to industrial laboratories as a consultant reinforced this belief. Furthermore, the author realized that he was teaching his students the way he had been taught. As a result, a course in separations was developed in which commercial products, natural samples, and official methods of analysis procedures would be used as much as possible. All of the laboratory techniques that the author has learned are being passed on to students by incorporating them into the directions of the experiments.

Students should know the basic principles and have actual practice with the operational techniques of a wide variety of separation methods. In addition, they should be familiar with a great many other methods of separation that may be useful in the future. This book includes far more experiments than any one student can be expected to do in a one-semester class. The purposes of this are

1. Students can select a wide variety of experiments to do based on their past experience and interests.
2. Students can readily learn something about a separation they are unfamiliar with and have an experiment to try if they desire.
3. Because students can learn many ways to separate a mixture, they are less likely to force an inefficient method of separation as a solution to a problem just because it is the only one they know.

Two points of caution:

1. The student is using real samples, strange equipment, and many new techniques all at the same time. It is unrealistic to expect that the analytical results on first try will be perfect, but they usually will be acceptable.
2. Separation is not not spelled sep<u>e</u>ration.

An experiment with detailed directions accompanies each separation scheme. Instructors must know what apparatus is needed, how it is set up, what solutions are required, strength of solutions, and how much to prepare for each student, particularly if teaching assistants are involved.

The format for each experiment is to provide first a brief summary of the principles involved. A problem is presented that requires a separation to be made. A list of all of the equipment needed for one complete setup is provided. A list of all chemicals and samples needed is provided with the amount for each student (S) listed at the end such as 15 mL/S. In the author's laboratory, each experiment is available on audio tape so students can proceed at their own pace. It is imperative that the apparatus be set up to follow the directions. That is one of the purposes of the apparatus diagram. The second reason is so that students can recognize the experiment setup without being embarrassed because they didn't know what it looked like.

A one paragraph summary of what is to be done gives the student a broad view of what is expected. It also provides the instructor with a quick review of the separation so if a question is asked, the instructor can quickly provide an answer or a suggestion.

The directions are detailed. It is cookbook, and intentionally so. This is not Chem I with beakers and test tubes where "discovery" is paramount and the apparatus is inexpensive. These separations involve a limited amount of expensive equipment, and good techniques are to be learned. In addition, when real-world samples are used, the experiments can take considerable time, so directions are necessary. Unless students are shown what good technique is, they will seldom learn it on their own, and unless they are shown how to take care of an instrument, it will be a "machine" and treated as roughly.

If an instrument such as an infrared is used, general directions are given to show what is needed. It is expected that the local instructors will make modifications for their particular instruments. An example calculation is provided at the end. The samples often smell, won't dissolve, are messy, and have many interfering substances present. However, that is reality after graduation, so it is best to get used to it now.

This text has been used with both undergraduate and graduate classes. Undergraduate students should do those portions assigned by their instructor and as many others as their interest dictates. Graduate students should begin with new material and continue on.

ACKNOWLEDGMENTS

Eileen Schofield, Senior Editor, Kansas State Agricultural Experiment Station, for editing the manuscript. Justin Vardeman, Manhattan, Kansas, Vocational Technical Institute, for the following drawings: Figures 2-3, 2-11, 2-14, 2-25, 3-8, 7-1, 7-19, 8-14, 11-1, 12-7, 13-4, 20-5 and 20-8. Kelly Johnson, Kansas State University, Architectural Engineering, for the following drawings: Figures 1-4, 5-1, 7-6, 7-9, 7-10, 7-14, 9-3, 9-7, 10-10, 11-3, 12-2, 21-4, 23-11, 26-15, 27-8, 29-1, 29-3, 29-8, 29-12, 29-14, 29-19, 29-21, 30-2, 34-1, 36-20, 43-1, 53-3, 56-3, and 56-15.

Clifton E. Meloan
Kansas State University

CONTENTS

Introduction   1

SEPARATIONS INVOLVING PHASE CHANGES

1	Volatilization	5
2	Zone Melting	13
3	Distillation: General Information	21
4	Azeotropic and Extractive Distillations	43
5	Steam and Immiscible Solvents Distillation	49
6	Vacuum Distillation	57
7	Molecular Distillation and Sublimation	71
8	Lyophilization (Freeze Drying)	85

SEPARATIONS INVOLVING EXTRACTION

9	Extraction: General Concepts	93
10	Continuous Extraction	107
11	Countercurrent (Extraction) Chromatography: The Ito Coil-Planet Centrifugal Extractor	117
12	Solid Phase Extraction	129
13	Supercritical Fluid Extraction	137

SEPARATIONS INVOLVING CHROMATOGRAPHY

14	Chromatography: General Theory	149
15	Displacement and Multiple Column Partition Chromatography	155
16	Affinity Chromatography	165
17	Size Exclusion Chromatography (Gel Filtration; Gel Permeation)	171
18	Flash Chromatography	179
19	High Performance Liquid Chromatography (HPLC)	183
20	Gas–Liquid Chromatography (GLC)	211
21	Paper Chromatography	249
22	Thin Layer Chromatography	255

SEPARATIONS INVOLVING ION EXCHANGE RESINS

23	Ion Exchange	269
24	Ion Chromatography	277
25	Ion Retardation, Ion Exclusion, and Ligand Exchange	289

SEPARATIONS INVOLVING ELECTRIC FIELDS

26	Electrodeposition	299
27	Electrophoresis—General	315
28	Immunoelectrophoresis	339
29	Polyacrylamide Gel Disc Electrophoresis	345
30	Ion Focusing, SDS, and Two-Dimensional Electrophoresis	351
31	Capillary Zone Electrophoresis	359
32	Field Flow Fractionation	371

SEPARATIONS INVOLVING FLOTATION

33	Purge and Trap and Dynamic Headspace Analysis	385
34	Foam Fractionation, Gas-, and Liquid-Assisted Flotation	395

SEPARATIONS INVOLVING MEMBRANES

35	Osmosis and Reverse Osmosis	413
36	Dialysis and Electrodialysis	423
37	Filtering and Sieving	431

SEPARATIONS INVOLVING MISCELLANEOUS TECHNIQUES

38	Density Gradients	449
39	Centrifugation	455
40	Masking and Sequestering Agents	469
41	Solubility	473
42	Gas Analysis by Portable Orsat	489

APPENDIX A 499

Experiment 1	Volatilization: The Determination of CO_2 in Either Limestone or Baking Powder	500
Experiment 2	Volatilization: The Determination of Mercury in Hair by Flameless Atomic Absorption Spectroscopy	504
Experiment 3	Zone Refining: The Separation of Traces of Methyl Red from Naphthalene	507
Experiment 4	Azeotropic Distillation: The Production of Absolute Ethanol from 95% Ethanol–Water	509
Experiment 5	Extractive Distillation: Separation of Cyclohexane from Benzene Using Aniline	516
Experiment 6	Steam Distillation: The Separation of Oil of Clove from the Whole Cloves	519
Experiment 7	Immiscible Solvents Distillation: The Use of Toluene to Remove Water from Fruits, Vegetables, and Meats	521
Experiment 8	Vacuum Distillation: A Vacuum Distillation to Purify the Solvent Dimethylformamide	523
Experiment 9	Molecular Distillation: The Determination of Vitamin E (Tocopherol) in Margarine	527
Experiment 10	Sublimation: The Separation of Benzoic Acid from Saccharin	532
Experiment 11	Entrainer Sublimation: The Separation of Caffeine from Coffee by Entrainer Sublimation	535
Experiment 12	Lyophilization: Preparing Freeze-Dried Beverages	538
Experiment 13	Continuous Extraction—Solvent Heavier Than Water: The Determination of Caffeine in Cola Drinks Using a Continuous Extractor; Solvent Heavier Type	540
Experiment 14	Continuous Extraction—Solvent Lighter Than Water: Separating the Anti-amoebic Alkaloids from Ipecac Syrup	544
Experiment 15	Continuous Extraction—The Soxhlet Extractor: Extracting Oil from Nutmeats	548
Experiment 16	Countercurrent Extraction: Separation of the Phenolic Flavor Components, Carvacrol and Thymol, from Oregano Leaves	550
Experiment 17	Solid Phase Extraction: Atrazine, Simazine, and Propazine in Ponds, Lakes, and Rivers (Supelco Applications)	554
Experiment 18	Supercritical Fluid Extraction: The Separation of Oil from Pecans	556

Experiment 19	Displacement Column Chromatography: The Separation of *cis-trans*-Azobenzene	561
Experiment 20	Multiple Column Partition Chromatography: The Separation of Codeine, Chlorpheniramine, and Phenylephrine in a Cough Syrup	564
Experiment 21	Affinity Chromatography: Separation of a Lectin from Soybeans	567
Experiment 22	Size Exclusion Chromatography: Gel Filtration Chromatography—The Separation of Vitamin B_{12} from Dextrans by Means of Sephadex G-100	572
Experiment 23	Flash Chromatography: Separation of Vanillin from Vanilla Extract	575
Experiment 24	High Performance Liquid Chromatography: Detecting Explosive Residues	579
Experiment 25	High Performance Liquid Chromatography: Separation of Aspirin, Acetaminophen, and Caffeine in Excedrin	584
Experiment 26	Gas–Liquid Chromatography—Arson Investigation: Detection of Accelerants in Debris by Head Space Analysis	586
Experiment 27	Gas–Liquid Chromatography—Determination of Antioxidants: The Determination of BHA and BHT Antioxidants in Breakfast Cereals	591
Experiment 28	Paper Chromatography: The Determination of Sulfamethazine, Sulfametrazine, and Sulfadiazine in Trisulfa Tablets	595
Experiment 29	Thin Layer Chromatography: Ascending TLC Separations Requiring Color Development—Vasodilator Drugs; the Aliphatic Nitrates	599
Experiment 30	"Instant" Thin Layer Chromatography: The Determination of Antihistamines in Drug Preparations	601
Experiment 31	Two-Dimensional Chromatography: Separation of Sulfamethazine or Carbadox from Medicated Feeds	606
Experiment 32	Ion Exchange: Separations of Several Cations in Mineral Water, Pond Water, or Low Acid Foods	608
Experiment 33	Ion Chromatography: The Separation of Corrosion Inhibitors and Contaminants in Engine Coolants	614
Experiment 34	Ion Retardation: Separating Salt and Urea from Urine	617
Experiment 35	Ion Exclusion: Determination of Total Vitamin C in Applesauce	618
Experiment 36	Ligand Exchange: Determination of Hydrazine, Methylhydrazine, and Dimethylhydrazine in Liquid Rocket Fuels	621
Experiment 37	Electrodeposition: The Determination of Copper and Nickel in a U.S. 5-Cent Piece	623
Experiment 38	Agar Horizontal Strip Electrophoresis: The Separation of the Isoenzymes of Lactic Dehydrogenase (LDH) in Blood Serum	628
Experiment 39	Agarose Gel Horizontal Strip Electrophoresis: DNA Fingerprinting—A Paternity Case	634
Experiment 40	Immunoelectrophoresis: Immunoelectrophoresis of Human Proteins in Urine or Serum	642
Experiment 41	Disc Electrophoresis: Identification of Fish Species by Their Amino Acid Pattern	646
Experiment 42	Ion Focusing Electrophoresis: Ion Focusing of Ferric Leghemoglobins from Soybean Root Nodules	652
Experiment 43	Capillary Zone Electrophoresis: Micellular Capillary Electrophoresis Separation of Aspartame Breakdown Products in Soft Drinks	657
Experiment 44	Field Flow Fractionation: Determination of the Size/Mass of Latex Spheres Using Sedimentation FFF	662
Experiment 45	Purge and Trap: Separation of Aroma and Flavor Compounds in Fresh and Aged Beer	665
Experiment 46	Foam Fractionation: Separation of Detergents from Waste Water	667
Experiment 47	Gas-Assisted Flotation: Separating Suspended Solids from River Water	671
Experiment 48	Liquid-Assisted Flotation—Tomatoes: Separating Fly Eggs and Maggots from Canned Tomatoes	673
Experiment 49	Liquid-Assisted Flotation—Cornmeal and Flour: Separating Insect Fragments and Rodent Hairs from Cornmeal and Flour	676
Experiment 50	Liquid-Assisted Flotation—Leafy Vegetables: Separating Mites, Aphids, and Thrips from Leafy Vegetables	681
Experiment 51	Osmosis: Determining the Osmotic Pressure of a Potato	685
Experiment 52	Dialysis: The Determination of Phosphatase in Milk to Determine Proper Pasteurization	687
Experiment 53	Air Filtration: Separating Pollen and Molds from Air	691
Experiment 54	Density Gradients For Soils: Comparing Soils at a Hit-and-Run Accident	694
Experiment 55	Centrifugation: The Separation of Fat from Milk and Ice Cream	696
Experiment 56	Masking and Sequestering Agents: The Determination of Zn Accelerators in Rubber Products	700

Experiment 57 Solubility of Plastic Materials: Separation and Identification of Thermoplastic Polymers   703
Experiment 58 Solubility of Fibers: Identification of Textile Fibers by Solubility Measurements   704
Experiment 59 Gas Analysis: Separation of CO_2, O_2, CO, and N_2 from Car Exhaust Gas Fumes   707

APPENDIX B

Answers to the Review Questions   711

INDEX   747

ACRONYMS

AOAC	Association of Official Analytical Chemists, International	MBAS	Methylene blue active substance
		MDFF	2-Methoxy-2,4-diphenyl-3-(2H)-furanone
AUFS	Absorbance units full scale	MECC	Micellar electrokinetic capillary chromatography
BHA	Butylated hydroxy anisole		
BHT	Butylated hydroxy toluene	MFFF	Magnetic field flow fractionation
BIS	N,N-methylene bis acrylamide	NBT	Nitro blue tetrazolium
BSA	Bovine serum albumin	NHE	Normal hydrogen electrode
CHEF	Crossed clamped homogeneous electric fields	NIST	National Institute of Standards and Technology
CCC	Counter current centrifuge	PAM	Pesticide Analytical Manual
CPC	Coil planet centrifuge	PAT	Purge and trap
CZE	Capillary zone electrophoresis	PCR	Polymerase chain reaction
DMF	Dimethylformamide	PFGGE	Pulsed field gradient gel electrophoresis
DNA	Deoxyribonucleic acid	PMD	Programmed multiple development
DPN	Diphosphopyridine	PTFE	Polytetrafluoroethylene
DTA	Differential thermal analysis	PVDF	Polyvinylidene difluoride
EDTA	Ethylenediaminetetraacetic acid	PVP	Polyvinylpyrrolidinone
FDA	The Food and Drug Administration	RAM	Relative apparent mobilities
EFFF	Electric field flow fractionation	RFLP	Restriction fragment length polymorphisms
FFF	Field flow fractionation		
FFFF	Flow field flow fractionation	SDS	Sodium dodecylsulfonate
FSOT	Fused silica open tubular	SFE	Supercritical fluid extraction
GLC	Gas–liquid chromatography	SFFF	Sedimentation field flow fractionation
GPC	Gel permeation chromatography	SI	Système International
HDES	Hydrodynamic equilibrium system	SPE	Solid phase extraction
HEPA	High efficiency particulate air	SRM	Standard reference material
HETP	Height equivalent to a theoretical plate	Steric FFF	Steric field flow fractionation
HPLC	High performance liquid chromatography	TAFE	Transverse alternating field electrophoresis
HSES	Hydrostatic equilibrium system	TEMED	N,N,N',N'-tetramethylethylene diamine
IEF	Isoelectric focusing	TFFF	Thermal field flow fractionation
IEP	Immunoelectrophoresis	TLC	Thin layer chromatography
ITLC	Instant thin layer chromatography	TRIS	tris(Hydroxymethyl)aminomethane
LDH	Lactate dehydrogenase	VOC	Volatile organic compounds

INTRODUCTION

With the possible exception of sampling, the largest sources of error in an analytical determination are those processes that are known as *separations*. A separation is a process whereby the compounds of interest are removed from the other compounds in the sample that may react similarly and interfere with a quantitative determination.

As an example of a separation, the presence of rotten eggs in powdered eggs can be determined by titrating the lactic acid formed in the rotten eggs. However, acetic, propionic, and butyric acids are also present in both fresh and rotten eggs and would also be titrated, causing an error. The lactic acid is "separated" (liquid-solid extraction) by adding diethyl ether, which will selectively dissolve the lactic acid and leave the other acids in the egg where they will not interfere.

Separations are also important in the preparation of pure compounds. The separation of penicillin from mold is an example. In fact, every chemical that you have ever used has been purified, at least to some extent, by a separation of some type.

Separations are often complex and may require several different methods and much time before the final determinative step. The Food and Drug Administration's total diet study of foods examines over 250 foods for over 350 pesticides and industrial chemicals. It requires 600 to 700 hours of analytical effort of which between 300 to 400 hours are involved in separations.

TERMS COMMON TO CHEMICAL SEPARATIONS

Cleanup: A practical term used to describe the handling of a sample before the measuring step. For example, after the pesticides are extracted from a vegetable, the extract may contain fats and plant waxes that will interfere with the determination of the pesticides. These fats and waxes are removed by solid phase extraction, column chromatography, or gel permeation chromatography, and the extract then is said to be *cleaned up,*

Spiking: The addition of a known amount of a standard to a sample so that recoveries can be determined and to calibrate the signal for quantitation.

Standard reference material (SRM): Materials of known concentration such as those prepared by the National Institute of Standards and Technology (NIST); for example, orchard leaves, bovine liver, oyster tissue, lobster pancreas, steel.

Incurred residue: The residue originally in the sample.

Percent recovery: The amount of a spike that can be recovered when processed through the entire analysis procedure. Values of 95 - 105 % are desired. If the value is below 80 %, then the method should be modified to improve the recovery.

Collaborative study: A study by others to validate the proposed method. Several national agencies set the standards. In the U.S. these are the *Association of Official Analytical Chemists, International* (AOAC Intl.), the *United States Pharmacopeia* (USP) and the American Society for Testing and Materials (ASTM).

The AOAC Intl. requirements are:
1. An intra-lab collaboration first.
2. At least eight laboratories should test the method as written at five spike levels. If the results look reasonable, they are presented at an AOAC meeting for *Official First Action*.
3. After any recommended modifications, which are then tested, the method may be accepted as *Official Final Action*. The method then can be used for court cases.
4. In selected cases where specialized apparatus is involved or only a few laboratories would use the method, then three laboratories and three spike levels are acceptable, but this must be noted in the method.

The need for such methods is extensive in commerce and in litigation. If your company is buying a raw material of a specified purity, you need to know that it meets the specifications. The company selling it to you needs to know what it is selling. If there is a difference in the analysis, and a third company is brought in to settle the argument, what method is to be used? A sample that has been analyzed by an official method will settle most arguments, and will stand up in court.

Remember - cover the material you have time for, do the experiments selected by your instructor for which you have the apparatus and leave the rest of the material for reference.

IF IT EXISTS - CHEMISTRY IS INVOLVED
IF YOU CAN BUY IT - A CHEMIST WAS INVOLVED SOMEWHERE

SEPARATIONS INVOLVING PHASE CHANGES

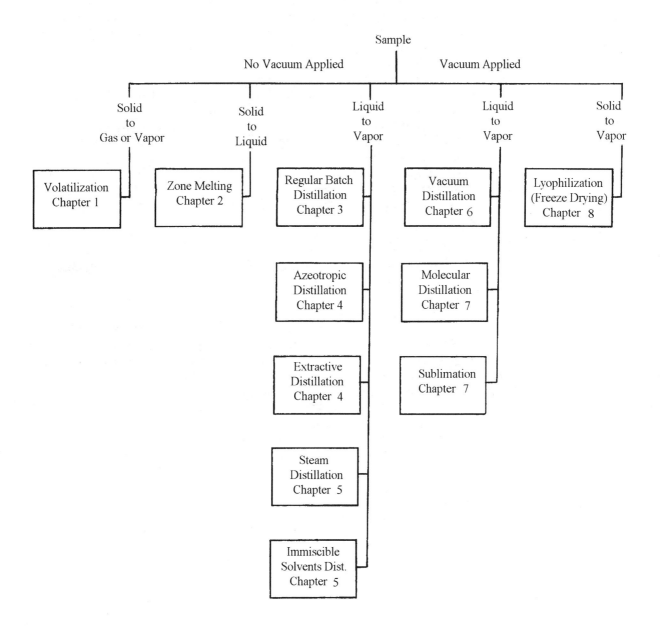

The *phases of matter* considered are *solid, liquid,* and *gas*. These are not to be confused with the *states of matter*. A glass of water is in the liquid phase, but its state may be 20 °C and 740 torr. Remember, if a liquid changes into a gas, this gas is called *a vapor*, since it did not exist naturally as a gas. If, in a mixture of compounds, the amount of one or more of the compounds changes if part of the mixture undergoes a phase change, then it is often possible to take advantage of this and effect a separation. The simplest example is the removal of water from a piece of lumber. When the lumber is heated, the water (liquid) changes to vapor and separates from the solid compounds in the lumber. This is called *volatilization* and is discussed in Chapter 1. Mercury in fish is determined in this manner.

If a portion of a solid mixture, such as an impure drug, is melted, the impurities are set free in the liquid portion. If this portion is then partially resolidified, the impurities will collect preferentially in the remaining liquid portion. This change, from solid to liquid to solid, is called *zone melting* and is discussed in Chapter 2. Computer chips are *zone refined* using this technique to evenly distribute the impurities,

The change from liquid to vapor back to liquid is given the term *distillation*, and there are several variations. For example; a mixture of benzene and toluene is liquid. The vapor coming from its surface contains more benzene than toluene, and if this vapor is condensed back to a liquid, then an enrichment in benzene occurs. The rate of this process can be increased by applying heat for the first phase change and applying cooling for the change back to liquid. Most laboratory-scale distillations are *batch distillations*, in that a single charge is distilled without adding more material during the distillation. The general theory, the apparatus used, and the techniques involved are discussed in Chapter 3.

There are times when a mixture of two or more compounds forms a constant boiling mixture and will not separate any further during distillation. These mixtures are called *azeotropes* and are discussed in Chapter 4. A common azeotrope is 95% ethanol-5% water; 100% ethanol, or absolute ethanol, is required for gasohol. How can this 5% of water be removed? You will learn that you form two additional azeotropes and use azeotropes as an advantage, a process called an *azeotropic distillation,* in Chapter 4.

There are times when there is a component in a mixture that is chemically different from the rest of the system, such as one component having an aromatic ring, while the others do not. If a compound is added to the mixture that can pi bond to the aromatic compound, it can raise its boiling point sufficiently to cause a separation to occur during this *extractive distillation*, discussed also in Chapter 4.

Raoult's law is the basis for *steam distillation*, Chapter 5, and its reverse, *immiscible solvents distillation*, also in Chapter 5. When the sum of the individual vapor pressures reaches 760 torr at sea level, the mixture will boil. By adding steam to a liquid, the vapor pressure of the water is quite high compared to the other components, and they will distill at a much lower temperature. This can minimize destruction caused by overheating, and is usually applied to liquids immiscible with water. The reverse, using a liquid such as toluene, can be used to remove large amounts of water from something like a watermelon section prior to additional chemical analysis.

By reducing the pressure of the system during the distillation, the boiling points decrease about 15 °C for each halving of the pressure. At these reduced temperatures it is possible to reduce decomposition, polymerization, azeotropes, and increase boiling point differences. Distillations at reduced pressures are called *vacuum distillations*, discussed in Chapter 6. Modifications of the application of a vacuum are discussed in Chapter 7. If a very high vacuum is applied, such that the length of the mean free path of the desired compound is greater than the distance from the liquid's surface to the surface of the collecting vessel, then there is no rebound. This is called a *molecular distillation*, Chapter 7, and is the commercial method for obtaining vitamin E.

If the phase change is solid to vapor, then this is a sublimation but the compound is lost. The preferred technique is solid to vapor back to solid, so the product can be collected. If the vapor is carried away from the sample surface as soon as it forms, so as to reduce the vapor pressure immediately above the sample, then this is called an *entrainer sublimation*, also in Chapter 7. If a liquid is solidified, changed to a vapor, and then to a solid, this is called *freeze drying*. This is discussed in Chapter 8.

1
VOLATILIZATION

PRINCIPLES

Volatilization is the conversion of all or part of a solid or a liquid to a gas. This evolved gas may be recovered and measured or the residue remaining may be measured. The gas may be produced by several methods: (1) by direct heating, such as heating NH_4NO_2 to form N_2 and H_2O; (2) by applying the principle that strong acids displace weaker acids and strong bases displace weaker bases; for example, the evolution of gaseous CO_2, a weak acid, from solid $CaCO_3$ by adding HCl, a strong acid, or the removal of gaseous NH_3, a weak base, from solid $(NH_4)_2SO_4$ by adding NaOH; (3) by oxidation, such as burning a sulfide in air to produce SO_2; or (4) by reduction, or converting the elements to hydrides such as AsH_3 or H_2Se. The most common method to determine Hg, As, and Se is to use volatilization procedures and detect the volatile components by atomic absorption spectroscopy.

In all cases a gaseous substance is formed. If this is to be collected, then it is passed through an absorption train to remove unwanted gaseous impurities. The desired gas then is either absorbed on a solid absorber and weighed or is passed into a solution where it reacts quantitatively with another chemical and the excess chemical is determined. Alternatively, the evolved gas mixture could be trapped and then passed through a gas chromatograph or the vapors can be examined spectrophotometrically.

If the evolved gas is not to be collected, then the residue is examined. One example of this approach is to determine the moisture in lumber by driving off the water and weighing the residue. Another example is to examine polymer compositions by heating the polymer in an N_2 stream and reducing the polymer to carbon, which then is weighed.

The point is that volatilization as a means of separation can be simple and can be applied to many problems, far more than most chemists realize. Although almost any element can be made volatile under stringent conditions, about 30 elements can be volatilized in one form or another by using reasonably mild conditions and simple apparatus. A brief summary of how several of these can be obtained in a volatile form is shown below for the inorganic compounds.

Carbon (inorganic)
    Carbonates (except the alkali metals) --------> $MO + CO_2$
    Carbonates and bicarbonates + strong acid (HCl) --------> CO_2
    Cyanides + strong acid (H_2SO_4) --------> HCN
Sulfur
    Sulfites and bisulfites + strong acid ($HClO_4$) --------> SO_2
    Acid soluble sulfides + strong acid ($HClO_4$) --------> H_2S
Nitrogen
    Nitrates and nitrites + strong acid (H_2SO_4, $HClO_4$) --------> NO_2
    NH_4^+ + strong base (KOH) --------> NH_3
Fluorine
    F^- + strong acid (H_2SO_4, $HClO_4$) --------> HF
    F^- + Si + H_2SO_4 --------> $SiF_4 + H_2SiF_6$
    F^- + B + $HClO_4$ --------> BF_3
Boron
    $H_3BO_3 + HF + HClO_4$ --------> BF_3
    $H_3BO_3 + CH_3OH + HCl$ --------> $(CH_3)_3BO_3$

Selenium and Tellurium
  -ates, ites, or ides + H· (NaBH$_4$)   --------> H$_2$Se or H$_2$Te
  -ates, ites, or ides + Cl$_2$ --------> SeCl$_2$ or TeCl$_2$ + TeCl$_4$

Arsenic
  AsO$_2^{-3}$, AsO$_3^{-3}$, AsO$_4^{-3}$ + Zn + H$_2$SO$_4$ (or NaBH$_4$)   --------> AsH$_3$
  + HCl + H$_2$SO$_4$ --------> AsCl$_3$

Antimony
  SbO$_3^{-3}$, SbO$_4^{-3}$ + HCl + HClO$_4$ --------> SbCl$_3$
  + Zn + H$_2$SO$_4$ --------> SbH$_3$

Silicon
  Silicates + HF + H$_2$SO$_4$ --------> SiF$_4$

Chromium
  Chromates, chromites + HCl + HClO$_4$ --------> CrO$_2$Cl$_2$

Mercury
  Hg^{++} + SnCl$_2$ --------> Hg + SnCl$_4$
  HgCl$_2$ + H$_2$SO$_4$ --300°C HCl--> HgCl$_2$

Halogens (F, Cl, Br, I)
  X$_2$ + H$_2$SO$_4$ --------> HX

Phosphorus
  PO$_3^{-3}$, PO$_4^{-3}$ + Zn + H$_2$SO$_4$ --------> PH$_3$

Osmium
  Osmates + HNO$_3$, H$_2$SO$_4$ --------> OsO$_4$

Rhenium
  Rhenates and perrhenates + H$_2$SO$_4$, HClO$_4$ --------> Re$_2$O$_7$

Gold
  Aurates + HCl, HNO$_3$, H$_2$SO$_4$ --200°C--> AuCl$_3$

Oxidation
  Sulfides + O$_2$ --265°C--> SO$_2$

Reduction by hydrogen
  Oxides of inactive metals + H$_2$ --------> H$_2$O (steam) + M or weigh the metal residue

N$_2$, O$_2$, F$_2$, Cl$_2$, Br$_2$, I$_2$, He, Ne, Ar, Kr, and Xe all can be volatilized from their surroundings by gentle heating.

Organic compounds
  Convert to CO$_2$ + H$_2$O and absorb each one; CO$_2$ on NaOH coated on asbestos (*Ascarite*) and H$_2$O on Mg(ClO$_4$)$_2$ (*Anhydrone*).
  C$_{10}$H$_8$ + 12 O$_2$ --------> CO$_2$ + H$_2$O

TECHNIQUES

MOISTURE DETERMINATION

Figure 1-1 (p. 8) shows a photograph of a computer-controlled microwave oven for determining moisture and any other volatiles that come off when a food sample is dried. A piece of paper is placed on the balance pan and automatically tared. A gram or so of homogenized sample is spread out on the paper and again placed on the balance pan. It is weighed automatically and recorded. The microwave oven is turned on and the sample is dried to constant weight, which takes about 1 minute. The loss in weight is obtained, and the computer calculates the loss in weight as percent moisture. A 9-pin printer prints out the results.

EXAMPLE CALCULATION

An aluminum dish weighed 1.9593 g empty and 2.8410 with sample. After the sample was brought to constant weight, the combination weighed 2.7555 g. What is the % moisture in this sample on an *as received basis* and on an *oven-dried basis*?

ANSWER

"As received" sample   Sample weight   1.0135 g
"Oven-dried" sample   Al dish + sample   2.8410 g
                        Al dish   1.9593 g
                        Sample   0.8817 g
                        Al dish + sample   2.8410 g
                        Weight after drying   2.7655 g
                        Moisture loss   0.0755 g

$$\% \text{ Moisture loss} = \frac{0.0755 \text{ g} \times 100}{0.8817 \text{ g}} = 8.56\%$$

        U tube H   U tube I
After   77.8624 g   After   74.2916 g
Before   77.5167 g   Before   74.2749 g
Wt gain   0.3457 g   Wt gain   0.0167 g

Total weight gained = 0.3457 g + 0.0167 g = 0.3624 g

"As received" calculations:

$$\% CO_2 = \frac{Wt \text{ of } CO_2 \times 100}{Sample \text{ weight}} = \frac{0.3624 \text{ g} \times 100}{1.0135} = 35.7\% \quad (1\text{-}1)$$

$$\% CaCO_3 = \frac{Wt \text{ of } CO_2 \times 100/44 \times 100}{Sample \text{ weight}} = \frac{0.3624 \text{ g} \times 100/44 \times 100}{1.0135 \text{ g}} = 81.3\%$$

"Oven-dried" calculations:

$$\% \text{ Oven-dried} = \frac{\% \text{ As received} \times 100}{100 - \% \text{ Oven-dried loss}} \quad (1\text{-}2)$$

% CO_2

$$= \frac{35.7\% \times 100}{100 - 8.56\%} = 39.0\%$$

% $CaCO_3$

$$= \frac{81.3\% \times 100}{100 - 8.56\%} = 88.9\%$$

Figure 1-1. A CEM Corp. model AVC-80 computer-controlled microwave oven for drying foods.
(Courtesy - CEM Corp., Mathews, NC)

EVOLVED GAS PURIFICATION

There is more to doing a volatilization determination quantitatively than you might believe. The desired gas often must be separated from other gases and it must be dried. The Knorr alkalimeter shown in Figure 1-2 for the determination of CO_2 in limestone is a good example of the precautions that must be taken to ensure an accurate analysis. CO_2 from the air must be removed as well as other possible impurities; HCl, SO_2, H_2S, and water vapor.

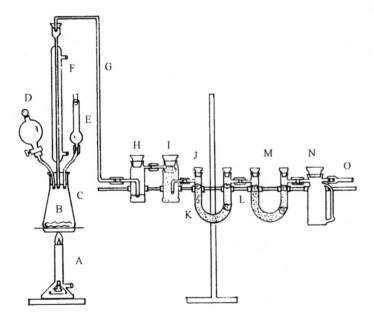

Figure 1-2. Knorr alkalimeter for the determination of CO_2.

A gentle suction is applied to draw the gas in the desired direction. (A) Bunsen burner. (B) Sample. (C) Wide mouth Erlenmeyer reaction flask. (D) HCl reservoir. (E) Drying tube filled with Ascarite (NaOH on asbestos) to remove CO_2. (F) Condenser to remove large amounts of water. (G) Z-shaped connecting tube. (H) H_2SO_4 scrubber to remove water, but let the weaker acid gases pass through. (I) $CuSO_4$ to remove HCl. (J) U tube for CO_2 absorption. Caution; soft glass breaks easily. (K) Ascarite to remove CO_2. (L) Anhydrone [$Mg(ClO_4)_2$] to absorb water.

$$CO_2 \text{ (f.w.} = 44) + 2\,NaOH \longrightarrow Na_2CO_3 + H_2O \text{ (f.w.} = 18).$$

(M) Backup absorption tube. (N) Drying tower to remove any water vapor backing up from the aspirator. (O) To a water aspirator to draw the CO_2 through the system.

DETERMINATION OF MERCURY IN THE ENVIRONMENT BY VOLATILIZATION

Excessive mercury in the environment is of concern because many organomercury compounds have the ability to pass through the blood-brain barrier and cause severe neurological disorders. Mercury used to get into the environment in large amounts from the production of chlorine, from slimicides and fungicides used in the paper industry, and from fumigants. Mercury is one of the electrodes for the electrolysis of salt water to produce chlorine and sodium hydroxide. The contaminated mercury used to be dumped into lakes and rivers (0.45 lb/ton of Cl_2 produced) because it was believed to be so heavy that it would just settle to the bottom. However, it was found that certain microorganisms could convert the mercury to organic mercury compounds, almost exclusively methyl mercury, which can be quite toxic. Methyl and phenyl mercuries were used

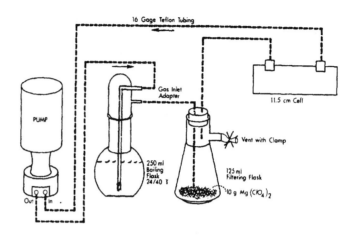

Figure 1-3. Apparatus for flameless atomic absorption analysis. (Courtesy - *Official Methods of Analysis,* 16th Ed., **1**, chapter 9, p. 20, 1995, Arlington, VA)

by the paper industry to spray on logs to keep them from getting a slime mold on them while they were waiting to be processed. Fish accumulate these compounds and at certain concentrations may cause a problem if eaten. Pike can concentrate Hg as much as 3,000-fold. Wheat used to be treated with organomercury compounds (pink wheat) to protect it during transportation and storage.

The determination of Hg in fish is done routinely by a volatilization method. Figure 1-3 shows a diagram of the apparatus used. The same apparatus can be used to determine Hg in hair as illustrated in Experiment 2, in Appendix A. Inorganic and organic Hg compounds are converted to Hg^{++} by acid digestion. $SnCl_2$ is added to reduce the Hg^{++} to Hg, which is volatile.

$$Hg^{++} + SnCl_2 \longrightarrow SnCl_4 + Hg \text{ (volatile)}$$

Air is passed through the solution, and the Hg vapor transfers to the air bubbles (2nd Law of Thermo-dynamics - high concentration goes to low concentration). Water vapor is removed in a drying tower, and the Hg vapor is passed through the beam of an atomic absorption spectrometer. The tolerance is 0.5 ppm.

DETERMINATION OF AS AND SE IN FOODS BY HYDRIDE GENERATION

Arsenic compounds are used widely in industry for the manufacture of glass, paint pigments, dyes, weed killers, and tanning hides to mention a few. Most are toxic, and several are suspected carcinogens.

Selenium compounds are used widely in glass manufacture, photography, electrodes for arc lights, TV sets, photocells, semiconductors, catalysts, and vulcanizing agents. Selenium is necessary in humans and cattle to prevent muscle degenerative diseases. However, too much can cause problems in animals (loco weed); several compounds are toxic, and surprisingly, some compounds may be anticarcinogens.

In 1972, R.S. Braman, L.L. Justen and C.C. Forbank (*Anal. Chem.*, **44**, 2195) described a method for determining As, Bi, Ge, Pb, Sb, Se, Sn, and Te by forming the volatile hydrides using $NaBH_4$. A 0.3 % solution in 0.9 % NaOH is used to keep it from decomposing.

Figure 1-4 (p. 10) is a diagram of the hydride generator. The screening levels for As and Se in the environment are so low that the only way to determine these elements is to use atomic absorption (AA) spectroscopy. The problem is that their normal wavelengths have many flame interferences. One way to get around this is to convert As to AsH_3 and Se to H_2Se, both volatile compounds, and pass these through AA without using a flame - *flameless AA*. This process is also called *hydride generation*.

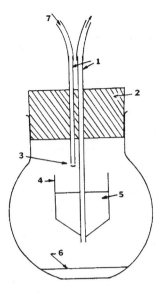

Figure 1-4. Hydride generator. (1) Polyethylene tubing; (2) rubber stopper; (3) flame-sealed polyethylene tubing with holes punched in one end; (4) reagent cup; (5) sodium borohydride solution; (6) sample solution; (7) nitrogen inlet.
(Courtesy - *Official Methods of Analysis*, 16th Ed., **1**, chapter 9, p. 2, 1995, Arlington, VA)

The hydrogen can be generated either by adding acid to Zn metal or the newer method of using a sodium borohydride solution. The borohydride is preferred because it is easier to handle, and the reaction appears to go better.

$$As^{+3}, As^{+5} + Zn + H_2SO_4 \longrightarrow AsH_3 + ZnSO_4$$
$$Se^{+4}, Se^{+6} + Zn + H_2SO_4 \longrightarrow H_2Se + ZnSO_4$$

or

$$As^{+3}, As^{+5} + NaBH_4(NaOH) + HCl \longrightarrow AsH_3 + Na_3BO_3 + ?$$
$$Se^{+4}, Se^{+6} + NaBH_4(NaOH) + HCl \longrightarrow H_2Se + Na_3BO_3 + ?$$

EXAMPLE CALCULATION

A 1.1385 g sample of hair was dissolved in 10 mL of concentrated HNO_3. The sample then was diluted to 50 mL with deionized water, and the Hg was determined as described in Experiment 2. An absorbance reading of 0.037 was obtained. A set of five standards produced the results shown in Figure 1-5. These values were plotted to produce the calibration curve shown in Figure 1-6. Use these data to determine the Hg content in ppb.

ANSWER

Referring to Figure 1-6, an absorbance of 0.037 gives a concentration of 6.5 ng/mL. Remember, the sample was diluted to 50 mL. Also, 1 ng/mL or 1 ng/g (if water) = 1 ppb.

$$6.5 \frac{ng}{mL} \times 50.0 \; mL \times \frac{1}{1.1385 \; g} = 285 \frac{ng}{g} = 285 \; ppb \; or \; 0.29 \; ppm.$$

EXAMPLE CALCULATION

A 10 ng/mL Hg standard produced an absorbance of 0.060. When 10 ng/mL was added to 1.5004 g of a fish sample with no incurred residue, an absorbance of 0.056 was obtained. When 10 ng/mL was added to 1.5970 g of a fish sample obtained near a logging plant, a value of 0.125 was obtained. (A) What is the % recovery with this method? (B) What is the concentration of Hg in the fish?

ANSWER

A. % Recovery = (actual/expected) x 100;   0.056/0.60 x 100 = 93 %.
B. 0.056 of the absorbance was due to the spike, leaving 0.125 - 0.056 = 0.069 for the Hg in the fish. Referring to Figure 1-6, this represents 11.3 ng/mL. Calculating as in the previous example gives a value of 354 ppb or 0.35 ppm in the fish. 0.50 ppm is the legal limit.

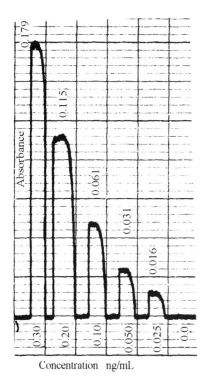

Figure 1-5. Flameless atomic absorption data for mercury standards.

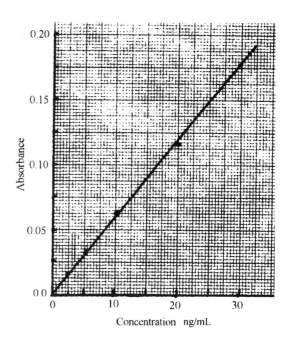

Figure 1-6. Plot of absorbance vs. concentration of mercury.

REVIEW QUESTIONS

1. Name at least two ways to produce a volatile compound from a liquid or solid.
2. A compound is suspected to be NH_4NO_3. How might you volatilize the NH_3?
3. How is the element H determined in an organic compound?
4. How is the Hg in a food sample made volatile?
5. What is meant by *flameless* atomic absorption spectroscopy?
6. What would be the effect on the CO_2 results if a rock sample evolved SO_2 or H_2S?
7. What effect on the results would occur if a weak organic acid such as acetic acid were evolved from organic impurities in the ore sample?
8. If all of the water displaced from the Ascarite were lost, how much lower would the final results be on a % basis?
9. A student filled U-tube H with Ascarite and no Anhydrone and then filled U-tube I with only Anhydrone. Would this have any effect on the results? If not, then why not do it this way?
10. There are areas in Kansas where limestone samples contain over 100% $CaCO_3$. How can this be? The analyses are correct.
11. Use the following data and determine the % $CaCO_3$ in the sample.
    Sample weight as received   0.9565 g
    Moisture loss after drying   0.0087 g
    Weight increase of U tube 1   0.3306 g
    Weight increase of U tube 2   0.0059 g

12. Use the following data and determine the % CO_2 in the sample on both an "as received" and an "oven dried" basis.
    Sample weight as received   1.1593 g
    Moisture loss after drying   0.0096 g
    Weight increase of U tube 1   0.1535 g
    Weight increase of U tube 2   0.0042 g

13. A 1.4420 g sample of a fingernail was found to contain 0.85 µg of Hg. What is the Hg concentration in ppm?

Review Questions

14. A 1.6732 g sample of the same fingernail spiked with 0.6 µg Hg was found to have a value of 1.55 µg of Hg. What is the % recovery?

UNDERGRADUATES - COVER THOSE PORTIONS ASSIGNED BY YOUR INSTRUCTOR, PLUS ANY OTHER PORTIONS YOU ARE INTERESTED IN.
GRADUATE STUDENTS - START WITH NEW MATERIAL AND CONTINUE ON.

IF IT EXISTS - CHEMISTRY IS INVOLVED
IF YOU CAN BUY IT - A CHEMIST WAS INVOLVED SOMEWHERE

2
ZONE MELTING

PRINCIPLES

Zone melting encompasses a group of related techniques to control the distribution of soluble impurities or solutes in normally crystalline solids. It was first introduced in 1952 by W.G. Pfann (*Trans. AIME*, **194**, 747) to produce ultrapure semiconductor materials. Although normal freezing is used to describe the underlying principle, the main areas are *zone refining* and *zone leveling*. As a separation scheme, it needs no solvents or other reagents, complete recovery is possible, it is simple and can be automated. However, it is a slow process, and not all of the original charge can be purified. In some cases, volume changes can be a problem, but usually these can be handled by tilting the apparatus. The main applications are to produce very pure metals or organic compounds or to provide a very uniform level of intentional impurity in a metal such as a semiconductor used in transistors.

What can be done? With a k of 0.1 (defined later), the ultimate purification is about 10^{-14} of the original value after 20 passes and a charge to zone length of 10. Three passes purified 99.91 % benzoic acid to 99.997 mole %. This is equivalent to 11 recrystallizations from benzene or 25 from water.

NORMAL FREEZING

Those of you who live in the northern climates may have observed that when muddy rivers, lakes, or farm ponds freeze in the winter, the ice is clean and not muddy at all!! Why is this so? Consider the diagram in Figure 2-1 illustrating what happens during normal freezing. Systems tend to the lowest energy. As a result, a crystal would "prefer" to be uniform and have no irregularities that would set up a point of stress. It is also known that all systems are in a continuous state of equilibrium, and if the system is a liquid-solid mixture, some ions or molecules are leaving the surface of the solid and others are adding to the surface of the solid. The areas around the impurities (points of stress) are slightly more soluble than the main body, and the impurity will leave the crystal if given enough time. Rivers and lakes are usually so deep and the layer of ice so thin by comparison that the layer of ice is quite pure. However, if you freeze a shallow pond or a bucket of river water, the ice on top will be pure, but that deeper down and on the bottom will still be muddy. The reason is that the impurities are now so concentrated in the liquid that they have no place to go and reenter the ice. *This is normal freezing - only one surface is involved.* How much material is removed depends upon the relative solubilities of the solute in the original material and in the melted portion. The ratio of C, the concentration of solute in the newly frozen solid, and C_o, the concentration in the original material, is k. If the solute lowers the m.p. of the solid, then k will be < 1 and if it raises the m.p. it will be > 1 Equation 2-1 assumes that k remains constant throughout the process. In actual practice, this seldom occurs, and k changes when the concentration changes are extreme or if a eutectic or peritectic composition is attained. For a single step, the concentration of the purified system can be represented by:

$$C = kC_o(1-g)^{k-1} \qquad (2\text{-}1)$$

where C = solute concentration in the solid; C_o = original average concentration of the solute; g = fraction of the original solution; k = distribution coefficient; C_{solid}/C_{liquid} > 1 if solute rises that is frozen; the m.p of the solvent; <1 if it lowers the m.p.

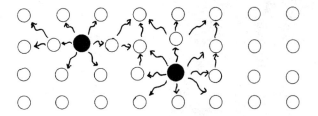

Figure 2-1. Diagram of stress caused by impurities.

Figure 2-2 shows the variation for nine zones in a 10-zone system after one pass.

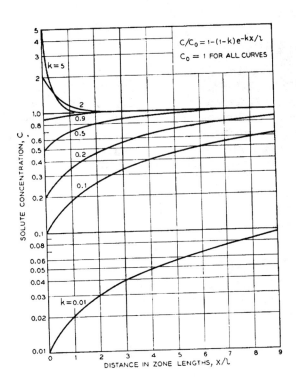

Figure 2-2. Relative solute concentration as a function of charge length.
(Courtesy - Pfann, W. G., *Zone Melting*, 2nd Ed., McGraw-Hill, New York, NY 1958)

EXAMPLE CALCULATION
What is the concentration of dissolved solids in the frozen portion of a glass of orange juice, if the liquid is 10 % frozen? Assume $k = 0.05$ and $C_o = 0.044$ M.

ANSWER
Use equation 2-1.
$$C = 0.05 \times 0.044 M (1 - 0.1)^{0.05 - 1}; \quad = 0.0026 \, M$$

ZONE REFINING
The same concept shown in Figure 2-3 works with a solid as the starting impure material. If you melt it, the impurities are set free. If you then slowly freeze it, the impurities will concentrate in the liquid. One such step is called *zone melting. The difference between normal freezing and zone melting is that two surfaces are involved in zone melting, one melting and one freezing.* If, in addition, you place the solid in a long column and heat a small section of it starting at one end and moving the heated zone slowly to the other end, you will drive the impurities ahead of the refreezing

solid toward the opposite end of the column. This is known as *zone refining* and was developed by W.G. Pfann in 1952 (*Trans. AIME*, **194**, 747).

Zone refining was developed originally to produce very pure germanium for the semiconductor industry. It was successful in that Bell Laboratories produced 99.99999999 % pure germanium! Since then, many elements have been prepared in high purity, as well as many organic compounds.

Refer to Figure 2-4 to help explain how zone refining works. If the composition of the solid that freezes out at any temperature is different from the composition of the liquid then zone refining is possible. Consider a melted section of composition A, and lower it C, its freezing point (T_1). The solid that freezes out (B), will have a higher concentration of X than the liquid at the same temperature. This exclusion of the solute Y will raise the concentration of Y in the liquid (D). If the sample is now cooled to the freezing point at this composition (T_2), the solid that forms (E) will be richer in X than the corresponding liquid. This in turn raises the concentration of Y in the liquid.

Referring to Figure 2-5, three regions are observed. Assume k is constant and less than one, the solid sample is of uniform composition, and the molten zone is of constant length (l) and cross-sectional area. Berg (1963) explains this as: "In the first part the solute concentration has been reduced, in the central part the solute concentration is unchanged, and in the final part the solute concentration has been increased.

"The concentration of the solute in the molten zone at the very beginning of the pass will be C_o, the concentration of solute in the unmelted ingot, because no segregation of solute occurs at the melting interface. After the zone has advanced a short distance, a freezing interface forms at x = 0. The solute concentration in the newly frozen solid will be kC_o. Simultaneous with the deposition of solute in the freezing melt, the melting interface at x = l is taking in a solute concentration of C_o. Thus, as the zone advances through the ingot, the concentration in the zone builds up, resulting in the freezing out of higher concentrations of solute. This enrichment of the zone continues at a decreasing rate until the concentration of solute in the zone is equivalent to C_o/k. From this point on, the amount of solute entering and leaving the molten zone at the two interfaces is equal, and the concentration of solute in the newly frozen ingot remains constant until the melting interface reaches the end of the ingot. Further movement of the zone merely decreases the length of the zone, causing the solute concentration to rise sharply in the melt and the solid. This latter stage is completely analogous to a normal freezing technique.

"The variation of solute concentration with ingot length for a single pass of the molten zone can be described by a single equation up to x = L - l, where L is the length of the ingot, and *x* is the distance the zone has traveled. The equation as presented by Pfann (*Zone Melting*, 2nd Ed, McGraw-Hill, New York, NY, 1958) is:

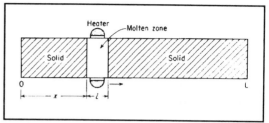

Figure 2-3. Creating and moving a molten zone through a solid.
(Courtesy - Berg, E.W., *Physical and Chemical Methods of Separation*, McGraw-Hill, New York, NY, 1963)

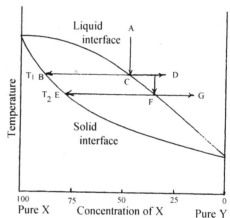

Figure 2-4. Phase diagram of an ideal two-component system.

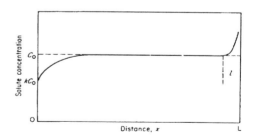

Figure 2-5. Solute distribution after one pass of the molten zone.
(Courtesy - Berg, E.W., *Physical and Chemical Methods of Separation*, McGraw-Hill, New York, NY, 1963)

$$\frac{C}{C_o} = 1 - (1 - k)e^{-kx/l}$$ (2-2)

EXAMPLE CALCULATION
What is the concentration of anthracene in a bar of naphthalene at a distance of 20 cm if l is 1 cm, k is 0.2, and C_o is 0.30 M, and this is the first pass?

ANSWER
Use equation 2-2.

$$\frac{C}{0.30} = 1 - (1 - 0.20)2.71^{-0.2 \times 20/1} = 0.29 M$$

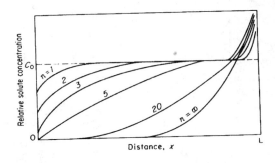

Figure 2-6. Relative solute concentration as a function of length of charge after n Passes of the molten zone.
(Courtesy - Berg, E.W., *Physical and Chemical Methods of Separation*, McGraw-Hill, New York, NY, 1963)

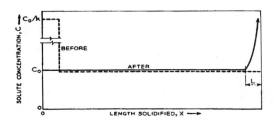

Figure 2-7. Zone leveling using a starting charge to eliminate the initial transition region.
(Courtesy - Pfann, W.G., *Zone Melting*, 2nd Ed., McGraw-Hill, New York, NY, 1958)

What happens as the number of passes increases? Figure 2-5 shows what happens. Notice that the concentration at the beginning approaches zero, and that the impurities collect at the end of the ingot.

Look at Figure 2-6 after 20 passes. Notice that the concentration approaches 0 and stays there for quite a distance. This is *zone refining*.

ZONE LEVELING

Suppose you wanted to intentionally dope a metal with 1 ppm of an impurity. One way is to add a small amount of the impurity to one end of a pure ingot and make several passes. The impurity will be spread evenly throughout the ingot for a considerable length. This is called *zone leveling* and is excellent for making semiconductor materials where the level of impurity is critical. Refer to Figure 2-7.

Ideally, what is needed is an ingot with the same amount of impurity in it from end to end. However, zone refining slowly pushes the impurity to one end. Suppose a small charge of the impurity were added at the start to just overcome the initial refining and when the melted zone reached the end, it were reversed and went back in the opposite direction. That might eliminate one of the zones and greatly reduce the other. If multiple passes are made (repeated pass zone melting), nearly all deviations from uniformity can be eliminated. This leveling is particularly effective if 1-k is large. Equation 2-3 is one equation to represent this. C_f is the final concentration of the uniform charge.

$$\frac{C_f}{C_o} = \frac{1}{1 + \frac{l}{L}\left(\frac{1}{k} - 1\right)}$$ (2-3)

EXAMPLE CALCULATION
What final concentration would be predicted if a 30 cm bar with an 8 mm molten zone had a 15 ppm of As as an initial concentration and *k* is believed to average 0.1?

ANSWER
Use equation 2-3.

$$\frac{C_f}{15\ ppm} = \frac{1}{1 + \frac{0.8\ cm}{30\ cm}\left(\frac{1}{0.1} - 1\right)}; = 12.1\ ppm$$

TECHNIQUES

Zone melting techniques are slow, very slow. Rates of zone movement vary from 1 to 20 cm/h. Figure 2-8 shows a more efficient arrangement. Once the heated zone has refrozen there is no reason why it can't be remelted. In this case, three heaters are used, and the charge is moved. Figure 2-9 shows a helical type where you can get many molten zones. This is more appropriate for organic compounds, which are lower melting than metals.

Figure 2-10, p. 18, shows a commercial apparatus. There is a waiting time of 3 months for this apparatus, because none are stocked. However, it is reasonably priced ($2,500 in 1997 + $150.00 for the educational kit) and can be used for both student and research purposes. Its capacity is 20-160 g of material that melts between -20 and 400 °C

The least expensive way to do zone refining it to place the sample in a quartz, ceramic, or graphite tube, and place the tube in a metal lathe. Move the heater with the lathe.

Figures 2-11 and 2-12, p. 18, show two inexpensive zone refiners that can be made locally and work well. In Figure 2-11, A and B are microswitches for reversing the stepping motor C; D is heating elements; and E is the wire to the controller box. The heaters are several turns of nichrome wire, and the tube is 8 mm Pyrex closed at the bottom. The heaters are moved up the tube by a screw attached to a stepping motor. At the top of the tube, a microswitch turns off the heaters, and the stepping motor is returned automatically to the bottom of the tube. The wiring diagram, which is not all that difficult, is presented in *J. Chem. Ed.*, **59** (1), 63, 1982 by G.F. Needham, G. Boehme, R.D. Willitt, and D.W. Swank.

Air gaps are removed by heating the entire length with a hot air gun. If the solid expands on melting and a vertical tube is used, then the heater must start at the top. The design in Figure 2-12 is clever in that it moves the tube rather than moving the heaters, but it floats the tube on a rising column of water.

A circle of 1/4" plywood acts as a float which moves upward at a constant rate as water fills the desiccator. A coat of paint keeps it from becoming water logged. The overflow tube acts the same as that in a Soxhlet extractor (Chapter 10) and the water level lowers. The cycle can be repeated as necessary, two or three passes being all that is usually necessary. A cycle rate of about 1 hour will clearly show the principle. The sample tube is 1 cm in diameter and about 20 cm long. The heaters are made from nichrome wire wound around a 3 mm thick glass ring slightly larger in diameter than the sample tube. A variable transformer is used to control the heat and temperatures in excess of 300 °C are easy to obtain. The asbestos shield is needed to keep the heated zone as small as possible. A large graduated cylinder can be used to collect the overflow water to measure the rate.

Experiment No. 3 in Appendix A can be used to show what happens when a small amount of methyl red is added to naphthalene.

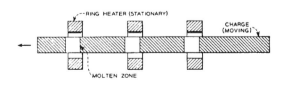

Figure 2-8. A 3-stage fixed heater with moveable charge.
(Courtesy - Pfann, W.G., *Zone Melting*, 2nd Ed., McGraw-Hill, New York, NY, 1958)

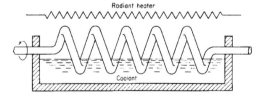

Figure 2-9. A rotating helical tube refiner.
(Courtesy - Berg, E.W., *Physical and Chemical Methods of Separation*, McGraw-Hill, New York, NY, 1963)

18 Principles

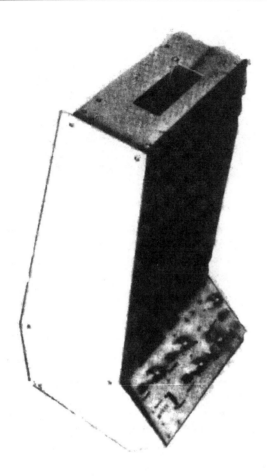

Figure 2-10. A model 200 zone refiner.
(Courtesy - Atomergic Chemetals Corp., Farmingdale, NY)

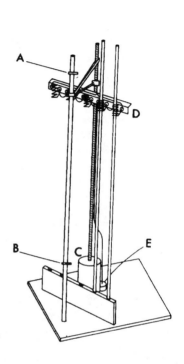

Figure 2-11. Schematic diagram of an inexpensive zone refining apparatus.
(Courtesy - Needham, G.F., Boehme, G., Willett, R.D., and Swank, D.W., *J. Chem. Ed.*, **59** (1), 63, 1982)

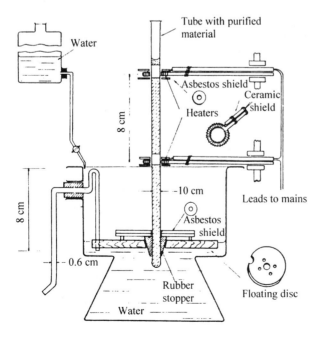

Figure 2-12. Water-controlled automatic zone refiner.
(Courtesy - Knypl, E.T., and Zielenski, K., *J. Chem. Ed.* **40** (7), 352, 1963)

FLAT TRAY ZONE REFINING

This is a new approach, developed in the late 1990's that overcomes many of the problems associated with tube type zone refining. Most of the description are those of Dr. Dean M. Ball, *"Zone Refining of Organic Chemicals: A New Approach", American Laboratory*, **31**, (5) 66, (1999).

The primary reason for the slowness of conventional zone refining is that the transfer of heat into and out of the sample is very slow. The thermal conductivity of both glass and organic chemicals is relatively low. This limits the speed with which the zone can be moved through the specimen to rates of 0.1 to 1 cm/h. Another complicating factor that requires special experimental techniques is that the coefficient of thermal expansion of organic chemicals is about 50 times higher than glass. If the heating is too vigorous or too fast, the glass tube will burst.

Instead of a round glass tube, the sample is placed on a flat metal tray having a very thin (0.025 mm) bottom, Figure 2-13. This configuration allows the transfer of heat into and out of the sample at least 1,000 times faster than the glass tube design. This, in turn, allows the molten zones to be closer together and to be moved faster. The commercial apparatus is shown in Figure 2-14.

With this type of zone refiner, a sample is prepared for purification by loading between 1 and 30 g of sample into the metal tray. The tray is then placed in the zone refiner and covered with the tray lid provided. A small positive pressure (0.03 atm) is applied by a built-in air pump and pressure regulator. This pressure has the benefit of forcing the thin metal bottom of the tray against the hot and cold surfaces so that thermal contact is near optimal levels. Without this small positive pressure, the bottom of the metal tray will warp and move away from the hot and cold surfaces, making the melting and freezing processes ineffective. Provision is made for an inert gas to be substituted for the air pump to allow the purification to be performed under an inert atmosphere. The temperature difference between the hot and cold zones is normally between 60 and 80 °C. The hot temperature is usually from 0 and 150 °C and the freezing temperature is from -10 to 80 °C.

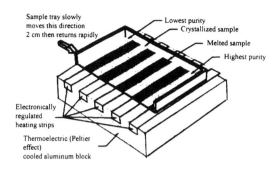

Figure 2-13. Purifier 10 zone refiner mechanism.
(Courtesy - Design Scientific, Inc., Gainesville, GA)

Figure 2-14. Purifier 10 zone refiner.
(Courtesy - Design Scientific, Inc., Gainesville, GA)

Referring to Table 2-1, p. 20, the most critical factors that determine purification time are the wall heat conduction ratio and the form factor or ratio of surface-area-to-volume of the sample. The combination of these improved factors enable sample purification to be accomplished in hours instead of days.

The use of a solvent

Recent work has shown that by adding a solvent, compounds that would decompose during normal zone refining can be refined as well. This is essentially a multiple recrystallization. The solvents can be any that the compound would normally dissolve in.

Table 2-1. Comparison of traditional and Purifier 10 zone refiners.
(Courtesy - Design Scientific, Inc., Gainsville, GA)

Sample container	Traditional	Purifier 10
Shape	Cylinder	Flat tray
Dimensions	2.5 x 50 cm	10 cm^2
Material of construction	Glass	Stainless steel
Wall thickness	1 mm	0.025 mm
Wall heat conduction ratio	1	560
Ratio of surface area to volume of sample	0.4 cm^2/mL	10 cm^2/mL
Container breakage during operation	Probable	Unlikely
Sample removal after purification	Difficult	Easy
Rate of movement of melted zone	0.1 - 1 cm/hr	1-50 cm/hr
Time required for typical purification	Several days	4-8 hrs
Amount of sample required	50-200 g	1-30 g

REVIEW QUESTIONS

1. What are the main uses of zone melting?
2. What forces are operating to aid a solid to purify itself if it is in contact with a liquid?
3. What is the concentration of dissolved solids in the frozen portion of a glass of grapefruit juice, if the liquid is 5 % frozen? Assume k = 0.07 and C_o = 0.040 M.
4. What is the difference between zone melting and normal freezing?
5. Why does zone refining work?
6. What are the three main areas in a zone refining operation?
7. What is the concentration of anthracene after one pass in a bar of naphthalene at a distance of 15 cm, if l is 1 cm, k is 0.15 and C_o is 0.25 M?
8. What final concentration would be predicted if a 40 cm bar with a 9 mm molten zone had 12 ppm of Sb as an initial concentration and k is believed to average 0.12?
9. What has been done to speed up the zone melting process?
10. What is the primary cause for the slowness of conventional glass tube type zone refining?
11. How does the thermal expansion of glass compare to that of most organic compounds?
12. Why is a small positive pressure applied to the tray?
13. About how far does the tray move before it is returned?
14. What provides the cooling under the tray?
15. What might be tried if you wanted to zone refine an organic compound that readily decomposes?

UNDERGRADUATES - COVER THOSE PORTIONS ASSIGNED BY YOUR INSTRUCTOR, PLUS ANY OTHER PORTIONS YOU ARE INTERESTED IN.
GRADUATE STUDENTS - START WITH NEW MATERIAL AND CONTINUE ON.

IF IT EXISTS - CHEMISTRY IS INVOLVED
IF YOU CAN BUY IT - A CHEMIST WAS INVOLVED SOMEWHERE

3
DISTILLATION: GENERAL INFORMATION

PRINCIPLES

One of the easiest ways to separate large quantities of heat stable compounds is by a process called *distillation*. Distillation is the process of producing a vapor from a liquid by heating the liquid in a vessel, then condensing the vapors and collecting them in another vessel. In Chapters 3 through 8 you will learn how several types of distillations work, the principles behind them, and the basic techniques that are necessary,

The types of distillation processes to be examined will be simple, fractional, steam, immiscible solvent, azeotropic, extractive, vacuum, molecular, entrainer sublimation, and freeze drying.

Simple and fractional distillations will be discussed in this chapter, but no experiments are provided because most students will have done experiments of this type previously. Such a discussion will permit the major areas of distillation to be put in proper perspective.

VAPOR PRESSURE

Molecules are in continuous motion, and in the gas phase they are moving several hundred miles an hour. For example, water molecules at room temperature are traveling at about 1340 mph and ethanol molecules at about 840 mph. They make between 3 to 5 billion collisions with each other or against a surface each second. These collisions with a surface are what we recognize as *pressure*. The pressure exerted by our atmosphere pressing down against the earth's surface at sea level is called an *atmosphere* (at 0 °C). Table 3-1 indicates several ways of expressing an atmosphere plus a few other common conversion factors.

Table 3-1. Pressure conversion factors

The Systeme Internationale (SI) unit accepted worldwide for pressure is the *Pascal*, Pa:
    1 Pascal   = 1 Newton/m^2
    1 Atmosphere = 1.0139 x 10^5 Newtons/m^2 = 1.0139 x 10^5 Pascals
Older units more commonly used in the U.S.:
    1 Atmosphere = 76 cm of Hg
                     = 760 mm of Hg (1 mm of Hg = 1 torr = 133 Pascals)
                     = 29.92 in of Hg
                     = 33.90 ft of water
                     = 14.7 lb/in^2

If molecules are constantly colliding with each other, it seems reasonable to assume that some hit each other head on, whereas others hit with more glancing blows. This should certainly affect the energies remaining in each molecule. Figure 3-1, p. 22, shows a typical energy plot for a box of molecules at different temperatures. We see that, in fact, they do have a variety of energies and that more of the molecules have higher energies as the temperature increases.

The size and shape of molecules differ from compound to compound, and because of that, some of them have a nonuniform electrical charge distribution within the molecule. For example, carbon disulfide, CS$_2$, is a very symmetrical and straight molecule and is nonpolar whereas water is slightly bent and is polar. Ethanol is a linear molecule, but because of the oxygen at one end, it is also polar. In addition, attractions can occur between molecules by what are known as *hydrogen bonds*. Molecules that have O, N, and F atoms located near one end can readily form hydrogen bonds. Some of these possibilities are shown in Figure 3-2, p. 22.

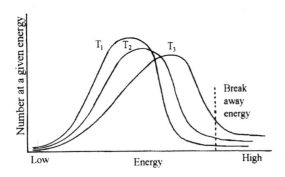

Figure 3-1. The energy distribution of molecules (approximate) at various temperatures ($T_1 > T_2 . T_3$).

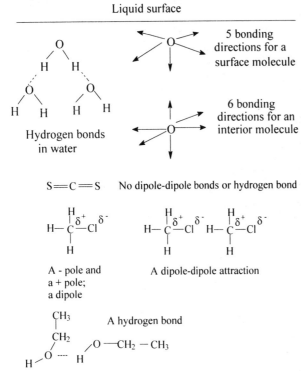

Figure 3-2. Surface effects on a liquid and several types of bonding.

These differences in bonding mean that the molecules in some liquids are held more tightly together than other liquids. Let us look at a diagram of the surface of a container of water (Figure 3-2) and see what happens because of this.

At the surface only five bonds are possible for a given molecule whereas six are possible in the interior. These five bonds at the surface can be stronger, because there is no sixth molecule to squeeze in and weaken them and there is no force to pull in the upward direction in this diagram. The net result is that the surface of a liquid has a tension on it, which is called *surface tension*.

For a compound like water, which can have rather strong hydrogen bonds in addition to a dipole-dipole attraction, the surface tension can be fairly strong. Water bugs take advantage of surface tension to "walk on water." Unfortunately, it is not strong enough to support professors, although many have tried. Let us again look at the surface of a liquid.

Because the molecules are in constant motion, some will have enough energy to overcome the surface tension and break away from the surface of a liquid. This breakaway energy will vary from compound to compound depending upon the strengths of the bonds holding the molecules together as well as the overall mass of the molecules.

The pressure exerted by those molecules that have escaped from a liquid's surface is called *vapor pressure*. For water at room temperature this pressure is only 23 mm of Hg (3059 Pa). This is not very much, but it does permit water to slowly evaporate. In contrast, hexane, a component of gasoline, has a vapor pressure of about 220 mm Hg (29,260 Pa), and ethanol, about 63 mm Hg (8379 Pa) at room temperature.

Figure 3-1 indicates that if the temperature of the system is increased, a larger portion of molecules reach the breakaway energy. This means that as the temperature increases, the vapor pressure increases. Figure 3-3, p. 23 is a diagram showing the dynamic equilibrium of molecules at a surface that produces vapor pressure, and Figure 3-4, p. 23, shows how vapor pressure changes with temperature for two compounds, water and ethanol.

SIMPLE DISTILLATION

A simple distillation involves applying heat to vaporize a liquid and then cooling the vapor until it condenses as a liquid. The separation of water from the salts in sea water is one example. See Figure 3-5, p. 24. The sea water is placed in a container called a *still pot*, or simply a *pot*, and heat is applied. The water will evaporate easily, but the salt will not and is left behind. However, the water will be lost unless something is done to collect it. A *condenser* is added to the system to cool the water vapor and condense it back to a liquid. The pure water then is collected in a container called a *receiver*.

FRACTIONAL DISTILLATION

In the simple distillation just discussed, only one compound of the sea water had sufficient vapor pressure to be distilled. What do you do if the mixture has two or more compounds that have appreciable vapor pressure? Rather than do several simple distillations, with a partial separation occurring at each step, the same effect can be obtained in a single column if the vapors of the volatile components are condensed, brought into contact with part of the condensate flowing down the column, and then boiled out of the descending liquid. A large surface area is needed so the process can be repeated several times along the length of the column. A column for this purpose is called a *fractionating column*, and the process is called *fractional distillation*.

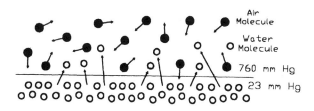

Figure 3-3. Diagrammatic form of a liquid surface.

Vapor Pressures of Mixtures of Compounds
Dalton's Law

Dalton's law states that each gas in a system behaves as if it were by itself, and therefore the total pressure of a system is the sum of all of the pressures added together.

$$P_{total} = P_1 + P_2 + P_3 + \cdots \quad (3\text{-}1)$$

P_1 then is called a *partial pressure*.

Refer to Figure 3-4 to see what this means in terms of boiling points. We see that when the SUM of the pressures of each component in the mixture equals the external pressure, the system will boil. In this case, at about 68 °C, the vapor pressure of water is about 230 torr and that of ethanol about 530 torr, giving a total of 760 torr.

Gibb's Phase Rule

The Gibb's phase rule tells how many degrees of freedom (variables you can change) you have in any system if you know the number of components (*C*), the number of phases (*P*), and the temperature and pressure.

$$F = C - P + 2 \quad (3\text{-}2)$$

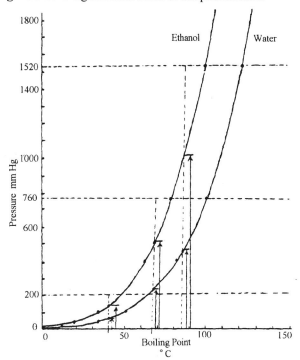

Figure 3-4. A plot of vapor pressure vs. temperature for water and ethanol.

EXAMPLE CALCULATION
How many degrees of freedom would you have if you had a beaker of benzene and toluene sitting on a bench top?

ANSWER
$C = 2$ (benzene and toluene); $P = 2$ vapor and liquid; $2 =$ temperature and pressure.
$F = 2 - 2 + 2 = 2$ Pressure is constant; therefore, the composition will vary with temperature.

Raoult's Law

What happens to the boiling point if there is 10 times more of one component than the other? Raoult's law corrects for this. The partial vapor pressure of one component is directly proportional to the number of molecules in the mixture. In practice, we refer to this as the *mole fraction* (*X*) of molecules.

$$P_{solvent} = X_{solvent} \times P^o_{solvent} \quad (3\text{-}3)$$

where P = the partial pressure; P^o = the vapor pressure of the pure material; X = the mole fraction of the material.

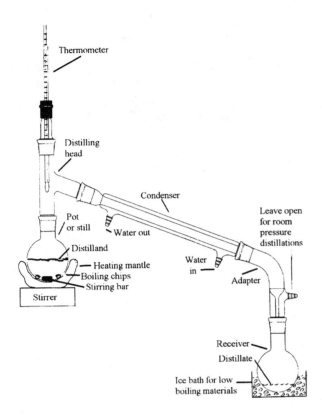

Figure 3-5. Apparatus for a simple distillation.

EXAMPLE CALCULATION

What would be the vapor pressure for a mixture of 15 mL of benzene (f.w. = 78.1, d = 0.878 g/mL, P^o = 86 torr at 25 °C) and 25 mL of toluene (f.w. = 92.1, d = 0.866 g/mL, P^o = 32 torr at 25 °C)?

ANSWER

$$N\ (\ benzene\ ) = \frac{15\ mL \times 0.878\ g/mL}{78.1\ g/mole} = 0.169\ moles$$

$$N\ (\ toluene\ ) = \frac{25\ mL \times 0.866\ g/mL}{92.1\ g/mole} = 0.235\ moles$$

$$X\ (\ benzene\ ) = \frac{0.169}{0.169 + 0.235} = 0.418 \qquad X\ (\ toluene\ ) = 1.000 - 0.418 = 0.582$$

$$P\ (\ benzene\ ) = 0.418 \times 86\ torr = 36\ torr \qquad P\ (\ toluene) = 0.582 \times 32\ torr = 19\ torr$$

P_t = 36 torr + 19 torr = 55 torr

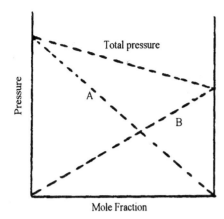

Figure 3-6. Theoretical vapor pressure curve at constant temperature.

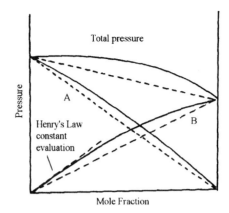

Figure 3-7. Actual vapor pressure curve at constant temperature.

The apparatus for a fractional distillation is more complex in that a *column* and a *distilling head* are placed between the pot and the condenser. The purpose of the column is to provide a large surface so that many simple distillations can take place at the same time, and the distilling head allows one to remove only a small portion of the material that reaches the top of the column, sending the rest back down the column to be redistilled, a process known as *refluxing*.

Consider a mixture of two compounds, A and B, at constant temperature. If Raoult's law holds, then we should get a curve something like that shown in Figure 3-6. Actually, there is some deviation from Raoult's law due to molecular interaction, and a vapor pressure curve like that shown in Figure 3-7 is more likely to occur. If both the liquid and the vapor are considered, then a curve like that shown in Figure 3-8 can be obtained. If the pressure is held constant and the temperature is changed, then a similar set of curves can be obtained.

Figure 3-9, p. 26, shows a plot of what happens to a mixture of benzene (b.p. = 80.0 °C) and toluene (b.p. = 110.6 °C) as the temperature changes. If a mixture of 80 mole % toluene and 20 mole % benzene (point D) is heated to boiling at constant pressure, the vapor will have the composition at point C, which is about 40 mole % benzene and now only 60 mole % toluene. This vapor could be condensed (point C) and collected (simple distillation), but benzene would not have been completely separated from toluene. The mixture would have to be redistilled several times, a time-consuming process. Refer to Figure 3-10, p. 26, which is a diagram of what might happen to this mixture in a distillation column. Most columns contain a packing of some sort to increase surface area. These are discussed later. The purpose of this packing is to provide a large condensing and re-evaporation surface. The objective is to do multiple simple distillations in the same column at the same time. Referring to Figure 3-9, let us condense the vapor at C. It is now liquid at E and is at 95 °C. If we were to continue to apply heat, this liquid at 95 °C would vaporize to form a vapor a little higher up the column at F. This vapor, in turn, is condensed to a liquid (point G), which upon heating at 89 °C forms a vapor at H even farther up the column. The more heat that is applied to the pot, the higher up the column the process will continue. The highest place up the column where the vapor condenses is called the *reflux line*.

In the case just described, six major equilibrium steps were involved. These are called *plates*. Although plates actually exist in large

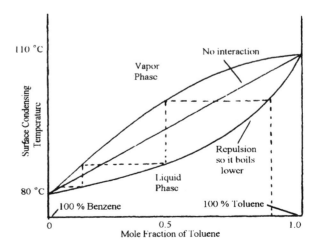

Figure 3-8. A combined liquid-vapor curve.

26 Principles

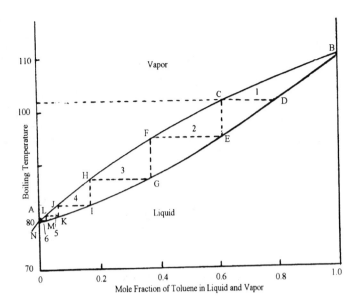

Figure 3-9. Boiling point - composition diagram for a mixture of benzene and toluene at 760 torr pressure.

commercial distilling columns to help hold the column packing, they are just there in theory in most laboratory columns, so they are called *theoretical plates*.

Suppose that the distance from where the product is removed at the top of the column to the surface of the liquid in the pot is 60 cm. The average distance between each theoretical plate in the example shown in Figure 3-9 (6) would be 10 cm. This is known as the *Height Equivalent to a Theoretical Plate* or *HETP*. HETP is a measure of the efficiency of distilling columns, and a standard equimolar mixture of hexane (b.p. = 69 °C) and methyl cyclohexane (b.p. = 100.4 °C) often is used to determine HETP. Figure 3-11, p. 27, shows the effect of the number of plates on a separation.

Table 3-2 shows the relationship between boiling point difference and the number of theoretical plates required to obtain a good separation.

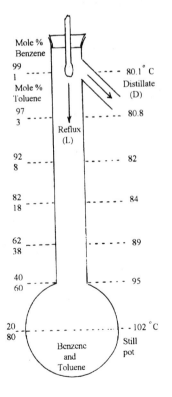

Figure 3-10. Composition changes in a distillation column.

Table 3-2. Relationship between the number of theoretical plates and the difference in boiling point for a good separation (Courtesy- K. Wiberg- *Laboratory Technique in Organic Chemistry*, McGraw-Hill, NY, 1960)

Number of plates	Difference in b.p. °C
0	215
1	108
2	72
3	54
4	43
5	36
10	20
15	13
20	10
30	7
50	4
100	2

DETERMINING THE NUMBER OF PLATES REQUIRED FOR A SEPARATION

If you had a mixture of two liquids, how many plates would be needed to separate them? This is important, because if you do not use enough plates, the mixture will not be adequately separated, and if you use too many, you are wasting time and energy, both of which are expensive. In this section, several equations will be reproduced that have been developed by others to determine the number of plates for a system at total reflux, for a system at partial reflux, and for a graphical method to do it quicker and with reasonable accuracy. The equations developed below were from the following original papers:

Fenske, M.R., *Ind. Eng. Chem.* **24**, 482 (1932).
Underwood, A.J.V., *Trans Inst. Chem. Engrs., London*, **10**, 112 (1932). Rose, A., *Ind. Eng. Chem.*, **33**, 594 (1941).
Sorel E., *Compt. Rend.* **108**, 1128, 1204, 1317, (1889).
Lewis, W.K., *Ind. Eng. Chem.* **1**, 522 (1909), 14, 492.

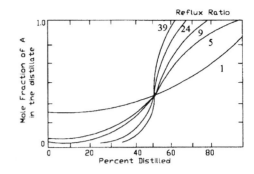

Figure 3-11. The effect of the number of plates on a separation of an equimolar mixture.

Volatility

Volatility, V, is the ratio of the mole fraction of a component in the vapor phase to its mole fraction in the liquid phase.

$$V = \frac{y \ (mole \ fraction \ in \ the \ vapor \ phase)}{x \ (mole \ fraction \ in \ the \ liquid \ phase)} \tag{3-4}$$

$V = 1$ for a pure compound because the mole fraction is the same for both the vapor and the liquid.

For a binary mixture. V_1 is for the more volatile component, V_2 for the less volatile component.

$$V_1 = \frac{y_1}{y_2} \qquad V_2 = \frac{x_1}{x_2} \tag{3-5}$$

$$Relative \ volatility = \alpha = \frac{V_1}{V_2} = \frac{y_1 \, x_2}{y_2 \, x_1} \qquad or \qquad \frac{y_1}{y_2} = \alpha \frac{x_1}{x_2}$$

Since
$$x_2 = 1 - x_1 \quad and \quad y_2 = 1 - y_1$$

$$\frac{y_1}{1 - y_1} = \alpha \frac{x_1}{1 - x_1} \tag{3-6}$$

If Raoult's law holds for the solution and Dalton's law of partial pressures holds in the vapor

$$\alpha = \frac{P_1^0}{P_2^0} \qquad P^o = \text{vapor pressure of pure materials.}$$

Raoult's law:

$$P_1 = X_1 P_1^0$$

For ideal gases $V_1 = P_1$; X = mole fraction.

$$\alpha = \frac{P_1^0 X_2}{P_2^0 X_1} \tag{3-7}$$

If pressures are not available, but boiling points are, you can combine **Trouton's rule**

$$\frac{\Delta H}{T} = 21 \; cal \; deg^{-1} \; mole^{-1} (or \; 87.8 \; J \; deg^{-1} \; mole^{-1}) \tag{3-8}$$

and the **Clapyron equation:**

$$\log \frac{P_1}{P_2} = \frac{-\Delta H (T_2 - T_1)}{2.3 \; R \; (T_2 \times T_1)} \tag{3-9}$$

to get **Rose's equation:**

$$\log \alpha = \frac{8.9 (T_2 - T_1)}{(T_2 + T_1)} \tag{3-10}$$

EXAMPLE CALCULATION

The b.p.'s of 1-hexanol and 2-hexanol are 157 and 140 °C, respectively. What is the predicted α for this mixture based on Rose's equation?

ANSWER

$$\log \alpha = \frac{8.9 \; [(273 + 157) - (273 + 140)]}{[(273 + 157) + (273 + 140)]}$$

$$\log \alpha = \frac{151.3}{843} = 0.1795 \qquad \alpha = 1.51$$

CONTINUOUS DISTILLATION - TOTAL REFLUX CONDITIONS

In a distillation, the process of returning part of the condensed vapors back down the column to be redistilled and improve the separation is called *refluxing*. With glass apparatus, the *reflux line* can be seen easily as a line of liquid somewhere up the column where the vapors are condensing. Unless some product is being removed, the process is called *total reflux*. This condition is used primarily to understand what is going on in the column.

Let X_s and Y_s be for the more volatile component; α must be constant over the range of concentration considered. Use equation 3-6.

$$\frac{Y_1}{1 - Y_1} = \alpha \frac{X_1}{1 - X_1} \tag{3-11}$$

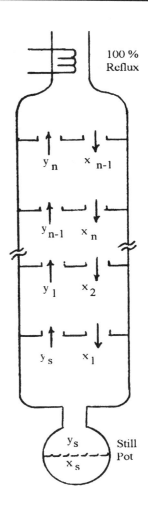

Figure 3-12. Diagram of a distillation column for total reflux. (Courtesy - E.W. Berg, *Physical and Chemical Methods of Separation*, McGraw-Hill, New York, NY, 1963)

Refer to Figure 3-12. X represents the mole fraction of the liquid returning to the pot, and Y represents the mole fraction of vapor heading for the top of the column.

$$\frac{Y_1}{1 - Y_1} = \alpha \frac{Y_s}{1 - Y_s} = \alpha^2 \frac{X_s}{1 - X_s}$$

Assuming that the vapor (X_1) entering the first plate of the column is condensed and in equilibrium with the vapor Y_1 entering the second plate of the column, continue the pattern.

$$\frac{Y_2}{1 - Y_2} = \alpha^3 \frac{X_s}{1 - X_s}$$

In general:

$$\frac{Y_n}{1 - Y_n} = \alpha^{n+1} \frac{X_s}{1 - X_s} \tag{3-12}$$

The n + 1 term = the enrichment factor.

With this equation you can
1. Calculate the efficiency of a column.
2. Calculate the number of plates required to separate two components
3. Calculate the number of theoretical plates.

n-Hexane (b.p. = 69 °C) and methylcyclohexane (b.p. = 101 °C) are used to determine the number of plates in a column since very accurate vapor-liquid data are available.

EXAMPLE CALCULATION

Chlorobenzene and bromobenzene have $P°_1$ and $P°_2$ = 861.5 and 455.8 torr, respectively. How many theoretical plates are required to produce a vapor containing 99.9 mole % chlorobenzene at the top of the column under total reflux if the pot contains a 50-50 mole ratio mixture?

ANSWER

$Y_n = 0.999 \quad X_s = 0.50$

$$\alpha = \frac{861.5}{455.8} = 1.89 \qquad \frac{0.999}{1 - 0.999} = (1.89)^{(n+1)} \frac{0.50}{1 - 0.50}$$

(n+1) log 1.89 = log 999   2.99956/0.27646 = 10.85 = n+1
n = 9.8 or 10 plates.

CONTINUOUS DISTILLATION, PARTIAL REFLUX

With total reflux no product is obtained. This is an unsatisfactory condition for all parties concerned. What is desired is to be able to determine prior to distillation just how many plates are needed and to prepare a column accordingly. Too many plates waste energy and time, and too few plates produce an inferior product. The equations developed below show how the number of plates required for any system can be determined. Refer to Figure 3-13.

Assume

    (1) The column is working adiabatically.
    (2) There is no heat of mixing.
    (3) The heat necessary for vaporization of each plate is provided by the condensation of the descending vapors.

L = Moles of reflux
V = Total number of moles leaving the top plate, therefore, also entering the bottom plate
D = Moles of distillate
L/V = Reflux ratio = R_D

$$V = L + D \tag{3-13}$$

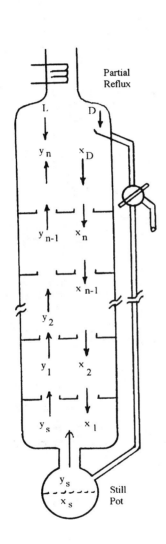

Figure 3-13. Diagram of a distillation column for partial reflux.
(Courtesy - E.W. Berg, *Physical and Chemical Methods of Separation*, McGraw-Hill, New York, NY, 1963)

Leaving the top plate	Entering the base of the column
DX_D =	$VY_s - LX_1$

Divide by V:

$$Y_s = \frac{L X_1}{V} + \frac{D X_D}{V}$$

$$= \frac{L X_1}{L + D} + \frac{D X_D}{L + D}$$

Therefore, between any two plates:

$$Y_1 = \frac{L X_2}{V} + \frac{D X_D}{V} \qquad Y_2 = \frac{L X_3}{V} + \frac{D X_D}{V}$$

In general:

$$Y_{n-1} = \frac{L X_n}{V} + \frac{D X_D}{V} \tag{3-14}$$

The foregoing is called the *operating line equation*.
$VY_{n-1} = LX_n + DX_D$.

    VY_{n-1} = moles of component leaving the n-1 plate.
    LX_n = moles returning by reflux.
    DX_D = moles leaving as distillate.

At total reflux: D = 0   L/V = 1
so:
$$Y_{n-1} = X_n \text{ which is a 45° line.}$$
(needed for the McCabe-Thiele diagram described later)

Distillation

EXAMPLE CALCULATION

Calculate the number of theoretical plates required to produce a distillate containing 99.0 mole % chlorobenzene from a 1:1 mole ratio of chlorobenzene and bromobenzene, if the reflux ratio is 9.

ANSWER

$V = L + D = 9 + 1 = 10$

$Y_{n-1} = 0.9 X_n + 0.1 X_D$

$Y_n = X_D$ because the more volatile compound is nearly pure at the top of the column. 0.990 is nearly 1.0

If $X_D = Y_n = 0.990$

$$\frac{Y_n}{1 - Y_n} = \alpha \frac{X_n}{1 - X_n} \; ; \qquad \frac{0.990}{1 - 0.990} = 1.89 \frac{X_n}{1 - X_n}$$

$X_n = 0.9812; \quad Y_{n-1} = 0.9(0.9812) + 0.1(0.99) = 0.9821$

Now:

$$\frac{Y_{n-1}}{1 - Y_{n-1}} = \alpha \frac{X_{n-1}}{1 - X_{n-1}} \; ; \qquad \frac{0.9821}{1 - 0.9821} = 1.89 \frac{X_{n-1}}{1 - X_{n-1}}$$

$X_{n-1} = 0.966 \quad Y_{n-2} = 0.9(0.9666) + 0.1(0.990) = 0.9689$

Continue the process until X n-? is equal to or less than 0.5 (a 1:1 ratio in the pot).

$$\frac{Y_{n-2}}{1 - Y_{n-2}} = \alpha \frac{X_{n-2}}{1 - X_{n-2}} \; ; \qquad \frac{0.9689}{1 - 0.9689} = 1.89 \frac{X_{n-2}}{1 - X_{n-2}}$$

$X_{n-2} = 0.9428 \quad Y_{n-3} = 0.9(0.9428) + 0.1(0.990) = 0.9475$

$$\frac{Y_{n-3}}{1 - Y_{n-3}} = \alpha \frac{X_{n-3}}{1 - X_{n-3}} \; ; \qquad \frac{0.9475}{1 - 0.9475} = 1.89 \frac{X_{n-3}}{1 - X_{n-3}}$$

$X_{n-3} = 0.9052 \quad Y_{n-4} = 0.9(0.9052) + 0.1(0.990) = 0.9137$

$$\frac{Y_{n-4}}{1 - Y_{n-4}} = \alpha \frac{X_{n-4}}{1 - X_{n-4}} \; ; \qquad \frac{0.9137}{1 - 0.9137} = 1.89 \frac{X_{n-4}}{1 - X_{n-4}}$$

$X_{n-4} = 0.8485 \quad Y_{n-5} = 0.9(0.8485) + 0.1(0.990) = 0.8626$

$$\frac{Y_{n-5}}{1-Y_{n-5}} = \alpha \frac{X_{n-5}}{1-X_{n-5}} \; ; \qquad \frac{0.8626}{1-0.8626} = 1.89 \frac{X_{n-5}}{1-X_{n-5}}$$

$$X_{n-5} = 0.7686 \qquad Y_{n-6} = 0.9(0.7686) + 0.1(0.990) = 0.7907$$

$$\frac{Y_{n-6}}{1-Y_{n-6}} = \alpha \frac{X_{n-6}}{1-X_{n-6}} \; ; \qquad \frac{0.7907}{1-0.7907} = 1.89 \frac{X_{n-6}}{1-X_{n-6}}$$

$$X_{n-6} = 0.6664 \qquad Y_{n-7} = 0.9(0.6664) + 0.1(0.990) = 0.6987$$

$$\frac{Y_{n-7}}{1-Y_{n-7}} = \alpha \frac{X_{n-7}}{1-X_{n-7}} \; ; \qquad \frac{0.6987}{1-0.6987} = 1.89 \frac{X_{n-7}}{1-X_{n-7}}$$

$$X_{n-7} = 0.5509 \qquad Y_{n-8} = 0.9(5509) + 0.1(0.990) = 0.5948$$

$$\frac{Y_{n-8}}{1-Y_{n-8}} = \alpha \frac{X_{n-8}}{1-X_{n-8}} \; ; \qquad \frac{0.5948}{1-0.5948} = 1.89 \frac{X_{n-8}}{1-X_{n-8}}$$

$$X_{n-8} = 0.4389 \qquad Y_{n-9} = 0.9(0.4389) + 0.1(0.990) = 0.4948$$

But $X_{n-8} = 0.4389$ which is less than what is in the still pot ($X_s = 0.50$); therefore, this is the end. The number of plates equals the number of steps, or 9. Use Y; it is easier, because the n-number gives you the number of plates.

Figure 3-14 is a plot of the above example calculation. Table 3-3 shows the number of plates required to get a 99 % mole fraction separation depending upon the boiling point difference between two liquids. A computer can be programmed to do the above calculation.

With extractive distillation, discussed later, a 0.3 °C difference can be separated 90-95% by <10 plates.

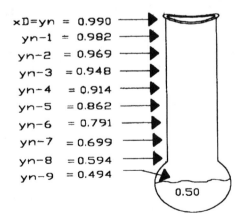

Figure 3-14. The plate - mole fraction distribution for the example calculation.

Table 3-3. The number of plates required to get a 99 % mole fraction separation depending upon the boiling point difference between two liquids

Boiling Point Difference °C	Number of Plates
15-20	10
8-12	25
5-10	45
3-5	80
2	150

MCCABE-THIELE EQUATION

There must be a quicker way to determine the number of plates required in a practical situation. In an industrial situation, the quality of the feedstock can change daily or weekly depending upon the supplier. If the product distilled is to be used for a variety of applications, then the purity of the product must be changed from day to day. Suppose your current supplier is sending you a feedstock that is 20 mole % pure and you have a customer that needs it to be 80 mole % pure, what do you do if you only have one high quality still?

If you put the feedstock in the pot and recover the product from the top of the column, then you are wasting energy and time at considerable extra expense, because you are redistilling some feedstock that has already been partially distilled. Therefore, you add the feedstock to the column where its composition matches that of the column, and you take it off at the desired composition. Many years ago, McCabe and Thiele developed an equation and a diagram technique to solve the above type of a problem graphically. This is shown in Figure 3-15.

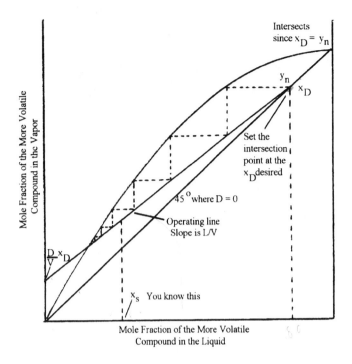

Figure 3-15. A plot showing the McCabe-Thiele equation and how it can be used.

$$Y_s = \frac{L X_1}{V} + \frac{D X_1}{V} \qquad \text{is a straight line of the form } Y = mX + B \qquad (3\text{-}15)$$

$$Slope = \frac{L}{V} \qquad\qquad Intercept \quad B = \frac{D X_D}{V}$$

EXAMPLE CALCULATION

Your current supplier is sending you a feedstock that is 20 mole % pure and you have a customer that needs it to be 80 mole % pure, how many plates are needed to meet these conditions?

ANSWER

Plot a diagram like that in Figure 3-15 with a 45° line beginning at 0 % mole fraction. The $D X_D/V$ are the various reflux ratios you might want to use. These are plotted on the Y axis. The vapor curve is plotted based on previously obtained vapor pressure data of the mixture.

Place a point on the 45° line at the composition of the desired final product (X_D) desired (0.80). Draw a vertical line upward at the feedstock composition (X_s, 0.20). Place a straightedge so it crosses the X_D point and extends across the axis. Move the left side of the straight edge up and down until it crosses the vertical X_s line somewhere between the vapor curve and the 45° line. Draw a line between the two points. Starting at X_D, step off the plates (dashed line) until you cross the vertical line. This is the number of plates required. Where the line crosses the Y axis is the reflux ratio required. As D decreases, the slope of L/V increases and fewer plates are needed. If L/V intersects at X_s, an infinite number of plates are required. If it intersects to the right of X_s, MORE than an infinite number of plates are required.

TECHNIQUES

HEATING DEVICES

Hot Plates

Hot plates, heating mantles, and steam baths are commonly used as heaters. Burners are not used unless nonflammable materials are involved or unless rapid heating is required. Hot plates, Figure 3-16, are relatively inexpensive and simple to operate, but they do not always provide even heat over the entire bottom of a round-bottom flask.

Figure 3-16. Various types of hot plates: non-stirring, left, stirring, right.
(Courtesy - Fisher Scientific Co., Pittsburgh, PA)

Heating Mantles

A heating mantle, Figure 3-17, which consists of layers of wire covered with insulation, provides much more uniform heating over a large surface area of the pot than a hot plate and its temperature is easily controlled by a variable transformer. Mantles come in all sizes to fit any size flask.

The coil of wire on the outside of a heating mantle should not be broken or pulled off. It is connected to a thermocouple (temperature measuring device) inside of the mantle and can be used to measure the temperature. The disadvantage of a heating mantle is that it takes a long time to reach the desired temperature compared to heating with a burner.

Figure 3-17. Various types of heating mantles.
(Courtesy - Ace Glass Co., Vineland, NJ)

Variable Transformers

A *variable transformer* (trade names Variac and Powerstat) is a device for changing the voltage applied to the heating wires in the heating mantle.

Figure 3-18. A variable transformer (Variac or Powerstat).
(Courtesy - Ace Glass Co., Vineland, NJ)

CAUTION: The cord connecting a heating mantle to a variable transformer has a *twist-on* plug, which must be turned 1/8 of a turn before it can be removed.

NOTE: The scale on the dial normally does not read directly in volts. Common scales are 100, 240, and 280. These were made for European systems. For a 220 volt line, the 280 represents the *peak-to-peak* voltage, and the 240 represents the *root mean square* voltage. For example, if 55 volts were needed in a heating mantle, a Variac with a 240 scale plugged into a 110 volt line would have to be adjusted to 120 (half of 240).

Boiling Chips

A *boiling stone* or *boiling chip* is a small piece of porous material (unglazed pottery, silicon carbide, brick chips) that is added to provide a continuous stream of air bubbles to prevent a buildup of gas in the bottom of the pot, which if allowed to collect can cause the solution to *bump* when it finally escapes. Bumping occurs with most materials, particularly those that are viscous or when the pot is filled half to two-thirds full. The air trapped in the boiling stone escapes when the pot is heated, and the stream of bubbles given off provides an escape path for the gases formed on the very bottom of the pot. Boiling stones usually are not used in vacuum distillations described in a later chapter.

CAUTION: *A boiling chip should never be added to a hot solution.* A violent eruption often takes place, which can break the apparatus, and you are likely to be severely burned from the hot solution. If you forgot to add boiling chips and the solution is now hot, the proper technique is to cool the system back to nearly room temperature before boiling chips are added. In an emergency, you can do the following, but be very careful. Cool the system several degrees and remove the thermometer. Add the boiling chips into this opening *one at a time.*

Steam Baths

A *steam bath* (or ring bath) provides a constant source of heat at 100°C and is quite useful for distillations that involve heat-sensitive or low-boiling compounds. These heaters are preferred for low-boiling materials or when large quantities of material are to be distilled, because they can be left unattended for considerable periods of time.

If small quantities of material are to be distilled and the compound has a medium to high boiling point, then a small burner is quite acceptable. It provides rapid heat and, if a thermometer is placed in the pot, then it is easy to control the temperature rather closely. **CAUTION:** If a burner is used, it should *never* be left unattended.

Figure 3-19. Various types of steam baths. Left: Multiple port, water heater attached. Right: Single port, variable size, requires steam source.
(Courtesy - Fisher Scientific Co., Pittsburgh, PA)

MAGNETIC STIRRERS

A magnetic stirrer is much preferred to a boiling chip. A small, plastic-covered, bar magnet is placed in the pot. Another bar magnet, attached to the shaft of an electric motor, is placed under the pot. When the motor turns, the magnet in the pot turns also, stirring the solution. The speed of rotation is controlled by using a variable transformer to vary the voltage to the stirring motor.

Magnetic stirrers are available that are either air- or water-driven. They are very simple and rugged. All that is necessary is to connect the inlet to either a stream of air or a water line. Figure 3-20 shows this type.

Figure 3-20. An air- or water-driven magnetic stirrer.
(Courtesy - G F S Chemical Co., Columbus, OH)

ANTIFOAMS

Several materials, particularly those that contain protein, have a tendency to foam. The foam contains impurities, and it must not be allowed to get into the condenser and be collected. If foaming occurs, then either the rate of distillation must be slowed or an antifoaming agent added. A drop or two of benzene is often effective. Silicone oils usually have much higher boiling points than the materials being distilled, and one or two drops of oil is also effective. The structure of a methyl silicone that can be used as an antifoam is shown to the left.

Antifoams are in many foods. For example, when making strawberry preserves, the system foams excessively, and a slower, more expensive heating process is required. However, if 10 ppm of antifoam is added, the cooking process is much faster and less expensive. Silicone antifoams are nontoxic.

COLUMNS

The distillation column is where the separation actually takes place.

Holdup is the amount of liquid actually in the column as vapor or reflux. You want holdups to be as small as possible yet still provide a good separation. If you were distilling a 5 mL sample, you would need a very small column or the sample would be lost in the column.

Throughput is the rate at which vapor is passing up the column. How many mL/min are you distilling? You want this to be as large as possible yet get the desired separation.

The desire is to get as much surface area for condensation and redistillation as possible yet provide sufficient openings for the vapor to pass.

Column Packings

To increase the surface area inside of a column, usually it is packed with small curved ceramic pieces that look like Pringle's potato chips (*Berl saddles*); sections of small diameter ceramic, glass, or metal pipe (*Raschig rings*); glass beads; or protruded metal clips made from either stainless steel, nickel, or Monel. The purpose of this packing is to provide a large condensing and re-evaporation surface. The objective is to do multiple simple distillations in the same column at the same time. Figure 3-21, p.37, shows several types of packings.

Figure 3-21. Column packing materials. From left to right: Glass beads. Rashig rings. Berl saddles. Protruded metal.

Packed Columns

Most packed columns are homemade types. A glass wool plug or a piece of screen wire is used to hold the packing in the column. Figure 3-22 shows a Hemple column that usually is packed with glass beads. The HETP for a 10-mm-diameter column is about 4 cm.

Fixed Internal Surface Columns
Vigreux Column

A *Vigreux* (Vig-row) column contains glass indentations on the side of the column. The indentations are made by heating a spot on the glass column and pushing either the point of a sharp pencil or the end of a small file into the hot glass. The indentations are usually pointed downward at about 45°. This is an inexpensive column that does a good job.

Snyder Column

This is a more elaborate column with extensive glass working on the inside to increase the surface area. The glass bulbs are free to move and move up and down randomly during a distillation. This is a noisy, but very effective, column. *Pour a few milliliters of solvent down the top of the column before you start the distillation and the bulbs will move freely. Otherwise, one sometimes sticks and the internal pressure blows the column off of the pot.*

Perforated Plate

These columns are more expensive, but they have essentially one plate per step, so it is easy to determine if you have a sufficiently good column to do the separation you want to do.

Spinning Band Columns

The most efficient distillation column at the present time is the *spinning band column* (Figure 3-23). It consists of either a spiral of wire mesh or a Teflon band being turned by a motor at the top of the column.

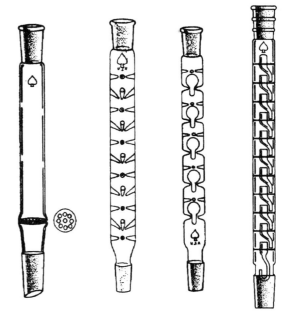

Figure 3-22. Several distilling columns shown from left to right: Hemple. Vigreux. Snyder. Perforated plate.
(Courtesy - Ace Glass Co., Vineland, NJ)

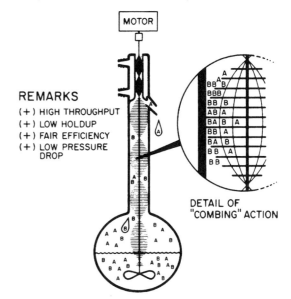

Figure 3-23. Diagram illustrating the principle of a metal mesh spinning band column.
(Courtesy - R.W. Yost, Distillation Primer, *American Laboratory*, Jan., 1974)

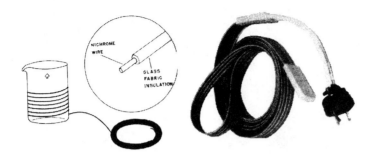

Figure 3-24. Column heating devices: Left, Insulated nichrome wire. Right, Heating tape.
(Courtesy - Ace Glass Co., Vineland, NJ)

These columns can produce up to 200 plates in an 18-to-20-inch column.

In the spiral wire mesh type, the edges of the wire scrape against the sides of the column. This causes a thin film of liquid to form on the side wall and, in addition, increases the surface area considerably as the wires move across it. The wire is spun so as to throw the vapors *back down* into the column. If spun the opposite way, it can suck the system dry in a few minutes. The spinning spiral band operates in the same manner. These are expensive columns, but they have a low holdup and a high throughput.

HEATING TAPES

To retain heat, the column should be wrapped with heating tape (Figure 3-24) or covered with a layer of glass wool held in place with tape. Aluminum foil can be wrapped around the column. Insulated nichrome wire connected to a variable transformer also can be wrapped around the column or any object to heat it.

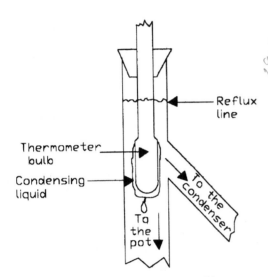

Figure 3-25. Proper thermometer position

THERMOMETER PLACEMENT

The thermometer should be placed so the bottom bulb is centered by the outlet to the condenser (Figure 3-25). Refer back to Figure 3-10, p. 26, and notice how the composition changes dramatically at the top of the column and how a small difference in thermometer placement can make a large difference in the purity of the final product. The measured temperature of the system is accurate only if the thermometer bulb is totally bathed in the condensing liquid. This is achieved by having the reflux line just slightly above the top of the bulb.

CONDENSERS

For distillations done at atmospheric pressure in apparatus with a ground glass joint (see Figure 3-5, p. 24), no joint grease is needed at either end of the condenser unless the distilling head or the adapter has a lip. The sections of the joint are pressed firmly together, and the condensing vapors seep up the joint and provide lubrication. A lip makes this seeping process difficult, and grease must be added. Any grease on the joints of horizontal condensers that dissolves or melts will contaminate the product, because it can no longer be distilled.

The *inlet water* tubing is attached to the *lower end* of the condenser so that the condenser can be easily filled and kept full of water. It also prevents very cold water from coming in contact with the hottest condensing liquid thereby which reducing the possibility of cracking the condenser.

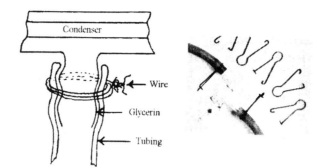

Figure 3-26. Connecting tubing to a condenser. Wire (left). Commercial clamp (right).
(Courtesy - Ace Glass Co., Vineland, NJ)

When the inlet and outlet tubing are being connected to the condenser, a drop or two of glycerine is placed on the glass connection part so that the tubing can be slipped off easily later. The tubing then is wired on with a few turns of copper wire, or a tubing clamp is used. See Figure 3-26 for examples.

One of the largest sources of damage in a laboratory is condenser tubing that has either broken or slipped off. Several gallons of water are then on the floor in a few minutes. This usually soaks through the floor at the pipe chases and can cause damage several floors below.

CAUTION: *Do not use old tubing.* If the tubing has surface cracks or feels stiff, discard it. It is less expensive to use good tubing than to repair the water damage done when the tubing breaks. Figure 3-27 shows several types of condensers. The Liebig and West condensers generally are used tilted about 30° from a horizontal position as shown in Figure 3-5, p. 24. The outer tube of a West condenser fits closer to the inner tube than the Liebig condenser, so the cooling water must be passed through at a faster rate. The Allihn, coiled, Graham, and Friedrichs condensers are best used in a vertical position. The Allihn condenser is the most commonly used type; the bulbs provide more surface area than those in either the West or Liebig condensers. The coiled and Graham condensers are used where very high efficiency is needed, and the Friedrichs condenser, which is made to have the condensate drip off the bottom center, is used most in continuous extractions. Look carefully to see the difference between the coiled and the Graham condensers. The cooling water goes *inside* of the coil in a coiled condenser, whereas it goes *outside* of the coil in a Graham condenser.

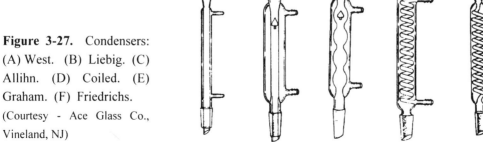

Figure 3-27. Condensers: (A) West. (B) Liebig. (C) Allihn. (D) Coiled. (E) Graham. (F) Friedrichs.
(Courtesy - Ace Glass Co., Vineland, NJ)

RECEIVERS

For a simple distillation at atmospheric pressure, almost any type container can be used to collect the *distillate* (the material that distills over). If the distillate has a low boiling point, then the receiver is placed in an ice bath.

CAUTION: *The system must be open to the atmosphere at one place.* This opening is usually between the adapter and the receiver. When using equipment with ground glass joints, it is very easy to accidentally close the system. The apparatus can blow apart if this happens. More will be said about receivers in the vacuum distillation section.

DISTILLING HEADS

If a column is not doing a sufficiently good job, a part of the material that reaches the top of the column can be sent back down the column to be redistilled. This is called *refluxing*. The ratio of the amount of material sent back down the column to the amount collected per unit of time is called the *reflux ratio*.

In a simple distillation, the distilling head holds the thermometer and connects the pot to the condenser. In fractional distillation, the distilling head must provide some means to alter the reflux ratio, and some even contain a condenser.

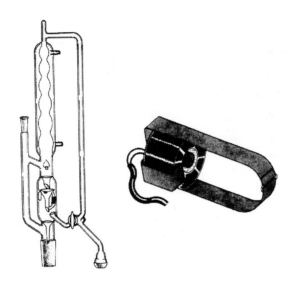

Figure 3-28. Swing funnel type distilling head.
(Courtesy - Ace Glass Co., Vineland, NJ)

Figure 3-29. Stopcock control type distilling head.
(Courtesy - Ace Glass Co., Vineland, NJ)

Swing Funnel Type

Figure 3-28 shows a quality distilling head. This is a swinging funnel, magnetic controlled type that is quite common. The coil of wire is placed alongside of the swinging funnel, and a pulse of electricity is sent through it by a relay. This causes the funnel to tilt. The speed of the tilting and, therefore, the reflux ratio are controlled by a timer in the circuit.

The condenser is usually an Allihn type, because of its large surface area, and usually it is placed vertically so all of the condensed vapors will drip into the reflux control.

Stopcock Control Type

Figure 3-29 shows a simpler, stopcock-type reflux controller. The reflux ratio is controlled by how much the stopcock is opened. This type does not work well with viscous materials, and you cannot measure the reflux ratio.

EXAMPLE CALCULATION

During a 5-minute period, 10.0 mL of material reach the top of the column, 1.0 mL is collected, and 9.0 mL are sent back. What is the reflux ratio?

ANSWER

9.0 mL sent back/ 1.0 mL collected = a reflux ratio of 9.

Distillation 41

REVIEW QUESTIONS

1. What causes "pressure"?
2. What is a Pascal?
3. What is meant by a "polar" molecule?
4. Why are the water molecules present in our atmosphere called "vapor" rather than a "gas"?
5. What is a "still pot" or "pot" when referring to a distillation?
6. What is a fractionating column?
7. What does the Gibb's phase rule tell you about a system?
8. $FeCl_4^-$ in a 6 M aqueous solution of HCl is to be extracted with diethyl ether at room temperature and atmospheric pressure. How many degrees of freedom do you have?
9. One fluid ounce (29.6 mL) of ethanol (f.w. = 46.1, d = 0.789 g/mL, v.p. at r.t. = 78.4 torr) is mixed with 4 fluid ounces (118.4 mL) of water (f.w. = 18, d = 1.00 g/mL, v.p. at r.t. = 100 torr), an olive, and a few drops of vermouth to make a Martini. What is the vapor pressure just above the surface of this combination, if you assume the contribution from the olive and the vermouth is negligible?
10. What does HETP stand for?
11. What does HETP mean and of what value is it?
12. Ethyl isobutyrate (b.p. = 111 °C) is used in the manufacture of flavors and essences, as is ethyl isovalerate (b.p. = 135 °C). How many plates would a distilling column have to have to effect a good separation of these two compounds?
13. What is meant by the term "volatility"?
14. What would you estimate the ΔH of vaporization to be for ethyl formate, a flavoring agent for lemonade, if it boils at 54 °C?
15. Butyl acetate is used as a solvent for fingernail polish. What is the difference in vapor pressure if a woman polishes her nails on a 100 °F (38 °C) day as compared to polishing them outside on a 40 °F (5 °C) day? ΔH_{vap} = 43.9 kJ/mole, v.p. at 38 °C = 37 torr.
16. What is α and what is its significance?
17. 1-Butanethiol (b.p. 98.5 °C) and 2-butanethiol (b.p. = 85 °C) are believed to be two of the major components of the "essence of skunk". What is the predicted α for this mixture based on Rose's equation?
18. What is meant by refluxing and why would it be used?
19. Cyclohexanone (b.p. = 155 °C, v.p. = 760 torr) is used as a solvent for resins and natural rubber and is prepared by the catalytic dehydrogenation of cyclohexanol (b.p. = 161 °C, v.p. = 508 torr at 155 °C). Assuming total reflux, how many theoretical plates are required to produce a vapor containing 99.0 mole % cyclohexanone at the top of the column, if the pot contains a 50-50 mole ratio mixture?
20. Repeat question 19, but use a reflux ratio of 9. Notice how many more plates it takes than the example problem when essentially the only thing that changed was that α was decreased from 1.89 to 1.49.
21. What is an advantage of the McCabe-Thiel diagram?
22. Why is a heating mantle usually preferred over a hot plate?
23. What is the purpose of the small coil of wire on the outside of a heating mantle?
24. What is the purpose of a boiling chip?
25. Why is it dangerous to add a boiling chip to a hot solution?
26. Why are silicone antifoams preferred for food processing?
27. What is "holdup" when referring to a distillation column?
28. Ceramic pieces that look like small Pringles' potato chips are called what?
29. What spins in a spinning band column?
30. What are some advantages of a spinning band column?

31. Where is the proper place to put a thermometer during a distillation?
32. Why is the inlet water always added at the bottom of a condenser?
33. What is the penalty for not properly wiring condenser tubing to a condenser?
34. What is the purpose of a distilling head?

UNDERGRADUATE STUDENTS - DO THOSE PORTIONS ASSIGNED BY YOUR INSTRUCTOR,
PLUS ANY OTHERS THAT INTEREST YOU.
GRADUATE STUDENTS - START WITH NEW MATERIAL AND CONTINUE ON.

IF IT EXISTS - CHEMISTRY IS INVOLVED
IF YOU CAN BUT IT - A CHEMIST WAS INVOLVED SOMEWHERE

4
AZEOTROPIC AND EXTRACTIVE DISTILLATIONS

AZEOTROPIC DISTILLATION
PRINCIPLES

If a mixture of ethanol and water is distilled, eventually it will form a solution that is 95 % ethanol regardless of the starting composition. Hydrochloric acid and water will form a 20.22 % HCl solution, and chloroform and acetone will form a 65.5 % $CHCl_3$ solution. These solutions are called *azeotropes* (Gr: a + zeo *to boil*; trope *more at*). Webster's dictionary defines an azeotrope as *a liquid mixture that is characterized by a constant maximum or minimum boiling point which is lower or higher than that of any of the components and that distills without change in composition*. An *azeotropic distillation* involves the formation of an azeotrope with at least one of the components of a liquid mixture, which then can be separated more readily because of the resulting increase in the difference between the volatilities of the components of the mixture. Figure 4-1, p. 44, shows the water-ethanol system (A) and the HCl-water system (B).

Notice that each line (A and B) crosses over the 45° line. At the point where it crosses over the line, the compositions are the same in both the liquid and the vapor. What does this mean in practice? Figures 4-2 and 4-3, p. 44, show the resulting temperature-mole fraction curves (constant pressure) for these situations.

To use an azeotrope to effect a separation a solvent is added to mixture that will form a constant boiling mixture with one or more of the components in the mixture. There are two requirements:
 1. The added component reduces the partial pressure of one of the original components.
 2. The formed azeotrope is removed more easily than anything else in the pot.

For example; You have a mixture AB. Add excess C to it. AC is distilled off leaving B and C, which then are separated by distillation. Refer to Figure 4-4, p. 45, as an example of what happens during a distillation. Suppose a mixture of A and B had a liquid composition at point 6. The vapor in equilibrium with this liquid has the composition at point 5. If the temperature is raised, then the concentration of the material in the pot will slowly increase in A from points 6 to 8 to 10 etc., and the vapor will be very rich in component B and move to the left along the vapor curve. The end result will be that B will distill away, and the liquid composition will change until point 18 is reached. At that time, the composition will remain the same, and the temperature will not change. The material distilling will have the composition indicated by point 18.

If the mixture initially is at point 22, then compound A would distill away, and the concentration of the material in the pot will move on the line from 22 to 20 to 18. Again, the end result is a liquid composition at point 18. Regardless of the starting composition, a mixture of constant composition is the end result.

Azeotropes are much more common than expected; in fact, there are entire books that list two- and three-component azeotropes. Some references are:

1. Horsley, L.K., *Anal. Chem.* **19**, 508-600 (1947). 2. Ibid. **21**, 831-865 (1949). 3. Advances in Chemistry Series, No. 6. *Azeotropic Data*, American Chemical Society, Washington D.C. 1952.

Example of an Azeotrope Distillation
Preparation of absolute ethanol is as follows:
 Water + EtOH forms an azeotrope (95% EtOH) b.p. = 78.15 °C.
 Add benzene to form another azeotrope (74% benzene, 18.5% alcohol, 7.5% water) b.p. = 65 °C.
 Distill at 65 °C. This removes the water.
 Now benzene and EtOH form an azeotrope (67.6% benzene, 32.4 % EtOH) at 68 °C.
 Distill at 68 °C until the temperature rises. Pure EtOH can be obtained at 78.5 °C.

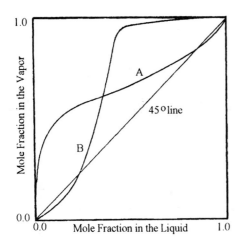

Figure 4-1. X-Y diagrams for the more volatile components of typical azeotropes. (A) Minimum b.p. (B) Maximum b.p.

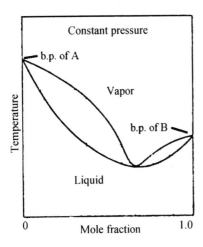

Figure 4-2. A minimum boiling point azeotrope.

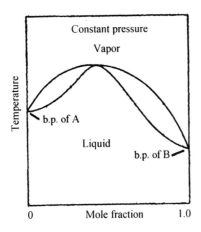

Figure 4-3. A maximum boiling point azeotrope.

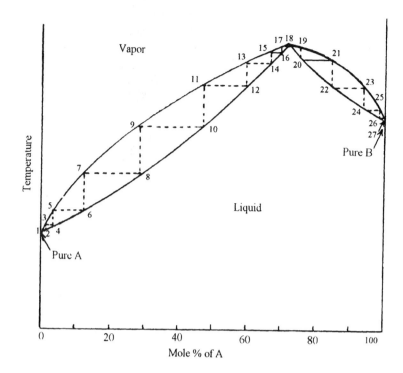

Figure 4-4. A general temperature-composition diagram for compounds A and B.

Tables 4-1, 4-2, and 4-3 (L.H. Horsley, *Anal. Chem.* **21**, 831, 1949) list some representative azeotropic mixtures. As you can see, an azeotrope is not a rare occurrence. As a chemist, *you cannot assume that you have a pure liquid just because it distills at a constant temperature.* Recall that an *azeotropic distillation* involves the formation of an azeotrope with at least one of the components of a liquid mixture which, thereby can be separated more readily because of the resulting increase in the difference between the volatilities of the components of the mixture. The third component added is often called an *entrainer* (see p. 47).

Table 4-1. Some binary azeotropic systems containing water

A component	B component	B.P.°C	B.P.°C	Weight % A
Water		100		
	Trichloroethylene	86.4	73.6	5.4
	2-Chloroethanol	128.7	97.8	57.7
	Iodoethane	70.	66.	3.5
	Methylallylamine	78.7	78.4	4.1
	Nitrobenzene	211.	98.6	88.
	Butyl vinyl ether	93.8	76.7	11.5
	Diisopropylamine	84.	74.1	9.2
	Triethylamine	89.4	75.	10.
	Styrene	145.	93.	
	m-Xylene	139.	92.	35.8
	Butyl ether	142.6	92.9	28.
	Ethyl hexyl ether	143.	92.9	29.

Table 4-2. Some binary system azeotropes (weight %; b.p. °C)

A Component	Weight %	B.P.	B Component	Weight %	B.P.	B.P. Azeotrope
Ethanol	53.	78.3	1,3-Butadiene	47.	-4.5	74.5
Nitromethane	56.5	101.1	Dioxane	43.5	101.3	100.5
Methanol	22.3	64.7	2-Methylfuran	77.7	63.1	51.5
Acetic acid	30.4	118.5	3-Picoline	69.6	144.	152.5
Acetamide	23.	221.2	Camphor	77.	209.1	199.8
Acetone	61.	56.3	2-Propanol	39.	69.0	54.2
2-Butanone	30.	79.0	Butyl nitrite	70.	78.2	76.7
Ethyl acetate	71.	77.2	Butyl nitrite	29.	78.2	76.2
Butyl amine	60.	77.8	Cyclohexane	40.	80.7	76.5
Pyridine	45.	115.5	3-Pentanol	55.	116.0	117.4
Cyclohexanone	65.	156.7	Cumene	35.	152.8	152.

Table 4-3. Some ternary azeotropic systems

A	B	C	Weight % A	Weight % B	Weight % C	B.P. °C
HBr	Water	Chlorobenzene	10.4	11.0	78.6	105.
HF	Water	Ethanol	30.	10.	60.	103.
Water	CCl$_4$	t-Butanol	3.1	85.	11.9	64.7
Water	CHCl$_3$	Acetone	40.	57.6	38.4	60.4
Water	Formic acid	m-Xylene	10.6	40.4	49.	97.5
Water	Trichloroethylene	Acetonitrile	6.4	73.1	20.5	67.
Water	Acetonitrile	Benzene	8.2	23.3	68.5	66.
Water	Ethanol	Ethyl chloroacetate	17.5	61.7	20.8	81.3
Water	Ethanol	Acetal	11.4	27.6	61.0	77.8
Water	Ethanol	Triethylamine	9.	13.	78.	74.7
Water	Propanol	Ethoxypropoxymethane	17.6	22.9	59.6	83.8
Water	Butanol	Butyl chloroacetate	41.8	50.3	7.9	93.1
Water	2-Butanol	Isobutyl chloroacetate	33.6	53.1	13.3	90.2
Water	Isoamyl alcohol	Isoamyl chloroacetate	46.2	47.3	6.5	95.4
CHCl$_3$	Methanol	Acetone	47.	23.	30.	57.5

TECHNIQUES

Ways to Break an Azeotrope

1. Form a new azeotrope. Experiment 4 is an example of this technique.
2. Add a salt.

    Anhydrous potassium carbonate was added to break the ethanol-water azeotrope over 150 years ago. This gave way to quicklime, which was cheaper. The difficulty with adding salts in commercial separations is that the salt must be accounted for where ever it is because of environmental considerations.
3. Changing pressure.

    Figure 4-5 shows the effect of changing the pressure on the acetone-water system. The ethanol-water system shifts away from the azeotrope if a vacuum is applied. This is often a reason why vacuum distillations (Chapter 6) are done.

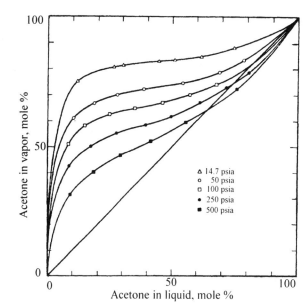

Figure 4-5. Vapor-liquid equilibria for the acetone-water system at several pressures.
(Courtesy - R.E. Kirk and D.F. Othmer, *Encyclopedia of Chemical Technology*, Vol. 3, John Wiley & Sons, NY 1980)

EXTRACTIVE DISTILLATION

PRINCIPLES

In extractive distillation, a third component is added to extract one of the other components. For example: when cyclohexane (b.p. = 80.8 °C) is formed by hydrogenating benzene (b.p. 80.1 °C), the desired product cannot be separated by an ordinary distillation. Aniline (b.p. = 184 °C) is added to form a complex with benzene, probably by a pi bond interaction. This complex boils at a much higher temperature than benzene, and the cyclohexane then can be separated by distillation.

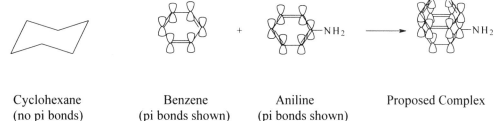

| Cyclohexane | Benzene | Aniline | Proposed Complex |
| (no pi bonds) | (pi bonds shown) | (pi bonds shown) | |

A third component, called an *entrainer* in azeotropic distillation and a *solvent* in extractive distillation, is added to increase the difference in volatility between the key components.

In an extractive distillation the *attraction* between the solvent and one or more of the components in the mixture is necessary whereas a *repulsion* between the entrainer and one or more components in the mixture is required for an azeotropic distillation. For example: advantage is taken of the highly repulsive forces that develop between benzene and water, when benzene is used as the entrainer to separate a mixture of ethanol-water. This repulsion results in an immiscible pair. However, when the same mixture is separated by an extractive distillation, ethylene glycol is used as the solvent. Water is attracted to the solvent and is removed with it from the bottom of the tower.

Hydrogen, dipole-dipole, ion-dipole, and pi bonds are common methods that chemists can select to create a successful extractive distillation. In general, extractive distillation offers a wider range of possibilities than does azeotropic distillation because (1) the phase relationships are not as critical so more solvents are available and (2) less heat input is required because the solvents are not taken overhead. Complete separations with extractive distillations are difficult to obtain and, as a result, extractive distillation is used when large quantities of reasonably pure materials are needed. Azeotropic distillations are used when higher purity is required.

Some commercial applications are:
1. The separation of acetone from methanol by adding water as the solvent to form a hydrate with methanol.
2. The removal of water from methyl ethyl ketone with pentane as the solvent.
3. The removal of isoprene from ethylene by using either dimethylformamide or acetonitrile.
4. The preparation of 98-99 % HNO_3 from 50-70 % HNO_3 by adding 93 % H_2SO_4 as a dehydrating solvent. If small amounts of high quality HNO_3 are needed and must be sulfate free, then an aqueous solution of $Mg(NO_3)_2$ can be used as the solvent.

REVIEW QUESTIONS

1. What is an azeotrope?
2. What is an azeotropic distillation?
3. What are the requirements for a successful azeotropic distillation?
4. What is the % water in a water-nitrobenzene azeotrope?
5. What is an *entrainer* when discussing azeotropic distillations?
6. How can azeotropes be broken?
7. Most wines are 12.5% alcohol by volume. What "proof" is this? See Experiment No. 4 in Appendix A.

Refer to the examples in Experiment 1 on refractive index to answer the next set of questions.

8. What is refractive index?
9. What does each of the parts of the symbol n_D^{20} mean? See Experiment No. 4 in Appendix A.
10. Acetone and diethyl ether should not be used to clean the prisms of a refractometer. Can you think of two reasons why this might be so?
11. Room temperatures are seldom at exactly 20 °C or 25 °C. How can you make a set of refractive index measurements, say at 23 °C, and be sure they are correct? (Hint: use water as a standard)
12. Pick any of the binary system azeotropes listed in Tables 4-1 and 4-2 and calculate an expected refractive index. Assume ideal solution behavior.
13. Pick any of the ternary system azeotropes listed in Table 4-3 and calculate the expected refractive index. Assume ideal solution behavior.
14. m-Xylene (f.w. = 106) has a refractive index of 1.4973 at 20 °C. It forms an azeotrope with 35.8 % water. What is the calculated refractive index of this system?
15. Allyl amine has a refractive index of 1.4190. A solution of this compound and water was distilled and when it was believed that pure allyl amine was being distilled, a few drops were collected and the refractive index was 1.4185 at 20 °C. How much water (mole %) is in the allyl amine?
16. What is the difference between an *entrainer* and a *solvent*?
17. What is the difference in operating principle between an azeotropic distillation and extractive distillation?
18. What are the advantages of an extractive distillation compared with an azeotropic distillation?
19. If an extraction solvent is added to the still pot at the beginning of the distillation, why is it necessary to add additional solvent into the top of the column?

UNDERGRADUATES - DO THOSE PORTIONS ASSIGNED BY YOUR INSTRUCTOR PLUS ANY ADDITIONAL PORTIONS YOU HAVE AN INTEREST IN.
GRADUATE STUDENTS - START WITH NEW MATERIAL AND CONTINUE ON.

IF IT EXISTS - CHEMISTRY IS INVOLVED
IF YOU CAN BUY IT - A CHEMIST WAS INVOLVED SOMEWHERE

5
STEAM AND IMMISCIBLE SOLVENTS DISTILLATION

STEAM DISTILLATION

PRINCIPLES

If the components of a solution decompose in some manner at its usual boiling point, then it must be distilled under less drastic conditions. One such method is a vacuum distillation discussed in Chapter 6. A second method that is much simpler, but is limited to those compounds that are immiscible with water, such as fats and oils, is a *steam distillation*.

When immiscible liquids are heated, each exerts its own vapor pressure irrespective of the other. When the sum of the vapor pressures of these liquids becomes equal to the atmospheric pressure, they *distill over together*. Water is a preferred as one of the liquids, because it has a high vapor pressure, a low molecular weight, and is quite inexpensive.

Figure 5-1, p. 50, shows the simple apparatus required. It consists of a steam generator, safety valve, distillation flask, condenser, and receiver.

Earlier you had some exposure to the concept of vapor pressure. When the surface molecules of a liquid escape, the liquid is said to *evaporate*. The gas molecules given off strike other molecules above the surface and push on them. The pressure created by this pushing is called *vapor pressure*. For example, at room temperature, water has a vapor pressure of about 24 mm Hg and you know that water will slowly evaporate. If the water is heated to 70 °C, the vapor pressure increases to 234 mm Hg and the water will evaporate much faster. At 100 °C the vapor pressure is 760 mm Hg, which is equal to the pressure of the air (at sea level) in the atmosphere pressing down on it. This state of equilibrium - the temperature at which the vapor pressure of the liquid equals the atmospheric pressure (at sea level), is said to be the *standard boiling point* of the liquid. At any other room pressure, it is called simply the *uncorrected boiling point*. Refer to Figure 5-2, p. 50, which shows how the vapor pressure varies with temperature for three compounds.

Suppose some carbon tetrachloride is mixed with the water, and the mixture is heated. We find that at about 68 °C, the CCl_4 has a vapor pressure of 550 torr and water has a vapor pressure of 210 torr. Together they have a vapor pressure of 760 torr, and the system will boil. The water will boil at 68 °C rather than 100 °C. It is the *total pressure* that is important.

The preceding is in reality a statement of Dalton's Law of partial pressures, which states that the total pressure is equal to the sum of the partial pressures of the individual vapors. Therefore, the molar concentration is proportional to the number of moles present or:

$$\frac{N_a}{N_b} = \frac{P_a}{P_b} \qquad (5\text{-}1)$$

50   Principles

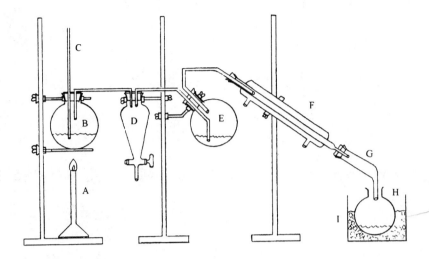

Figure 5-1. A steam distillation apparatus.

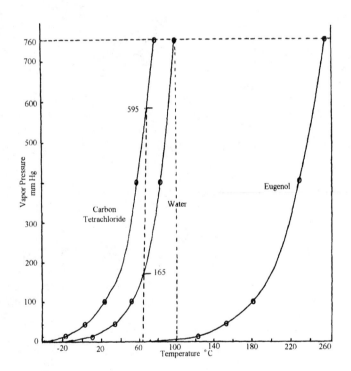

Figure 5-2. Vapor pressure-temperature diagram for water, carbon tetrachloride, and eugenol.

in which $P_a + P_b = P_{atmospheric}$

The weight of each material is obtained by multiplying by the corresponding molecular weights.

$$\frac{Weight\ of\ a}{Weight\ of\ b} = \frac{P_a \times M_a}{P_b \times M_b} \tag{5-2}$$

EXAMPLE CALCULATION

What is the weight ratio of water (f.w. = 18) to CCl_4 (f.w. = 154) that will be collected during a steam distillation? Refer to Figure 5-2, p. 50. Move along the X axis until the sum of the pressures = 760 torr.

ANSWER

$$\frac{\text{Weight of } H_2O}{\text{Weight of } CCl_4} = \frac{210 \times 18}{550 \times 154} = 0.0446 = \frac{1}{22.4}$$

The distillate will be about 95.5% CCl_4 by weight.

EXAMPLE CALCULATION

During a steam distillation of cloves, the eugenol (f.w. = 164) is collected. If 10.0 grams of water also is collected, how many grams of eugenol (oil of cloves) would you expect if the distillation was 100% efficient?

ANSWER

Refer to Figure 5-2, p 50, for vapor pressure data. Move along the X axis until the sum of the pressures = 760 torr.

$$\frac{10.0 \text{ g of } H_2O}{X \text{ g of Eugenol}} = \frac{755 \times 18}{5 \times 164} \qquad X = 0.60 \text{ g Eugenol}$$

In order to distill a compound below its normal boiling point, steam is generated and passed through the system. Condensed water plus low concentrations of materials in the sample that are not miscible with water will be collected.

Steam distillations are used commonly in many areas of industry and research. The fat in fish meal is determined by passing steam into a suspension of the fish meal. The fat is separated easily and floats to the top of the water in the receiver, where its amount can be determined. This is one of the ways that "oils" are prepared, such as oil of lemon, oil of cinnamon, oil of pine, or oil of peppermint.

TECHNIQUES

Emulsions

An *emulsion*, a dispersion of two immiscible liquids in each other, is a common occurrence in steam distillations and is recognized by the formation of a milky condensate in the receiver. An emulsion usually is formed when the densities of the two liquids are nearly the same. In Experiment No. 6 in Appendix A, the density of clove oil is about 1.04 g/mL and that of water is 0.99 g/mL. There are several techniques for breaking emulsions, and each is tried until one is found that works most easily with that system.

1. Place a piece of glass wool in the adapter or filter the condensate through a glass wool plug. The glass wool provides a surface on which the droplets can coalesce.

2. Centrifuge. This is useful if small amounts are being collected.

3. Add a salt such as sodium sulfate or sodium chloride to the condensate. This changes the density of the water and also "salts out" the organic compounds by lowering the activity of the water.

4. Make the system acid. In most cases, the very fine droplets, those approaching colloidal size, carry a negative charge. By adding acid, a positive counter ion is provided that neutralizes the surface charge and allows the drops to coalesce. Avoid making the system basic, because this provides negative surface charges and makes the emulsion harder to break.

5. Filter the condensate through a piece of *phase separation paper* such as Whatman PS-1. This paper is treated with silicones that then repel water but permit organic liquids to pass through.

6. Pray.

IMMISCIBLE SOLVENTS DISTILLATION

PRINCIPLES

In the previous section, we discussed how steam could be used to distill immiscible organic compounds. The process can be reversed, and hot organic compounds can be used to remove water from materials. This method can be used to determine the amount of water in an apple, an orange, or a piece of watermelon without ruining the fruit for other analyses. This technique also is used to determine the water in natural rubber that affects the curing or the water in coal products that affects the burning rate.

The technique involves placing the sample in a flask and then adding benzene, toluene, or some other low-boiling solvent that is *immiscible* with water. Heat is applied, and the solvent and the water in the sample both distill. The material is collected in a Dean-Stark, Bidwell-Stirling, or Barrett trap, or some modification of them. The water settles to the bottom, and the solvent drains back into the sample flask to be reused. The water can be drained off and its amount measured. The traces of solvent can be evaporated from the sample, and the dry sample is ready for further analyses. This is a separation method for large amounts of water and is particularly useful for food samples that would be quite large unless the water is removed.

Figure 5-3 shows several types of traps that can be used to collect and measure the water. The equation for calculating the results of the distillation is the same as equation 5-2, p. 50.

A    B    C    D

Figure 5-3. Types of traps used to collect water: (A) Dean-Stark. (B) Bidwell-Sterling. (C) Barrett. (D) Solvent heavier than water.
(Courtesy - Ace Glass Co., Vineland, NJ)

EXAMPLE CALCULATION
A section of peeled orange is the sample. What is the percent of water in this sample?

ANSWER

Step 1.
Weight of Al dish and orange section   18.063 g
Weight of Al dish   2.014 g
\-\-\-\-\-\-\-\-\-
Weight of orange section   16.049 g

Step 2.
Weight of weighing bottle, lid and water   34.469 g
Weight of weighing bottle and lid   20.637 g
\-\-\-\-\-\-\-\-\-
Weight of water   13.832 g

Step 3.
$$\% \ H_2O = \frac{13.832 \times 100}{16.049 \ g} = 86.2 \ \%$$

> Step 4. (optional)
>   Actual weight of the dried orange section 2.170 g
>   Expected weight: 16.049 - 13.832 = 2.217 g
> The expected weight is usually higher than the actual weight because other components also distill.

EXAMPLE CALCULATION

Using the equation 5-2, p. 50, calculate how many mL of toluene (m.w. = 92, density = 0.866 g/mL) probably would have been distilled in order to remove the water from your sample?

ANSWER

Step 1.   Atmospheric pressure in Manhattan, Kansas is about 740 torr. Plotting the data from step 30 of Experiment No. 7 in Appendix A, values of 325 torr for toluene and 415 torr for water produced a total of 740 torr at 83 °C.

Step 2.

$$\frac{Weight\ of\ toluene}{13.83\ g} = \frac{325 \times 92}{415 \times 18} \qquad Weight\ of\ toluene =\ 55.4\ g$$

Step 3.

$$\frac{55.4 \times 1.0\ mL}{0.866\ g} = 64\ mL$$

TECHNIQUES

A wide-mouth flask is used if foods are being examined, so that something like an orange section can be added without squeezing the water out of it when adding it to the flask.

Benzene (b.p. 80 °C) is a favorite solvent, because it boils at a lower temperature than water and there is less damage to the food sample. However, it is a suspected weak carcinogen. If this bothers you, then switch to toluene (b.p. = 110 °C). If the distillation is done in a hood there is no danger of exposure.

Erlenmeyer flasks with wide mouths are easier to handle and less likely to spill than round-bottom flasks with wide mouths. Because of the flat bottom on the Erlenmeyer, a burner is used to supply heat rather than a heating mantle. However, because of safety precautions with flames, a round bottom flask and a heating mantle are used here.

The Barrett trap is preferred because the water can be drained from it and weighed, giving a more accurate measurement of the amount of water. A typical apparatus is shown in Figure 5-4.

Table 5-1 lists the water content found in several foods. The original source, the USDA Handbook No. 8, *Composition of Foods*, lists 2,250 additional foods.

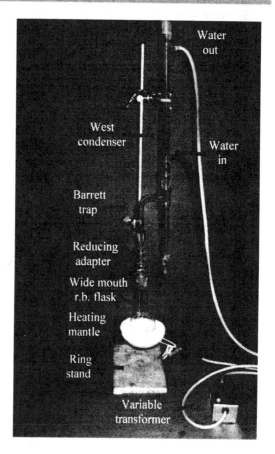

Figure 5-4. Apparatus for an immiscible solvent distillation.

Table 5-1. Water content of various foods

Food	% Water	Food	% Water
Abalone, raw	75.8	Crappie, raw	81.9
Almonds, dried	4.7	Cream puffs with custard	58.3
Apples, raw	84.4	Cucumbers, raw	95.1
Apricots, raw	85.3	Dandelion greens, raw	85.6
Artichokes, raw	85.5	Dates	22.5
Asparagus, raw	91.7	Doughnuts, cake	23.7
Avocados, raw	74.0	Duck, flesh, raw	68.8
Bacon, raw	19.3	Duck, wild, flesh	70.8
Bacon, Canadian, cold	61.7	Eel, raw	64.6
Bamboo shoots, raw	91.0	Eel, smoked	50.2
Bananas, raw	75.7	Eggs, chicken, raw	73.7
Barbecue sauce	80.9	Eggs, dried, whole	4.1
Bass, black sea, raw	79.3	Eggplant, raw	92.4
Bass, largemouth	77.3	Elderberries, raw	79.8
Beans, red, raw	10.4	Endive, raw	93.1
Beans, lima, raw	67.5	Fat, vegetable	0.0
Beans, green, raw	90.1	Fennel, raw	90.0
Beef, T-bone steak, raw	47.5	Figs, raw	77.5
Beef, choice sirloin raw	55.7	Figs, dried	23.0
Beef, 6-12th ribs, raw	47.2	Frog legs, raw	81.9
Beef, hamburger, raw	60.2	Garlic, raw	61.3
Beets, raw	87.3	Goose, raw flash	51.1
Blackberries, raw	84.5	Gooseberries, raw	88.9
Blueberries, raw	83.2	Grapefruit, raw	88.4
Boston brown bread	45.0	Grapefruit peel, candied	17.4
Brains, all types	78.9	Grapes, raw	81.6
Bread, white	35.8	Haddock, raw	80.5
Broccoli, raw	89.1	Hake, raw	81.8
Butter	15.5	Halibut, raw	76.5
Bullhead, raw	81.3	Herring, smoked	64.0
Cabbage, raw	92.4	Herring, kippered	61.0
Cake, angel food	34.0	Honey	17.2
Cake, icing, chocolate	14.3	Horseradish, raw	74.6
Candy, caramel	7.6	Ice cream	63.2
Candy, sweet chocolate	0.9	Ice milk	66.7
Candy, chocolate fudge	8.2	Jams and preserves	29.0
Carp, raw	77.8	Jellies	29.0
Carrots, raw	88.2	Kale, raw	82.7
Cauliflower, raw	91.0	Kohlrabi, raw	90.3
Caviar, sturgeon	46.0	Kumquats, raw	81.3
Celery, raw	94.1	Lake trout, raw	70.6
Cheese, cottage	78.3	Lamb leg, raw	60.8
Cheese, blue	40.0	Lard	0.0
Cheese, American	40.0	Lemons, raw	87.4
Cheese, Swiss	40.0	Lentils, raw	11.1
Cherries, raw	83.7	Lettuce, raw	95.1
Chewing gum	3.5	Limes, raw	89.3
Chicken, light meat, raw	73.7	Liver, beef, raw	69.7
Chicken, dark meat, raw	73.7	Lobster, raw	78.5
Chickpeas, (garbanzos)	10.7	Loganberries, raw	83.0
Chocolate syrup	31.6	Lychees, dried	22.3
Clams, soft raw	85.8	Macaroni, dry	10.4
Coconut meat, fresh	50.9	Mackerel, raw	67.2
Cod, raw	81.2	Mackerel, salted	43.0
Coffee, instant	2.6	Mackerel, smoked	59.4
Collards, raw	85.3	Malt, dry	5.2
Cookies, brownies	9.8	Mangos, raw	81.7
Cookies, oatmeal	2.3	Margarine	15.5
Crabapples, raw	81.1	Milk, whole, cow	87.4
Crackers, Graham	6.4	Milk, human	85.2
Crackers, saltine	4.3	Milk, reindeer	64.1
Cranberries, raw	87.9	Molasses	24.0

Table 5-1 Continued

Food	% Water	Food	% Water
Muffins, plain	38.0	Radishes, raw	94.5
Mushrooms, raw	90.4	Raisins, raw	18.0
Muskmelons, raw	91.2	Raspberries, black	80.8
Muskrat, cooked	67.3	Raspberries, red	84.2
Mussels, meat	78.6	Rhubarb, raw	94.8
Mustard greens, raw	89.5	Rice, brown or white	12.0
Nectarines, raw	81.8	Rutabagas, raw	87.0
Octopus, raw	82.2	Rye flour	11.0
Okra, raw	88.9	Salad dressings	
Olives, green	78.2	French	38.8
Onions, raw	89.1	Thousand Island	32.0
Onions, green	89.4	Mayonnaise	15.1
Opossum, cooked	57.3	Salmon, raw	63.6
Oranges, peeled, raw	86.0	smoked	58.9
Orange peel	72.5	Sardines, raw	70.7
Oysters, raw	84.6	Sausage, bologna	56.2
Pancakes, enriched flour	50.1	Frankfurters	55.6
Papaws, raw	76.6	Scallops, raw	79.8
Papayas, raw	88.7	Seaweeds, kelp	21.7
Parsley, raw	85.1	Shrimp, raw	78.2
Parsnips, raw	79.1	Snail, raw	79.2
Pate de foie gras	37.0	Spaghetti, cooked	63.6
Peaches, raw	89.1	Spinach, raw	90.7
Peanuts, raw	5.6	Squab, raw	58.0
Peanuts, roasted	1.8	Squash, raw	94.0
Peanut butter	1.8	Strawberries, raw	89.9
Pears, raw	83.2	Sweet potatoes, raw	70.6
Peas, raw	83.3	Tangerines, raw	88.9
Pecans	3.4	Tapioca, dry	12.6
Peppers, hot, chili	88.8	Taros, raw	73.0
Perch	75.7	Tea, dry powder	3.8
Persimmons, raw	78.6	Tomatoes, ripe, raw	93.5
Pheasant, raw	69.2	Tomato catsup	68.6
Pickles, dill	93.3	Tuna, raw	70.5
Pie, apple	47.6	Turkey, raw	64.2
Cherry	46.6	Turnips, raw	91.5
Chocolate	48.4	Turnip greens, raw	90.3
Pumpkin	59.2	Turtle, raw	78.5
Pig's feet, pickled	66.9	Waffles	41.4
Pineapple, raw	85.3	Walnuts, black	3.1
Pizza, frozen	48.9	Watermelon, raw	92.6
Plums, raw	81.1	Whale meat, raw	70.9
Popcorn, popped	4.0	Wheat, hard red winter	13.0
Pork, country style	42.0	Yams, raw	73.5
Potatoes, raw	79.8	Yoghurt	89.0
Potato chips	1.8	Zwieback	5.0
Pretzels	4.5		
Prunes, dried	28.0		
Pumpkin, raw	91.0		
Quail, raw	65.9		

REVIEW QUESTIONS

1. What are some limitations of a steam distillation?
2. Why is water preferred for a "steam" distillation?
3. What would you call the point where the vapor pressure of a solid equals the vapor pressure of its same liquid?
4. At about 99 °C, water (f.w. = 18) has a vapor pressure of 544 torr and diethylaniline (f.w. = 149) has a value of 15 torr. How many grams of diethylaniline would steam distill if 100 grams of water were collected?

56   Techniques

5. Go to Figure 5-2, p. 50, and plot the following data on it for nitrobenzene (f.w. = 123).

Pressure torr	Temperature °C	Pressure torr	Temperature °C
1	45	100	140
10	85	400	185
40	115	550	211

   a. At what temperature would you predict a mixture of nitrobenzene and water would steam distill?
   b. If 10.0 grams of nitrobenzene is desired, how many mL of water would also distill?

6. What might happen if the safety valve tube were omitted?
7. Why not fill the steam generator 3/4 or more full of water?
8. Suppose you connected your steam distillation apparatus to a water aspirator that reduced the pressure to 250 mm Hg:
   a. What would happen?
   b. What would be the ratio of eugenol (f.w. = 154) to water under these conditions?
9. Why is the sample flask tilted 30-40°?
10. What do you believe would happen if the burner is removed from the steam generator and the stopcock on the trap is left closed?
11. Sometimes solid materials are steam distilled. If large amounts are used, the solids may collect in the colder parts of the condenser. If the condenser gets plugged, a dangerous situation can exist. Can you suggest a simple way to unplug the condenser without taking the system apart or shutting it down?
12. For breaking an emulsion, acid usually does a much better job than a base. How can you account for this?
13. Phase separation paper is filter paper treated with a silicone. Water will not pass through, but organic liquids will. How do you explain this?
14. What happens when anhydrous Na_2SO_4 is used to remove water from an organic liquid?
15. What is the difference between a Barrett trap and a Dean-Stark trap?
16. How many grams of a potato sample will you need to obtain 10.0 grams of water? Refer to Table 5-1, p. 54.
17. Why do we weigh these water-containing samples only to ± 0.001 g when most of you will have a balance that will weigh to ± 0.0001 g?
18. Why is stopcock grease not used on the joints during this distillation?
19. Is there a situation in which the use of stopcock grease on the joints would be acceptable?
20. What is the difference between a West condenser and a Liebig condenser?
21. Emulsion formation is generally never a problem in immiscible solvents distillation. Can you explain why this is the case?
22. Determine the % water in the following chocolate iced cake doughnut:

Weight of doughnut and Al dish	66.217 g
Weight of Al dish	2.186
Weight of weighing bottle, lid, and water	34.755
Weight of weighing bottle and lid	20.660

23. Plot the following data for benzene along with the data for water found in step 30 of the directions in Experiment No. 7.

Pressure	Temp. °C
760 torr	80.1
400	60.6
100	26.1
40	7.6
10	-11.5 (s)

    Determine the partial pressures of water and benzene that will produce a boiling mixture and estimate the boiling temperature at 760 torr total pressure.

24. Use the data obtained in problem 23 and determine how much benzene (m.w. = 78, density = 0.879 g/mL) would be distilled if 14.095 g of water was distilled.
25. In the example calculations in step 4 of the example on p. 52, the expected weight of the dried orange section was 2.217 g, yet it actually weighed only 2.170 g. Was this an analytical error or could there be another reason for the difference?

6
VACUUM DISTILLATION

PRINCIPLES

There are many times when distilling compounds that it becomes impossible at atmospheric pressure because they must be heated to such a high temperature that they may polymerize, react with other compounds in the mixture, explode, decompose to an unidentifiable mass or contain an azeotrope. A way to reduce the temperature of a distillation is to reduce the atmospheric pressure of the system. Any distillation below atmospheric pressure is called a *vacuum distillation*. By reducing the pressure, you minimize all of the above difficulties and may well break up any azeotrope. In addition, although the boiling points may be too close together to get a good separation at atmospheric pressure, the difference in boiling points often increases at reduced pressures. See Figure 6-1, p. 58; compare horizontally at any pressure.

This chapter will describe briefly what a vacuum is, how it is produced, and how it can be controlled to produce a quality separation.

According to Aristotle some 2,000 years ago, "That in which the pressure of body, though not actual, is possible [is] void." A void was a place with nothing in it, but into which something could be put. The *Random House Dictionary* defines vacuum as "an enclosed space from which matter, especially air, has been partially removed so that the matter or gas remaining in the space exerts less pressure than the atmosphere."

PRESSURE NOMENCLATURE

The international unit of pressure, the *Pascal*, Pa, is seldom used in this country at this time, mainly because it has inconvenient numbers that are not related to the common measuring devices. A person can visualize 10 mm of mercury, but not 1330 Pascals. Normally, mm Hg is used with coarse gauges, such as the U-tube manometer. For pressures obtained with mechanical pumps, the term *torr* is common (1 torr = 1 mm Hg), and the term *micron* (1×10^{-6} atmosphere) is used with diffusion pumps. We must eventually switch to Pascals, so the pressure in Pascals or kilo Pascals (kPa) will be included, where convenient, in parentheses. A person's lungs produce a vacuum of about 300 torr, and the tentacles of an octopus can attain 100 torr.

An example of how the boiling point decreases with a decrease in pressure is shown in Table 6-1, p. 58, for *m*-nitrotoluene. By lowering the pressure to 1 torr (0.13 kPa), which you will find is very easy to do, the boiling point was lowered 162 °C. As a general rule, the boiling point will decrease about 15 °C each time the pressure is reduced by one-half.

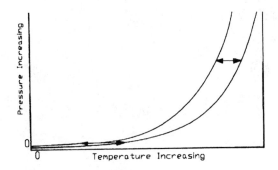

Figure 6-1. Changes in boiling point differences with reduced pressure.

Table 6-1. Boiling point changes with pressure for m-nitrotoluene

Pressure, torr	(kPa)	B.P. °C
760	101.3	231.9
100	13.3	156.9
40	5.3	130.7
20	2.66	112.6
10	1.33	96.0
5	0.67	81.0
1	0.133	50.2

Vacuum distillation can be thought of as fractional distillation under reduced pressure. Figure 6-2 shows a block diagram of the equipment needed for distillation down to about 0.1 torr (0.0133 kPa).

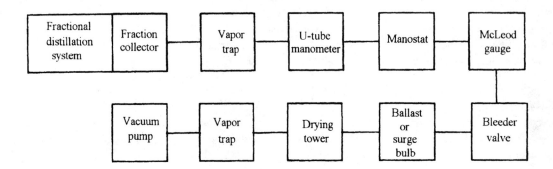

Figure 6-2. The components for a vacuum distillation. The surge bulb is optional unless working at low pressures.

Figure 6-3, p. 59, is called a *nomograph.* A nomograph is used for quick calculations that need not have great accuracy. The nomograph in Figure 6-3 is used to get an approximate idea of where a compound will boil at reduced pressure. All that is needed to use a nomograph is a ruler or the edge of a piece of paper.

EXAMPLE CALCULATION
   Column A of Figure 6-3 tells what the new boiling point will be. Column B lists the boiling temperatures of our material at some pressure (usually 760 torr, 101.3 kPa). Column C lists the various pressures that you might want to use. Dibenzyl ether, a solvent in perfumery, readily decomposes at its normal boiling point of 298 °C. What would be its boiling point at 20 torr (2.66 kPa), if we assume ideal behavior?

ANSWER
  Place a ruler so it lies across column B at 298 °C and column C at 20 torr (2.66 kPa). Read the expected boiling point in column A, which is about 168 °C. The sample actually boils at 173-174 °C at 21 torr (2.67 kPa).

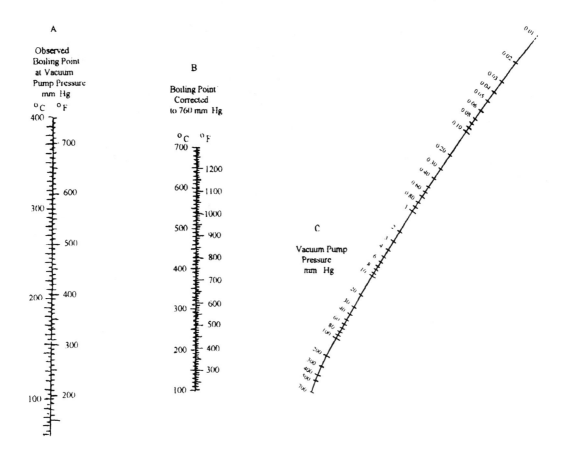

Figure 6-3. A pressure-temperature nomograph.
(Courtesy - Matheson Gas Products, Inc., Montgomeryville, PA)

EXAMPLE CALCULATION
Tiglic acid (trans-2-methyl-2-butanoic acid), formed during the charcoaling of maple syrup, is used to break emulsions. It normally boils at 200 °C with some decomposition. You decide to reduce its boiling point to 50 °C. About what pressure would you have to use to distill this material?

ANSWER
Set the ruler at 50 °C on column A and 200 °C on column B. Read the desired pressure on column C, which is about 2 torr (0.266 kPa).

TECHNIQUES

GENERAL RULES AROUND VACUUM LINES
1. Always wear safety glasses.
2. Wear a lab coat or a long-sleeved shirt or blouse
3. Never make sudden movements.
4. Never turn a stopcock until you determine what will happen when it is turned.
5. Always turn stopcocks slowly.

Pyrex glass tubing up to 25 mm in diameter can stand internally applied positive pressures of 3-5 atmospheres (atm), maybe up to 10. Therefore, a negative pressure of 1 atmosphere, which is the best possible vacuum, is generally quite safe. Big flasks usually are enclosed in a screen wire cage or wrapped with tape just in case of an accident.

VACUUM PUMPS

Two types of vacuum pumps commonly are found in a laboratory. These are the *mechanical pumps*, often called *forepumps*, and *diffusion pumps*. There are several more types of pumps, but they are used for specialized high vacuum work.

Mechanical Pumps (Forepumps)

The basic concept of a mechanical pump is to produce a chamber at a lower pressure than the vacuum line, let it fill with molecules from the line by molecular collision, then force the molecules out of the chamber into the atmosphere.

Two common types of mechanical pumps are the Duoseal and the Hyvac. These are shown in cross section in Figures 6-4 and 6-5. Each works by trapping some of the gas from the system to be evacuated, compressing it, and forcing out the compressed gas.

The seal between the vane, rotor, and the frame is maintained by a thin film of oil. This is shown in Figures 6-4 and 6-5 to the left. This oil must be kept clean, if the pump is to work at its best.

The oil should be changed every 3 to 4 months if the pump is used daily. Most pumps have a small window on one side to serve as an oil level gauge. Do not overfill the pump with oil. The extra oil is just thrown out. *If you shut off a vacuum pump, be sure to first open the system to atmospheric pressure or the pump oil may be drawn out of the pump into the system.*

There are three main purposes of the oil. Lubrication and sealing are obvious, but the oil also is necessary to help provide the very high compression ratios required at low inlet pressures. At low inlet pressures the exhaust valve may not open because there is dead volume between the valve and the rotating vane. However, if there is oil in the compression/exhaust chamber the dead volume is reduced, sometimes eliminated, just before the gas reaches exhaust pressure. Continued movement of the piston then forces the oil and the small volume of gas through the exhaust valve. Too little oil and the compression ratio is not obtained, too much oil and the oil takes up space that would otherwise be gas.

Diffusion Pumps

Both Hg and oil diffusion pumps will be discussed in Chapter 7, Molecular Distillation.

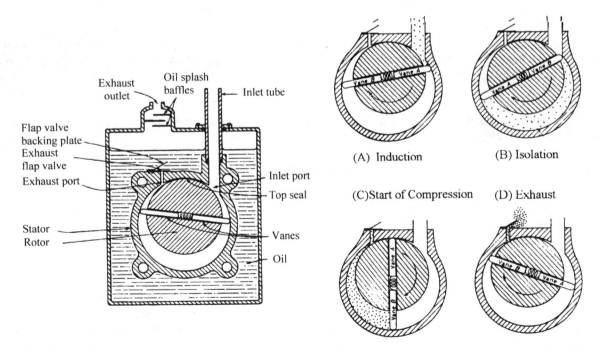

Figure 6-4. Cross section of a rotating vane vacuum pump and how it operates.
(Courtesy - L. Ward and J. Bunn, *Introduction to the Theory and Practice of High Vacuum Technology*, Butterworths, London, 1967)

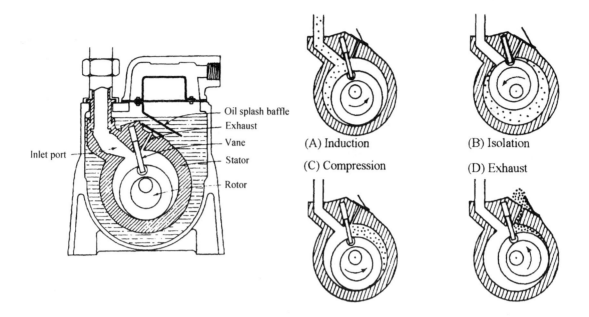

Figure 6-5. Cross section of a sliding vane vacuum pump and how it operates.
(Courtesy - L. Ward and J. Bunn, *Introduction to the Theory and Practice of High Vacuum Technology*. Butterworths, London, 1967)

TRAPS

A pump is only as good as the purity of the oil in it. A new mechanical pump should be able to reduce the pressure in a system to about 0.1 torr (0133 kPa). If solvents or other low-boiling compounds get into the pump oil during the pumping operation, then the pump may be able to pump down only to 20-25 torr (2.6 - 3.3 kPa). If corrosive chemicals get into the pump, then the metal parts can be damaged. In order to protect the pump and the pump oil, a trap is placed in the line. Figure 6-6, p. 62, shows the most common type.

Notice that these traps appear to be connected backwards compared to ordinary water-cooled traps. The reason for this is as follows. Material coming from the system will condense along the sides of the trap (B-E) and, if not solidified, will run down and collect at A. If the dry ice - isopropanol level drops from C to D, as it can during a long distillation, then the material at B can evaporate. If the trap is connected properly, this vapor must go farther down the trap before it can get to the pump. When it does this, it refreezes and is again trapped out of the system. If the trap is connected incorrectly, then the material at B can get into the pump because there is no place for refreezing. These traps must be checked after each distillation, and if there is an appreciable amount of material at A, then it must be removed.

CAUTION: Never turn on a vacuum pump just to see if it works, unless cooling material is present in the trap. This can cause immediate contamination of the oil.

DEWAR FLASKS

These are "Thermos" bottles made out of heavier glass than your coffee Thermos and are evacuated to about 10^{-7} atmospheres. Household type vacuum bottles should not be used around vacuum lines, because the temperature differences encountered are sufficient to cause most to crack in a short while. The common cooling material is dry ice, and the cooling solvents are isopropanol (-89.5 °C) and acetone (-94 °C), isopropanol being preferred because it is not as combustible and is less likely to dissolve the synthetic materials from which clothes are frequently made.

Liquid nitrogen (b.p. -209 °C) also is used, but liquid air should not be used. The nitrogen in liquid air boils away first, leaving the oxygen behind. If any organic vapors get into this liquid oxygen, an explosion can result.

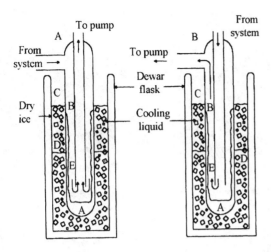

Figure 6-6. A vapor trap. (A) Connected correctly. (B) Connected incorrectly.

Polystyrene Containers as "Dewars"

Acids are shipped in containers containing six 5-pint bottles surrounded by molded polystyrene. This container can be sawed into six pieces to make six quite suitable "Dewar flasks". Acetone is not used as the liquid, because *acetone dissolves polystyrene*. Isopropanol and dry ice work quite well in these "Dewars".

Filling Technique

The proper way to fill a Dewar is to add the powdered dry ice first and then the solvent. The reverse order, which produces large gas bubbles in the liquid, often causes the liquid to overflow. A part of Experiment 8 in Appendix A will have you test the difference.

Handling Dry Ice

You can pick up small pieces of dry ice with your fingers, if the pickup and release is done quickly. The film of air between your fingers and the dry ice will protect you for a few seconds. If dry ice is held for more than a few seconds with bare fingers, then freezing of the fingers can occur. It is recommended that a towel or gloves be used when you handle dry ice. About 3/4 lb (340 g) is required for a one pint Dewar. The dry ice is wrapped in a towel and crushed to marble-size pieces with a hammer before it is placed in the Dewar.

DRYING TOWERS

A drying tower, if used, is placed between the bleeding valve and the pump to remove any water vapor from the air during the initial pressure adjustments when the pump is first turned on. Potassium hydroxide was the old-time favorite desiccant, but today, indicating Drierite ($CaSO_4$ with $CoCl_2 \cdot H_2O$) or Anhydrone [$Mg(ClO_4)_2 \cdot 2H_2O$] is used more often because these desiccants do not liquify when moist.

BALLAST OR SURGE TANK

This is sometimes omitted. Its purpose is to keep the pump side of the manostat below the system pressure even when the pump is disconnected for a short time, such as when you change fractions. The tank is a large bulb, usually 2-3 liters in size. It should be taped to keep glass fragments from flying around if it breaks under vacuum. It also can be enclosed in a screen wire cage as further protection. Figure 6-7 shows how this is done. It also can be coated with a "paint-on transparent plastic" which gives excellent protection. The photo at the extreme right shows how well the tape worked on a 4 L bulb whose contents exploded.

Remember - you are our nation's most valuable resource - we want to take good care of you.

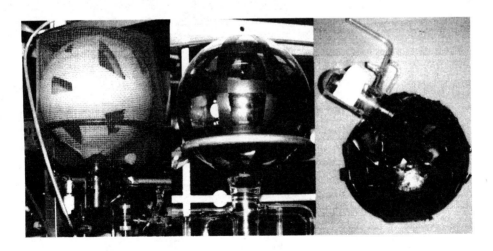

Figure 6-7. A screen wire cage, a taped bulb, and the effect of taping.

BLEEDING VALVES

A bleeding valve is used to release the vacuum in the system quickly and in a controlled manner (1) during the early stages of pumping so that the system doesn't pump out too fast, (2) when a new fraction is to be collected, and (3) when the distillation is over. If it pumps out too fast, the dissolved gases in the sample can cause severe bumping.

The fraction collector probably will have a bleeding valve on it. If it does not, then the bottom of a Bunsen burner or any natural gas burner can be inserted in the line.

PRESSURE MEASURING DEVICES

The three most common methods for measuring pressure in systems in the region of 0.1 torr (0.013 kPa) or above are the U-tube manometer, and the Zimmerli and McLeod gauges. Digital gauges are becoming available.

U-Tube Manometers

A *U-tube manometer*, shown in Figure 6-8, is quite easy to operate. The pressure is measured in the *open tube type* by measuring the differences in height between the two columns of Hg. In this case, the height difference *increases* as the pressure *decreases*. The *closed end type* requires that the Hg be boiled under reduced pressure until the dissolved gases are removed before you fill the tube. If this is not done, then an air bubble eventually forms at the top of the sealed end and ruins the measurements. In this type, the differences in mercury height *decreases* with a *decrease* in pressure. This type of manometer can be used to measure pressures from atmospheric down to about 1 to 2 torr (0.13 - 0.26 kPa), but the accuracy decreases below 5 to 10 torr (0.67-1.3 kPa).

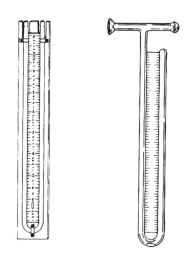

Figure 6-8. U-Tube Manometers. Open end (left) and closed end (right).
(Courtesy - Ace Glass Co., Vineland, NJ)

Zimmerli Gauge

Figure 6-9 shows a *Zimmerli gauge* (Zimmerli, A., *Anal. Chem.*, **10**, 283, 1938) which allows you to easily remove a trapped air bubble. This type of gauge is used for pressures from about 20 torr (2.6 kPa) down to about 1-2 torr (0.13 - 0.26 kPa) where dissolved gases are more likely to be a problem. Larger gauges require too much mercury to be convenient to operate. Figure 6-9A shows it filled with mercury, and Figure 6-9B shows how you read it. When used the first time, the gauge should be filled as in A, then the mercury "boiled" by heating it with a hair dryer or placing the gauge in boiling water. In all subsequent measurements, the gauge simply needs to be tilted clockwise to remove any trapped air bubbles through the capillary. Neither the U-tube manometer nor the Zimmerli gauge is truly satisfactory below 5 torr (0.67 kPa) because of the large error in reading the pressure.

Tilting Type McLeod Gauge

The original gauge was developed in 1884 (McLeod, H., *Phil. Mag.*, **48**, 110) and the tilting modification was developed in 1938 (Flosdorf, E.W., *Ind. Eng. Chem.*, **10**, 534). For pressures between 5 torr (0.67 kPa) and 0.01 torr (1.33 Pa), a *tilting McLeod gauge*, Figure 6-10, is used. The gauge is normally in position A. Notice that all portions above the mercury are at the same pressure as the system being measured (P_1V_1). When a pressure is to be measured, the gauge is turned clockwise until the mercury level in tube A is just to the top of tube B or a calibration mark near there. This seals a volume of gas in the bulb and capillary on the right (P_2V_2). The pressure is read on the scale at C, which has been calibrated based on $P_1V_1 = P_2V_2$. This is an *absolute* gauge, meaning that it can be calibrated by measuring the volume

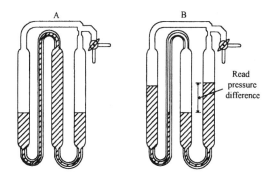

Figure 6-9. A Zimmerli gauge. (A) Hg position with no vacuum applied. (B) Position with a vacuum applied.

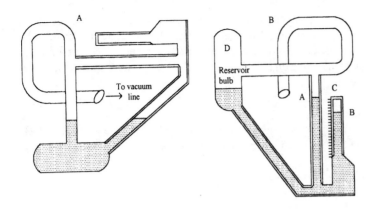

Figure 6-10. Tilting McLeod gauge. (A) Starting position. (B) Reading position. (Courtesy - Ace Glass Co., Vineland, NJ, Photograph)

of the bulb and needs no other gauge to calibrate it. These gauges can measure down to 10^{-4} torr and are used with diffusion pumps.

Most McLeod gauges are calibrated with a quadratic scale. Tilt the gauge until the level of the Hg (A in B above) so it is just to the top of arm C (in B above) and read the pressure at B. This is a quadratic calibration and follows equation 6-1.

$$P = \frac{\pi r^2 L^2}{V} \tag{6-1}$$

where: P = pressure in cm of Hg; r = radius of the capillary in cm; L = length of the air space above the Hg (V_2); and V = the volume of the bulb and capillary in cm³ (V_1).

EXAMPLE CALCULATION

A student made a McLeod gauge in glass blowing class. The volume was measured to be 6.6 cm³. The diameter of the capillary was 2 mm. What pressure would be read if the Hg was 9.2 cm from the top of the capillary?

ANSWER

Using equation 6-1

$$P = \frac{3.14 \times (0.1)^2 \times (9.2)^2}{6.6}$$

$P = 0.4$ cm $= 4$ mm Hg or 4 torr

DIFFERENTIAL PRESSURE REGULATORS - MANOSTATS

A *manostat* (one pressure) is a device for maintaining any pressure desired from 760 to 2 torr with an accuracy of 1 to 2 torr. Suppose you had prepared *m*-nitrotoluene and wanted to purify it. At 20 torr, the compound boils at 112.8 °C. How can you maintain a pressure of 20 torr in your system during the entire distillation so that you can be sure that you collect only the pure material? A manostat is a device that allows you to do this. The two most common types are the Cartesian diver and the Lewis-Nester.

Cartesian Diver

The Cartesian type (Figure 6-11) operates as follows. Mercury is placed in (A) until the diver (B) floats but does not quite touch the capillary tip (C). Stopcocks D and E are opened, and the system is pumped close to the pressure desired as determined by a manometer. Stopcock E then is closed, which slows down the pumping rate. When you get about 1 to 2 torr above the pressure you

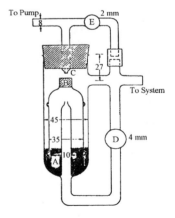

Figure 6-11. Cartesian diver manostat.

want, stopcock D is closed, trapping this pressure in the diver. If the pressure in the system is less than in the diver, the diver rises and closes off the capillary. If the pressure in the system is greater than in the diver, the diver sinks and the vacuum pump removes the excess gas. This type requires only 10-15 mL of Hg. It is critical that the top of the diver be flat and clean or the manostat will leak.

Lewis-Nester Manostat

Because the diver-capillary interface is so critical in the Cartesian diver manostats, a simpler method was developed by Lewis and improved by Nester (Figure 6-12). No rubber is used, and the large area of the glass frit permits a fine control (± 0.1 torr). In this case, the mercury at A acts as the diver and closes off the glass frit. The manostat is filled with clean Hg, so that the Hg comes as close to the frit as possible without touching it. All stopcocks are opened. Suppose it is desired to hold a pressure of 20 torr in the vacuum line. The vacuum pump is turned on, and the pressure adjusted to 20 torr with the bleeding valve. Stopcock D is closed, trapping a volume of gas in B at 20 torr. Stopcock E, for faster pumping, is closed also. If the pressure at C lowers to 16 torr, the Hg rises and closes off the frit. If the pressure at C raises to 22 torr, it pushes the Hg away from the frit and the gas is pumped out. This type of manostat is less troublesome than the diver, but it requires a large amount of mercury (about 1,800 g). Often a vapor trap is placed between the line and the manostat to keep the mercury clean. Another safety recommendation is to place a small tray under the manostat to catch the Hg, in case the manostat should break.

MERCURY

Mercury Poisoning

Mercury is a very useful chemical, and if handled properly, can be used routinely without fear. There is no need to shut down a laboratory because of a mercury spill. Clean it up immediately and continue. However, if spillage is not cleaned up, the vapors can cause chronic metal poisoning after prolonged exposure (Table 6-2). An extreme example is the old-time Hatter's disease. This occurred when hat makers treated beaver fur with $Hg(NO_3)_2$ to permit the fur to kink into felt. After continued exposure, the hatters often got the shakes. Anyone who shook was *mad as a hatter*. Mercurous compounds are much less soluble than mercuric compounds. In fact, a spoonful of calomel (Hg_2Cl_2) often was given

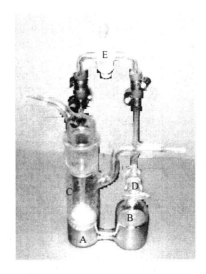

Figure 6-12. A Lewis-Nester type manostat.

Table 6-2. Symptoms of metallic Hg poisoning

Chronic	Acute
Inflammation of mouth and gums	Vomiting
Excessive salivation	Abdominal pain
Loosening of teeth	Bloody diarrhea
Jerky gait	Kidney damage
Personality changes	Death in 10 days
Depression	
Irritability	
Loss of memory	

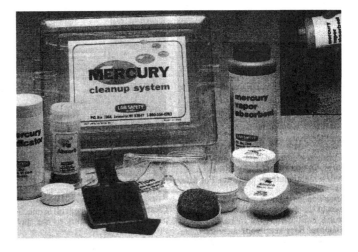

Figure 6-13. Mercury cleanup kit. (Courtesy - Lab Safety Supply, Janesville, WI)

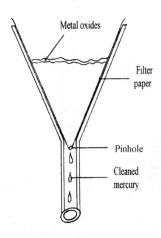

Figure 6-14. Pinholing mercury.

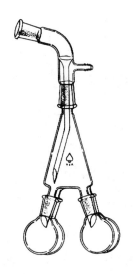

Figure 6-15. Distilling receiver, udder type.
(Courtesy - Ace Glass Co., Vineland NJ)

Figure 6-16. Distillation head, fraction cutter.
(Courtesy - Ace Glass Co., Vineland, NJ)

as a *dose of salts* to cure whatever ailed you.

Organic mercury compounds are much more toxic, because some have the ability to pass through the blood-brain barrier and react with brain cells. *These must be handled with care.*

The recommended limit of mercury vapor in air where people are continuously exposed is 0.1 mg/m^3. The equilibrium concentration in air from spilled mercury is about 200 times this amount, and reducing it to safe limits only by opening the doors and windows is difficult. It has to be cleaned up, but panic is unwarranted.

Cleaning Up Mercury Spills

Mercury scoops should be used to clean up larger amounts. If mercury spills into cracks in the floor, then a disposable pipet connected to a vacuum pump can be used. Nitric acid will dissolve it, and powdered sulfur will slowly react with it to reduce the vapor concentration in the room. Many commercial kits are now available to handle Hg spills. The one shown in Figure 6-13, p. 65, contains a scoop, a powder to amalgamate the Hg, four jars of absorbent for small drops, a vapor absorbent for areas not reachable, and an indicator to indicate places that may have been missed.

Cleaning Metallic Mercury

None of the pressure measuring or regulating devices described above will work correctly unless clean mercury is used. Metallic mercury has become very expensive, so it is best to recover and recycle as much as possible. Although only freshly distilled mercury is recommended, usually mercury recycled by pinholing is used. Before mercury is distilled it should have a preliminary cleanup to keep the still from getting dirty too quickly. Oils and liquids float to the top and can be absorbed with paper toweling. Many metals will dissolve in mercury and are slowly converted to oxides. These oxides are lighter than mercury and float to the top. This is the scum you see. It can be removed by what is known as "pinholing" (Figure 6-14). A small funnel is used, and a piece of filter paper is folded in the normal manner. A large pin, or end of a paper clip, is pushed through the bottom of the paper to form a small hole. The mercury is poured through this system, and the metal oxides stick to the paper. The paper also removes small quantities of oil and water.

DISTILLING RECEIVERS

During a distillation, it is desirable to collect liquids that boil at different temperature. These are called *fractions*. At atmospheric pressure, this is no problem, because the receiver is changed easily. However, when the system is under vacuum, the receiver cannot be removed without first shutting off the vacuum. This is very inconvenient, so multiple receivers, called *fraction collectors* are used. The one shown in Figure 6-15 is the *udder* type, often called a *cow*. It is quite useful, if only three or four fractions are needed. The collector is turned, and a new receiver is moved into position for each fraction without breaking the vacuum. If more than a few fractions are to be collected or the fractions are quite large, then the distilling head, fraction cutter shown in Figure 6-16 is used. It is designed so that the rest of the system can be kept under vacuum and the receiver raised to atmospheric pressure, removed, a new one put in place, and then re-evacuated. A bleeding stopcock is also included.

VACUUM LINE JOINTS

Ball and socket joints are preferred for vacuum systems because (1) they don't freeze as easily as standard taper joints, and (2) they can be twisted and turned so the apparatus is less likely to break if bumped into. A ball and socket joint, a joint clamp, and the joint greasing technique are shown in Figure 6-17.

Figure 6-17. Ball and socket joints and clamps. (Courtesy - Ace Glass Co., Vineland NJ)

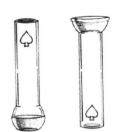
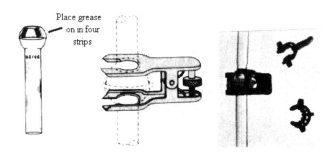

Ball and socket joints ground glass joints will leak unless sealed with a vacuum grease. Place four strips of grease on the ball lengthwise rather than around the ball. When the ball is fitted into the socket less air is likely to be trapped and the desired pressure can be attained quicker. A technique that is common for joints that cannot be greased is to place an O-ring between the ball and socket sections of the joint. The i.d. of the O-ring must be somewhat larger than the orifice diameter of the joint. A spring clamp is used to hold the sections together.

Figure 6-18. Tesla coil leak detector. (Courtesy - Fisher Scientific Co., Pittsburgh, PA)

CHECKING FOR LEAKS

Glass tubing often gets pinhole leaks in it, and these can be hard to find. About 10,000 volts is required for a spark to jump across 1 cm of air at atmospheric pressure. A Tesla coil (Figure 6-18) can produce 20 to 30,000 volts at a few microamperes and is used to test for leaks of this type. The system is pumped down, the Tesla coil is turned on by adjusting the knob at the back of it, and the spark-producing tip is passed slowly over the glass. If a leak is found, the diverging sparks will come together and point in the direction of the leak. A luminous area will form inside the glass tube. This is easier to see if the lights are out. The holes then are either sealed with Apiezon black wax or closed by glass blowing techniques.

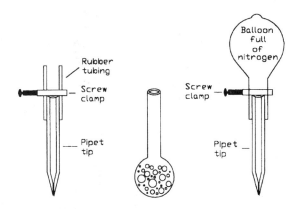

Figure 6-19. Air bubblers and providing an inert atmosphere.

AIR BUBBLERS

A means of providing stirring and also introducing an inert atmosphere into the system is a *bubbler,* two types of which are shown in Figure 6-19.

Disposable pipets also can be used for these purposes. Be sure to close the pinch clamp at the end of the distillation *before* the pump is shut off, or the material in the flask can come out of the bubbler onto the floor, heating mantle, and stirrer.

Figure 6-20, p. 68, shows a minimum component vacuum system for doing distillations. It is mounted on a cart on wheels so it can be moved out of the way when not in use or from room to room.

REVIEW QUESTIONS

1. N, N-dimethylaniline boils with decomposition at 192 °C. If you wanted to distill it at 100 °C, what pressure would you use?
2. When benzaldehyde is distilled, some of it reacts with the oxygen in the air and is converted to benzoic acid. Benzaldehyde boils at 179 °C. What pressure should it be distilled at in order to have it boil at 60 °C?

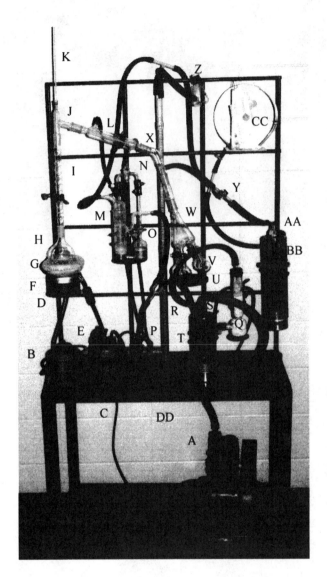

Figure 6-20. Minimum-component vacuum distillation apparatus. (A) pump, (B) and (C) Variacs, (D) support plate, (E) terminal strip, (F) stirring motor, (G) heating mantle, (H) r. b. flask, (I) Vigreux distilling column, (J) distilling head, (K) thermometer, (L) Liebig condenser, (M) manostat, (N) stopcock, (O) stopcock, (P) bleeding valve, (Q) drying tower, (R) U-tube manometer, (S) trap, (T) Dewar flask, (U) support plate, (V) r. b. flasks, (W) cow, (X) vacuum adapter, (Y) stopcock, (Z) Hg overflow, (AA) trap, (BB) Dewar flask, (CC) McLeod gauge, (DD) cork rings.

3. Styrene has a boiling point of 146 °C. If it is heated at this temperature for short periods, it polymerizes into polystyrene. The pump you have will produce a pressure of 9 torr (1.2 kPa). What boiling point would you expect for styrene at this pressure?
4. Strychnine was found to have a boiling point of 270 °C at 5 torr (0.67 kPa). What would you expect its boiling point to be at room pressure?
5. Table 6-1, p. 58, shows the m-nitrotoluene has boiling points of 231.9 °C at 760 torr and 130.7 °C at 40 torr (5.3 kPa). At what temperature does your nomograph say it should boil when it is distilled at 40 torr?
6. Why are vacuum distillations sometimes necessary?
7. What is the purpose of oil in a mechanical vacuum pump?
8. Why does a volatile organic solvent ruin pump oil if it gets into the oil?
9. What is a Dewar flask and how does it function?
10. Why should liquid air be avoided as a cooling agent?
11. Does it make any difference whether the solvent or the dry ice is added first to a Dewar flask? Please explain.
12. What is the chemical reaction that causes dry-indicating Drierite (blue) to change to pink when it gets wet?
13. Why tape a ballast bulb?

14. What is an implosion? How does it differ from an explosion? This is one you have to research.
15. Mercury boils at 357 °C at 760 torr (101.3 kPa). If you want to boil out any trapped gases and you wanted to use a steam bath at 100 °C, what pressure must the system be pumped to?
16. Sketch a Zimmerli gauge and explain how it works.
17. Sketch a Nester-type manostat and explain how it works.
18. Normally, a manostat will hold a pressure to ± 1-2 torr. How might you modify it to make it as sensitive as possible, say ± 0.5 torr?
19. What do we mean by "pinholing" mercury?
20. Why place four strips of grease lengthwise on a ball joint rather than one strip around it?
21. What is the technique for greasing a ball and socket joint?
22. Look up Nicola Tesla in an encyclopedia and learn something about who he was and what he did. I believe you will be pleasantly surprised.
23. The name "Apiezon" is a trade name for all types of vacuum waxes and greases. One of these waxes is known as black wax. It is applied by heating it with a burner and then applying it to the glass where the leak is. Find out something about what black wax is made of and what it will dissolve in so it can be removed.
24. Why not use boiling chips in a vacuum distillation?
25. Which is correct; desiccant or dessicant?
26. A McLeod gauge is to be designed so that its highest pressure to be read is 8 torr. If a 2 mm capillary is used and the length from the top of the bulb to the top of the capillary is 10 mm, what must the volume of the system be?

IF IT EXISTS - CHEMISTRY IS INVOLVED
IF YOU CAN BUY IT - A CHEMIST WAS INVOLVED SOMEWHERE

7
MOLECULAR DISTILLATION AND SUBLIMATION

MOLECULAR DISTILLATION

PRINCIPLES

Many liquids are too viscous to be distilled in an ordinary vacuum distillation apparatus, because they will condense and plug the column. They also may decompose thermally at the high temperatures required for their distillation. In cases like this, a *molecular distillation* is performed. A true molecular distillation employs a sufficiently high vacuum so that the *mean free path of the molecules to be distilled is greater than the distance from the sample surface to the condenser surface.* Ideally, this means that molecular distillation has no refluxing. In practice, this goal is seldom reached, but even then, molecular distillations can be useful. Because of the low pressures, boiling points are usually 200 - 300 °C lower than normal, so thermal decomposition is at a minimum. A molecular distillation is about as good as a batch distillation with a 50 °C difference in boiling points. Molecular weights are limited to between 1000 and 1500. J.N. Bronsted and G. Hevesy (*Philos Mag.* **43**, 31, 1922) are given credit for making the first molecular still. They used it to separate isotopes of Hg.

One type of laboratory-scale molecular still is called *a batch* or *pot* still. Batch stills have too low a capacity and are too slow for most industrial needs. A continuous operation is required. Industrial models are usually of the centrifugal or the cylindrical flowing film type depicted later.

The batch still shown in Figure 7-1 (Wiberg type) operates as follows. The sample is placed in A. The collecting tubes are placed in B and connected to C. Cold water is pumped in at D and flows out at E. A good vacuum is applied at F. A metal heating block or a heating mantle (with a hole punched in the bottom) is placed under the sample. The sample distills and condenses on the sides of the condenser (called a *cold finger*) at G, runs down the sides and collects at H, where it finally drips into I. Solids simply stay on the surface at G.

Equation 7-1 is Langmuir's method to determine the maximum amount of material that can distill (*Phys. Rev.*, **8**, 149, 1916):

$$N = \frac{P}{\sqrt{2 \pi R M T}} = moles/second/cm^2 \tag{7-1}$$

However, the pressure is given in dynes/cm^2 which is inconvenient. Equation 7-2 revises Langmuir's equation to have the pressure in torr and the amount distilled in grams:

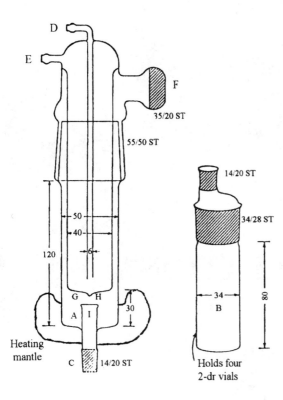

Figure 7-1. A combination molecular still - sublimator. (Courtesy - Wiberg, K., *Laboratory Techniques in Organic Chemistry*, McGraw-Hill, NY, 1960)

$$W = \frac{p \times 1.013 \times 10^6 \times \sqrt{M}}{760\ 2\ 8.314 \times 10^7 \times \sqrt{T}} = grams/second/cm^2 \qquad (7-2)$$

where W = amount distilled in grams/sec/cm^2; p = vapor pressure (torr) of the compound being distilled; M = m.w. of the compound being distilled (g); and T = temperature (°K) of the distillation.

The above equation expresses what should happen theoretically. The problem is that few molecules go directly from the evaporating surface to the condenser in a straight line and at a 90° angle. A correction factor, f, called the *evaporative coefficient*, developed by Burrows (*Molecular Distillation*, Oxford Clarendon Press, 1960) considers three factors: (1) the fraction of molecules reaching the condenser without collision, (2) the fraction that collides first and how many reach the condenser, and (3) those that have collided and reach the condenser by random motion.

By combining the constants in equation 7-2 and adding f, the following equation is obtained:

$$W = 0.0583\ p\ \sqrt{\frac{M}{T}}\ f\quad grams/second/cm^2 \qquad (7-3)$$

A nomograph to estimate the rate of molecular distillation (Gokarn, A.N. and Paul, R.N., NCL Communication 1686, 1973) is shown in Figure 7-2, p. 73. An example of how it is used follows. Notice the magnitude of the f values in particular and that the temperature is in °K.

EXAMPLE CALCULATION

What will be the molecular distillation rate of a substance at 600 °K, when M = 700, p = 10^{-3}, and f = 10^{-1}? (From Gokarn and Paul)

ANSWER

Connect 700 on the M-scale to 600 on the T-scale with a broken line that intersects the α-line at a point. Then connect 10^{-3} on the p-scale with 10^{-1} on the f-scale with a broken line that intersects the β-line at a point. Connect these points on the α and β lines with a broken line that intersects the W_e-scale at 6.3×10^{-6}. Thus, the required molecular distillation rate is 6.3×10^{-6} gm/sec cm².

Referring to equation 7-1, p. 71, you can see that the relative amount of materials distilled varies according to $P/\sqrt{M}$ and not directly with vapor pressure as in other distillations.

$$\frac{p_1}{\sqrt{M_1}} : \frac{p_2}{\sqrt{M_2}} : \frac{p_3}{\sqrt{M_3}} \tag{7-4}$$

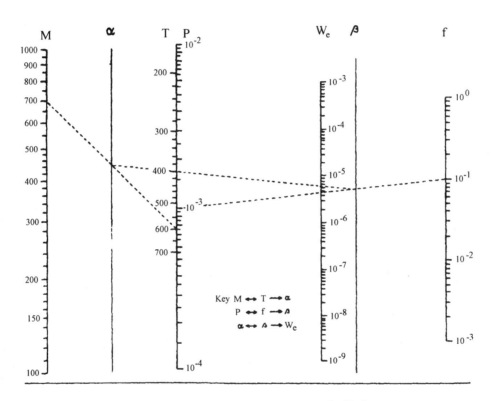

Figure 7-2. Nomograph to determine the rate of molecular distillation.
(Courtesy - A.N. Gokan, and R.N. Paul, National Chemical Laboratory, Poona, India, Communication No. 1685, 1973)

EXAMPLE CALCULATION

Consider an equimolar mixture of the following three compounds at 100 °C. In a regular distillation, they would distill in the ratio $P_1:P_2:P_3$ or 1:1:1. As a molecular distillation, how would they distill?

Compound	b.p. °C	Vapor pressure (torr) at 100 °C
2-Heptanone	114	100
2-Methyl propanylbromide	150	100
1,3-Diethoxytetramethyl siloxane	222	100

ANSWER

$$\frac{P_1}{\sqrt{M_1}} : \frac{P_2}{\sqrt{M_2}} : \frac{P_3}{\sqrt{M_3}} \quad or \quad 1.4 : 1.2 : 1.0$$

The basic differences between a molecular distillation and a regular vacuum distillation are (1) the liquid condensed in the column is not allowed to flow back into the sample, (2) the pressure in the system is reduced to such a low value that a molecule of the sample has a good chance of reaching the condenser surface without hitting another molecule first, and (3) the ratios at which different component molecules distill is proportional to the partial pressure divided by the square root of the molecular weight.

An entirely different equation to more closely represent what actually takes place was developed based on work by Luchak and Langstroth (*Canad. J. Res.* A., **28,** 574, 1950).

$$W = \frac{1.6 \times 10^{-5} \, p \, M \, D}{T \, d} \tag{7-5}$$

where D = diffusivity of the organic molecules in an air atmosphere. Values from 0.01 to 0.1 are typical. The lower the applied vacuum, the higher the D. d = distance in cm from the evaporation surface to the condensing surface.

EXAMPLE CALCULATION

Compare the results between the Langmuir and the Luchak-Langstroth equations. Lauric acid, $C_{12}H_{24}O_2$, has a m.w. of 200 and a m.p. of 48 °C. At 200 °C, it has a vapor pressure of 40 torr. Assume D = 0.05 and d = 3.5 cm. How much material would you expect to distill per cm^2 of surface?

ANSWER

Modified Langmuir equation, equation 7-3, no f term applied:

$$W = 0.0583 \times 40 \; \sqrt{200/473}$$
$$= 1.5 \text{ g/sec/cm}^2$$

Luchak-Langstroth equation:

$$W = \frac{1.60 \times 10^{-5} \times 40 \times 200 \times 0.05}{473 \times 3.5}$$

$$= 3.86 \times 10^{-6} \text{ g/sec/cm}^2, \text{ a considerably lower value}$$

The rate of distillation depends on how fast material from below the surface can get to the surface. Therefore, a device to keep a fresh surface exposed, a *flowing film still* is used in commercial apparatus. Flowing films can either fall by gravity or move sideways by centrifugal force, just as long as a thin film is produced.

A molecular distillation can occur at any temperature, in fact, many organic compounds are purified in the laboratory using no more than a water aspirator to reduce the pressure. However, if the molecular weight gets large and high efficiency is desired, then the pressure must be reduced so that the mean free path of the molecule to be distilled will be longer than the distance between the evaporator and the condenser. This prevents the distilling molecule from striking another molecule on the way to the condenser and, thus, not being distilled. The colder the temperature of the condenser surface, the less likely the molecule is to rebound back to the evaporator.

Mean Free Path

Air molecules have a mean free path of about 10^{-5} cm at room temperature and atmospheric pressure. Molecular stills have distances of 10 to 12 mm. Therefore, some changes in both temperature and pressure must be made to have a mean free path that long. The mean free path, L, is calculated by the following equation:

$$L = \frac{1}{\sqrt{2}\ \pi\ \sigma^2\ n} \tag{7-6}$$

where σ = Distance separating the centers of two molecules (cm). σ for several molecules is:

N_2	= 3.5 x 10^{-8} cm	NH_3	= 3.09 x 10^{-8}
CO_2	= 4.2 x 10^{-8}	CO	= 3.16 x 10^{-8}
N_2O_5	= 8.5 x 10^{-8}	C_2H_2	= 3.44 x 10^{-8}
H_2	= 2.1 x 10^{-8}	O_2	= 2.93 x 10^{-8}
Tocopherol	= 22 x 10^{-8}	He	= 2.38 x 10^{-8}

n = Number of molecules/mL. n is calculated by the following equation:

$$n = \frac{6.02\ x\ 10^{23}}{22,400\ x\ T_2/T_1\ x\ P_1/P_2} \tag{7-7}$$

EXAMPLE CALCULATION

Calculate the mean free path at the following conditions: 25 °C (298 °K) and 760 torr, n = 2.46 x 10^{19}; 25 °C (298 °K) and 150 torr, n = 4.86 x 10^{18}; 125 °C (398 °K) and 0.1 torr, n = 2.42 x 10^{15} and at 200 °C (473 °K) and 0.0001 torr, n = 2.04 x 10^{11}

ANSWER

At 25 °C (298 °K) and 760 torr, n = 2.46 x 10^{19}

$$L = \frac{1}{2\ (3.5\ x\ 10^{-8})^2\ x\ 2.46\ x\ 10^{19}} = 1.32\ x\ 10^{-5}\ cm$$

At 25 °C (298 °K) and 150 torr, n = 4.86 x 10^{18}
L = 6.7 x 10^{-5} cm
At 125 °C (398 °K) and 0.1 torr, n = 2.42 x 10^{15}
L = 0.004 cm
At 200 °C (473 °K) and 0.0001 torr, n = 2.04 x 10^{11}
L = 49.8 cm

COMMERCIAL STILLS

The main thrust of a commercial still is to provide a large evaporating surface and a continuous process. Batch stills first were rotated to increase the surface area, but several developments since then have occurred. The falling film method, the centrifugal method, the wiped film method, and the fractionating method are all now common.

Falling Film Stills

The distilland (sample) is added to the top and allowed to flow by the force of gravity down the surface in a thin film. The still usually consists of two vertical concentric cylinders, one being the evaporator and the other the condenser. Most of the sample is stored at low temperature and only the portion that is in immediate contact with the evaporator is heated and then only for a few seconds (10-50). Most falling films are from 0.1 to 2.0 mm thick. The efficiency is better than that of all but the smallest batch apparatus with f values approaching 1 and having one theoretical plate between the evaporator and the condenser. A value of 5-6 g/sec/m^2 is reasonable with small units. From 5-10 % of the sample is distilled in one pass. Such a still is shown in Figure 7-3.

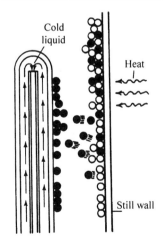

Figure 7-3. Falling film stills. (Courtesy - Pope Scientific Co., Menomonee Falls, WI)

Figure 7-4. A multistage wiped-film processing plant. (Courtesy - Pope Scientific Inc., Saukville, WI)

An improvement in increasing the surface area is the *wiped-film still*. Figure 7-4 shows the Pope Scientific Inc. multi-stage, turnkey wiped film processing plant. It includes a degasser/devolatilizer (1st stage), 12" molecular still (2nd stage), liquid and vacuum pumps, multiple cold traps, heat recovery exchangers and a complete control system. This system is capable of continuous operation to 1 millitorr, 375 °C and up to 200 kg/h feed rate. The basic process is shown as a schematic in Figure 7-5. Their system employs a wiper blade that moves each plug of material downward as shown in Figure 7-6. Thin films are preferred for a variety of reasons. Some listed by the manufacturer are

1. Turbulence created by a rapidly moving wiper or controlled clearance blade assists in heat transmission, thus lowering the temperature needed on the inside evaporator wall for a given system pressure.

2. A maximum surface area per unit volume of flow is generated facilitating rapid evaporation.

3. The liquid exposure time to the elevated wall temperature can be controlled within seconds or less. This minimizes product degradation of heat sensitive materials by controlling the wiper assembly speed.

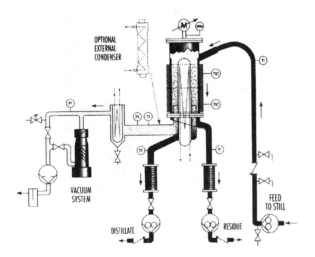

Figure 7-5. Schematic of the process of a wiped-film still. (Courtesy - Pope Scientific Inc., Saukville, WI)

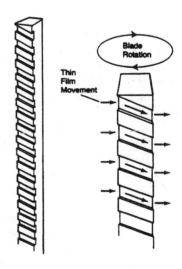

Figure 7-6. Fluid flow through a wiper blade. (Courtesy - Pope Scientific Inc., Saukville, WI)

4. High viscosity materials can be transported through the system for distillation or solvent stripping.
5. Pope slotted wiper blades promote plug flow with little back mixing. This minimizes dwell time distribution, ensuring that material flowing through the system has a uniform exposure to process conditions.

Figure 7-7 is a photograph of a 2" laboratory size wiped film still and is shown as a diagram in Figure 7-8.

Centrifugal Stills

The centrifugal still is the most refined (and expensive) of the commercial stills. A cross-sectional diagram is shown in Figure 7-9, p. 78, and a bank of them is shown in Figure 7-10, p. 78. The following excerpt from Barrows explains the operation of the centrifugal molecular still. " The degassed liquid to be distilled is introduced continuously at the bottom of the inner surface of the rotor, a rotating conical shaped surface. The rotor may be as large as 1.5 m in diameter at the top and may revolve at speeds of 400 to 500 r/min. A thin layer of liquid to be distilled, 0.05 to 0.1 mm thick, then spreads over the inner surface and travels rapidly to the upper periphery under the action of centrifugal force. Heat is supplied to the liquid through the rotor by radiant electrical heaters, and the vaporized material is condensed

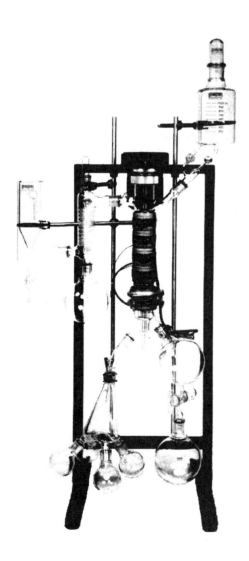

Figure 7-7. A 2 inch wiped film still.
(Courtesy - Pope Scientific Inc., Saukville, WI)

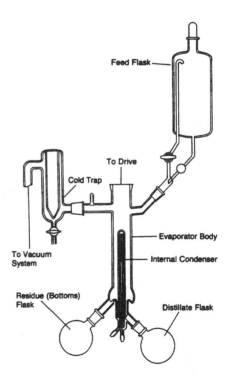

Figure 7-8. Diagram of the commercial molecular still shown in Figure 7-7.
(Courtesy - Pope Scientific Inc., Saukville, WI)

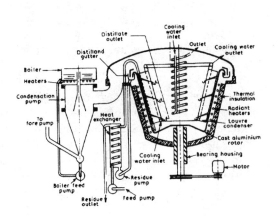

Figure 7-9. Schematic section of a Hickmam style centrifugal molecular still.
(Courtesy - *Molecular Distillation*, G. Burrows, Oxford Clarendon Press, 1960)

Figure 7-10. Commercial 36 inch molecular distillation units.
(Courtesy - *Molecular Stills*, P.R. Watt, Reinhold Publishing Co., New York, NY, 1963).

upon the water-cooled louver-shaped condenser. This is maintained at temperatures sufficiently low to prevent re-evaporation or reflection of the vaporized molecules. The residue liquid is caught in the collection gutter of the top of the rotor, and the distillate is drained from the collection troughs on the condenser. Each product is pumped from the still body, which is evacuated to the low pressures necessary for molecular distillation, and the time of residence of the substances in the still may be as low as 1 second or less. Such a device is capable of handling 5×10^{-5} to 25×10^{-5} m^3/s (50 to 250 gal/h) of liquid to be distilled and gives a separation of 80 to 95 percent."

Some Commercial Applications of Molecular Distillations

Removal of odors and colors from plasticizers
Vitamin recovery
Isolation of natural oils
Purification of drugs
Distillation of waxes and fatty acids
Isolation of perfumes
Deodorizing oils
Solvent stripping
Remove color bodies from materials of high molecular weight
Stripping monomer from polymer
Stripping free fatty acids from fats and oils
Distilling fats and oils
Concentrating heat-sensitive pharmaceuticals from their substrates
Concentrating fruit juices
Isolation of aromatic compounds

TECHNIQUES

Diffusion Pumps

The vacuum pumps discussed in Chapter 6 are capable of reducing the pressure to about 0.1 torr when they are new, are well trapped, and have clean oil in them. This is not a sufficiently low pressure to permit a high efficiency molecular distillation. A previous example calculation indicated that 10^{-4} torr or less is necessary. The most common method for reducing pressures from 10^{-1} to 10^{-4} torr (13.3 - 0.013 Pa) is the *diffusion pump*. Multiple stage diffusion pumps are capable of reducing pressures to 10^{-8} torr. A diagram of a down jet diffusion pump is shown in Figure 7-11, p. 79.

Mercury (10^{-6} torr) or a low vapor pressure oil (Myvoil, 10^{-7} torr; Octoil, 10^{-8} torr) is used as the liquid. This liquid is heated, boils up the left side (Figure 7-11) and comes down the right side. These heavy gas molecules produce a jet effect as they pass through the opening (A). Gas molecules from the line to be evacuated (B) get trapped in this jet stream and are carried into region C. There can be about a 1000-fold difference in pressure between points B and C before the pumping action stops. A mechanical pump, like a Hyvac or a Duoseal, is connected at point D to remove the evacuated gas so the pressure does not build up. The mechanical pump used this way is called a *fore pump*. The mercury or oil is condensed by a water condenser at E to be recycled. The vapor pressure of the diffusion pump liquid sets the lower pressure limit that the pump can reach.

Figure 7-12 shows a three stage oil diffusion pump which can reach a pressure of 7×10^{-8} torr without a cold trap. It costs about $ 2,250. Figure 7-13, p. 80, is a photo of a combination fore pump and diffusion pump made of metal for rugged use.

Devices for Measuring Low Pressure

When pressures get below about 10^{-4} torr (0.013 Pa), McLeod gauges are not practical. There are several types of gauges that can be used. Two of the most common gauges that cover the range from 10^{-1} to 10^{-6} torr are (1) the Pirani gauge, and (2) a thermocouple gauge. These are both reliable and can be obtained for about $500.00. Figure 7-14, p. 80, is a diagram of the detector portion of a Pirani gauge, Figure 7-15, p. 80, is a diagram of a thermocouple gauge, and Figure 7-16, p. 80, is a photo of a commercial thermocouple gauge.

A current is passed through a small filament of wire to heat it. A voltage is obtained based upon Ohm's law (E = IR). If the surrounding pressure is decreased, there are fewer gas molecules striking the wire and it is cooled less. The wire heats, the resistance increases, and the voltage increases. This voltage increase is related to pressure by calibration with a McLeod gauge. The description of the operating principle of the thermopile gauge (Figure 7-15) as provided by Matheson Gas Co. is: "Operation is based on a low voltage ac bridge noble metal thermopile circuit. A change in pressure in the tube creates a change in the thermal conductivity of the gas cooling the thermopile, which in turn causes a temperature shift of ac thermocouples A and B. This shift results in deviating the dc output from the two thermocouples.

"The dc thermocouple, C, is unheated and in series with the indicating meter circuit. Ambient temperature variations will develop voltages in all thermocouples; however, the transient effects in heated and unheated elements are equal and opposite; therefore the unheated couple compensates for transient temperature changes."

Figure 7-16 shows a commercial gauge connected to a vacuum line.

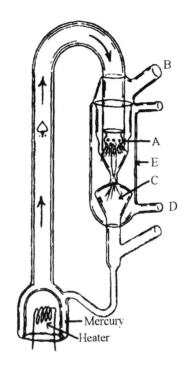

Figure 7-11. A down jet Hg diffusion pump.
(Courtesy - Ace Glass Co., Vineland, NJ)

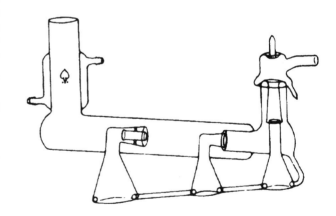

Figure 7-12. A three stage oil diffusion pump.
(Courtesy - Ace Glass Co., Vineland, NJ)

Figure 7-13. A fore pump - diffusion pump combination
(Courtesy - Van Waters & Rogers Scientific Co., McGaw Park, IL)

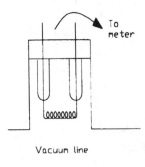

Figure 7-14. Diagram of a Pirani gauge.

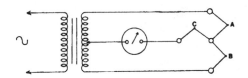

Figure 7-15. A thermopile circuit.
(Courtesy - Matheson Gas Co., Montgomeryville, PA)

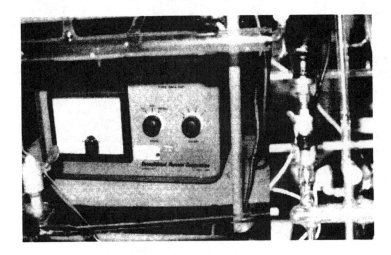

Figure 7-16. A commercial gauge connected to a vacuum line.

SUBLIMATION AND ENTRAINER SUBLIMATION

PRINCIPLES OF SUBLIMATION

Sublimation is a process in which a solid becomes a gas without first becoming a liquid. A solid will sublime if its vapor pressure reaches atmospheric pressure below its melting point. Figure 7-17, a temperature-pressure diagram of a general substance, will be used to illustrate how this can happen.

Normally, most compounds behave as shown starting at point A. As the temperature increases, the compound melts to a liquid and then vaporizes to a gas. However, if the pressure is lowered below the triple point pressure, to point B, the compound can go directly to a gas when the temperature is raised. If the pressure is lowered from A or B to point C, the compound will vaporize without a change in the temperature. Theoretically, all compounds can be sublimed. However, for most compounds, the triple point is at such a low pressure and temperature that sublimation is not practical. Fortunately, there are still several hundred compounds with characteristics such that they can go easily from the solid to the gas form. If the gas is then cooled, it resolidifies and can be collected. This material is usually of very high purity, because few compounds sublime under the same conditions. The material to be sublimed is called the *sublimand,* and the product is called the *sublimate.*

Iodine, dry ice, camphor, and arsenic readily sublime at room temperatures and pressures. Sulfur, benzoin, and NH_4NO_3 are obtained readily in pure form by sublimation. Saccharin, quinine, cholesterol, and atropine are additional examples of compounds that are conveniently separated by sublimation.

The subliming temperature at reduced pressure is usually several degrees below the melting point of the compound, so less compound destruction occurs during a sublimation than during a distillation.

Examples:
Naphthalene   m.p. 79 °C   Sublimes at 25 °C at 1 torr (0.13 kPa).
Urea   m.p. 132 °C   Sublimes at 50 °C at 1 torr.

The apparatus for a sublimation can be the same as that shown in Figure 7-1, p. 72. All of the material collects at surface G, and the collection chamber for liquids (B) simply is not used.

Entrainer Sublimation

A simple sublimation can be speeded up in many cases by passing an inert gas called a *carrier* or an *entrainer* over the sublimate. The purpose of this gas is to reduce the partial pressure of the sublimate below the triple point pressure. This is similar to a breeze blowing over you in the summertime when you are sweating. The breeze feels cool because it reduces the partial pressure of the water droplets on your body, and they evaporate faster. Compared to a normal sublimation, therefore, a compound will sublime faster at the same temperature or at the same rate at a lower temperature.

Figure 7-18, p. 82, is a diagram of a simple entrainer sublimator. If the tube that collects the sublimate is cooled to different temperatures at successive regions along the length, then it is possible to fractionally sublime a mixture of compounds. The separation in this case is usually not clean, but it can be used in some situations. Commercially, coffee is decaffeinated by entrainer sublimation.

The basic differences between a molecular distillation and a sublimation at low pressures are that (1) the molecular distillation goes from solid to liquid to gas, which usually requires more energy than a sublimation that goes from the solid directly to the gas, and (2) the surface of the subliming compound is continuously renewed by evaporation, convection, or diffusion. As a result, a sublimation in many cases is faster and more economical.

Figure 7-19, p. 82, shows the components required for a regular sublimation and Figure 7-20, p. 82, shows the components needed for the simplest type of entrainer sublimations.

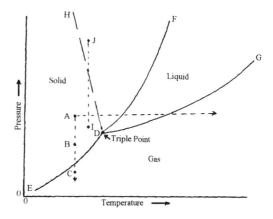

Figure 7-17. A temperature-pressure diagram of a compound

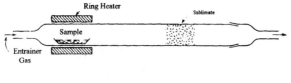

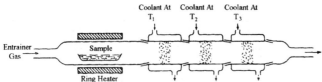

Figure 7-18. Entrainer sublimators: Top, air cooled. Bottom, liquid cooled multiple condensers.
(Courtesy - E.W. Berg, *Physical and Chemical Methods of Separation*, McGraw-Hill, New York, N.Y. 1963)

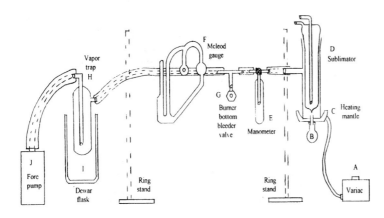

Figure 7-19. Apparatus arrangement for a regular sublimation.

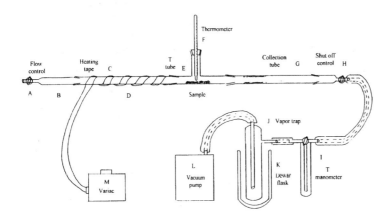

Figure 7-20. Diagram of an entrainer sublimation apparatus.

REVIEW QUESTIONS

1. How does a molecular distillation differ from what would be considered a regular vacuum distillation?
2. Why do you not want a gas bubble to form under the still (C of Figure 7-13, p. 80)?
3. What are three factors considered when evaluating the evaporative coefficient?
4. Use the nomograph (Figure 7-2, p. 73) and determine the distillation rate of a substance of m.w. 350 if the temperature is 500 °K, the pressure is 5×10^{-4}, and f is estimated at 0.05.

5. The following three compounds have boiling points of 100 °C at 1 torr. In what ratio would they distill in a molecular still?

    | 2-nitrophenyl acetate | $C_8H_7NO_4$ |
    | 2,3,4,6-tetrachlorophenol | $C_6H_2Cl_4O$ |
    | Champacol | $C_{15}H_{26}O$ |

6. The following three compounds have boiling points of 130 ° at 40 torr.
    a. Triethyl hexylsilane $C_{12}H_{22}Si$ m.w. = 200
    b. Pivalophenone $C_{11}H_{14}O$ m.w. = 162
    c. Unknown
    They distilled in the ratio 1.00 : 1.15 : 1.28, respectively. Estimate the m.w. of the unknown compound.
7. How many molecules are there in 1.0 cm³ of air at 220 °C and 10^{-3} torr?
8. Calculate the mean free path for alpha tocopherol at 220 °C and 10^{-3} torr.
9. What is the main thrust of a commercial molecular still?
10. What is an average film thickness in an unwiped falling film still?
11. What is the purpose of brushes or scrapers in a wiped film still?
12. What is the purpose of spinning a centrifugal still?
13. About how fast does a centrifugal still turn and what is a typical capacity?
14. Draw a single-stage diffusion pump and explain how it operates.
15. The Pirani gauge was discussed as a device for measuring low pressures. Discuss another type of gauge that can be used.
16. What is sublimation and how does it differ from a molecular distillation?
17. Refer to Figure 7-17, p. 81. Change the line D-F for water to line D-H. Suppose you have an ice surface at point I. If the pressure is increased at constant T to point J, what will happen? What happens under the runners of a pair of ice skates?
18. You have just received a new T-type manometer. How do you fill it with mercury?
19. Is there anything you need to do to adjust a T-type manometer when making a reading?
20. What is a heating tape?
21. Would the entrainer apparatus work for tea leaves?
22. What it is the difference between sublimate and sublimand?
23. What it is the purpose of using an entrainer gas during a sublimation?

IF IT EXISTS - CHEMISTRY IS INVOLVED
IF YOU CAN BUY IT - A CHEMIST WAS INVOLVED SOMEWHERE

8
LYOPHILIZATION (FREEZE DRYING)

PRINCIPLES

Lyophilization is a process of drying materials by subliming the water from a frozen sample. This works well for materials that are heat sensitive and would be wholly or partially destroyed if dried at atmospheric drying conditions. The basic process is to first freeze the sample, then apply a vacuum so that the ice will sublime without melting - hence, the more common name - *freeze drying*. Figure 8-1, p. 86, is a diagram of the phases of water. From this it can be seen that water will sublime at any combination of temperatures below 0°C and pressures below 4.57 torr. Commercially, freeze drying is considered to be any process of this type operating below 100 Pa (0.75 torr).

A photograph of a commercial laboratory apparatus is shown in Figure 8-2, p. 86. (A) is the sample holder. (B) is a special holder so that a sample can be added or removed without breaking the vacuum on the other sample chambers. (C) is a large trap to collect the sublimed water, (D) is a McLeod gauge to measure the pressure, and (E) is the temperature gauge. The pump (F) may be a simple mechanical pump, but is more likely a mechanical-diffusion pump combination.

HISTORY

It is known that the Incas prepared their food for long-term storage by drying it on mountain tops. This provided a cold temperature, low pressure, and radiant heat. They were "freeze drying", although they probably didn't call it that. Many of you have freeze dried foods unintentionally without knowing it. *Freezer burn* is caused by a combination of the slow sublimation of water from foods stored for long times in freezers or the freezing compartment of a refrigerator and the destruction of the cells by the large ice crystals formed by slow freezing.

R. Altman was the first to report microscopic investigations of freeze-dried tissue in 1890. Arrhenius, and later Becquerel, proposed that life originated by freeze-dried organisms that arrived on this planet from outer space and then were reconstituted by water. Becquerel calculated that if a dehydrated organism could live for one year at 10 °C, then it could live for 70 billion years at -270 °C, the temperature of outer space.

The process of freeze drying was developed on a large scale during WWII (1942) for processing blood plasma. This made it easier to ship overseas because the water was removed, and the material would last much longer without spoiling. Freeze drying was proposed for foods in 1949 but really didn't get started until 1964. It has evolved into a process having important commercial applications in the preparation of vaccines; antibiotics; pharmaceuticals; virus cultures; and foodstuffs such as meat, fish, milk, fruits and vegetables, and coffee.

A freeze-dried material undergoes fewer adverse changes than those preserved by pickling, salting, canning, or normal heat dehydration. The product requires simpler storage and transportation systems, and if sealed under vacuum or an inert gas, usually retains most of its biological and physical characteristics indefinitely.

PROCESS

The material to be freeze dried is first frozen to a temperature just below its lowest eutectic. A *eutectic* is the lowest temperature at which a mixture of two or more components will melt. This eutectic temperature can be determined in an unknown system by measuring the electrical resistance between two probes in the sample or by

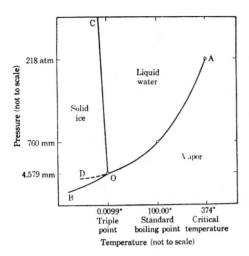

Figure 8-1. A temperature-pressure diagram of water.
(Courtesy - F. Daniels, *Outlines of Physical Chemistry*, J. Wiley & Sons, New York, NY)

Figure 8-2. A bench-top freeze-drying apparatus.
(Courtesy - Labconco Co., Kansas City, MO)

performing a *differential thermal analysis* (DTA). When the system undergoes a phase change, the resistance and the temperature change significantly. Figure 8-3, p. 87, shows a DTA for orange juice at several concentrations.

Bx = Brix, a measure of sugar concentration based on density. A hydrometer is used. *One degree Brix equals the % by weight of sucrose in a water-sugar solution*.

A vacuum of from 4-6 torr (0.53 - 0.80 kPa) is common, and the heat loss by the subliming water usually will keep the material frozen. During small-scale laboratory operations the sample holders are left in the open air, but in large-scale commercial applications, heat must be applied to provide the sublimation energy at the rate just below that required to keep the material frozen. To sublime 1 g of ice at 0 °C requires 666 calories (2.78 kJ). This heat is provided by warm-water trays placed under the sample trays, by radiation from heated walls surrounding the sample trays, or by microwave warming. Sublimation from commercial warm water-heated trays occurs at the rate of 0.1-1.0 kg H_2O/hr/m^2. Few systems are cooled below -30 °F, because the vapor pressure is then too low for rapid sublimation.

The process at this stage will have reduced the water content about 90 %, but the 5-10 % of bound water is difficult to remove. If it is necessary to remove this bound water as well, then the temperature is raised to 30-35 °C and the vacuum reduced to a few tenths of a torr. This combination of temperature and pressure also will remove O_2 and this aids in preserving many materials. To remove this last bit of water is expensive in that it takes 25-30 % of the total time.

FREEZING RATE EFFECTS

Slow freezing produces large ice crystals that may rupture the cells, but this may be desirable if fast rehydration is necessary. Fast freezing produces tiny crystals, and, in some cases, this produces undesirable color and texture changes in the final product. Therefore, it is necessary to carefully control the freezing rate if the product must have eye appeal. Figure 8-4, p. 87, shows representative types of crystals.

If cells undergo slow cooling, ice crystallization starts in the external medium and there is a gradual dehydration of the cells. However, more rapid cooling results in intracellular crystallization. With very rapid cooling, the crystals thus formed are usually quite small and ultra rapid cooling can produce crystals so small that the intracellular water appears glass like. The *combination of the supply of water molecules* from the liquid phase and the rate of *removal of the available free latent heat determines how the ice crystal grows once a nucleus has formed.* Molecular motion plays a major role for the conversion from a liquid into a solid, particularly so as the temperature is lowered and the viscosity decreases.

Several methods can be used to rapidly freeze foods. The most efficient is to immerse the product in a very cold liquid. This permits a liquid-solid interface for rapid heat transfer. Liquid nitrogen (LN_2) and liquid Freon freezant (LFF) are the most commonly used, the latter being the most popular because it is easier to reclaim. LFF is dichlorofluoromethane (Freon-12) which boils at -30 °C. It was approved for use in 1967 and can still be used because it can be reclaimed, but it is being phased out because of potential upper air ozone reactions. Peas can be frozen solid in 20-30 seconds. Strawberries freeze so fast that they will crack, so they must be sprayed rather than dipped. Examples of the use of this process are: shrimp, diced cooked chicken, fried onion rings, sliced beans, breaded scampi, hamburgers, and corn on the cob. The Freon rapidly evaporates from the food at room temperature and is not considered to be a contaminant. Porous products, which can absorb large amounts of liquid, such as bakery goods, are not suitable for this process.

TECHNIQUES

COMPONENT PARTS

Sample Holder

Figure 8-5 is a photo of one type of sample holder. They are usually of two-piece construction and have either on O-ring or grease seal between the parts. The top end of the lower part is usually a wide opening to facilitate the easy removal of the dried product. They come in sizes from 25 to 1,000 mL.

The technique of filling the sample holder is to fill it 1/3 to 1/2 full of liquid. Place the lower end into the dry ice-acetone or isopropanol container, and tilt the holder to about a 60° angle or as flat as possible without spilling the liquid. Slowly turn the holder so a film of solid sample freezes around the inside of the sample holder. Freeze the sample for a few minutes longer, connect the top section and then connect it to the condenser before it begins to melt. This rotation technique increases the surface area, and the water in this thin film, about 1 cm thick at most, can be sublimed rapidly.

Connecting Valves

This valve (Figure 8-6, p. 88) attaches to the main condenser. It permits attachment of a sample to the condenser without breaking the vacuum seal on the other sample holders. Samples also can be removed in the same way. The white valve stem on the front is turned to control the operation. The valve usually is made of a rather hard rubber with a plastic inner stem. It also includes an internal baffle to prevent cross contamination from one sample to another. It can be taken apart for cleaning.

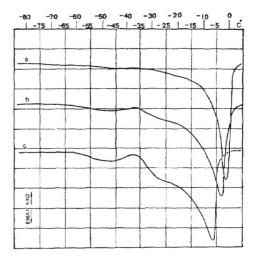

Figure 8-3. DTA of orange juice: (A) 10 °Bx, (B) 20 °Bx, (C) 30 °Bx. Freezing rate 6 °C/min to -120 °C. Heating rate 1 °C/min.
(Courtesy - *Freeze Drying and Advanced Food Technology*, S.A. Goldblith, L. Rey, and W.W. Rothmayr, Eds, Academic Press, New York, NY 1975)

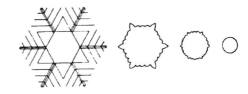

Figure 8-4. Free growth of ice crystals in water. increasing supercooling from left to right.
(Courtesy - *Freeze Drying and Advanced Food Technology*, S.A. Goldblith, L. Rey, and W.W. Rothmayr, Eds, Academic Press, New York, NY, 1975)

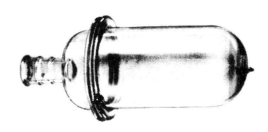

Figure 8-5. Freeze-drying sample holder.
(Courtesy - Labconco Co., Kansas City, MO)

Figure 8-6. Freeze drying connecting valve.
(Courtesy - Labconco Co., Kansas City, MO)

Figure 8-7. Multiport condenser.
(Courtesy - Labconco Co., Kansas City, MO)

Figure 8-8 Backup condenser.
(Courtesy - Labconco Co., Kansas City, MO)

Multiport Condenser

Shown in Figure 8-7 is a 12-port drying chamber. It is made of stainless steel, usually weld free to minimize vacuum leaks, and contains an inner chamber about 7-10 cm smaller in diameter. The inner chamber is filled with dry ice and acetone or isopropanol and is the chamber usually used to freeze the sample when it is first prepared. It is about 30 cm high and 22-25 cm in diameter. The progress of the sublimation can be followed roughly by watching the dry ice-acetone (isopropanol) mixture. As the ice condenses, the coolant will boil off the CO_2 in it - sometimes so vigorously early in the process that the pressure may have to be increased a bit to slow down the sublimation.

A second *backup condenser* often is used as a backup for the primary condenser and to protect the vacuum pump in case the primary condenser exhausts its refrigerant. It is about 20 cm tall and 22 cm in diameter. This is shown in Figure 8-8. The pressure is measured with a tilting *McLeod gauge* having a range from 5 to 5,000 millitorrs. The pump is usually a combination of a mechanical fore pump and an oil diffusion pump.

Figure 8-9 shows a simple noncommercial arrangement that works very well for small scale laboratory preparations.

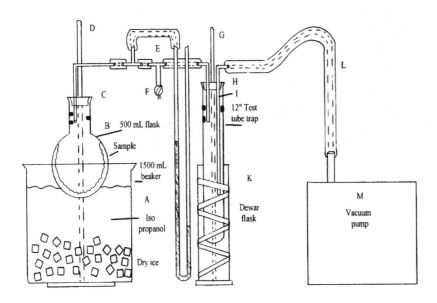

Figure 8-9. Simple freeze-drying apparatus.

REVIEW QUESTIONS

1. Compare dry ice acetone and dry ice isopropanol in the following categories:
   a. Temperature attained.
   b. Flammability at the dry ice temperature.
   c. Solubility of clothing in each mixture.
   d. Solubility of polystyrene containers.
2. What are some reasons why some samples should be lyophilized?
3. What is the purpose of the backup condenser?
4. Why is the sample frozen to just below the solution eutectic rather than freezing it to a steady low temperature?
5. Commercial freeze driers use a refrigerator unit rather than dry ice acetone or dry ice isopropanol. Why is this? (Hint - refer to question 4)
6. What is the effect of slow freezing?
7. A sample that is freeze drying will usually stay frozen once the vacuum is applied. Why is this?
8. How much water is usually removed by normal freeze drying?
9. What is the technique used to freeze the sample in the sample tubes for laboratory preparations?
10. How can you determine the maximum eutectic temperature of an unknown material?
11. It requires 666 calories (2.78 kJ) to sublime 1.0 g of ice. How many calories (and kilojoules) does it take to evaporate 1.0 g of ice by normal heat application? Assume the ice is at -10 °C initially.
12. In the process of freeze drying coffee, it is said that the original 20-25 % material is concentrated to 30-40 % by freezing. How does this process work? (Hint - what happens if you freeze a bucket of muddy water?).
13. Several years ago, the author thought it would be nice to freeze dry a bottle of wine and take it on a hike. The sample froze as expected, but after a few hours, the sample melted and was never satisfactorily refrozen. While it was drinkable it was disappointing. What happened?
14. What is the principle of operation of a DTA?
15. The first attempt at a commercial freeze drying process was in Florida in the late 50's to store fruit juices. The process was a failure at that time, because the product became quite gummy. Can you suggest a possible explanation for this, knowing that fruit juices contain sugars?
16. Why do people place foods in plastic bags filled with water before they store them for a long time in a freezer?
17. What does 10 °Bx mean?

IF IT EXISTS - CHEMISTRY IS INVOLVED
IF YOU CAN BUY IT - A CHEMIST WAS INVOLVED SOMEWHERE

SEPARATIONS INVOLVING EXTRACTION

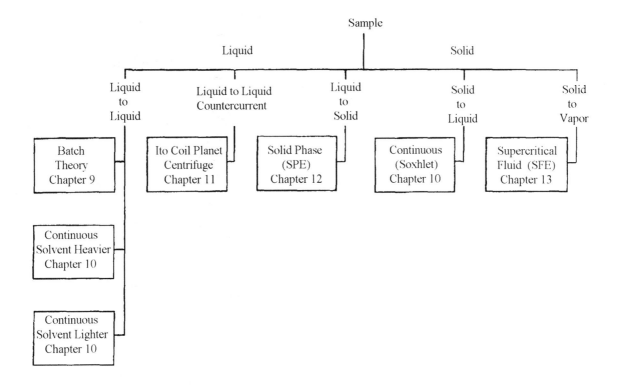

Liquid-liquid extraction is a process by which one or more components of a mixture, usually in water, are selectively transferred to another, usually organic liquid. *Liquid-solid extraction* involves the use of a liquid to selectively remove components from a solid. The majority of laboratory scale extractions are *batch extractions*, involving separatory funnels. The theory and general techniques are discussed in Chapter 9.

The ratio of the amount extracted to the amount remaining is called the *distribution ratio*. There are many situations in which the distribution ratio is low, which would require large amounts of solvent if separatory funnels were used exclusively. This can be very expensive in both chemical and labor costs, and the additional cost of solvent disposal is now often prohibitive. *Continuous extractors*, in which a small volume of solvent is used to extract a portion of the compound, then evaporated, condensed, and used again, are an ideal solution. This process can be repeated for days if necessary; and at the end, there is only a small volume of solvent to remove and dispose of. Continuous extractors that involve solvents both heavier and lighter than water are discussed in Chapter 10. A widely used apparatus for continuously extracting components from solids is the batch extractor developed by Soxhlet. The sample is placed in a porous paper thimble and then placed in a horizontal tube with a closed bottom. The extraction solvent is dripped onto the top of the solid, percolates through it, and siphons off after a short time, the process is repeated as often as necessary. *Soxhlet extraction* is covered in Chapter 10.

A further improvement is to move one solvent past the other in the opposite direction. This *countercurrent* technique can not only separate several compounds from a sample but can separate them from each other at the same time. This is then a combination of an extraction and a chromatographic separation. The *coil planet centrifuge*, developed by Ito, is discussed in Chapter 11.

A somewhat recent advancement is to use the same types of surface coatings as are used on high performance liquid chromatography column packings but to make the particles larger. These are then placed in a porous matrix in the form of a disc from dime to half dollar size, and in a 1-3 cm long column, 1-2 cm in diameter. This is known as the solid phase, and when a liquid containing the desired components passes through it, they are extracted from the liquid. These components can then be stripped away later by using a different polarity solvent. This is known as *solid phase extraction* (SPE), and the details are presented in Chapter 12.

One of the problems with using a liquid as the extraction solvent is its removal when the extraction is finished. The most recent way to eliminate this problem is to use a *supercritical* gas, CO_2 being the gas of choice at the moment. A gas in the supercritical state has solvent properties comparable to a liquid; but it is less viscous, so it can penetrate the sample faster. When the extraction is complete, the pressure is released, and the gas evaporates away from the extracted components. CO_2 is nonpolar so more polar compounds such as methanol are sometimes added in small amounts. This excellent *supercritical fluid extraction* (SFE) technique is described in Chapter 13.

9
EXTRACTION: GENERAL CONCEPTS

PRINCIPLES

An *extraction* is the selective transfer of a compound or compounds from one liquid (usually water) to another immiscible liquid (usually organic) or from a solid to a liquid. The former process is called a *liquid-liquid* extraction and the latter is called a *liquid-solid* extraction. More recently, with the development of reactive solid phases, materials are extracted from liquids by a solid. This is called *solid phase extraction* and is discussed in Chapter 12.

There are various types of extractions. They range from food, drug, and cosmetic extractions, which separate several compounds at one time, to the more sophisticated analytical extractions, which are designed to cleanly separate a selected compound or compounds from a complex mixture.

A food extract, such as a lemon extract, is really not a chemical extraction but consists of the essential oils that are mechanically pressed out of lemon skin and then kept in a 45% solution of alcohol. A drug extract is called a *tincture. Tinctures are usually alcohol solutions* and are liquid-solid extractions. Examples are the tincture of iodine, oil of wintergreen, oil of clove, and oil of peppermint. Cosmetic extracts such as ambergris, musk, castor, and benzoin, are called *essences* and are used to make perfumes.

BASIC IDEA

Refer to Figure 9-1, p. 94. Suppose you had four compounds (A, B, C, and D) dissolved in water, and you desired to remove A and leave the others behind. If a suitable immiscible organic solvent could be found to dissolve A, and you could either adjust the chemistry of B, C, and D so they would not dissolve in the organic layer or adjust A so it would, (A(x)), then to perform a liquid-liquid extraction you would just have to shake the two layers together and allow them to separate

Figure 9-2, p. 94, illustrates how an extraction can be used. One method for determining the wear on engine rings is to measure the amount of vanadium in the crankcase oil by visible spectroscopy. However, iron interferes. The oil first is digested with $HClO_4$. When HCl and diethylether are added, the iron transfers to the ether layer, leaving the vanadium in the water layer. The water layer can be drained and analyzed. Notice that two layers are formed that can be separated easily.

The principle that controls a liquid-liquid extraction is the *Gibb's phase rule:*

$$F = C - P + 2 \qquad (9\text{-}1)$$

where F = degrees of freedom. These are the variables you can change. This is usually the composition of each material being extracted. C = number of components. P = number of phases. 2 = temperature and pressure. At constant T and P, which is usual for room temperature extractions, then $F = C - P$

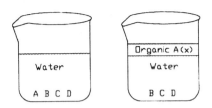

Figure 9-1. Diagram illustrating the basic principle of liquid-liquid extraction.

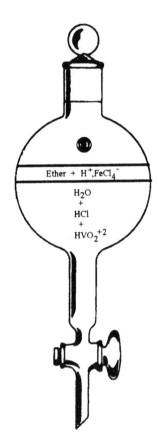

Figure 9-2. Separation of Fe from V by a liquid-liquid extraction in a separatory funnel.

EXAMPLE CALCULATION
Referring to the problem described in Figure 9-2, how many degrees of freedom do you have and what might they be?

ANSWER
$C = 5$ (water, ether, HCl, Fe^{+3} and HVO_2^{+2}).
$P = 2$ (both liquids; water and ether).
Pressure is constant during the short time of the extraction. Generally, there is little heat of mixing so *temperature is considered to be constant* as well.
$F = 5 - 2 = 3$. This is a change in composition, Fe^{+3}, HVO_2^{+2}, and HCl. If you specify the concentration in one phase, it sets the concentration in the other phase. What determines this concentration is the *distribution law* discussed below.

DISTRIBUTION LAW

Based upon the properties of the compound and how it interacts with each liquid phase and assuming the system is not saturated, then each compound will distribute itself between the water and organic layer in a definite ratio, which will remain constant until one phase becomes saturated.

$$K_D = \frac{[X_2]}{[X_1]} \quad \text{from } X_1 \leftrightarrow X_2 \tag{9-2}$$

For example; if you had 100 molecules and 90 went into the organic layer, then if you had 1000 molecules, 900 would go into the organic layer, and this ratio would remain constant until one layer becomes saturated.

DISTRIBUTION RATIO

From the above examples, it can be seen that an extraction is the transfer of a compound from one liquid to another or from a solid to a liquid. This transfer is seldom ever complete in a single operation. Generally, a small amount of the compound is always left behind. Because the compound divides between the two layers, it is called a *partition*. The usual way to express the extent of this partition is the *partition coefficient, P*, or the *distribution ratio, D*.

$$D = \frac{g/mL \text{ solute in the organic (upper) phase}}{g/mL \text{ solute in the water (lower) phase}} \tag{9-3}$$

EXAMPLE CALCULATION

700 mg of antimony (V) was dissolved in 100 mL of HCl, an equal volume of isopropyl ether was added, and the system thoroughly shaken. The layers were separated; and the ether, evaporated. The residue was found to contain 685 mg of antimony. What is the distribution ratio?

ANSWER

$$D = \frac{g/mL \ in \ organic}{g/mL \ in \ water} = \frac{0.685/100}{(0.700 - 0.685)/100} = 45.7$$

The distribution ratio shows greater restraint than the phase rule, because now the ratio of concentration is constant (provided it has the same m.w. in each solvent). This takes into account all reactions that take place. If no reactions take place, then $D = K_D$.

EXAMPLE CALCULATION

(This example was developed by Prof. Henry Freiser, Univ. of Arizona.) Given the following data, develop an equation to determine D for the distribution of oxine (8-hydroxyquinoline) between $CHCl_3$ and H_2O.

Ionization equilibrium

$$H_2Ox^+ \rightarrow HOx + H^+ \qquad K_1 = \frac{[HOx][H^+]}{[H_2Ox^+]} = 8 \times 10^{-6}$$

$$HOx \rightarrow Ox^- + H^+ \qquad K_2 = \frac{[Ox^-][H^+]}{[HOx]} = 1.4 \times 10^{-10}$$

Distribution

$$[HOx]_w \rightarrow [HOx]_o \qquad K_D = \frac{[HOx]_o}{[HOx]_w} = 720$$

ANSWER

$$D = \frac{[HOx]_o}{[HOx]_w} = \frac{Total \ organic}{Total \ water} = \frac{[HOx]_o}{[H_2Ox^+] + [HOx]_w = [Ox^-]}$$

divide by $[HOx]_w$

$$D = \frac{\frac{[HOx]_o}{[HOx]_w}}{\frac{[H_2Ox^+]}{[HOx]} + \frac{[HOx]}{[HOx]} + \frac{[Ox^-]}{[HOx]}} = \frac{K_D}{\frac{[H^+]}{K_1} + 1 + \frac{K_2}{[H^+]}}$$

EXTRACTION REQUIREMENTS

A quantitative prediction of what a *D* will be for any given system is not possible unless all of the various equilibrium constants are known, but for a substance to be extracted, it *must be neutral*. There are three basic ways to prepare a neutral molecule: (1) chelate formation, (2) ion association, and (3) micelle formation.

Chelates (Key-lates)

If a metal ion reacts with another ionic compound (ligand) to form one or more rings, the resulting compound is called a *chelate* (Gr. *Kelos* - claw). Two such chelates are shown below.

Extracts into $CHCl_3$. The two negative ligands neutralize the +2 charge of the central ion.

This chelate extracts into 1-hexanol to separate uranium from thorium and cerium.

Ion Association

In this case, two or more ions associate to neutralize each other's charge.

a. Incorporation of a large ion.

$$ZnCl_4^{-2} + 2(C_6H_5CH_2)_3NH^+ \xrightarrow{\text{xylene}}$$

b. Oxonium

The solvent oxygen atom takes part in the coordination sphere.

$$FeCl_4[(C_2H_5)_2O]_2^-, (C_2H_5)_2O:H^+$$

c. Mixed

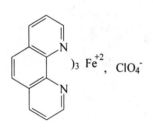

The unpaired electrons of the N on 1,10-phenanthroline (diagram to the left partially neutralize the iron's charge, and the ClO_4^- completes the process. This complex will extract into nitrobenzene.

Micelles

A *micelle* (diagram, lower left) is a complex molecule that may be colloidal in size. Metal ions are incorporated into salts of high molecular weight that dissolve in organic solvents the same as soap in water, that is, by forming colloidal aggregates. The organic part is turned out, and the ionic parts are hidden. In the case shown at the left, the polar amine end with its nonbonding electrons surrounds the Cu^{+2} ion and effectively isolates the charge. Also: U, Mo by amines in kerosene. Cu^{+2} caproate (C_{10}-COOH) in $CHCl_3$.

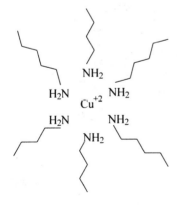

Research done in the author's laboratory indicates that the processes described nearly, but not completely, neutralize the metal ion's charge. Polar water molecules are added as adducts around the chelate, the ion association complex or the micelle as necessary to complete the charge neutralization process as shown in the diagram at the top of the next page. This produces a neutral molecule that is hydrophilic and tends to stay in the aqueous layer. If a suitable organic liquid is added, these molecules usually can

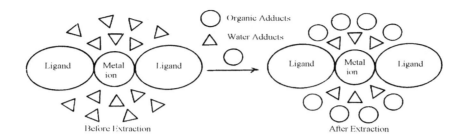
Before Extraction → After Extraction

displace a portion of the water molecules, thus making the total species hydrophobic and extractable. The distribution ratio depends on the relative amounts of water displaced. This lack of organic solvent adduct formation may be why charged particles do not extract. Charged particles hold more water molecules and tend to hold them so tightly by ion-dipole bonding that they are not replaced by the weaker dipole-dipole bonds between water and the organic phase molecules.

DISTRIBUTION OF THE EXTRACTABLE SPECIES

One of the surprises in predicting an extraction is that you cannot go by solubilities alone. This is believed to be due to (1) a change in activity coefficient of the solute in each phase (when saturated then solubilities hold) and (2) the effect of the second solvent upon the first. Because the layers are immiscible, you generally assume there would be no effect due to the second solvent. However, since the concentration of the extractable species is quite low, even a small solubility of one solvent in the other provides sufficient molecules to alter the distribution.

EXAMPLE
$AgClO_4$ = 85% in water   Expect a D nearly 1, yet you get a D of 100
       = 50% in toluene
Mix together = 100% in water   (May be due to a hydrated Ag salt insoluble in toluene.)

Electrostatic bonds

These bonds involve electrostatic forces, such as those between two ions, an ion and a dipole, or two dipoles. The ratio of the cation (r_c) to the anion (r_a) is called the radius ratio. As the ratio of the cation to anion decreases below 1, the lattice energy remains about the same at first and then decreases rapidly. Because the sum of the magnitudes of the solvation energies of the gaseous ions rise steadily (smaller r), a salt will tend to be more soluble as its radius ratio becomes smaller than 1.

$$\frac{r_c}{r_a} = < 1$$

H-bonds

Generally, the solubility of a substance in water or alcohol is governed more by the ability of the substance to form H bonds than by its "polarity" as measured by dipole moment.

Chemical bonds.

Chemical bonds include acid-base interactions. I_2 extracts into mesitylene > xylene > benzene because of pi bonds and sigma bonds.

Synergism

If the effect of two components in a mixture is greater than the sum of each component individually, the effect is called *synergism*. For example: One solvent extracts a certain amount of a compound, and another solvent extracts a certain amount. But a mixture of the solvents may cause a 10^2 to 10^3 increase. Synergism is very important with tributylphosphate extractions.

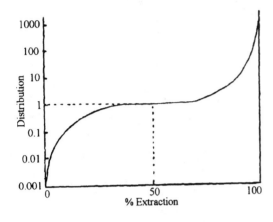

Figure 9-3. A plot of D vs % extraction

PERCENT EXTRACTION

Equation 9-4 is used to calculate the percent of a compound that can be extracted if the D and the volume of each phase are known.

$$\% \ Extraction = \frac{100 \ D}{D + V_w/V_o} \quad (9\text{-}4)$$

where: V_w = volume of water layer (lower). V_o = volume of organic layer (upper).

Figure 9-3 shows how the percent extraction varies with the distribution ratio. Notice that this is a log plot on the Y axis. The effect of varying the amount of extraction solvent is presented in the following example calculation.

EXAMPLE CALCULATION

Iron forms a chelate that will extract into nitrobenzene with a D of 3. Ten mL of this solution is extracted with 5 mL of nitrobenzene. A second 10 mL aliquot is extracted with 15 mL of nitrobenzene. Does it make any difference what volume of extracting solvent is used?

ANSWER

Part A

$$\% = \frac{100 \times 3}{3 + 10/5} = 60 \ \%$$

Part B

$$\% = \frac{100 \times 3}{3 + 10/15} = 82 \ \%$$

Yes, it makes a difference, as you would expect. How advantage is taken of this is explained later involving equation 9-5, p. 99.

In order to rapidly separate two components, not only must they have different distribution ratios, but the distribution ratios must have the correct range. Referring to Figure 9-3, it can be seen that for the best separations between systems with similar D's, values between 1 to 10 are best. The following example calculation will illustrate this observation.

EXAMPLE CALCULATION

Both systems described here have a 10-fold difference in D's, yet in practice we find that system 1 is far easier to separate than system 2. Why? System 1: Cpd A has a D of 1 and cpd B has a D of 10. System 2: Cpd X has a D of 100 and cpd Y has a D of 1000. Assume that V_w = 10 mL and V_o = 10 mL.

ANSWER

$$\%\ A = \frac{100 \times 1}{1 + 10/10} = 50\%$$

$$\%\ X = \frac{100 \times 100}{100 + 10/10} = 99.0\%$$

$$\%\ B = \frac{100 \times 10}{10 + 10/10} = 90.9\%$$

$$\%\ Y = \frac{100 \times 1000}{1000 + 10/10} = 99.9\%$$

B is not completely extracted, but it is separating from A. After three or four extractions, it will be fairly well separated.

Both are almost completely extracted, and no separation takes place. Even after a dozen extractions, little separation will take place.

BATCH EXTRACTIONS

Most laboratory extractions are done with a separatory funnel and done one or more times, changing the extracting solvent each time. These are called *batch extractions*. For the following equations, let D = distribution ratio; A = amount of solute originally present; X = amount not extracted; V_w = volume of water; and V_o = volume of organic.

$$D = \frac{\frac{A - X}{V_o}}{\frac{X}{V_w}} = \frac{V_w(A - X)}{X V_o}$$

Then:
$$DXV_o = AV_w - V_w X$$
$$DXV_o + V_w X = AV_w$$
$$X(DV_o + V_w) = AV_w$$

$$X = \frac{A V_w}{D V_o + V_w} \tag{9-5}$$

For a second extraction, X of the first = A of the second.

$$X_2 = \frac{V_w}{D V_o + V_w} \times \frac{A X_1 V_w}{D V_o + V_w} = A\left(\frac{V_w}{D V_o + V_w}\right)^n \tag{9-6}$$

If you use a volume twice as large

$$X = A\left(\frac{V_w}{2 D V_o + V_w}\right)^n \tag{9-7}$$

MULTIPLE EXTRACTIONS

A common mistake is made when an analyst gets in a hurry. The directions call for three 10 mL extractions, but time is short, so one 30 mL extraction is made. That is assumed to be the same; *it is not*. Equation 9-6 provides the basis for the answer.

EXAMPLE CALCULATION

A fat is to be removed from a meat sample by an ether extraction. Three 30 mL extractions are recommended to remove 0.10 g of fat from 1.0 g of meat dispersed in 30 mL of water. Assume $D = 2$. Which is best, one 90-mL extraction or three 30-mL extractions?

ANSWER

For a single 90 mL extraction:

$$X = 0.1 \left(\frac{30}{[2 \times 90] + 30} \right)^1 = 0.014 \text{ g}$$

For three 30 mL extractions:

$$X = 0.1 \left(\frac{30}{[2 \times 30] + 30} \right)^3 = 0.0037 \text{ g}$$

almost a 4-fold improvement

TECHNIQUES

The simplest type of extraction vessel is a separatory funnel; several types are shown in Figure 9-4. They are usually either cylindrical (A), globe shaped (B), or pear shaped (C, D). They may have either glass or plastic stoppers and stopcocks and may have a standard taper joint at the bottom. They usually are made of borosilicate glass, but the wine red colored "low actinic" also can be obtained for use with light-sensitive compounds.

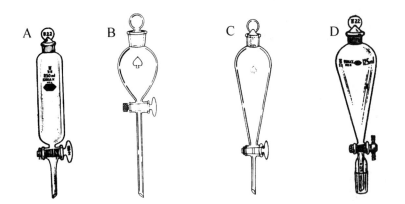

Figure 9-4. Various types of separatory funnels. (A) Cylindrical. (B) Globe. (C) Pear. (D) Standard taper bottom.
(Courtesy - Ace Glass Co., Vineland, NJ)

The more common sizes can be obtained made from plastic (usually polypropylene). The usual capacities are from 30 mL to 2 L. Figure 9-5, p. 101, shows the proper way to hold a separatory funnel.

1. Separatory funnels should be rinsed with the extracting solvent before use.

2. The funnel should be *inverted by rotating your wrist. Vigorous shaking is not necessary* to attain equilibrium and, in fact, may cause the formation of an emulsion, which will delay the final separation.

3. When two immiscible liquids are mixed, there is still a small amount of each liquid that dissolves in the other, and the heat of solution that may develop can cause a pressure buildup. Figure 9-5 shows how this is vented safely.

EXTRACTION TECHNIQUES

Choice of solvent

a. Watch the density. Large density differences are desired.

b. Because of cost, toxicity, flammability, or density you may want to mix solvents.

Stripping

Removal of the extracted solute from the organic layer is called *stripping*. It occurs if $D < 1$. Add a few mL of water to hold the solute and then evaporate off the solvent. You may need to add H_2SO_4 or $HClO_4$ to remove the organic solvent in the water layer by evaporation. If the solvent is nonvolatile (boils higher than water), shake with strong acid solution to destroy the chelate, and the metal will go into the aqueous layer.

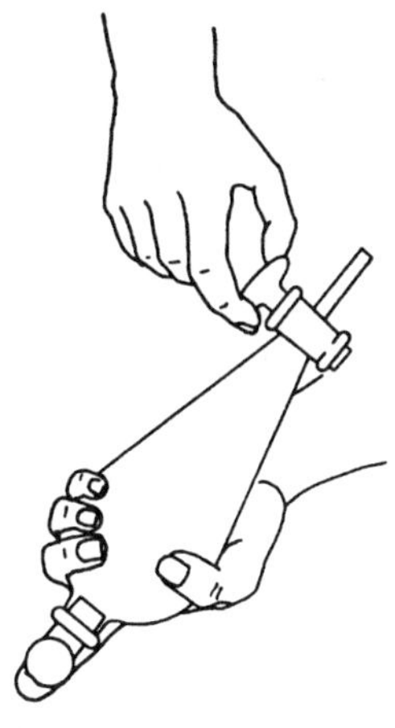

Backwashing

Backwashing is used with batch extractions. The combined phases of several extractions are washed to remove impurities. Because the D of the impurities is usually much less than that of the metal extracted, this removes impurities much the same as reprecipitation.

Emulsions

Emulsion formation depends on viscosity, density difference, and drop size. To prevent, reduce, or break emulsions refer to Chapter 5, p. 51.
   a. Invert the layers slowly; do not shake vigorously.
   b. Use centrifugation for small volumes. It is the best and fastest way.
   c. Use mixed solvents.
   d. Add neutral salts, which increase the density and change the surface tension.
   e. Use only a small amount of water with a large amount of organic solvent.
   f. Filter through a porous substance.
   g. Filter through phase separation paper (silicone treated).

Figure 9-5. Separatory funnel venting technique. (Courtesy - R.J. Oullette, C.A. Bonn, J.S. Swenton, S. Marcus, *Introductory Experimental Chemistry*, Harper & Row, New York, 1975)

Use of masking and sequestering agents

A *masking agent* forms a *complex* (no rings) with the material to be extracted; a *sequestering agent* forms a *chelate*. Masking agents are charged complexes that do not extract. (See Chapter 40.)

A. Several cation masking agents that are used include: CN^-, tartrate, citrate, F^-, EDTA, SCN^-, $S_2O_3^{-2}$. These are generally used with chelate extractions, because the more acidic pH's used with ion association systems will not allow the association to form.

   Al + Fe + oxine -------> $Fe(CN)_6^{-3}$ and does not interfere.
        CN^-, KOH
   Ni + Co + Dimethylglyoxime
        Add CN^-, then H_2O_2, $Ni(CN)_4^{-2}$ $Co(CN)_6^{-3}$

The Ni complex is destroyed by H_2O_2 or formaldehyde, but the Co complex is not.

B. For anion masking, EDTA forms anionic chelates/complexes with quite a number of metal ions, including the alkaline earths. F^- interferes with UO_2^{+2} extractions and can be removed by adding B or Al to mask the fluoride.

Salting-out agents

   a. Use ammonium salts, if possible, because these can be removed from the aqueous layer for work later on.
   b. Al and Fe salts are strong salting-out agents because of their high charge, but they are difficult to remove.
   c. As charge increases, the salting-out effect increases, probably because more solvent molecules are involved.

CONCENTRATORS

When trace quantities of materials are extracted, it is necessary to evaporate most of the solvent to concentrate the desired compound for further analysis. This can be difficult to do without losing the desired compound as well. The two most common ways of removing large volumes of solvent without appreciable loss of the desired compound are a Kuderna-Danish concentrator and a rotary evaporator. The former is preferred for low-boiling solvents (< 100 °C, acetonitrile, acetone, methylene chloride), and the latter for higher-boiling solvents. A *Kuderna-Danish concentrator* is shown in Figure 9-6.

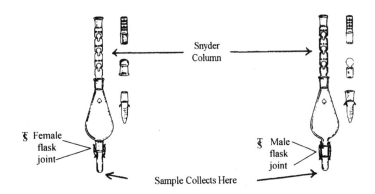

Figure 9-6. A Kuderna-Danish concentrator.
(Courtesy - Ace Glass Co., Vineland, NJ)

The bottom collector is held in place by two small springs. The extract is placed in the flask, and the *Snyder column* attached. The apparatus is placed in a steam bath, so that the collector is in the steam and hot water. If the solvent boils at a low temperature, do not increase the heat at a later time or it may bump vigorously and come apart. If you add fresh solvent to wash the Snyder column, add a fresh boiling chip. The boiling chip is not a problem later on. The solvent is evaporated to very near dryness, and then any solvent desired is added by a microliter syringe to the volume desired.

With a fatty compound and/or a large amount of solvent, use two Snyder columns placed one on top of the other, because the sample may foam out the top of a single column.

A *rotary evaporator* is shown in Figure 9-7, and several arrangements are shown in Figure 9-8, p. 103. The complete apparatus consists of the rotary evaporator, a heating bath, a cooling bath, a refrigeration unit or cooling water, a pressure gauge, and a vacuum pump. The rotary evaporator itself consists of two round bottom (r.b.) flasks, a hollow rotating tube, a motor, and a condenser. The extracted sample is placed in the r.b. flask on the right, which is then attached to the rotating tube. The condenser and collecting flask are attached and refrigerant (cooling water) is passed through the condenser. The entire apparatus is lowered until the r.b. flask containing the sample is immersed in a water bath with the water heated to some low temperature (35 °C for toluene). Vacuum is applied; and the apparatus, turned on. The flask in the water bath slowly rotates, and a thin film of the solvent and sample coats the inside of the flask. This coating provides a thin layer and a large evaporating surface. The vacuum reduces the pressure to enhance the evaporation rate. The condenser liquifies the vapors that are collected in the collector flask, which can be immersed in a refrigerant as well. Apply the vacuum slowly at first until the sample is degassed and low boilers are removed. Increase the vacuum until a steady evaporation is taking place. For reproducible results, a pressure gauge is recommended.

Figure 9-7. A Buchi type rotoevaporator.
(Courtesy - Thomas Scientific Inc., Swedesboro, NJ)

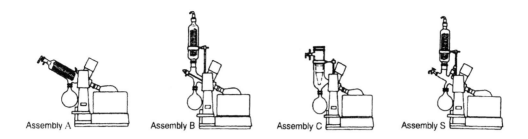

Figure 9-8. Several configurations for a Buchi/Brinkman rotoevaporator. (A) Diagonal condenser. (B) Vertical condenser. (C) Two-piece cold-trap condenser. (S) Special distribution head.
(Courtesy - Fisher Scientific Co., Pittsburgh, PA)

Referring to Figure 9-8, Assembly A has a sturdy, diagonally positioned, tap water-cooled condenser. This is the most versatile arrangement and ideal for simple distillation of solvents. Assembly B has a vertical tap, water-cooled condenser, ideal for higher boiling solvents. It is more compact than assembly A. Assembly C features a two-piece, cold-trap condenser that uses coolant (dry ice/acetone or mechanical refrigeration). It is particularly well suited for low boiling solvents and for conditions in which a tap water-cooled condenser is insufficient. Assembly S is equipped with a vertical tap water-cooled condenser and special stationary distribution head with a polytetrafluoroethylene (PTFE) valve.

EXTRACTION P-VALUES

This technique was developed primarily to help identify pesticides in foods. There is no reason why it could not be extended to other compounds found in foods or even other materials.

Sometimes, the gas chromatographic retention times of two or more materials are so close that an identification is difficult, even if two different columns have been used. A thin layer chromatography separation (Chapter 22) then can be used, but this generally requires a lengthy cleanup and concentration because of the small (nanogram) samples involved. A much quicker method is the use of *P-values*. This method is based on the distribution of the pesticides between two immiscible phases.

The procedure of Beroza and Bowman, Anal. Chem. **37**, 291, 1965, which follows, shows how it is done. Assume a hexane-acetonitrile system. A 5.0 mL aliquot of the upper phase containing a given pesticide is analyzed by gas-liquid chromatography. To a second 5.0 mL aliquot in a graduated, glass-stoppered, 10.0 mL centrifuge tube is added an equal volume of lower phase, the tube is shaken for about 1 minute, and the upper phase is analyzed exactly like the first 5.0 mL aliquot. The ratio of the quantitative results of the second analysis to those of the first (peak areas or heights) is the *P-value*. This is the amount of the pesticide in the upper phase (second analysis) divided by the total amount of pesticide (first analysis). Table 9-1, p. 104, shows how valuable this can be; o,o'-DDT and TDE have identical retention times.

Aliquots of the equilibrated phases are measured with volumetric pipets, and the phase volume is noted before and after equilibration. Emulsions are separated by centrifugation. After the distribution, the upper phase is passed through a 2.5 cm layer of anhydrous Na_2SO_4 to dry the solvent.

The P-value is quite reproducible (± 0.02), providing the two phases are saturated with each other before distribution. Table 9-2 shows P-values of several insecticides (Beroza and Bowman).

Table 9-1. P-values of two insecticides having indistinguishable retention times TDE = tetrachlorodiphenylethane)

Compound	Extraction Systems		
	Cyclohexane Methanol	Hexane 90% Dimethyl Sulfoxide	Isooctane 80% Dimethyl Formamide
o,o'-DDT	0.61	0.55	0.44
TDE	0.37	0.08	0.16

Table 9-2. P-values of several insecticides

Pesticide	Hexane Acetonitrile	Hexane 90 % Aq. Dimethyl Sulfoxide	Isooctane 85 % Aq. Dimethyl Formamide	Isooctane Dimethyl Formamide
Aldrin	0.73	0.89	0.86	0.38
Carbophenothion	0.21	0.35	0.27	0.04
Gamma Chlordane	0.40	0.45	0.48	0.14
p,p'-DDE	0.56	0.73	0.65	0.16
o,o-DDT	0.45	0.53	0.42	0.10
p,p'-DDT	0.38	0.40	0.36	0.08
Dieldrin	0.33	0.45	0.46	0.12
Endosulfan I	0.39	0.55	0.52	0.16
Endosulfan II	0.13	0.09	0.14	0.06
Endrin	0.35	0.52	0.51	0.15
Heptachlor	0.55	0.77	0.73	0.21
Heptachlor epoxide	0.29	0.35	0.39	0.10
1-Hydroxychlordane	0.07	0.03	0.06	0.03
Lindane	0.02	0.09	0.14	0.05
TDE	0.17	0.08	0.15	0.04
Telodrin	0.48	0.65	0.63	0.17

EXTRACTION MODIFICATIONS

Many variations and special apparatus have been developed over the decades to solve specific problems and to do extractions more efficiently. Several of these are solvent heavier than water (Chapter 10); solvent lighter than water (Chapter 10); continuous countercurrent (Chapter 11); solid phase extraction (Chapter 12); liquid-solid extraction, microwave heated solvents (Chapter 10); and supercritical fluid extraction (Chapter 13).

REVIEW QUESTIONS

1. What is a tincture?
2. What is C in the Gibb's phase rule?
3. What is a partition coefficient or the distribution ratio?
4. The Aldrin in a sample of strawberries was extracted with hexane and concentrated to 5.0 mL. Five mL of 90% dimethylsulfoxide (DMSO) (consider as the water layer) was added, and it was found that 83% of the Aldrin was in the hexane layer. What is the distribution ratio of this compound between these solvents?
5. What is the one basic requirement for a compound to be extracted?
6. What is the difference between a chelate and a complex?
7. Is it possible to predict the D of a system if you know the solubilities of each component?
8. What is synergism?
9. Iron forms a chelate that will extract into nitrobenzene with a D of 3. What % of the iron will be extracted from a 25 mL sample if 10 mL of nitrobenzene is used?
10. In problem 4, it was found that 83% of the Aldrin was extracted in one extraction when the volumes were equal. Assume you wanted 98% of the Aldrin to be extracted into the hexane in one extraction. How many total mL of hexane must be used? (DMSO is the water layer.)
11. Suppose you know that cholesterol in egg noodles can be extracted with ethyl acetate. The D is 0.3 (assumed). How many 50 mL extractions will it take to remove 95% of the cholesterol in a batch of egg noodles (20 mL H_2O), if they contain 2.0% cholesterol? Assume a 10.0 g sample.
12. The iodine residue from a water purification compound can be detected qualitatively by extraction into CCl_4. If you have 10.0 mL of water and I_2 has a D of 4, what would be the volume of CCl_4 necessary for a 50% extraction? What would be the volume of water necessary for a 98 % extraction if 5.0 mL of CCl_4 were used? Is this practical?
13. Iron interferes with most analyses for vanadium, chromium, molybdenum, and titanium and must be removed. Ferric iron can be extracted into ether from a 6 N HCl solution with a D of 110. If 0.3 g of iron were present in 150 mL of HCl solution (water layer), how many mL of ether would it take to ensure that no more than 1.0 mg of Fe remained after one extraction?
14. Arsenic compounds are very good as sprays for apple trees. It is desired to perform a quantitative extraction (1.0 mg left) of a 0.2 g sample of arsenic as the dibenzoylmethane chelate. What must be the partition coefficient, if not more than three extractions can be performed and the volume of $CHCl_3$ is 25 mL for each 50 mL of H_2O?
15. Bismuth alloyed with tin or cadmium is used as a low melting alloy in automatic sprinkler systems for fire prevention. A typical alloy contains 30% Bi. If an 0.80 g sample is dissolved in 50 mL of acid and a chelate formed, how many extractions employing 20.0 mL of benzene will be required to reduce the Bi concentration to 0.0001 g? The chelate has a D of 1.4.
16. How much more effective will four extractions using 5.0 mL of 1-hexanol be than using one 20.0 mL portion? Assume a 10.0 mL volume of water, a D of 2, and a sample of 0.5 g.
17. What is stripping? What value of D do you want to do this?
18. What is an inversion when referring to an extraction?
19. What is the difference between a masking agent and a sequestering agent?
20. What is the purpose of a Kuderna-Danish apparatus?

UNDERGRADUATES - DO THOSE PORTIONS ASSIGNED BY YOUR INSTRUCTOR,
PLUS ANY OTHERS YOU HAVE AN INTEREST IN.
GRADUATE STUDENTS - START WITH NEW MATERIAL AND CONTINUE ON.

10
CONTINUOUS EXTRACTION
SOLVENT HEAVIER THAN WATER

PRINCIPLES

Many materials have such low distribution ratios that neither separatory funnels nor the older, no longer commercially available, Craig apparatus are practical to use. What is needed is a continuous extraction with an apparatus that can be left unattended for long periods of time. Several different approaches are described in this chapter.

To have a high efficiency in a continuous extraction, the following factors are beneficial:
    a. A large surface area of contact between the two phases.
    b. A long contact time.
    c. A high relative volume of extractant to solvent.
    d. A high *D*.

With low *D*'s, a large volume of extracting solvent is needed. However, it can be the same solvent used over and over, if it is evaporated from the extract between uses. A 50 mL sample may have 50 mL of extraction solvent initially employed for an actual ratio of 1:1. If however, the solvent is redistilled and reused 20 times, then the relative volume is 1,000 to 50 or a 20:1 ratio. A high D is desirable, but if you had a high *D*, then a continuous extraction probably would not be necessary.

Continuous Extractor Efficiencies

The efficiency of continuous extractors is measured by the *half extraction volume*, determined as shown in Figure 10-1, p. 108. This is the volume of extracting solvent required to reduce the amount of the desired compound in the sample to one-half of its original concentration. A distribution factor, *k*, can be used to evaluate individual systems.

$$k = \frac{0.693\ W}{V} \tag{10-1}$$

where $k = \dfrac{\text{Concentration of solute in the extracting solvent}}{\text{Concentration of solute in the original solution}}$

W = the original solution volume and V = the half extraction volume.

Although *k* often is close to *D*, *k* is seldom equal to *D* because the extraction conditions for continuous extractions are not as easily defined or controlled as with a single extraction system, and in many cases, true equilibrium is not attained. *k* more often approaches *D* when *D* is small.

EXAMPLE CALCULATION

A comparison of two extractors is to be made. 50 mL of sample solution containing 5.0 mg of caffeine requires 900 mL of CCl_4 to reduce the concentration to 2.5 mg in extractor A. Extractor B requires 1275 mL of CCl_4 to reduce 6.0 mg to 3.0 mg in a 75 mL sample. Which extractor is best and what is the k for each system?

ANSWER

$$\text{Extractor A} \quad \frac{900 \text{ mL of extraction solvent}}{50 \text{ mL of original sample}} = 18 \text{ mL/mL}$$

$$\text{Extractor B} \quad \frac{1{,}275 \text{ mL of extraction solvent}}{75 \text{ mL of original sample}} = 17 \text{ mL/mL} \qquad \text{B is slightly better.}$$

$$\text{Extractor A} \quad k = \frac{0.693 \times 50 \text{ mL}}{900 \text{ mL}} = 0.0385; \qquad \text{Extractor B} \quad k = \frac{0.693 \times 75 \text{ mL}}{1{,}275 \text{ mL}} = 0.0407$$

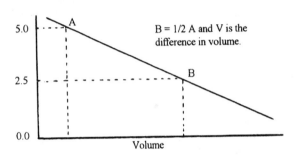

Figure 10-1. Diagram showing how the half extraction volume is determined.

TECHNIQUE

Commercial Extractors

A true continuous extraction requires that fresh solvent continuously flow through the sample and that fresh sample be continuously flowing in the opposite direction. If the system is designed properly, the compound of interest will be completely extracted from the sample just as the sample reaches the end of the apparatus, and the extracting solvent will be saturated with the extracted compound when it reaches the end of the apparatus. Several such *continuous countercurrent extractors* are shown in Figure 10-2.

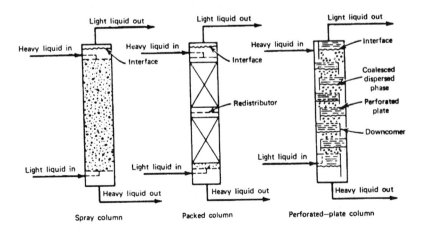

Figure 10-2. Continuous countercurrent extractors - unagitated.
(Courtesy - R.E. Kirk and D.F. Othmer, *Encyclopedia of Chemical Technology*, John Wiley & Sons, New York, NY, 1980)

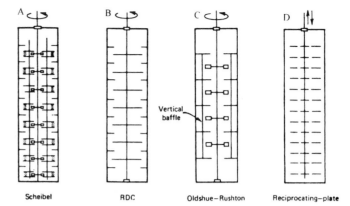

Figure 10-3. Continuous countercurrent extractors - mechanically agitated.
(Courtesy - R.E. Kirk and D.F. Othmer, *Encyclopedia of Chemical Technology*, John Wiley & Sons, New York, NY, 1980)

The following diagrams are hopefully self-explanatory. To improve the efficiency of the extraction, mechanical agitation often is added. Figure 10-3 shows several arrangements of these types of extractors. These are divided into rotary and reciprocating types.

According to the authors in the *Encyclopedia of Chemical Technology*, "The Scheibel column (A) was developed in 1948. Alternate compartments are agitated with impellers, whereas the others are packed with an open woven wire mesh. Columns up to 2.6 m in diameter are in service.

"The rotating disc contactor (B), developed in the Netherlands in 1951, uses the shearing action of a rapidly rotating disc to interdisperse the phases. These have been used in the petrochemical industry for furfural and SO_2 extraction, propane deasphalting, sulfolane extraction for the separation of aromatics from aliphatics, and caprolactam purification. Columns up to 4.3 m in diameter are in service.

"The Oldshue-Rushton column (C) was developed in the early 1950s and has been widely used in the chemical industry. It consists essentially of a number of compartments separated by horizontal stator-ring baffles, each fitted with vertical baffles and a turbine-type impeller mounted in a central shaft. Columns up to 2.7 m in diameter have been reported.

"The reciprocating plate extractor (D) is a vibrating type developed in the 1940's. The open perforated reciprocating plate columns consist of a stack of perforated and baffle plates with a free area of about 58%. The central shaft supporting the plates is reciprocated by means of a drive mechanism located at the top of the column. These columns are used in the pharmaceutical, petrochemical and waste water treatment industries and columns up to 1 m in diameter are in service."

Laboratory-size Extractors

The industrial-scale apparatus shown in Figures 10-2, p. 108, and 10-3 are excellent for applications involving large volumes of sample. However, most laboratory extraction problems involve extractions of a limited amount of sample - *a batch extraction*. The apparatus described in this chapter are simple, efficient, and economical for laboratory use and require small volumes of solvent. Small solvent volumes are advantageous because of (1) lower initial solvent costs, (2) less time spent removing the solvent when the extraction is finished, and (3) less solvent to worry about disposing of according to EPA regulations.

With low *D*'s, a large volume of extracting solvent is needed. However, it can be the same solvent used over and over. A typical laboratory-scale extractor for use with solvents heavier than water is shown in Figure 10-4.

An extractor that operates on the same principle, but for microscale extractions, is shown in Figure 10-5, p. 110.

Referring to Figure 10-4, the solvent is placed in the flask, A, and heated. The vapor rises to B and then up to C, where it is condensed. The liquid cannot get back in the flask because of the seal at B and runs into

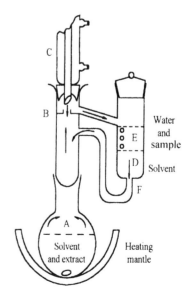

Figure 10-4. A continuous solvent-heavier-than-water extractor.

Figure 10-5. Microscale extractor.
(Courtesy - Ace Glass Co., Vineland, NJ)

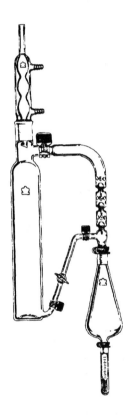

Figure 10-6. Continuous heavier-than-water extractor with built-in concentrator.
(Courtesy - Kontes Glass Inc., Vineland, NJ)

the extractor, D. The liquid drips through the water layer (sample), E, and collects at F. The excess solvent containing the extract runs out the bottom at F and back into the flask. This is a continuous process with the extract collecting in the flask. If chloroform is used, caffeine can be extracted easily from cola beverages in about 45 minutes to an hour.

Figure 10-6 is a photograph of a continuous heavier-than-water extractor with a built-in concentrator.

SOLVENT LIGHTER THAN WATER

PRINCIPLES

Continuous extractors using solvents lighter than water follow the same principles as described in the above section, except that the apparatus must be modified to handle the less dense extraction solvent. Figure 10-7, p. 111, shows several types of apparatus: Refer to the right apparatus. The sample, usually dissolved in water, is placed in the reservoir at A to fill it to less than half full. The extraction solvent, usually diethyl ether, is placed in flask B. Flask B is heated by a heating mantle with the temperature controlled by a variable transformer. When the solvent is heated, it evaporates and goes up through the apparatus to the condenser (C), where it is condensed. The condensed solvent drips into the collecting funnel (D) and gradually fills the tube. When the solvent level gets sufficiently high to build up enough pressure, it will force the solvent out of holes in the bottom of the collecting tube (E). The solvent, being lighter than water, will slowly rise up through the sample, extracting compounds as it rises. To prolong the rise, the drops of solvent are forced around the spiral on the collecting tube. The solvent and extract collect on top of the water layer. Eventually, enough solvent will collect to fill the tube, and it will drain (F) back into flask B, where the process continues. There are two problems. One, if too much sample is placed in A, the weight of the solvent might not be enough to force droplets out the bottom of the collector tube. Two, the refractive index of diethyl ether is about the same as that of water, so it is hard to determine if the extraction is actually working. Look closely. If the solvent starts to run over the funnel at D, then remove some of the sample.

TECHNIQUES

Some applications of separations involving these extractors are the separation of henna from hair dyes with ethyl acetate, the isolation of lactic acid from rotten eggs with diethyl ether, the removal of excess iron from steel samples as $FeCl_4^-,H^+$ with diisopropyl ether, and the removal of the acids from the hydrolysis of rosins with diethyl ether.

Figure 10-8, p. 111, is a diagram of a microscale continuous extractor. Figure 10-9, p. 111, is a diagram of an extractor that can serve as both a solvent-heavier or a solvent-lighter continuous extractor by merely switching the solvents in the left and right-hand flasks. For a solvent-lighter extraction, "An aqueous mixture is placed in a two liter flask on the lower side arm. A lighter than water solvent is placed in a one liter flask attached to the upper arm. Both liquids are heated to reflux, vapors are condensed and passed through the capillary tube. The solute is transferred to the solvent and both liquids are returned to their original flask. In some cases it may be

necessary to wrap the side arms to prevent premature condensation. If a heavier than water solvent is used, you need only reverse the position of the flasks (Ace glass)."

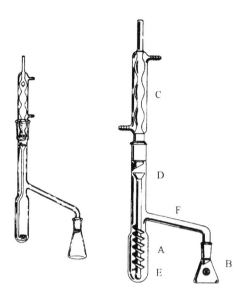

Figure 10-7. Continuous extractors using solvents lighter than water.
(Courtesy - Ace Glass Co., Vineland, NJ and Lab Glass Inc., Vineland, NJ)

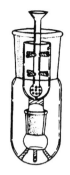

Figure 10-8. Microscale continuous extractor for solvents lighter than water.
(Courtesy - Lab Glass Inc., Vineland, NJ)

Figure 10-9. Solvent heavier - solvent lighter continuous extractor.
(Courtesy - Ace Glass Co., Vineland NJ)

THE SOXHLET EXTRACTOR

PRINCIPLES

In ancient times, water was used to remove alkalies (KOH, NaOH) from wood ashes, a process known as *lixiviation*. Later on, the process for partially dissolving a solid material to remove a desired component was called *leaching*, a term still used by industry. Chemists prefer the term *liquid-solid extraction* to distinguish the process from *liquid-liquid extraction*.

The major difficulties with a liquid-solid extraction were that the distribution ratios of the desired products were low; the process was slow; and large amounts of solvent were required, which then had to be removed. What was desired was an apparatus that could (1) hold finely divided solid particles so a large surface area could be exposed, (2) continuously pass solvent over those particles, and (3) somehow use a small amount of solvent over and over.

The most convenient laboratory-size apparatus for liquid-solid extraction in the one developed by Franz Ritter von Soxhlet (socks-let), a German agricultural chemist (1848-1926). This is shown in Figure 10-10.

This apparatus can be made in several sizes from 125 to 5000 mL. It is commercially available, costing from $100 to $800 for a complete glass portion of the apparatus.

The apparatus consists of a heating device (A) and a solvent reservoir and extract flask (B), which will be a round-bottomed flask if a heating mantle is used, or a flat-bottomed Florence flask if a hot plate is used to supply the heat. The heating mantle's temperature is controlled by a variable transformer (C). An extraction chamber (D) is placed on top of B and contains a porous paper extraction thimble (E), Figure 10-11. Porous ceramic extraction thimbles are also available. The sample (F) is placed inside of the thimble and covered with a small amount of glass wool (G) to spread out the extraction solvent and reduce channeling during the extraction. A condenser, usually an Allihn type (H), is connected to the top of the extraction chamber.

When the solvent is heated, it will go up the side arm (I), be condensed, and drop onto the sample. When the solvent level reaches (J), the solvent and the materials it has extracted then will siphon back into (A), leaving a few milliliters at the bottom of (D). The process then repeats itself. The solvent is distilled from the extracted material, so fresh solvent is always dripping onto the sample. After each cycle, the extract becomes a little more concentrated in the flask. Figure 10-12 is a photo of a commercial multiple unit for the extraction of fat.

TECHNIQUES
1. Never fill the reservoir flask more than two-thirds full at the start. The expansion and boiling can cause spillover problems.
2. Place a plug of glass wool over the top of the sample after it has been placed in the thimble. This spreads out the dripping solvent for a more uniform extraction and also prevents some samples from floating.
3. Do the ether evaporation in a hood, so the dense ether fumes are drawn away from the hot plate. Rarely if high concentrations of the fumes come in contact with the heating coils, a fire may start. Have a fire extinguisher handy - just in case.

MICROWAVE-ASSISTED EXTRACTION

PRINCIPLES
If you have many liquid-solid extractions to perform and can afford more expensive apparatus, then microwave-assisted extractions should be seriously considered. The sample is placed in a sealed Teflon container and heated with microwave energy. The *increased temperature and pressure permit a vary rapid extraction.* I.J. Barnabas, J.R. Dean, I.A. Fowlis, and S.P. Owen,

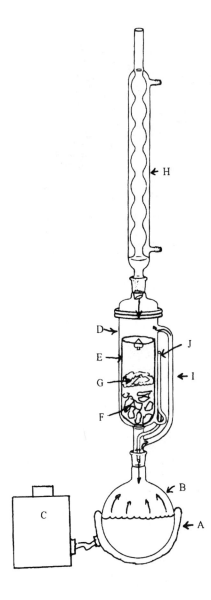

Figure 10-10. A Soxhlet extractor.
(Courtesy - Ace Glass Co., Vineland, NJ)

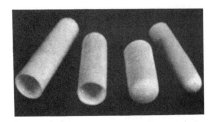

Figure 10-11. Extraction thimbles.
(Courtesy - Ace Glass Co., Vineland, NJ)

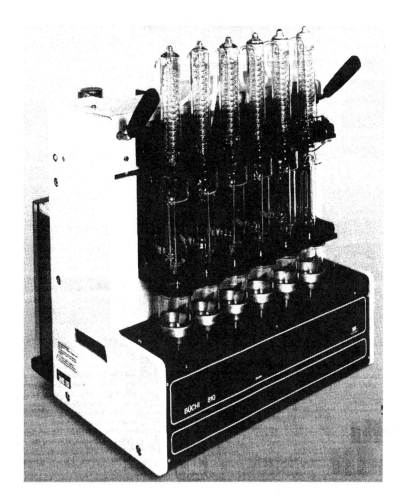

Figure 10-12. A Buchi model 810 multiple-unit Soxhlet extractor.
(Courtesy - Brinkman Instruments, Westbury, NY)

Analyst, **120**, 1897 (1995) state, "Microwave energy is a non-ionizing radiation that causes molecular motion by migration of ions and rotation of dipoles, but does not increase changes in molecular structure. It has a frequency range of 300 - 300,000 MHz with four frequencies used for industrial and scientific purposes, the most common being 2,450 MHz which is used in all domestic microwave units.

"In organic microwave sample preparation, dipole rotation refers to the alignment, due to the electric field, of molecules in the solvent and sample that have permanent dipole moments. As the field decreases, thermally induced disorder is restored which results in thermal energy being released. At 2,450 MHz, the alignment of the molecules followed by their return to disorder occurs 4.9×10^9/second, which results in rapid heating. The polarizability of the solvent molecules obviously depends on the nature of the solvent and its relative permittivity. Therefore the greater the relative permittivity the more thermal energy is released and the more rapid the heating is for a given frequency. Non-polar solvents, such as hexane and toluene, with low relative permittivities are not affected by microwave energy and therefore require polar additives if they are to be used as solvents in microwave extraction.

"*Polarizability:* Some molecules in the presence of electric fields (E) can have a dipole moment (m) induced.

$$m = \alpha E \tag{10-2}$$

"The proportionality constant, α, is called the *molecular polarizability*. E has the dimensions q (charge) x cm^{-2} and m is $q \times$ cm so α is a volume measurement and is of the order of 10^{-24} cm^3.

"*Permittivity*: In Coulombs Law relating the force of repulsion between two charges:

$$F = \frac{q_1 \, q_2}{\varepsilon \, r^2} \tag{10-3}$$

ϵ is the absolute permittivity of the medium and the ratio ϵ/ϵ_o = K or the *dielectric constant*."

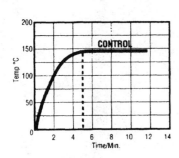

Figure 10-13. Microwave heating of CH_2Cl_2 in a closed vessel.
(Courtesy - CEM Co., Mathews, NC)

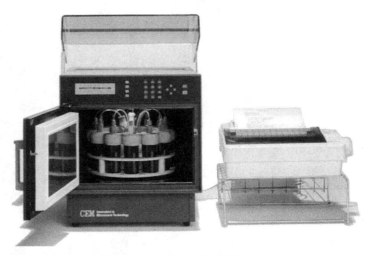

Figure 10-14. A CEM Co. model MES 1000 microwave extraction unit.
(Courtesy - CEM Co., Mathews, NC)

"A typical Soxhlet extraction, by conductive heating will be completed in 5-6 hours. Alternatively, closed vessel extractions by microwave heating can be completed in 15 minutes. The difference is due to the sample heating method. In conductivity heating, vaporization at the liquid surface causes a thermal gradient to be established by convection currents and only a small portion of the liquid is at the temperature of the heat that is applied to the outside of the vessel. However, microwaves heat all of the sample simultaneously without heating the vessel. Therefore, with microwave heating the solution reaches its b.p. very rapidly. Also, because the solvent is in a sealed system, it is capable of reaching a far greater b.p. than at atmospheric pressure. Polar solvents, such as acetone and dichloromethane, are heated to approximately 100 °C above their normal b.p.s. It is these high extraction solvent temperatures combined with rapid heating which increase extraction efficiency and therefore greatly reduce extraction time."

Figure 10-13 shows how fast dichloromethane (b.p. 40 °C) can be heated and its final temperature at 200 psig.

TECHNIQUES

Figure 10-14 shows a 12-position unit. Figure 10-15 shows the units in more detail, Figure 10-16 shows a cross-sectional diagram of a pressure-and-temperature sensitive cell, and Figure 10-17 shows the details of the infrared optical temperature probe. The vessels are normally Teflon lined, of 100 mL volume, and capable of operating at 200 °C and 200 psig. Typical solvent systems are hexane-acetone (1:1), hexane-methanol (1:1), and dichloromethane. The fiberoptic temperature control is constructed of high-purity fused silica, is microwave transparent, and is thermally and electrically nonconductive. The temperatures can be set over a 0-200 °C range with an accuracy of 1% full scale. It is immersed directly into the reaction process for quick response. Either the temperature or the pressure is set as a limit, and the heating will shut off when these levels are attained. In addition, the unit has a rupture membrane as an additional safety feature.

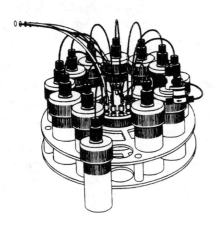

Figure 10-15. Twelve lined vessels, containment holder, and temperature and pressure probes.
(Courtesy - V. Lopez-Avla, R. Young, and W.F. Beckert, *Anal. Chem.*, **66** (7), 1097, 1994)

Microwave heating requires less solvent to purchase, concentrate, and discard. Tests have shown that 530 microwave extractions can be performed with the same amount of solvent as 32 Soxhlet extractions. Using 16 L of acetone at a cost of $125, 32 Soxhlet extractions cost $3.90 each, whereas the 530 microwave extractions cost $0.24 each.

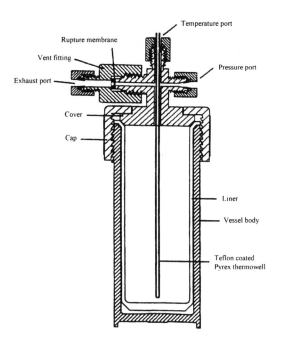

Figure 10-16. Lined vessel with temperature and pressure control.
(Courtesy - V. Lopez-Avila, R. Young, and W.F. Beckert, *Anal. Chem.*, **66** (7), 1097, 1994)

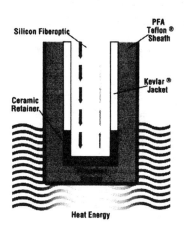

Figure 10-17. The CEM fiberoptic probe.
(Courtesy - CEM Co., Mathews, NC)

REVIEW QUESTIONS

1. Explain how any one of the agitating extractors operates.
2. What factors are necessary to have a highly efficient continuous extraction?
3. If a continuous extractor operates for 2 hours at a distillation rate of 2 drops/second of fresh extracting solvent, what is the half extraction volume, if we assume the sample is half extracted in 50 minutes? Assume 1 drop = 0.05 mL.
4. If the original sample volume in problem 3 was 50 mL, what is the k value?
5. Refer to the cola nut analysis in Experiment 13, Appendix A, and determine what the concentration of caffeine is in ppm.
6. List three solvents other than chloroform that could be used in a heavier-than-water extractor.
7. What is the solubility of caffeine in water? (*Merck Index* or *CRC Handbook*)
8. What is the solubility of caffeine in chloroform? (*Merck Index* or *CRC Handbook*)
9. The absorbance of 5.0 mg of caffeine standard/25 mL of acetone had an absorbance of 0.225. The absorbance for the extracted caffeine from 50 mL of a cola soft drink diluted to 25 mL is 0.532. What is the concentration of caffeine in ppm, if you assume 1 mL of the soft drink = 1 g?
10. A mocha raw coffee is estimated to contain 1.1% caffeine. Fifty mL of hot water is used to extract the caffeine from the coffee along with several other compounds from a 10.0 g sample. If 5.0 mg caffeine/25 mL of acetone produces an absorbance of 0.225, what size aliquot of the sample extract should be taken to provide a final absorbance in the 0.225 range?
11. A common problem with this type of apparatus is that too much sample is placed in the tube. If this is the case, what will happen when the solvent is distilled?
12. What is the purpose of the spiral glass on the one type of apparatus shown in Figure 10-7, p. 111?
13. What are three other liquids lighter than water that could be used with this type of extractor?
14. Examine the right extractor in Figure 10-7, p. 111, and explain how it works.
15. Examine the extractor in Figure 10-9, p. 111, and explain how it works.
16. In the early days of anesthesia, diethyl ether was a favorite. Can you suggest an explanation as to why people in the operating room had to wear shoes that could be grounded?

17. The following has happened enough times in the evening to make it worth discussing. Proper laboratory safety procedures will eliminate this problem, but it still seems to happen occasionally. Diethyl ether was used during the day and the excess poured down the sink. Sometime later, a cigarette butt was thrown into the sink, often by janitors, and there was quite an explosion. Explain how this could happen? Usually no one gets hurt badly, but hair is sometimes singed, and the noise scares the ---- out of you.
18. A 10 mL sample of Ipecac was treated as in Experiment 15. For the titration, 5.00 mL of 0.1500 N H_2SO_4 was added, which then required 31.60 mL of 0.025 N NaOH to reach the endpoint. The blank was 2.20 mL. How many mg of Ipecac alkaloids, calculated as emetine, were present? This bottle had the same label as that in the example calculation. What are your conclusions?
19. What is leaching?
20. Why is a piece of glass wool placed over the sample during the extraction?
21. What causes "siphoning"?
22. What are two characteristics that make microwave-heated extraction faster than an ordinary solvent-heated extraction?
23. What is the most common frequency used in most domestic microwave units?
24. How fast do the average molecules restore themselves after being disordered in an electric field?
25. Why is acetone or methanol added to hexane as an extraction solvent in microwave-heated extraction?
26. What is meant by polarization?
27. What is meant by permittivity?
28. What prevents a microwave-heated vessel from exploding?

IF IT EXISTS - CHEMISTRY IS INVOLVED
IF YOU CAN BUY IT - A CHEMIST WAS INVOLVED SOMEWHERE

11
COUNTERCURRENT (EXTRACTION) CHROMATOGRAPHY
THE ITO COIL-PLANET CENTRIFUGAL EXTRACTOR

PRINCIPLES

Note: The diagrams and text for the following discussion were abstracted from the many papers of Dr. Y. Ito; from the text by Y. Ito and N. Mandava, *Countercurrent Chromatography, Theory and Practice* (Marcel Dekker, New York, 1988); and from discussions with Dr. Walter Conway of State University of New York at Buffalo.

Countercurrent extraction is a liquid-liquid partition system in which no solid matrix is required to hold the stationary phase. R.E. Cornish, Archibald, R.C., Murphy, E.A. and Evans, H.M., *Ind. Eng. Chem.*, **26**, 397 (1934) are credited with the first successful apparatus. In 1966, Ito, Y., Harada, R., Weinstein, M.A., and Aoki, I. in *Ikakikaigakuzaski*, **36**, 1 (1966) and in *Nature* **212**, 985 (1966) developed the first coil-planet centrifuge. This was followed in 1981 by a multilayer arrangement (Ito, Y., *J. Chromatogr.*, **214**, 122), which is the centerpiece of today's apparatus. The coil-planet arrangement (Figure 11-1) sets up an unsymmetrical centrifugal force, which can cause immiscible liquids in the coil to counterflow and produce extensive mixing in sections of the coil. While it may have only 500 plates compared to the several thousand of a high performance liquid chromatograph, it has a much larger capacity; and because the stationary phase is so large, it can provide a sharp separation with a wide variety of partition coefficients.

This technique usually is not as good as high performance liquid chromatography, but if you need mg to g quantities rather than nanograms, then this is a technique to seriously consider. It has been applied to all types of compounds. To quote Dr. Walter Conway, "It is good for both polar and non-polar organic materials as well as inorganic mixtures such as the rare earths. It has been applied to many classes of compounds, including agricultural chemicals, alkaloids, amino acids, peptides, proteins, antibiotics, drug metabolites, dyes, food products, flavonoids, glycosides, herbicides, pesticides, pharmaceuticals, optical isomers, saponins, tannins, metals and other inorganic materials.

"Basically, the method retains a stationary phase in a coiled column by centrifugal means while a mobile phase is passed through it. Highly efficient mixing is accomplished by rotating the column within the centrifugal force field. In other words, there are *two distinct motions, revolutionary and planetary,* working in synchronization. The configuration of the multilayer coil creates force field gradients that allow faster flow rates and rapid separation of larger volume samples. A seal-free flow-through system facilitates a simple leak-free continuous elution process.

"The instrument consists of a motor and an off-axis column that rotates in a particular mode of planetary motion. In this mode, the two liquid phases establish a hydrodynamic equilibrium wherein the stationary phase dominates the total volume of the coil, often exceeding 80% of the total column volume. Under this equilibrium, the introduced mobile phase moves with relatively high speed through the column.

"The apparatus can rotate up to 1200 rpm, but speeds in the region of 800-1000 rpm are sufficient to maintain good retention of the stationary phase. The column is wound in layers to provide a centrifugal gradient that further enhances stationary phase retention. The multilayer coil consists of one continuous winding of polytetrafluoroethylene (PTFE) tubing. In some apparatus, the spool contains more than one coil and these may be connected in series to provide larger capacity. The columns are easily interchangeable and balancing is easily accomplished with a set of aluminum and brass discs provided with the system.

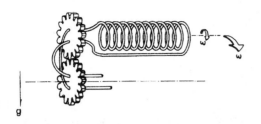

Figure 11-1. Diagram of a horizontal flow-through gear-driven coil-planet centrifuge.
(Courtesy - Mandava, N, and Ito, Y., *Countercurrent Chromatography, Theory and Practice,* Marcel Dekker, New York, 1988)

"Typically, the two phases of the solvent system are mutually saturated by shaking in a separatory funnel. The chosen stationary phase is loaded into the coil in the absence of rotation. The sample is then injected as a solution in either one, or a mixture of both phases. Rotation is then started and the mobile phase is pumped into the coil. Some stationary phase will be displaced and the sample constituents will be separated and eluted in the order of their partition coefficients. Mass transfer of the solute is promoted by a train of dynamic mixing zones, generated by the planetary motion of the coil as it rotates on its own axis while traversing a planetary orbit."

On each half orbit as a planet, the stationary phase is forced across the tube. This provides for substantial mixing of

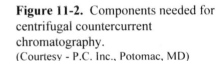

Figure 11-2. Components needed for centrifugal countercurrent chromatography.
(Courtesy - P.C. Inc., Potomac, MD)

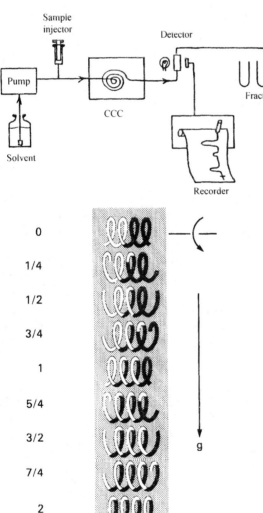

Figure 11-3. Countercurrent process of two immiscible liquids in a rotating helical coil.
(Courtesy - Mandava, N, and Ito, Y., *Countercurrent Chromatography, Theory and Practice,* Marcel Dekker, New York, 1988)

the two phases and rapid partitioning. During the next half orbit the stationary phase returns to its original position.

Figure 11-1 is a diagram to illustrate the principle of the flow-through coil planet centrifuge. "The coil holder revolves twice on its own axis for each circuit of its orbit at the same speed. This planetary motion allows continuous elution through the rotating column and the anti-twisting motion eliminates the use of rotary seals."

Figure 11-2 is a diagram of the component parts needed for an analysis. There are two main instrument systems in countercurrent extraction: (1) the *hydrostatic equilibrium system* and (2) the *hydrodynamic equilibrium system*. In the hydrostatic equilibrium system (HSES), a stationary coil is placed in a horizontal position. In the hydrodynamic equilibrium system (HDES), the coil rotates around its own axis These will be discussed individually in more detail, but first, what happens when two immiscible liquids are placed in a tube and the tube rotated slowly will be examined.

Refer to Figure 11-3. "The coiled tube at the top is filled with the upper phase (clear) at the head end and the lower phase (shaded) in the tail end forming an interface at the bottom of the coiled tube. When the tube is subjected to a slow rotation under a gravitational field as indicated, the two phases

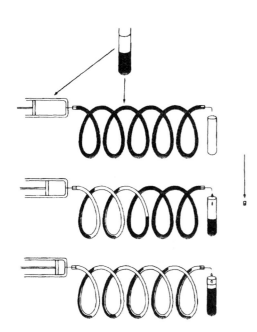

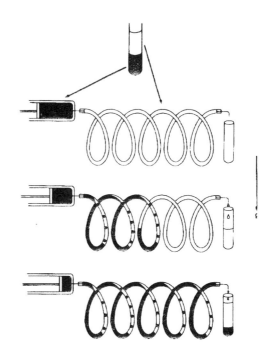

Figure 11-4. The HSES system with a high wall affinity.
(Courtesy - Mandava, N, and Ito, Y., *Countercurrent Chromatography, Theory and Practice*, Marcel Dekker, New York, 1988)

Figure 11-5. The HSES system with low wall affinity.
(Courtesy - Mandava, N, and Ito, Y., *Countercurrent Chromatography, Theory and Practice*, Marcel Dekker, New York, 1988)

begin to interchange through the interface. In each half rotational cycle of the coil, a new interface is formed at the bottom of the adjacent coil loop through which the countercurrent process continues. The process occurs four times during two complete rotational cycles when the two phases finally establish the hydrodynamic equilibrium state. At this stage each coil unit is occupied by approximately equal volumes of the two phases as illustrated, and the whole length of the tube contains alternating segments, each about half a coil unit in length, of the two phases.

"If the sample solution is introduced beforehand at the interface of the two phases the solutes are subjected to an efficient continuous partition process and eventually distributed throughout the length of alternating segments of the two phases in the coil according to their relative partition coefficients."

THE HYDROSTATIC EQUILIBRIUM SYSTEM

"A model of this basic system is shown in Figures 11-4 and 11-5. Figure 11-4 shows a system of five coils. The stationary coil is first filled with the lower (heavier) phase of an equilibrated two-phase solvent system and the upper (lighter) phase is slowly introduced through the end of the coil. The upper phase then pushes the lower phase downward in the first coil unit until the interface reaches the bottom of the coil, where the upper phase starts to percolate upward through the lower phase affected by the gravity. This retains the lower phase in one half of the first coil unit. This process is repeated in each coil unit to produce alternating segments of the upper and the lower phases throughout the coil. Once this so-called hydrostatic equilibrium state is established, continued flow of the upper mobile phase results in the replacement of the upper phase only, leaving the lower phase stationary in each coil unit.

"Figure 11-5 illustrates the similar process when the lower phase is used as the mobile phase. The stationary coil is first filled with the upper phase and the lower phase is slowly introduced at one end of the coil. In this case, the lower phase immediately starts to percolate downward through the upper phase by the effect of gravity and quickly reaches the bottom of the first coil unit. The lower phase then pushes the upper phase upward until the interface reaches the top of the coil. This process is repeated in each coil unit to produce alternating segments of the two phases throughout the coil. After this hydrostatic equilibrium state is established, the mobile lower phase displaces the lower phase while retaining the stationary upper phase in each coil.

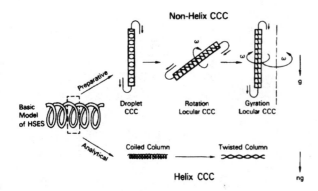

Figure 11-6. Several schemes based on the HSES system.
(Courtesy - Mandava, N, and Ito, Y., *Countercurrent Chromatography, Theory and Practice,* Marcel Dekker, New York, 1988)

"Solutes locally introduced at the inlet of the coil are subjected to a partition process between the flowing mobile phase and the retained stationary phase in each partition unit and finally eluted out with the mobile phase according to the order determined by their relative partition coefficients. If the partition process in each partition unit is highly efficient, this model is expected to yield a partition efficiency close to five theoretical plates.

"Figures 11-4 and 11-5 also illustrate the effects of liquid-wall interaction. When the mobile phase has a strong wall surface affinity, it flows smoothly along the wall of the tube to form a continuous stream (Figure 11-4). When the mobile phase lacks the wall surface affinity, it forms multiple droplets into the stationary phase segments (Figure 11-5)."

Figure 11-6 illustrates several arrangements based on the HSES system.

THE HYDRODYNAMIC EQUILIBRIUM SYSTEM

This system is the same as the HSES system, except that the coil is slowly rotating around its own axis. This simple rotation introduces a new feature to the system that involves complex hydrodynamic interactions of the two solvent phases in the coil. Refer to Figure 11-7. (a) "Consider a coil filled with water into which some glass beads and a few air bubbles are added. A rotating coil exhibits an Archimedean screw effect. Any object, either lighter (air bubbles) or heavier (glass beads) than the water, moves toward one end of the coil. This end is called the *head* and the other end, the *tail* of the coil."

(b) "A similar phenomenon is observed with the two immiscible solvent phases, such as chloroform and water. A small amount of the lower (heavier) phase suspended in the upper (lighter) phase travels toward the head (top) while a small amount of the upper phase suspended in the lower phase also travels toward the head of the coil (bottom). This suggests that the relative volumes of the two phases govern their behavior in the rotating coil."

(c) "A two-phase solvent system introduced in a slowly rotating coil soon establishes a hydrodynamic equilibrium state whereby each solvent phase occupies nearly equal space in each helical turn and any excess of either phase remains at the tail of the coil."

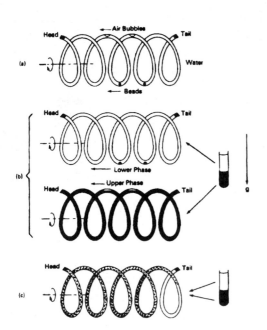

Figure 11-7. Motion of various objects in a rotating coil.
(Courtesy - Mandava, N, and Ito, Y., *Countercurrent Chromatography, Theory and Practice,* Marcel Dekker, New York, 1988)

Now refer to Figures 11-8 and 11-9, p. 121. "In 11-8 the coil is first filled with the heavier stationary phase or the lighter stationary phase (11-9) and the mobile phase is introduced at the head of the coil while the coil is slowly rotated around its own axis (*top*). As soon as the mobile phase meets the stationary phase in the coil, a hydrodynamic equilibrium is established between the two phases (*middle*). This process continues until the mobile phase reaches the tail of the coil. Thereafter, the continued elution results in the displacement of

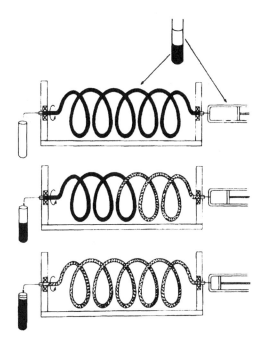 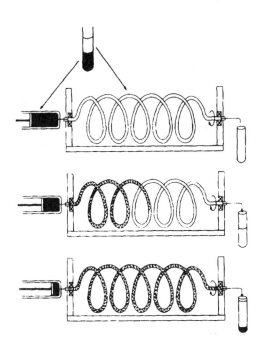

Figure 11-8. Coil first filled with lower phase. (Courtesy - Mandava, N, and Ito, Y., *Countercurrent Chromatography, Theory and Practice,* Marcel Dekker, New York, 1988)

Figure 11-9. Coil first filled with upper phase. (Courtesy - Mandava, N, and Ito, Y., *Countercurrent Chromatography, Theory and Practice,* Marcel Dekker, New York, 1988)

the mobile phase only and a large amount of stationary phase is permanently retained in each helical turn of the coil (*bottom*). Consequently, solutes introduced locally at the head of the coil are subjected to an efficient partition process between the mobile and the stationary phases and separated according to their partition coefficients.

"As soon as the mobile phase reaches the first coil unit, the two phases interact to establish the hydrodynamic equilibrium where one phase with less wall surface affinity is split into multiple droplets which oscillate synchronously with the coil rotation. While continuous pumping of the mobile phase keeps breaking this equilibrium state at the head end of the coil, the two phases can quickly react to restore the equilibrium by readjusting their relative volumes in each coil unit. Thus the stationary phase pushes newly introduced excess amounts of the mobile phase toward the tail end of the coil.

"The hydrodynamic system has an advantage over the hydrostatic system in that the rotating coil has no inefficient free space which is completely occupied by the mobile phase and therefore every portion of the coil is considered to be efficient column space. At any portion of the coil, one of the phases forms droplets regardless of the choice of mobile phase. Each droplet becomes an efficient stirrer of the other phase by the violent local oscillation."

THEORY

Resolution

Resolution is a measure of how well two components are separated. In later chapters, you will learn that it involves primarily three factors: k, the *capacity factor*, α, the *selectivity*, and N, the *number of plates*. In countercurrent extraction, another term is involved, S_f, the *fraction of stationary phase in the column.* Equation 11-1 shows how these are related. Equation 11-2 is easier to use in practice and gives nearly the same results.

$$R_s = 0.25\,(\alpha - 1)\sqrt{N}\,\frac{K_1}{0.5\,K_1\,(\alpha + 1) + \dfrac{(1 - S_f)}{S_f}} \tag{11-1}$$

where $\alpha = k_2/k_1$ (peak 2 and peak 1), $K = (V_R - V_m)/(V_c - V_m)$. V_R = retention volume in mL. V_m = volume of the mobile phase in the column in mL. V_c = known column volume in mL, $N = 16(V_R/w)^2$, w = baseline peak width in mL (4 σ), $S_f = V_s/V_c$. V_s = volume of the stationary phase in the column in mL.

EXAMPLE CALCULATION

What is the resolution between two components in the following experiment? A 300 mL column filled with 200 mL of stationary phase was used. 2 mL fractions were collected, and the solutes determined with a uv spectrophotometer. The first component came through with a band width of 12 mL, after 130 mL of mobile phase had passed through the column. The second component came through with a band width of 18 mL, after 144 mL of mobile phase had passed through the column.

ANSWER

$$k_1 = \frac{(130 \ mL - 100 \ mL)}{(300 \ mL - 100 \ mL)} = 0.15; \quad k_2 = \frac{(144 \ mL - 100 \ mL)}{(300 \ mL - 100 \ mL)} = 0.22; \quad \alpha = \frac{0.22}{0.15} = 1.47$$

$$N_1 = 16\left(\frac{130 \ mL}{12 \ mL}\right)^2 = 1878 \ plates; \quad N_2 = 16\left(\frac{144 \ mL}{18 \ mL}\right)^2 = 1024 \ plates; \quad Avg = 1,451; \quad \sqrt{1,451} = 38.1$$

Use an average N when using Equation 11-1.

$$R_s = 0.25 \ (1.47 - 1)(38.1) \ x \left[\frac{0.15}{0.5 \ x \ 0.15 \ x \ 2.47 + \left(\frac{1 - 0.67}{0.67}\right)}\right] = 0.98$$

This is < 1. This means that the two peaks are not baseline resolved. 130 + 6 = 136 and 144 - 9 = 135. There is a slight overlap, and 0.98 is about right.

$$R_s = \frac{2(V_{R_2} - V_{R_1})}{W_1 + W_2} \tag{11-2}$$

where Vr_1 and Vr_2 = the retention time of the first and second peaks, respectively. W = 4 σ base width of the peaks.

EXAMPLE CALCULATION
Using the same data as in the above example, calculate the resolution.

ANSWER

$$R_s = \frac{2(144 - 130)}{18 + 12} = 0.93$$

According to Ito, "CCC is governed by the same equations that apply to all forms of chromatography but, because the phase volumes are known in CCC, it is simple to express the equations in terms of the partition coefficient $K = C_s/C_m$, where C represents solute concentration in the stationary phase, s, and mobile, m, phases, rather than the capacity factor $k' = K(V_s/V_m)$, where V represents the volume of the stationary, s, and mobile, m, phases. Because there is no supporting matrix, and the column volume, V_c (column) is known, the stationary phase volume can be obtained from

$$V_s = V_c - V_m = V_c - V_{co} \qquad (11\text{-}3)$$

where the initially displaced stationary phase, V_{co}, is used to estimate V_m. The retention equation is

$$V_R = V_m + K V_s \qquad (11\text{-}4)$$

where V_R is the retention volume.

"The relationship between K and the cc chromatogram can be discerned visually from Figure 11-10. The universal chromatogram shows the elution of hypothetical solutes with $K =$ 0, 1, 2, and 3. When $K = 0$, the solute spends no time in the stationary phase and emerges as the mobile phase front, V_m. Any solute with $K = 1$ emerges at the column volume, V_c. Note that this is the focal point of the countercurrent chromatogram and is independent of the phase volume ratio. This follows directly from the retention equation; when $K = 1$, $V_R = V_m + V_s = V_c$. Solutes with higher K emerge at multiples of V_s.

"Because V_c will be known, its position on the chromatogram can be determined from the flow rate or fraction volumes. It will be a fixed point, but the $K = 0$ point will depend on the mobile phase volume, which can be estimated from either V_{co} or the emergence of an unretained solute or as simply the first peak or baseline aberration on the chromatogram. The distance between $K = 0$ and $K = 1$ gives V_s, which can be laid off beyond $K = 1$ to give the elution volumes for higher K values. K can be calculated with the equation:

$$K = \frac{V_R - V_m}{V_c - V_m} \qquad (11\text{-}5)$$

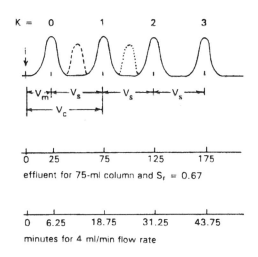

Figure 11-10. Universal cc chromatogram with volume and time scales for a 75-mL column and $S_F = 0.67$.
(Courtesy - Conway, W., Sharma, M., and Chen, L., Dept. of Pharmacy, Univ. of Buffalo, Amherst, NY, 1998)

A good visual estimate can be made by linearly interpolating between the above index points. Thus, the peak illustrated by the dashed line on Figure 11-10 has a K of about 0.5 whereas the dotted line peak is about 1.5.

"It is interesting to note that as V_m increases, V_s will decrease, causing all of the peaks to emerge at the $K = 1$ peak when $V_m = V_s$. This illustrates that resolution depends on the volume of stationary phase retained in the column and S_f tends to be high in CCC, which partially explains the good resolution often obtained."

EXAMPLE CALCULATION

V_m for a 12.2 mL coil was found to be 2.1 mL. The retention volumes for thymol and carvacrol under conditions described in Experiment 16 were 14.8 mL and 13.1 mL, respectively. Calculate the partition coefficients, K_p, for each compound. The sub p refers to the partition coefficient in the polar phase. What is the stationary phase fraction?

ANSWER

$$\text{Carvacrol: } K_p = \frac{13.1 \ mL - 2.1 \ mL}{12.2 \ mL - 2.1 \ mL} = 1.10 \qquad \text{Thymol: } K_p = \frac{14.8 \ mL - 2.1 \ mL}{12.2 \ mL - 2.1 \ mL} = 1.28$$

The stationary phase fraction is calculated as follows:

$$S_f = \frac{(V_c - V_m)}{V_c} \tag{11-6}$$

$$S_f = \frac{(12.2 \text{ mL} - 2.1 \text{ mL})}{12.2 \text{ mL}} = 0.83 \text{ or } 83\%$$

TECHNIQUE

SEAL-FREE CENTRIFUGE SYSTEMS

How can you add sample, mobile phase, or pump these materials into a rotating system without having a rotating seal? Why would you want to anyway? Ito lists seven reasons why eliminating a rotating seal would be beneficial.

1. The system is leak free under a high pressure of several hundred psi.
2. It readily facilitates the use of multiple-flow channels.
3. Samples are free from mechanical damage at the site of the seal.
4. It eliminates the heat dissipation problem at the seal.
5. It eliminates the contamination problem.
6. It permits the use of corrosive solvents.
7. It minimizes the dead space at the seal, which would cause unnecessary sample band broadening.

Figure 11-11, p. 125, is a diagram of several seal-free designs. They increase in complexity from top left to bottom right and are divided into three categories based upon their mode of planetary motion. They are called *synchronous, nonplanetary,* and *nonsynchronous*. The arrows are not clear, but note that in I (upper left) the angular rotation of the planet is opposite that of the orbit, whereas in J (lower left), the planet rotates in the same sense as the orbit because this planet is now inverted. "In the nonplanetary scheme, the coil holder simply rotates around the central axis of the centrifuge as in the conventional centrifuge system. On the other hand, both the synchronous and nonsynchronous schemes produce planetary motions of the coil holder; hence they are called *coil planet centrifuges* (CPC). In the synchronous schemes the rotation and revolution of the holder are synchronized to give rotation-revolution ratios of 1:1 as in I or 2:1 as in J, while in the nonsynchronous schemes the rotational rate of the holder is independently adjustable with respect to the revolution.

"In scheme I, the holder undergoes the simplest mode of synchronous planetary motion by keeping its axis always parallel to and at a given distance from the central axis of the centrifuge. In order to avoid the flow tubes being twisted, the holder synchronously counterrotates about its own axis while revolving around the central axis of the centrifuge. As a result; the motion of the holder becomes similar to the circular motion of a beaker observed when a chemist gently mixes its contents by hand. This is called the *flow-through CPC* and is used for both analytical and preparative separations. It is also analogous to a Ferris wheel motion, since it is exactly like the motion of the cars on the Ferris wheel. It is also the same motion generated by a vortex mixer.

Scheme I-L has a tilted orientation of the holder that creates a very complex pattern. Each coil unit experiences two effective forces: the first component acts to trap the stationary phase like the gravity in the basic HSES model, and the second component acts to produce efficient mixing of the two solvent phases as in the basic HDES model. Relative strength of these two force components is determined by the inclination of the holder. This type is called the *angle rotor flow-through CPC* and is used mostly for analytical scale separations.

Scheme L is perpendicular to the centrifuge axis and has two components. The second can be minimized by proper location along the holder. This is called the *elution centrifuge* and is used primarily for analytical scale separations.

Scheme J is an inverted orientation of the holder; the holder rotates about its own axis and revolves around the central axis of the centrifuge at the same angular velocity in the same direction. This produces a quite different pattern of the centrifugal force field compared to scheme I, displaying either a rotating or an oscillating pattern depending on the location of the point on the holder.

FACTORS THAT INFLUENCE SEPARATION

Mandava and Ito list 11 factors that influence the separation:
1. Number of coil units in which the countercurrent process takes place.
2. Internal diameter of the tube used.
3. Rotational speed of the coil or "contact time" (the time required for the media to pass through a given number of the coil units).
4. Centrifugal force applied.

5. Helical diameter of the coil.
6. Tubing material.
7. Sample size.
8. Molecular weight and shape of the solute.
9. Interactions between two phases and inner surface of the tube.
10. Viscosity of the media.
11. Vibration applied to the coiled tube during partitioning.

Number of Coil Units

Figure 11-12 shows the relationship between the number of coils and the inner diameter of the tube in a coil-planet centrifuge. Clearly, the *efficiency increases as the tubing diameter gets smaller and the number of coils increases.*

Internal Diameter of the Tube Used

Figure 11-13, p. 126, shows the effect of coil diameter. In a large-diameter coil, the droplets tend to accumulate on the side of the coil, whereas with a smaller-diameter coil, the droplet can occupy the space across the diameter of the coil. Therefore, the *efficiency rises in inverse proportion to the internal diameter.*

Contact Time of the Media

Contact time is related to the rotational speed of the coil unit. As a general rule, the *efficiency rises nearly in proportion to the square root of the contact time* for a given number of turns of the tube.

Centrifugal Force

The centrifugal force (50 *x* g to 350 *x* g) influences the volume ratio between the upper and lower phases in the coil unit. Under a given centrifugal force and rotational speed, *the volume of the advancing phase in a coil unit decreases as the tube becomes smaller and/or the viscosity of the media increases.*

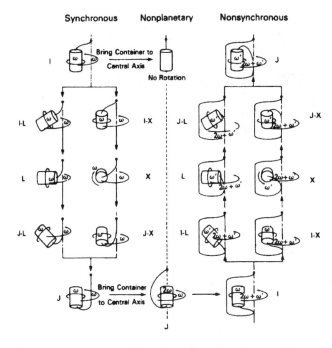

Figure 11-11. A series of seal-free flow through centrifuge schemes and their mutual relationship.
(Courtesy - Mandava, N, and Ito, Y., *Countercurrent Chromatography, Theory and Practice,* Marcel Dekker, New York, 1988)

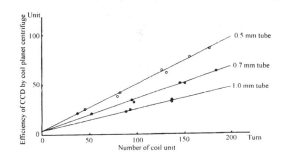

Figure 11-12. Effects of number of helical turns and internal diameter of the tube on partition efficiency.
(Courtesy - Mandava, N, and Ito, Y., *Countercurrent Chromatography, Theory and Practice,* Marcel Dekker, New York, 1988)

Helix Diameter

An increase in the helix diameter provides a larger capacity, so that larger samples can be run. *For a given length of tube, the partition efficiency decreases with the increase of the helix diameter.*

Tubing Material

The harder the tube, the better the expected result is. Polytetrafluoroethylene (PTFE) is preferred for most organic solvents because few of these solvents will dissolve in it.

A common coil consists of a winding of 130 m of polytetrafluoroethylene tubing, 1.7 mm i.d. x 2.5 mm o.d. on

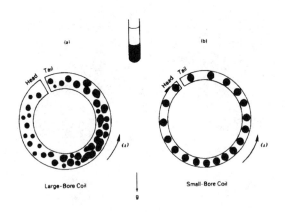

Figure 11-13. Motion of a two-phase solvent system as a function of coil diameter.
(Courtesy - Mandava, N, and Ito, Y., *Countercurrent Chromatography, Theory and Practice*, Marcel Dekker, New York, 1988)

a 5 cm wide spool with a 10 cm core and an o.d. of 17.8 cm, giving a column volume of about 300 mL. This is used with an orbital radius of 10 cm. A 15 mL column can elute a solute with a $k = 1$ in about 10 minutes. This is good for a preliminary cleanup. A 75 mL column is used for most applications, and separations take 1-2 hours. The 300 mL column is used for fine separations and can require several hours, depending on the partition coefficients.

Figure 11-14 is a photograph of a commercial coil-planet centrifuge.

Sample Size

As a general rule, the sample size can be up to nearly 5 % of the total volume of the media without altering the partition coefficient. The sample is dissolved in one of the phases and loaded by a sample loop.

Figure 11-14. A multilayer coil-planet countercurrent chromatograph, model DP-950.
(Courtesy - P.C. Inc., Potomac, MD)

Figure 11-15 is a photograph of a multilayer coil containing three windings: 15 mL, 75 mL, and 225 mL. Figure 11-16 is a photograph of a sample injection apparatus used to force the mobile phase through the column. It is simply a sample loop injector device where the syringe is used to load the loop, which then is injected in the mobile phase stream using a Rheodyne 6-port injection valve.

Figure 11-15. A triple multi-layer coil.
(Courtesy - P.C. Inc., Potomac, MD)

Figure 11-16. Sample injection system.
(Courtesy - P.C. Inc., Potomac, MD)

REVIEW QUESTIONS

1. How does the Ito system differ from the older Craig system? Do not try to answer this unless you have a diagram of a Craig apparatus.
2. What are the distinct motions in a coil-planet centrifuge?
3. Why is the column wound in layers?
4. How is the fixed orientation of the helix sample coil maintained?
5. What is the difference between the hydrostatic equilibrium system and the hydrodynamic equilibrium system?
6. What causes the mobile phase to break into droplets in the hydrostatic equilibrium system?
7. Considering the rotating coil, what is the advantage of the hydrodynamic system over the hydrostatic system?
8. What would be the resolution between two components in this case? A 75 mL column was filled with 60 mL of stationary phase; the first compound came through with a band width of 3 mL after 25 mL of mobile phase had passed through the column, and the second component had a band width of 4 mL after 40 mL of mobile phase had passed through.
9. When using orbital turns per plate for comparing systems, what term is actually used?
10. As V_m increases, V_s decreases. What does this indicate?
11. V_m for a 12.2 mL coil was found to be 2.8 mL. The retention volumes for two spice components were 13.8 and 16.5 mL, respectively. What is the partition coefficient, K_p, of each compound?
12. What is the basis for the synchronous scheme of rotation?
13. What is the main use for scheme L in Figure 11-11, p. 125?
14. How is efficiency related to the internal coil diameter?
15. What is the effect of helix diameter?
16. What is the maximum size sample that can be used without altering the partition ratio?

IF IT EXISTS - CHEMISTRY IS INVOLVED
IF YOU CAN BUY IT - A CHEMIST WAS INVOLVED SOMEWHERE

12
SOLID PHASE EXTRACTION

PRINCIPLES

Solid phase extraction (SPE) is a liquid-solid separation. It was developed commercially by the Waters Co. in 1978 and is sold as Sep-Pak. Since then, others have entered the field. The material is either a small porous disc, about the size of a dime, or a small plug, about 1 cm in diameter, of silica coated with various compounds. The liquid containing the desired compounds is poured through the disc or plug. Usually the desired compounds are retained and the impurities pass through. The desired compounds, now highly concentrated and in a much purer form, are eluted with just a few milliliters of solvent. The type of compounds retained is determined by the choice of disc material. The disc materials are usually the same as those used for high performance liquid chromatography (HPLC), but of larger particle size. Particle sizes for HPLC are from 3 to 10 µm in diameter while those for SPE separations are from 40 to 80 µm. The compounds on the particles have various end caps to provide selectivity, and gravity flow is common with either low pressure or low vacuum usually being applied. Columns are typically about 10 mm in diameter and about 75 mm long. Figure 12-1, p. 130, summarizes the process.

Solid phase extraction has several advantages over conventional liquid-liquid extraction when trace components are of interest. It is faster, requires less solvent, reduces the need for large concentration steps, and is easily automatable. Several liters, if necessary, can be poured through the column, and the trace components collected. These can be eluted with only a few mL of solvent, so the solvent removal for further concentration is nearly or completely eliminated. However, an HPLC column of 30 cm can provide 15,000 plates, whereas a solid phase disc can provide only 10 to 50 plates. What this means is that a solid phase disc can separate classes of compounds, but HPLC can then separate the compounds within that class. In fact, one of the major uses for SPE is as a clean-up for HPLC. The sequence is (1) select the proper size SPE tube, (2) condition the tube, (3) add the sample, (4) wash the packing, and (5) elute the compounds of interest.

An excellent summary of the technique and several representative methods are presented in the Supelco *Guide to Solid Phase Extraction*, which can be obtained from Supelco, Inc. Bellefonte, PA 16823-0048 (Telephone 1-800-247-6628).

PARTICLE COATINGS

The starting material is usually a glass bead made with a high boron content. This boron then is leached away, leaving a highly porous particle. The surface of this particle (silica) contains many -OH groups, which make the particle surface quite polar. This type of bead can be used directly to retain polar compounds. However, most organic compounds are weakly polar or nonpolar and are not retained. Advantage is taken of the -OH groups to chemically bond organic molecules to the silica bead, thus making a nonpolar surface. The final reaction is called *end capping* and is used to alter the selectivity of the surface. Table 12-1, p. 131, illustrates some combinations.

130   Principles

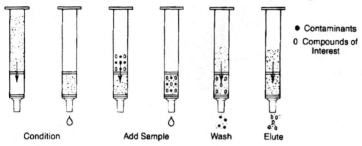

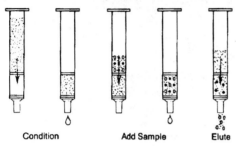

Figure 12-1. A summary of the solid phase extraction process. (Courtesy - Supelco, Inc., Bellefonte, PA)

The following material is abstracted from an article by Edouard Bouvier of Waters Corp. with modifications to fit this chapter. "Can a Sep-Pak cartridge be used as an inexpensive analytical column? The answer can be found by examining the fundamental HPLC equation (see Chapter 19, p. 190, for more details) for the resolution between two components:

$$R_s = \frac{1}{4}\left(\frac{\alpha - 1}{\alpha}\right)\left(\frac{k}{k + 1}\right)\sqrt{N} \qquad (12\text{-}1)$$

where α is *selectivity*, k is the *retention factor* for the more retained component, and N is the *number of plates*. This selectivity term is a measure of how the two different components behave on the chromatographic site. The higher the α, the greater the difference in chromatographic behavior. The particle surface material and end cap influence α. The retention factor is a measure of how strongly retained the component is. The solution pH and ionic strength are important here. N is a measure of the efficiency of the chromatographic system. The thickness of the disc and the size of the particles are important here. If one compared a Sep-Pak cartridge to a µBondapak analytical column, one would find a similar α and k for both. The difference between the two systems resides in the N, the number of plates. Typically a µBondapak analytical column (3.9 x 150 mm) will have approximately 7500 theoretical plates, while a Sep-Pak short-body cartridge will have less than 50 plates. Using the resolution equation, a minimum α can be determined for baseline resolution.

"Assuming a resolution of 1.2 (typical value for baseline resolution), one can solve the above equation for α and find that $k = 10$; a minimum α of 1.06 is required for baseline resolution on the HPLC system, whereas an α of 3.9 is necessary with a Sep-Pak cartridge. The HPLC column will be able to separate components within a class, whereas the Sep-Pak cartridge can separate one class from another. This explains why Sep-Pak cartridges are used most often in a step-gradient mode, because one can elute components with high α's in relatively small volumes.

Table 12-1. Solid phase extraction phases and recommended uses
(Courtesy - Supelco, Inc., Bellefonte, PA)

Tube Group	Tube Type	Phase	Polarity Classification of Compound for Extraction
1	LC-8 LC-18 LC-Ph LC-CN	Octyl-bonded silica Octadecyl-bonded silica Phenyl-bonded silica Cyanopropyl-bonded silica	Nonpolar to moderately polar compounds
2	LC-CN LC-Si LC-Diol LC-Florisil LC-NH$_2$	Silica gel (no phase) Diol-bonded silica Magnesium silicate Aminopropyl-bonded silica	Polar compounds
3	LC-CN LC-NH$_2$ LC-SCX LC-SAX LC-WCX	 Sulfonic acid-bonded silica (strong cation exchanger) Quaternary amine-bonded silica (strong anion exchanger) Weak cation exchanger (proprietary)	Carbohydrates, cations Carbohydrates, weak anions, organic acids Strong cations, organic bases Strong anions, organic acids Weak cations

EFFECT OF k

"A low value of k (2 or less) means the component is not retained. A high value of k (>10) means the compound is difficult to elute. For HPLC separations (Chapter 19, p. 183), the chemist tries to alter the conditions to provide a k between 4 to 8. A k of 4 means that four column volumes must be passed through the column packing to elute that component.

"Most SPE applications consist of five basic steps: (1) cartridge conditioning, (2) cartridge equilibration, (3) sample loading, (4) intermediate elutions or rinses, and (5) product elution. The first step is of greatest importance with reversed-phase adsorbents, as they are not easily wetted by water. Typically, acetonitrile or methanol is used to completely wet the solid phase support as well as to solvate the hydrocarbonaceous ligand. The equilibrium step simply removes the organic solvent from the cartridge and prepares it for the sample. Care must be taken to prevent the support from drying at this point in order to keep it completely wetted. For this equilibration and the sample load step, typically one wants to use conditions that will make the analytes of interest adsorb strongly to the sorbent, that is, to drive $k \rightarrow \infty$. (Note that in some SPE applications, loading solvents are chosen to let the analytes of interest pass through the column unretained, while the interferences are adsorbed. In this case, one wants to drive $k \rightarrow 0$.)

"For the case of strong adsorption, this typically means equilibrating the cartridge with neat water or aqueous buffer for a reversed phase Sep-Pak cartridge, or using a nonpolar solvent such as hexane for a silica cartridge. In practice, k values are finite, say 100-500. This means that for trace enrichment applications in which large sample volumes are passed through a Sep-Pak cartridge, one needs to be concerned about sample breakthrough, especially of the least retained analyte. Consider the case in which $k = 500$. The retention volume of an injected peak can be calculated from the relation shown in Equation 12-2, p. 132. If $V_o = 1$ mL, then $V_r = 501$ mL. However, because of peak dispersion, the component will start to elute well before this volume. If $V_o = 1$ mL, then $V_r = 501$ mL. However, because of peak dispersion, the component will start to elute well before this volume.

$$k = \frac{V_r - V_o}{V_o} \quad (12\text{-}2)$$

This is shown graphically in Figure 12-2. Note that the more efficient the Sep-Pak cartridge, the greater the volume one can load before breakthrough.

"If interferences are present, they can be minimized or eliminated by implementing a cartridge rinse procedure and/or choosing an appropriate elution solvent. The cartridge rinse step is important to remove more weakly retained interferences from the components of interest by using a rinse solvent that is of sufficient polarity or solvent strength to elute undesired contaminants without co-elution of the desired analytes. Typically, this means using a mixture of water (or buffer) and a miscible organic solvent (such as methanol or acetonitrile) for a reversed phase cartridge or moderately polar solvent with a silica sorbent, such as a mixture of hexane and ethyl acetate.

"For the elution step, the objective is to elute the analytes of interest in as small a volume as possible without also co-eluting more strongly adsorbed contaminants. However, keep in mind that the more retained the analyte is, the greater is the elution volume required for complete recovery. As a result, the analyte will be more dilute and must be further concentrated, or the sensitivity of the analytical will be less.

"Figure 12-3 shows the effect that k has on the volume required for 100% recovery. When highly adsorbed interferences are not an issue, neat acetonitrile or methanol can be used to elute the analytes from reversed-phase adsorbents. Also, nonpolar solvents such as methylene chloride can be used on these adsorbents for very hydro-phobic analytes, as long as a cartridge drying step is first employed to remove all water. For silica adsorbents, highly polar eluents such as neat ethyl acetate, acetonitrile, or methanol can be used.

EFFECT OF FLOW RATE

"One key parameter is the flow rate. This is because mass transfer resistance affects the chromatographic performance of the Sep-Pak cartridges. At high flow rates (or more precisely high linear velocities), the number of plates decreases. Typical flow rates on a Sep-Pak Plus Cartridge are 1-10 mL/min. Figure 12-4 shows the effect that flow rate has on cartridge performance for two different Sep-Pak cartridges, one a C_{18}, the other a †C_{18}. One difference between the two adsorbents is the particle size. The C_{18} material contains 55-105 µm bonded silica,

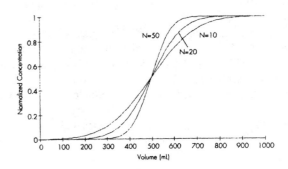

Figure 12-2. A plot of the breakthrough profile for frontal loading of a sample onto a Sep-Pak with k = 500 at different cartridge efficiencies.
(Courtesy - *Waters Column*, **5** (1), 2, 1994, Milford, MA)

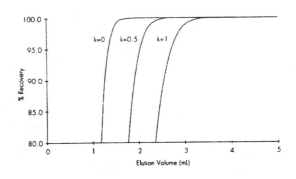

Figure 12-3. A plot of recovery as a function of elution volume for different k's with $N = 20$.
(Courtesy - *Waters Column*, **5** (1), 2, 1994 Milford, MA)

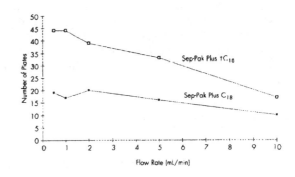

Figure 12-4. A plot of the number of plates vs flow rate for Sep-Pak cartridges.
(Courtesy - *Waters Column*, **5** (1), 2, 1994 Milford, MA)

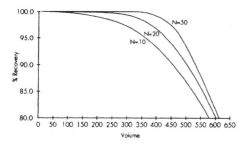
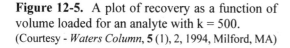

Figure 12-5. A plot of recovery as a function of volume loaded for an analyte with k = 500.
(Courtesy - *Waters Column*, **5** (1), 2, 1994, Milford, MA)

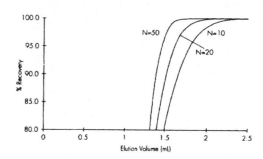

Figure 12-6. A plot of recovery as a function of elution volume.
(Courtesy - *Waters Column*, **5** (1), 2, 1994, Milford, MA)

while the †C_{18} material contains 37-55 μm particles. The figure shows two major points. First, the particle size affects chromatographic performance as one would expect. The smaller diameter particle material results in a higher efficiency sorbent bed. Secondly, and most important for the SPE method development, the flow rate affects performance. One can see from Figure 12-2 that changes in the loading flow rate can lead to poorer reproducibility. The same is the case for elution flow rates. Figure 12-5 shows recoveries one may expect at different cartridge efficiencies for the loading, and Figure 12-6 the elution steps. In order to make a method rugged, it is necessary to either implement some control on the flow rate through the Sep-Pak cartridge or develop conditions assuming a maximum loading and elution flow rate. Figure 12-7 provides a guide on how to solve problems in general. The letters A, B, etc. refer to specific recommendations given in the Supelco publication referred to previously.

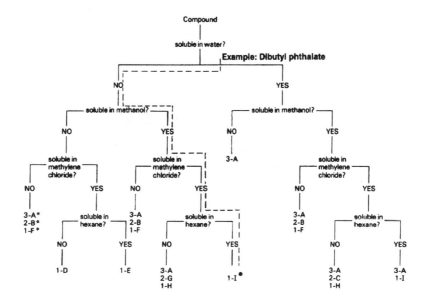

Figure 12-7. Methods development flow chart based upon compound solubility.
(Courtesy - Supelco Inc., Bellefonte, PA)

EFFECT OF CARTRIDGE SIZE

"The linear velocity will change at a given flow rate due to differences in inner diameters. The linear velocity is proportional to the cross-sectional area of a cartridge or proportional to the square of the cartridge inner diameter. One must take this into account and scale the flow rate appropriately. For example, the dimensions of the Sep-Pak light cartridge are 5.44 *x* 12 mm. The i.d. is ~1.9 *x* less than that of the Sep-Pak Plus cartridge, so flow rates must be reduced by a factor of 3.6 to achieve comparable linear velocities."

TECHNIQUES

The simplest arrangement for doing a SPE is shown in Figure 12-8. A 50 mL plastic syringe is fitted into the top of the Sep-Pak cartridge, which drains into a 10 mL graduated cylinder or Kaderna-Danish bottom. The sample is placed in the syringe and allowed to gravity flow through the sorbent or is forced through at a measured rate by adding the plunger and applying pressure. The syringe is rinsed, and eluting solvent then is placed in the syringe and forced through the sorbent. This arrangement is not efficient if several samples are being run. Figure 12-9 shows two arrangements, one for 12 tubes and one for 24 tubes on a vacuum manifold. The Supelco description is: "The newly designed vacuum bleed valve offers better control and sealing via a screw-type mechanism fitted with a Teflon seal. The new, stand-alone cover prevents solvent guides from resting on the work surface and becoming damaged or contaminated.

"Precise flow control is provided through each SPE tube by rotating the independent, screw-type valves built into the cover. This precise control ensures the packing will not dry out in some tubes while others finish drying."

"A larger apparatus especially designed for the SPE of oil and grease is shown in Figure 12-10, p. 135. "This apparatus is unique in two ways. First, the collection vessel is located outside of the filter flask or manifold and uses a vacuum bridge to connect it to the vacuum source. Second, a Luer adaptor is located in the solvent collection arm that allows the use of an in-line Na_2SO_4 drying cartridge. These two features allow the analyst to collect the solvent with eluted oil and grease directly into a pre-weighed collection flask without any disassembly or reassembly of the glassware. The eluent is already dried with Na_2SO_4 and can be directly evaporated to obtain the weight of the extracted oil and grease. It is designed to use 47 mm discs."

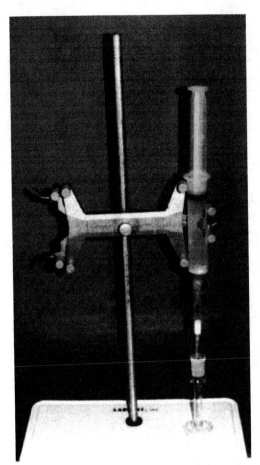

Figure 12-8. A solid phase extraction cleanup apparatus.
(Courtesy - FDA Total Diet Center, Lenexa, KS)

 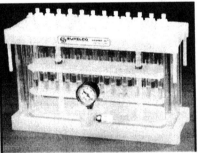

Figure 12-9. Vacuum manifolds for solid phase extraction. Left, 12 port; right, 24 port.
(Courtesy - Supelco Inc., Bellefonte, PA)

SOLID PHASE MICROEXTRACTION

A microscale SPE technique was developed in 1992 by Janusz Pawliszyn at the University of Waterloo, Ontario Canada. According to J. Berg of Varian Assoc., Walnut Creek, CA, the device shown in Figure 12-11, p. 135, "consists of a holder and a replaceable fiber assembly. The assembled unit looks much like a syringe, but in place of the hollow needle is a fiber inside a protective sheath. The fiber is attached to the holder plunger, so that it may be exposed by moving it out of the sheath. The fiber itself consists of a piece of fused silica rod coated with an adsorbent.

"In use, the fiber is exposed to a liquid sample or to the headspace above the sample [Figure 12-12, p. 136]. If, for example, the coating is hydrophobic and the fiber is immersed in an aqueous solution, then organic compounds will

Figure 12-10. A StepSaver solid phase extraction apparatus.
(Courtesy - Environmental Express, Mt. Pleasant, SC)

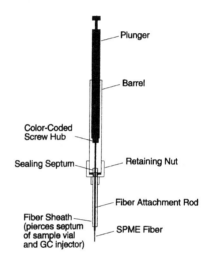

Figure 12-11. Fiber holder and SPME fiber for an Autosampler.
(Courtesy - J.R. Berg, *American Laboratory*, Nov, 1993)

migrate from the solution to the coating. If the fiber is immersed for a long enough period of time, then migration will occur until equilibrium is reached. The length of time required to reach equilibrium depends upon the partition coefficient of a particular solute and whether the solution is agitated or not.

"The actual sampling procedure, whether manual or automated, is as follows: (1) The fiber sheath pierces the septum of the sample vial. (2) The fiber is extended from the sheath into either liquid sample or the headspace above liquid, solid, or semisolid samples. (3) Equilibration of analytes between the fiber and liquid or headspace occurs over a period of time. (4) The fiber is withdrawn into its sheath. (5) The sheath is withdrawn from the vial and inserted into a hot or cold injection port of a gas chromatograph. (6) The sample desorbs into the carrier gas stream." Store opened, but unused, cartridges in a plastic bag in a desiccator.

REVIEW QUESTIONS

1. What is meant by "solid phase extraction"?
2. What is the main difference between HPLC particles and SPE particles?
3. What are some advantages of SPE over liquid-liquid extraction?
4. About how many theoretical plates does a typical cartridge provide?
5. What is a major purpose of the "end cap" on a particle coating?
6. The α term is called selectivity. What does this mean?
7. What does k measure?

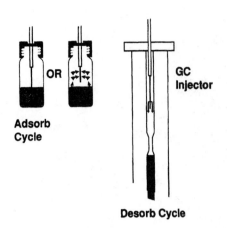

Figure 12-12. Solid phase microextraction. (Courtesy - J. R. Berg, *American Laboratory*, Nov. 1993)

8. If a compound has a $k = 1$, what does this mean?
9. If you wanted to change k, what would you alter in the system first?
10. What is the surface coating on an LC-8 particle?
11. Which tube type would be preferred to separate a mixture of carboxylic acids in a fruit juice; LC-CN, LC-Si, or LC-SAX?
12. A cartridge has a void volume of 0.3 mL. If a compound with a $k = 10$ is retained, how many column volumes are required to elute this compound?
13. Why should a cartridge be conditioned before being used?
14. A compound required 2.0 mL to elute from a column with a void volume of 0.2 mL. What is its k?
15. What is the purpose of the equilibration step?
16. What is the purpose of the rinse step?
17. What does Figure 12-4, p. 132, illustrate in general terms?
18. An analyst switched from using a 1 cm diameter cartridge to a 2 cm diameter cartridge. What effect will this have on the flow rate?
19. Refer to Figure 12-7, p.133. What type of cartridge would you consider using if the compounds of interest were water soluble, soluble in methanol, or soluble in methylene chloride, but not soluble in hexane?
20. What is the fiber in a microextraction?
21. How should opened, but unused, cartridges be stored?

UNDERGRADUATES - DO THOSE PORTIONS ASSIGNED BY YOUR INSTRUCTOR
PLUS ANY OTHERS THAT INTEREST YOU.
GRADUATE STUDENTS - START WITH NEW MATERIAL AND CONTINUE ON.

IF IT EXISTS - CHEMISTRY IS INVOLVED
IF YOU CAN BUY IT - A CHEMIST WAS INVOLVED SOMEWHERE

13
SUPERCRITICAL FLUID EXTRACTION

PRINCIPLES

A supercritical fluid extraction is a gas-solid extraction in which the extracting agent is a gas under supercritical conditions. The low viscosity and near zero surface tension permit this gas to penetrate solids far better and quicker than most liquids, and when the extraction is complete, the gas is removed from the extract by simply lowering the pressure.

Normally, if you want to dissolve a compound, you first select a solvent that has a similar polarity (like dissolves like); then you try to get as much contact (large surface area) as possible between the molecules of solvent and the compound to be dissolved and heat the solvent to further increase the number of collisions. One reason liquids are preferred to gases as solvents is that a high concentration of molecules attacks the surface of the compound to be dissolved, and, therefore, more solvent-solute bonds can be formed to compete with the solute-solute bonds of the solid or liquid. The difficulty with liquid solvents is that once the compound has dissolved, it must now be separated from the solvent. If the compound would dissolve in a gas, then separating the solvent from the solute would be much easier. However, the number of molecules of gas attacking the surface of the compound is quite small compared to that of a liquid, and for that reason, few solids dissolve in gases. What must be done is to make a gas behave like a liquid. If you could compress a gas so it would approach the density of a liquid, then you should get increased solubility and be able to evaporate the gas once the compound was extracted. Under normal conditions, if you compress a gas too much, it may collapse into a liquid.

In 1822, Baron C. Cagnaird de la Tour (*Ann. Chim.*, **22**, 410) discovered that if you heat a gas above a certain temperature, no matter how much pressure you apply, you cannot compress it into a liquid. It was called the *Cagnaird de la Tour point*. Now that temperature is called the *critical temperature*, T_c, and the lowest applied pressure at that temperature is the *critical pressure*, P_c. Any gas above these conditions is known as *supercritical*. This is shown in general in Figure 13-1, p. 138.

In 1879, J.B. Hannay and J. Hogarth *(Proc. R. Soc. London*, Ser A **29**, 324) reported to the Royal Society of London on their observations of what happened to several inorganic salts (KI, KBr, $CoCl_2$), not normally soluble in ethanol, when they were placed in ethanol and heated to a high temperature with sufficient pressure to keep the ethanol from evaporating. They noticed that at high temperatures and pressures, the compounds dissolved, and when the pressure was reduced, the compounds precipitated. They said this was a new phenomenon. Others thought it was just salt dissolving in a "hot liquid". This was the first recorded observation of what is now known as supercritical fluid extraction (SFE). It is interesting how high pressures were obtained in these early studies. Pumps were not available nor were "pressure gauges". Mercury in heavy-walled tubes was used and provided both a high pressure and a measure of it at the same time. In 1879 E.G. Amagat (*Comptes Rendus des Seances de L'Academie des Sciences*, **88** (I), 336) attained 400 atmospheres by using a Hg column extending into the bottom of a mine shaft. L. Cailletet (*Comptes Rendus des Seances de L'Acadamie des Sciences* **112**, 764, 1891) used the Eiffel tower similarly after it was built. Even though the pipes were heavywalled metal (2 in. o.d, and 1/4 in. i.d.), they apparently had many breaks. Can you imagine what the environmentalists would do today?

138   Principles

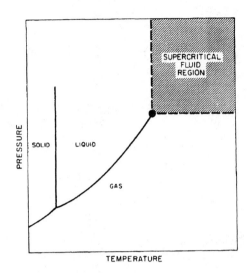

Figure 13-1. Phase diagram showing the supercritical region.
(Courtesy - McHugh, M., and Krukunis, V., *Supercritical Fluid Extractions - Principles and Practice*, Butterworths, Boston, 1986).

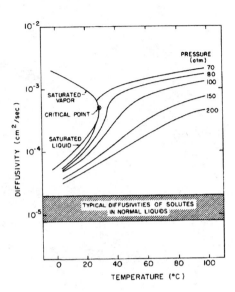

Figure 13-2. Diffusivity behavior of carbon dioxide.
(Courtesy - McHugh, M. and Krukunis, V., *Supercritical Fluid Extractions - Principles and Practice*, Butterworths, Boston, 1986).

What is gained by using supercritical fluids? They have gas-like transport properties and their diffusivities are 1-2 orders of magnitude higher than those of liquids (Figure 13-2); therefore, extractions from solids can be fast. They have very high "liquid-like" densities (Figure 13-3), which increase their dissolving power to that approaching liquids. They *have viscosities nearly as low as gases* (Figure 13-4), and they *have essentially zero surface tension*, which allows them to easily penetrate into microporous structures.

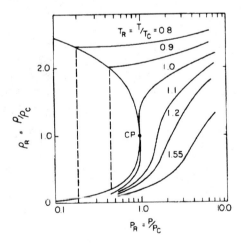

Figure 13-3. Variation of the reduced density of a pure component in the vicinity of its critical point.
(Courtesy - McHugh M. and Krukunis, V., *Supercritical Fluid Extractions - Principles and Practice*, Butterworths, Boston, 1986).

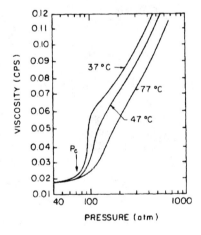

Figure 13-4. Viscosity behavior of carbon dioxide.
(Courtesy - McHugh M. and Krukunis, V., *Supercritical Fluid Extractions - Principles and Practice*, Butterworths, Boston, 1986).

Supercritical fluid extraction

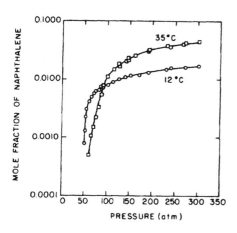

Figure 13-5. Solubility behavior of solid naphthalene in supercritical ethylene.
(Courtesy - Diepen, G. and Sheffer, F. *J. Am. Chem. Soc.* **70**, 4085, 1948).

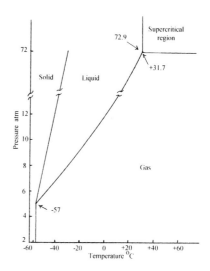

Figure 13-6. Phase diagram for carbon dioxide showing the critical conditions.

After extraction, their high volatility at regular pressures enables them to be separated readily from the solute with little or no residue. This is energy efficient and safety efficient, particularly with foods.

Cochran et al. (*Sepn. Sci. Technol.*, **25** (13-5), 2017, 1990) report that at supercritical conditions, the solvent actually forms clusters averaging 100 molecules surrounding the solute molecule and that the solute molecules cluster as well. This may explain the increased dissolving power of the solvents at the higher pressures. Figure 13-5 shows how the solubility of naphthalene changes in ethylene at various pressures. Advantage often is taken of this ability to change the degree of solubility as the pressure changes to more selectively separate one compound from another or to separate classes of compounds from each other.

Notice that CO_2 has easy to attain critical conditions. It is also inexpensive, nonpolar, nontoxic, nonflammable, and easy to obtain in high purity. Therefore, it is a favorite for SFE. Figure 13-6, is a phase diagram of CO_2.

Table 13-1 shows the T_c and P_c for several compounds. Table 13-2, p. 140, shows the type of pesticide residue recoveries that can be obtained from foods.

Table 13-1. Some selected critical temperatures and pressures

Compound	T_c (°C)	P_c (Atm)	Compound	T_c (°C)	P_c (Atm)
Carbon dioxide, CO_2	31.7	72.9	Freon 13, $CClF_3$	28.8	38.2
Water, H_2O	374.1	218.3	Freon 14, CF_4	-45.7	39.9
Sulfur dioxide, SO_2	157.8	77.7	Chloroform, $CHCl_3$	263.	47.7
Acetone, C_3H_6O	235.5	47.	Fluoroform, CHF_3	25.7	46.9
Methanol, CH_3OH	240.	78.5	Methane, CH_4	-82.1	45.8
Acetonitrile, C_3H_5N	290.8	41.3	Ethane, C_2H_6	32.2	48.2
Chlorodifluoromethane, $CHClF_2$	96.	49.12	Propane, C_3H_8	96.8	42
Freon 11, CCl_3F	198.	43.2	Butane, C_4H_{10}	152.	37.5
Freon 12, CCl_2F_2	111.5	39.6	Hexane, C_6H_{14}	234.2	29.9

Notice that most of the critical pressures are below 60 atm or below about 880 p.s.i., pressures which are easily obtainable. Therefore, if you have a pump that can produce several thousand p.s.i., you can alter the pressure to increase the density and the solubility of the material to be extracted should increase.

For those who worry about adding CO_2 to the environment, if you drive your car once around the block, you will add more CO_2 to the environment than from one extraction, and the CO_2 used for the extraction was originally obtained from the atmosphere.

Table 13-2. Pesticide recoveries from selected foods using large-scale supercritical fluid extraction with CO_2

(Courtesy - M. Hopper, FDA Laboratory Kansas City MO.)

Food	Pesticide	Fortification ppm	Recovery	Mean	%RSD
Butter	cis-Chlordane	0.10	0.095-0.101	0.099	2.2
	Malathion	0.10	0.099-0.107	0.103	2.6
Cheddar cheese	Heptachlor epoxide	0.001	0.0012-0.0013	0.0012	4.3
	Dieldrin	0.001	0.0022-0.0029	0.0025	10.2
Saltine crackers	Chlorpyrifos	0.0230	0.0275-0.0381	0.0291	5.7
	Malathion	0.0190	0.0300-0.0342	0.0316	4.5
Sandwich	Malathion	0.0310	0.0350-0.0390	0.0372	3.6
Ground beef cooked	p.p'-DDE	0.0070	0.0087-0.0093	0.0090	2.8

Table 13-3, p. 141, shows a comparison of SFE (1.5 hrs), the AOAC Soxhlet method (overnight extraction), and the Pesticide Analytical Manual (PAM) method (4 hrs) for extracting fats from several fatty foods. These two tables show that SFE in 1.5 hours can extract pesticides from the most difficult samples with excellent recoveries and that fat can be removed so additional analyses can be made free of complications from fat.

Table 13-3. Comparison among fat extraction procedures (%fat extracted)

(Courtesy - M. Hopper, FDA Laboratory, Lenexa, KS)

Item	Soxhlet (4 g sample) AOAC 15th ed., II 960.39 Mean (N=6)	%RSD	PAM 1 (25 g sample) 211.13 F,K & C Mean (N=6)	%RSD	SFE (CO_2) (15 g sample) Mean (N=6)	%RSD
Pork sausage	30.55	4.75	29.83	1.55	29.83	1.32
Peanut butter creamy	50.32	0.39	49.29	0.60	49.51	0.44
Cheddar cheese sharp/mild	33.85	3.13	33.94	3.46	33.26	1.14
Corn chips	31.32	0.62	31.80	0.76	31.51	0.46

SOME RECENT APPLICATIONS

The first large-scale industrial application was to decaffeinate coffee (Zosel, K., U.S. Patent 3,969,196, 1976) and was exploited by the Hag AG Corp. in Germany. It now processes over 50,000,000 kg/yr.

Figure 13-7 is a list of the major industries that use SFE. American industries were late on the scene. Maxwell House (1988) was the first-large scale user, putting through about 25,000,000 kg/yr. Phillip Morris uses SFE with CO_2 to remove nicotine from tobacco. Metals can be extracted, if they are first chelated with fluorinated chelating agents. Fat is extracted from bones, potato chips, and other snack foods. SFE has been tried as a cleanser for fine electrical parts.

Several examples of laboratory-scale extractions follow. (1) SFE was used in a comparison of extraction methods for the determination of the apparent tannins in common beans. (2) Homogenates of shrimp meat were extracted with ethyl acetate followed by precolumn cleanup on a silica Sep-Pak cartridge to determine 4-hexylresorcinol, a processing aid. Two extraction procedures were compared to extract oil from olive foot cake with acidic hexane. An open-air method and exhaustive Soxhlet extraction with the latter removed about two times more oil. (3) The fractionation of butter oil with supercritical CO_2 at 40 °C and pressures of 125 or 350 bar gave initial fractions that were colorless and consisted primarily of glycerides enriched in short chain fatty acids. (4) Postharvest retention of the red color of litchi fruit pericarp was investigated; results showed that MeOH extracts of the red pericarp absorbed strongly at 525 nm, but extracts of brown pericarp had a low peak absorbance at 525 nm even after acidification. (5) The viscosities of milk-type fatty acids and methylated fatty acids saturated with supercritical CO_2 were studied at 40 and 60 °C and 85 to 350 bar. (6) Pressures from 3,000 to 7,000 p.s.i. and temperatures from 40-55 °C were used to study the extraction of annatto seed pigments by supercritical CO_2. Maximum solubilities were 0.003 for pure bixin and 0.26 mg/g CO_2 for annatto seed pigment. (7) SFE with CO_2 was used to fractionate oat lipids into polar and nonpolar components, so aqueous phase interactions could be studied.

Year	Operator	Materials Processed
1982	SKW/Trotsberg	Hops
1984	Fuji Flavor Co.	Tobacco
	Barth and Co.	Hops
	Natural Care Byproducts	Hops, Red pepper
1986	SKW/Trotsberg	Hops
	Fuji Flavor Co.	Tobacco
	CEA	Aromas, Pharmaceuticals
1987	Barth and Co.	Hops
	Messer Griesheim	Various
1988	Nippon	Tobacco
	Takeda	Acetone residue from antibiotics
	CAL-Pfizer	Aromas
1989	Clean Harbors	Waste waters
	Ensco, Inc.	Solid wastes
1990	Jacobs Suchard	Coffee
	Raps and Co.	Spices
	Pitt-Des Moines	Hops
1991	Texaco	Refinery wastes
1993	Agrisana	Pharmaceuticals from botanicals
	Bioland	Bone
	U.S. Air Force	Aircraft gyrosopic components
1994	AT&T	Fiber optics rods

Figure 13-7. Some industrial uses of SFE. (Courtesy - Phelps, C.L., Smart, N.G. and Wai, C.M., *J. Chem. Ed.*, **73** (12), 1163, 1996)

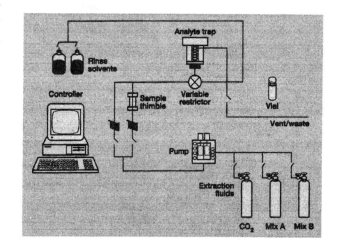

Figure 13-8. Components needed for a supercritical fluid extraction. (Courtesy - Taylor, L., *R&D Magazine*, Feb., 1992)

TECHNIQUES

A supercritical fluid extraction involves placing the sample in a heavy-walled metal chamber, which is heated above the T_c. A gas above its supercritical temperature and pressure then is passed through it to extract the desired

142 Principles

compounds. The emerging gas cools, and the pressure reduces to atmospheric in a collecting vessel. The gas vaporizes from the extract, leaving a deposit of the material behind.

Figure 13-8 is a block diagram of the component parts needed for a typical laboratory-scale SFE. The most common system at the moment is supercritical CO_2 which is nonpolar. Figure 13-9 is a photograph of the ISCO Model SFX-2-10 supercritical fluid extractor showing two pumps. Figure 13-10, p. 143, is an isometric drawing of a commercial apparatus. Pumps are usually syringe type so that a constant pressure can be maintained.

MODIFIERS

You can alter the selectivity of the extraction by changing the pressure, or you can add a "modifier" that increases the polarity of the system. Modifiers are added in small amounts, usually less than 5 %. Some common ones are methanol or acetone to make the system more polar or propane and octane for nonpolar effects. Experiments by L. Taylor of Virginia Polytechnic Institute have shown that recoveries of polychlorinated biphenyls can be increased from 50 % to 80 % by adding 0.8 % toluene and to 100 % by adding 5 % methanol. Just how these modifiers work at the molecular level is still under investigation. Some believe they just alter the polarity of CO_2 because CO_2 is slightly polarizable. Others believe they are involved in the solvent sphere around the solute.

Figure 13-9. Front view of the ISCO model SFX-2-10 supercritical fluid extractor.

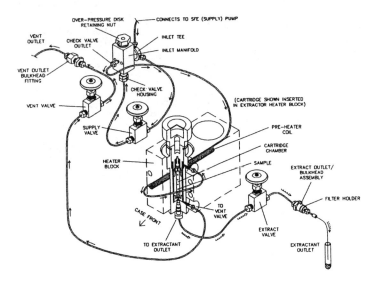

Figure 13-10. Fluid flow diagram of a supercritical fluid extractor.
(Courtesy - Isco, Inc., Lincoln, NE)

DRYING AGENTS

More than 10% water in the sample can impede CO_2 interaction with the sample, thus diminishing the extraction. Water in the sample acts as a modifier and can ruin the extraction in some cases. If it's available and time permits, freeze drying can be used to remove water from the sample. Rather than dry the sample before it is placed in the extraction tube, an absorbent is added to retain and disperse the water which sometimes interferes with the extraction. At the moment, the compound of choice is pelletized diatomaceous earth called *Hydromatrix*. It does not cake.

For samples containing more than 70% water, a value of 1.6 g of Hydromatrix/mL of water in the sample is recommended. Therefore, a 15 g sample would require about 16 g of Hydromatrix. For samples containing less than 70% water, a ratio of about 1:1 is recommended.

Another workable drying agent is anhydrous magnesium sulfate.

CARTRIDGE AND CARTRIDGE HOLDERS

The cartridge holder must be quite strong, but the cell itself usually need be only finger tightened. Figure 13-11 shows a cartridge holder that has been tested to 40,000 p.s.i. Figure 13-12, shows several different sizes of cartridges and the filter placement.

Mark one side of each filter frit. This is to ensure that filters are placed in the cartridge in the same direction each time. Their purpose is to filter any particles coming from the solvent or the connecting lines, and very small particles can become trapped inside of the pores with the solvent going in one direction. If the frit is turned over for the next extraction, the particles can be dislodged.

Keep the extraction cartridge full. Void volume should be minimal. If you only have 1 gram of sample in a 10 mL cartridge, then fill the empty space with Hydromatrix, glass wool, glass beads, clean sand, or Celite.

Figure 13-11. Double-ended SFE vessel. (Courtesy - VICI Valco Instruments Co. Inc., Houston, TX)

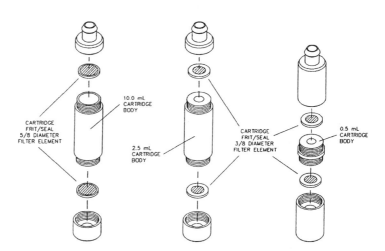

Figure 13-12. Cartridge assemblies. (Courtesy - ISCO, Inc., Lincoln, NE)

Have no void volume on the downstream side of the cartridge. If the cartridge sits upright in the holder and the extractant goes in at the bottom, be sure to fill the void space at the top of the cell.

Always use a dispersing agent. Dispersing agents are the same as those used to fill the void volume. They are mixed in with the sample as homogeneously as possible. This is to prevent channeling. Use good column packing techniques discussed in the chromatography chapters, and you will get good recoveries.

PUMPS

A pump capable of several thousand p.s.i. commonly is used. Not only is the pump needed to maintain supercritical conditions, but the solubilizing power of the system varies greatly with pressure, usually dissolving more solutes as the pressure increases. For example, CO_2 at 1.23 g/cm^3 will dissolve compounds with Hildebrand's solubility parameter (Chapter 41, p. 479) from 7-10, about the same as benzene, chloroform, ethyl acetate, acetone, cyclohexane, carbon tetrachloride, toluene, ethyl ether, and pentane. If the pressure is reduced so that the CO_2 is about 0.9 g/cm^3, then it will dissolve compounds with parameters from 7-9 (solvents like cyclohexane, carbon tetrachloride, toluene, ethyl ether, and pentane) and if further lowered to 0.6 g/cm^3, it will dissolve only compounds with parameters of 7-8 (ethyl ether and pentane).

Pressures of 6,000 to 10,000 p.s.i. are common. At these pressures, the pump heads must be kept cool to ensure that CO_2 stays liquid while the pump is filled. Ordinary HPLC pumps (Chapter 19, p. 194) can be used, but they must be cooled. This has been done by circulating cold water over the head or by passing the gas from expanding CO_2 over the head. The ISCO syringe pump has been used up to 10,000 p.s.i. with no cooling.

Metering Devices

The flow of extracting solvent needs to be known and monitored for reproducible results. Some pumps have meters attached to measure the volume displaced at that T and P. Others have a movement guide to measure the distance the piston travels. This is then a measure of the volume.

RESTRICTORS

The purpose of the restrictor is to *regulate the flow rate and the pressure*. A 30 cm length of fused silica capillary column, 50 μL i.d., used in gas chromatography works well. If methanol is used as a modifier, this restrictor sometimes breaks. A better arrangement is the stainless steel restrictor sold by ISCO Inc., Lincoln, NE. A flow rate of about 1.5 mL/min is a good starting place. Better control can be obtained with micro metering valves, but they are more costly. Manual operation can be obtained, but the best is the computer-controlled automatic type. The restrictor *must be heated or it will plug*. As a general rule; *1 mL of liquid CO_2 = 500 mL of gaseous CO_2*. Also, as a general rule; *use five to 10 column void volumes of extractant over a period of 10-20 minutes to attain 100% extraction*.

DECOMPRESSION

The extracted sample emerges in decompression. When the gas emerges from the restrictor and encounters atmospheric pressure, adiabatic decompression and cooling occurs. The solute now becomes frozen particles, and the rapid expansion of the solvent can cause some of the solute to blow away. If the distance from the restrictor is greater than 1.5 cm and there is a collecting surface, then little of the solute is lost.

REVIEW QUESTIONS

1. What are some characteristics that make supercritical fluid extraction attractive?
2. What is a supercritical gas?
3. If you placed a thick-walled pipe, with a 1 cm^2 hole in it lengthwise, vertically in the stairwell of a 10-story building (100 ft), what pressure could you attain on a piston at the bottom end? (Hg = 13.6 g/cm^3)

4. What is a suggested mechanism for how the solvent molecules behave under supercritical conditions?
5. Freon 13 becomes supercritical more easily than methanol and it is polar (Table 13-1, p. 139). Why is methanol usually used?
6. How is solubility affected as you increase the pressure in SFE?
7. Refer to Table 13-3, p. 140. Comparing the time of analysis and the results, what advantages does SFE have, if any?
8. What was the first large-scale industrial application of SFE?
9. What does Phillip Morris use SFE for?
10. How can metals be extracted using SFE?
11. What is the purpose of a modifier?
12. What is Hydromatrix?
13. A 10 g sample of steak (40% water) is to have the fat extracted. How much Hydromatrix would you use?
14. Why should one side of the filter frit be marked?
15. Why is the void volume kept small?
16. What are some substances that can be used to fill the large voids in the cartridge?
17. What is the purpose of a dispersing agent? Why not add just the sample by itself?
18. How is flow rate commonly determined?
19. How much liquid is needed and how long would it to take to get a good extraction?
20. You have a 30 cm long restrictor, and the pressure is 3000 p.s.i. If you want to try 4000 p.s.i., do you open the CO_2 tank valve a little more?
21. What is a common flow rate for SFE?
22. Why is the restrictor heated?
23. If the flow rate is 1.0 mL/min of liquid CO_2, how many mL of gaseous CO_2 come out the vent?

IF IT EXISTS - CHEMISTRY IS INVOLVED

IF YOU CAN BUY IT - A CHEMIST WAS INVOLVED SOMEWHERE

SEPARATIONS INVOLVING CHROMATOGRAPHY

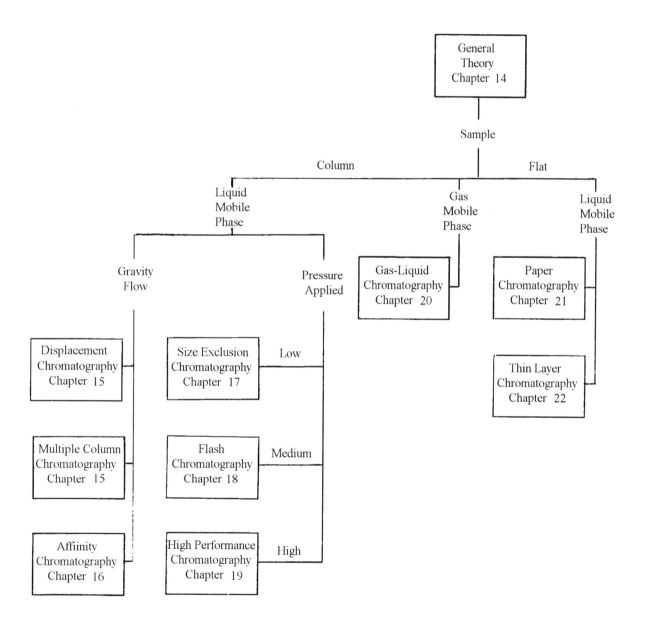

The separation techniques which collectively are called *chromatography* are probably the most widely used chemical separations in the laboratory. They have had a tremendous impact on the chemical sciences and finally upon society. The general theories, Chapter 14, and nine variations, Chapters 15 to 22, will be discussed in this subsection.

The order of presentation is not based entirely on historical development. The first division is between the use of a column or no column. The "yes" column division is based on the type of mobile phase, liquid or gas. The techniques involving a liquid mobile phase passing through a column are then categorized based on the amount of pressure applied, beginning with gravity and continuing to several hundred pounds per square inch. The two main "flat" chromatographies, or "no" column used, are discussed last, although the last one, thin layer chromatography, is not last in use but very popular.

The original chromatographic system was a glass column containing a packing of fine particles. To the top of this was added a narrow band of sample, which was then washed down the column (eluted) with a liquid. This is described in Chapter 15, *displacement chromatography*. In traditional column chromatography, the analyst was at the mercy of the polarity of the column packing. This packing can be coated with a liquid so partitioning can take place. The last part of Chapter 15, *multiple column partition chromatography*, discusses how more control over the separation can be obtained, and how individual compounds can be retained entirely on one column, eluting the others to additional columns. The next refinement was to synthesize groups onto the surface of the column, packing particles that were specific for a single compound or a class of compounds with a particular functional group. This is known as *affinity chromatography*, and is presented in Chapter 16.

The above mentioned techniques usually all employ gravity as the driving force, although it is not mandatory. Column packings can be made that separate compounds based on their size, the very large ones not being retained at all, the very small ones not being eluted in a reasonable time, but those in between being separated based on their size. The newer term for this is *size exclusion chromatography*, but the time-honored names are either *gel filtration* or *gel permeation chromatography* (GPC). The pressures employed are very low; a few inches of water is common. See Chapter 17.

Pressures of a few pounds are used to force the liquid through the column in *flash chromatography*, Chapter 18. This is a good technique to use if the number of impurities is small, such as the side reaction components of a synthesis.

The original columns were glass so the separation could be observed. However, most compounds are not sufficiently colored to view, and if too much pressure is applied to force the separation, the glass will crack. To avoid the breakage problem, and to increase the rate of the separation, the glass was replaced with a stainless steel column. Now, pressures of several thousand pounds per square inch could be employed and the separations were excellent and fast. This was originally known as high *pressure* liquid chromatography, but it was found that very good separations could be obtained on most systems with only several hundred pounds of pressure, so the name was changed to *high performance liquid chromatography* (HPLC). This is a superb technique and is described in Chapter 19.

In general, up to a point, as long as equilibrium is maintained, the faster the separation the better the separation, because the molecules do not have time to diffuse. To increase the speed, and reduce the pressure, a gas can be used as the eluent. This is known as *gas-liquid chromatography* and described in Chapter 20. This is also a superb technique, and while it is limited to about 15% of the available compounds, those compounds are the ones of major interest.

After nearly 40 years of column chromatography, a bold step was taken, which was to get rid of the column and flatten out the packing. The best way to do this was to use a piece of paper and place the sample as a spot at one end. The liquid could then be washed down or up the paper. This is known as *paper chromatography* (Chapter 21), and a Nobel prize was awarded for its conception.

The use of paper was a real advance over a column in regard to reducing analysis time, but you were limited by the polarity of the paper, and the paper was not uniform. The solution was to make a "paper" out of particles, usually silica gel, alumina, or cellulose. Chapter 22, on *thin layer chromatography* (TLC) provides the latest techniques.

14
CHROMATOGRAPHY: GENERAL THEORY

PRINCIPLES

Chromatography *(chromatus* = color; *graphein* = to write) is a multistage separation technique based on differences between compounds in adsorbing on a surface or dissolving in a thin film of liquid. Although Brunschwig, a Strasbourg surgeon (1512); purified ethanol by a chromatographic technique, and Day, an American geochemist, separated crude oils on Fuller's earth (1898-1903), it was the work of Michael Tswett, a Russian botanist (*Ber. Deut. Botan. Ges.*, **24**, 384, 1906), that was the first systematic study and is recognized as the beginning of chromatography.

What they did is what we now call *column chromatography*. It consists of a glass column 1-5 cm in diameter and 20-100 cm long, packed with a solid material (called the inert phase, although it seldom is) such as silica gel, diatomaceous earth, or powdered sugar. The sample is placed as a thin band on top of the packing and washed down the column with a liquid (mobile phase). As the compounds move down the column, small differences in their shape, charge distribution, and polarity allow them to become separated. Extremely good separations can be achieved. For example: chlorophyll *a* (m.w. = 893) differs from chlorophyll *b* by only one methyl group, yet these two compounds are easily separated in the simplest apparatus.

Many variations have been developed since the original discovery. The more common chromatographies are paper, thin layer, high performance, gas, and gel permeation. The techniques of each of these will be discussed in subsequent chapters. However, regardless of the type of chromatography involving neutral compounds, the mechanisms involved in the separation are basically the same. The two major mechanisms at work during a chromatographic separation are *displacement* and *partition*. Few instances involve only one mechanism, but certain types of separations emphasize one mechanism over the other. This chapter will describe these mechanisms, and the specific manipulative techniques will be described in the following chapters when appropriate.

DISPLACEMENT CHROMATOGRAPHY

Refer to Figure 14-1A, p. 150. A glass column is packed tightly with powder-size dry particles, the *inert phase*. A thin layer of a sample mixture of two compounds (W = weakly adsorbed, S = strongly adsorbed) is added to the top of the inert phase. A liquid, which is more strongly adsorbed to the inert phase than any of the compounds in the sample, is now poured into the top of the column. This is called the *mobile phase* (M) and serves to wash the sample down the column, a process called *elution*.

The "inert" particles are not actually totally inert. Remember, they are mainly silicates and aluminates, and these groups have many nonbonded electrons. When these particles are magnified, they can be seen to have corners, edges, holes, projections, and flat surfaces. Electrical charge distribution tends to be *higher on corners than on edges and on edges than on flat surfaces.* These places of activity are called *active sites*. Suppose compound W can be held to the inert phase only at the highly charged corners and that compound S can be held not only at the corners, but at the edges as well. If W is the only compound present, then molecules of it will attach themselves to the corners of the particles and go down the column as far as necessary until all of the molecules are adsorbed. *An equilibrium will be established - some molecules will be leaving corners, while other molecules are being adsorbed on corners.*

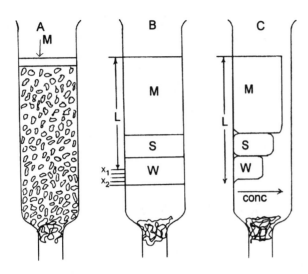

Figure 14-1. Diagram of the stages of displacement chromatography.

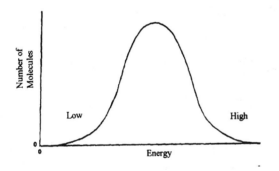

Figure 14-2. Energy distribution of a group of molecules.

If a second compound, S, is now present, then there is competition for the available reactive sites. Because S is more strongly adsorbed, it will displace the W molecules as soon as they leave during their equilibrium process. True, the S molecules are also in equilibrium and adsorbing to an active site and leaving it, but because their attraction is stronger, they win the battle. It is like a game of musical chairs, with S molecules usually winning. The only place for the W molecules to go is on down the column until they can find a vacant corner - they have been displaced by an adsorption process.

The above sequence will stop very quickly unless a mobile phase is added to force the process to continue. The molecules of the mobile phase are selected so as to be more strongly adsorbed to the active sites than either W or S. When the mobile phase molecules reach an active site, they have two distinct advantages: (1) they are more strongly adsorbed to the site than either W or S, and (2) they are much more concentrated. The net result is that they displace the S molecules, and the S molecules in turn displace the W molecules. This displacement sequence continues along the length of the column as shown in Figure 14-1B. In order to complete the separation, M is added until W and S are eluted from the column.

In order to understand the problems that often arise with adsorption chromatography and how to correct them, as well as being able to compare displacement to partition chromatography, a closer examination must be made. Figure 14-1C is a plot of the concentrations of W, S, and M in various sections of the column. These data can be obtained by pushing the packing out of the column, slicing it into 1-2 mm sections, and determining the concentration of each compound in that section. This concentration then is plotted along the X axis. For each section; x_1, x_2 ---.

You can see that W has a small leading edge, rises to a certain concentration, essentially maintains that concentration, then has a small trailing edge. The same general pattern is repeated for each compound; however, S is present in its region of the column in higher concentration than is W in its region of the column. The reason for this is that S can adsorb to corners as well as edges, so there are more places in a given section of column where the S molecules can adsorb. The same reason explains the even higher concentration of M, because it can adsorb to edges, corners, and flat surfaces.

Leading and Trailing Edges

Why are there leading and trailing edges rather than a sharp dividing line between the compounds? One reason is the fact that molecules, even of the same kind, have different energies. Figure 14-2 shows a typical energy distribution curve for a group of molecules. As you can see, some molecules have lower energy than the main body. When they arrive at an active site, they do not have enough energy to react and proceed down the column a bit farther than the main body. This produces the leading edge. On the other hand, some very active molecules can be adsorbed more strongly than those of the main body. This produces the trailing edge. This is normal and usually will be present.

A mixture of compounds in a displacement mechanism causes a problem, because the compounds always have a small region where they are mixed. It is for this reason that displacement chromatography is not used for quantitative separations, but can be quite useful for "preparative" separations, when large amounts of material are to be separated. The mixed fraction simply is collected and recycled.

How do you remove M? This is done be heating the inert phase and desorbing the mobile phase. A few hours at 110 °C is usually sufficient to reactivate the "inert" phase.

Why don't some molecules "fall out of the column" because of very low energy? This can be explained by the *Langmuir adsorption isotherm*, Figure 14-3.

Langmuir did his original work with oxygen and tungsten. A common experiment to illustrate the same principle is to adsorb acetic acid on charcoal. A known amount of dilute acetic acid is added to a known amount of charcoal placed in a flask. After thorough shaking of the mixture, the charcoal is removed by filtration, and the acid remaining in the solution is titrated to determine the amount adsorbed to the charcoal. This experiment is repeated several times with increasingly concentrated acetic acid. The data, when plotted, resemble the curve shown in Figure 14-3.

The significance is that at low concentrations of acid, the acid is adsorbed very strongly to the charcoal (steep slope), but as the acid concentration increases, the acid does not adsorb as strongly (shallow slope). The low energy molecules do not "fall out of the column" because once they get ahead of the main body, their concentration is low and they are more strongly adsorbed.

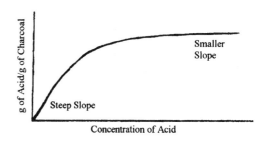

Figure 14-3. Langmuir adsorption isotherm for acetic acid on charcoal.

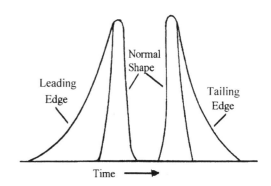

Figure 14-4. Illustrating leading and trailing (tailing).

Leading and Tailing

Sometimes the peaks are not symmetrical, being distorted on the front side (leading) or on the back side (tailing) more than can be explained by normal energy differences (Figure 14-4). *Leading* usually is caused by too loose of a packing or channeling allowing some of the sample to get ahead of the main body before equilibrium is established. It can be cured by packing the column tighter. *Tailing* is more of a problem. It can ruin a separation unless eliminated. It occurs because the compound following the first compound down the column is close to it in polarity and does not do a proper job in the displacement mechanism. What must be done is either to make the "inert" surface more inert or to use a more polar mobile phase to desorb the adsorbing molecules. The former can be done by *acid washing* or *silinizing* the column to cover the active sites. The latter can be done by a process called *gradient elution* (explained in Chapter 19, p. 197) whereby you start out with a mobile phase of low polarity and then slowly add to it another solvent of higher polarity.

Acid Washing and Silinizing

Figure 14-5, p. 152, is a diagram of a small section of the surface of a clay particle. Because it has been in an oxygen environment for millions of years it can have places where extra oxygen is present. This oxygen has nonbonding electrons that make it an active site and can adsorb molecules. There are two common ways to remove this problem: (1) washing the material with an acid, so the protons react with the nonbonding electrons and neutralize them or (2) reacting the material with dichlorodimethylsilane to react with the oxygen and neutralize the nonbonding electrons.

$(CH_3)_2SiCl_2$ ------> $2\ CH_3Cl$ + Si atoms

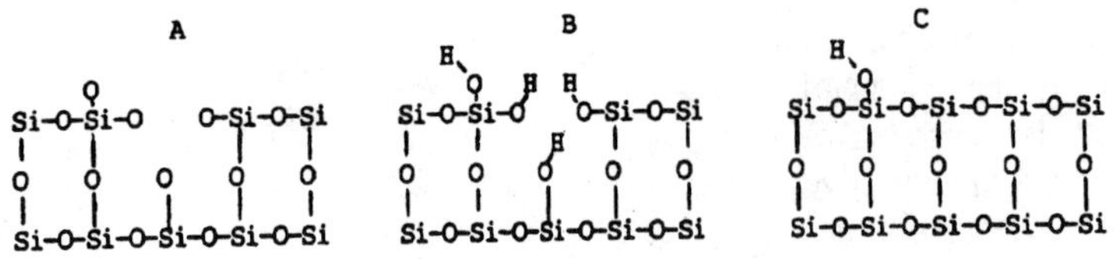

Figure 14-5. (A) A clay particle surface. (B) Acid washed. (C) Acid washed and silinized.

PARTITION CHROMATOGRAPHY

The same equipment is used here as with displacement chromatography, with one exception. The inert phase is coated with a thin film of liquid, which is strongly adsorbed to the surface of the inert phase particles. Because this liquid does not move, it is called a *stationary phase*. A mobile phase is selected that (1) will not dissolve or mix appreciably with the stationary phase and (2) will not adsorb to the inert phase as did the stationary phase. Refer to Figure 14-6A.

The chemical property of the sample molecules of importance now is their difference in ability to dissolve in the stationary phase. Those compounds that are more soluble in the stationary phase are the ones retained the longest. Figure 14-6A shows several particles coated with a stationary phase. To make it easier to understand the partition process, the diagram is divided vertically, the progress of one compound being shown on one side and that of another compound on the other side. A sample consisting of two compounds is dissolved in a small amount of mobile phase and placed in a narrow band at the top of the column. One of the compounds, MS, is more soluble in the stationary phase ($P = 1$) than the other compound, S ($P = 0.67$). To keep the numbers small, assume 1000 molecules of each are in the mixture and begin moving down the column when the mobile phase is added.

Consider the MS molecules first. As they travel down the column, they come in contact with the stationary phase. The *Second Law of Thermodynamics* dictates that *systems tend to maximum disorder*. This means that the "high concentration - high order" MS molecules can gain disorder if some dissolve in the stationary phase where the initial concentration of MS is zero - "low concentration - high disorder".

This separation between phases is usually called the *distribution ratio* in extractions and *partition coefficient* in chromatography. In this case, 500 molecules dissolve and 500 stay in the mobile phase. However, fresh mobile phase continues to flow past the particle. Those MS molecules remaining in the mobile phase move on to the next particle to be further partitioned, 250 to 250. The molecules in the stationary phase - again following the second law of thermodynamics - partition themselves between the stationary and mobile phases, 250 and 250.

The same process is taking place with the other compound. In this case, 600 molecules stay in the mobile phase, and 400 dissolve in the stationary phase. The 600 molecules that move on partition themselves 240 in the stationary phase and 360 in the mobile phase. The 400 molecules in the stationary phase partition themselves in the fresh mobile phase, 240 in the mobile phase and 160 remaining in the stationary phase. Diffusion in other directions can occur. The diffusion causes the bands to broaden or spots to widen, the amount decreasing as the speed of the mobile phase increases, as long as equilibrium conditions are established.

Figure 14-6B shows how the compounds distribute themselves down the column. Figure 14-6C shows the concentration profile for this type of separation. Notice that the compounds can be separated completely, but that each is mixed with the mobile phase. Partitioning chromatography is used when a quantitative separation is required.

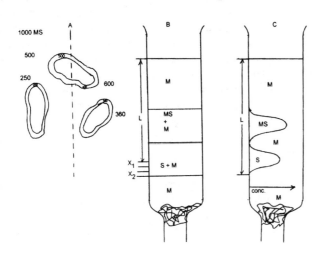

Figure 14-6. Diagram of the stages of partition chromatography.

Practical Significance of the Partition Coefficient

$$Partition\ Coefficient = P = \frac{Concentration\ of\ solute\ in\ the\ stationary\ phase}{Concentration\ of\ solute\ in\ the\ mobile\ phase} \qquad (14\text{-}1)$$

If a compound has a partition coefficient of 5, this means that 5 column volumes of mobile phase will be needed to completely elute the compound from the column. *Therefore, make the volume of the column packing as small as necessary to just separate the mixture.* The less mobile phase that is required, the less solvent that has to be removed during the concentration step which usually follows, and the less overall time is required.

REVIEW QUESTIONS

1. What are the two major mechanisms believed to be involved in chromatographic separations?
2. What does an inert phase look like in general?
3. Where does electrical charge tend to congregate on particles?
4. What are two general properties of a mobile phase?
5. What is meant by a leading edge and what causes it?
6. What is the basic conclusion of the Langmuir adsorption isotherm?
7. What does acid washing do to a silica particle?
8. What are some properties of a stationary phase in general terms?
9. What directs "high concentration goes to low concentration"?
10. If equilibrium can be established, why are faster separations preferred?
11. What is the practical significance of a $P = 10$?
12. A person has two compounds to separate; one has a $P = 8$ and the other $P = 2$. He has found a column 60 cm long and 4 cm in diameter. He decides to fill it to a height of 45 cm with Florisil to be sure he gets a good separation. What would you do if you saw this?

UNDERGRADUATE STUDENTS - DO THOSE PORTIONS ASSIGNED BY YOUR INSTRUCTOR
PLUS ANY OTHERS YOU HAVE AN INTEREST IN.
GRADUATE STUDENTS - START WITH NEW MATERIAL AND CONTINUE ON.

IF IT EXISTS - CHEMISTRY IS INVOLVED
IF YOU CAN BUY IT - A CHEMIST WAS INVOLVED SOMEWHERE

15
DISPLACEMENT AND MULTIPLE COLUMN PARTITION CHROMATOGRAPHY

DISPLACEMENT COLUMN CHROMATOGRAPHY

PRINCIPLES

When chemists refer to column chromatography, they usually mean a glass column of 1-4 cm in diameter and 15-60 cm in length with a mobile phase passing through the inert phase by the force of gravity, mild suction, or at most a few pounds of pressure. Figure 15-1, p. 156, shows several types of commercially available columns. The type shown in A is a plain column with a glass frit for a packing support. B is the same with a stopcock to control the flow rate. C has an expanded solvent reservoir at the top. D is the same as B with an additional standard taper joint at the top to hold a solvent reservoir of any capacity. E is a simple column with a removable bottom (ball and socket joint).

Stopcocks are used to control the flow rate and should be wide bore to prevent plugging with viscous mobile phases. Removable bottoms are used for those separations that require long times. The bottom can be removed and the column packing pushed out with a plunger (CARE: this is not as easy as it sounds). The column packing then can be sectioned, and the desired compounds extracted or washed from the inert phase for further use.

Adsorbents

Table 15-1 lists several adsorbents used for column chromatographic separations.

Table 15-1. Adsorbents

In Order from Most Active to Least Active	
1. Fuller's earth	9. Calcium phosphate
2. Charcoal	10. Calcium carbonate
3. Activated alumina	11. Potassium carbonate
4. Magnesium silicate	12. Sodium carbonate
5. Silica gel	13. Talc
6. Calcium oxide	14. Inulin
7. Magnesium oxide	15. Starch
8. Calcium carbonate	16. Powdered sugar (sucrose)

Mobile Phases

Table 15-2 lists several liquids that are good mobile phases for column chromatography.

Table 15-2. Mobile phases

In Order from Least to Greatest Adsorption by the Column Adsorbent	
1. Petroleum ether 30-50°	11. Esters of organic acids
2. Petroleum ether 50-70°	12. 1,2-Dichloroethane, dichloro-
3. Petroleum ether 50-100°	methane, chloroform
4. Carbon tetrachloride	13. Alcohols
5. Cyclohexane	14. Water (varies with pH and
6. Carbon disulfide	salt concentration)
7. Ether	15. Pyridine
8. Acetone	16. Organic acids
9. Benzene	17. Mixtures of acids and bases
10. Toluene	with water, alcohol or pyridine

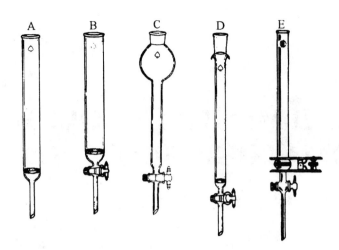

Figure 15-1. Various types of columns for column chromatography.
(Courtesy - Ace Glass Co., Vineland, NJ)

Calculations

The main equations required to solve several column chromatography problems are presented below and in the example calculations. These equations are based upon a liquid-liquid partition process rather than an adsorption process and are usually valid only for processes involving very dilute systems.

For the most part, the mathematical problems involved here concern (1) how much mobile phase is necessary for a separation, (2) when a mixture is actually separated, and (3) when each component comes through the column. The basic terminology used in this connection is:

A_i = cross-sectional area of the inert phase in cm^2, A_s = cross-sectional area of the stationary phase in cm^2, A_m = cross-sectional area of the mobile phase in cm^2, and A = cross-sectional area of the column in cm^2.

$$A = A_i + A_s + A_m$$

$$P = \frac{\text{Concentration of solute in the stationary phase}}{\text{Concentration of solute in the mobile phase}} \tag{15-1}$$

h = height equivalent to a theoretical plate, HETP, in cm. V_h = effective plate volume in cm^3;

$$= h(A_m + PA_s) \tag{15-2}$$

V = volume of solvent required to elute the component from the column in cm^3. r = plate number, starting at the top of the column. $f_{(n,r)}$ = fraction of solute in plate r after n extractions. R_f = ratio of the distance the solvent moves to the distance the mobile phase moves.

$$R_f = \frac{A_m h}{V_b} = \frac{A_m}{A_m + PA_s} \tag{15-3}$$

One of the main problems of analytical interest is trying to determine how long a column is required to effect a separation and what the percent contamination of the emerging bands is. Consider the separation of two compounds, A and B, shown in Figure 15-2, p. 158, as a concentration profile.

As A and B are eluted down the column, they will separate. The problem is, will they have separated completely by the time they get to the end of the column and if not, how much contamination is present? In addition, because the vast majority of chromatographic separations involve colorless substances, it is desirable to have some idea how much eluent is necessary to wash the compound through the column, thus predicting to some extent when to start collecting the fractions of interest.

The various equations involved in the solution of a problem of this type are illustrated in the second example.

EXAMPLE CALCULATION

The following system was used to separate a mixture of acetyl proline and acetyl phenylalanine. If the experimentally determined partition coefficients are 9.5 and 1.3 for acetyl proline and acetyl phenylalanine, respectively, calculate the R_f values for these two compounds using the following system.

Column: glass, 1 cm x 20 cm containing 5 g of silica gel, 3.5 mL of water, and 10 mL of chloroform. Mobile phase: chloroform:1-butanol (99:1). Density of the silica gel: 2.27 g/mL.

ANSWER

A = cross sectional area of the column = $\pi r^2 = 3.14 \times (0.5)^2$.

$$\text{Volume of silica gel} = \frac{5.0 \text{ g silica gel}}{2.37 \text{ g/mL}} = 2.20 \text{ mL}$$

$$A_m \text{ (chloroform)} = \frac{10.0 \text{ cm}^3}{20 \text{ cm}} = 0.500 \text{ cm}^2$$

$$A_i \text{ (silica gel)} = \frac{2.20 \text{ cm}^3}{20 \text{ cm}} = 0.110 \text{ cm}^2$$

$$A_s \text{ (water)} = \frac{3.5 \text{ cm}^3}{20 \text{ cm}} = 0.175 \text{ cm}^2$$

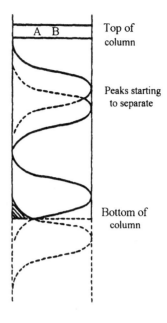

Figure 15-2. Concentration profile of two incompletely separated compounds.

$A = 0.785 \text{ cm}^2$ (This is the cross-sectional area looking down on the top of the column.)

Acetyl proline:

$$R_f = \frac{A_m}{A_m + PA_s} = \frac{0.500}{0.500 + (9.5 \times 0.175)} = 0.23$$

Acetyl phenylalanine:

$$R_f = \frac{A_m}{A_m + PA_s} = \frac{0.500}{0.500 + (1.3 \times 0.175)} = 0.69$$

EXAMPLE CALCULATION

A 15.0 cm long column has an A_m of 0.30 and an A_s of 0.15. A mixture of compounds, A and B, is placed at the top of the column. What is the fraction of component A left in the column if the eluent is cut midway between the peaks of the bands at the end of the column? $h = 0.003$ cm; $P_A = 1.5$ and $P_B = 1.7$. The answer involves a calculation of the amount of in the shaded area of Figure 15-2.

ANSWER

Step 1. Determine the relative rates of motion of each compound down the column. From equation 15-3 we get

$$\frac{(R_f)_A}{(R_f)_B} = \frac{A_m + P_B A_s}{A_m + P_A A_s} = \frac{0.30 + (1.7 \times 0.15)}{0.30 + (1.5 \times 0.15)} = 1.058 \tag{15-4}$$

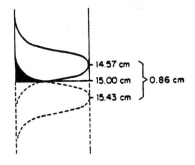

The 1.058 means that compound A moves down the column 1.058 times faster than compound B. When the main concentration peak of compound B reaches the 15.0 cm point, the corresponding peak for compound A would be at 15.86 cm, a 0.86 cm difference. However, the midpoint between the peaks is to be at 15.0 cm. Figure 15-3 shows how this adjustment is made.

When the midpoint of the peaks is placed at the end of the column, the main peaks are 0.43 cm in each direction from it.

Figure 15-3. The midpoint of the peaks.

Step 2. Find how much eluent it will take to wash the main peak of compound A to 15.43 cm. The total volume of the mobile phase required will equal the effective plate volume times the number of theoretical plates.

$$V = N (V_h)_A \tag{15-5}$$

where V = volume of mobile phase in mL, N = plate number, and $(V_h)_A$ = effective plate volume of compound A.

$$N = \frac{\text{Column length}}{\text{HETP}} \tag{15-6}$$

$$N = \frac{15.43}{0.003 \ cm/plate} = 5143 \ plates$$

Notice that the column length used is 15.43 cm, even though the actual column is only 15.0 cm long. This is necessary because the midpoint of the peaks is to be moved to 15.0 cm, and this will take a volume of mobile phase equivalent to a 15.43 cm column. Using equation 15-2:

$V_b = h(A_m + PA_s)$
   $= 0.003 (0.30 + 1.5 \times 0.15)$
   $= 0.00158 \ cm^3/plate$

Therefore, $V = 5143$ plates x $0.00158 \ cm^3/plate = 8.10 \ cm^3$

Step 3. Calculate the area of the small shaded section in Figure 15-3. The area under the curve is related to the number of theoretical plates and the volume of the mobile phase by

$$t = \frac{V}{V_b \sqrt{N}} - \sqrt{N} \tag{15-7}$$

Figure 15-4 is a Gaussian distribution curve showing t values where t is the number of standard deviations away from the center of the curve. Statisticians have worked out the standard deviation to area relationship, and the results are shown in part in Table 15-3.

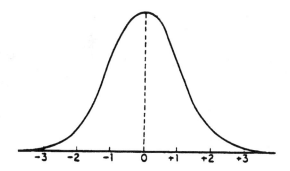

Figure 15-4. A Gaussian distribution curve.

Table 15-3. t values and % area

t	%	t	%	t	%	t	%
0.0	0.00	1.0	68.26	2.0	95.44	3.0	99.730
0.1	7.96	1.1	72.86	2.1	96.42	3.1	99.806
0.2	15.86	1.2	76.98	2.2	97.22	3.2	99.862
0.3	23.58	1.3	80.64	2.3	97.86	3.3	99.902
0.4	31.08	1.4	83.94	2.4	98.36	3.4	99.932
0.5	39.30	1.5	86.64	2.5	98.76	3.5	99.952
0.6	45.14	1.6	89.04	2.6	99.06	3.6	99.968
0.7	51.60	1.7	91.08	2.7	99.30	3.7	99.978
0.8	57.62	1.8	92.82	2.8	99.48	3.8	99.984
0.9	63.18	1.9	94.26	2.9	99.62	4.0	99.992

Although Table 15-3 is based upon an infinite number of plates, the 5000 plates in this column are sufficient to be in error by only a small fraction of a percent. The N used now is based on the actual column length because the desired t is at the end of the column and not at the peak concentration.

$$t = \frac{8.10}{0.00158 \sqrt{5{,}000}} - \sqrt{5{,}000} = 2.02$$

From Table 15-3 this corresponds to an area of 95.5%. This leaves 2.3% of compound A in the column (½ of 4.5%).

Unfortunately, there are no formulas for calculating the proper solvent or adsorbent for the separation of a given compound. Tables 15-1 and 15-2 show some of the materials that have been found to work.

TECHNIQUES

Never let air get into a packing once a column is packed. These air bubbles alter the flow of the mobile phase, skewing the dividing lines between compounds and making quantitative recovery nearly impossible. In severe cases, the bubbles may even stop the flow of mobile phase. To prevent air from getting into the packing, always keep a small volume of mobile phase on top of the packing and never let it go dry during a separation. This can be helped by using columns with large reservoirs for the mobile phase.

Pack columns firmly. If the packing is too loose, channeling and wall effects can occur. Both cause uneven fronts between compounds, *leading* (p. 151), and poor resolution. *Channeling* is caused by openings between the additions of packing that permit the mobile phase to move faster there than through the regular packing. *Wall effects* are present if there is excessive space between the column wall and the inert phase. The mobile phase then moves faster along the wall than in the main body of the packing and skews the shape of the solvent front. However, don't overdo it. A too tightly packed column will have a slow mobile phase flow, resulting in an increase in diffusion and band broadening. Experience is the best teacher. The following is a technique that has been found to work well.

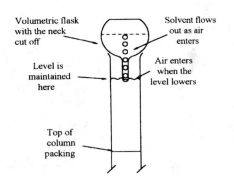

Figure 15-5. Technique for obtaining constant flow and constant level of mobile phase.

When dry packing a column, add the packing in increments of a few grams and tap the end of the column gently on a notebook or pad after each addition. This provides the proper firmness of packing and reduces the boundary lines between additions. Once the packing has been added, then press it firmly in place with a tamper to make the top flat. The tamper can be made by placing a glass or wood rod on a single-hole rubber stopper so that the wide end of the stopper makes contact with the packing. If very fine powders are used as the inert phase, you can rough up the surface of each increment before the next addition is made. This reduces the formation of a smooth boundary and a channeling effect. As a rule, you should not add all of the packing at once in a large column. When pressure is applied to pack it, the top few centimeters are packed much more tightly than the material below, and the mobile phase will move at a variable rate. This makes duplicating results from day to day difficult.

To maintain a constant flow rate and a continuous flow of mobile phase, invert a modified volumetric flask filled with the solvent in the top of the column. This is shown in Figure 15-5. The top is cut off of an appropriate sized volumetric flask, and the desired amount of mobile phase is added. Place the flask in the top of the column, and after a few minutes, the liquid level will stay at the level of the bottom of the neck. When the solvent level drops below the neck, air enters the flask and allows a small amount of solvent to come out. This process continues until all of the solvent is used. Because the solvent level is constant, the flow rate of the mobile phase is constant, and the analyst does not have to pay constant attention to the column.

Use dry mobile phases. Water is quite polar and will elute most compounds adsorbed on a column packing. The easiest way to ensure that your solvent is dry is to place an adsorbent on top of the column packing. Either place a second small column containing 1-2 cm of anhydrous Na_2SO_4 on top of the main column or place a small section of glass wool on top of the column packing and then add a few grams of *anhydrous* sodium sulfate. As the solvent passes through, any water will immediately form $Na_2SO_4 \cdot 10H_2O$ (Glauber's salt).

Gradient Elution

When compounds of widely different adsorptive capacities are present in the same mixture, then it is difficult to remove the more polar compounds and still separate the easy eluters with the same mobile phase. What is done is to start off the separation with a weakly polar mobile phase, and then slowly add another liquid of higher polarity. This produces a polarity gradient, and as the polarity increases, the more highly polar compounds begin to move down the column. With a bit of research, it is often possible to develop a gradient that will provide the same degree of separation between each component of the mixture. Figure 15-6 shows a diagram of one type of gradient producer.

In the simplest case, a liquid of low polarity (N) is placed in tube 1. A liquid of higher polarity (P) is placed in tubes 2 and 3. As liquid is drained from tube 1, liquid from tube 2 flows in and is mixed by the stirrer. As this process continues, the polarity of the mobile phase slowly increases.

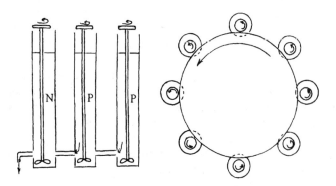

Figure 15-6. Apparatus for producing liquid gradients. Side view (left). Top view (right).

One commercial apparatus consists of 11 tubes in a circular arrangement. Each is fitted with a stirrer on a vertical shaft with a gear on the top. Each stirrer is connected to a sun gear, so all turn at the same time.

Fraction Collectors

A *fraction collector* (Figure 15-7) is an apparatus that contains dozens of small test tubes or vials on a moveable rack. Modern collectors can be programmed to collect fractions by volume, counting drops or by time. They can be programmed further to collect only peaks of interest. The one shown can hold up to 174 12-13 mm tubes, and an LCD displays the tube number, drop count, and help messages.

Figure 15-7. A SpectraChrom CF-1 fraction collector.
(Courtesy - Fisher Scientific Co., Pittsburgh, PA)

MULTIPLE-COLUMN PARTITION CHROMATOGRAPHY

PRINCIPLES

With one exception all of the chromatographic, ion exchange, and electrophoretic methods of separation require that the sample be in a very narrow band or small-diameter spot at the beginning of the separation. The exception occurs when only one component of a mixture is desired. In that case, the desired compound is retained on the column, and the other compounds are washed away. The desired compound then can be washed off the column and collected. Doing this requires that the surface of the column packing be altered to retain the desired compound. With conventional adsorption column chromatography, the analyst is at the mercy of the chemical nature of the inert phase. The following experiment, which is typical of those used in the drug industry and by the FDA to detect adulteration and dose levels in commercial products, shows how an analyst can alter the packing to the desired conditions.

This separation depends on the fact that most compounds are acids or bases to some extent. A column packing that is initially neutral is selected, and it is made acidic to retain basic compounds or basic to retain acidic compounds. One of the best neutral packing materials is called "Celite" (pronounced cell ite, not sea light). It is a finely powdered mixture of the skeletons of small sea animals. Figure 15-8 is a photomicrograph of some of the particles.

Figure 15-8. Photomicrograph of some particles of "Celite".
(Courtesy - Johns-Manville Co., Denver, CO)

TECHNIQUES

Columns

The columns used for this type of separation are very simple, as shown in Figure 15-9. They are made of glass and are 2.5 cm x 15 cm.

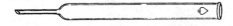

Figure 15-9. Glass columns for single-compound separations.
(Courtesy - Ace Glass Co., Vineland, NJ)

Weighing Technique

Celite is a dry powder and difficult to wet with either the sample or acids or bases. It is also messy to handle and weigh. However, because only an approximate weight of the material is necessary, a simple technique has been developed to obtain 2-5 g of material within ± 0.1 g with no mess. The apparatus is shown in Figure 15-10.

It is made by screwing the bottoms of two 1-ounce plastic bottles onto the ends of a 1 cm x 20 cm dowel rod. A razor blade is used to trim the bottles to the appropriate size. The proper size is determined by pushing the open end of the scoop into the Celite, withdrawing it, and then by simple pressure, emptying it onto a paper for weighing. One end is adjusted to 3 g and the other end to 2 g. Because most columns require 3-5 g of Celite, this combination can be used for all occasions.

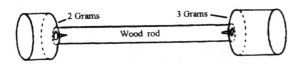

Figure 15-10. A homemade Celite "scoop."

Coating Technique

Liquids can be added to Celite's surface at *a rate of no more than 1 g of liquid to 1 g of Celite*. The best ratio is in the range of 1 g of liquid to 3 g of Celite. To prepare a basic Celite, for example, place 3 g of Celite in a 100 mL beaker and add 1 mL of 1 N NaOH solution. The liquid will immediately form balls and not disperse. Use a glass rod or a flat spatula and mix the liquid *thoroughly* into the solid until no balls of liquid are visible and the mixture appears homogeneous. *This step is very important.* If all of the particles are not coated uniformly, then channeling will occur and sample recoveries will be low. The sample is placed on the Celite in the same manner.

Packing Technique

Place a small piece of glass wool in the bottom of the column. Transfer the coated Celite to the column, tapping it gently on a paper pad after each addition to ensure a firm packing. When all of the sample is transferred, use a tamping rod to scrape the sidewalls of the column clean and tamp the packing firmly. Use a small piece of glass wool to wipe the inside of the beaker clean. Place this piece of glass wool on top of the column and push it down on top of the Celite with the tamping rod. This helps clean the surface of the column's side wall.

For multiple-compound separations, two techniques are common. The first, which is more common, is to prepare a column for each compound desired and stack the columns vertically on a tall ring stand (Figure 15-11). The second method is to use a larger column and place the packings on top of each other. About 2 mm of dry Celite is packed firmly between each section, so they can be separated easily.

The first method is preferred, because it is faster. Once the compounds have been separated, the top column can be removed, and the compound it contains can be eluted without waiting for the other columns to empty.

Elution Technique

Refer to Figure 15-5, p. 160, for the elution technique. Drugs usually require less than 250 mL of eluent, so the bottom of a 250 mL round bottom is adequate.

EXAMPLE CALCULATION

To pass FDA regulations, the product should contain between 95% to 105% of what is declared on the label. Two mL of a sample of cough syrup is pipeted onto 3 g of Celite with a Mohr pipet. The sample is examined by the procedure described in Experiment 20. The sample, examined for codeine, has an absorbance of 1.152 and the standard curve had an absorbance of 1.20 at 80 µg/mL. What is the concentration of the codeine and does it fall within the 95-105% limit?

ANSWER
Step 1: Determine the µg codeine/mL in the test solution.
Step 2: Determine the µg codeine in the sample.

$$\frac{80 \ \mu g/mL}{1.20} = \frac{X \ \mu g/mL}{1.152} \qquad X = 76.8 \ \mu g/mL$$

76.8 µg/mL x 25 mL (dilution) = 1920 µg in 2.0 mL of sample.

Step 3: Determine mg codeine/5 mL and verify with the label.

$$\frac{1920 \ \mu g}{2 \ mL} \times \frac{1 \ mg}{1000 \ \mu g} \times 5 \ mL = 4.8 \ mg$$

95% of 5 mg = 4.75 mg; 105% of 5 mg = 5.25 mg. Therefore the sample meets specifications.

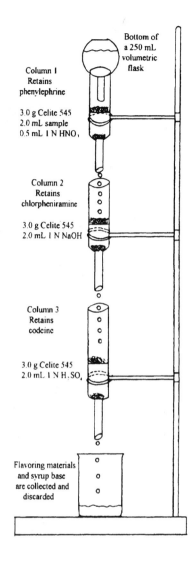

Figure 15-11. Stacked column arrangement for cough syrup analysis.

REVIEW QUESTIONS

1. Why is it essential that the top of the column packing be flat?
2. Why is it good technique to use dry mobile phases unless experience indicates that a little water is acceptable?
3. How do you dry a solvent on a column?
4. Why do you want the sample to be in as narrow a band as possible at the beginning of the separation?
5. What is one cause of "leading"? See p. 151 for help.
6. What causes the wall effect and what does it do to a separation?
7. What is a gradient elution and how can a gradient be established?
8. Why is air a problem in a column?
9. How can you store a packed column for days without it going dry?
10. Why are removable bottoms placed on chromatography columns?
11. What is the technique used to supply several hundred mL of mobile phase to a column without danger of an overflow?
12. Determine the R_f values for phenol and salicylic acid if a column 2 cm in diameter and 30 cm long, contains 83 g of alumina ($d = 3.6$ g/cm^3), 9.6 mL of water, and 61 mL of a butanol-pyridine mixture. The partition coefficients are 0.194 and 3.42, respectively.

13. Using the column described in the previous problem, what are the partition coefficients of gallic acid and protocatechuic acid, if their R_f's are 0.43 and 0.74, respectively?
14. A column having an inside diameter of 14 mm is filled with silica gel ($d = 0.61$ g/cm^3) containing 10% water by weight, 15 g being required and filling the tube to the 25 cm level. Calculate A_m. The R_f values for pyrogallol and phenol were 0.51 and 0.97, respectively. What are the partition coefficients?
15. Using the previous column, calculate the R_f values if the partition coefficients of citric acid and lactic acid are 8.3 and 3.9, respectively.
16. In a particular partition chromatography column, the volumes of mobile phase, stationary phase, and inert phase are in the ratio $A_m : A_s : A_i$ of 0.20 : 0.05 : 0.75 and the HETP is 0.0050 cm. Two substances having partition coefficients of 1.50 and 1.55 are to be separated. Calculate (a) the R_f values for the two substances, (b) the volume of eluent required to bring each of the bands in turn to a point 10.0 cm down a column of total area 1.0 cm^2, (c) the volume of eluent required to wash all but 0.13% of the leading component from a 30 cm column, and (d) the % of the lagging component that has been removed under the conditions of part c.
17. A flask of chloroform is inverted and placed in the top of a column packed with Celite. The chloroform does not all drain out and overflows the column. Why is this?
18. What is "dry washing" of the beaker in which the sample was mixed?
19. A beginner mixed a measured amount of cough syrup into Celite and packed a column. The recoveries were low. What might have caused this?
20. What is the preferred solid to liquid ratio range for Celite columns?
21. Why is it not necessary to have the sample in a narrow band at the top of the column in this case?
22. Why is air in the column not as serious a problem in this type of separation?
23. Earlier it was mentioned that dry mobile phases usually were preferred, yet in this case the eluent is saturated with water. Why?
24. What is an easy way to clean these columns?
25. It is recommended that the plug of glass wool placed in the bottom of the column not be too small. What is the reason for this?
26. Assume that the absorbance for the sample had been 1.225 for the example problem. What would the results have been then?

IF IT EXISTS - CHEMISTRY IS INVOLVED
IF YOU CAN BUY IT - A CHEMIST WAS INVOLVED SOMEWHERE

16
AFFINITY CHROMATOGRAPHY

PRINCIPLES

Affinity chromatography, developed by R. Axen, J. Porath, and S. Ernback in 1967 and named by P. Cuatrecasas in 1968 is a separation in which the surface of the inert phase has been modified so that it selectively binds compounds having a specific functional group. The binding force is strong enough to effect a separation, but weak enough so that the desired compound can be washed off when desired. For example, Affi Gel 601 contains a boronic acid group that selectively retards compounds having 1,2-dihydroxy groups such as glycols and some sugars.

That some biological molecules had affinities for other biological molecules was first observed by Starkenstein in 1919, who noticed that amalyase binds tightly to insoluble starch. In 1953, Lerman used an azo dye immobilized on cellulose to separate mushroom tyrosinase from other proteins.

This technique became known as *dye binding* and is still a valid method of separation. What is now known as *affinity chromatography* had its beginnings in 1967 with the introduction by Axen et al. of activated agarose as the inert matrix. The name *affinity chromatography* was proposed as a general name to include both organic and inorganic applications. However, the applications have been related almost entirely to biological systems, so the name *bioselective chromatography* has been proposed. It is unlikely that the name will change quickly, because there are currently over 20,000 publications and several books referring to *affinity chromatography*.

Several commercial *beads* are available. Of more importance is the fact that by using a few well-established reactions, any one of hundreds of ligands can be added to the spacer arm. A few of these reactions will be discussed later.

INERT MATRIX

The matrix should (1) be stable to the eluant, (2) be mechanical and chemically stable, (3) have a large surface area, (4) be easily derivatized, and (5) have good flow characteristics. Currently used matrix materials are agarose,

166   Principles

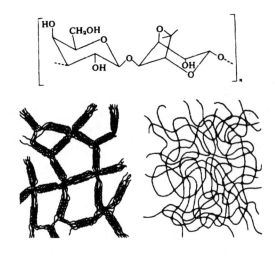

Figure 16-1. Molecular agarose (top). Systematic comparison of Sephadex (lower left) to agarose (lower right).
(Courtesy - Arnott et al., *J. Mol. Biol.* **90**, 269, 1974)

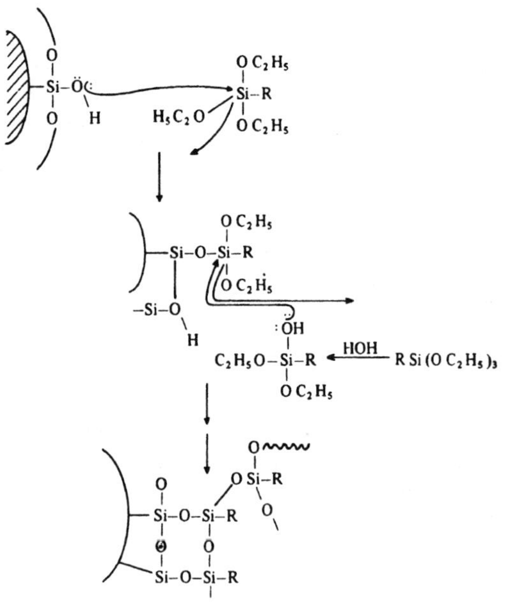

porous glass, cellulose, glycidoxy-coated glass, Ultragel (agarose coated polyacrylamide), and Enzacryl (polyacrylic-coated iron particles). About 90 % of the separations are with agarose and porous glass.

Agarose is obtained from sea kelp and is a linear polysaccharide consisting of alternating residues of α-1,3 and β-1,4 linkages plus a few carboxylate and sulfate ionic residues. The ionic residues must be removed, because they interfere with most separations. This can be done by a $NaBH_4$ reduction. The unpurified material is commonly called *agar*.

Agarose consists of pentagonal pores created by the bridging of triple helical agarose chains. This is a rather fragile substance and is often strengthened by additional cross linking with epichlorohydrin. Figure 16-1 is a diagram comparing the structures of Sephadex and agarose.

Controlled pore glass was developed because agarose lacks sufficient mechanical strength for high pressure, fast flow rate applications; it swells and shrinks excessively with changes in ionic strength; dissolves in too many solvents; is difficult to use with organic solvents; and is difficult to dry easily.

Controlled pore glass is prepared by heating certain types of borosilicate glass to 500-800 °C until it separates into silicate-rich and boron-rich phases. (Figure 16-2). The borate phase is then dissolved with acid, producing a sponge like arrangement with pores from 2.5-7.0 nm.

If the alkali is dissolved, then pores from 4.5 to 400 nm can be obtained. This form of glass will readily adsorb proteins, usually without altering their biological function. This glass does possess considerable ionic surface character, which is often undesirable. This can be removed by refluxing with a 10 % aqueous solution of aminopropyl triethoxysilane. The amino propyl group also can act as a spacer arm if desired.

SPACER ARMS

The spacer arm is used to move the active group well away from the bead, so that steric hindrances are at a minimum. This is shown in diagrammatic form in Figure 16-3. The effectiveness of spacer arms depends upon (1) their length, (2) stability of the attachment to the bead, (3) hydrophobic nature, (4) the presence of fixed charges, and (5) their concentration.

Most commercially available materials can be purchased with the spacer arm already attached and, in some cases, the ligand as well. Table 16-1 shows several spacer arms available from Bio-Rad Laboratories. Ligands with free carboxyl groups can be attached to Affi-Gel 101 and 102 by the EDAC method described later. Acid chlorides, N-hydroxysuccinimide esters, aldehydes, anhydrides and alkyl halides react directly.

Figure 16-2. Mechanism of silinization of porous glass.
(Courtesy - W.H. Scouten, *Affinity Chromatography*, Wiley-Interscience, 1981)

Ligands with free amino groups can be *immobilized* on Affi-Gel 201 and 202 by the EDAC method. Ligands can be attached to Affi-Gel 401 by disulfide, thioester, or thioether formation. Affi-Gel 501 has a high capacity for selectively purifying SH-containing proteins.

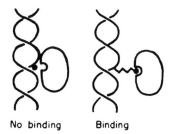

Figure 16-3. Effect of spacer groups on the accessibility of binding sites.
(Courtesy - Gelb, W.G., *American Laboratory*, Oct., 1973)

Table 16-1. Affi-Gel supports for affinity chromatography.
(Courtesy - Bio-Rad Laboratories, Richmond, CA)

Compound	Spacer Arm
Affi-Gel 10	I—O—$CH_2CH_2CH_2NHCCH_2CH_2C$—O—N(succinimidyl)
Affi-Gel 101	I-O-$(CH_2)_3NH_2$
Affi-Gel 102	I-O-$CH_2CONH(CH_2)_3NH(CH_2)_3NH_2$
Affi-Gel 201	I-O-$(CH_2)_3NHCO(CH_2)_2COOH$
Affi-Gel 202	I-O-$CH_2CONH(CH_2)_3NH(CH_2)_3NHCO(CH_2)_2COOH$
CM Bio-Gel A	I-O-CH_2COOH
Affi-Gel 401	I—O—$CH_2CONH(CH_2)_3NH(CH_2)_3NHCOCH$(NHCOCH_3)($(CH_2)_2SH$)
Affi-Gel 501	I—O—$(CH_2)_3NHCO$—C$_6$H$_4$—HgCl

ACTIVATING METHODS

Cyanogen Bromide

One of the common, but somewhat hazardous, methods of activating agarose is to use cyanogen bromide, CNBr.

$$\text{diol} \xrightarrow{\text{BrCN}} \text{cyclic imidocarbonate (C=NH)} \xrightarrow{H_2N-R} \text{isourea derivative}$$

Acid Chloride

This is a proven method to activate carboxylic acid groups for further reaction with amines.

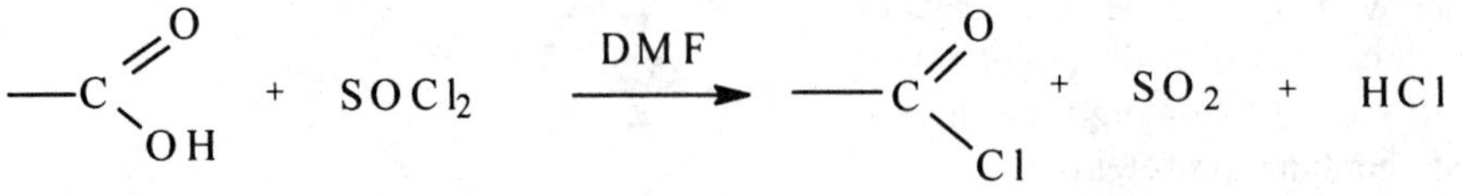

$$\text{—COOH} + SOCl_2 \xrightarrow{DMF} \text{—COCl} + SO_2 + HCl$$

ATTACHMENT OF LIGANDS

There are many ways to attach ligands to the spacer arm. A few are shown below.

Direct Attachment

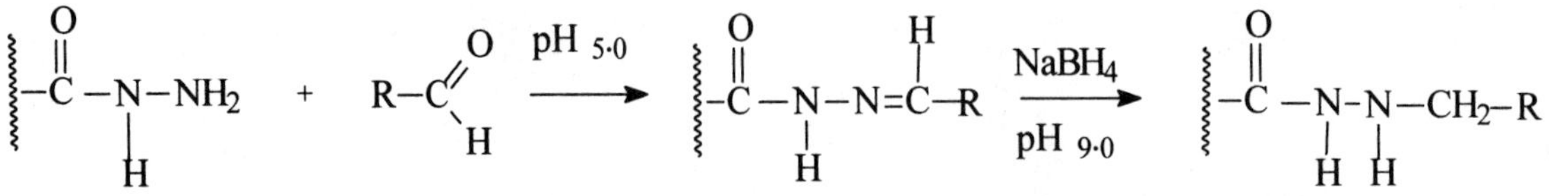

$$\text{−C(=O)−N(H)−NH}_2 + R\text{−CHO} \xrightarrow{pH\ 5.0} \text{−C(=O)−N(H)−N=CH−R} \xrightarrow[pH\ 9.0]{NaBH_4} \text{−C(=O)−N(H)−N(H)−CH}_2\text{−R}$$

Reactive Intermediate

N-Hydroxysuccinimide

Ligands with free alkyl or aryl amino groups will couple spontaneously within hours in aqueous solution.

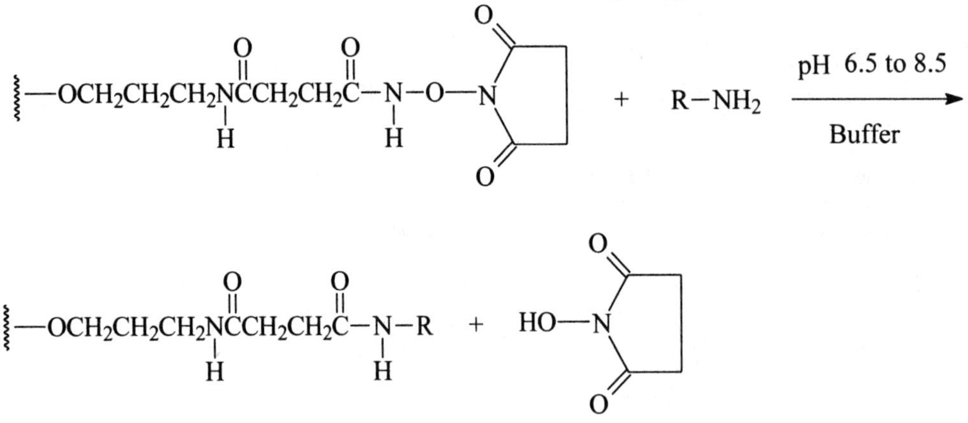

$$\text{—OCH}_2\text{CH}_2\text{CH}_2\text{N(H)C(=O)CH}_2\text{CH}_2\text{C(=O)−N−O−NHS} + R\text{−NH}_2 \xrightarrow[\text{Buffer}]{pH\ 6.5\ to\ 8.5}$$

$$\text{—OCH}_2\text{CH}_2\text{CH}_2\text{N(H)C(=O)CH}_2\text{CH}_2\text{C(=O)−N(H)−R} + HO\text{−NHS}$$

EDAC (1-ethyl-3-(3-dimethylaminopropyl) carbodiimide HCl)

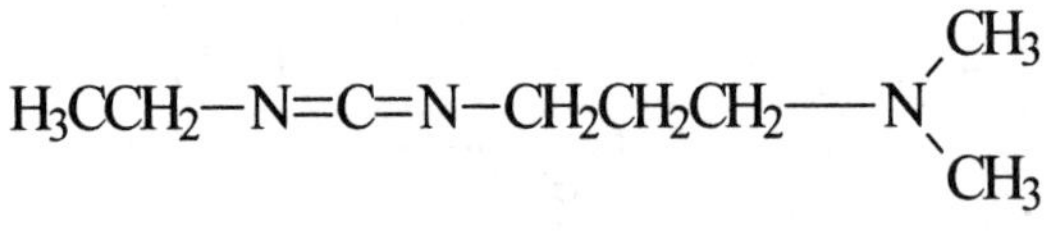

$$H_3CCH_2\text{−N=C=N−CH}_2\text{CH}_2\text{CH}_2\text{−N(CH}_3)_2$$

For the preparation of amides, peptides, thioesters and esters in either aqueous or anhydrous solutions in about 2 hours. The following scheme, p. 169, from Lowe and Dean, *Affinity Chromatography*, J. Wiley & Sons, 1974, illustrates the reaction sequence for carbodiimides in general.

[Reaction scheme: R-COO⁻ + carbodiimide (C=N-R₁, N-R₂) → at pH 4-5 → O-acylisourea intermediate, which reacts with (i) R₃-NH₂ to give amide R-C(=O)-NH-R₃ + urea; (ii) rearrangement; (iii) R₃-SH to give thioester + urea]

Table 16-2 lists several ligands and the types of compounds for which they have an affinity. Affinity chromatography offers the research chemist an almost unlimited opportunity to develop rather selective separations. However, the preparation of the bed matrix usually is not easy, requiring much technique.

Table 16-2. Selected ligands and their affinity compounds

Ligand	Affinity Compounds
Diazo-NAD⁺	Dehydrogenases
AMP analogues	NADP⁺ binding proteins
AMP, AEP, AMP	Kinases
Blue dextran 2000	Yeast phosphofructokinase
Coenzyme A	3-Hydroxy-3-methyl glutaryl coenzyme A reductase
Methotrexate	Dihydrofolate reductase
B$_{12}$	Transcobalamin I and II
Thiamine	Pyruvate oxidase
Pyridoxyl phosphate	Tyrosine aminotransferase

REVIEW QUESTIONS

1. What is the purpose of the "spacer arm"?
2. What is dye binding?
3. What are some advantages of controlled pore glass and how is it prepared?
4. How is porous glass silinized?
5. Why use EDAC first rather than couple the ligand directly to the spacer arm?
6. Why are lectins considered to be hazardous unless properly handled? Refer to Appendix A, Experiment No. 21.
7. What is the difference between lectin and lecithin? Same reference as 6, but you will need more.

IF IT EXISTS - CHEMISTRY IS INVOLVED
IF YOU CAN BUY IT - A CHEMIST WAS INVOLVED SOMEWHERE

17
SIZE EXCLUSION CHROMATOGRAPHY
(Gel Filtration; Gel Permeation)

PRINCIPLES

Size exclusion chromatography is a chromatographic technique in which the separation is based on differences in the size of the sample molecules. The column packing is made from *beads* of a porous gel shown as a photomicrograph in Figure 17-1, p. 172.

Molecules of various size are drawn on the photo and show that the large molecules cannot fit at all and would be washed out of the column. The smallest m. w. molecule that does not enter the gel is called the *exclusion limit,* and the volume required to elute the large molecules is called the *void volume, V.* The smaller molecules can permeate the gel; the smaller the molecule, the farther the penetration and the greater the retardation. This results in a separation because it requires more solvent to wash the smaller molecules out of the gel. This has a limit. If the molecules become too small, then their molecular size compared to the pores is so small that they behave the same and require essentially the same amount of solvent to elute them. This volume is called the *total permeation volume, V_t.* Figure 17-2, p. 172 illustrates these terms. V_i is the *interstitial volume*.

The size of the pores determines the molecular weight range of the compounds that can be separated. Size exclusion often is used to clean up an extracted solution. For example, it is used to remove the fats and waxes from the pesticide residues that are extracted from a food composite.

A general rule is that *compounds that differ by 10% in size can be separated in the same column.* A series of columns, each fitted with a different exclusion limit gel, can be used to separate a multiple-component mixture.

GEL FILTRATION

Parts of the material below were abstracted from Pharmacia-Biotech Products literature.

Gel filtration is a rapid and efficient alternative to dialysis for desalting and buffer exchange. It is used, for example, in the preparation of protein samples prior to freeze drying or fractionating by ion exchange chromatography. Applications include:

Determining the m.w. distribution of polymers.
Preparative fractionation of polymers.
Purification of biological samples.
Determining complex equilibrium constants.
Desalting biological materials.
Exchanging buffers.

Sephadex

The concept of what is now called *size exclusion* was developed by J. Porath and P. Flodin in 1959 and was called gel filtration. The gels they prepared were made from dextran that had been polymerized with epichlorohydrin and are known as *Sephadex*. They are prepared by what is now the Pharmacia Biotech Products Co., Uppsala, Sweden. Their structure is shown on p. 172.

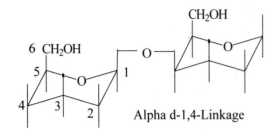

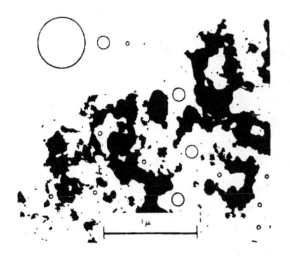

Figure 17-1. An electron micrograph of a gel particle matrix.
(Courtesy - Cantow, M.J.R., *Polymer Fractionation*, Academic Press, Orlando, FL, 1967)

Sephadex is the standard choice for fractionating mixtures of proteins, peptides, and the smaller nucleic acids and polysaccharides. By varying the degrees of cross linking, gels of different porosities and different fractionation ranges are obtained. Sephadex G-10 to G-25 is stable from pH 2 to 13 and G-50 to G-200 from pH 2 to 10. Sephadex can be used with buffers containing dissociating agents such as urea and detergents and can be autoclave sterilized.

Dextran is obtained from starch. Starch is a polysaccharide consisting primarily of α-d-1,4 and α-d-1,6 linkages. If the starch is treated with either of the enzymes *Leuconostoc mesenteroides* or *Lactobacteriaceae dextranicum*, the α-d-1,4 linkages are broken, leaving a residue of α-d-1,6 linkages, which is isolated by precipitation with methanol. This residue is known as *dextran*. Typical commercial dextrans have molecular weights from 40,000 to 75,000.

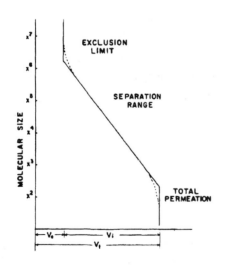

Figure 17-2. The volume terms in size exclusion chromatography.
(Courtesy - Waters Corp., Milford, MA)

These materials use aqueous salt solutions for the mobile phase; they swell considerably and collapse rather easily. In fact, the large pore gels (greater than G-25) must have the mobile phase enter from the bottom of the column to prevent the gel from collapsing and plugging the column. *Glass frits* should not be used to support the bead column, because the sharp edges cut into the beads, causing them to collapse and plug the column.

Gel filtration usually involves low pressures (solvent heights of a few inches), and the flow rates are slow, from 0.1 to 1 mL/min. Figure 17-3 is a diagram of some commercially available columns. The Kontes columns are glass with plastic frits and are designed to have a low dead volume below the packing. The Pharmacia Biotech columns are plastic and work very well with the soft types of Sephadex.

Lipophilic Sephadex

Lipophilic materials (Sephadex LH-20 and LE-60) are used when organic solvents are required. They are prepared from Sephadex G-25 and G-50 by hydroxypropylation. They are designed for use in aqueous buffer systems, polar organic solvents, and aqueous solvent mixtures. In mixed solvents, the gel preferentially takes up the more polar component. They have wide applications in the fractionation of lipids, steroids, fatty acids, hormones, vitamins, and other small molecules.

The ideal gel should be in the form of porous superfine beads with high mechanical strength to allow good resolution of large molecules at conveniently high flow rates. Sephacryl is prepared by covalently cross-linking allyl dextran with N,N'-methylene bisacrylamide to give a highly stable matrix. By carefully controlling the exact conditions of the synthesis, gels are produced with exclusion limits ranging from approximately 250,000 (Sephacryl S-200) to several hundred million (Sephacryl S-1000). The rigidity of the matrix allows beads of Superfine grade to be prepared that still provide excellent flow rates. Columns can be packed and equilibrated quickly, and cycle times reduced.

Sephacryl can be used in aqueous buffer systems, pH 2-11, in concentrated urea or guanidine HCl, and in a number of organic solvents. The fractionation ranges of the five types of Sephacryl (Table 17-1, p. 174) cover molecules from small polypeptides to particles of diameter 300-400 nm. Sephacryl S-200 and S-300 are useful for most proteins.

Sephacryl

Sephacryl S-400 and S-500 are excellent for polysaccharides and other macromolecules with an extended structure. Sephacryl S-1000 fractionates restriction fragments of DNA, very large polysaccharides, proteoglycans, and membrane-bound vesicles up to 300-400 nm in diameter.

The bead structure of Sephacryl is allyl dextran and N,N'-methylene bisacrylamide.

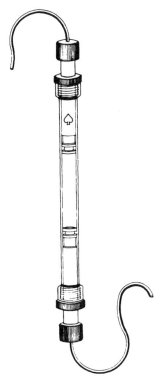

Figure 17-3. Diagram of a column and plunger arrangement.
(Courtesy - Ace Glass Co., Vineland, NJ)

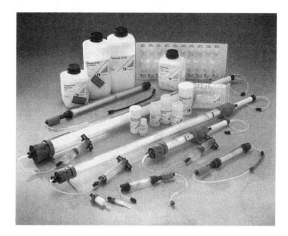

Figure 17-4. Columns and bed volume control plungers.
(Courtesy - Pharmacia Biotech Co., Piscataway, NJ)

Table 17-1 lists several gel filtration compounds and some of their characteristics.

Table 17-1. Selected data on gel filtration products
(Courtesy - Pharmacia Co., Uppsala, Sweden)

Sephadex		Useful Fractionating Range (mw)		Dry Bead
		Peptides/Globular Proteins	Dextrans	Diameter (μm)
G-10		-700	-700	40-120
G-15		-1,500	-1,500	40-120
G-25	Coarse	1,000-5,000	100-5,000	100-300
	Medium			50-150
	Fine			20-80
	Superfine			20-50
G-50	Coarse	1,500-30,000	500-10,000	100-300
	Medium			50-150
	Fine			20-80
	Superfine			20-50
G-75		3,000-80,000	1,000-50,000	40-120
	Superfine	3,000-70,000		20-50
G-100		4,000-150,000	1,000-100,000	40-120
	Superfine	4,000-100,000		20-50
G-150		5,000-300,000	1,000-150,000	40-120
	Superfine	5,000-150,000		20-50
G-200		5,000-600,000	1,000-200,000	40-120
	Superfine	5,000-250,000		20-50
Sepharose				
6B/C1-6B		$10,000 - 4 \times 10^6$	$10,000 - 1 \times 10^6$	45-165
4B/C1-4B		$60,000 - 2 \times 10^7$	$30,000 - 5 \times 10^6$	45-165
2B/C1-2B		$70,000 - 4 \times 10^7$	$100,000 - 2 \times 10^7$	60-200
Sephacryl				
S-200 Superfine		5,000-250,000	1,000-80,000	40-105
S-300 Superfine		$10,000 - 1.5 \times 10^6$	2,000-400,000	40-105
S-400 Superfine		$20,000 - 8 \times 10^6$	$10,000 - 2 \times 10^6$	40-105
S-500 Superfine			$40,000 - 2 \times 10^7$	40-105
S-1000 Superfine			$500,000 \rightarrow 10^8$	40-105

Molecular Weights

Table 17-2 lists several compounds that can be used to calibrate a column for the determination of molecular weights.

Table 17-2. Reference compounds for molecular weight determinations
(Courtesy - Pharmacia Biotech Products, Uppsala, Sweden)

Protein	m.w.	Source
Ribonuclease	13,700	Bovine pancreas
Chymotrypsinogen A	25,000	Bovine pancreas
Ovalbumin	43,000	Hen egg
Albumin	67,000	Bovine serum
Aldoase	158,000	Rabbit muscle
Catalase	232,000	Bovine liver
Ferritin	440,000	Mouse spleen
Thyroglobulin	669,000	Bovine thyroid

Blue dextran 2000, m.w. 2,000,000 is used to determine void volume.

GEL PERMEATION

One of the initial disadvantages of the Sephadex polymers was being limited to use with aqueous systems. In 1964, John Moore of Dow Chemical patented a polymer that could be used for size exclusion separations involving organic solvents. These initially were highly crosslinked styrene-divinyl benzene (Styragel, Poragel, Bio-gel-S). Pharmacia had the patent on the gel filtration process, so this system was called *gel permeation*.

These beads can be used with higher pressures. Some, like hydrogel, can withstand pressures up to 3000 p.s.i.. Organic solvents can be used; therefore, a wide variety of compounds can be separated. Typical solvents are benzene, toluene, xylene, carbon tetrachloride, dimethylformamide, methylene chloride, ketones, dimethyl sulfoxide, dichlorobenzene, ethylene dichloride, perchlorethylene, tetrahydrofuran, trichlorobenzene, and chloroform.

Bio-gel-P

Bio-gel-P is a polyacrylamide polymer cross-linked with methylene bisacrylamide. Bio-gel-P normally is not employed with water-miscible organic solvents because the beads contract, causing reduction in pore size. It is compatible with dilute organic acids, 8M urea, 6M guanidine HCl, chaotropic agents, and detergents.

Enzacryl

The enzacryls are polymers of nitrogen containing compounds crosslinked with acrylamide. Most are used for immobilizing enzymes. Immobilized enzymes are very important, because chemical reactions can be performed on a column and the enzyme will retain its activity.

Enzacryl AA   R = -NHC$_6$H$_4$NH$_2$
         AH   R = -NH-NH$_2$ (activate with HNO$_2$)
         Polythiol   R = -NHCH(CO$_2$H)CH$_2$SH   (good for enzyme binding)

Polythio R =   —NHCH—C=O   (enzyme binding - need not be activated)
                         | |
Lactone   CH$_2$ -S

Activating and Binding Process
Gel + HNO$_2$ ----------> Acid azide
Acid azide + Enzyme ----------> Bound enzyme (at the NH$_2$ of lysine on the enzyme)

Table 17-3, p. 176, shows the exclusion limits for the Bio-gel-P's and Styragels.

TECHNIQUES

The following techniques apply to gel permeation as well as gel filtration.

Preswell the beads in the solvent to be used. The bead volume varies considerably from solvent to solvent and with ionic strength when using aqueous solvents. If end plungers are not used, the bed volume tends to form channels, and if plungers are used before pre-swelling, pressure buildup can blow out the end plugs. Beads usually take about 4 hours to reach maximum size during preswelling.

Slurry pack the column. Keep the column full of solvent when the swollen beads are added. The beads are light and settle slowly. Be patient, or channeling will occur.

Table 17-3. Fractionation range for several Bio-gel-P and Styragel polymers
(Courtesy - Bio-Rad Laboratories, Richmond, CA; Waters Associates, Milford, MA)

Polymer		Fractionating Range	Packed Bed Volume
Bio-gel	P2	100-1,800	3.5
	P4	800-4,000	5
	P6	1,000-6,000	8
	P10	1,500-20,000	9
	P30	2,500-40,000	11
	P60	3,000-60,000	14
	P100	5,000-100,000	15
	P150	15,000-150,000	18
	P200	30,000-200,000	25
	P300	60,000-400,000	30
Styragel	10^7 Å	500,000-500,000,000	
	10^6	100,000-50,000,000	
	10^5	50,000-2,000,000	
	10^4	1,000-700,000	
	10^3	500-50,000	
	10^2	100-8,000	

Bio-beads SX-1 through SK-12 have exclusion limits from 400 to 14,000.

Use a two-plunger apparatus, if possible. The plungers are used to hold the column packing in place. The beads are very flexible and tend to collapse when pressure is applied to the plunger. The beads do not relieve applied pressure evenly, and the plunger end is more compact than the opposite end. This causes channeling. The use of two plungers minimizes this effect. Adjust the plungers slowly. The beads tend to stick to the sidewall of the column and get squeezed if the plunger is moved too fast.

Test the uniformity of the column packing with a colored compound. For example, 0.5 g of butter dissolved in 1:1 methylene chloride and hexane can be used to check SX-3 columns. The band should stay narrow, and the front should be straight. If it is not, then repack the column with the same material, only use a little more care. Some columns have to be repacked 3-4 times before one is obtained that has no channeling.

Optimum pressures are 7-11 psi. The plungers are adjusted until 7-11 psi of pressure is acquired to force liquid through the column. Less pressure leads to channeling; more pressure leads to bead collapse.

Figure 17-5, p. 177, is a photograph of a complete automated apparatus setup, and Figure 17-6, p. 177, shows a setup in diagrammatic form of a nonautomated system. Figure 17-7, p. 177, is a photograph of a gradient mixer, and Figure 17-8, p. 177, is a diagram of how the gradient mixer is set up. Figure 17-9, p. 178, shows the apparatus setup for use with either gel filtration or gel permeation.

Figure 17-9 is a photograph of an automated system that incorporates 23 columns and is used by the FDA to remove fats and waxes from plant extracts so pesticide residues can be determined. It uses a Milton-Roy pump and is set up on a 5 mL/min flow rate. The flow rate is adjusted by the pressure on the plungers. It has "dump", "collect", and "wash" cycles that can be varied from 0.1 to 10 minutes and from 1 to 99 times. For example; the FDA procedure uses the standard 5 mL/min flow rate, 11 cycles on dump, and 20 cycles on collect.

Figure 17-9. An automated gel filtration system.
(Courtesy - O.I. Analytical Laboratories, Columbia, MO)

REVIEW QUESTIONS

1. Why are plastic columns and support frits preferred over glass columns for gel filtration?
2. Why is there so much volume change when different ionic strength liquids are used with these compounds?
3. Why should preswelling be done?
4. What is the purpose of plungers, and why are two preferred?
5. What do we mean by size exclusion?
6. How do you prepare Sephadex?
7. How do you immobilize an enzyme?

IF IT EXISTS - CHEMISTRY IS INVOLVED
IF YOU CAN BUY IT - A CHEMIST WAS INVOLVED SOMEWHERE

Size exclusion chromatography 177

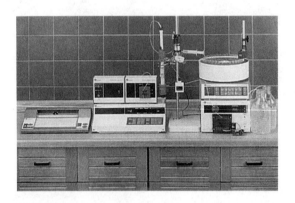

Figure 17-5. An automated flexible low pressure chromatography system.
(Courtesy - Pharmacia Biotech Co., Piscataway, NJ)

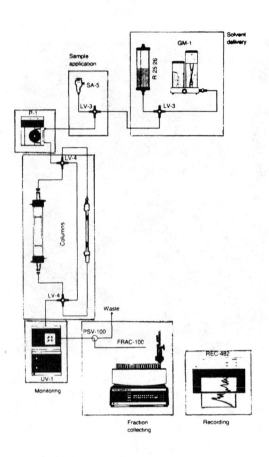

Figure 17-6. Diagram of a gel filtration apparatus.
(Courtesy - Pharmacia Biotech Co., Piscataway, NJ)

Figure 17-7. Gradient mixer.
(Courtesy - Pharmacia Biotech Co., Piscataway, NJ)

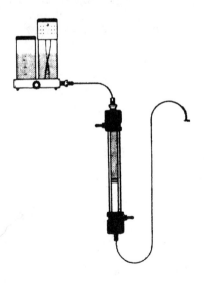

Figure 17-8. Arrangement for a gradient mixer.
Courtesy - Pharmacia Biotech Co., Piscataway, NJ)

18
FLASH CHROMATOGRAPHY

PRINCIPLES

Flash chromatography is simply regular chromatography with a low pressure (usually < 20 p.s.i.) applied to force the eluting solvent through the column faster. This produces medium quality, but fast (10-15 minutes) separations. It is not appropriate for separating a mixture of a dozen components, but it is excellent for separating the few side reactants from the main component in an organic synthesis. Depending upon the column size, several grams of sample can be purified at one time. The method was developed by Still, W.C., Kahn, M., and Mitra, A. (*J. Org. Chem.*, **43** (14), 2923, 1978). Figure 18-1, p. 180, shows a common arrangement for pressures less than 20 p.s.i. with a manual flow controller. Figure 18-2, p. 180, shows a unit for higher pressures (>50 p.s.i.) with a pressure gauge attached to regulate flow and a pressure relief valve.

Typical columns are from 30 to 45 cm long and from 1 to 15 cm in diameter. Reservoirs vary from 250 mL to 3000 mL and for the higher pressure units are epoxy coated for safety. The column packing is usually silica gel. The components are held together by clamps or are screwed together. In any case, the separation should be done behind a safety shield. Glass frits are not used on the column bottoms because of too much dead volume below them. Glass wool and fine sand are used instead.

TECHNIQUES

COLUMN PACKING

The most common packing is silica gel. However, recently some reversed phase packings have been employed. Still et al. found that the best particle size was 40-63 mesh (Figure 18-3 *top*, p. 180). Resolution is defined as retention time (r) divided by the peak width (w).

Dry Packing

The method of Still et al. for a 2 cm diameter column is as follows. It is important that the column packing does not fragment, or channeling can occur.

"Place a small piece of glass wool in the bottom of the column. Add 2 mm of 50-100 mesh sand and level it. Add 40-63 mesh silica gel to a depth of 14-15 cm in a single addition. With the stopcock open, the column is gently tapped vertically on a bench top to settle the packing. Add 2 mm of 50-100 mesh sand to the top of the now flat gel bed. The column is clamped for pressure packing and elution. The solvent is carefully poured over the sand to fill the column completely. The needle valve of the flow controller is opened all of the way, fitted tightly to the top of the column and secured with strong rubber bands. The main air line valve leading to the flow controller is opened slightly and a finger placed fairly tightly over the bleed port. This will cause the pressure above the adsorbent bed to climb rapidly and compress the silica gel as solvent is rapidly forced through the column. It is important to maintain the pressure until all of the air is expelled and the lower part of the column is cool, otherwise the column will fragment and should be re-packed."

180 Principles

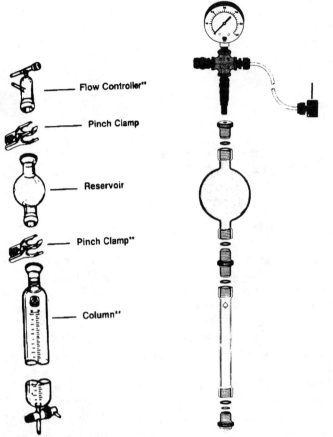

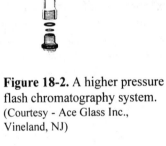

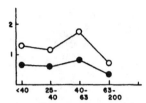

Silica gel particle size[6] (μm): (●) r/w; (○) r/(w/2).

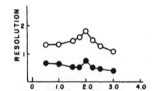

Eluant flow rate (in./min).

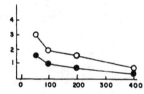

Sample size (mg).

Figure 18-1. Components needed for flash chromatography.
(Courtesy - Supelco, Bellefonte, PA)

Figure 18-2. A higher pressure flash chromatography system.
(Courtesy - Ace Glass Inc., Vineland, NJ)

Figure 18-3. The effect on resolution by (*top*) particle size, (*middle*) flow rate, and (*bottom*) sample size.
(Courtesy - Still, W.C., Kahn, M. and Mitra, A., *J. Org. Chem.*, **43** (14), 2923, 1978)

Wet Packing

Most people have trouble getting all of the air out of the column using the dry packing method, so a wet slurry packing method is used.

In this method, the silica gel is made into a slurry and poured into the column with the outlet open. Pressure is applied to force the solvent out the bottom and also pack the column tight without air being trapped. The pressure is released, the stopcock closed, the sample added, and then the regular elution procedure is followed.

PUMPS

For most applications, the air from a compressed air line is used, and a pump is not needed. This may be from a cylinder of air or the lab air supply. However, for educational purposes or for inexpensive occasional separations, two low-pressure providers have been tested. The lowest pressure "pump" is a double balloon arrangement (Bell, W.L. and Edmondson, R.D., *J. Chem. Ed.*, **63** (4), 361, 1986). A large balloon is placed inside of another balloon, blown up, and attached to the top of the column. Pressures of 2-3 p.s.i. are achieved. The pressure is fairly steady because the balloon volume is much larger than the volume of solvent used.

For somewhat higher pressures, depending upon the size, and a more even flow rate, a fish aquarium pump, available at any pet store, can be used (Jacobson, B.M., *J. Chem. Ed.*, **65** (5), 459, 1988). The air controller obtained there also will work well.

PRESSURIZING GASES

Compressed air normally is used unless the compounds are air (oxygen) sensitive, then compressed nitrogen is used. These gases are compressed by pumps that have oil seals, so the gas always contains a small amount of oil vapor. A trap between the supply and the column can be used, but is required only if very high product purity is desired. Keep the system as simple as possible.

SAMPLE ADDITION

Figure 18-3 (*bottom*) indicates that there is a best sample size for a given column diameter. This is summarized in Table 18-1 for several columns. Sample addition can be a problem. Even though this is medium resolution chromatography, it is still necessary to apply the sample in as thin a layer as possible. To do this without mixing it in with the gel bed takes some practice. One way to make it easier is to use the sample-addition funnel shown in Figure 18-4. It is a regular long-stem funnel with the end bent and a 4 mm hole in the side at the bottom. This deflects the sample to the column sidewall, and it spreads out evenly when it reaches the top of the column bed without disturbing the top of the bed.

FLOW CONTROL

The purpose of the flow controller is to reduce the pressure of the incoming gas to that required for the proper solvent flow rate and to let the excess gas escape through a vent.

Figure 18-3 (*middle*) shows that for a *2 cm column, an eluant flow rate of 5 cm/min through the column is optimal.*

Commercial controllers give fine control and can be varied as needed for different column sizes and changes in the viscosity of the solvents. However, if a simple system is needed for separations that are routine, then the controller proposed by A. Feigenbaum (Figure 18-5) can be considered: "Air is admitted into the system through a restrictor. The restrictor is simply an 8 mm o.d. glass tube, constricted at one end to 2 mm. The Hg height is adjusted to provide a solvent flow rate of 5 cm/min. The excess is vented to a hood. The pressure reservoir is a 3 L flask, wrapped with screen wire for safety."

ELUTING SOLVENTS

Experience has shown that the best separations take place if the desired component has an R_f of 0.35 on a silica gel thin layer plate and the impurities are ± 0.15 R_f values away. (See Chapter 22 for thin layer chromatography, TLC.) The chemist must try several solvents or solvent combinations on a TLC plate to get the right separation. Once this has been done, then the solvent system is simply transferred to the column. Solvent systems that have been found to work well include 10-50% mixtures of ethyl acetate/30-60°C with petroleum ether (hexanes).

Table 18-1. Recommendations for various size columns
(Courtesy - Still, W.C., Kahn, M. And Mitra, A., *J. Org. Chem.*, **43** (14), 2924, 1978)

column diameter, mm	vol of eluant,[a] mL	sample: typical loading (mg) $\Delta R_f \geq 0.2$	$\Delta R_f \geq 0.1$	typical fraction size, mL
10	100	100	40	5
20	200	400	160	10
30	400	900	360	20
40	600	1600	600	30
50	1000	2500	1000	50

[a] Typical volume of eluant required for packing and elution.

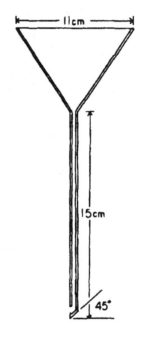

Figure 18-4. A sample-addition funnel.
(Courtesy - Thompson, W.J. and Hanson, B.L., *J. Chem. Ed.*, **61** (7), 645, 1984)

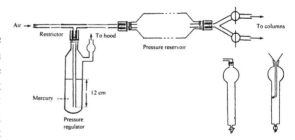

Figure 18-5. A low cost flow controller for student laboratories.
(Courtesy - Feigenbaum, A., *J. Chem. Ed.*, **61** (7), 649, 1984)

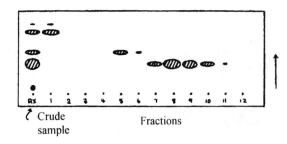

↑ Crude sample   Fractions

Figure 18-6. The use of thin layer chromatography to monitor the collected fractions.
(Courtesy - Still, W.C., Kahn, M. And Mitra, A., *J. Chem. Ed.*, **43** (14), 2923, 1978)

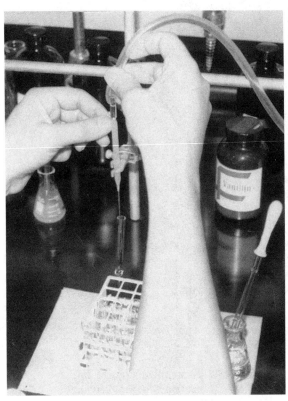

Figure 18-7. Micro flash chromatography apparatus.
(Courtesy - Dr. Keith Buzsek, Kansas State University, Manhattan, KS)

COLLECTION AND DETECTION

The sample comes through so fast that an easy way must be found to collect the fractions. Table 18-1 provides recommendations on the size of the fraction to collect based on the sample size and column volume.

The effluent can be monitored by a uv detector, and the desired fraction collected. What has been found to work well is to use a test tube rack containing several 20 x 150 mm test tubes and collect a fraction in each one. Refer to Figure 18-6 to see a convenient method of detection. The crude sample is spotted along one side of the TLC plate, 5 μL from each fraction, then is spotted along the edge of the plate. When the plate is developed, the fractions that contain the desired compound are identified, and the degree of separation achieved is also known.

GENERAL CONSIDERATIONS

Taking all of the above into consideration, Still et al. make some general recommendations. These are summarized in Table 18-1.

SMALL-SCALE SYSTEMS

The apparatus illustrated in Figure 18-7 is a good example of how common laboratory apparatus can be used for microscale separations. The apparatus shown can easily separate 10 mg or less of sample, with an upper limit of 25-30 mg, and if the apparatus is already set up, a separation can be done in less than an hour. A 5.5 inch disposable pipet is packed and held in place with a small 3-pronged clamp mounted on a ring stand. A 25 mL Erlenmeyer flask for eluting solvent is placed nearby. A series of 10 x 75 mm test tubes in a test tube rack is placed by the outlet, and one is shown in place. The air line is held in place as needed. A flask for rinse solvent is to the upper left.

REVIEW QUESTIONS

1. What is meant by "flash chromatography"?
2. Why are glass frits usually not used on the bottom of the columns?
3. In the dry packing method of Still et al., why is it important to maintain the pressure until all of the air is expelled and the lower part of the column is cool?
4. What types of inexpensive pumps can be used for flash chromatography?
5. What is the purpose of the flow controller?
6. What is the best R_f compared to that of any impurities?
7. What are some good solvents or solvent mixtures to use with flash chromatography?
8. If it is desired to separate the major component with a ΔR_f of ≥ 0.2, what is the maximum recommended sample size when using a 30 mm diameter column?

19
HIGH PERFORMANCE LIQUID CHROMATOGRAPHY (HPLC)

PRINCIPLES

What is now known as high pressure liquid chromatography or high performance liquid chromatography (HPLC) was first presented by Huber, J.F.K., and Hulsman, J.A.G., in 1967.

We have learned that the first chromatographic technique was column chromatography. Since then, paper, thin layer, and gas chromatography were developed. Now, nearly 100 years later, we are going back to column chromatography! This seems like a strange sequence of events, but there are good reasons for it. The original column chromatographic technique employed glass columns and either gravity flow or a slight vacuum to move the mobile phase through the column. This was also *slow* chromatography and *hard to reproduce* chromatography. However, it was *extremely flexible* chromatography in that an almost unlimited variety of solvents and column packings could be used. It was because of this recognized flexibility that scientists reexamined column chromatography.

Use of steel columns and high pressures showed that column chromatography could be "fast" and "reproducible" as well as flexible. A basic instrument, shown in diagrammatic form in Figure 19-1, p. 184, consists of a solvent reservoir, a pump, a gradient chamber, an injection port, a column, a detector, a fraction collector, and a recorder. Depending upon the quality of the individual components and the number of components actually used, the cost of such a combination can vary from $4,000 to $35,000.

Figure 19-2, p. 184, is a photograph of a quality high pressure liquid chromatograph.

TYPES OF LIQUID CHROMATOGRAPHY

There are many ways to classify the types of liquid chromatography. One of these is discussed below. Four types of high pressure liquid chromatography to be discussed here are: *liquid-solid, bonded reversed phase, ion-exchange, and paired-ion*. These are all based on the differences in chemical properties of the materials to be separated.

Liquid-Solid Chromatography
(This is used to separate polar compounds.)

The title liquid-solid aptly describes the column conditions. True, the packing is a solid, but of more importance, the outer layer of the packing material that comes in contact with the mobile phase and sample compounds is a solid. This is shown in diagrammatic form in Figure 19-3.

Most solid column packings are clays or silica type materials, which means that their surfaces are aluminates or silicates and consist of large numbers of terminal -OH groups that are highly polar. Usually a nonpolar solvent such as hexane is used as the mobile phase. When moderately polar compounds are dissolved in the mobile phase and passed over the column packing, the more polar compounds are retained longer than the less polar compounds, and a separation results.

Do not use solvents above pH 8 or the silica particles may dissolve.

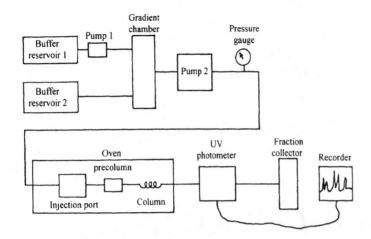

Figure 19-1. Component diagram of a Varian LCS 100 high pressure liquid chromatograph.
(Courtesy - Varian Aerograph Co., Walnut Creek, CA)

Figure 19-2. A Spectraphysics model 8800 HPLC with a Spectrascan FL2000 fluorescence detector and a linear LC304 fluorescence detector.

Bonded Phase (Usually Reversed Phase) Liquid-Solid Chromatography
(This is used to separate nonpolar compounds.)

The solid surfaces of untreated particles are polar and are not very useful for separating nonpolar compounds, which have little affinity. To solve this problem, a **reversed phase bonded packing** is now available.

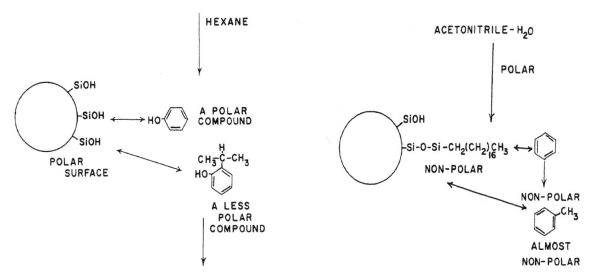

Figure 19-3. A liquid-solid system.
(Courtesy - Waters Corp., Milford, MA)

Figure 19-4. A reversed phase bonded packing.
(Courtesy - Waters Corp., Milford, MA)

In this situation, shown in Figure 19-4, the reactive surface of the particle is changed to a nonpolar compound and it is chemically bonded to the -OH group so it cannot be stripped off. A common system is octadecyl silane (ODS). One way to attach these bonded phases is shown on page 184.

The final reaction is called *end capping* and is used to add various groups other than a methyl group to the -OH so the reactivity can be varied, usually to a less polar group. Several reversed phase systems are shown in Table 19-1.

Table 19-1. Several common bonded phase surfaces

ODS, C-18	$-O-\underset{CH_3}{\overset{\|}{Si}}-(CH_2)_{17}-CH_3$	Amino, NH_2	$-O-\underset{O}{\overset{\|}{Si}}-(CH_2)_3-NH_2$
Octyl, C-8	$-O-\underset{CH_3}{\overset{CH_3}{Si}}-(CH_2)_7-CH_3$	Cyano, CN	$-O-\underset{CH_3}{\overset{CH_3}{Si}}-(CH_2)_3-CN$
Hexyl, C-6	$-O-\underset{CH_3}{\overset{CH_3}{Si}}-(CH_2)_5-CH_3$	Phenyl	$-O-\underset{CH_3}{\overset{CH_3}{Si}}-Phenyl$
Butyl, C-4	$-O-\underset{CH_3}{\overset{CH_3}{Si}}-(CH_2)_3-CH_3$	Anion	$-O-\underset{O}{\overset{O}{Si}}-(CH_2)_3-\underset{CH_3}{\overset{CH_3}{N^+}}Cl$
Methyl, C-1	$-O-\underset{CH_3}{\overset{CH_3}{Si}}-CH_3$	Cation	$-O-\underset{O}{\overset{O}{Si}}-(CH_2)_3-\underset{O}{\overset{O}{S}}-O^-$

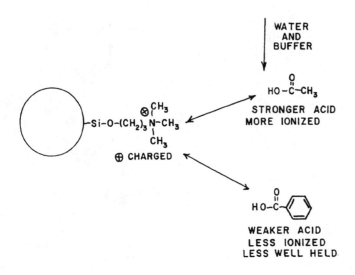

Figure 19-5. A bonded ion exchange packing.
(Courtesy - Waters Corp., Milford, MA)

Table 19-2. Ion exchange groups for HPLC

Strongly basic	$-CH_2N(CH_3)_3$, Cl^-
Medium basic	$-N(CH_3)_3$, Cl^-
Weakly basic	$-NH(R)_2$, Cl^-
Strong acid	$-SO_3^-$, H^+
Medium acid	$-HPO_3^=$, Na^+

Ion-Exchange Liquid Chromatography
(This is used to separate ionic compounds.)

Normal ion-exchange resins, which are made from styrene and divinyl benzene, usually are not used in high performance liquid chromatography because the beads are too soft, compress, and may plug up the column. However, some newer polymers of styrene divinyl benzene with higher cross linking are being used. What is more commonly used is a bonded packing of typical ion-exchange functional groups on a hard silica particle. This is shown in Figure 19-5. Others are shown in Table 19-2.

Generally, the compounds to be separated are placed in an aqueous system buffered about 1.5 pK_a units above the highest pK_a in the mixture. This ensures that the compounds are completely ionized and retained. They then are eluted by:

1. Passing the solvent buffer through the column, the order of elution being the compound with the highest pK_a eluting first.
2. Using an ionic strength gradient, with a low ionic strength being used first.
3. Changing the pH. For anions, go more acidic. This changes the sample compounds back to neutral molecules, and they then will elute. The reverse is done with cation exchangers. Table 19-2 shows several ion-exchange groups that are used on commercial columns.

Paired-Ion Chromatography (PIC)
(This is used to make a reversed phase column polar so polar compounds can be separated.)

This technique, in its most popular application, is a modification of reversed phase liquid-solid chromatography. It is based entirely on concentration equilibrium and can be used to separate highly polar materials with a nonpolar surface. A counter ion to the ion desired to be separated is added to the mobile phase along with a buffer to maintain ionic strength and pH. A "paired ion" is formed that is neutral and can be separated from other similar compounds by a normal reversed phase column. A diagram of how this is done is shown in Figure 19-6.

If the right system can be found, PIC usually provides better separation efficiencies than ion exchange. The ion pair reagents for cations are organic sulfonic acids like $CH_3(CH_2)_6$-SO_3^-, H^+.

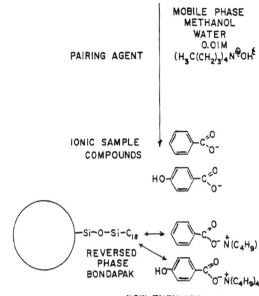

Figure 19-6. The paired ion chromatography concept.
(Courtesy - Waters Corp., Milford MA)

The ion pair reagents for anions are usually quaternary ammonium salts such as phosphates and hydrogen sulfates. Several are shown in Table 19-3.

Table 19-3. Ion pairing reagents

For Anions	For Cations
Tetrabutylammonium phosphate	Butane sulfonic acid
Tetramethylammonium hydrogen sulfate	Pentane sulfonic acid
Cetyltrimethylammonium hydrogen sulfate	Hexane sulfonic acid
	Octane sulfonic acid
	Dodecane sulfonic acid
	1-Pentane sulfonate, sodium
	1-Octane sulfonate, sodium
	1-Dodecyl sulfate, sodium

Figure 19-7, p. 188, is a chart that can be used to determine which of the above column packings to use.

THEORY

Although many have contributed to the theory of HPLC, only the summary by Kaizuma, H., Myers, M.N. and Giddings, J.C., *J. of Chromatog. Sci.*, **8**, 630 (1970) will be discussed now. Refer to the van Deemter section in Chapter 20, p. 217, to help you compare the theories of gas-liquid chromatography (GLC) and HPLC. In order to get the best separations, band broadening must be held to a minimum. The major factors that contribute to this are summarized in equation 19-1.

$$H = H_P + H_D + H_S + H_M \tag{19-1}$$

where H_P = flow inequalities along the column, similar to eddy diffusion in gas-liquid chromatography, Chapter 20; H_D = longitudinal diffusion along the column, similar to gas diffusion in gas-liquid chromatography, Chapter 20; H_S = resistance to mass transfer due to the stationary phase; and H_M = resistance to mass transfer due to the mobile phase.

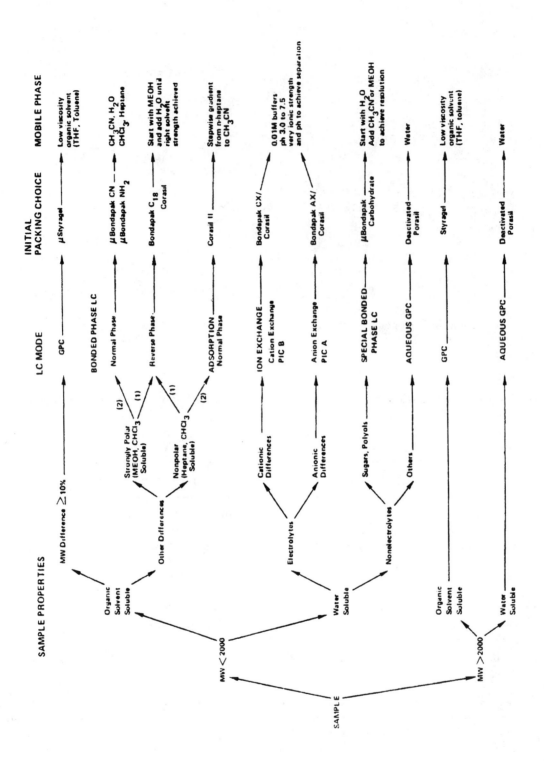

Figure 19-7. Recommendations for the choice of column packing. (Courtesy - Waters Corp., Milford, MA)

A plot of HETP vs. flow rate is shown in Figure 19-8 to illustrate the relative effects of each of the above factors. As with the van Deemter equation, there is an *optimum flow rate*. It is difficult to see, but the optimum flow rate for HPLC is *10,000 times slower* than that for GLC. Based on theory alone, it would seem that HPLC should never be used. However, the $H_s + H_m$ combination tends to level off at higher flow rates, so a flow rate that is reasonable is selected and used, knowing full well that better flow rates exist.

Flow Irregularities

When water flows around a partially submerged rock, you usually can see a small whirlpool just behind it downstream. This causes a partial remixing of any separated components. Small rocks provide different size whirlpools than big rocks, and so it is with the particles in an HPLC column. *To get the minimum broadening, keep the particle size small and as uniform as possible and use columns of 1-5 mm in diameter to reduce channeling and wall effects.*

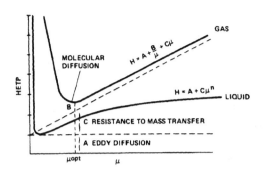

Figure 19-8. HETP vs. flow rate for an HPLC column.
(Courtesy - Waters Corp., Milford, MA)

$$H_p = 2 \lambda d_p \tag{19-2}$$

where λ = a constant related to packing and particle size; d_p = particle diameter.

Molecular Diffusion

Molecular diffusion is the normal motion of molecules due to diffusion in the mobile phase.

$$H_D = \frac{2 \gamma D_M}{\mu} \tag{19-3}$$

where γ = a constant related to how diffusion is restricted by the column packing. The value is usually < 1. D_M = the diffusion coefficient of the solute in the mobile phase; μ = the flow velocity.

The slower the flow rate, the more time the sample molecules have to diffuse. The more viscous the mobile phase, the less diffusion occurs. Lateral diffusion does not affect band broadening as much as does longitudinal diffusion.

Mass Transfer - Stationary Phase

When sample molecules are moving down a column in the mobile phase and some are absorbed in the stationary phase, those that are absorbed fall behind those not absorbed. This causes band broadening. The factors contributing to this are related below.

$$H_S = \frac{q \, r \, d_f^2 \, \mu}{D_S} \tag{19-4}$$

where q = a constant related to the shape of the packing particles; r = a constant related to the relative migration rates of the solute and the mobile phase; d_f = thickness of the stationary phase; D_S = the diffusion coefficient of the solute in the stationary phase.

To reduce band broadening use a thin film of stationary phase, one that has a high D_S and a low flow rate. Notice that the film thickness is a *square term* so a small change is significant.

Mass Transfer - Mobile Phase

$$H_M = \frac{\omega\, d_p^2\, \mu}{D_M} \tag{19-5}$$

where ω = a constant combining column diameter, shape and the packing arrangement; D_M = diffusion coefficient of the solute in the mobile phase. Values range from 2.43×10^{-5} cm^2/sec for KCl in water to 7×10^{-7} cm^2/sec for albumin in water.

To reduce band broadening, pack the column tightly and uniformly with small particles and use a slow flow rate and a mobile phase with a high D_M.

Factors Controlling Resolution

The previous section discussed several factors that cause band broadening. Those major factors that control how the bands can be resolved from each other are the *capacity factor*, *selectivity*, and the *number of plates*. Together, these are called *resolution*. Figure 19-9 will be used to illustrate these concepts.

Resolution (R_S) is the distance between the centers of two peaks divided by the average base width of the peaks.

$$R_S = \frac{V_2 - V_1}{1/2\,(w_1 + w_2)} \tag{19-6}$$

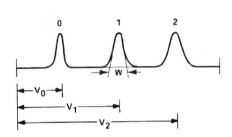

Figure 19-9. Chromatogram to illustrate resolution components.

V_0 is the volume required for a nonretained compound to pass through the column. For resolution, a value of 1 means that the end of the first peak just reaches the baseline before the second peak starts. Any value less than 1 means that the peaks are not separated. Any value much greater than 1 means that time and solvent have been wasted.

EXAMPLE CALCULATION

Referring to Figure 19-9, if V_1 is 4.92 mL, V_2 is 6.95 mL, w_1 is 0.43 mL, and w_2 is 0.51 mL, what is the resolution between peaks 1 and 2?

ANSWER

$$R_S = \frac{6.95 - 4.92}{0.5\,(0.43 + 0.51)} = \frac{2.03}{0.47} = 4.32$$

Capacity Factor

The capacity factor, k, is a measure of the number of column volumes required to retain a compound. It is defined mathematically as:

$$k = \frac{V_1 - V_0}{V_0} \tag{19-10}$$

where V_0 = is the retention volume of a nonretained peak.

Small numbers (< 1) mean the compound is not retained and most likely not separated from other compounds. Large numbers (> 8) mean the compound stays on the column a long time. Values between 2-6 are preferred, and it is the job of the analyst to alter the system accordingly. For a complex mixture, the spread should be kept between 1-10 to obtain maximum separation in the minimum time.

EXAMPLE CALCULATION
Refer to Figure 19-9, p. 190. If V_0 is 1.50 mL, V_1 is 4.92 mL, and V_2 is 9.15 mL, what is the capacity factor range of this separation?

ANSWER

$$k_1 = \frac{4.92 - 1.50}{1.50} = 2.28 \qquad k_2 = \frac{9.15 - 1.50}{1.50} = 5.1$$

Selectivity

Selectivity, α, is a ratio of the net retention volumes of adjacent compounds. It also may be thought of as a ratio of the distribution coefficients. The desire is to get α to be large, so a short column can be used.

$$\alpha = \frac{V_2 - V_0}{V_1 - V_0} = \frac{k_2}{k_1} \tag{19-11}$$

EXAMPLE CALCULATION
If $V_1 = 4.92$ mL, $V_2 = 6.95$ mL, $V_3 = 8.17$ mL, $V_4 = 9.15$ mL, and $V_0 = 1.5$ mL, what is the selectivity between compounds 1 and 2 and compounds 3 and 4?

ANSWER

$$\alpha_{1,2} = \frac{6.95 - 1.50}{4.92 - 1.50} = 1.60 \qquad \alpha_{3,4} = \frac{9.15 - 1.50}{8.17 - 1.50} = 1.15$$

Number of Theoretical Plates

The number of theoretical plates discussed here is the same concept as described in earlier chapters. The equation used is the same as in all column separations and is shown below. Remember - *the number of theoretical plates required to separate two compounds in a chromatographic column is approximately the square of the number required by distillation*, because only a small portion of the column is being used at any one time.

$$N = 16 \left(\frac{t_R}{w} \right)^2 \tag{19-12}$$

EXAMPLE CALCULATION
Determine the number of theoretical plates for a compound with $V = 4.92$ mL and $w = 0.43$ mL. What is the HETP if the column is 150 cm long?

ANSWER

$$N = 16 \left(\frac{4.92 \ mL}{0.43 \ mL} \right)^2 = 2095 \ plates; \quad HETP = \frac{Length}{No. \ of \ plates} = \frac{1500 \ mm}{2095 \ plates} = 0.72 \ mm/plate$$

Resolution Revisited

Earlier, resolution was defined. Combining the factors that play a major role in resolution yields equation 19-13.

$$R_S = 1/4 \left(\frac{\alpha - 1}{\alpha} \right) (\sqrt{N}) \left(\frac{k}{k + 1} \right) \tag{19-13}$$

EXAMPLE CALCULATION
Using the values obtained in previous examples for compounds 1 and 2 determine R_S using only N and k for compound 1. (Answer will be somewhat different because averages are not used.)

ANSWER

$$R_S = 1/4 \left(\frac{1.6 - 1}{1.6} \right)^2 \times (2095)^2 \times \left(\frac{2.28}{2.28 + 1} \right) = 2.98$$

A suggested approach by Waters Corporation applications analysts to utilize the above equation is as follows:
1. Use one column length initially.
2. Use relatively rapid flow rates initially (2-3 mL/min).
3. Determine the k' range for all components (2-6).
4. Adjust for selectivity.
5. Resolve standards away from other components.
6. Adjust N for optimization.
7. Make easy changes - flow rate, column length.
8. Change only one variable at a time.

TECHNIQUES

BASIC COMPONENTS
The following is a brief discussion of the components necessary to make an HPLC instrument function.

Solvents
The range of solvents that can be used in high performance liquid chromatography is quite wide. Water and aqueous buffer solutions are common, as are any nonaqueous solvents of low viscosity. High viscosity solvents are avoided because they require longer time to pass through the column, which results in peak broadening and poorer resolution. This also requires a higher pressure to force the solvent through the column (see equation 19-14). Table 19-4, p. 193, lists several HPLC grade solvents.

Solvents for use in an HPLC *must be free of particles and dissolved air*. Light scattering measurements have shown that if you make one twist when taking a glass stopper off of a reagent bottle, 40,000 particles fall into the liquid! They are much too small to see, but they are filtered by the very small particles of the column packing and can plug a column quite quickly. Degassing also may be necessary if a reciprocating pump is used. This type of pump (discussed later) has a tendency to produce gas bubbles from any dissolved gases during the intake stroke. These bubbles do not always redissolve and can partially block a column or create noise at the detector. The solvent can be degassed by bubbling helium through it (*sparging*). Helium is used because it is quite soluble in most solvents and seldom forms bubbles in the detector.

Figure 19-10 shows a solvent filter arrangement. The filter, made from sintered stainless steel or titanium, is placed in the solvent reservoir, and ultrapure helium is passed through it at about 5-10 psi. Under such conditions, a cylinder of helium will last about 5 months in continuous operation.

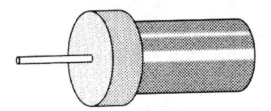

Figure 19-10. A solvent filter-degasser system.
(Courtesy - Alltech Associates Inc., Deerfield, IL)

Table 19-4. HPLC-grade solvents in order of increasing polarity index, P'
Snyder, L.R., J. Chromatogr., **92**, 223, 1974; (Courtesy - Burdick and Jackson Inc., Muskegon, MI)

Solvent	Polarity Index	Solvent	Polarity Index
Pentane	0.0	Isobutyl alcohol	4.0
Heptane	0.1	Tetrahydrofuran	4.0
Hexane	0.1	Chloroform	4.1
Petroleum ether	0.1	Methyl isobutyl ketone	4.2
Iso-octane	0.1	Ethyl acetate	4.4
Cyclohexane	0.2	Methyl n-propyl ketone	4.5
Hexadecane	0.5	Methyl ethyl ketone	4.7
Trichloroethylene	1.0	2-Ethoxyethanol	5.0
Carbon tetrachloride	1.6	β-Phenethylamine	5.0
Toluene	2.4	Acetone	5.1
Methyl t-butyl ether	2.5	Methanol	5.1
o-Xylene	2.5	Pyridine	5.3
Benzene	2.7	Diethyl carbonate	5.5
Chlorobenzene	2.7	2-Methoxyethanol	5.5
o-Dichlorobenzene	2.7	Acetonitrile	5.8
Diethyl ether	2.8	Propylene carbonate	6.1
Methylene chloride	3.1	Dimethyl formamide	6.4
Ethylene dichloride	3.5	Dimethyl acetamide	6.5
2-Propanol	3.9	Dimethyl sulfoxide	7.2
2-Butanol	4.0	Water	10.2

Figure 19-11 shows an inexpensive arrangement to filter a solvent and degas it as well. A disadvantage is that with a mixed solvent, the lower boiling component sometimes can be partially lost. Another low cost method to degas a solvent if helium and/or a vacuum is not available is to use an ultrasonic bath. Figure 19-12 shows a bath that can be used for degassing solvents. Several open bottles containing the solvent are placed into the 2.8 L bath, and water is added to surround the bottles. The high frequency pulses set up standing waves in the liquids, which then cause low pressure cavities to be formed. The gases form bubbles in these low pressure areas and drift to the top of the solvent and escape.

Figure 19-13, p. 194, shows a combination filtration-delivery system. This apparatus filters the mobile phase and provides a line for sparging with helium and a line for delivery to the pump. The dissolved sample is also a problem. You must be sure that there are no suspended or undissolved particles in the sample before it is injected onto the column.

Figure 19-11. A simple vacuum filter - degas apparatus.
(Courtesy - Alltech Associates Inc., Deerfield IL)

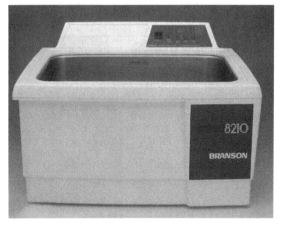

Figure 19-12. An ultrasonic bath for degassing solvents.
(Courtesy - Alltech Associates Inc., Deerfield, IL)

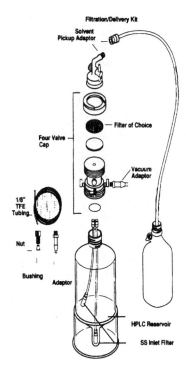

Figure 19-13. A filtration - delivery system.
(Courtesy - Alltech Associates Inc., Deerfield, IL)

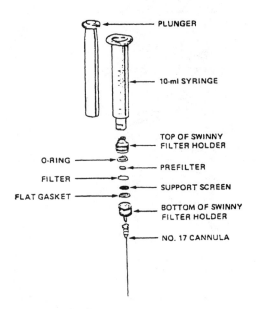

Figure 19-14. A sample clarification kit.
(Courtesy - Waters Corp., Milford, MA)

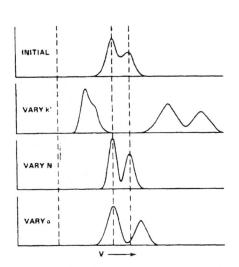

Figure 19-15. Summary of varying k', N, and α.
(Courtesy - Waters Corp., Milford, MA)

To ensure that the sample is free of particles, it is usually passed through a Millipore type filter such as the one shown in Figure 19-14.

The filter is placed on the end of a syringe with a 12 mm diameter membrane. The sample is placed in the barrel of the syringe, and when the plunger is depressed the sample is forced through the membrane. It usually is injected directly into the sample injector. Nylon disposable filters or nylon membranes are available and work well.

A Teflon filter of about 0.45 microns is recommended when organic solvents are used, and a cellulose acetate filter is used with aqueous systems. Figure 19-15 shows the effects of varying k', N, and α.

Pumps

The pump is one of the major components of a high pressure liquid chromatograph. Equation 19-14 relates several of the variables producing a pressure drop across a column.

$$P = \frac{L \, \eta \, \mu}{\theta \, d_p^2} \tag{19-14}$$

where L = column length; η = solvent viscosity; μ = flow rate; θ = a constant, and d_p = particle diameter.

EXAMPLE CALCULATION

When acetonitrile ($\eta = 0.37$) was passed through a 300 mm long column containing 10 μm C-18 beads at 0.70 mL/min., a pressure of 650 p.s.i.g. was required. What would be the pressure if a 5 μm bead column was used, the flow rate was kept the same, and θ did not change?

ANSWER

A. Determine the constant θ using equation 19-14.

$$650 = \frac{300 \times 0.37 \times 0.70}{\theta \times (10)^2} \qquad \theta = 0.0012$$

B. Use θ and 5 μm beads to determine the new pressure.

$$P = \frac{300 \times 0.37 \times 0.70}{0.0012 \times (5)^2} \qquad P = 2590 \; p.s.i.g.$$

In order to *obtain the very best separations, small particles must be packed into the column*, but from equation 19-14, you can see that this greatly affects the pressure required. Pressures greater than 500 psi and sometimes more than 5000 p.s.i. are required, with usual pressures being from 700 to 1500 p.s.i. To attain these pressures requires good pumps, and many different types of pumps have been developed, some capable of 10,000 psi. Two major types, the *reciprocating pump* and a *positive displacement pump,* will be discussed

Reciprocating Pumps

These pumps have the advantage of being able to pump solvent in unlimited quantities and are of use when long separations or preparative situations exist. Figure 19-16 shows a diagram of how one type functions. These pumps usually have a maximum operating pressure of 6,00 p.s.i.g. (42 MPa) and flow rates adjustable from 0.5 to 5.0 mL/min. The internal volumes are from 60-200 μL/stroke.

A disadvantage is that it produces a pulsating pressure, which affects most detectors. This makes it difficult to provide reproducible flow control over a range of solvent types; pump volumes vary as the solvents change due to the compressibility of the solvents, and they cavitate - that is, on the inlet stroke when the pressure is released, gas bubbles can form either from dissolved gases or from high vapor pressure solvents. This means that lower vapor pressure (usually also higher viscosity) solvents must be used, and the solvents should be degassed. Bubbles can cause detector noise, variable volume delivery, and "vapor lock", requiring the pump to be reprimed. The pulsating effect can be reduced greatly by using pulse suppressors and/or two pistons, one compressing while the other is releasing. Figure 19-17, p. 196, shows a dual-piston dual-cam reciprocating pump. Some designs place just one cam in the center, and it runs two pistons placed in opposite directions. Figure 19-18, p. 196, shows a pulse dampener and its effects.

Positive Displacement Pumps

Figures 19-19 and 19-20 show two variations of pumps of this type. These often are called *syringe pumps* because they act just like a syringe. Figure 19-19 shows a constant volume style, and Figure 19-20 shows a constant pressure style. The constant volume style uses a motor-driven screw to force the piston forward. This forces a constant volume through the column regardless of what pressure develops. This permits the retention time to stay the same from one run to another. The constant pressure type (older and seldom used now) uses compressed gas to drive the piston forward. To start with, the gas is added between the piston and the diaphragm to push the piston down in the diagram. This creates a reduced pressure on the other side of the piston, and the chamber can fill.

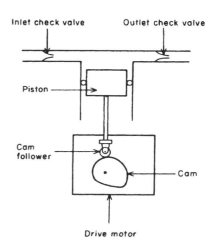

Figure 19-16. Diagram of a single reciprocating piston pump.
(Courtesy - R. Macrae, *HPLC in Food Analysis*, Academic Press, London, 1982)

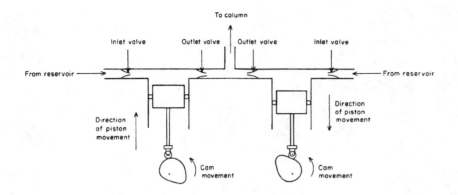

Figure 19-17. A dual piston two-cam reciprocating pump. (Courtesy - R. Macrae, *HPLC in Food Analysis*, Academic Press, London, 1982)

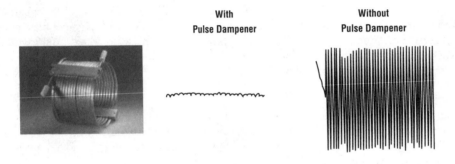

Figure 19-18. A pulse dampener and its effect at 1000 p.s.i.g. and 2.0 mL/min. (Courtesy - Alltech Associates Inc., Deerfield, IL)

The gas then is switched to the other side of the diaphragm to push the piston forward. Because of the relatively large area of the diaphragm compared to the area of the piston head, a gas pressure of 100 psig can force the piston forward with a pressure of 10,000 psig. The advantage is that the mobile phase flow is smooth, so there is no detector noise. They are also inexpensive and easy to repair. However, if anything is present to change the back pressure (column becoming plugged), then the retention times will change.

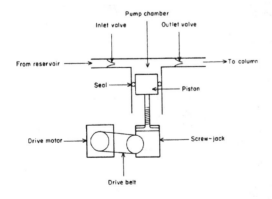

Figure 19-19. Diagram of a constant volume syringe pump. (Courtesy - R. Macrae, *HPLC in Food Analysis*, Academic Press, London, 1982)

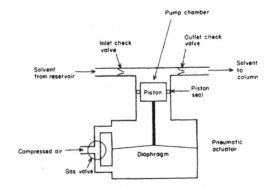

Figure 19-20. Diagram of a constant pressure syringe pump. (Courtesy - R. Macrae, *HPLC in Food Analysis*, Academic Press, London, 1982)

A syringe pump (Figure 19-21) has the disadvantage that only a single charge of solvents can be delivered, hopefully enough to complete the analysis. The advantages are: no cavitation, no pulsation, and changing solvents is easy. *Always prime the pump if it has not been used recently or solvent bottles are changed.*

A new pump must be primed to get it started. If you replace a seal or change solvents in a pump, you will have to reprime it to get it started.

Figure 19-21. A syringe for priming a pump. (Courtesy - Alltech Associates Inc., Deerfield, IL)

Air gets in the line and causes the pump to stop pumping for awhile until the air is pumped through. In addition, if the pump operates dry, the seals wear out more quickly. Prime the pump by placing a syringe into the prime/purge valve and withdrawing the air until solvent comes through.

Gradient Elution

If the *solvent composition is kept constant throughout the entire separation,* the process is called *isocratic elution.* However, if the *solvent composition is changed in any manner,* (pH, buffer strength, different solvent mixtures), then the process is called *gradient elution.*

EXAMPLE
Which of the following are considered isocratic, and which are considered gradient?
A. A system that begins with 20% acetonitrile/water and ends with 50 % acetonitrile/water.
B. A system that begins and ends with methanol.
C. A system that begins and ends with 50% methanol/water.

ANSWER
A. Gradient - the solvent changes from the beginning to the end.
B. Isocratic - the solvent remains the same throughout.
C. Isocratic - the solvent remains the same throughout even though it is a mixture.

Gradient elution is particularly useful when separating mixtures having widely varying characteristics. However, the solvent characteristics can be progressively changed, which, in turn, alters the behavior of the sample compounds. If done properly, a slow separation can be speeded up or groups of compounds that will not separate can be made to separate. This requires a mixing chamber and usually a second pump.

A less expensive method that can be used at low pressures is to use just one pump, a control valve, and a mixing chamber. Figures 19-22 to 24, p. 198, show the effect of gradient elution on a sample. The explanations are by Waters Corporation applications analysts.

Figure 19-22 shows a systematic approach for using a gradient to improve a separation where no components are baseline separated initially.
Step 1. Decrease the initial solvent strength. Make it more polar for reversed phase or less polar for normal phase.
Step 2. Increase the time of the gradient. This gives a better selectivity.
Step 3. Increase the flow rate. This is opposite to isocratic solvents. This effects the k' by forcing more of the weaker component of the solvent through the column.

Figure 19-23 shows how to handle a system that is poorly resolved at the beginning.
Step 1. Delay the rate at the start and increase it at the end.
Step 2. Increase the rate of change at the end.
Step 3. Decrease the initial conditions. This is called a *concave* gradient.

Figure 19-24 shows how to handle a system that is poorly resolved at the end.
Step 1. Increase the rate at the beginning and decrease it at the end.
Step 2. Vary the degree.
Step 3. Increase the rate at the beginning as needed. This is called a *convex* gradient.

198   Techniques

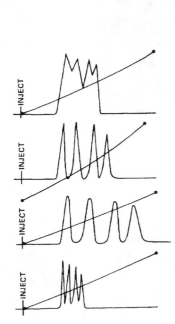

Figure 19-22. Correction for overall poor resolution.
(Courtesy - Waters Corp., Milford, MA)

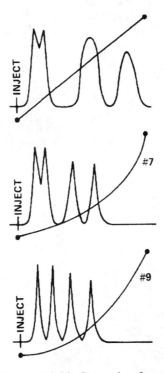

Figure 19-23. Correction for inadequate resolution at the beginning.
(Courtesy - Waters Corp., Milford, MA)

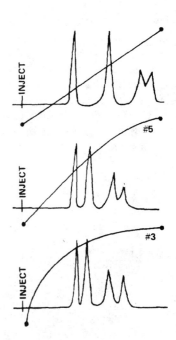

Figure 19-24. Correction for inadequate resolution at the end.
(Courtesy - Waters Corp., Milford, MA)

Sample Injection

Sample volumes of 5 to 50 µL are normal. One way of injecting a sample is shown in Figure 19-25. This injector allows sample introduction without interrupting the solvent flow. All of the parts are either stainless steel or Teflon.

Never use a gas chromatographic syringe in an HPLC injection block. Gas chromatographic syringe needles have *a tapered tip* whereas HPLC needles have a *square end* on the tip. The tapered tip scratches the entry port and causes the port to leak after a while.

More elaborate chromatographs contain several columns of different characteristics. A multiport sample injection device is shown in Figure 19-26. The sample size is determined by the size of the loop, 10 to 20 µL being common.

Precolumns

Columns are expensive and can get clogged easily if the sample, the solvent, or wear particles from the pump have not been filtered properly. There are two ways to protect the column. One is to use a *precolumn* and the other is to use another *in-line filter* between the injector and the column. The precolumn is a short section of column, about 5 cm long, packed with the same material as the regular column and attached immediately in front of it. This column acts as a filter and provides some separation as well.

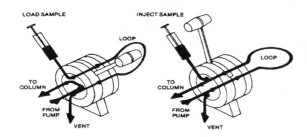

Figure 19-25. A syringe-loading injection block.
(Courtesy - Alltech Associates Inc., Deerfield, IL)

Figure 19-26. A 6-port loop injection system and some two-port sample loops.
(Courtesy - Supelco, Bellefonte, PA)

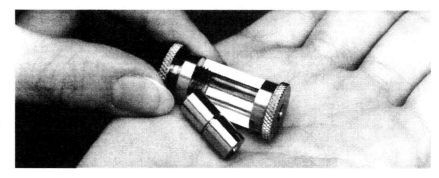

Figure 19-27. A cartridge-type guard column.
(Courtesy - Alltech Associates Inc., Deerfield, IL)

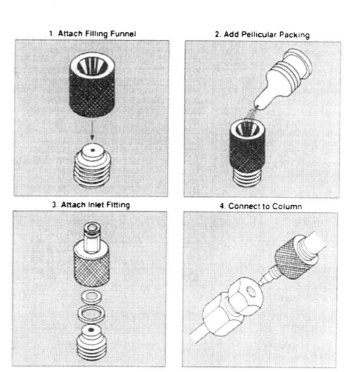

Figure 19-28. Steps to refill a refillable guard column.
(Courtesy - Alltech Associates Inc., Deerfield, IL)

If it begins to plug (the pressure must be increased to maintain the flow rate), it can be removed, cleaned out, and repacked. Figure 19-27 shows a cartridge guard column, which has advantages over the self packed. It is more efficient, and easier to replace, and the same cartridge holder can be used for a variety of different cartridges. Figure 19-28 shows a four-step process to refill a refillable type.

Figure 19-29 shows an expanded diagram of an in-line column prefilter. This is a direct connect type that can be connected to any fitting without tubing.

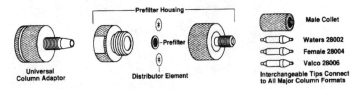

Figure 19-29. Expanded view of an in-line column prefilter. (Courtesy - Alltech Associates Inc., Deerfield, IL)

Columns and Column Packing

Columns are commonly 10 to 30 cm in length and from 3-10 mm in diameter, the larger columns being used for preparative work. They usually are made out of stainless steel. Several commercial columns are shown in Figure 19-30.

The high pressure employed requires a very hard packing material. Furthermore, if high efficiency is to be obtained, then uniform packing must be achieved. These requirements usually result in a packing being made from silica or alumina and consisting of round particles.

Figure 19-31 shows an expanded view of a typical HPLC column. They are generally made from 316 stainless steel tubing, fittings, and frits.

Currently three general types of particles are used: fully porous, pellicular, and microporous. These are shown in Figure 19-32.

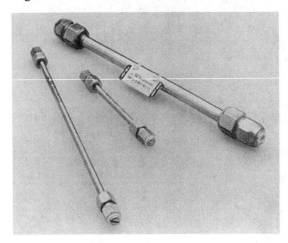

Figure 19-30. HPLC columns. (Courtesy - Alltech Associates Inc., Deerfield, IL)

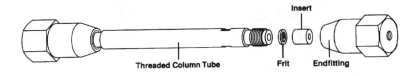

Figure 19-31. Expanded view of a conventional HPLC column. (Courtesy - Alltech Associates Inc., Deerfield, IL)

Fully Porous Particles

Beads are made of glass with a high boron content that then is leached away by acids to make them porous. These come in various sizes, but a popular size is about 50 μm in diameter. They have very large surface areas - 300 to 500 m^2/g. This large area means a high column capacity, so these materials are used for preparative separations. The retention times are longer, because the diffusion distances (19 μm in and 19 μm out) can be long.

Pellicular Particles

The pellicular particles consist of a solid core with a 2 to 3 μm crust etched onto the surface. They can also be about 50 μm in diameter, but their surface area is much less than that of the fully porous particles. Because of the smaller diffusion distances (2 μm in and 2 μm out), they are highly efficient and excellent for analytical separations.

Figure 19-32. Diagrams of the general types of column packings: Left: fully porous, Middle: pellicular, Right: microporous.
(Courtesy - Waters Corp., Milford, MA)

Microporous Particles

These originally were 10 µm in diameter and fully porous. More recently, sizes as small as 3 µm have been made. They provide a highly efficient and high speed packing and are used for the most efficient separations. Their high porosity means that heavier loading is possible than with the pellicular particles. In addition, because the diffusional distances are small, efficiency can be very good. An excellent analogy of the advantages of a microparticle over a regular size-particle is that given by the Waters Corporation. "Consider for a moment a sewer pipe which is packed with basketballs and filled with water. Now between each one of these basketballs, if you stop the flow of water, there is a certain volume of liquid; if there is anything dissolved in the water, each one of these volumes of liquid acts like a large mixing chamber. Consider now the same sewer pipe, only this time it is packed with baseballs. The volumes between the balls are much smaller; therefore the mixing chambers are much smaller. The total volume of the water in both cases is going to be nearly the same. But because the mixing chambers are smaller and the time between mixing and interaction between the liquid and the particles is smaller, the net result is that you end up with a lot less mixing with the smaller particles and therefore higher efficiencies."

Packing Columns

HPLC columns are much more difficult to pack correctly than GLC columns, because they must be packed under high pressure. Ideally, this is at the pressure limit of the chromatograph's pump, usually 6000 p.s.i.g. Because of the high cost of HPLC columns, some investigators prefer to pack their own. If a high pressure compressed air line (100 - 150 p.s.i.g.) is available, then a slurry-packing apparatus like that shown in Figure 19-33 can be used. It has an "air amplification pump" and can be used to repack a column.

Column Cleanups

Liquid chromatograph columns are expensive, $190 to over $1000 each. As a result, special care must be taken to protect them. They easily become inactive from surface contamination and must be regenerated. A good policy is to flush the column at a slow rate (0.1 mL/min) overnight. If a quicker cleanup is required, then the following can be tried. A silica column can be regenerated with 100 mL of isopropanol followed by about 100 mL each of successively less polar materials such as acetone, chloroform, and finally hexane, at a rate of 2-4 mL/min.

The reversed phase packings, such as the Bondapaks, are eluted first with the same concentration of organic and water as was used in the organic aqueous buffer, then with methanol or acetonitrile. Proteins are removed with 8M urea followed by water. The important concept is that *columns do go bad, but they can be regenerated and should not be thrown away too hastily.*

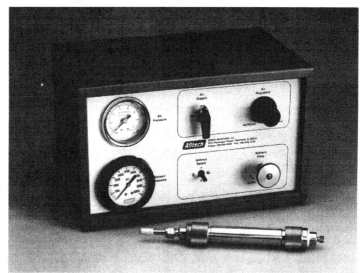

Figure 19-33. A slurry packing apparatus.
(Courtesy - Alltech Associates Inc., Deerfield, IL)

Cutting the Tubing

All of the above components, including the detectors described next, must be connected with tubing capable of withstanding pressures up to 10,000 p.s.i.g. Furthermore, the lines must have a minimum of dead volume and no mixing of the separated compounds. This means that 1.6 mm (1/16 inch) diameter (0.007 or 0.01 i.d.) stainless steel tubing is used in varying lengths. It must be cut so there are no irregular edges to introduce mixing. A tubing cutter that will cut both 1.6 and 3.2 mm o.d. tubing is shown in Figure 19-34.

Figure 19-34. A tubing cutter.
(Courtesy - Alltech Associates Inc., Deerfield, IL)

Detectors

The detector is another of the critical components of a high pressure liquid chromatograph, and in fact, the practical application of liquid chromatography had to await a good detector system. Many types of detectors are now on the market. The four most common, the ultraviolet absorption (uv), fluorescence, refractive index (RI), and electrochemical (EC) detectors, will be discussed as well as the newer light scattering mass sensitive detector.

The Ultraviolet Absorption Detector

Many ultraviolet absorption detectors use a low pressure mercury lamp as a source; the cells are about 1 cm in path length and have volumes of 8 to 30 µL. Most are double beam. Normally, they have absorbance ranges from 0.001 to 3 absorbance units; that corresponds to a minimum sample requirement of about 5×10^{-10} g/mL for a favorable compound. Figure 19-35 is a diagram of one such detector.

The uv region is particularly useful for many compounds such as food additives, pesticides, explosives, and drugs, because these compounds contain -C=C-, -C=O, -N=O, and -N=N- functional groups that readily absorb uv radiation. Aromatic rings absorb very strongly at the 254 nm wavelength of the mercury lamp emission. By using a filter, the wavelength at 280 nm is also available, although it is not as sensitive for most compounds.

Multiple wavelength detectors are available, but they usually have somewhat higher detection limits at each wavelength, because it is difficult to focus sufficient intensity of radiation through the cell. However, they are used commonly because of their flexibility and often are associated with a diode array for the detection of specific wavelengths.

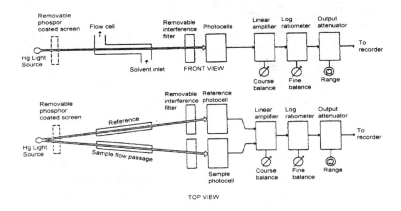

Figure 19-35. A diagram of a uv detector.
(Courtesy - Varian Associates, Walnut Creek, CA)

Making Sensitive Derivatives

Many compounds are not sensitive to uv detection. Currently, two main methods are used to improve this sensitivity, both requiring either a pre- or a postcolumn reaction. The first method is to chemically add a highly absorptive group to the compound to be detected, and the second is to add a compound that fluoresces to the compound being detected. Once the compounds are through the column and have been separated, they then are reacted with other reagents in small-volume reaction chambers before they get to the detector. One postcolumn apparatus to derivatize compounds is the Pickering apparatus discussed later.

Table 19-5 shows several reagents for the first approach, to make them more sensitive to the 254 nm wavelength of the uv detector. Basically, a benzene ring is added to the system.

Table 19-5. Selected derivatizing reagents

Type of Compound	Reagent
Carboxylic acids	p-Bromophenyl bromide (PBPB)
	O-p-Nitrobenzyl-N,N'- diisopropylisourea (PNBDI)
Aldehydes and ketones	p-Nitrobenzyloxyamine·HCl
Alcohols, phenols, I° and II° amines	3,5-Dinitrobenzoyl chloride
Amino acids	N-succinimidyl-p-nitrophenyl acetate
Isocyanates	p-Nitrobenzyl-N-n-propylamine·HCl

Many compounds either fluoresce or can be reacted with a fluorescing group. For example; compounds such as aliphatic amines, which do not have a good uv absorbance, can be detected easily by reacting them with fluorescamine, that will cause them to fluoresce. Table 19-6 lists several reagents for producing fluorescence.

Table 19-6. Selected fluorescent derivitizing agents

Type of Compound	Reagent
Thiols, I° and II° amines	7-Chloro-4-nitrobenzo-2-oxa-1,3-diazole
I° and II° Amines, phenols, peptides, amino acids	Dansyl chloride-1-dimethyl aminonaphthalene-5-sulfonyl chloride
Aldehydes, reducing sugars, ketones	Dansyl hydrazide-1-dimethyl aminonaphthalene-5-sulfonyl hydrazide
Carboxylic acids	4-Bromomethyl-7-methoxy coumarin
I° Amines, amino acids	Fluorescamine; o-Phthaldehyde (OPA)
Amino groups	4,4'-Diisothiocyanostilbene-2,2'-disulfonic acid (DIDS)
Sulfhydryl groups	Fluorescein-5-maleiamide; 5-Iodoacetamidofluorescein

Fluorescence Detection

A fluorescence detector is similar to a uv detector except that the fluorescent radiation is measured at right angles to the incident radiation. Figure 19-36 shows one type of design.

The detector shown in Figure 19-36 uses a xenon lamp pulsed at 20 Hz for normal operation and 100 Hz for traces. The standard excitation and emission ranges are 250 - 650 nm. An optional red-sensitive photomultiplier tube extends the range to 800 nm. The flow-through cell is 2 mm x 3.5 mm, with a 3 µL illuminated volume. This detector can detect as little as 100 fg of anthracene.

Pickering Postcolumn Reaction Apparatus

Figure 19-37, p. 204, shows a diagram of one type of postcolumn reaction apparatus, the Pickering apparatus.

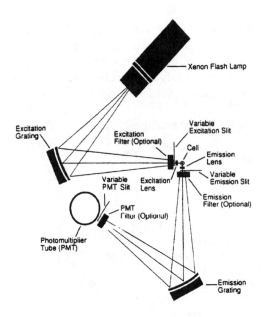

Figure 19-36. A model LC 304 fluorescence detector.
(Courtesy - Thermoseparation Products, Schaumburg, IL)

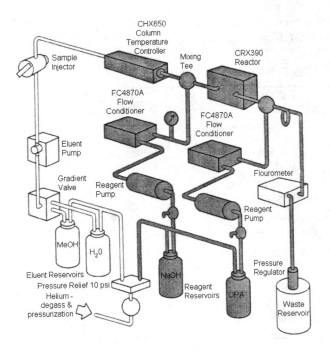

Figure 19-37. A diagram of a Pickering post column reaction apparatus.
(Courtesy - Pickering Laboratories, Mt. View, CA)

Photodiode Arrays (PDA's)

Scanning the wavelengths by passing the radiation past a single detector has been done for decades. However, it difficult to achieve sufficient intensity at all of the wavelengths to provide acceptable sensitivity, and the scan is usually too fast to provide adequate resolution. A new approach is to place an array of *photodiodes* side by side, each one detecting a small region of the spectrum. Arrays vary from 16 to 2048 and can detect an entire spectrum in a matter of milliseconds. Figure 19-38 shows the principle of operation of a photodiode. A typical wavelength vs. intensity spectrum can be obtained.

Doped silicon (n-type) is a common receptor covering the range from ~180 to 1100 nm. The incoming radiation ejects electrons, producing positively charged holes, the electrons migrating to the p-type surface where the charge is collected and measured after a fixed amount of time. The diodes are arranged as shown in Figure 19-39 in an integrated circuit with the electronic switches (field effect transistors, FET) built right into the "chip." The size of each diode varies, the one shown is 25 x 2500 nm, being elongated to match the instruments slit geometry. InGaAs diodes are used for the near infrared (800 - 1700 nm).

PDA's, compared to a charge coupled device (CCD), discussed next, are preferred over CCD's where high levels of light are available for absorbance, transmittance, reflectance, and radiometric measurements because they have a 10x higher signal/noise ratio than CCD's (10,000:1 to 900:1).

Charged Coupled Devices (CCD's)

A charge coupled device (CCD), Figure 19-40, is a one- or two-dimensional array of photosensors, called *pixels*, and like the PDA, comes in a semiconductor "chip" package. The readout mechanism for these devices differs greatly from the PDA's. On a CCD, each pixel is overlaid with a small voltage carrying element known as an electrode. During illumination of the chip, charge accumulates in the pixel. To collect the data, a sequence of voltages is applied across the electrodes to move the charge row by row down the vertical dimension of the chip and into a shift register at the bottom of the array (Figure 19-41). This charge is then moved similarly across the shift register to the output node where it is converted to a digital form for processing. The most significant benefit of this readout method is that the associated readout noise is very low. Typically, CCD's have a sensitivity similar to a photomultiplier tube; however, unlike photomultiplier tubes, a CCD is not damaged by overexposure to bright lights. If the data are plotted vs. time, much more information can be obtained as shown in Figure 19-42

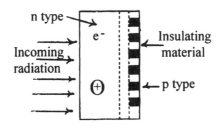

Figure 19-38. Illustrating the basic principles of a photodiode.

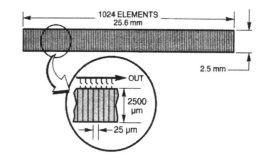

Figure 19-39. Layout of a photodiode array detector.
(Courtesy - Oriel Instruments, Stratford, CT)

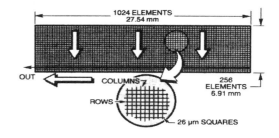

Figure 19-40. Layout of a spectroscopic charge coupled device (CCD).
(Courtesy - Oriel Instruments, Stratford, CT)

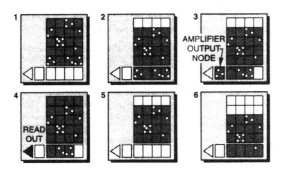

Figure 19-41. Readout pattern of two dimensional CCD's.
(Courtesy - Oriel Instruments, Stratford, CT)

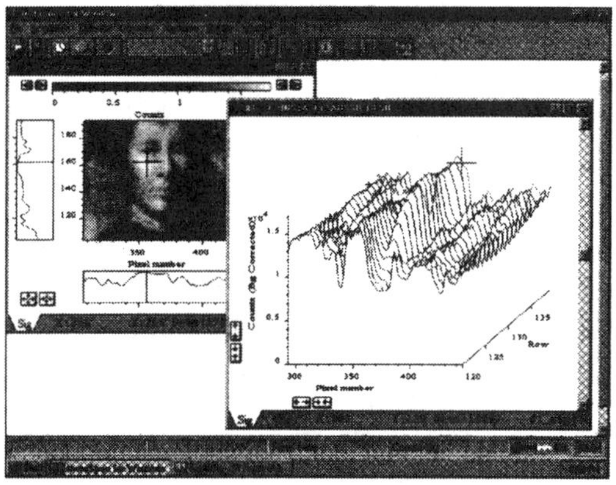

Figure 19-42. Image of fluorescing currency and relative intensity contour.
(Courtesy - Oriel Instruments, Stratford, CT)

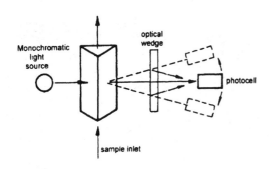

Figure 19-43. Deflection type RI detector.
(Courtesy - Waters, Corp., Milford, MA)

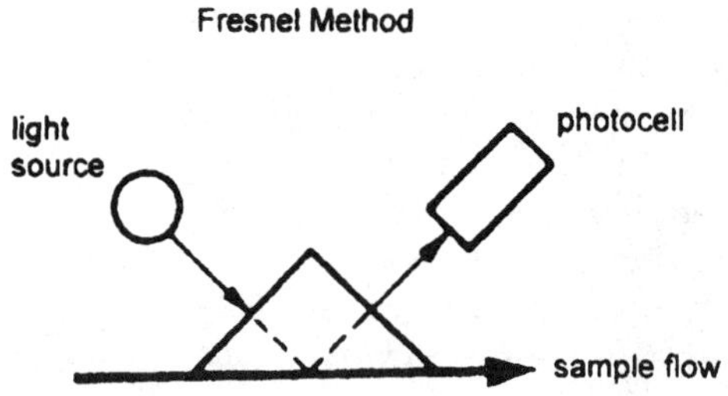

Figure 19-44. Fresnel reflection type RI detector.
(Courtesy - Varian Corp., Walnut Creek, CA)

Refractive Index Detector

The refractive index detector is applicable to all compounds, although it is not as sensitive as the uv detector. An RI detector can detect differences of about 10^{-5} RI units, which means that about 5×10^{-7} g/mL must pass through the detector for a favorable response. As a general rule, *the sensitivity in mg of sample is almost equal to the reciprocal of the differences in refractive index between the solvent and the sample.* Figures 19-43 and 19-44, show two basic methods of refractive index measurement.

Differential refractometers operate by utilizing one of the following principles. In the first type, the measurement is based on an optical displacement of the beam (Figure 19-43). A mirror controls the reflection of the light beam. The light passes through the divided cell the first time without being reflected, and this then is followed by a reflected beam. If there is a difference in the solutions, a deflection will occur. The deflection is the sum of the deflection in the two cells.

A second method utilizes the *Fresnel principle* (Figure 19-44), which relates the transmittance of a dielectric interface to the refractive indices of the interface materials. Such an interface may be formed between a glass prism of selected optical properties and the liquid whose refractive index is to be measured. These detectors are difficult to use with gradient elution systems, and temperature control of the solvents is critical.

Electrical Chemical Detector

This detector can detect compounds in the 10^{-6} to 10^{-8} mol/L range. Any ion can be detected, as can any compound that has a pK_a or pK_b greater than 7 (see Table 19-7, for examples). What is measured is the change in

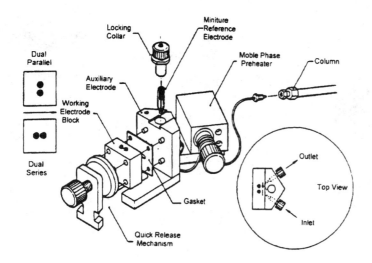

Figure 19-45. A model LC-17AT electrochemical detector.
(Courtesy - Bioanalytical Systems, Inc., West Lafayette, IN)

resistance of a solution as a solute passes between two electrodes. A pulsed a.c. potential is applied to reduce electrolysis and electroplating. The circuit is a Wheatstone bridge. Figure 19-45, p. 206, shows a diagram of the newest design, a thin-layer, sandwich type, microliter volume, flow-through cell.

Three electrodes are involved. A glassy carbon *working electrode*, with an Ag,AgCl *reference electrode* and a metal *auxiliary electrode* about 50 μm directly across. A constant voltage is applied, and quantitation depends upon the amount the current changes when compounds pass across the working electrode and are either oxidized or reduced. The more positive the potential, the stronger an oxidizing agent the electrode becomes. The phenyl urea herbicides are detected this way at the present time.

Table 19-7 lists several types of compounds that can be detected electrochemically. The table is most useful to indicate the functional groups that are electroactive.

Table 19-7. Selected types of compounds suitable for oxidative or reductive electrochemical detection
(Courtesy - Bioanalytical Systems Inc., West Lafayette, IN)

Class	Examples	Class	Examples
Oxidative mode			
Phenols	Phenol	Indoles	Tryptophan
	Pentachlorophenol		Serotonin
	Parabens	Ascorbic acid	Vitamin C
	Morphine	Xanthines	Uric acid
	Tyrosine		Theophylline
Hydroquinones	Catecholamines	Thiols	Cysteine
Vanillyl cpds	Homovanillic acid		Glutathione
	Vanilylmandelic acid	Phenothiazines	Chloropromazine
	Ferulic acid		
Aromatic amines	Aniline		
	Benzidine		
Reductive mode			
Quinones	Vitamin K_3	Aromatic nitro	Chloramphenicol
Aliphatic nitro	Nitroethane	Organometallics	Triphenyllead
	Nitroglycerin		*cis*-Platinum
N-oxides	Chlordiazepoxide	Azomethine	Diazepam
			Nitrazepam
Azo compounds	4-(2-pyridylazo)resorcin	Peroxides	Benzoyl peroxide
Nitrosamines	Dimethylnitrosamine	Thioamides	Prothionamide

Mass Sensitive Detector

This detector can detect any type of compound down to nanogram detection limits. Figure 19-46, p. 208, is a drawing of the VAREX "Universal" evaporative light scattering detector (ELSD). The sample is nebulized by a stream of nitrogen into a heated chamber, which causes the sample to condense into small particles, which pass out of the end of the jet. A laser diode light beam is focused onto the cloud of particles, and the amount of scattered radiation is measured by a silicon photodiode. The detection limits are 5 ng/20 μL of injected volume or 250 ppb at specified chromatographic conditions.

The advantages are that it can be used with multisolvent gradients without baseline drift, it responds to virtually all solutes, and it needs to be calibrated only once for extended periods of use to obtain mass directly from the peak area.

The Mass Spectrometer

The quadrupole-ion trap mass spectrometer combination provides the most information about a separated compound. It is by far the most expensive detector, but the ion trap portion, which essentially holds an ion in place so it can be scanned several times, permits selective and sensitive information about the compound of interest. Figure 19-47 is a block diagram of the components needed.

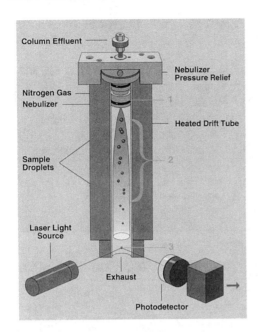

Figure 19-46. A light-scattering mass sensitive detector.
(Courtesy - Alltech Inc., Deerfield, IL)

Mass spectrometers operate at 10^{-6} torr, or less, and the sample coming from the LC outlet is at atmospheric pressure. Therefore, the interface is critical. The following description is abstracted from material supplied by the Finnegan Corp: Samples are injected into the LC inlet. The LC then separates the sample molecules by liquid chromatography. The separated constituents from the LC flow through the transfer line and are introduced into the atmospheric pressure ionization (API) source. Upon entering the API source, sample molecules are ionized by electrospray ionization (ESI), Figure 19-48. This process forms a fine mist of the liquid and places a charge on the droplets. This charged mist is then passed through an ion optics section, Figure 19-49, that evaporates the solvent, then focuses and accelerates the resulting sample ions into the mass analyzer. Here they are analyzed according to their mass-to-charge ratios, Figure 19-50. The sample ions are then detected by an ion detection system, Figure 19-51, that produces a signal proportional to the number of ions detected. The ion current signal from the ion detection system is received and amplified by the system electronics and is then passed on to the data system for further processing, storage, and display.

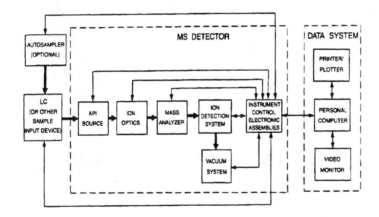

Figure 19-47. Block diagram of an LC-MS system.
(Courtesy - ThermoQuest Finnegan Corp., Austin, TX)

REVIEW QUESTIONS

1. What are the advantages of HPLC over low pressure column chromatography?
2. What is the principle of liquid-solid chromatography?
3. What is meant by "paired ion" chromatography?
4. Name two compounds that can be used to separate anions by paired ion chromatography?
5. Why is a reverse phase column packing desired in many cases?
6. How is a reversed phase column prepared?
7. What precautions must be taken to prepare an HPLC grade solvent?
8. Why is He commonly used for degassing solvents and how does it work?
9. What is a syringe pump?
10. Reciprocating pumps usually have two "check valves". How do they work and what is their purpose?

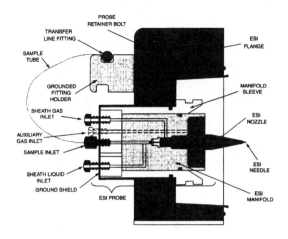

Figure 19-48. Cross sectional view of the ESI probe.
(Courtesy - ThermoQuest Finnegan Corp., Austin, TX)

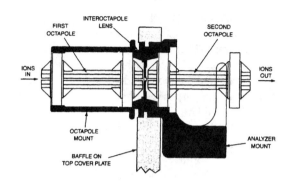

Figure 19-49. Cross sectional view of the ion optics.
(Courtesy - ThermoQuest Finnegan Corp., Austin, TX)

11. What is the purpose of derivatizing compounds? Name two such compounds and tell what groups they react with.
12. What are the structural formulas for fluorescamine and dansyl chloride?
13. What is meant by "2 pi steradians" in the description of the fluorescence detector?
14. How does an RI detector work?
15. How does an electroconductivity detector work and what types of compounds can it be used for?
16. What is the purpose of a precolumn?
17. An HPLC procedure required a pressure of 800 p.s.i. when using a 30 cm column packed with 50 μm beads. A column packed with 10 μm beads was to be purchased. The pump available had a 6000 p.s.i. capacity. Could a 30 cm column be used?
18. An analyst had been using a column packed with 5 μm beads and 10 cm long to separate a steroid composition, usually operating at 3200 p.s.i.. A salesman convinced him to use a 3 μm bead column. Would his 6000 p.s.i. pump be usable?

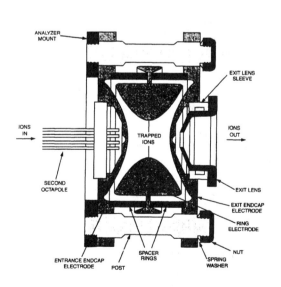

Figure 19-50. Cross sectional view of the mass analyzer.
(Courtesy - ThermoQuest Finnegan Corp., Austin, TX)

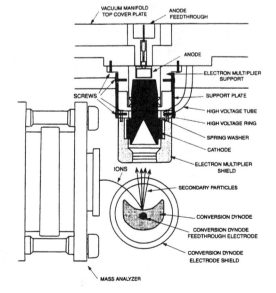

Figure 19-51. Cross sectional view of the ion optics detection system.
(Courtesy - ThermoQuest Finnegan Corp., Austin, TX)

19. A system that used a 30% methanol (η = 0.60 centipoises), 70% water (η = 1.0 centipoises) operated at 1600 p.s.i. It was decided that a solvent gradient would be needed that would end with 70% methanol-30% water. Would the pressure be expected to change, assuming all other parameters remained constant? Can you prove your conclusion by a calculation?
20. A separation that employed ethanol (η = 1.20 centipoises) operated at 5700 p.s.i. and was causing excessive seal failure. The analyst noticed that acetone, with a polarity index of 5.4, was close to that of ethanol whose polarity index was 5.2 yet had a viscosity of 0.32 centipoises. If acetone were substituted for ethanol, what pressure might be expected to be required if all other factors remained constant?
21. What type of compounds does a liquid-solid HPLC column separate?
22. What elution order would you predict if a polar, a nonpolar, and a medium polar compound were placed on a liquid-solid HPLC column?
23. What is meant by a reversed phase column?
24. What does ODS C-18 mean when discussing HPLC columns?
25. Why aren't ordinary ion exchange beads used with HPLC?
26. If you had a mixture of organic acids to be separated by an ion exchange HPLC column, what pH would you use as a starting point?
27. What is the purpose of a paired ion system?
28. What types of compound are used as pairing agents for cations in PIC?
29. What are some typical general classes of solvents used for HPLC?
30. What must be done to a solvent before it can be used for HPLC?
31. Why is dissolved air a problem?
32. Why is helium slowly bubbled through an HPLC solvent if no gas should be present?
33. Why must every sample be filtered before it is injected into an HPLC?
34. An acetonitrile-water solvent (η = 0.50) was passed through a 300 mm long column containing 10 µm beads at 0.60 mL/min. A pressure of 700 p.s.i.g. was required. What is θ?
35. What pressure would be required if 3 µm beads were now used in the above column, all else remaining the same?
36. What is an advantage and a disadvantage of a reciprocating pump?
37. What is an advantage and a disadvantage of a syringe pump?
38. When is it usually necessary to prime a pump?
39. What is an isocratic elution?
40. What is a gradient elution?
41. What is a normal size sample for an HPLC analysis?
42. Why is an injection valve necessary when adding a sample to an HPLC? Why not just do it the same as with GLC?
43. Why use either a precolumn or an in-line filter?
44. What is meant by a pellicular bead?
45. Why do smaller beads provide better resolution, all other items being the same?
46. An experienced HPLC chromatographer usually runs solvent through the column overnight at about 0.1 mL/min. Why is this done?
47. What is the source for a low pressure uv detector for HPLC?
48. What types of groups absorb in the uv region?
49. What is one nonfluorescing reagent that can be added to alcohols, phenols, and I° and II° amines to make them detectable with a uv detector?
50. What is a fluorescing reagent that can be added to carboxylic acids to make them fluoresce?
51. What is the source for a typical fluorescence HPLC detector?
52. The detection limit for the Model LC 304 fluorescence detector is listed as 100 fg of anthracene. What is fg?
53. What does the Pickering apparatus do in general, and what does it specifically do when determining carbamates?
54. How does the speed of obtaining a spectrum with a PDA compare to that of traditional methods?
55. Why is the diode shape in a PDA elongated?
56. What is a common material for the semiconductor in a photodiode for the uv/vis region?
57. Why is a PDA preferred to a CCD for most spectroscopic measurements even though the CCD is more sensitive?
58. If the refractive index detector is so insensitive and responds to pressure changes, why use it at all?
59. What types of compounds can be detected with the electrical chemical detector?

IF IT EXISTS - CHEMISTRY IS INVOLVED

IF YOU CAN BUY IT - A CHEMIST WAS INVOLVED SOMEWHERE

20
GAS-LIQUID CHROMATOGRAPHY (GLC)

PRINCIPLES

Gas-liquid chromatography is a partitioning chromatography in which the mobile phase is a gas. This technique was developed in England in 1952 by A.J.P. Martin and A.T. James (*Biochem. J.*, **50**, 679). The longer compounds remain in a column, the more time they have to diffuse and remix. Martin suggested that the best separations should occur if the stationary phase was very thin and the mobile phase was changed from a liquid to a gas. This would provide fewer "foreign" molecules for the sample molecules to hit, thus reducing the mixing by diffusion. The detectors available in 1952 had poor response, and in order to separate sufficient material to be detected, a relatively thick layer of stationary phase was required. However, the liquid mobile phase could be changed to a gas.

The first commercial instrument was sold in 1956, and by 1959 about 600 papers had been published. Now, thousands of papers are published annually on this technique, and it probably has done as much to influence the course of fundamental research, medical science, and industrial progress as any technique yet discovered. As a general rule, *any compound that boils without decomposition below 350°C can be separated by gas-liquid chromatography*. This represents about 15% of all of the known compounds. Although this seems like a small percentage, those compounds constitute a significant percentage of the compounds in common use.

The importance of gas-liquid chromatography lies in its powerful separating ability. Columns can have the equivalent of up to 1,000,000 theoretical plates. Compare this with an ordinary distillation, which normally operates in the range of 5 to 10 theoretical plates. What this means in practice is that a gas chromatograph can separate the odor components of coffee into over 400 compounds!

In the case of gas-liquid chromatography (GLC or just GC), the two phases are a flowing gas phase and a liquid phase held stationary on an inert "solid support". In practice, for the older style using packed columns, a small sample (1-5 µL) is injected by means of a syringe or sample loop into a heated injection port, where the liquid is vaporized into the flowing gas stream (usually N_2, He, or Ar). The carrier gas transports the sample through the column (glass, copper, or stainless steel, 3 to 6 mm dia, 1 to 2 m long), which is packed with a solid of large surface area. This *solid support*, which is commonly a diatomaceous earth or crushed firebrick, is impregnated with a *nonvolatile liquid phase*. Separation of the injected mixture occurs by partitioning between the gas stream and the liquid phase. A large number of partitions occur and even very small differences in physical and chemical properties allow the components to become separated.

The gas stream, containing the separated compounds, then is passed through a detector; the simplest type is one that measures the difference in thermal conductivity between the sample plus He or N_2 gas and a reference He or N_2 gas. The detector signal is monitored continuously by a recorder, which plots the components as a function of time, resulting in Gaussian-shaped peaks.

More recently, capillary columns, 0.05 to 0.53 mm in diameter and 10 to 100 m long, having as many as 4000 plates/m, are becoming common. Figure 20-1 is a diagram of the component parts, and Figure 20-2 is a photograph of a commercial instrument.

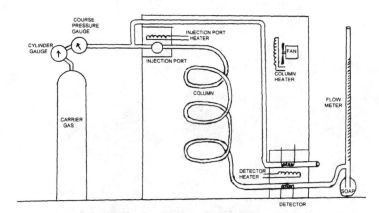

Figure 20-1. Diagram of the components of a gas-liquid chromatograph.

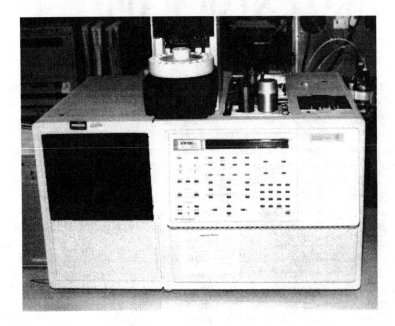

Figure 20-2. A Varian model 3600 gas-liquid chromatograph.
(Courtesy - Varian Associates, Walnut Creek, CA)

The elapsed time between the injection and the center of a peak is called the *retention time* of that compound. Note that although a gas chromatograph separates compounds, it DOES NOT IDENTIFY them. However, the retention time may be used as corroborating evidence in the identification of a compound by injecting a known sample under identical conditions and observing identical retention times. The unknown-known comparison then should be repeated on another column of *different polarity* to substantiate the previous agreement of retention times.

Failure to use columns of different polarity to identify compounds can be embarrassing. DDT was blamed for many problems caused by PCB's by some early investigators, who failed to use at least two columns.

Figure 20-3, p. 213, is used to illustrate some common terms for gas and liquid chromatography. The elapsed time between the injection and the center of a peak is called the *retention time*, t_R, of that compound. The baseline is a straight line drawn from where the peak begins on one side to where the peak ends on the other side. Several situations are shown. The *peak width*, w, is measured as shown. It is the distance on the baseline between the intersections of tangents to the inflection points on the sides of the peak. The *retention time*, t_R, is the time from the injection to the highest part of the peak. *Retention volume*, V_R, is the volume of carrier gas required to elute a compound off of a column. It is the product of flow rate (mL/min) and time (min).

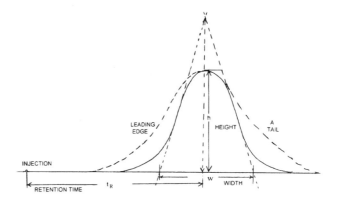

Figure 20-3. Diagram to illustrate some common gas and liquid chromatography terms.

THEORETICAL PLATES

The concept of HETP was described in Chapter 3, p. 26. The same concept applies here except that HETP for a distillation differs from an HETP for a chromatographic separation, because in a distillation, the *entire column is in use at the same time,* whereas in a chromatographic column, only a *small portion is being used at any one time.*

As a generalization: *the number of plates required to separate a mixture by a gas chromatographic procedure is the square of the number of plates required to separate the same mixture by distillation.*

The smaller the HETP is, the better the column. The older 6 mm diameter packed columns would have values from 1 to 3 mm/plate, whereas values of 0.25 mm/plate are common for capillary columns. One equation to determine the number of theoretical plates and the HETP is the *van Deemter* equation shown below:

The broadening of a band varies with the $\sqrt{N}$ (or r for column chromatography). N is the number of theoretical plates in the column. From the Poisson distribution, the band width = $4\sqrt{N}$.

If distance = L, then $L = kN$ and $w = k\, 4\sqrt{N}$
or

$$N = 16 \left(\frac{t_R}{w} \right)^2 \tag{20-1}$$

where w = distance between intersections of tangents to the inflection points on the baseline.

The height equivalent to one theoretical plate is defined by

$$\text{HETP} = L/N \tag{20-2}$$

where L = is the length of the column, usually in mm.

EXAMPLE CALCULATION
Calculate the number of theoretical plates and the HETP in a 2 m column whose major peak was 40 mm from the injection point and t_i and t_f are 37 and 43 mm, respectively.

ANSWER
A. From equation 20-1:

$$N = 16 \left(\frac{40}{6} \right)^2 = 711 \ plates$$

B. From equation 20-2:
2 m = 2000 mm; therefore, HETP = 2000 mm/711 plates = 2.8 mm/plate.

THEORY

Liquid-liquid theory can be expanded to cover gas-liquid systems by taking into account the compressibility of the mobile phase, which introduces a gradient down the column.

P_i = inlet pressure. P_o = outlet pressure. F_c = gas flow rate, mL/min (corrected to correspond to the column temperature and outlet pressure P_o). t_R = retention time (for the center of the peak to emerge). V_R = retention volume (column T and P_o) = $t_R F_c$. V_R^o = corrected retention volume - corrected for the compressibility of the gas. V_M is the volume of the mobile phase. A_m and A_s were defined in Chapter 15 as column cross sectional areas

If P is the pressure at a point along the column, the gas velocity can be calculated by

$$V = \frac{F_c P_c}{V_M P} \qquad (20\text{-}3)$$

The volume rate of flow of gas is proportional to the pressure gradient at that point.

$$A_m V = K \frac{dp}{dx} = \frac{F_c P_o}{P} \qquad (20\text{-}4)$$

K depends upon the viscosity of the gas and how tightly the column is packed; P is obtained by integration.

$$\frac{K P^2}{P_o} = 2 F_c X + K P_o \qquad (20\text{-}5)$$

$$\text{or} \qquad F_c = \frac{P_o K}{2 x} \left(\frac{P}{P_o}\right)^2 - 1 \qquad (20\text{-}6)$$

The time, dt, required for the solute to move a distance dx is given by

$$dt = dx/VR_f \qquad \text{and} \quad t_R = \int_0^1 dx/VR_f = \int_0^1 A_m P dx / R_f P_o F_c \qquad (20\text{-}7)$$

L = column length.

Because $dx = KPdP/F_c P_o$, then $t_R = \int_{P_o}^{P_i} A_m K P^2 dP / R_f P_o^2 F_c^2$

$$\text{In terms of pressures} \qquad t_R = \frac{K A_m (P_i^3 - P_o^3)}{3 R_f P_o^2 F_c^2} \qquad (20\text{-}8)$$

$$\text{The retention volume is} \qquad V_R = t_R = \frac{K A_m P_o}{3 R_f F_c} \left(\frac{P_i}{P_o}\right)^2 - 1 \qquad (20\text{-}9)$$

$$\text{or at a distance } L \qquad V_R = \frac{2 A_m L \ (P_i/P_o)^3 - 1}{3 R_f \ (P_i/P_o)^2 - 1} \qquad (20\text{-}10)$$

as $P_i/P_o \rightarrow 1$; $V_r \rightarrow A_m L/R_f$ which is V_R^o (corrected for pressure drop)

$$V_R^o = V_R \frac{3\,(P_i/P_o)^2 - 1}{2\,(P_i/P_o)^3 - 1} \qquad (20\text{-}11)$$

EXAMPLE CALCULATION
For the same pressure drop of 0.2 atm, how does the retention volume compare for outlet pressures of infinity, 1.0 and 0.1 atm?

ANSWER
$P_o = $ infinity, $V_R = V_R^o$

$P_o = 1.0$; $P_i = 1.2$; $V_R = 2/3 V_R^o \dfrac{(P_i/P_o)^3 - 1}{(P_i/P_o)^2 - 1} = 1.10\, V_R^o$

$P_o = 0.10$, $P_i = 0.30$ $V_R = 2.16\, V_R^o$

$V_G^o = A_m L$ = total gas volume in the column (corrected for T and P_o). $VL = A_s$ = total volume of the liquid phase. P = the partition coefficient in the equations below.

$$V_R^o = \frac{A_m L}{R_f} = \frac{V_G^o}{R_f} \qquad R_f = \frac{A_m}{A_m + P A_s} = \frac{V_G^o}{V_G^o + PVL} = \frac{V_G^o}{V_R^o}$$

For two substances

$$\frac{(V_R^o)_A}{(V_R^o)_B} = \frac{V_G^o + P_A VL}{V_G^o + P_B VL}$$

Note that they appear in the order of P values, because of the reciprocal V_R^o, and we have a difference from liquid-liquid column chromatography.

Get V_G^o by passing O_2 through the column because it has a P of 0.

$$\text{From} \quad \frac{V_G^o}{V_G^o + PVL} = \frac{V_G^o}{V_R^o}; \qquad R = \frac{V_R^o - V_G^o}{V}$$

If the activity coefficient varies with concentration, tailing occurs, both front and back, depending on the direction of the deviation.

Retention Volume
Retention volume is the volume of carrier gas required to elute a compound off of a column. It is the product of flow rate, μ, (mL/min) and time, t (min).

$$V_R = \mu t \qquad (20\text{-}12)$$

EXAMPLE CALCULATION

It required 7.8 minutes for the peak of a compound to emerge when the flow rate of N_2 was 35 mL/min. What is the retention volume?

ANSWER

V_R = 35 mL/min x 7.8 min = 273 mL.

KOVATS' RETENTION INDEX

The relative retention time of a compound compared to another compound on a given column is of more value than the absolute retention time, because the absolute retention time is hard to reproduce from one laboratory to another.

Several attempts have been made to simplify the comparisons of retention times from column to column and compound to compound. The first and simplest is by E. Kovats (*Helv. Chim. Acta*, **41**, 1915, 1958) involving the use of hydrocarbons. The next was by L.J. Rohrschneider (*J. Chromatog.*, **22**, 6, 1966), who used benzene, ethanol, methylethyl ketone, nitromethane, and pyridine to characterize liquid phases. W.O. McReynolds (*J. Chromatog. Sci.*, **8**, 685, 1970) improved upon this by using 10 compounds to relate over 200 liquid phases to squalene. The results indicated that many of the liquid phases behaved nearly the same and that really just a few were needed to separate all of the compounds. Kovats introduced a *retention index*, *I*, which is defined as:

$$I = 100 \frac{\log V_s - \log V_n}{\log V_{n+1} - \log V_n} + 100\, n \qquad (20\text{-}13)$$

where V_s = retention volume of the sample, V_n = retention volume of a reference compound of *n* carbons, V_{n+1} = retention volume of a reference compound of *n* + 1 carbons; n and *n* + 1 are compounds that bracket the sample.

Very little change occurs in the retention index with temperature, so the relative values obtained on a column at 100 °C will be nearly the same at 200 °C.

Figure 20-4 illustrates how a Kovats' index can be obtained. Basically, it is done this way. Two straight chain hydrocarbons are found that differ by one carbon and that bracket the sample compound. The retention volumes of the reference alkanes are converted to logs and are assigned values 100 times the number of the carbons (C_4 = 400; C_5 = 500). The sample then has a value in between.

The top scale (Figure 20-4) shows the net retention times of two samples and five straight chain alkanes, and the middle scale is a plot of log t_r. The lower scale shows the assigned values for the known hydrocarbons and the values determined for A and B. These are summarized in Table 20-1, p. 217.

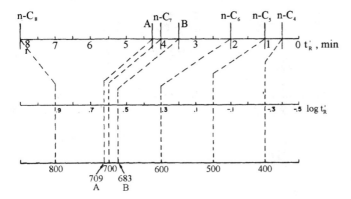

Figure 20-4. Diagram to Illustrate how to Determine Kovats' Retention Indices. (Courtesy - L.S. Ettre, *Anal. Chem.*, **36**, 31A, 1964)

Table 20-1. Summary of the values in Figure 20-4

Carbon No.	Retention Time min	Log t_r	Kovats Index, I
C4	0.5	-0.30	400
C5	1.0	0.0	500
C6	2.0	+0.30	600
B	3.5	+0.54	683
C7	4.0	+0.60	700
A	4.25	+0.63	709
C8	8.0	+0.90	800

EXAMPLE CALCULATION

What would be the Kovats' retention index for a compound that had a retention time of 5.0 minutes using the data in Figure 20-4?

ANSWER

$$I = 100 \frac{[0.70 - 0.60]}{[0.90 - 0.60]} + 100 \times 7 = 733$$

VAN DEEMTER EQUATION

The theoretical plate concept is useful in determining the efficiency of any given column, but it does not indicate the effects of various operational parameters. Here, however, the van Deemter rate theory proves valuable. The van Deemter equation is:

$$\text{HETP} = A + B/\mu + C\mu \tag{20-14}$$

where $A = 2\lambda d_p$. $B = 2\gamma D_G$. $C = 8kd_f^2\mu/\pi^2 D_L(1+k)^2$. μ = velocity (cm/sec), λ = a quantity characteristic of the column packing, d_p = average particle diameter, k = the fraction of the sample in the liquid phase divided by the fraction in the vapor phase, d_f = the average liquid film thickness, D_L = the diffusion coefficient in the liquid phase, γ = a correction factor for the interparticle spaces, and D_G = the diffusion coefficient in the gas phase.

The first term in equation 20-14 is due to turbulence as the gas goes around the column packing, the same as water swirls as it goes around a bridge support. This is called *eddy diffusion* and does not change as the flow rate increases. The second term is due to *molecular diffusion*. If there is no gas flow, then mixing can be complete, and HETP approaches infinity. As the flow rate increases, it is easier to keep two different molecules apart once they are separated, and HETP gets quite low. The third term is due to *resistance to mass transfer*. The higher the flow rate, the more likely it is for molecules to be remixed because of an increase in the number of collisions, and HETP increases with flow rate.

Figure 20-5 shows data for the variation of the HETP of a column prepared by coating Celite with tricresyl phosphate, as a function of the linear velocity of the carrier gas, in cm/sec.

The curve R-S is the actual variation of the HETP with velocity. The tangent to this curve, P-Q, allows the extrapolation to zero velocity. Thus the eddy diffusion is shown as the straight line, P-T. The resistance to mass transfer is represented by the contributions between lines P-T and P-Q for n-butane and P-T and P-U for air. The molecular diffusion contribution is represented by

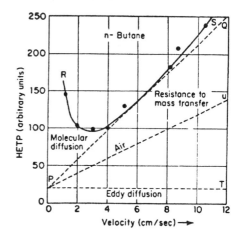

Figure 20-5. Variation of HETP as a Function of Carrier Gas Velocity.

the difference between P-Q and R-S. Thus, at very low velocities of carrier gas, the molecular diffusion is more important, whereas at high gas velocities, the mass transfer resistance becomes more important.

The optimum velocity of carrier gas to be employed is at that point where the HETP-velocity curve is at a minimum.

The consequences of the van Deemter equation are that: (1) small particles improve the HETP; (2) there is an optimum flow rate; (3) in general, a high m.w. carrier gas improves the efficiency; (4) the smaller the i.d. of the column is, the higher the efficiency; (5) the lower the inlet to outlet pressure ratio is, the better the efficiency; (6) low vapor pressure, low viscosity liquid phases are best; (7) the thinner the stationary phase is, the better and faster the separation; (8) the lowest temperature possible is used to obtain a separation.

TECHNIQUES

CARRIER GASES

The most common carrier gases (mobile phase) are nitrogen and helium. Hydrogen and air are used for the flame ionization detector and ultrapure hydrogen (99.9999%) and ultrapure nitrogen (99.999%) are required for the Hall detector. The type of carrier gas to be employed presents some choice, but it usually is dictated by the type of detector used.

Figure 20-6. Gas Cylinder Safety brackets. Left: Bench mount. Right: Wall mount.
(Courtesy - Chemical Research Supplies, Addison, IL)

Safety Note:
Always fasten cylinders to the lab bench with safety straps. A common type of gas cylinder safety strap is shown in Figure 20-6. The holder is screw clamped to the bench top and the cylinder is then strapped to it.

Small amounts of impurities in carrier gases can cause baseline problems (oil), reaction with the stationary phase (oxygen), and loss of sensitivity for several detectors (water), and hydrocarbons.

Usually place a gas purifier and/or an oxygen trap between the gas cylinder pressure regulator and the injection port, particularly if a capillary column is being used. Oxygen reacts with many stationary phases. This causes a change in the retention time, and with capillary columns, causes a noticeable peak broadening after just a few samples have been run.

Figure 20-7 shows some purifiers. The purifier for water, oil, and particles contains Drierite ($CaSO_4$), 5Å molecular sieves and a filter frit. If organics are suspected, purifiers containing activated charcoal can be obtained.

Figure 20-7. Gas purifiers. Left: O_2 purge. Right: O_2 trap.
(Courtesy - Alltech Associates Inc., Deerfield, IL)

PRESSURE REGULATORS

The gas velocity through the column is regulated by applying pressure at the inlet. Commonly, the outlet pressure is atmospheric, and the inlet pressure is maintained somewhere between 2 and 3 atm. For routine work, two-stage, stainless steel diaphragm regulators (Figure 20-8) are used on the gas cylinder. One gauge measures the main cylinder pressure, and the other gauge measures the pressure you are applying to the column. For finer control, an additional metering valve often is used.

Do not overtighten the control valve handle of the metering valve when closing it. This mars the end of the needle, and it will leak after a while.

Before you attach a pressure regulator to a gas cylinder, open the main tank valve for 1-2 seconds to blow away any grease or dirt. Cylinders are shipped in open trucks across the country. The dirt collected plus any oil from the compressors used by the manufacturers should be removed, or the regulator will be contaminated.

INJECTION PORT

The injection port is the place where the sample is entered into the chromatograph. The injection port is fitted with a small *septum* made of materials that permit passage of a needle and then reseal after the needle is withdrawn. This is usually silicone rubber, 1 mm thick and 6-8 mm in diameter. Figure 20-9 shows several septa. Typically, 20 to 30 injections can be made with common 26 gauge needles before they need to be replaced.

Leakage from a damaged septum can be detected by inaccuracies in quantitation, problems with the chromatography such as a change in retention time, or the deterioration of the column from the entry of air. The septa can bleed substances into the gas flow and also can become brittle with use. Shreds from the penetrated septa can break loose and block wide-bore columns and block gas flow. Tailing can be caused by septum bleed and can be checked for as follows: the noise level increases, the early eluters show decreased response, and the baseline drifts if programmed temperature is used.

In order for the best separations to occur, the sample should be vaporized as rapidly as possible after injection. Therefore, the septum is surrounded by a brass or steel block that can be heated and act as a heat reservoir. This rapid vaporization produces what is known as *plug injection.*

As a general rule, injection block heaters are set about 20 °C hotter than the column. This serves as a starting point; further adjustments may be necessary if the peaks tail and there is no other cause for the tailing.

Figure 20-8. A Gas Cylinder Pressure Regulator.
(Courtesy - Alltech Associates Inc., Deerfield, IL)

Figure 20-9. Septa.
Courtesy - Alltech Associates Inc., Deerfield, IL)

SYRINGES

Sample addition generally is done by means of a syringe needle being pushed through a rubber septum. Injection volumes of 1 to 10 µL for liquids and from 0.1 to 5 mL for gases are normal. Figure 20-10, p. 220, shows syringes for adding liquids and gases and the valves used for gas tight-syringes.

The plunger wire on small volume syringes is fragile and often bends even if care is taken. To prevent this, plunger guides are used. Figure 20-11 *top*, p. 220, shows a plunger guide, and 20-11 *bottom* shows the guide mounted on a syringe.

Reproducing the injection of an exact volume of liquid into the chromatograph is quite difficult. One method to improve the reproducibility to ± 1% is to use a *Chaney adapter*. This is a plunger guide with a depth gauge on one side that is held in place with a small set screw. Figure 20-12 *top*, p. 220, shows a Chaney adapter, and 20-12 *bottom* shows it attached to a syringe.

Cautions

1. *Never oil syringe plungers.* The solution in the barrel serves as a lubricant.
2. *Never touch the plunger with your fingers.* The plunger fits the barrel so tightly that even a fingerprint often can produce enough resistance to cause the barrel to break when the plunger is pushed in.
3. *Never interchange plungers between barrels.* They seldom fit well and will either leak or eventually jam.
4. *Never clean metal plungers with acids.* They will eventually leak. Clean the syringe barrel by flushing with solvent, then blowing it dry with a stream of filtered dry air. A liquid soap (dishwashing detergent) and water solution flushed through the syringe works best. *However, it must be rinsed thoroughly with distilled or deionized water, alcohol, and then acetone.*
5. *Withdraw the syringe immediately after the sample is added.* If you do not, then additional sample in the needle may evaporate. Because this amount is seldom the same from injection to injection, the final results will be erratic, and the peaks will tail.
6. *Always rinse the syringe with solvent before and immediately after using it.* This eliminates cross-contamination.

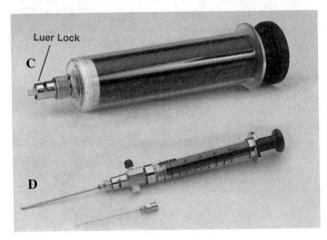

Figure 20-10. Gas chromatography sample addition syringes: (A) For liquids. (B) Micro syringe. (C) For gases. (D) Gas tight valves.
(Courtesy - Alltech Associates Inc., Deerfield, IL)

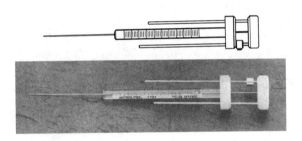

Figure 20-11. *Top*: Plunger guide. *Bottom*: Guide mounted on a syringe.
(Courtesy - Supelco, Bellefonte, PA)

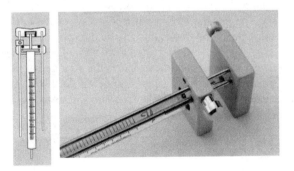

Figure 20-12. *Left*: Chaney adapter. *Right*: Adapter mounted on a syringe.
(Courtesy - Supelco, Bellefonte, PA)

Syringe Filling Technique

The solvent flush method is the best method for (1) sample mixtures that contain a wide range of boiling points such as insecticides and (2) analysts with little experience. Place 1-2 µL of solvent in the syringe. Add 1-2 µL of air, then add 1-5 µL of sample and another 1-2 L of air. Inject the entire system. Read the volume of the solvent plug only.

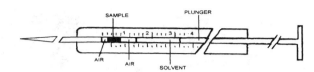

Figure 20-13. The solvent flush method.

Capillary Injection Systems

Capillary columns require addition of a very small sample. This requires a special injection system called a *splitter*. Most often, a larger sample than necessary is injected; a portion of this actually goes into the column, and the remainder is discarded. Figure 20-14 shows one type of a capillary injection system with a splitter.

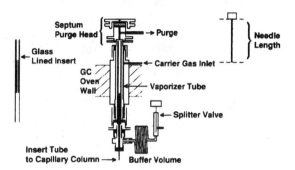

Figure 20-14. Diagram of a capillary injection system.
(Courtesy - Alltech Associates Inc., Deerfield, IL)

Solid Samples

Solids can be injected as solutions, but this often results in broad and tailing peaks. Solids, which are low melting, can be added directly to a gas chromatograph by using a solid sampler. This is essentially a syringe within a syringe as shown in Figure 20-15. It is not pushed through a septum because of its large diameter but is attached to a fitting that is placed on the injection port.

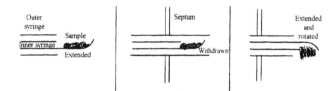

Figure 20-15. Solid Sampler Operation.

COLUMNS

The column, including the packing, is the heart of the chromatograph. Columns used for residue detection usually are made out of glass or fused silica. FSOT means *fused silica open tubular*. Table 20-2 shows current nomenclature for columns based on their *inner diameter* (i.d.). Figure 20-16, p. 222, shows several columns for GLC.

Table 20-2. Column nomenclature based on the inner diameter

Inner Diameter (mm)	Name
> 6	Preparative
2-6	Normal or regular
< 1	Capillary
0.32	Widebore (narrow)
0.25	Narrow (more narrow)
0.18	Minibore
0.05-0.10	Microbore

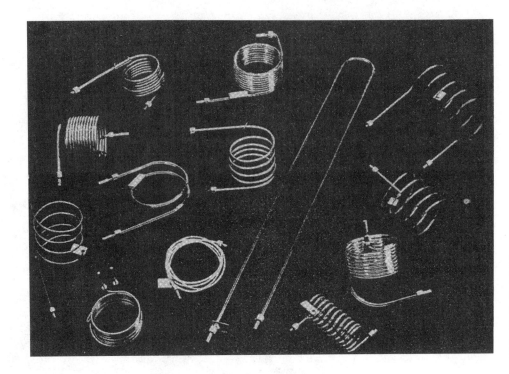

Figure 20-16. Representative columns for gas-liquid chromatography. (Courtesy - Varian Associates, Palo Alto, CA)

As a starting point, the column should be heated to two-thirds of the boiling point of the highest boiling compound in the mixture.

Columns of 6 mm o.d. should not be in coils smaller than 15 cm in diameter, and columns of 3 mm o.d. not in coils less than 8 cm in diameter. There are no restrictions on capillary columns. Surprising as it may seem, because of the comparatively fast flow rate, the molecules moving next to the wall around the outside of a coiled column do not move to the inside as might be expected based on their small sizes and their continual motion. This means that the outside molecules travel farther in the column than those on the inside, much like runners in a 400 meter race if they did not have a staggered start. To minimize or eliminate this effect, short columns are bent into a U shape, and longer columns are coiled in large diameters.

In order to maintain the sample in a gaseous state, the column is heated. To provide reproducible results, all parts of the column must be at the same temperature. To do this uniformly, the column is coiled. It is heated by a stream of heated air. The better instruments have a separate heater for the column chamber.

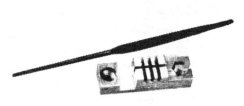

Figure 20-17. Glass tubing cutter, file type. (Courtesy - Alltech Associates Inc., Deerfield, IL)

Figure 20-18. A tubing cutter for metal. (Courtesy - Alltech Associates Inc., Deerfield, IL)

Tubing Cutter

Only glass or fused silica columns should be used for pesticide residue work, particularly if the organohalogen pesticides are to be detected. Hot metal tubing reacts with many of the halogenated pesticides to dehydrohalogenate them and change their composition. However, nickel tubing is used in the furnace in the Hall detector, and it must be cut and cut very smoothly. Figure 20-17 shows two cutters for glass tubing, and Figure 20-18 shows a cutter for copper, aluminum, and stainless steel tubing. Figure 20-19, p. 223, shows a cutter preferred for nickel tubing.

Glass can be cut by making a single scratch mark on it with a file or knife. The scratch is moistened with a drop of saliva, the scratch mark placed upward, and the tubing pulled apart with a *slight* downward bending motion. The bottom photograph in

Figure 20-17 is for larger tubing. The tubing is placed in the holder, then the cutter is pressed firmly on the glass and a scratch mark is made as the tubing is rotated slowly. The tubing is removed, the scratch mark is moistened, and pulled apart with a slight downward bending motion.

Metal tubing should not be sawed, because the rough edges make it difficult to attach tubing connectors, and the jagged inside causes flow turbulence. The easiest way is to use a tubing cutter like that shown in Figure 20-18. As it is twisted around the tubing, the handle is slowly screwed tighter, forcing the cutter wheel into the tubing. If the edge is rough, a *reamer* can be used to smooth out the inside of the cut.

Nickel tubing is difficult to cut without leaving a burr on the inside. The cutter shown in Figure 20-19 has been found to make a clean cut that requires no reaming. It is used just like a pair of pliers.

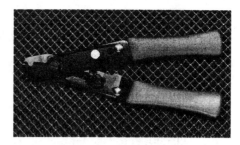

Figure 20-19. A tubing cutter for nickel metal.
(Courtesy - Alltech Associates Inc., Deerfield, IL)

Tubing Benders

Most chromatographers bend metal tubing by wrapping it around a gas cylinder, the cylinder head cover, or something similar. For a more professional look, a coiling machine like that shown in Figure 20-20 can be used to make bends of any diameter from 7 to 60 cm in diameter. There are additional bends, such as connecting the tubing to the chromatograph ports, that are often quite difficult and require a tool with flexibility. One such type is shown in Figure 20-21.

Although metal columns are not as popular as in the past, they still are used. In addition, short lengths of metal tubing are sometimes used to connect parts of the apparatus. Place the tubing on the wheel where the bend is desired. The handles are squeezed together to make the bend. This prevents crimping of the tubing.

Tubing Connectors

The column tubing must be connected to the chromatograph with a leak-free seal. Compression fittings of some type are used. Figure 20-22 shows the *Swagelok* method for connecting tubing. When the nut is tightened it forces the back ferrule into the front ferrule and crimps the front ferrule onto the tubing.

Figure 20-23, p. 224, shows an *O-ring* method for connecting glass tubing to metal connection ports. These devices can be used up to about 200 °C.

Ferrules

Ferrules are the tapered sections of a connection that clamp onto the tubing to make a gas-tight seal. They are used to make metal-metal and metal-glass connections. They are made from many materials: brass, Teflon, graphite, Vespel (polyimide), and a Vespel-graphite combination. Brass is used mainly for metal-metal connections. Teflon is used for temperatures up to 250 °C and is so soft that the connection can be *finger tightened.* Graphite can be used up to 450 °C and is softer than brass but not as soft as Teflon. Vespel does not cold flow, can be used up to 350 °C, and can be used many times. The Vespel-graphite combination can be used up to 400 °C. A *straight ferrule* connects glass of the same size, whereas a *reducer ferrule* connects glass of different sizes. Figure 20-24, p. 224, shows several ferrules. Vespel, Vespel-graphite, and graphite ferrules do not require a back ferrule.

Figure 20-20. A metal tubing coiler.
(Courtesy - Alltech Associates Inc., Deerfield, IL)

Figure 20-21. A hand-held tubing bender.
(Courtesy - Alltech Associates Inc., Deerfield, IL)

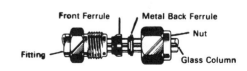

Figure 20-22. Swagelok tubing connections.
(Courtesy - Alltech Associates Inc., Deerfield, IL)

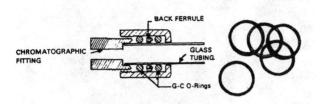

Figure 20-23. O-ring method for connecting glass tubing.
(Courtesy - Alltech Associates Inc., Deerfield, IL)

Figure 20-24. Graphite ferrules
(Courtesy - Alltech Associates Inc., Deerfield, IL)

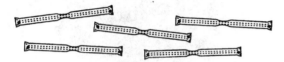

Figure 20-25. Direct-connect capillary column connectors.
(Courtesy - Alltech Associates Inc., Deerfield, IL)

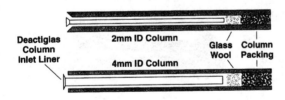

Figure 20-26. Column inlet liners.
(Courtesy - Alltech Associates Inc., Deerfield, IL)

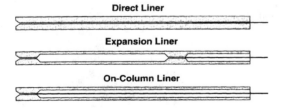

Figure 20-27. Direct injection liners.
(Courtesy - Alltech Associates Inc., Deerfield, IL)

A recent development is the direct-connect capillary connector shown in Figure 20-25. Two pieces of capillary tubing can be connected without tools or other fittings and be gas tight. Simply push the end of the tubing into the connector. *You must heat them to 150 °C to complete the seal. The ends must be cut square to get a good fit. If the column is old, dip the end into pentane before inserting it.*

Liners

Injection liners are short sections of glass tubing smaller in diameter than either the column or the injection port. They have two purposes: (1) to protect a column and (2) to serve as a quick connect that is not necessarily gas tight. To protect a column, they are inserted into the first 5-8 cm of a column to collect any nonvolatile or very low-volatile material that would slowly plug the column. After several injections, the liner with its contaminants is removed and replaced, thus saving an expensive column. Two are shown in Figure 20-26.

The injection port liner is used to minimize dead volume when a sample is injected. Several are shown in Figure 20-27. Those shown are 0.8 mm i.d. with an internal volume of 12 µL. The solvent expansion type prevents flashback when injecting large volumes of dilute samples.

SOLID SUPPORTS

The function of the solid support is to act as an inert platform for the liquid phase in the column. Solid supports most often used are made either from diatomaceous earth or firebrick (clay). The Chromosorbs are the most common and are of various compositions. There are several types: A, G, P, W, R, and T. Most can be *acid washed (A W)* and/or *silanized (S)*.

Several solid packing materials are used without a liquid phase. This is called *gas-solid chromatography*, but because the instruments used are the same, they are commonly grouped together and called gas chromatography. Several of those materials are the Porapaks, Celite, silica gel, alumina, carbon pellets, and molecular sieves.

Several Common Supports for Gas-Liquid Chromatography

Chromosorb P is a pink diatomaceous earth material that has been carefully size graded and calcined. The surface area is 4-6 m^2/g. This material is the least inert of the Chromosorb supports but offers the highest efficiency. It is used when hydrocarbons are to be separated.

Chromosorb W is a white diatomaceous earth material that has been flux-calcined with about 3% sodium carbonate and has a surface area of 1 to 3.5 m^2/g. This material is more inert but less efficient than Chromosorb P. It is used when polar compounds are to be separated.

Chromosorb G to a substantial degree combines the high column efficiency and good handling characteristics of Chromosorb P with the following advantages over present white diatomite supports: greater column efficiency, less surface adsorption, harder particles, and less breakdown in handling. Because Chromosorb G is about 2.4 times heavier than Chromosorb W, a 5% liquid loading on Chromosorb G is equal to 12% on Chromosorb W. No more than 5% liquid phase should be used on Chromosorb G.

Chromosorb A is used mainly for preparative columns. It can be loaded to 25%; therefore, large samples can be separated.

Chromosorb T is prepared from Teflon 6. It is used when highly polar or reactive compounds such as water, halogens, and interhalogen compounds are to be separated. If acid washing and silylation fail to reduce peak tailing, then this support should be used.

Chromosorb R is an ultrafine powder (1-4 μm dia.) used for coating the inside of capillary columns. *Acid washing* is used to reduce the surface activity of the inert phase. The mechanism of this procedure was described earlier.

Silanized supports are Chromosorb supports that have been reacted with dimethyldichlorosilane (DMCS) or some similar compound to reduce surface active sites of the diatomaceous earth material. Chromosorb supports deactivated with DMCS provide less tailing, minimized catalytic effects, and improved results with low liquid loadings. Hexamethyldisilane (HMDS) also has been used. The combination of acid washing and DMCS treating is particularly effective in reducing adsorption.

To silanize a column, the silanizing agent (5-10%) is dissolved in a solvent such as methylene chloride, hexane, or toluene. It then is added to the inert phase or placed in the column if its walls are to be treated. The system is heated to about 50 °C for 10 minutes. The material then is rinsed 4-5 times with dry methanol and dried by passing air over or through it, as the case may be.

Proposed reaction

$$\text{—Si—OH} + (CH_3)_2SiCl_2 \xrightarrow{-HCl} \text{—Si—O—Si(CH_3)_2—Cl} \xrightarrow[-HCl]{+CH_3OH} \text{—Si—O—Si(CH_3)_2—O—}$$

Material active site ... DCMS ... Addition of dry methanol ... Silyated material

Porapak is a polymer of ethylvinyl benzene cross-linked with various amounts of divinyl benzene to produce polymers of various porosities. The eight basic types are listed in order of increasing polarity: P, P-S, Q, Q-S, R, S, N, and T. These materials do not require a liquid phase, because the organic polymer apparently acts like a liquid phase. Porapak P and Q are the same, except that P has a larger pore size. The other Porapaks are modified by adding specific monomers.

Common Supports for Gas-Solid Chromatography

Porous silica replaces the former *Porasils*. They are uncoated porous silica beads. Porous silica X-075LS has a surface area of 185 m^2/g and a pore diameter of 150 angstroms. Porous silica X-200LS has a surface area of 100 m^2/g and a pore diameter of 300 angstroms.

Tenax-GC is a polymer of 2,6-diphenyl-4-phenylene oxide. It is suitable for the separation of high-boiling polar compounds.

Silica gel is a dehydrated form of silicon dioxide. It can be reactivated by heating to 100 °C for 30 minutes.

Florisil is a magnesia-silica gel, not as polar as silica gel.

Carbosphere is activated charcoal in pellet form of about 30 mesh. It can be reactivated by heating for 30 minutes at 100°C.

Molecular sieves are porous synthetic zeolites of some modified form of $Na_{12}(AlO_2)_{12}(SiO_2)_{12} \cdot 27H_2O$. The most common pore sizes are 0.5 and 1.3 nm. They separate based on size and can be reactivated by heating to 300 °C for 1 hour.

Stationary Phases

The stationary phase is a high-boiling liquid that is applied to the surface of the inert phase. Table 20-3, p. 226, lists several of the common stationary phases, their abbreviations, and their chemical compositions.

Coating a Support with a Stationary Phase

A weighed amount of the liquid phase is dissolved in an easily evaporable solvent such as acetone or methylene chloride and placed in a small evaporating dish. The amount of solvent used (mL) is about 1.5 times the weight of the solid support (g). Slowly, with gentle stirring, add the solid support to the dissolved liquid phase and stir until the solid material is wetted thoroughly. The mixture should be of "paste" consistency and not "runny". The mixture is either placed on a hot plate (low) or under an infrared lamp and occasionally stirred until the solvent is evaporated. Do not stir too often or too vigorously or the particles will break. These "fines" make good resolution difficult and are hard to separate by sieving.

Table 20-3. Some common stationary phases

Abbreviation	Chemical Composition
Aroclor	Chlorinated biphenyls
Carbowax	Polyethylene glycol
DC-200	Silicone
DEGA	Diethyl glycol adipate
DEGSE	Diethylene glycol sebacate
Dexsil	Carborane methyl silicone
EGA	Ethylene glycol adipate
EGS	Ethylene glycol succinate
IGEPAL	Nonyl phenoxy polyoxyethylene ethanol
OV-1	100% Methyl silicone
OV-3	Polysiloxane, 5% phenyl, 95% methyl
OV-17	50% Phenyl silicone, 50% methyl silicone
OV-210	Polysiloxane, 50% trifluoropropyl, 50 % methyl
OV-225	50% Cyanopropylphenyl, 50% methyl polysiloxane
OV1701	Polysiloxane, 14% cyanopropylphenyl, 86 % methyl
SE-30	Methyl silicone gum rubber
Superox	Polyethylene glycols
THEED	Tetrahydroxyethyl ethylene diamine
TCP	Tricresyl phosphate
TWEEN	Polyoxyethylene sorbitan mono-oleate
QF-1	Fluoro silicone

Rotary evaporators (Rotovaps) can be used, but they have a tendency to cause more particles to break than does occasional and gentle stirring. An infrared heat lamp can be used to hasten the evaporation.

Packing Columns

If a column is to produce sharp peaks, it must be packed uniformly and reasonably tightly. This is done either by adding the packing under pressure or by applying a vacuum and vibrating the column as the packing is being added. The gas in the carrier gas cylinder can be used to apply pressure and a water aspirator or forepump with a bleeding valve can be used to apply a vacuum.

A small piece of silanized glass wool is placed in the end of the coiled column, and this end attached to a water aspirator. A small funnel (Figure 20-28, p. 227) is attached to the other end of the column, and the packing added. As vacuum is applied, the packing is drawn into the column. The column is continuously vibrated with either a vibrator engraving tool or a column vibrator (Figure 20-29, p. 227). The funnel is removed, and a small amount of glass wool is placed in the end so the packing cannot move. End caps are put in place if the column is to be stored.

Marking Columns

As soon as a column is packed, it must be tagged, or you will be banished from the laboratory. Figure 20-30, p. 227, shows one type of tag. The information usually required is: *column length, solid support, stationary phase and percent, mesh, and the date packed*. The information is engraved on these tags with an engraving tool as shown in Figure 20-31, p. 227.

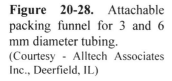

Figure 20-28. Attachable packing funnel for 3 and 6 mm diameter tubing.
(Courtesy - Alltech Associates Inc., Deerfield, IL)

Figure 20-29. A column vibrator.
(Courtesy - Alltech Associates Inc., Deerfield, IL)

Figure 20-30. A metal column identification tag.
(Courtesy - Alltech Associates Inc., Deerfield, IL)

Conditioning Columns

Before a newly packed column is used, it should be *conditioned*. The solid support may contain traces of moisture, organics, and adsorbed gases. The liquid phase may contain traces of impurities from the synthesis and usually will have traces of the solvent used to dissolve it. These impurities will produce an unstable baseline and must be removed.

To condition a packed column, first check the maximum temperature of the liquid phase - do not exceed! and disconnect it from the detector. With the carrier gas flow at about one-fifth normal, outgas at ambient temperature for 20-30 minutes, then heat the column starting at about 60 °C for 3 minutes, then program at 1 °C/ min to 20 °C above the intended use, and hold at this temperature overnight.

Capillary Columns

One of the limitations with *packed columns* is *eddy diffusion*. This is a *remixing of the separated compounds as they go around the particles in the column packing*, much like you get whirlpools when water flows around a rock in a stream. To completely eliminate this problem requires the elimination of the column packing! If you make the column diameter very small (< 1.0 mm), the inside wall of the column can act as the inert phase. A thin film (0.1 µm) of stationary phase is coated onto the inside of the tubing wall or even bonded to it. These capillary columns are made of fused silica and then coated with a polyimide polymer to reduce breakage. They are quite flexible and can be tied in a knot, but that is seldom done. Their natural tendency is to form a straight column, and they are hard to coil and keep

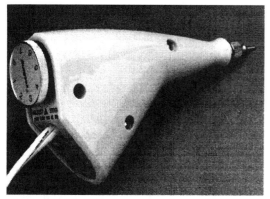

Figure 20-31. Vibrator engraving tool.
(Courtesy - Alltech Associates Inc., Deerfield, IL)

Figure 20-32. A cage and bracket for fused silica capillary columns.
(Courtesy - J & W Scientific, Folsom, CA)

coiled unless a *cage* is used. Figure 20-32, p. 227, shows a capillary column cage. They usually are made of stainless steel and either electropolished or coated with a polyimide polymer to reduce abrasion. A cage 5 cm high and 17.5 cm in diameter will hold a 30 m wide-bore column.

The following is a list of items associated with *fused silica open tubular* (FSOT) capillary columns.

Column

Usually, a wide-bore, 15 m or 30 m, fused silica capillary column, 0.53 mm i.d. with a cross-linked stationary phase and a film thickness of 0.1 or 1.5 µm is used.

Injection

Direct injection into a retention gap with an injection liner onto a 0.53 mm FSOT capillary column is used. This involves introduction of the sample into a hot vaporizing inlet with total transfer (no splitting) of the injected materials onto the analytical column. Volumes of 0.5-6.0 µL are used.

Retention Gap (Injection Liner)

Retention gaps (guard columns) are recommended for GLC capillary columns, especially those used in pesticide-residue determinations. A *retention gap* is a segment of deactivated fused silica tubing (without stationary phase) that is placed between the instrument inlet and the top of the capillary column; in effect, it serves as an extension of the column inlet. Deactivated fused silica tubing 1-5 m long with 0.53 mm i.d. is commonly used; a length of 5 m is recommended for pesticide-residue determinations. Capillary column connectors with low volume or zero dead volume are suitable for connecting analytical column and retention gap.

A retention gap serves two purposes: to provide space for the injected solution to vaporize and expand, thus permitting injection of solvent volumes (>1 µL) that could not otherwise be injected into capillary tubing and to provide surface area for the deposition of co-extractives, thereby protecting the analytical column from buildup of nonvolatiles that can cause loss of efficiency and analyte decomposition or adsorption. In this role, the retention gap often is called a "guard column." Properly installed, a retention gap will not noticeably reduce column efficiency.

With time and use, nonvolatile residues accumulate in all capillary columns, regardless of the use of retention gaps or other protective measures. Efforts to improve deteriorated chromatography should always begin with the removal of portions of a contaminated retention gap or replacement of the retention gap. If the analytical column is also contaminated and replacement of the retention gap is insufficient to improve the chromatography, a 15-20 cm portion of the inlet end of the analytical column can be removed by cutting the column as previously described. If a capillary column ≥ 5 m is used, removal of a relatively short segment does not significantly affect its overall length or behavior, even if segments are removed repeatedly.

Inlet Adapter

Use an inlet adapter for the column injection port that is an easily replaced disposable liner. See Figure 20-26.

Press Fit Connectors

Use press-in universal style connectors that connect tubing of any size. See Figure 20-25.

Gases

Use hydrogen or helium carrier gases. Operate in the "packed column mode", 10-25 mL/min. at 200 °C isothermic or capillary mode 2-5 mL/min. with programmed column temperature.

Cutting Capillary Columns

Proper cutting of the column and retention gap is critical. Square, clean cuts minimize flow disturbances. A cleaving tool for cutting polyimide-coated fused silica capillary tubing always should be used (Figure 20-33). Make the cuts after the ferrules have been installed onto the tubing.

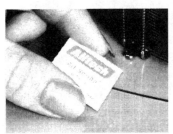

Figure 20-33. A quartz capillary column and retention gap cutter.
(Courtesy - Alltech Associates Inc., Deerfield, IL)

Testing for Leaks

After making a connection, check for leaks by placing a few drops of ethanol, an electronic leak detector, or a high boiling polypropylene glycol solution around the joint. If bubbles appear, then a leak is present. *Do not use a soap solution (phosphates) or other leak detectors (halogens).*

Conditioning Columns

Conditioning need not be extensive, only 1 hour at 20-30 °C above the maximum operating temperature, with the detector connected.

Rejuvenating Columns

Short segments of a contaminated capillary analytical column can be broken off repeatedly with little effect on column performance. If insufficient restoration is accomplished, backwashing of the analytical column and/or the retention gap should be attempted. Kits for backwashing capillary columns are commercially available. Instructions are contained with the kits.

Never allow a heated GLC column to be connected to a cold detector, or the detector will become contaminated.

FLOW METERS

In order to obtain reproducible results, it is necessary to determine the carrier gas flow rate. In addition, the flow rates of the gases used with several of the detectors must be controlled. Carrier gas flow rates usually are determined with a *soap film* flow meter, and detector gas flow rates are determined by *rotameters*. Figure 20-34 shows a soap film flow meter. It consists of a graduated tube (like a buret) with a pipet bulb, partially filled with a soap solution, attached to the bottom. To use, the flow meter is attached to the outlet of the column, and the bulb gently squeezed to force a soap film up into the gas stream. As the film rises, the time it takes to pass two points is measured. A drop of grease at the top will break the film, and it will run back into the bulb. Magnets can be attached to hold it to the chromatograph, or it can be placed on a ring stand. Capillary column flow rates of 3-25 mL/min are reasonable for 0.53 mm i.d. columns, 1-5 mL/min for 0.32 mm i.d. columns, and 0.5-0.8 mL/min for 0.25 mm i.d. columns. Packed columns vary from 10-80 mL/min depending upon the packing. Figure 20-35 shows a set of three (H_2, N_2, air) used for a hydrogen flame ionization detector. A rotameter is a vertical tube with a small ball inside. As gas rushes past the ball, the ball rises, the amount being proportional to the flow rate. *These are calibrated for specific gases and will read incorrectly if used with other gases.* Normal ranges are from 10-100 mL/min for hydrogen to 20-400 mL/min for air.

Figure 20-34. A soap film flowmeter. (Courtesy - Alltech Associates Inc., Deerfield, IL)

BACKFLUSHING

Backflushing is the technique of reversing the flow of carrier gas through the column once the early eluting desired compounds of a mixture have been eluted. The purpose is to speed up the analysis. For example: suppose a sample contains 10 components that require 50 minutes to completely elute, but the compound of interest comes out in 7 minutes. This means that you must wait 43 minutes before the next sample can be injected. The remaining compounds have gone down the column only a short distance, and it would be quicker to drive them backwards than to have them continue. A bypass loop is attached to the column by two three-way valves, one before the inlet and one after the outlet. After the desired compound emerges the valves are adjusted to reverse the carrier gas flow through the column. In this example, only 7 minutes would be required to flush out the compounds rather than 43 minutes. The column is kept at the same temperature, so no time must be wasted cooling it to the starting point.

Figure 20-35. Three rotameters in a Mounting. Courtesy - Alltech Associates Inc., Deerfield, IL)

TEMPERATURE PROGRAMMING

Real world samples usually contain a large number of components. If the column temperature is set low so the early eluters can be separated, then it may take a few hours for the higher boiling materials to be removed. If the temperature is set high initially to speed up the analysis, then the early eluters are seldom separated. The solution is to *temperature program*. This means setting a low initial temperature and then slowly increasing it at a desired rate. This is easy to do with the microprocessors common on most chromatographs. Many possibilities are available. For example: you may want to maintain a low initial temperature for 3 minutes. This is a *hold*. Then *ramp* at 8 °C/min to a set temperature, then "hold" again for 5 minutes. Figure 20-36, p. 230, shows a regular separation and how it can be improved by temperature programming.

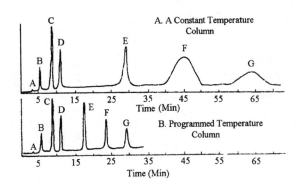

Figure 20-36. (A) A regular separation. (B) Temperature programming.

DETECTORS

The purpose of the detector is to determine when and how much of a compound has emerged from the column. Although the goal of all detectors is to be as sensitive as possible, many detectors are designed to be selective for certain classes of compounds. Dozens of different types of detectors have been developed, but only a few are used routinely. Those are: *thermal conductivity* (TC), *thermionic* (N/P), *electron capture* (ECD), *flame photometric* (FPD), *Hall electroconductivity detector* (Hall or ELCD), *hydrogen flame ionization detector* (FID), *argon ionization* (AI), *photoionization* (PID), *gas density balance* (GDB), and the mass spectrometer. Chemists usually select a detector by the following criteria, listed in priority:

1. Compounds to be detected.
2. The types of compounds that can be detected.
3. Sensitivity in *limit of detection (LOD)* and *limit of quantitation (LOQ)*.
4. Sample destruction: Whether the sample is destroyed during the measurement.
5. *Linear dynamic range (LDR):* The concentration range from the detection limit upwards until the response becomes nonlinear.
6. Cost.

Definitions of Terms

Detector Sensitivity

This is a determination to compare detectors. It is *the amount of a compound that must pass through the detector element in one second to produce a 1% increase in response above background*. For example, a thermal conductivity detector has a value of about 10^{-7} g/sec. This is not as good as a person's nose.

Limit of Detection (LOD)

The traditional limit is *three times the noise level*. Turn on the instrument, set every adjustment at the highest sensitivity, and record the signal with no sample present. The *random fluctuation is the noise level*. Three times this is taken as the minimum level of detection.

Limit of Quantitation (LOQ)

This is taken at *10 times the noise level*. This is a signal of such strength that it can be duplicated and quantitated with sufficient accuracy for regulatory work.

Trace

A value between the LOD and the LOQ. You know something is there, but you can't be sure how much.

Not Detected

Indicates that it was looked for, but not detected. You must specify the LOD.

Linear Dynamic Range (LDR)

The straight line portion of a calibration curve from the lowest detection level to the highest concentration before the line begins to bend. This usually is given in powers of 10. For example, the LDR of a hydrogen flame ionization detector is about 10^6.

Thermal Conductivity (TC) Detector

Use this detector for all compounds unless trace amounts are needed.
This was one of the first GLC detectors used. It can detect all types of compounds over a LDR of 10^5 with a

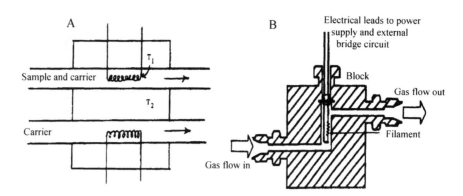

Figure 20-37. (A) Diagram of a TC detector (B) Diagram of a commercial detector. (Courtesy - Varian Associates, Walnut Creek, CA)

sensitivity of 10^{-7} g/sec. This detector is not as sensitive as your nose. With two sets of detector elements, the sensitivity can be improved to 10^{-9} g/sec. It is common knowledge that blowing air over a hot wire cools the wire. This principle along with Avogadro's and Ohm's laws are the basis of the thermal conductivity detector. Avogadro's law states that equal volumes of gases at the same temperature and pressure have the same kinetic energy.

$$1/2\, M_1 V_1^2 = 1/2\, M_2 V_2^2 \tag{20-15}$$

Notice that heavy molecules have a lower velocity than light molecules. If you use a light molecule for the carrier gas, then a heavier sample molecule will produce a different cooling effect on a hot wire, because it will strike the wire fewer times. Ohm's Law is:

$$E_{volts} = I_{amperes} \times R_{ohms} \tag{20-16}$$

Figure 20-37A is a diagram of a TC detector, and Figure 28-37B is a diagram of a commercial detector. A thin filament of wire is placed at the end of the column and heated by passing a constant current through it to a temperature, T_1, which is greater than the temperature of the wall of the detector, T_2. The carrier gas molecules strike the hot wire, and as each molecule hits the wire, it takes a bit of heat away. As the wire is cooled, its resistance changes, reaching equilibrium and producing a *baseline voltage*. When a sample passes over the hot wire, the sample molecules, because of different velocities and masses, will cool the wire in different amounts than the carrier gas does. This means that the resistance of the wire changes. Because the current is kept constant, the voltage changes, and a signal will be produced. The sensitivity of detection with these cells generally increases with a reduction in the wall temperatures - to ensure greater differences between T_1 and T_2. The wall temperature must not be lowered too greatly or condensation of the components in the cell will occur. The better instruments have a separate heater for the detector, and it is kept at column temperature or 20-30 °C hotter.

The greater the difference in thermal conductivities of the carrier gas and the components of the mixtures, the greater the sensitivity of the method will be. Table 20-4, p. 232, lists the thermal conductivities of several representative materials. Note that hydrogen and helium have thermal conductivities that are greatly different from those of the other compounds. Hydrogen would appear to be superior, but is seldom used because of its flammability. Helium is the second choice, but is expensive. As a result, nitrogen is used most often.

If one hot wire is used, then the detector is very unstable, because it is sensitive to all temperature, flow rate, and filament current changes. To provide stability, the Wheatstone bridge circuit is used (Figure 20-38), with one cell

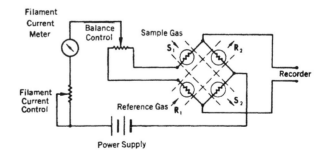

Figure 20-38. A diagram of a Wheatstone bridge circuit. (Courtesy - Varian Associates, Walnut Creek, CA)

Table 20-4. Thermal conductivities of selected gases and vapors x 10^5

Substance	K	Substance	K
Hydrogen	41.6	Diethyl ether	3.6
Helium	34.8	Propane	3.58
Methane	7.21	Ethanol	3.5
Oxygen	5.89	Water	3.5
Air	5.83	Methanol	3.45
Nitrogen	5.81	i-Butane	3.32
Nitric oxide	5.71	n-Butane	3.22
Carbon monoxide	5.63	n-Pentane	3.12
Acetylene	4.53	n-Hexane	2.96
Nitrous oxide	4.5	Acetone	2.37
Ethane	4.36	Methyl chloride	2.20
Argon	3.98	Freon-12	1.96

through which carrier gas passes and another through which carrier gas with components of the sample passes as two opposite legs of the bridge. The other two legs are two standard resistances.

This and the gas density balance are the only detectors that do not destroy the sample when making a measurement. In times of need, a detector of this type can be made from either the filaments of a light bulb or those from a model airplane glow plug.

Gas Density Balance (GDB) Detector

Use this detector for all compounds, especially those that are very corrosive.

The gas density balance was described by A.J.P. Martin and A.T. James in 1956 (*Biochem. J.*, **63**, 138). It can detect all compounds at a level of 10^{-8} g/sec, about the same as a thermal conductivity detector. Its LDR is about 1000. The main advantage of this detector is that it can be used with the most corrosive compounds, such as the halogens and interhalogen compounds. The carrier gas is either something lighter than the sample compounds (H_2) or something heavier than the sample compounds (SF_6). Referring to Figure 20-39, the carrier gas entering at point A divides and passes over the two detector filaments, B and C, then on to the outlet, D.

The filaments are connected in a Wheatstone bridge configuration as was the TC detector. Because the gas passes each filament at the same velocity, cooling each the same amount, there should be no signal as long as only carrier gas from the column enters at point F, and nothing changes. When the sample enters at point F, it can go up or down, depending on its density. If the sample is more dense than the carrier gas, more of the sample will tend to go toward point G than toward point H. At point G, the carrier gas coming over B is now slowed because of the sample gas coming past point G and blocking the way. At point H, the effect is much less. The result is that B heats up more than C, because the gas is not flowing past B as fast as it is past C, and there is an imbalance in the signal. Notice that, except for some minor diffusion, none of the corrosive sample came in contact with the detector elements. This detector can be made out of glass and two model airplane glow plugs for about $20.

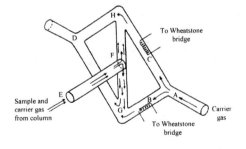

Figure 20-39. Diagram of a gas density balance. (Courtesy - From a design by Martin, A.J.P. and James, A.T., *Biochem. J.*, **63**, 138, 1956)

Hydrogen Flame Ionization Detector (FID)

Use this detector for all organic compounds except CO, CO_2, HCOOH, and HCO. It will not detect inorganic compounds.

The FID detects all compounds that contain C, except very small molecules where the C is in a high oxidation state. Thus, small carbon-containing compounds like CO_2, HCOOH, and CO cannot be detected, because the C is all highly oxidized. This detector has a linear dynamic range of 10^6 and a sensitivity of 10^{-11} g/sec. This detector is as good as a person's nose. Figure 20-40 is a diagram of a commercial detector.

This detector operates as follows. When the carrier gas and sample emerge from the column, hydrogen (equal parts of hydrogen and carrier gas) and air (10 parts air and 1 part carrier gas) are added to the carrier gas to produce a flame with a temperature of approximately 2100 °C. This flame produces positive and negative ions. The negative ions and electrons are collected by a positively charged wire ring surrounding the flame. The current produced is the background current. When a sample containing carbon enters the flame, the concentration of the ions increases to 10^{10} to 10^{12} ions/mL. It is believed that the carbon formed in this reducing atmosphere forms aggregates, such as (-C≡C-CH=C=CH-C≡C-). The associated electrons in these multiple bonds are easy to remove, having a work function of only 4.3 ev. The freed electrons are collected, and this produces a signal proportional to the concentration of carbon present. This explains why carbon compounds are about all that are detected. The sample passes through the flame so fast that oxidized carbon, such as CO and CO_2, does not have time to become reduced to the carbon aggregates and, therefore, is not detected.

This detector is insensitive to temperature change, so it is a favorite when programmed column temperatures are used. The detector temperature limit is about 400 °C. Nitrogen, helium, or argon are the usual carrier gases.

Because the hydrogen flame is colorless and about 2100 °C, a piece of paper stuck into the flame (it will burn) or a mirror held over it (water will condense on it) should be used rather than your finger to determine if the flame is on.

Thermionic Specific (TSD or N/P) Detector

Use this detector for organic compounds that contain nitrogen or phosphorus.

This detector (Figure 20-41) is quite sensitive (10^{-13} g/sec) and selective for compounds containing phosphorus. It is also good for compounds containing nitrogen (10^{-11} g/sec) but not nearly as good for other compounds. Its best use is for organophosphate pesticides. The original observation of this phenomenon

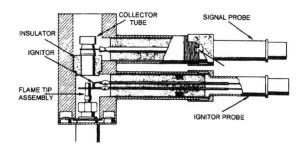

Figure 20-40. Cross section diagram of a Varian hydrogen flame ionization detector.
(Courtesy - Varian Associates, Walnut Creek, CA)

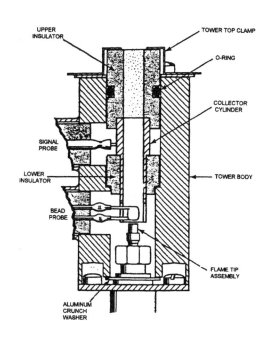

Figure 20-41. Cross section diagram of an N/P detector.
(Courtesy - Varian Associates, Walnut Creek, CA)

was by L. Guiffrida of the FDA in 1964 who found that if a small amount of sodium metal was added to a hydrogen flame, phosphorus-containing compounds responded about 500 times better than with an FID. However, the detector response was not stable. In 1966 Oaks, Dimick, and Hartman of Varian Instruments found that satisfactory stability could be attained in 5 minutes and showed no changes after 400 hours if a small piece of either NaCl or KCl was placed in the plasma portion of the flame.

The mechanism by which the sodium reacts with the phosphorus or the flame has not been established. However, a 5000/1 increase in selectivity is observed for P-containing compounds and about 50/1 for N compounds.

The LDR is 10^4. The disadvantages are that it is sensitive to detector temperature changes, and precise flow control of the hydrogen and air is needed. It is the same as an FID with the addition of a small ceramic bead containing KCl placed just above the end of the burner.

Possible Problems

1. Age of the alkali source - it is consumed over a period of time.

2. Gas flow stability - not a problem with newer instruments.
3. Position of column outlet.
4. Solvents and reagents - Acetonitrile.
5. Column stationary phases - cyano - such as the DB-225 capillary column.
6. Septa - Low temp type - not a problem if the correct ones are selected.
7. No smoking - nicotine.
8. Phosphate detergents used to clean glassware.
9. Phosphoric acid-treated glass wool.

Electron Capture Detector (ECD)

Use this detector primarily for halogen-containing compounds. It also can be used for compounds containing an electron-deficient center, such as a nitro group, an alpha dicarbonyl group, S, or P, but with reduced sensitivity.

In an attempt to make a more selective detector, the idea of *electron capture* was investigated. Assume that the electrons striking the sample have just enough energy to penetrate the electrical field of the molecule and be captured, but not enough energy to break the molecule into ions. This would cause the original electrical signal, based on free electrons, to *decrease* in proportion to the sample concentration. This was found to be true for certain atoms or groups, if the source of electrons was a β-particle. The result is that this detector is quite sensitive (5×10^{-14} g/sec) to organic compounds containing halogens, sulfur, nitrogen (nitriles, nitrates), and conjugated carbonyls.

The LDR for the older systems is only about 50, unless a linearizer is used. A modern detector has an LDR of about 10^4. The carrier gas is nitrogen or helium, which must be pure and dry. With packed columns, the detector is sensitive to column temperature changes and, therefore, is difficult to use with temperature programming. However, temperature programming can be done with capillary columns. The β-particle (electron) source for older instruments is 250 mC of tritium as titanium tritide, and this limits the detector temperature to 220 °C. *Note: A β-particle is the same as an electron, except that it originated from the decay of an atom.*

Nickel-63 Electron Capture Detector

This detector is quite sensitive (2×10^{-14} g/sec) to alkyl halides, metal organics, conjugated carbonyls, nitriles, nitrates, and sulfur-containing compounds.

Nickel-63 (12 mC) is less volatile than tritium, and the more recent EC detectors use this source up to detector temperatures of 400 °C. The carrier gas is nitrogen, which must be free of oxygen or 5% CH_4 in Ar. Figure 20-42 shows the ^{63}Ni detector.

NOTE: The β-particle source is 12 mCi of ^{63}Ni electroplated on an electrode. Even though β-particles penetrate only 8-10 cm of air and the detector cover completely shields the source, in the U.S. this is considered to be a *radioactive source* and *cannot be discarded when the instrument wears out*. It must be labeled as being radioactive, it must be tested twice a year by wipe tests, and it must be sent to a licensed operator if it needs cleaning. It must be disposed of separately as a radioactive source and a record kept for its entire lifetime. This is not considered trivial by the U.S. Nuclear Regulatory Commission (NRC), even though common sense indicates otherwise.

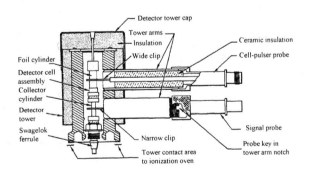

Figure 20-42. Cross section diagram of a ^{63}Ni EC detector. (Courtesy - Varian Associates Inc., Walnut Creek, CA)

Possible Problems: 1. Handling plastics, rubber products, hand lotions, and cleaning solutions. 2. Cabbage, radishes, lettuce, carrots, and onions (all Florisil eluates); a strong irregular baseline forms.

Argon Ionization Detector

Most materials, except the noble gases, some inorganic gases, and fluorocarbons, can be detected.

In 1961, J.E. Lovelock found that the cross-sectional process could be made much more efficient (10 to 100 times more sensitive than an FID) if DRY argon were used as the carrier gas. The β-particle would produce an excited argon atom that then would transfer its energy to the sample molecules very efficiently. The sequence of reactions is suggested as:

Ar + high energy β particles (15.7 ev) -----> $Ar^+ + 2e^-$

Ar + low energy β particles (11.6 ev) -----> Ar^* (11.6 ev, 10^{-4} sec half-life)

$Ar^* + 2 Ar$ -----> $3 Ar$

$Ar^* +$ impurities -----> $Ar + M^+ e^-$

$Ar^* +$ sample -----> $Ar + M^+ e^-$

The first four steps are the background, and the fifth is the signal. Materials that can be ionized by 11.6 ev are detectable. Moderate variations in carrier flow rate and temperature have negligible effects on the detector base current. The LDR is about 5×10^3.

The argon must contain < 0.003% H_2O and 0.01% O_2. If the water concentration is as much as 0.1%, then a loss of 10 in sensitivity occurs.

Flame Photometric Detector (FPD)

This detector is quite selective for compounds that contain phosphorus (10^{-12} g/sec) and sulfur (10^{-10} g/sec) or both, which makes it a favorite for pesticide analysis, air pollution, and petroleum compounds.

The sample is burned the same as in an FID detector. Phosphorus compounds form HPO, which emits radiation at 526 nm, and sulfur compounds form S_2 emitting at 394 nm. An interference filter with a half band width of about 7 nm is placed between the flame and a phototube such as EMI 9524B, and the transmitted radiation collected and measured. The LDR is 10^5 for P and 10^3 for S. Figure 20-43 shows a diagram of a two-flame arrangement FPD. One flame decomposes the sample, and the other flame is used to produce the light emission, thus providing for a better signal.

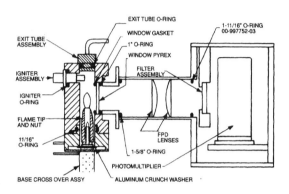

Figure 20-43. Cross sectional diagram of a dual-flame photometric detector.
(Courtesy - Varian Associates, Walnut Creek, CA)

Beilstein Detector

A *Beilstein detector* is a modified FPD detector in which a copper gauze is placed just above the flame. The halogen of halogen-containing compounds produces a green emission (540 nm) upon reacting with the copper.

Photoionization Detector (PID)

The photoionization detector can be used to detect essentially the same types of compounds as an FID detector as well as many inorganic compounds but at levels 10-50 times lower. The PID can detect S-containing compounds about 20 times lower than an FPD. Its LDR is 10^7.

Ions can be produced more efficiently by ultraviolet irradiation than by bombardment with electrons. The sample and carrier gas are passed in front of an ultraviolet radiation source (10.2 ev), where the sample components are ionized, but not the major components of air such as N_2, O_2, CO_2, H_2O, CO, and Ar.

$$RH + h\nu \longrightarrow RH^+ + e^-$$

A pair of electrodes with 200-220 v applied is

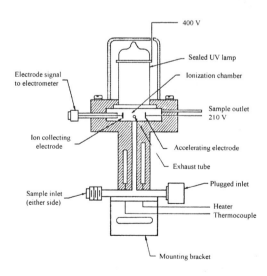

Figure 20-44. Diagram of a HNU model P151 photoionization detector.
(Courtesy - Varian Associates, Walnut Creek, CA)

placed in the cell (0.2 mL). The positive ions are driven to the negative collector plate, and the signal registered. Two picograms of benzene can be detected, and 1-2 picograms of most inorganic compounds can be detected, the only requirement being that the ionization energy be less than 10.2 ev. The carrier gas (He, N_2, Ar) flow rates are from 10-60 mL/min.

The detector can operate at temperatures up to 300 °C. Several lamp sources are available (9.5, 10.0, 10.2, 10.9, and 11.7 ev), which provides further selectivity. When used with the 10.2 ev source, such extraction solvents as chloroform, dichloroethane, acetonitrile, and carbon tetrachloride are not detected, but when operated with the 11.7 ev source, O_2, NH_3, H_2S, HI, ICl, Cl_2, I_2 and PH_3 can be detected. Figure 20-44 shows a diagram of one type of detector.

Hall Electrolytic Conductivity Detector (Hall or ELCD)

This is the most sensitive detector other than the mass spectrometer for nitrogen-containing compounds. It also can be used for compounds containing halogens and sulfur. However, it is somewhat temperamental and requires very pure hydrogen as well as operator experience.

The first successful electrolytic conductivity detector was described by D.M. Coulson in 1965 (*J. Gas Chromatogr.*, **3**, 134). The sample was combusted after passing through the column and dissolved in a liquid. The ions formed, and hence the change in conductivity, were detected by passing the solution past two metal plates. His system used direct current and was found to build up space charges and to electroplate metals on the plates' surfaces. This created baseline stability problems. Those problems were solved by R.C. Hall in 1974 (*J. Chromatogr. Sci.*, **12**, 152) when he switched to an alternating current system. Figure 20-45 shows the Tracor Model 1000. Figure 20-46 shows a cross-section of the furnace.

As the separated compound emerges from the column, it passes through a 20-cm length of tubing and is burned in a stream of *high purity hydrogen*. From the Tracor manual, "The gaseous reaction products formed in the reactor are directed to the cell and enter through the gas inlet. The conductivity cell is constructed of stainless steel and inert plastic. The differential conductivity cell contains the reference conductivity cell, the gas-liquid contactor, the gas-liquid separator, and the analytical conductivity cell.

"The conductivity solvent enters through the solvent inlet and flows through the reference conductivity cell that is formed by the top and outer electrode assemblies. The solvent then flows into the gas-liquid contactor where it is mixed with the gaseous reaction products entering through the gas inlet. The heterogeneous gas-liquid mixture formed in the gas-liquid contactor is separated into gas and liquid phases in the gas-liquid separator. The gas phase exits

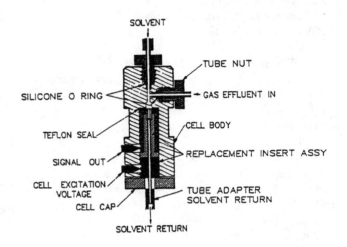

Figure 20-45. Cross sectional diagram of a Hall model 1000 detector.
(Courtesy - Tracor Instruments Analytical Division, Austin, TX)

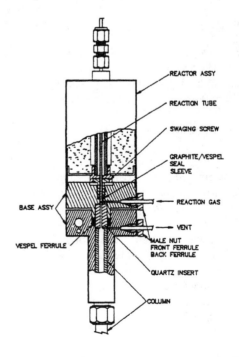

Figure 20-46. Cross sectional diagram of a Hall model 1000 detector combustion furnace.
(Courtesy - Tracor Instruments Analytical Division, Austin, TX)

through the hollow bottom electrode. The liquid phase flows between the outer wall of the bottom electrode and the inner wall of the gas-liquid separator. The cavity formed by these surfaces constitutes the analytical conductivity cell. The liquid phase exits into the hollow bottom electrode through a small hole in the wall of this electrode. At this point the gas and liquid phases are recombined and returned to the solvent reservoir."

Halogen mode (Assuming X, S, and N are also present)

$$\text{Org X} \xrightarrow[\text{Ni}]{\text{H}_2} NH_3 + CO_2 + H_2S + HX \text{ (Acid halide detected (e.g., HCl))}$$

Organic Catalytic
Halide Reduction

In the halogen mode, the alkali scrubber is removed from the outlet end of the nickel tube, and the electrolyte solvent is *n*-propanol. The conductivity due to H^+ of the HX is detected. The ionization of the other reaction products are suppressed by the *n*-propanol electrolyte.

Nitrogen mode (Assuming X and S are also present)
Step 1.

$$\text{Org N} \xrightarrow[\text{Ni}]{\text{H}_2} NH_3 + CO_2 + H_2S + HX$$

Organic Catalytic
Nitrogen Reduction

Step 2.
$NH_3 + H_2O \longrightarrow NH_4^+ + OH^-$
   Species Detected

Detector response in the nitrogen mode is due to the formation of NH_3 by the catalytic reduction of organic nitrogen in the nickel tube. When ammonia is dissolved in the 50% *n*-propanol/water solvent, it forms ammonium hydroxide, a weak base. The ammonium hydroxide dissociates to the electrolytically conducting species, NH_4^+ and OH^-. In order for the proper detector response to be obtained, the conductivity solvent must be slightly basic so that neutralization of the base does not occur.

A scrubber attached to the outlet end of the nickel tube containing $Sr(OH)_2$, a strong base, is used to remove the acid gases HX, H_2S and CO_2 yet pass the weak base NH_3. The ammonia dissolves in water, and the conductivity of the resulting NH_4^+ and OH^- ions formed is measured. The sensitivity is 10^{-15} g/sec.

Sulfur mode
Step 1.

$$\text{Org S} \xrightarrow{O_2} N_2 + CO_2 + SO_2 + HX \text{ (Assuming X and N are also present)}$$

Organic Oxidation
Sulfur

Step 2
$SO_2 + H_2O \longrightarrow H_2SO_3 \longrightarrow H^+ + HSO_3^-$

The compounds are combusted in oxygen rather than hydrogen and the gas stream bubbled through $AgNO_3$ to remove HX. The SO_2 forms H_2SO_3 in the water, and the H^+ is measured. The H_2CO_3 formed is too weakly ionized in 50% *n*-propanol to interfere.

Use only very high purity gases with this detector: 99.999% N_2 and 99.9999% H_2

If high purity hydrogen is not available, then two options exist: (1) a hydrogen gas generator or (2) a hydrogen purifier. Figure 20-47 shows a hydrogen generator. It is based on the electrolysis of either distilled or deionized water with a solid polymer electrolyte. Such a generator can

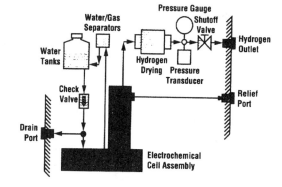

Figure 20-47. Packard Hydrogen Generator.
(Courtesy - Alltech Associates, Inc., Deerfield, IL)

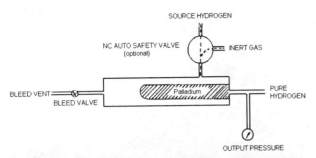

Figure 20-48. An AADCO model 560 hydrogen gas purifier.
(Courtesy - AADCO Instruments, Inc., Clearwater, FL)

produce up to 500 mL of 99.992% hydrogen/min with a maximum delivery pressure of 60 p.s.i.g. This is good enough for all detectors except the Hall detector, which requires a hydrogen purifier such as that shown in Figure 20-48. This shows the inner mechanism of the AADCO Model 560 and a photograph of the apparatus. When palladium is heated to 420 °C, it will pass only hydrogen. The impure gas is passed into the heated chamber, and the hydrogen passes through the Pd and into the chromatograph, while the impurities are allowed to bleed out an exhaust port. The inlet pressure is set at about 100 p.s.i.g., and the outlet pressure is about 45-50 p.s.i.g.. Care must be taken if the heater fails (power goes off) for more than 10 minutes. Hydrogen becomes trapped in the Pd, which becomes brittle and may crack when reheated. This can be detected immediately because the outlet pressure is about the same as the inlet pressure. A slow cooldown with N_2 passing through prevents this embrittlement from happening.

Mass Spectrometer

The GLC-mass spectrometer combination is one of most important combinations in use today. Figure 20-49 is a block diagram showing the components needed. What is important is how a gas chromatograph is interfaced to the mass spectrometer. The gas coming from a gas chromatograph is usually at 1 atmosphere, whereas a mass spectrometer operates at about 10^{-6} atmospheres. The problem is how to reduce the pressure with a minimum loss of sample. The current method is to use a molecular jet separator. Figure 20-50, p. 239, shows a diagram of how it operates.

The carrier gas and sample mixture are passed through a small jet. The carrier gas, being more mobile and having less forward momentum, can spread away from the straight forward direction, while most of the sample continues forward into the outlet jet. The carrier gas then is pumped out of the system. The inlet jet is 0.1 mm in diameter, the separation is 0.35 mm, and the outlet jet is 0.25 mm in diameter.

Figure 20-51, p. 239, shows overall transfer line connection between the GLC and the electron ionization (EI) and chemical ionization (CI) ion source of the mass spectrometer detector.

Refer to Figures 19-48 to 19-51, p. 209, for the details of the rest of the mass spectrometer configuration.

Always label each chromatogram as soon as it is obtained - or your days in the laboratory may

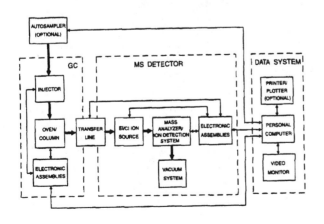

Figure 20-49. Block diagram of a GC/MS/MS system.
(Courtesy - ThermoQuest Finnegan Corp., Austin, TX)

be few. Next week, you will have forgotten most of the conditions used to obtain today's chromatogram, and by next month, it will be a total loss. A stamp can be obtained (Figure 20-52) that reminds you of the desired information.

EXAMPLES OF VARIOUS CALCULATIONS

This material was written previously by the author for the *Pesticide Laboratory Training Manual* published by the *Association of Official Analytical Chemists, International* (1996) and is quoted by permission.

The following examples are related to food samples, which are the hardest because of corrections for moisture, sugars, and fats.

1. *Calculation of the strength injected (mg/mL) for a typical fresh fruit or vegetable sample run by the Luke method* (Luke, M.A., Froberg, J.E., Doose, G.M., and Masumoto, H.T., *J. Assoc. Off. Anal. Chem.*, **64**, 1187, 1981).

$$\frac{mg \ sample}{\mu L \ injected} = \frac{mL \ (extract) \ x \ sample \ weight}{(200 + W - 10) \ (V)}$$

(20-17)

where sample weight = initial g of sample blended, W = mL water in 100 g sample, mL extract = mL taken for analysis, 10 = acetone/water concentration factor (shake out), V = mL of final injection solution, 200 = mL acetone used for blending.

Note: As a first approximation, $W = 85$ for most fruits and vegetables. The water content of raw produce may be found in Food Composition Tables, *USDA Bulletin No. 8*, Washington, DC, and several can be found in Table 5-1.

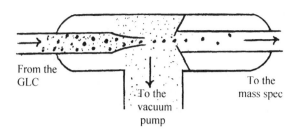

Figure 20-50. A diagram of a molecular jet separator.

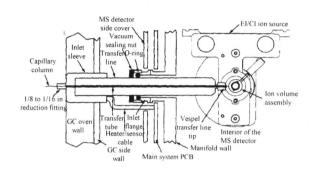

Figure 20-51. Transfer line connection between the GLC and the EI/CI ion source of the mass spectrometer detector. (Courtesy - ThermoQuest Finnegan Corp., Austin, TX)

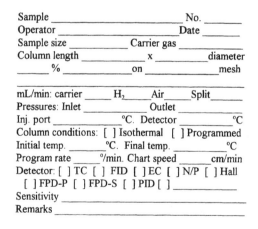

Figure 20-52. Chromatogram label.

EXAMPLE CALCULATION

A 100 g sample of prunes (44% moisture) was extracted with 200 mL of acetone. 80 mL of extract was recovered and concentrated to 5.0 mL before injection. What is the mg of sample/μL injected?

ANSWER

$$\frac{mg \ sample}{\mu g \ injected} = \frac{80 \ x \ 100}{(200 + 44 - 10) \ (5)} = 6.8 \ mg/\mu L$$

2. *Calculation of extraction volumes with a sugar correction.* These formulas can be used to calculate the extraction volume for pesticide-residue procedures that use a single extraction with acetone, acetonitrile, or methanol, with or without prior dilution of these solvents with water. These formulas account for the volume change on mixing and for the volume contributions from the soluble sugars, (sucrose, glucose, and fructose). Results using these equations are in agreement with the observed volumes of such mixtures within 0.3% throughout the following ranges, expressed as % water in the mixtures: 15 - 38% for acetone, 4.8 - 59% for acetonitrile, and 9.3 - 33% for methanol.

For acetone: $\quad V = 0.98 V_o + 0.92 V_w + 0.62 W \quad$ (20-18)

For acetonitrile $\quad V = 0.995 V_o + 0.96 V_w + 0.62 W \quad$ (20-19)

For methanol $\quad V = 0.99 V_o + 0.91 V_w + 0.62 W \quad$ (20-20)

where V = total volume in mL, V_o = volume of the organic solvent in mL, V_w = volume of water in mL, and W = weight of sugar in g.

EXAMPLE CALCULATION
A pure organic solvent is blended with the sample. A 100 g portion of frozen strawberries (75.7% water, 20% sugar) is blended with 200 mL of acetone by the method of Luke et al. What is the total volume to be used in equation 20-15?

ANSWER
$V = 0.98 \times 200$ mL $+ 0.92 \times 0.757 \times 100 + 0.62 \times 20; \quad V = 278$ mL

The usual method that does not correct for sugar gives a value of 206 mL.

EXAMPLE CALCULATION
A premixed water-organic solvent mixture is blended with the sample. A 25 g sample of immature lima beans (67.5% water) is blended with 350 mL of 35% water in acetonitrile. Sugar is too low to be significant. What is the total volume?

ANSWER
$V = 350 + 0.96 \times 0.675 \times 25 = 366$ mL   No factor is applied to the 350 mL of *premixed* solvent.

EXAMPLE CALCULATION
High moisture fatty products are blended with a sample charge. Acetonitrile is blended with a 50 g sample of milk, 20 g alumina, and sufficient extra water to give 20% water in 350 mL total volume. What volume of water and acetonitrile are to be added?

ANSWER
According to the specifications, the total amount of water is:
$0.20 \times 350 = 70$ mL. 44 mL of this is from the sample (milk is 88% water), and
$70 - 44 = 20$ mL is added.

From equation 20-19, the amount of acetonitrile is:
$(350 - 0.96 \times 70)/0.995 = 284$ mL

3. *Calculation for a dry low-fat sample by the dry Luke method, of the strength injected (mg/μL):* The equation is nearly the same as equation 20-17, but notice that the sample size is smaller and that a much larger volume of extractant is used. No correction volumes are needed, nor a moisture value.

$$\frac{mg\ sample}{\mu L\ injected} = \frac{mL\ (extract)\ \times\ sample\ weight}{350\ \times\ V} \quad (20\text{-}21)$$

EXAMPLE CALCULATION

A 15 g sample of millet (8.7% moisture) was extracted with 350 mL of a 35% acetone/water mixture. 80 mL of extract was concentrated to 4 mL. What is the mg of sample/µL injected?

ANSWER

$$\frac{mg\ sample}{\mu L\ injected} = \frac{80 \times 15}{350 \times 4} = 0.86\ mg/\mu L$$

4. *Calculation for a fat sample of strength injected (mg/µL):*
<u>On a fat basis</u>:

$$\frac{mg\ fat}{\mu L\ injected} = \frac{W}{V} \tag{20-22}$$

where W = g fat taken for analysis, and V = mL of final solution.

EXAMPLE CALCULATION

A 3.0 g sample of butter was extracted with petroleum ether. After a Florisil clean-up, the extract is concentrated to 7.0 mL. What is the mg of fat/ µL injected?

ANSWER

$$\frac{mg\ fat}{\mu L\ injected} = \frac{3.0}{7.0} = 0.43\ mg/\mu L$$

5. *Calculation for a fatty food of the strength injected (mg/µL):*
<u>On a whole food basis</u>:

$$\frac{mg\ fat}{\mu L\ injected} = \frac{W}{V \times \%} \tag{20-23}$$

where W = g of fat taken for analysis. V = mL of final solution. % = % fat in the sample expressed as a decimal (i.e., 12%fat = 0.12).

EXAMPLE CALCULATION

3.0 g of the fat extracted from raw Spanish peanuts (49.6% fat) was analyzed properly and concentrated to a final volume of 5.0 mL. What is the mg sample equivalent/µL injected?

ANSWER

$$\frac{mg\ fat}{\mu L\ injected} = \frac{3.0}{5.0 \times 0.496} = 12.2\ mg/\mu L$$

6. *Calculation of ppm residue.* (ppm = ng/mg)

$$\frac{Height\ sample\ peak}{Height\ standard\ peak} \times \frac{\mu L\ standard\ (ng/\mu L)}{\mu L\ sample\ (mg/\mu L)} = ppm \tag{20-24}$$

EXAMPLE CALCULATION

A 2.4 µL injection of a carrot solution representing 2.84 mg/µL strength produced a peak of 44 mm height. The peak matched the retention time of Dieldrin. An amount, 2.6 µL of a Dieldrin standard solution of 0.2 ng/µL strength, was injected and produced a 40 mm peak height. How many ppm of Dieldrin are on the carrot sample?

ANSWER

$$\frac{44 \text{ mm sample}}{40 \text{ mm standard}} \times \frac{2.6 \text{ µL sample} \times 0.2 \text{ ng/µL standard}}{2.4 \text{ µL sample} \times 2.84 \text{ mg/µL sample}} = 0.084 \text{ ppm Dieldrin}$$

7. *Recoveries.*

Sample fortification recoveries may be calculated as:

$$\frac{Fortification}{Recovery} = \frac{(ppm\ spk - ppm\ spl) \times 100}{ppm\ spike\ level} \qquad (20\text{-}25)$$

where ppm *spk* = concentration as ppm found in the spiked sample, ppm *spl* = concentration as ppm found in the sample.

EXAMPLE CALCULATION

To determine the recovery of Dieldrin, the previous sample of carrots (containing 0.084 ppm Dieldrin) was spiked with an additional 0.10 ppm of Dieldrin and analyzed. A 2.5 µL injection of the final solution (representing 2.84 mg sample/µL) was made. The peak found had the retention time of Dieldrin and did not exhibit significant broadening, (further indicating the original peak was Dieldrin). The peak height found was 96 mm. An injection of 5.5 µL was made of the Dieldrin standard solution (0.2 ng/µL). The recovery peak had a height of 85 mm. What is the percentage recovery of Dieldrin?

ANSWER

Use equation 20-21 to calculate the total ppm in the spiked (or fortification) sample. Then use equation 20-20 to calculate the % recovery:

$$\frac{96}{85} \times \frac{5.5 \times 0.2}{2.5 \times 2.84} = 0.175 \text{ ppm}; \qquad \frac{(0.175 - 0.84)(100)}{0.10} = 91\%$$

PEAK AREA MEASURING TECHNIQUES

This material was written previously by the author for the *Pesticide Laboratory Training Manual* published by the *Association of Official Analytical Chemists, International* (1996) and is quoted by permission.

Quantitations by manual techniques (Condal-Bosch, L., *J. Chem. Ed.*, **41** (4), A235, 1964) commonly rely on measurements of peak heights and peak areas. Peak heights usually are recommended for early eluters whose height to base ratios would be fairly large. It has been suggested that *peak height* measurements are appropriate for very small peaks and peak widths of 10 mm or less. *Peak area* is determined on other peaks and is accomplished by triangulation. Both techniques are demonstrated and explained in Figure 20-53. Whether the detector response (peak) is measured manually or electronically, proper positioning of the baseline below the peak is critical. Accuracy of the measurement depends in part on how well the detector's response to the residue can be distinguished from its response to sample co-extractives and co-eluting residues. Typically, a residue peak in a sample chromatogram may occur on a sloping baseline, on top of another peak, or incompletely separated from another peak. To measure the peaks manually, an analyst must literally draw the baseline on the chromatogram before measuring; to use automated measurement, the analyst must configure the system to include only that part of the signal that can reasonably be assumed to represent a residue. Appropriate setting of the baseline is integral to the directions below for measuring the peaks.

Keep the standard concentration as close to the sample residue concentration as possible, and the errors in the area methods tend to cancel.

Symmetrical Peaks
Measurement of Peak Height

To measure peak height, construct a baseline beneath the peak and measure the length of the perpendicular from peak apex to midpoint of the constructed baseline. In Figure 20-53, this is represented by line A-B on peak 1.

Measurement of Area by Triangulation

Measurement by triangulation involves drawing a triangle that approximates a peak's dimensions and calculating the area of the triangle. This method requires extreme care in construction of the triangle and in measuring its dimensions. Special treatment is required for peaks on sloping baselines and for skewed (asymmetrical) peaks. The technique is subject to error when the peak is narrow, but is preferred over measurement for peak height when the peak is >10 mm wide at the base. Figure 20-53 represents chromatograms with a flat baseline.

Method 1: Draw a baseline from one side of the bottom of the peak to the other (C-D). Draw straight lines tangent to each side of the peak, extending them so they cross at the top (E) and intersect the baseline (F, G). Draw a line from C down to the baseline and meeting it at a right angle (H). The distance C-H is the height (h) and the distance F-G is the width (w). The area is:

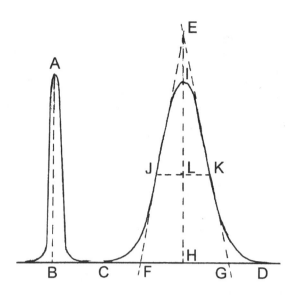

Figure 20-53. Techniques for measuring peak areas. (Courtesy - L. Condal-Bosch, *J. Chem. Ed.*, **41**, (4), A235, 1964)

$$Area = \tfrac{1}{2}\, w \times h \qquad (20\text{-}26)$$

The area calculated this way is 97 % of the actual area.

EXAMPLE CALCULATION

The height of a peak was found to be 88 mm and the width 35 mm. (A) What is the area of this peak? (B) If the peak just measured was for a 10 ppm standard, what is the concentration of a peak having an area of 1400 mm^2?

ANSWER

A. Use equation 20-26. $A = \tfrac{1}{2} \times 88 \times 35$ $A = 1540$ mm^2
B. Concentration is proportional to the areas involved.

$$\frac{1{,}540 \; mm^2}{1{,}400 \; mm^2} = \frac{10 \; ppm}{X \; ppm}; \qquad X = 9.1 \; ppm$$

Method 2: The half width method. Refer to Figure 20-53. Measure the height (*h*) from the top of the peak (I) to the baseline (I-H). Divide this distance by 2 and at that point (L) measure the width (J-K). Use equation 20-27. The area calculated this way is 90 % of the actual area.

$$A = h \times w \qquad (20\text{-}27)$$

EXAMPLE CALCULATION

The same peak as used in method 1 was measured using method 2. The height was found to be 69 mm, and the width at half height was 22 mm. What is the area of this peak?

ANSWER

Use equation 20-27. $A = 69$ mm $\times$ 22 mm $A = 1518$ mm^2
It is slightly smaller as predicted (90% to 97%).

Method 3: Refer to Figure 20-53. Measure the height as in method 2 (I-H) and the width as in method 1, (F-G). Use equation 20-20 to determine the area. The area calculated this way is 80% of the actual area.

EXAMPLE CALCULATION
The height from method 2 was 69 mm, and the width from method 1 was 35 mm. What is the area of this peak?

ANSWER
Use equation 20-24. $A = 69 \text{ mm} \times 35 \text{ mm} = 1208 \text{ mm}^2$

There must be a place where the area calculated is 100% of the actual. This magic place is found by measuring the width at 45% of the height as measured in Method 2. It is hard to determine the 45% point, so it is seldom done.

Skewed Peaks

Skewed peaks present another challenge to the validity of area measurement by triangulation. *The preferred solution to quantitation of skewed peaks is to improve the chromatography sufficiently to cause peaks to be symmetrical.* Use of a more polar column, changing column or inlet temperature, or cleaning and optimizing the injection system may effect an improvement. Figure 20-54 represents a chromatogram with a sloping baseline. The triangulation technique is as follows:

Draw a baseline from one shoulder to the other (B-C). Follow method 1 above and draw line A-D to determine the height (*h*). From point C, draw a line to the other side of the peak so it makes a right angle at (E). The distance C-E is the width (*w*). Use equation 20-26 as before.

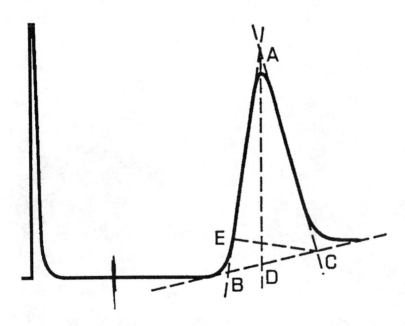

Figure 20-54. Triangulation method to measure peak areas of unsymmetrical peaks.
(Courtesy - FDA Pesticide Analytical Manual Vol. 1, 3rd Ed. 504B)

EXAMPLE CALCULATION
A peak shaped like that in Figure 20-54 had the following measurements: B-C = 29 mm, C-E = 23 mm, A-B = 68 mm, and A-D = 65 mm. What is the area of this peak if measured as recommended?

ANSWER
Line A-B is the base, and line C-E is the height. Therefore, $A = \frac{1}{2} 68 \text{ mm} \times 23 \text{ mm}$ or 782 mm^2

The most accurate manual method to determine the area of a skewed peak is to construct a trapezoid. This is easy if you use a template (included), and harder if you do not. Refer to Figure 20-55. Determine the height as shown. You need two widths, one at 15% and the other at 85% of the height.

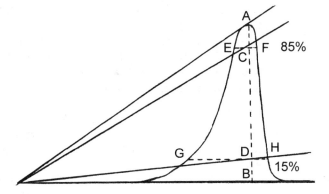

Figure 20-55. The Trapezoid Method to Determine Peak Areas Showing How the Template Is Used.
(Courtesy - Condal-Bosch, L., *J. Chem. Ed.* **41**, (4), A235, 1964)

$$A = \frac{(W_{0.15} + W_{0.85}) \times H}{2} \tag{20-28}$$

Make your own template by tracing the one in Figure 28-55, but extending the lines so the top line would be at the top of your largest peak. Then make a transparency of it. Draw a peak height line (A-B) as in method 2. Place the base of the template over the baseline of the peak and move the template left or right until the top line intersects at point A. Point C is the 85% intersection and point D is the 15% intersection. Mark these points, then draw a line parallel to the base through each point (E-F and G-H). Use equation 20-28 to determine the area.

EXAMPLE CALCULATION
A peak had a height (A-B) of 67 mm. The 15% width was 34 mm, and the 85% width was 10 mm. What is the peak area?

ANSWER

$$A = \frac{(10 \text{ mm} + 34 \text{ mm}) \times 67 \text{ mm}}{2} = 1474 \text{ mm}^2$$

Note: An inexpensive way that has been used for years is to photocopy the chromatogram, then cut out the peak for the residue, and weigh the piece of paper. Do the same for a standard and compare the two. This is easy and reasonably accurate.

Electronic Integration
Electronic peak area integration is commonplace and is utilized most often in large volume operations. It should be used if possible. It is almost always included in the software package on computer-controlled instruments.

REVIEW QUESTIONS

1. What is a general rule to determine if a compound can be separated by gas-liquid chromatography?
2. What is the retention time in a GLC analysis?
3. Calculate the number of theoretical plates and the HETP in a 3 m column whose major peak was 50 mm from the injection point and t_i and t_f are 38 and 42 mm, respectively.
4. How does V_r^o compare to V_r for a pressure drop of 0.5 atm, if the outlet pressure is 1.0 atm?
5. It required 4.2 minutes for the peak to emerge when the flow rate of He was 10 mL in 20 seconds. What is the retention volume?
6. What would be the Kovats' retention index for a compound that had a retention time of 5.3 minutes using the data in Table 20-4, p. 232?
7. What are the *A, B,* and *C* terms of the van Deemter equation due to?
8. What does the general shape of a plot of the van Deemter equation show?

9. Referring to the van Deemter equation, to improve efficiency, would you change to (a) a larger or smaller diameter column, (b) a higher or lower m.w. carrier gas, and (c) a smaller or larger diameter of the inert phase particles?
10. What are the usual carrier gases used for pesticide-residue analysis?
11. Why are gas-purifying traps placed just past the regulator valve?
12. What is a good technique to follow when you connect a pressure gauge to a gas cylinder?
13. How can you tell if a septum is leaking?
14. What is an advantage of having a gas-tight syringe?
15. When would you want to use a Chaney adapter?
16. Why should a syringe be withdrawn immediately after the sample is injected?
17. What is an advantage for using the solvent flush method of sample injection?
18. What does FSOT mean?
19. Why should 1/8 inch (3 mm) diameter packed columns not be coiled tighter than 3 in (7.5 cm) in diameter?
20. Why should metal columns be avoided when doing pesticide residue work, especially with chlorinated pesticides?
21. Why are tubing benders used when bending large diameter tubing?
22. What are ferrules and what is their purpose?
23. What type of ferrule would you use if you wanted to work at 370 °C?
24. What must you do to ensure a tight seal when using direct-connect capillary connectors?
25. What is the purpose of an inlet liner?
26. What do AW and S stand for when placed on the label of a solid support?
27. How would you prepare a silanizing agent?
28. What capillary column packing would you select to separate organosulfur pesticides?
29. What is the coating on a DB-210 capillary open tubular column? (Table 20-2).
30. Why are Rotovaps usually not recommended when coating a support?
31. What is the easiest way to pack an 1/8 inch (3 mm) diameter column?
32. What information should be put on a column marking tag?
33. What is the purpose of conditioning a column?
34. Why do you disconnect a column from the detector when it is being conditioned?
35. Why is a capillary column placed in a cage?
36. What is the purpose of a retention gap when using a capillary column?
37. Why is a quartz (ceramic) cutter used to cut capillary columns?
38. Why should a soap solution not be used to test for leaks?
39. Why should you never connect a heated column to a cold detector?
40. How do you test a capillary column for proper installation?
41. Capillary columns are conditioned differently than packed columns. What is the difference?
42. Can a rotameter calibrated for hydrogen be used for air?
43. What are the most common GLC detectors used for pesticide-residue work?
44. How do you determine noise level?
45. What does LOD mean, and how is it determined?
46. What does LOQ mean, and how is it determined?
47. When is "trace" reported?
48. Which of the following compounds can be detected by an FID detector? CH_4, $C_4H_9O_2$, H_2O, NH_3, $C_8Cl_4N_2$, CO.
49. How do you test to see if the flame is lit on an FID?
50. What is the basic difference between an FID and an N/P detector?
51. What are three items that might cause a N interference with a N/P detector?
52. What types of compounds can be separated best with an ECD?
53. What is a β-particle?
54. What is the most common source of β-particles used in the ECD?
55. Suppose you have a GLC with an ECD and it is worn out. How must you dispose of it?
56. What is measured in an FPD-P detector?
57. What does each flame do in a two-flame FPD?
58. What should you do if the flame doesn't light on an FPD?
59. What is the basic principle of operation of a photoionization detector?
60. What modification did Hall make on the Coulson detector to make it work better?
61. What is the basis of operation of a Hall detector?

62. Why are the NH_4^+, OH^-, H^+ of H_2CO_3, HCO_3^-, and SH^- ions not measured when the H^+ and Cl^- ions are detected in the halogen mode of the Hall detector?
63. What is the purpose of the $Sr(OH)_2$ scrubber when using the Hall detector in the N mode?
64. The noise level gets quite high unless ultrahigh purity gases are used with the Hall detector. What is meant by "ultrahigh" purity?
65. What type of interface is used to connect a GLC to a mass spectrometer?
66. A 100 g sample of red, raw, hot chili peppers (87.7% moisture) was extracted with 200 mL of acetone. 80 mL of extract was recovered and concentrated to 6.0 mL before injection. What is the mg sample/μL injected?
67. A 25 g sample of dried apricots (7.5% moisture) was extracted with 350 mL of an acetone/water mixture by the dry Luke method. 80 mL of extract was collected and reduced to 5 mL. What is the mg of sample/μL injected?
68. A 3.0 g sample of fat from a peanut butter (50% fat, 1.14% water) extract was concentrated to 6.0 mL. What is the mg of fat/μL injected?
69. Using the same sample above, calculate the mg of sample/μL injected on a whole food basis.
70. A 3.0 μL injection of a turnip solution representing 2.5 mg/μL produced a peak of 56 mm height. The peak matched the t_r of Carbaryl. An amount, 3.0 μL, of a Carbaryl standard solution of 0.25 ng/μL was injected and produced a 45 mm peak height. How many ppm of Carbaryl are on the turnips?
71. What are some guidelines for when to use peak height for quantitation rather than peak area?
72. Suppose you devise your own system to measure the area of a peak, and you have no idea how accurate it is. How can you eliminate this problem to make an accurate calculation?
73. If you wanted to be 100% accurate and wanted to use equation 20-28 to determine the area of a peak, at what height would you measure w?
74. What is the most accurate manual method to determine the area of a chromatographic peak?
75. A 50 g sample of children's suckers (high sugar, 35% moisture) was extracted with 100 mL of water and 200 mL of acetone. 80 mL of filtrate was recovered. How many grams were recovered?
76. A 50 g sample of shiitake dried mushrooms (9.5% water) was extracted with 350 mL of a water/acetone solvent, and 80 mL of filtrate recovered. How many grams were recovered?
77. An analyst turned up the sensitivity on the chromatograph's recorder to its most sensitive setting. It was noted that the normal fluctuations were about 4 mm. What are the LOD and the LOQ?
78. To determine the recovery of DDT, a sample of pineapple (containing 0.075 ppm DDT) was spiked with an additional 0.10 ppm of DDT and analyzed. A 2.0 μL injection of the final solution (representing 3.41 mg sample/μL) was made. The peak height found was 93 mm. An injection of 4.5 μL was made of the DDT standard solution (0.2 ng/μL). The recovery peak had a height of 74 mm. What is the percentage recovery of DDT?
79. A 100 g sample of dates (22.5% water, 64.2% sugar) was examined by the method of Luke et al. and blended with 200 mL of acetone. What is V total?

IF IT EXISTS - CHEMISTRY IS INVOLVED
IF YOU CAN BUY IT - A CHEMIST WAS INVOLVED SOMEWHERE

21
PAPER CHROMATOGRAPHY

PRINCIPLES

When analysts speak of *paper chromatography,* they mean that a piece of moist paper is used as the inert and stationary phases, the sample is placed as a small diameter spot at one end, and the components are eluted by having the mobile phase pass through the paper by capillary action.

This concept probably started with capillary analysis in 1861 by Schonbein, C.F. (*Ann. Chem. Liebigs*, **114**, 275, 1861). He showed that when a solution was allowed to rise through paper by capillary action, the solvent migrated faster than the solute and different solutes rose to different heights. His system was like our present-day *frontal analysis*.

In 1941, Martin, A.J.P. and Synge, R.L.M. *(Biochem J.*, **35**, 1358, 1941) developed *partition chromatography*. Martin's background was distillation. He was also a chemical engineer and worked with countercurrent extractions. He worked at the Wool Industries Research Institute trying to separate amino acids, and his very complex apparatus took 1 week to set up. He decided to fix one liquid and move the other. Water was fixed on silica gel with $CHCl_3$ and 1% ethanol was the mobile phase. It was his distillation background that led to the HETP concept in chromatography. In 1944, Consden, R., Gordon, A.H., and Martin, A.J.P., *(Biochem. J.*, **38**, 224, 1944) developed *paper chromatography*. Martin was trying to separate several amino acids on a cellulose column, and it was taking 3 weeks to elute them all. He considered paper because it was pure cellulose with no lignin, Cu, or other impurities. A small percent of water is tightly bound, so the stationary phase is already there. Other water molecules are adsorbed by the cellulose and fill the interspaces of the fibers. Paper can hold as much as 20% water. Other characteristics of paper are that cellulose is electronegative in water. Paper acts as a reducing agent (upon prolonged contact, it will reduce $KMnO_4$, Fe^{+3}, diphenylamines). Oxidation of the hydroxyl groups during manufacture produces some -COOH, -CHO, and R-C(=O)-R groups.

When paper chromatography was developed, it was so highly regarded that a Nobel prize was awarded to Martin in 1952 for its discovery. Mixtures that had taken several days to weeks to separate could now be separated in a day or even hours. Today, very few separations are made by paper chromatography. The methods have been converted largely to thin layer chromatography, which can perform the same separations in minutes. This chapter is included not just for historical value, but because there are still a few separations that work best by this method and there are some good techniques involved, which if mastered with paper systems, will work even more easily with thin layer chromatography. Many, many separations are done by paper chromatography in biochemistry laboratories, because larger samples can be more easily handled.

The principles of paper chromatography are similar to those of column chromatography discussed in Chapters 14 to 18. The main difference is that a piece of paper is used for the inert phase. The solution to be examined is deposited as a small spot on the paper. The mobile phase is allowed to move over the spot in a definite direction, and the substances are separated by their differences in solubility in the moving solvent and the stationary phase, which generally is considered to be the water normally present in paper. Isaac Asimov, the well-known science fiction writer, likened paper chromatography to a schizophrenic dilemma that each molecule of a mixture had to face: Shall it move or stand still? Shall it ignore the mobile phase or shall it go along? Each substance makes its individual compromise.

TECHNIQUES

There are several types of paper chromatography, and they are named according to the direction the mobile phase flows through the paper. These are *ascending, descending, horizontal, and two-dimensional*, and any of these with a *reversed phase*.

ASCENDING CHROMATOGRAPHY

The sample is spotted 1-2 cm from the end of a paper strip. The paper is suspended vertically in a chromatographic chamber with the spotted end down and allowed to dip about 5 mm into the solvent. The solvent rises up the paper by a capillary action, effecting a separation of the components as it ascends (Figure 21-1). The disadvantage is that the solvent will rise only about 20 cm. The main advantage is the very simple equipment required. The author uses empty 5 lb reagent jars, and adjusts the paper height with a wire pushed through a cork placed in a 1 cm diameter hole in the lid.

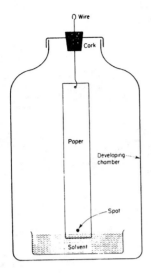

Figure 21-1. Apparatus required for ascending paper chromatography.

DESCENDING CHROMATOGRAPHY

The sample is spotted 5-8 cm from the end of a 30-45 cm long strip of paper. The strip is suspended vertically in a chromatographic chamber (Figure 21-2) with the spotted end up. The 5-8 cm "wick" is bent and placed in a solvent tank suspended from the top of the chamber. The strip is held in place by a small length of glass rod. After the solvent rises up the "wick", it descends across the spot and down the strip. An antisiphon bar usually is attached to the solvent tank to prevent the solvent from siphoning out of the tank. The siphoning will cause the solvent to flow too fast for complete equilibrium to be obtained, and poorer separations are obtained. The main advantage of the descending technique is that the paper can be of any length; therefore, much more difficult separations can be made.

HORIZONTAL CHROMATOGRAPHY

This also is known as the *Rutter technique*. The spot is placed in the center of a filter paper disc containing a slit from the center to the perimeter. By means of this "wick", solvent is fed to the paper by capillary action, and resolution occurs as circles rather than spots. The chamber can be a petri dish or a desiccator (Figure 21-3).

MICROHORIZONTAL CHROMATOGRAPHY

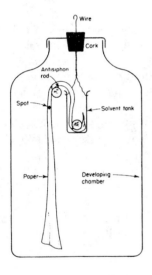

Figure 21-2. Apparatus required for descending paper chromatography.

This technique, known as the *Weisz ring oven technique*, is used primarily for the separation of inorganic ions when the sample is microsize. It provides an analyst with a way to divide a single drop of sample into at least 10 equal portions. It has been used to identify the metals in the paints of suspected forged artwork, the sample being obtained with a flattened-end syringe needle. Air pollutants can be concentrated by passing air through a paper filter for several days if necessary, removing a small sample, and detecting the metals.

To call the ring oven technique *horizontal chromatography* is probably a misnomer, although it can be used as such. The initial stages of this separation do produce a chromatographic separation, and if the process is stopped at that time, a typical chromatogram is produced with rings being formed rather than spots or lines. However, the ability of this method to concentrate the sample so it can be divided into several parts has been found to be of more value.

The apparatus consists of an aluminum doughnut-shaped ring, 65 mm o.d. x 33 mm i.d. This ring is heated electrically up to 120°C. A 5 cm diameter piece of filter paper is placed on the ring. The sample, either as a small drop, a small particle, or a small (2-3 mm) circle of paper, is placed in the middle of the paper. A micropipet filled with mobile phase is placed vertically over the sample. The ring is heated sufficiently to evaporate the solvent rapidly when it reaches the ring. Depending upon the solvent used, portions of the sample are washed outward along with the solvent. When the solvent approaches the heated ring, the solvent evaporates, leaving a very thin ring of one component of the sample. As the process continues, successive components are washed to the outer ring and recombine. This ring is so thin that the concentration of the sample is often 50 times greater than in the original area. The circumference of the ring is about 70 mm. The paper can be cut into sections, and each section then tested as desired for any ion; in this case, each section contains the same components.

TWO-DIMENSIONAL CHROMATOGRAPHY

Two solvents are employed, one after the other. The sample is spotted in the corner of a square or rectangular piece of paper. One solvent is used to develop the chromatogram in one direction. The paper is dried and turned 90°, and a second solvent of different polarity is used. An example of the high degree of separation that can be obtained is shown in Figure 21-4.

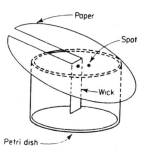

Figure 21-3. Apparatus for horizontal paper chromatography.

REVERSED-PHASE CHROMATOGRAPHY

The main functional groups on paper are hydroxyls with acids and aldehydes next in amount. These, along with the adsorbed water, make the system polar and not particularly good to separate nonpolar compounds. To obtain a nonpolar stationary phase (reversed phase), the paper is first dried thoroughly. It is then dipped in a nonpolar liquid and dabbed dry with paper towels. This produces a "reversed phase" suitable for separating water-insoluble substances like steroids, long chain fatty acids, or chlorinated insecticides. Figure 21-5 shows a commercial "dipping tank". This is made of stainless steel and is used for the uniform application of immobile solvents to paper sheets.

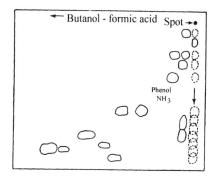

COMPONENTS

Paper

Commercially obtained chromatographic paper is a high quality filter paper. Ideally, it contains pure cellulose and no lignins, copper, or other impurities. Examination of cellulose under an electron microscope reveals the crystallinity of the macromolecules forming closely attached fiber bundles, held together by hydrogen bonding. A small percentage of water is tightly bound. Other water molecules are adsorbed by the cellulose and fill the interspaces of the fibers. It is not certain that this water takes part in the partition process. The fact that the interspaces can be filled readily with a nonpolar solvent may exclude the adsorbed water as a factor in the partition process. Chromatographic paper is usually available in sheets 45 x 55 cm or in strips 2.5 cm x 100 m and often contains organic impurities (lignin). Caution should be used when interpreting infrared or ultraviolet spectra of eluted spots of "unknown" compounds.

Do not handle chromatographic paper with your bare fingers. The oils from your fingerprints cause the mobile phase to alter its path, producing a skewed solvent front and altered R_f values. Use tweezers or wear gloves when handling chromatographic paper.

Figure 21-4. A two-dimensional paper chromatogram.

Figure 21-5. Dipping tank. (Courtesy - A.H. Thomas Scientific Co., Swedesboro, NJ)

It is important that the paper come to equilibrium with the mobile phase before the mobile phase is passed over the sample spot. The reason is that the paper will adsorb some of the mobile phase, particularly if it contains water, and this will slowly change the composition of the mobile phase. This will result in a *streak* or *long tail* forming for each component when the sample is developed, and a clean separation is not made. Equilibrium is attained by placing the spotted paper in the chamber, but *not letting it dip into the mobile phase*, and maintaining that position for 1-2 hours depending on the size of the chamber. The paper then is lowered into the mobile phase, and the development can take place.

Quicker attainment of equilibrium in the paper can be obtained by placing a paper towel, soaked with the mobile phase, around the inside of the chamber. Leave a 4-5 cm opening so the sample paper can be viewed.

When doing two-dimensional paper chromatography, pass the first mobile phase through the paper in the machine direction. When paper is prepared, it is passed through rollers, and this continual pulling tends to align the fibers in one direction. If the first mobile phase is allowed to move parallel to these fibers, less broadening of the sample spots occurs than if the mobile phase had to move across the grain. When the second mobile phase then is added, the spots are small, and good resolution can be obtained. If the first solvent is used against the grain, then the spots become quite large, making resolution with the second mobile phase difficult. The *long direction* on rectangular cut paper is the *machine direction*.

Preparation of the Sample

Conventional extraction procedures and evaporation techniques are applied. Depending on the nature of the compounds, different clean-up procedures are recommended: ion exchange for amino acids and organic acids, pyridine solutions for sugars. Because the final volume of the solution must be small, evaporations must be carried out in *vacuo* (Rinco evaporators) or by lyophilization. Sometimes inorganic ions (Na^+, Mg^{+2}) interfere in chromatography, and desalting may have to be carried out. Usually, commercial electrolytic desalters are used for this purpose.

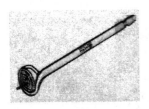

Spotting

The principle is to spot a small amount of the sample solution by successive applications of small volumes of about 1-2 µL. During the development, some diffusion occurs, and the original spot becomes larger during the process. Spots can be kept small by applying heat or a draft of warm air, as from a hair drier, while spotting. Micropipets are generally available and will empty when they touch the paper. Other pipets require a syringe control. Automatic spotting devices have been used. The simplest method is to use a very small loop of platinum wire sealed into a glass rod.

Figure 21-6. Streaking pipet.
(Courtesy - Fisher Scientific Co., Pittsburgh, PA)

Figure 21-6 shows a device for *streaking* a paper, that is, placing a line of sample on the paper. It is a horizontal spiral capillary with a 0.5 mL capacity.

Chromatographic Chambers

The simplest arrangement can be made for the ascending technique. Here, the upper edge of the paper can be fastened with a paper clip to a support rod and the bottom edge just immersed in the developing solvent. Test tubes, aquaria, biological specimen jars, and pickle or large reagent jars serve as excellent chambers. However, commercially you can obtain chrome-plated, insulated, explosion-proof equipment, which can be quite expensive, but very good.

Figure 21-7 shows some commercially available, equilibrium attainment chambers for paper chromatographic separations. Some are vented to a hood system, so that easily combustible or explosive mixtures are not formed.

The left chamber is a glass jar 308 x 308 x 610 mm deep with a ground top edge for a vapor-tight seal.

Figure 21-7. Equilibrium chambers for paper chromatography. Left, glass jar type. Right, Stainless steel type.
(Courtesy - A. H. Thomas Scientific Company, Swedesboro, NJ)

The inner rack and troughs can be adapted for either ascending or descending methods. The stainless steel chamber on the right is for precision work using either ascending or descending methods. Its outer dimensions are 762 mm wide x 813 mm high and 521 mm deep. It can hold 10 sheets and has 5 solvent entry ports for descending methods.

Mobile Phase

By knowing the partition coefficient of the solute in two partially miscible solvents (e.g., phenol-water), you can predict the movement of the solute or suggest a better mixture of solvents. Some general rules:

If the substance moves too slowly, increase the solvent constituent favoring solubility of the solute. If the solute moves near the solvent front, increase the other solvent.

Table 21-1 lists several common liquids used as mobile phases, either singly or in mixtures, for separations on paper. The solvents listed first are those that are either donors or acceptors of electron pairs and have a greater ability to form intermolecular H-bridges than those listed later on.

Table 21-1. Mixotropic Solvent Series
(Courtesy - E. Heftmann, *Chromatography*, 2nd ed. Reinhold Co., New York, NY, 1967)

Water	*t*-Butanol	Chloroform
Formamide	Phenol	Benzene
Methanol	*n*-Butanol	Toluene
Acetic acid	*n*-Amyl alcohol	Cyclohexane
Ethanol	Ethyl acetate	Petroleum ether
Isopropanol	Ether	Petroleum
Acetone	*n*-Butyl acetate	Paraffin oil
n-Propanol		

When using a two-phase system, the chromatographic chamber should be saturated with the stationary phase (e.g., water when using phenol). Some useful solvents are:

Phenol-saturated with water.
Butanol-ammonium hydroxide.
Acetone-water.
A "universal" solvent is:
    1-butanol-acetic acid- water   (shake and use the upper phase)
      40%   10%   50%

Sample Detection

After the separation, the chromatogram is removed from the chamber, the solvent front is marked with a soft pencil, and the paper is dried. Unless the substances are themselves colored, the spots must be visualized by a chemical reaction. This usually is suggested by the chemistry of the substances that are separated; for example:

Organic acids --------------- acid-base indicators
Reducing sugars ----------- silver nitrate NaOH
Amino acids ---------------- ninhydrin (triketohydrindene)

A variety of techniques have been developed to make the sample components visible. These are discussed in detail in Chapter 22, p. 262, on thin layer chromatography, and you are advised to read that section.

Evaluation of the Spots

The most commonly used factor is R_f.

$$R_f = \frac{\textit{Distance the solute front moves}}{\textit{Distance the mobile phase moves}} \qquad (21\text{-}1)$$

This is not an absolute value, unless all conditions are controlled. For example, the R_f values reported in the British literature for the amino acids may be slightly different from those obtained in Kansas during the summer. The reason is the high humidity in England, and the resultant higher water content of the paper. It is always best to spot knowns and unknowns on the same strip or sheet of paper. R_f tables are of value in judging the extent of resolution and

EXAMPLE CALCULATION

A sample of estrogens was separated using *o*-dichlorobenzene as the mobile phase. Two components were found when the spots were formed. The solvent fronts and solute fronts are shown below for both standards and unknowns. Determine the possible estrogens based on the experiments listed below.

	Solvent front mm	Solute front mm
Estriol	80	0
Estradiol	78	8
Equilenin	79	32
Estrone	77	69
Unknown 1	79	34
Unknown 2	79	68

ANSWER

Determine the R_f values for each spot.

Estriol	0/80	= 0
Estradiol	8/78	= 0.10
Equilenin	32/79	= 0.40
Estrone	69/77	= 0.90
Unknown 1	34/79	= 0.43
Unknown 2	68/79	= 0.86

The unknowns are most likely equilenin and estrone, but it would be wise to repeat the separation with a different mobile phase.

REVIEW QUESTIONS

1. What is meant by the "machine direction" on a piece of paper?
2. Why is HCl rather than NaOH used to dissolve sulfa drugs from a tablet?
3. What is one advantage of descending over ascending chromatography?
4. What is reversed-phase paper chromatography?
5. How do you reverse the phases for reversed-phase chromatography?
6. Why is a fingerprint on the paper a potential problem?
7. Why should standards be run at the same time as unknowns?
8. What causes tailing?
9. An analyst has a very low concentration extract. What technique can be used to concentrate the sample on a small spot?
10. Why mark the solvent front with a pencil mark before the mobile phase dries?
11. When using a mobile phase that formed two layers when it was prepared, it is considered good technique to use the nonmobile phase layer to establish equilibrium with the paper in descending chromatography. Why is this?
12. Why do some analysts place paper towels around the inside of equilibrium chambers?
13. Streaking, rather than spotting, is often done. Why would you want to do this?
14. What is the purpose of the small bar on the outside of the solvent tank in descending chromatography?
15. A separation is being performed, and the spots are moving slowly. What changes would you make to speed up the movement of the spots?
16. Why would you want to speed up the movement of the spots described above anyway?
17. What is R_f?
18. What technique errors might cause the developed spots to be not round?

IF IT EXISTS - CHEMISTRY IS INVOLVED
IF YOU CAN BUY IT - A CHEMIST WAS INVOLVED SOMEWHERE

22
THIN LAYER CHROMATOGRAPHY

PRINCIPLES

Column chromatography made separations possible that were otherwise impossible at the time, but complete separation usually required from 1 day to several weeks. In 1944, A.J.P. Martin, an English biochemist, showed that a piece of filter paper could be used instead of the column. The glass tube and the tightly packed columns could be eliminated, and the separation times were reduced, usually from a day or 2 to a few hours. One of the difficulties was that only cellulose could be used as an inert phase. Another difficulty was that the paper was made of fibers that were thicker in some parts than others, which caused irregular flow rates and poorer separations. The paper was destroyed easily by strong acids or bases, so the color-developing reagents had to be in mild solutions. These problems severely limited the types of compounds that could be separated. What was needed was a "paper" made from uniform-sized particles, regularly spaced, and capable of withstanding strong acids and bases.

The principle of what we now know as *thin layer chromatography* (TLC) was proposed first by N. Izmailov and M. Shaiber in 1938. They dusted Al_2O_3 (alumina) on glass plates. The particles were uniform and inert to acids, but the dust would blow or wash away unless extreme care was taken, so a photograph of the results had to be obtained. There was no way to store the plates.

TLC did not receive attention as an analytical technique until 1956 when Stahl, E., Schroter, G., Kraft, G., and Rentz, R. (*Pharmazie*, **11**, 633) described a method to produce uniform-thickness, thin layers of silica gel (SiO_2) held onto glass plates with 7-8% of plaster of Paris ($CaSO_4$). Their concepts formed the basis of our current procedures and reduced separation times to 10-30 minutes. Figure 22-1, p. 256, is a diagram of the apparatus commonly used. It is a small glass jar with a screw-cap lid and may or may not have filter paper placed around the walls on the inside. The sample, 1-2 drops, is placed on a single spot about 2 cm from one end of the plate, and the spot is indicated with a pencil mark. A scratch mark is made on the coating 12 cm from the bottom end. The developing solvent is added to the developing jar to a height of about 1 cm. The plate is inserted, sample end down, and the jar closed to help maintain a solvent-saturated environment. When the solvent has risen to the 12 cm mark by capillary action (10-30 min) the plate is removed. The solvent is allowed to evaporate or is dried with a hair dryer. The plate then is sprayed with a reagent that will form a color with the desired compounds. After spraying, colored spots become visible, and the distance they have moved compared to the solvent front is measured. Figure 22-2, p. 256, is a photograph of a more elaborate system including the developing tank, plate rack, and plate.

The mechanism for the separation is most likely a combination of adsorption and partition. If essentially round spots form that are separated from each other, then the separation is primarily partitioning. If tailing occurs, then adsorption is a major factor.

TECHNIQUES

(The contributions of Mr. Donald Meyers, FDA Drug Analyst, concerning the practical techniques and the experiments in Appendix A are appreciated.)

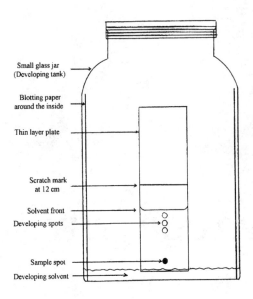

Figure 22-1. A diagram of a TLC apparatus.

Figure 22-2. A commercial TLC apparatus.
(Courtesy - Bodman Inc., Aston, PA)

PLATES

Most commercially prepared plates are 20 cm x 20 cm and are coated with a 0.25-mm-thick layer of adsorbent. The plate backing is usually glass, plastic, or aluminum. The coating is usually silica gel (SiO_2) or alumina (Al_2O_3); cellulose, modified cellulose, kieselghur, ion exchange resins, and octyldecylsilane on glass beads are used less often.

Glass plates are about 1-2 mm thick. They can be purchased prescored (cut with a glass cutter) every 2.5 or 5 cm so they can be separated easily into strips, if desired. Glass is easier to wet than aluminum or plastic and is preferred for systems that require a developing solvent with high water content and for systems that require being charred with sulfuric acid. The coating is held onto the plate with 7-8% of $CaSO_4$.

Aluminum plates are cut easily with a pair of scissors into any size, and alumina appears to bind well to this backing.

Plastic plates are about 0.2 mm thick and are made of inert and flexible polyethylene terephthalate. They are easy to cut, and modern sheets can be bent considerably without the coating flaking off. They are resistant to nearly all organic solvents but will dissolve in H_2SO_4 after long exposure and *tend to warp* if placed on a hot plate to dry the sample spots. Polyvinyl alcohol often is used as a binder.

To reduce edge effects, scrape 2-3 mm of coating from each side of the plate. Refer to Figure 22-3, p. 257. This should be done before using a plate, regardless of the size. This provides for a symmetrical and more reproducible solvent front and makes it easier to reproduce R_f values. Lay a ruler on the plate and with a spatula, razor blade, or your thumbnail, scrape off the rough, flaked edge.

To ensure a uniform final solvent front, make a scratch mark across the width of the plate 12 cm from the bottom end of the plate. Refer to Figure 22-3. Use the end of a spatula to do this. This mark will stop the solvent from rising beyond that distance. This allows you to be lax in watching the plate during development and yet produces an even final solvent front. You can be gone 15-20 minutes after the development has stopped, and the spots will not broaden significantly nor will the R_f change if the developing tank is sealed tightly.

Wash old plates with fresh developing solvent and dry them before using them. They often become brown around the edges from contaminants in the air. To salvage the plates, wash them with your developing solvent or methanol once a box of plates has been opened, dry them, and they can be used. Figure 22-4, p. 257, shows a plate carrier and drying rack.

Activating plates. If you prepare your own plates, they must be placed in an oven for 3-4 hours at 120 °C to remove the excess moisture. This is called *activation*. There is disagreement on whether activating is necessary or not with commercially prepared plates. The argument is that by the time you spot the sample and place it in the developing tank, the effects of activation have been lost. Most analysts will activate commercially prepared plates for 30 minutes at 110 °C before they use them. As a general rule, *if you are only going to do a few analyses, then activate the plates*. If you are going to repeat the analysis dozens or hundreds of times, then activate until you learn enough about the system to decide if activation is necessary or not.

NOTE: Plastic-backed plates will warp beyond repair if heated too hot or for too long. Try to use such plates without activation.

Thin layer chromatography 257

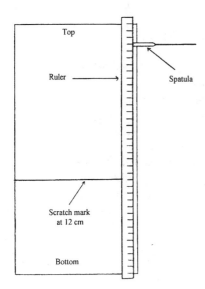

Figure 22-3. Preparing a TLC plate.

Figure 22-4. Plate carrying and drying rack.
(Courtesy - Bodman Inc., Aston, PA)

Figure 22-5 shows a drying oven that will hold one rack of plates. Figure 22-6 shows a glass-type desiccator, and Figure 22-7, p. 258, a cabinet-type desiccator used to keep the plates dry.

COATINGS

The coating consists of a layer of adsorbent usually 0.25 mm thick, although you can purchase plates with the layers 0.10 to 0.50 mm thick.

Silica gel. Silica gel ($SiO_2 \cdot X\ H_2O$) is prepared by gelling sodium silicate in a mineral acid. This produces small particles with a high surface area - 750-800 m^2/g with pore sizes of 2.2-2.6 nm. Silica gel is the favorite for most separations (see Table 22-1, p. 259). The resulting spots are more intense and do not fade as fast as they do on alumina. Most coatings use particles of 0.04 mm diameter.

Figure 22-5. Drying oven.
(Courtesy - Bodman Inc., Aston, PA)

Figure 22-6. A glass type desiccator cabinet.
(Courtesy - Bodman Inc., Aston, PA)

Figure 22-7. A cabinet type desiccator.
(Courtesy - Bodman Inc., Aston, PA)

Alumina. Alumina has a general formula Al_2O_3, although it usually exists as the trihydrate. Activated aluminas are prepared by dehydrating aluminum hydroxide (Gibbsite) at 370-400 °C in a current of air. The surface areas are 180 m^2/g (type E) and 90 m^2/g (type T), and the pore sizes range up to 10 μm.

Kieselghur. Another name for diatomaceous earth is kieselghur. Diatomaceous earth was formed by the slow accumulation of the shells of very small algae called diatoms. Their shells consist mainly of silica in the opaline form. Figure 15-8 in Chapter 15 showed several types of diatoms. The product looks like chalk and consists of fine-grained, porous material. Individual diatoms range in size from 0.002 to 2 mm, with the average being about 0.2 mm. This material is used primarily for "preabsorbent" coatings described later.

Cellulose. Cellulose is a long chain polymer of saccharides connected by β-1,4-linkages.

The OH groups can be reacted with other groups such as acetyl and carboxymethyl to produce modified celluloses and vary the activity. A 0.1 mm layer on a plate usually will expand to 0.25 mm when wetted. Binders usually are not required with cellulose. Microcrystalline cellulose has been cut, so there are no long fibers to form a preferred direction of flow.

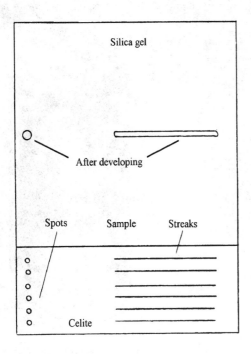

Figure 22-8. A preadsorbent plate coated with Celite and silica gel G.

Reverse phase, C-18. This coating consists of 0.05 mm diameter porous glass beads with a layer of octadecylsilane bonded to the surface. The purpose of the coating is to duplicate as closely as possible the column conditions for high pressure liquid chromatography (HPLC) regarding the solvent selection. TLC is much faster than HPLC but is difficult to quantitate. These plates contain 19 rows of 1 cm x 20 cm.

Preadsorbent Plate. This type of plate has two adsorbents, one covering a small portion of the plate, and the second covering the rest of the plate. Figure 22-8 shows one arrangement for a Celite-silica gel plate. This arrangement is used for concentrating dilute solutions or for obtaining large amounts of material (preparative TLC).

Celite. Celite has very little retention capacity, and any compound placed on it will move with the solvent front.

The left side of Figure 22-8 shows how such a combination is used for preconcentration. If a solution is very dilute and it would either take too long to concentrate or could be destroyed by heat during an evaporation step, then several drops of the sample are placed in a row on the Celite and thoroughly dried When the developing solvent passes over these spots, the sample moves with the solvent front and concentrates the sample. Only one spot will appear on the silica gel portion for each compound, regardless of the number of spots placed on the Celite.

The right side of Figure 22-8 shows how to use this plate for prep scale work. If it is desired to separate large amounts of a particular compound for further analysis (prep scale), then the technique is to place several streaks of the sample on the Celite and develop as usual. Only one line will appear on the silica gel for each compound. It may be wide because of overloading of the plate, but it will be much narrower than the original streaks combined.

Table 22-1 lists the types of compounds for which the various coatings are used.

Table 22-1. Plate coatings and the types of compounds separated
(Courtesy: J.T. Baker Co., Phillipsburg, NJ)

Coating	For the Separation of
Silica gel	Alcohols, aldehydes, alkaloids, amines, amino acids, amphetamines, antibiotics, antioxidants, barbiturates, carbohydrates, flavonoids, herbicides, heterocyclic compounds, hydrocarbons, indoles, insecticides, ketones, lipids, nitro compounds, organic acids, peroxides, pesticides, phenols, plasticizers, polypeptides, steroids, terpenes, unsaturated compounds, vitamins.
Aluminum oxide	Alcohols, alkaloids, amines (strongly basic), antioxidants, herbicides, heterocyclics, peptides, polyalcohols, steroids, terpenes, vitamins.
Celluloses	Amines, amino acids, antibiotics, carbohydrates, glycosides, hydrocarbons, inorganic ions, nucleic acids, organic acids, peptides, urea derivatives, vitamins
Modified Celluloses	
AC-10	Antioxidants
CM	Inorganics, metal ions
DEAE	Nucleoside mono-, di-, and tri-phosphates
ECTEOLA	Purines, pyrimidines, nucleosides
PEI	Monophosphate nucleosides

SPOTTING TECHNIQUES

Samples usually are placed on the plates in the form of small drops. These are called *spots*. Figure 22-9 shows one way to spot a plate with a capillary tube.

Microliter pipets (called lambda pipets; 1 lambda, λ = 1 µL) are preferred to syringes for adding samples to the plates, because they are less likely to scratch the coating off of the plate.

Figure 22-10, p. 260, shows a microliter pipet and syringe. *Use a volatile solvent in which the compound of interest is readily soluble.* This allows the solvent to be evaporated rapidly when the sample solution is applied to the plate. The resultant spot will have a smaller diameter. *Thoroughly dry each drop of sample before another drop is placed on top of it.* This keeps the original spot small while increasing the concentration, and much better separations can be obtained. A hair dryer is preferred for the drying step. Laying the plate on a warm hotplate is not recommended, because plastic plates tend to warp and some samples decompose.

Concentrate the sample as much as possible (just before precipitation occurs) before you spot the sample on the plate. Use 1-5 µL per spot.

Figure 22-9. Spotting a plate using a capillary tube.
(Courtesy - Bodman Inc., Aston, PA).

Always spot 2 cm from the bottom of the plate and place a small dot with a pencil alongside the spot. Figure 22-11 shows a spotting guide that can be used. *Be sure the sample spot is completely dry before development with the solvent is begun.* Figure 22-12 shows what happens if developing solvent is added before the sample spot is dry. The solvent tends to go around the sample spot initially, and you often get a skewed solvent front and will have trouble obtaining a good R_f value.

Preparative TLC

A common technique for organic chemists is to streak the plate rather than spot it. A strip of sample is placed on the plate, and the plate developed as before. As much as 30 mg can be cleaned up at one time by streaking. A syringe is used commonly, but the streaking pipet shown earlier in Figure 21-6 provides a more uniform streak. Once the compound is separated, the coating is scraped off with a spatula, collected, and extracted from the coating. This technique is so common that an automatic streaking apparatus is available. One type is shown in Figure 22-13, p. 261.

OVERLOADING

Overloading is caused by placing too much sample on the coating. This causes the coating particles to become saturated and results in poor separation. It can be detected as shown in Figure 22-14, p. 261. If overloading is present, the developed spot will be elongated and of fairly uniform concentration. If the spot is elongated with a sharp front and a gradual decrease in concentration, this is called *tailing* and is due to a lack of equilibrium, not excessive concentration.

DEVELOPING TANKS

The purpose of the developing tank is to cause the environment around the plate and the coating to become saturated with the vapors from the developing solvent. This results in less tailing and produces a more even solvent front, so the solvent moves faster up the plate. A developing tank for a 20 x 20 cm plate is shown in Figure 22-15, p. 261. A small screw-cap jar may also be used for smaller plates.

Place only enough solvent in the tank to cover the bottom 1 cm of the plate.

Wrap absorbent paper around the sides of the tank. When this absorbs the developing solvent, it will help saturate the inside of the tank. Leave a 2 to 4 cm opening so you can see the plate inside.

Never let the plate touch the absorbent paper, or the solvent in the paper may begin to siphon down the plate.

Wait at least 1 hour for the tank to become saturated before you begin a development.

Wipe away any drops of condensate from the tank cover or top of the side walls. If they fall onto the plate, they can ruin an analysis.

Keep the tank cover on tightly, particularly if you have scratched the coating at the 12 cm mark. This keeps the solvent from evaporating at the scratch mark and stops the capillary action. As a result, the R_f does not change.

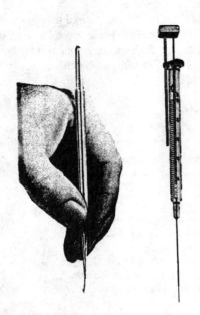

Figure 22-10. Left, a microliter pipet. Right, a microliter syringe.
(Courtesy - Bodman Inc., Aston, PA).

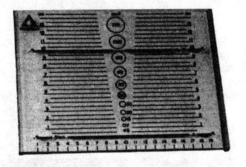

Figure 22-11. A commercial spotting guide.
(Courtesy - Bodman Inc., Aston, PA)

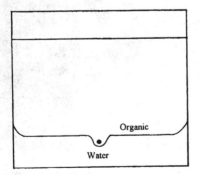

Figure 22-12. The effect of developing a wet sample spot.

Figure 22-13. Preparative TLC sample streaker. (Courtesy - Alltech Inc., Deerfield, IL)

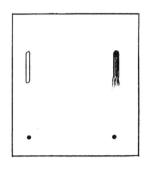

Figure 22-14. Left, Overloading. Right, Tailing.

DEVELOPING SOLVENTS

Most developing solvents are either mixtures of organic liquids or organic liquids saturated with water. The composition is varied to alter the polarity of the solvent.

Some solvents used singly or in combination are methanol, ethanol, butanol, benzene, toluene, hexane, heptane, chloroform, acetonitrile, acetic acid, diethylether, and water.

As examples; the separation of aspirin, caffeine, and phenacetin (APC tablets) is accomplished by a mobile phase of methanol-acetic acid-ether-benzene (1 : 18 : 60 : 120) and nitroglycerine, a vasodilator, present in heart medication capsules can be separated from the other capsule materials by benzene.

For the best results, never use a mixed solvent for more than two successive plates. The composition of the developing solvent changes and causes the R_f value to vary.

For very close separations, use freshly prepared solvents and change with each plate.

In two-dimensional separations, use the most polar solvent combination first and thoroughly dry the plate before the second solvent is used.

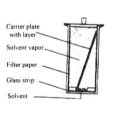

Figure 22-15. A commercial developing tank. (Courtesy - Bodman Inc., Aston, PA)

Table 22-2 lists most of the solvents used for TLC according to their eluting power.

Table 22-2. Elutropic (elution strength) solvent series
(Courtesy: Merck Co., Rahway, NJ)

Weakest	n-Heptane	Diethyl ether	
	n-Hexane	i-Butanol	
	n-Pentane	Acetonitrile	
	Cyclohexane	i-*Butyl* methyl ketone	
	Carbon disulfide	2-Propanol	
	Carbon tetrachloride	Ethyl acetate	
	Trichloroethylene	1-Propanol	
	Xylene	Ethyl methyl ketone	
	Toluene	Acetone	
	Benzene	Ethanol	
	Chloroform	Dioxane	
	Dichloromethane	Tetrahydrofuran	
	Di-*iso*-propyl ether	Methanol	
	t-Butanol	Pyridine	Strongest

SAMPLE DETECTION

When a sample is separated into its component compounds on a TLC plate, each compound exists as a round or oval spot. With few exceptions, such as the dyes in inks, lipsticks, foods, or gasolines, these spots are invisible. Three major methods are used to make these spots visible: spray reagents, fluorescence, and charring.

Note: Several compounds may occur on the coating, *but only those that react with the color reagent will become visible.*

Spray Reagents

Some reagents will react with the functional groups on compounds to produce a color or a fluorescence. For example, ninhydrin will form a blue-violet color with the amine group. The reagent is dissolved in a solvent and applied to the coating by spraying a fine mist onto it, hence the term spray reagents. Table 22-3 lists several reagents and the classes of compounds with which they react.

Table 22-3. Color developing reagents for paper and thin layer chromatography

Reagent	Compounds Detected
Ninhydrin	Amino acids, amines
4-Dimethylaminobenzaldehyde	Amino sugars, indole derivatives, ergot alkaloids.
Phosphomolybdic acid	Any easily reducible substance
Bromcresol green	Acids and bases
Rhodamine B	Insecticides
2,7-Dichlorofluorescein	Lipids
Aniline phthalate	Reducing sugars
2,4-Dinitrophenylhydrazine	Aldehydes and ketones
Iodoplatinate	Most drugs

Figure 22-16 shows two types of sprayers. The first is an atomizer operated by a hand-held squeeze bulb. The second operates on a propellent stored in a pressurized can.

Figure 22-17 shows a cardboard disposable-type spray chamber, and Figure 22-18, p. 263, shows a commercial spray chamber used to contain the excess mist.

When spraying plates, spray past the edge before you return. Figure 22-19, p. 263, shows both the correct and incorrect ways to spray a plate. Always spray past the sides of the plate before you reverse direction. This prevents a doubling of the reagent concentration along the edges and reduces the probability of the reagent dripping.

Figure 22-16. Two types of sprayers for TLC. Left, squeeze bulb type. Right, pressure can type.
(Courtesy - Bodman Inc., Aston, PA)

Figure 22-17. Spray chambers for TLC.
(Courtesy - Analtech Inc., Newark, DE)

Fluorescence

Two methods exist: natural fluorescence and fluorescent-coated adsorbents.

Natural fluorescence. The plate is placed in a small box (Figure 22-20) and irradiated with either short wavelength (254 nm) or long wavelength (366 nm) ultraviolet radiation. Those compounds that fluoresce produce a variety of colors. Many other compounds can be made to fluoresce by spraying the coating with acids, bases, or the above-mentioned reagents. There are still many compounds that cannot be detected by this technique. A way to detect almost any compound by fluorescence is described below.

Fluorescent plates. Inorganic phosphors are added to the TLC adsorbent. These include manganese- activated zinc silicate (254 nm), zinc-cadmium sulfide (254/366 nm), and lead-manganese activated calcium silicate (254 nm). Occasionally, organic compounds are added as phosphors, but they do not withstand treatment with strong acids and bases as well as the inorganic compounds. When these compounds are exposed to ultraviolet radiation, the entire plate glows with a blue-green fluorescence. Those sample compounds that do not fluoresce cover this background fluorescence and appear as dark spots. Those compounds that fluoresce naturally add their fluorescence to that on the coating and produce a colored spot.

Charring

Charring is to be used as a last resort. Concentrated sulfuric acid (H_2SO_4) is sprayed over the plate. Sulfuric acid reacts with organic compounds and dehydrates them to a charcoal black residue - a char. Nearly all organic compounds are detectable, and this is an advantage. The disadvantage is that the compounds are destroyed in the process. This technique should be used with care on plastic-backed plates, and the spraying should be done in a hood to contain the corrosive spray droplets.

NOTE: The detection limits for TLC are usually in the 0.1 to 1 µg region.

QUANTITATION

Semiquantitative estimation can be made by comparing the intensity of developed spots with a series of standards spotted on the same plate. Alternatively, the spot can be scraped from the plate, the compound extracted, and the amount determined by uv/vis spectroscopy. A better method is to place the plate in a *scanner*, which measures the reflectance of the spot, usually by uv/vis spectroscopy.

MULTIPLE DEVELOPMENT

An improved technique has been developed to separate those compounds that are separated only slightly even when using the best technique. It is called *multiple development*. Although it requires more time, a much better separation can be obtained and the sample spot is more uniform, so quantitative measurements are easier to make.

Figure 22-18. A spray chamber for TLC.
(Courtesy - Bodman Inc, Aston, PA)

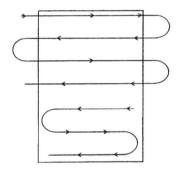

Figure 22-19. The correct way (top) and the incorrect way (bottom) to spray a TLC plate.

Figure 22-20. Irradiation chamber for viewing fluorescence on TLC plates.
(Courtesy - Brinkman Instruments Co., Westbury, N.Y.)

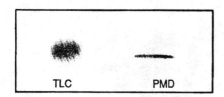

Figure 22-21. Left, regular development. Right, multiple development.

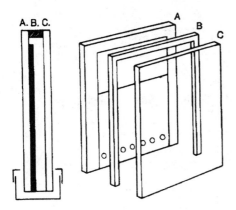

Figure 22-22. Plate arrangement: (A) TLC plate. (B) Spacer. (C) Cover plate.
(Courtesy - Regis Chemical Co., Morton Grove IL)

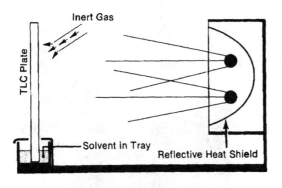

Figure 22-23. Diagram of the heating arrangement.
(Courtesy - Regis Chemical Co., Morton, IL)

The sample is either spotted or streaked as before. The developing solvent is allowed to pass over the sample until the front is about 2 cm past the original spot. The plate is removed from the solvent and dried with a stream of warm air (hair dryer). The plate is reinserted into the tank and developed again, this time allowing the solvent to rise about 4 cm above the spot. Again the plate is removed and dried. This is repeated 3 to 5 times, allowing the solvent to rise farther each time.

The result is a series of elongated ellipses rather than round spots. Figure 22-21 shows the difference between a regular TLC and a multiple-developed TLC.

The reason for the elongation is that once the solvent is removed by drying, the sample molecules do not move until more solvent is added. The second time the plate is dipped in the solvent, the solvent will start the lower sample molecules moving upward, but the top molecules do not move because there is no solvent to cause them to move. This in effect brings the trailing molecules up to the same place as the faster-moving molecules. If the process is repeated several times, the result is an ellipse, rounded at the top, but more concentrated and better separated from the other compounds. Separation is almost always improved if the R_f's of the components to be separated are below 0.5. If the R_f's are above 0.7, then little improvement is obtained, because the sample tends to run with the solvent front.

PROGRAMMED MULTIPLE-DEVELOPMENT TLC

Multiple-development TLC can be automated to remove much of the work and provide a nearly elliptical sample spot. This is called *programmed multiple-development* TLC. The following description is through the courtesy of the Regis Chemical Co.

In PMD, the plate is not removed from the solvent. The solvent is evaporated by a stream of hot air and/or radiant heat. Figures 22-22 and 22-23 show the plate arrangement and the recycling apparatus. The apparatus can be set for any number of cycles from 1 to 99 with each cycle varying from 10 to 100 seconds. Following each development, controlled evaporation causes the solvent front to recede, usually to or below the point of initial spotting. Because the solvent is dried by radiant heat or gas flow while the plate is still in contact with the solvent reservoir, the solvent always flows forward, even while the solvent front is receding.

As the receding front passes through each of the chromatogram spots in turn, the lower molecules of the spot are swept forward after the leading molecules have stopped. "Spot reconcentration" thus occurs twice during each cycle of a PMD run. After the last preset cycle, continued controlled evaporation prevents further development of the plate.

TWO-DIMENSIONAL THIN LAYER CHROMATOGRAPHY

The process of two-dimensional TLC involves first using a polar developing solvent, then after drying the plate, turning it 90° and using a second developing solvent that is of different polarity than the first. Two-dimensional chromatography is useful in two situations: (1) when the mixture is so complex that the components are not all separated by one solvent, and (2) when a more positive identification is required for one or more of the components. Figure 22-24 shows how the plate is scratched and where the sample and standards are spotted.

The sample is spotted 2 cm in and up from one corner, and the first standards are spotted to the right of the vertical scratch mark, usually 1 cm apart and 2 cm from the bottom of the plate. The plate is developed as usual and, in this case, produces the separation indicated by the O's.

Note: the plate is not sprayed to develop the sample spots. The plate is dried thoroughly and a second set of standards is spotted in what is now the top left quadrant of the plate. The plate is turned 90° so the left side of the plate becomes the bottom and the plate is developed with a second solvent. The plate is dried and sprayed with color developing reagents. The results of using the second solvent are shown in Figure 22-24 by ●'s. An identification is obtained if the dashed lines from the standards intersect with a spot of the unknown.

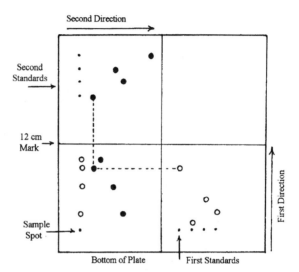

Figure 22-24. A diagram of the two-dimensional TLC process. Sample and standard spots: ○ after the first development; ● after the second development.

REVIEW QUESTIONS

1. What are the advantages of using 200-mesh silica gel or alumina for an inert phase rather than paper?
2. How do you keep the inert phase on a glass plate?
3. Why is the developing tank usually covered?
4. What may be the cause of spots that smear to the rear?
5. Why should you scrape the edge of the coating on a new plate?
6. Why are plates activated?
7. Of what value is a spotting guide and what are the circles in the center for?
8. How can you tell if you developed a wet spot?
9. How can you tell if you overloaded the coating?
10. Why is it good technique never to use a mixed developing solvent for more than two separations?
11. Table 22-2, p. 261, lists several solvents. What is meant by "increased eluting power"?
12. If you had a choice to use either a squeeze bulb type sprayer or a pressurized can type, which would you choose and why?
13. Why should you spray past the edge of a plate before you spray in the opposite direction?
14. Explain how multiple development changes a round spot into an ellipse.
15. When should you consider using two-dimensional TLC?
16. Why should spraying with sulfuric acid be considered a last resort?
17. What is the advantage of making a scratch mark in the coating 10-12 cm above the spot?
18. Look in a current chemical catalog and determine how much it would cost to prepare 200 mL of the iodoplatinate reagent.
19. Name two phosphors that are added to coatings to make them fluoresce.
20. As a general rule, what are the detection limits for TLC?

IF IT EXISTS - CHEMISTRY IS INVOLVED
IF YOU CAN BUY IT - A CHEMIST WAS INVOLVED SOMEWHERE

SEPARATIONS INVOLVING ION EXCHANGE RESINS

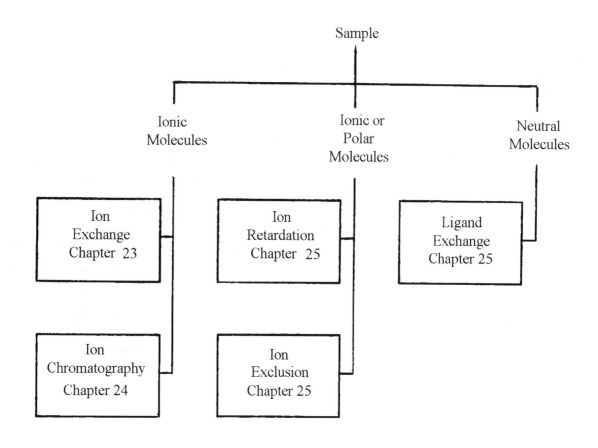

The observation that some of the compounds in soil could exchange ions, particularly cations, based on their size and charge, was made about 1850. Those compounds were the natural zeolites in the soil. They were inorganic compounds, and synthetic zeolites were eventually made. Their capacity was low; they were heat and pH sensitive, and did not work well with organic solvents. In 1935, organic polymers were prepared that had the same ion exchange ability, but were more rugged. In 1946, the preparation of the polymers in bead form provided the means for what is now known as *ion exchange*.

If you place a quantity of resin beads in the H^+ form in a beaker, add a solution of K^+, Ca^{+2} and Fe^{+3} ions, and stir them together, the K^+, Ca^{+2} and Fe^{+3} ions will exchange with the H^+ ions. This is known as *ion exchange* (see Chapter 23). If the beads are placed in a narrow diameter column, and the solution of K^+, Ca^{+2} and Fe^{+3} ions is passed down the column, the H^+ will again be exchanged. If then, a solution containing H^+ ions in a higher concentration is passed down the column, the K^+ will be exchanged first, the Ca^{+2} second, and the Fe^{+3} last, coming off of the column in that order. This is known as *ion exchange chromatography*. It is hard to detect the cations in the presence of the acid, so this technique was seldom used. However, if a second column (anion exchanger in the OH^- form) is placed below the first column, the H^+ is neutralized, and the other cations can be detected electrochemically. This is called *ion chromatography* and is presented in Chapter 24.

If chains of a cation exchange resin and chains of an anion exchange resin are captured in a porous polymer bead they tend to line up with their charges opposite to each other. A mixture of ions or polar compounds passing through the beads will be retarded. This *ion retardation* (Chapter 25) is an excellent way to separate ionic from nonionic compounds without having an exchange take place.

Nonionic compounds permeate ion exchange resin beads easier than do ionic compounds because of surface effects. Therefore, if repeated additions of sample and water are passed down the column there is an *ion exclusion*, discussed in Chapter 25.

Perhaps the most innovative use of ion exchange resins is to separate nonionic compounds by what is known as *ligand exchange* (see Chapter 25). As an example, a resin in the -COOH form is exchanged with $Ni(H_2O)_6^{+2}$. The Ni^{+2} coordinates with the $-COO^-$ octahedrally, leaving 4 H_2O molecules as ligands. These can then be exchanged with other ligands, such as NH_3, $H_2NCH_2CH_2NH_2$, etc.

23
ION EXCHANGE

PRINCIPLES

Ion exchange is a process in which *one type of ion in a compound is exchanged for a different type; cation for cation and anion for anion.* For example, calcium, iron, and magnesium ions can be removed from water by being exchanged for sodium ion. This can be done in a bucket or a bag, much like a home water softener

Ion exchange chromatography is an ion exchange process in which *the desired ions are exchanged in sequence and are eluted from a column much the same as compounds are eluted from a column in column chromatography.* Figure 23-1, p. 270, illustrates the basic process.

Ion chromatography is the same as ion exchange chromatography, except that it includes a provision for removing the ions in the eluting agent, so sensitive electrochemical detection is possible.

The equipment used in ion exchange separations is similar to that used in column chromatography and can be identical for most separations. Figure 23-2, p. 270, is a diagram of an apparatus with a water jacket and a means to prevent the column from going dry. This is an elaborate apparatus. Most of the time, a broken buret is used.

Professor W. Reinman of Rutgers University has stated that probably the first published account of what we now know as ion exchange was the biblical account of Moses at Marah, who "found a suitable tree which when he had cast it into the water the waters were made sweet." The explanation is that the oxidized cellulose exchanged with the salty tasting ions in the water.

Records as far back as Aristotle (384-322 B.C.) indicate that soils and sand filters were used to treat impure drinking water. Sir Roger Bacon (1561-1626), an English philosopher, observed in the early 1600's that "salt water passed through earth through 10 vessels, one within another, and yet it hath not lost its saltiness as to become potable, but when drayned through 20 vessels hath become fresh." The first systematic studies were made by Thompson, H.S.M., (*J. R. Agr. Soc. Engl.*, **11**, 313, 1850) and Way, J.T. (*J. R. Agr. Soc. Engl.*, **11**, 197, 1850; **13**, 123, 1852). H.S.M. Thompson, an English agricultural chemist, wanted to know why $(NH_4)_2SO_4$ and KCl did not wash out of soil. He observed that if these compounds were passed through a column of soil, $CaSO_4$ and $CaCl_2$ would be obtained in the filtrate. Two years later, he mentioned this to J. T. Way, a consulting chemist for the Royal Agricultural Society. Way began systematic studies of the process, which was called *base exchange* because of the basic character of the exchanged elements.

Way's observations, listed below from his paper on "The Ability of Soils to Absorb Manure," are still valid today.
1. The exchange of calcium and ammonia in soils noted by Thompson was verified.
2. Exchange of compounds in soils involved the exchange of equivalent quantities.
3. Certain compounds were exchanged more readily than others.
4. The extent of the exchange increased with concentration, reaching a leveling off value.
5. The temperature coefficient for the rate of exchange was lower than that of a true chemical reaction.
6. The aluminum silicates present in soils were responsible for the exchange.

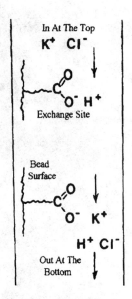

Figure 23-1. Diagram of the ion exchange process.

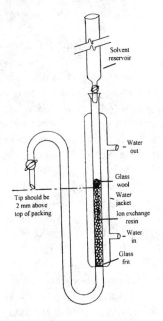

Figure 23-2. Diagram of an elaborate ion exchange column.

7. Heat treatment destroyed the soil.
8. Exchange materials could be synthesized from soluble silicates and alums.
9. Exchange of materials differed from true physical adsorption.

E. Lemberg, a German chemist, clearly established the reversibility and stoichiometry of the process in 1876 (*Z. Deut. Geol. Ges.*, **28**, 519); when he passed NaCl over *leucite* ($K_2O \cdot Al_2O_3 \cdot 4SiO_4$) and converted it to *analcite* ($Na_2O \cdot Al_2O_3 \cdot 4SiO_2 \cdot 2H_2O$) and then reversed the process when he added KCl.

The first attempt to employ ion exchange for commercial purposes was in 1896 by F. Harm, a German agricultural chemist, to remove Na^+ and K^+ from sugar beet juice. The first synthetic ion exchange compounds were aluminum silicates called *zeolites* and were available commercially about 1903.

The first truly successful use of ion exchange was in 1905, when R. Gans used a natural zeolite for water softening (Ger. Patent 174,097, 1906).

$$Na_2Al_2Si_3O_{10} + CaCO_3 = CaAl_2Si_3O_{10} + Na_2CO_3 \qquad (23\text{-}1)$$

Figure 23-3. p. 271, shows a diagram of a portion of a natural inorganic ion exchange compound, sodalite.

Inorganic type exchangers had low capacity and were quite sensitive to temperature and pH. In 1935, the English water chemist, Basil A. Adams suggested to Eric Leighton Holmes, an organic chemist, the idea of making an organic polymer type of ion exchanger. They condensed polyhydric phenols or phenol sulfonic acids with formaldehyde (*J. Soc. Chem. Ind. London*, **54**, 1 T, 1935).

These compounds had both increased capacity and stability. The next two most important developments were due to D'Alelio and Mark. In 1944, Gaetano F. D'Alelio, of General Electric Co., prepared the first styrene-divinyl benzene polymers (U.S. Patents 2,340,110 and 2,340,111). These allowed various porosities of material to be prepared and provided a backbone for various types of cation and anion exchange groups to be added. (See the reaction on p. 271.)

Until the work of W.P. Hohenstein and Herman Mark (*J. Polymer Sci.*, **1**, 127) at the University of Michigan in 1946, polymers were obtained in large chunks and had to be crushed or ground to obtain small particles and large surface areas. They found that small round beads of resin (Figure 23-4, p. 271) could be obtained if a dispersing agent such as polyvinyl alcohol was added during the polymerization. The resulting white beads were like small pearls, and the process is known as *pearl polymerization*.

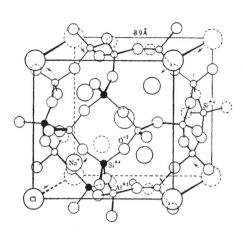

Figure 23-3. Portion of the aluminum silicate structure of sodalite present in chabazite.
(Courtesy - Hendricks, S.B., *Ind. Eng. Chem.*, **37**, 625, 1945)

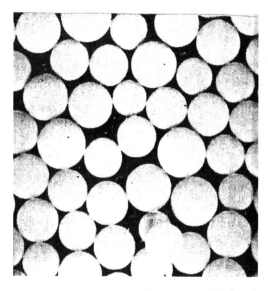

Figure 23-4. Cation ion exchange resin beads (20-40 mesh).
(Courtesy - Rohm and Haas, Philadelphia, PA)

X [phenol] + Y [phenol-SO₃⁻H⁺] + H–C(=O)–H ⟶ HO–[benzene]–CH₂–[benzene]–OH (with SO₃⁻H⁺)

The following represents the basic process for making a strong acid cation exchange resin. A 12:1 water to monomer ratio is common. The catalyst is 1 %, and 100 mL of 1% dispersant is used per 250 mL of monomer. A 2 L mixture would be stirred at 740 rpm for 5 hours at 80 °C.

Styrene (backbone) + Divinyl Benzene (cross-linker) + Benzoyl Peroxide (catalyst) + Polyvinyl Alcohol (dispersant) ⟶

[cross-linked polystyrene structure] $\xrightarrow{H_2SO_4}$

[sulfonated polystyrene structure with SO₃H groups]

Table 23-1 shows several types of functional groups that can be added to the basic polymer backbone arranged in their order of activity.

Table 23-1. Exchange functional groups

	Cation Exchangers	Anion Exchangers
Strong	$-SO_3^-H^+$	$-N^+(CH_3)_3, Cl^-$
	$-COO^-H^+$	$-N^+-(CH_3)_2 CH_2OH, Cl^-$
	$-CH_2SO_3^-H^+$	
	$-O^-H^+$	$-N^+R_2H, Cl^-$
	$-S^-H^+$	$-N^+RH_2, Cl^-$
Weak		$-N^+H_3, Cl^-$
	$-HPO_2^-H^+$	

LABEL INFORMATION

The label on a bottle of commercial ion exchange resin generally contains the following information: strongly acidic, sulfonic acid, Na^+, 20-50 mesh, medium porosity or 8X, 4.4 meq/g min dry. What does this mean?

This is a cation exchange resin in the sulfonic acid form. The cation already on the resin is sodium. The beads will all pass through a 20 mesh sieve, but not a 50 mesh sieve. 8X refers to the degree of cross-linking. It means that 8% divinyl benzene was added to the starting mixture, not that it is 8% cross linked. This provides a medium porosity resin. One gram of the dry resin has an exchange capacity of at least 4.4 meq.

Table 23-2 relates cross-linking, porosity, and moisture holding capacity.

Table 23-2. Approximate relations among crosslinking, porosity, and moisture holding capacity

Cross-linking	Porosity	Moisture Holding Capacity %
2X	High	85-95
4X	Intermediate high	58-65
8X	Medium	44-48
12X	Intermediate low	40-44
16X	Low	37-41

GENERAL RULES FOR ION EXCHANGE AFFINITY

1. At low concentrations and ordinary temperatures, the extent of exchange increases with increasing valency of the exchanging ion.
$$Na^+ < Ca^{+2} < Al^{+3} < Th^{+4}$$

2. At low concentrations, ordinary temperatures, and constant valence, the extent of exchange increases with increasing atomic number of the exchanging ion.
$$Li^+ < Na^+ < K^+ = NH_4^+ < Rb^+ < Cs^+ < Ag^+ < Be^{+2} < Mn^{+2} < Mg^{+2}$$
$$= Zn^{+2} < Cu^{+2} = Ni^{+2} < Co^{+2} < Ca^{+2} < Sr^{+2} < Ba^{+2}$$

3. At high concentrations, the differences in the exchange potential of ions of different valences (Na^+, Ca^{+2}) diminishes, and in some cases reverses. This is how resins can be regenerated.

4. At high temperature, in nonaqueous media, or at high concentration, the exchange potentials of ions of similar charge become quite similar and even reverse.

5. The relative exchange potentials of various ions can be approximated from their activity coefficients; the higher the activity coefficient, the greater the exchange is.

6. The exchange potentials of the H^+ and OH^- ions varys considerably with the nature of the functional group and depend upon the strength of the acid or base formed between the functional group and either the hydrogen or hydroxyl ion. The stronger the acid or base, the lower the exchange potential is.

7. For a weak base exchanger, the order of exchange usually is:
   $OH^- > SO_4^= > CrO_4^= > Citrate > Tartrate > NO_3^- > AsO_4^= > PO_4^= > MoO_4^= > Acetate = I^- = Br^- > Cl^- > F^-$
8. For strong bases, it is generally the same except for OH^-.

Olof Samuelson (*Z. Anal. Chem.*, **116**, 328, 1939) was the first to systematically examine ion exchange as a technique for chemical analysis.

ION EXCHANGE KINETICS

It is believed that ion exchange occurs within the bead because of the small outer surface and rather high capacity (2-10 meq/g). The ion exchange process has been broken into five possible steps (Figure 23-5) in an attempt to determine which step controls the rate of the exchange.

Steps 1 and 5 are not rate determining if the solution is agitated, such as happens when the solvent moves through the column. Step 3, the actual exchange, is believed to be instantaneous and, therefore, is not a limiting factor.

Step 4, the migration through the bead to the exchange site, depends on the degree of cross-linking and the concentration of the solution. If the solution concentration is greater than 0.1 M, then this step controls the rate of exchange.

Step 2, the passage from the outer solution into the bead matrix is believed to control the rate, if the solution concentration is less than 0.001 M. For concentration ranges between 0.001 and 0.1 M, a combination of steps 2 and 4 controls the rate.

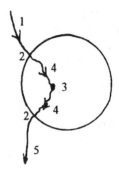

Figure 23-5. The steps in ion exchange kinetics.

THE DONNAN MEMBRANE THEORY

The Donnan membrane theory was developed in 1911 to explain the unequal diffusion of ions across a semipermeable membrane and is accepted as a reasonable explanation of the exchange mechanism for the organic type resins. The solution-bead surface interface is deemed to be similar to the membrane in the Donnan theory.

Remember - the driving force is the Second Law of Thermodynamics that states that systems tend to maximum disorder (entropy), or ions at high concentration migrate to regions of lower concentration.

The following material was abstracted from E.W. Berg, *Physical and Chemical Methods of Separation*, McGraw-Hill, New York, NY, 1963.

If solutions of NaCl and NaR are separated by a membrane permeable to sodium and chloride ions but not to the anion R^- of the other sodium salt, it is possible to derive an exchange expression. When equilibrium has been established, a certain amount of sodium and chloride will have diffused through the membrane and the chemical potential of the substances on both sides of the membrane must be equal:

$$\mu NaCl_I = \mu NaCl_{II} \tag{23-2}$$

where I and II refer to the two solutions. But since the chemical potential of the electrolyte can be taken as the sum of the potentials of the ions,

and
$$\mu°Na^+ + RT \ln \alpha Na^+_I + RT \ln \alpha Cl^-_I = \mu°Na^+ + RT \ln \alpha Na^+_{II} + RT \ln \alpha Cl^-_{II} \tag{23-3}$$

$$\alpha Na^+_I \, \alpha Cl^-_I = \alpha Na^+_{II} \, \alpha Cl^-_{II} \tag{23-4}$$

where α represents the activity of the indicated ions. If it is assumed that the solutions are dilute, activities can be replaced with concentrations and

$$[Na^+]_I [Cl^-]_I = [Cl^-]_I^2 = [Na^+]_{II}[Cl^-]_{II} = \{[Cl^-]_{II} + [R^-]_{II}\}[Cl^-]_{II} \tag{23-5}$$

provided the conditions of electroneutrality apply.

and
$$[Cl^-]_I^2 = [Cl^-]_{II}^2 + [Cl^-]_{II}[R^-]_{II} \tag{23-6}$$

$$[Cl^-]_I > [Cl^-]_{II} \tag{23-7}$$

The implication is that at equilibrium the concentration of diffusible electrolyte is greater in the solution free of nondiffusible ion.

If a second diffusible cation, H^+, is added to the system, then

and
$$[Na^+]_I[Cl^-]_I = [Na^+]_{II}[Cl^-]_{II} \qquad (23\text{-}8)$$
$$[H^+]_I[Cl^-]_I = [H^+]_{II}[Cl^-]_{II} \qquad (23\text{-}9)$$

Dividing one equation by the other results in

$$\frac{[Na^+]_I}{[H^+]_I} = \frac{[Na^+]_{II}}{[H^+]_{II}} \qquad (23\text{-}10)$$

According to this equation an exchange of ions must take place until the concentration ratios are equal on both sides of the membrane. *Note:* This does not mean that the concentrations must be equal, but the *ratios must be equal*. If the ratio of Na/H on one side of the membrane is 10:1, the ratio on the other side is also 10:1, even though one side may be many times more concentrated than the other.

COMPLEX AND CHELATE RESINS

In an effort to improve the selectivity of ion exchange resins, various complexing and chelating agents have been built into the resin structure. Many ion selective electrodes use resins of this type. If the cross-linking is < 1.0%, the resin is usually a liquid. Below are some examples. The left resin is selective for potassium ion. The middle resin, 8-hydroxyquinoline, is used for several ions, the selectivity based upon the pH of the eluting phase. The right resin, developed by Professor J.L. Lambert at Kansas State University, is of tremendous importance. It is used to purify water. It has been found to kill all gram - and gram + bacteria on contact and every virus it has been tested with thus far. All you do is pour the water in the top of the container and when it passes out the bottom, it will look the same and taste the same, but all the bacteria and most viruses will be dead. Its only drawback for municipal water systems is that it has no residual effect as does chlorination.

TECHNIQUE

Do not throw away used resin beads. They can be regenerated many times. Collect them in a sealed container to keep them moist; and when a good supply has been collected, regenerate them to the ionic form desired by batch exchange.

Pre-swell the beads before packing a column. Dry beads swell when placed in water, especially the 2X and 4X beads. The higher the cross-linking, the less the swelling will be. Soak them in water for several minutes or until their volume appears to stop increasing. This also will remove some trapped air and permit a quality column packing.

For the best ion exchange chromatography separations, use small beads of the same diameter. The smaller the bead, the faster the exchange can take place and the smaller the mixing chambers between the beads can be. This results in less time for diffusion. Small beads and beads of the same size keep step 4 (Figure 23-5, p. 273) fast and uniform, resulting is less band spread. Beads of any size can be used for the batch exchange of groups of ions, because neither time nor separation is a factor.

When a fresh bottle of resin is obtained, it should be cleaned very carefully or it will not produce reproducible results. Analytical grade resins have had the treatment described below and can be used immediately. Place the resin in a Soxhlet extractor and extract with 2-3 N HCl until the extract is clear. Follow this with 2-3 N NaOH until the extract is clear, and then extract with 95% ethanol until the extract is clear. Rinse with water until the rinsings are neutral and salt free. Keep the resin moist.

SETTING UP AN ION EXCHANGE COLUMN

Place the column in a buret holder on a ring stand. This keeps the column vertical. Place a small piece of glass wool in the bottom, unless a frit is present. Fill the column about half full with water. Place a powder funnel on top of the column and slowly add the resin beads. Attach a water line to the bottom of the column and slowly turn on the water. This back flow should remove any trapped air bubbles. Turn off the water and allow the resin to settle (optional). Place a small piece of glass wool on top of the beads. This stops any stirring and is used if a fine separation is being attempted. Drain the column of excess water until it is about 2 mm above the top of the beads. Check the flow rate as you do this and note the stopcock position for the flow rate desired.

POROUS PARTICLES AND MEMBRANES

In an effort to increase the speed of the exchange and to minimize the distance of travel within the bead, two approaches have been successful. One is to make a porous particle (30 μ), as shown in Figure 23-6, or to use a membrane (Figure 23-7). Figure 23-8 shows a diagram of a FASTCHROM unit. Notice that in both cases, the distance the ions must travel is greatly reduced compared to ordinary resin beads.

The separation module is the cylinder between the two end plates. Figure 23-9, p. 276, shows the results of a fractionation of a 1 Kb DNA ladder of 22 fragments ranging from 75 to 12,000 base pairs. A 2.5 cm diameter x 2.5 cm long cylinder was equilibrated with a 25 mM phosphate pH 7.0 buffer and a chloride gradient from 40 to 70% 0.5 M NaCl in the buffer.

Figure 23-10, p. 276, shows a laboratory-size ion exchanger to prepare deionized water. It can produce up to 2.0 L/min and up to 18.3 megohm-cm water with less than 10 ppb total organic carbon content. Use a three-module system for tap feed water, and a four-module system for high purity quality tap water.

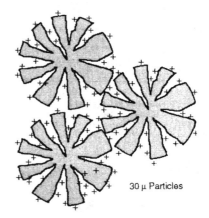

Figure 23-6. Porous particulate ion exchange medium.
(Courtesy - Kontes, Vineland, NJ)

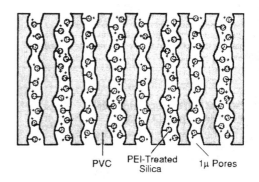

Figure 23-7. Microporous membrane ion exchange medium.
(Courtesy - Kontes, Vineland, NJ)

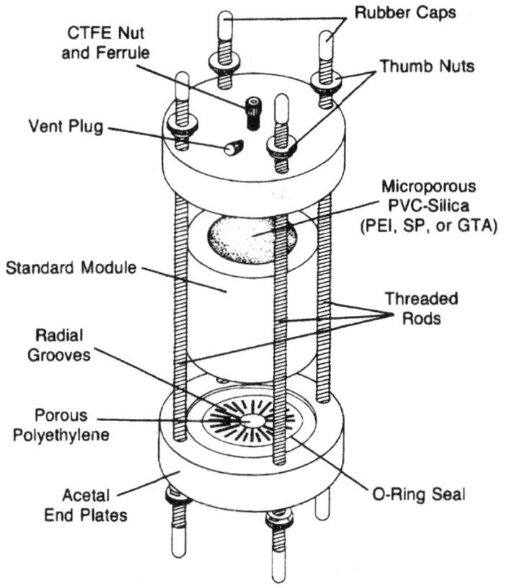

Figure 23-8. FASTCHROM standard module and end plates.
(Courtesy - Kontes, Vineland, NJ)

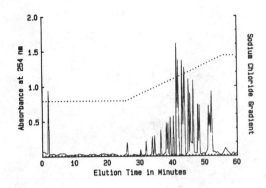

Figure 23-9. FASTCHROM PEI chromatography separation of a 1 kb ladder.
(Courtesy - Kontes, Vineland, NJ)

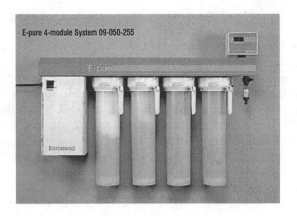

Figure 23-10. Ion exchange resin deionizer.
(Courtesy - A.H. Thomas Scientific, Swedesboro, NJ)

EXAMPLE CALCULATION

A 10.0 mL sample of pond water required 0.36 mL of 0.0107 N NaOH to reach the end point when a mixed indicator was used and 8.28 mL after a similar size sample had been passed through a cation exchange resin. (A) How many meq of ions (as H^+) are present initially per liter of sample? (B) How many meq of ions (as H^+) are present after exchange per liter of sample? (C) What is the total hardness (as $CaCO_3$) in ppm, and (D) what is the hardness in grains/gallon? 1 gal = 3.78 L.

ANSWER

A. (mL) (N) = No. of meq.
   (0.36)(0.0107) = 0.00385/10.0 mL or 0.385/L.

B. $[(mL_s - mL_b)(N)]$ = No. of meq.
   [(8.28 - 0.36)(0.0107)] = 0.0847/10 mL or 8.47/L.

C. We assume 1 mL = 1 g.
   (No. of meq.)(The meq. wht.) = g
   For $CaCO_3$, f.w. = 100, the meq wht = 100/2000 = 0.050
   (8.47)(0.050) = 0.42 g; This is 420 mg/L or 420 ppm

D.
$$\frac{1\ Grain}{17.12\ ppm} \times \frac{420\ ppm}{1\ L} \times \frac{3.78\ L}{1\ Gallon} = 92.7\ grains/gal$$

REVIEW QUESTIONS

1. How does the overflow control on the apparatus shown in Figure 23-2, p. 270, work?
2. What is the difference between ion exchange and ion exchange chromatography?
3. What are some limitations of inorganic ion exchanger materials?
4. What is "pearl polymerization"?
5. What does weakly acidic, carboxylic acid, H^+, 30-60 mesh, 4X, 3.6 meq/g min dry mean?
6. The following mixture of ions is poured into the top of a cation exchange column, what is the order of elution? Na^+, NO_3^-, Ba^{+2}, $FeCl_4^-$, Ag^+, Co^{+2}, and NH_4^+.
7. A 0.0004 M solution of Cs^+ is passed through a cation exchange column. What is most likely to control the rate of exchange?
8. Why should you use small beads of uniform size for the best ion exchange chromatography separations?

IF IT EXISTS - CHEMISTRY IS INVOLVED
IF YOU CAN BUY IT - A CHEMIST WAS INVOLVED SOMEWHERE

24
ION CHROMATOGRAPHY

PRINCIPLES

Although ion exchange chromatography had been known for decades, it never had been used widely, for two main reasons: (1) it was slow and (2) no simple, sensitive, and widely applicable detector was available. Neither refractive index nor electrical conductivity detectors were sufficiently sensitive because of large background signals. In 1971, at Dow Chemical Co., Small, H., Stevens, T.S. and Bauman W.C. found a way to eliminate these difficulties with the electrical conductivity detector by adding a second *suppressor* column and developed what is now known as *ion chromatography*. The first instrument was sold commercially in 1975 (Small, H., Stevens, T.S. and Bauman W.C. *Anal. Chem.* **47**, 1801). Since then, ways have been found to essentially combine the two columns into one column and to use thin membranes rather than column beads. More recently, electro-regeneration of the suppressor has been developed. In addition, ways have been found to use uv/vis, fluorescence, and pulsed amperometric detectors for some ions. *This technique is capable of separating and detecting essentially all anions from compounds that have pK_a's less than 7 and all cations from compounds that have pK_a's greater than 8.* It can be used to separate Fe^{+2} from Fe^{+3} and NO_2^- from NO_3^-. Column efficiencies of 25,000 plates/m and detection limits in the ppb region are now common. Although it can be used to separate cations, this technique has now developed into the best way we have to detect mixtures of anions in commercial and natural products.

Figure 24-1, p. 278, shows an example of determining the anionic components in red wine. The carboxylic acid components influence the flavor and color.

Ion chromatography can be thought of as a high performance ion exchange separation. Originally, it was a high performance liquid chromatography (HPLC) system with an ion exchange column to provide speed, an additional suppressor column to remove the background conductivity, and a conductivity detector. By properly selecting the eluting agent, the suppressor column can sometimes be eliminated, a technique called *single column ion chromatography* (SCIC). More recent techniques use electro-regenerated membranes to do all of the above operations.

Of the major methods of detection (refractive index, spectrophotometry, fluorescence, and electrochemical), the electrical conductivity method of detection is a favorite, because conductivity is a universal property of ionic species and, at low concentrations of solute, is linear with concentration. The difficulty previously was that the eluting solvents usually contained H^+ or OH^- ions whose conductance overwhelmed that of the ions to be measured. H. Small, T. Stevens, and W. Bauman eliminated this problem by adding a second ion exchange column, the *suppressor column*, following the separation column, that removes the ions from the eluting solution. Only the ions of interest are then present in a background of essentially deionized water. Figure 24-2, p. 278, shows a commercial instrument. How the apparatus functions at the molecular level for a suppressed system is shown in Figure 24-3, p. 279.

SUPPRESSED SYSTEMS

Refer to Figure 24-3, p. 279. As an example of how an ion chromatograph functions, consider the analysis of a sample containing LiCl, NaCl, and KCl using HCl as the eluent. The sample is added to a strong acid (H^+ form) cation exchange resin, with the Li^+, Na^+, and K^+ ions displacing H^+ ions where the sample is added. The Cl^- anions from the

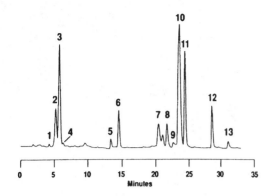

Figure 24-1. Anionic components of red wine: 1, α-hydroxybutyric acid; 2, acetic acid; 3, lactic acid; 4, propionic acid; 5, Cl$^-$; 6, galacturonic acid; 7, succinic acid; 8, malic acid; 9, maleic acid; 10, tartaric acid; 11, SO$_4^{-2}$; 12, PO$_4^{-3}$; 13, citric acid.
(Courtesy R.D. Rocklin, A. Henshall, and R.B. Rubin, *American Laboratory*, Mar. 1990)

sample move along as counterions for the exchanged H$^+$ ions. When HCl eluent is added, the H$^+$ exchanges with the Li$^+$, Na$^+$, and K$^+$ in a normal ion exchange process, and the ions elute from the bottom of the column in the order, Li$^+$ first, Na$^+$ next, and K$^+$ last based on size as described in Chapter 23, p. 272. Equivalent amounts of H$^+$ are removed from the stream, and the Cl$^-$ from the HCl moves down the column.

When the eluent ions reach the suppressor column, which is a strong base anion column, the Cl$^-$ ions exchange with the OH$^-$ ions. The freed OH$^-$ ions then react with the H$^+$ ions from the eluent to form water, which has no conductivity. All of the highly conducting eluent is neutralized. The Cl$^-$ counterions for the sample ions also exchange with the OH$^-$ ions on the suppressor equivalent to the amount of each cation present.

What the detector measures is the conductance of each cation plus the conductance of the associated OH$^-$, but without the highly concentrated and very highly conductive H$^+$ being present. An analogous scheme for anion analysis can be envisioned with NaOH as the eluent, an anion exchange resin in the separating column, and a strong acid resin in the H$^+$ form in the suppressor column. However, NaOH does not work well with organic anions, because the selectivity ratio is not favorable. Therefore, more neutralizing solution is required and the suppressor column has a correspondingly shorter life.

According to Professor J. S. Fritz at Iowa State University, "Carbonate has very good elution characteristics because of the valence state (-2) and it can be used at relatively low concentrations. Buffered solutions, prepared by mixing carbonate and hydrogen carbonate, are the most commonly used eluents. Changes in selectivity can be made simply by varying the ratio of hydrogen carbonate to carbonate salt, i.e., by varying the pH value of the eluent. Also, the eluent concentration can be varied so that faster or slower elution is obtained without affecting the elution order of analyte ions."

The phenate ion has been found to be quite satisfactory. Not only does it have a more favorable selectivity coefficient on Dowex-2 (0.14 for phenate *vs*. 1.5 for OH$^-$) but an acid resin as the suppressor would convert it to a phenol, which is a very weak acid that would be dissociated only feebly and contribute little to the conductivity of the effluent from the suppressor. A disadvantage of the phenate ion is that it forms oxidation products that tend to poison the column and shorten its life. Table 24-1 shows several eluents for anions.

Figure 24-2. An Odyssey high performance ion chromatograph.
(Courtesy - Alltech Associates Inc., Deerfield, IL)

Table 24-1. Some eluents for anion analysis by ion chromatography

Eluent	Eluting Ion	Suppressor Reaction Product
$Na_2B_4O_7$	$B_4O_7^{-2}$	H_3BO_3
NaOH	OH^-	H_2O
$NaHCO_3$	HCO_3^-	H_2CO_3
Na_2CO_3	CO_3^{-2}	H_2CO_3
Phenol	$C_6H_5O^-$	C_6H_5OH

SINGLE-COLUMN ION CHROMATOGRAPHY

By choosing an eluent that does not dissociate strongly it is possible to eliminate the suppressor column. Such compounds as phenol, $pK_a = 1.3 \times 10^{-10}$, or phthalic acid, $K_{a1} = 1.3 \times 10^{-3}$, $K_{a2} = 3.9 \times 10^{-6}$, are examples. When compounds of this type are used the system is called single column ion chromatography (SCIC). Table 24-2 lists some of the other common eluents for anions.

Table 24-2. Some eluents for anion analysis by ion chromatography

Eluent	Eluting Ion	Suppressor Reaction Product
Kbenzoate	$C_6H_5COO^-$	C_6H_5COOH
KHphthalate	$C_6H_4(COOH,COO^-)$	$C_6H_4(COOH,COOH)$
K_2phthalate	$C_6H_4(COO^-)_2$	$C_6H_4(COOH)_2$
Phenol	$C_6H_5O^-$	C_6H_5OH
Ammonium o-sulfobenzoate	$C_6H_4COO^-,SO_3^-$	$C_6H_4COONH_4,SO_3H$

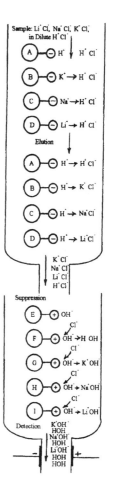

Figure 24-3. Diagram of an ion chromatograph's principle of operation.

TECHNIQUE

A severely fronted peak usually means too much sample.
A dip at the start (negative peak) is usually due to water.
Wear disposable gloves as much as possible when doing trace analysis.
Use presoaked polycarbonate containers rather than glass or plastic when doing trace analysis of cations.
The sample and eluents must be filtered through at least a 0.45 µm filter, and the eluents degassed.

COLUMNS

The columns are from 5 cm to 25 cm long and 3-5 mm i.d. packed with particles of 10-50 µm diameter. Table 24-3, p. 280, lists the characteristics of five favorite columns, and Figure 24-4, p. 280, shows the chromatograms obtained for each one using the manufacturers' specifications.

Caution: Ordinary ion exchange columns prepared for HPLC should not be used for ion chromatography. Their capacity is too large; thus they require large amounts of eluent with a corresponding reduction in the active lifetime of the suppressor column, and this also slows down the separation. Ion exchange columns made specifically for the separation columns in ion chromatography have very low capacities compared to those of regular columns. For example, 0.04 meq/g vs. 1-2 meq/g.

Silica-Based Particles

The silica-based particles are of the pellicular type. They have good packing rigidity and can stand high pressures and high flow rates. However, they tend to dissolve in eluents of pH greater than 8, and they adsorb F^-. This can be a problem, because it is often necessary to go above pH 6 to get rid of system peaks.

Table 24-3. Description and specifications of selected low capacity anion exchange columns
(Courtesy - Haddad, P.R., Jackson, P.E. and Heckenberg, A.L., *J. Chromatogr.*, **346**, 139, 1985)

Column	Size (mm)	pH Range	Capacity (μequiv/g)	Particle Size (μm)	Type of Packing Material
Vydac 242 IC 4.6	250 x 4.6	2-6	100	20	Spherical silica with bonded quaternary groups.
Interaction ION-100	50 x 3.2	0-14	100	10	Neutral hydrophilic macroporous resin with covalently bound quaternary ammonium groups.
Hamilton PRP-X100	150 x 4.1	1-13	200	10	Highly cross-linked polystyrene divinyl benzene coated with quaternary ammonium groups.
Bio-Gel TSX Anion PW	50 x 4.6	1-12	30	10	Polymethacrylate gel coated with quaternary ammonium groups
Waters IC Pak A	50 x 4.6	1-12	30	10	Same as above.

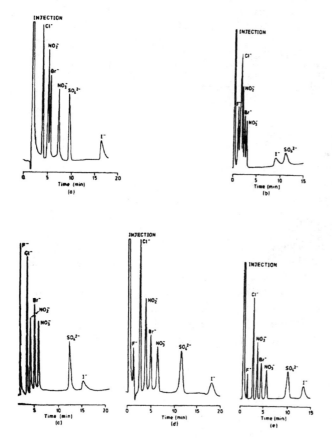

Figure 24-4. Chromatograms obtained with optimized phthalate eluents. A, Vydac 242 IC 4.6; B, Interaction ION-100; C, Hamilton PRP-X100; D, Bio-Gel TSK IC-Anion-PW; E, Waters IC Pak A.
(Courtesy - Haddad, P.R., Jackson, P.E. and Heckenberg, A.L, *J. Chromatogr.*, **346**, 139, 1985)

Resin-Based Particles

Resin-based particles have wide pH ranges, usually 1-12. They tend to compact at higher pressures, and they are temperature sensitive, so temperatures higher than ambient must be used with caution.

MEMBRANES

Figure 24-5 and the following discussion is that of R.D. Rocklin, A. Henshall, and R.B. Rubin *(American Laboratory*, Mar. 1990). "The mechanism of chemical suppression, using sodium hydroxide as the mobile phase for anion-exchange separation, is illustrated in [Figure 24-5]. Analyte anions elute from the column with sodium counterions. They enter the suppressor between two cation exchange membranes. On the other side of these membranes, a dilute solution of sulfuric acid (the regenerant) flows in the opposite direction. The membranes have fixed negative sulfonic acid functional groups and therefore exclude anions. Positive ions can pass through the membrane freely. Therefore, sodium ions from the mobile phase flow across the membrane into the regenerant side. An equal number of hydrogen ions from the regenerant flow across the membrane and enter the mobile phase side, thus neutralizing the highly conductive hydroxide to water. The resulting background conductivity is near zero, considerably lower than before suppression. Also, the counter ion in the anion analytes is now hydrogen ion, which has seven times higher conductivity than the original sodium counterion.

"For cation systems, the suppressor membranes are anion exchange polymers, which allow anions to pass freely, but exclude cations. Dilute acids such as HCl are used in the mobile phase. In the suppressor, chloride counterions are replaced by highly conductive OH⁻ ions, neutralizing the acidic mobile phase. Tetrabutylammonium hydroxide is used as the regenerant because it is excluded from the anion exchange membranes."

ELECTROCHEMICAL REGENERATED ION SUPPRESSION

Electrical regenerated ion suppression is a continuous electrochemical regeneration of the suppressor. Figure 24-6 shows the components. The Alltech ERIS 1000 Autosuppressor consists of a 10 port valve, two solid phase electrochemical suppressor cells packed with either cation (for anion analysis) or anion (for cation analysis) exchange resins, and a constant current power supply.

For anion analysis refer to Figures 24-7 and 24-8, p. 282. "The cell is packed with strong cation exchange resin in the hydrogen form. During suppression, the high conductivity mobile phase is converted to low conductivity acid or water by ion exchange of the mobile phase sodium ions with the hydrogen form cation exchange resin. At the same time, the low conductivity analyte ions are converted to their high conductivity acids by exchanging counter cations with the hydrogen form cation exchange resin. The background signal is reduced, while the analyte signal is enhanced, significantly improving detection.

"During electrochemical regeneration [Figure 24-8, p. 282], the detector effluent undergoes electrolysis when current is applied

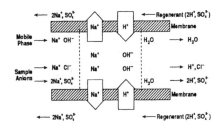

Figure 24-5. Schematic of suppression with membranes.
(Courtesy - Rocklin, R.D, Henshall A. and Rubin, R.B., *American Laboratory*, Mar. 1990)

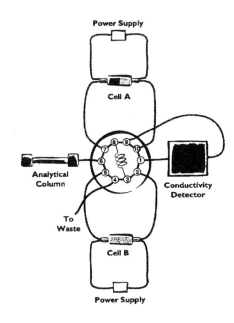

Figure 24-6. The Alltech model 1000 electrochemical regenerated ion suppression system.
(Courtesy - Alltech Assoc. Inc., Deerfield, IL)

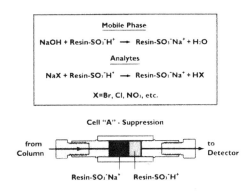

Figure 24-7. Suppression during anion analysis.
(Courtesy - Alltech Associates Inc., Deerfield, IL)

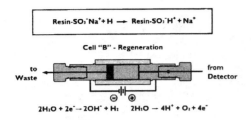

Figure 24-8. Electrochemical regeneration during anion analysis.
(Courtesy - Alltech Associates Inc., Deerfield, IL)

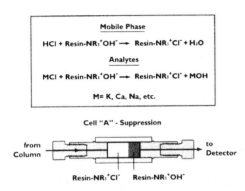

Figure 24-9. Suppression during cation analysis.
(Courtesy - Alltech Associates Inc., Deerfield, IL)

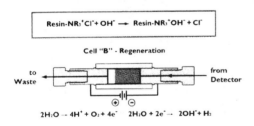

Figure 24-10. Electrochemical regeneration during cation analysis.
(Courtesy - Alltech Associates Inc., Deerfield, IL)

across the cell. Hydrogen ions and oxygen gas are generated at the cell inlet (anode), while hydroxide ions and hydrogen gas are generated at the cell outlet (cathode). The hydrogen ions are carried by the detector effluent across the exhausted sodium form cation exchange resin, converting it back to the original hydrogen form.

"For cation analysis, the cell is packed with strong anion exchange resin in the hydroxide form [Figure 24-9]. During suppression, the high conductivity mobile phase is converted to low conductivity water by ion exchange of the mobile phase chloride ions with the hydroxide form anion exchange resin. At the same time, the low conductivity analyte cations are converted to their high conductivity hydroxides by exchanging their counter anions with the hydroxide form anion exchange resin. The background signal is reduced, while the analyte signal is enhanced, significantly improving detection.

"During the electrochemical regeneration [Figure 24-10], the detector effluent undergoes electrolysis when current is applied across the cell. The cell polarity is reversed, compared to anion analysis, placing the cathode at the cell inlet. Hydroxide ions and hydrogen gas are generated at the cell inlet (cathode), while hydrogen ions and oxygen gas are generated at the cell outlet (anode). The hydroxide ions are carried by the detector effluent across the exhausted chloride form anion exchange resin, converting it back to the original hydroxide form."

DETECTORS

Traditionally, the detectors for the older ion exchange chromatography were refractive index, conductivity, and uv/vis spectroscopy in isolated cases. Today a wider variety is available. There are three types of electroconductivity detectors (d.c. conductivity, d.c. amperometry, pulsed amperometry): direct uv/vis with postcolumn reactants, "vacancy" uv/vis, and fluorescence.

Conductivity

In conductivity, no oxidation or reduction takes place on the surface of the electrodes; only the resistance changes due to the ions in the solution is measured. To understand why problems with the conductivity detector occurred prior to using the suppressor column, you need to examine the equivalent conductances of several ions. If you place two metal plates 1.00 cm apart with sufficient area that 1.00 L of a 1.00 N solution will exactly fit, the conductance across the plates is the *equivalent conductance*. It is the conductance of 1 mole of ions. That is why the values listed in Table 24-4, p. 283, are divided by 2 or 3 as the case may be. The table lists the equivalent conductances for several ions at infinite dilution. The old unit for this was the *mho*, but the SI recommended unit is the *Seimen* (S), $10^{-4} m^2\ mol^{-1}$. Refer to Chapter 19, p. 206, on HPLC detectors for a description of electrochemical detectors.

Notice that with the exception of H^+ and OH^-, the equivalent conductances do not vary greatly. However, if H^+ is present, it swamps the signal from the other ions. This was the problem with conductivity detection before the suppressor column was added. The conductance of the eluting agent was so high because of the H^+ ion, and also its presence in very high concentration compared to the eluted ion made detection of the sample ion difficult. It was like weighing a large ship and its captain, then weighing the ship to determine the weight of the captain. The suppressor column not only removes the H^+, but it replaces a lower conducting counteranion with a higher conducting OH^- ion, which increases the sensitivity.

Table 24-4. Ionic equivalent conductances at infinite dilution

Cations	S	Cations	S	Anions	S
H^+	349.7	$1/2\ Ni^{+2}$	50.0	OH^-	198.0
Li^+	38.7	$1/2\ Mg^{+2}$	53.1	Cl^-	76.3
Na^+	50.1	$1/2\ Ca^{+2}$	59.5	Br^-	78.4
K^+	73.5	$1/2\ Ba^{+2}$	63.6	I^-	76.8
NH_4^+	73.4	$½\ Co^{+2}$	45.0	NO_3^-	71.4
Ag^+	61.9	$1/3\ Cr^{+3}$	67.0	HCO_3^-	44.5
$½\ Fe^{+2}$	54.0	$1/3\ Fe^{+3}$	68.0	$½\ SO_4^{-2}$	79.0

This type of detector is useful for inorganic strong acid anions; inorganic strong base cations; sulfates; sulfonates; fatty acids; I°, II°, III°, and IV° amines; and carboxylic acids.

D.C. Amperometry

In the d.c. amperometry method, a voltage is applied that is sufficient to cause oxidation-reduction at the electrode surfaces. The advantage is that by carefully selecting the voltage, the selectivity can be increased. This method can be two orders of magnitude more sensitive for some aromatic compounds than uv detection. This can be difficult even for experienced personnel. Beginners should be extra careful. The preferred working electrode is a platinum electrode covered with platinum black. The reference electrode is Ag/AgCl. This method is particularly good for sulfite, sulfide, iodide, phenols, catecholamines, and cyanide. Figure 24-11 is a diagram of the Dionex detector for the 4000 series instrument, that can be either amperometric or conductive.

The silver electrode works best with anions that form insoluble compounds with silver such as Cl^-, Br^-, I^-, CN^-, HS^-, and $S_2O_3^{-2}$. A K_{sp} of $\leq 10^{-13}$ would give good sensitivity. The potential range is usually from 0.0 V to 0.4 V, and is pH dependent, shifting positively as the pH decreases. Insoluble silver oxide salts on the electrode surface decrease its sensitivity, but they can be removed by polishing the electrode. The reference electrode is Ag/AgCl. Table 24-5, p. 284, shows typical ranges for several electrodes.

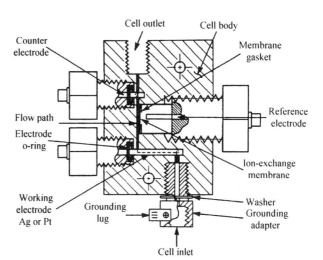

Figure 24-11. Cross section of an amperometric detector.
(Courtesy - Dionex Corp., Sunnyvale, CA)

Table 24-5. Potential limits in acidic and basic solutions *vs.* an Ag/AgCl reference electrode
(Courtesy - Dionex Corp., Sunnyvale, CA)

Working Electrode	Solution (0.1 N)	Negative Limit (volts)	Positive Limit (volts)
Gold	KOH	-1.25	+0.75
	HClO$_4$	-0.35	+1.1
Silver	KOH	-1.2	+0.1
	HClO$_4$	-0.55	+0.4
Platinum	KOH	-0.90	+0.65
	HClO$_4$	-0.20	+1.24
Glassy Carbon	KOH	~-1.50	~+0.6
	HClO$_4$	~-0.8	~+1.3
Mercury	KOH	-1.9	-0.05
	HClO$_4$	-1.1	+0.60

EXAMPLE CALCULATION

Explain how a Ag electrode can be used as a potentiometric detector to detect CN⁻ in a plating solution? (Dionex Corp., Sunnyvale, CA.)

ANSWER

The electrode reaction:

$$Ag + 2\ CN^- \rightarrow Ag(CN)_2^- + e^-$$

The Nernst equation for silver ion:

$$E = E^o_{Ag^+} + 0.059 \log \frac{[Ag^+]}{[Ag]}$$

Reaction of Ag⁺ with CN⁻:

$$Ag^+ + 2\ CN^- \rightarrow Ag(CN)_2^-$$

Formation constant K_f is:

$$K_f = \frac{[Ag(CN)_2^-]}{[Ag][CN^-]^2} = 10^{21.1}$$

Solving for [Ag⁺] and substitution into the Nernst equation, where $0.77 = E^o_{Ag^+}$ vs. NHE. (normal hydrogen electrode)

Since $\log K_f = -21.1$

$$E = 0.77 + 0.059 \log \frac{[Ag(CN)_2^-]}{K_f [CN^-]^2};$$

$$E = 0.77 - 0.059(21.1) + 0.059 \log \frac{[Ag(CN)_2^-]}{[CN^-]^2}$$

$$E = -0.47 + 0.059 \log \frac{[Ag(CN)_2^-]}{[CN^-]^2}$$

Where -0.47 represents the calculated $E^o_{Ag(CN)2^-}$ vs. NHE. Because $E^o_{Ag(CN)2^-}$ is lower than $E^o_{Ag^+}$ and E_{app} is held constant between these potentials by the potentiostat, oxidation of silver must occur to maintain equilibrium, generating a peak as the CN⁻ passes through the cell.

EXAMPLE CALCULATION

Explain how a Pt electrode can be used to detect ClO⁻ in a bleaching solution? (Dionex Corp., Sunnyvale, CA).

ANSWER

$$ClO^- + H_2O + 2e^- = Cl^- + 2OH^-; \qquad E^o = 0.89 \text{ V vs. NHE.}$$

$$E = E^o_{ClO^-} + \frac{0.059}{2} \log \frac{[ClO^-]}{[Cl^-][OH^-]^2} \qquad E = 0.89 + 0.059 \log \frac{1}{[OH^-]} + 0.030 \log \frac{[ClO^-]}{[Cl^-]}$$

$$\text{Because } \log [OH^-] = \log \frac{K_w}{[H^+]}; \qquad E = 0.89 - (0.059) \log \frac{K_w}{[H^+]} + 0.030 \log \frac{[ClO^-]}{[Cl^-]}$$

$$\text{Since } \log \frac{1}{[OH^-]} = -\log [OH^-]; \qquad E = 0.89 - (0.059) \log [OH^-] + 0.030 \log \frac{[ClO^-]}{[Cl^-]}$$

Because $K_w = 10^{-14}$ and $\log 1/[H^+] = -\log [H^+] = pH$, the second term of this equation may be expanded to give:

$$E = 0.89 - 0.059(\log 10^{-14} - \log [H^+]) + 0.030 \log \frac{[ClO^-]}{[Cl^-]}$$

$$= 0.89 - 0.059(-14 + pH) + 0.030 \log \frac{[ClO^-]}{[Cl^-]}$$

Which finally gives:

$$E = 1.72 - 0.059 \, pH + 0.030 \log \frac{[ClO^-]}{[Cl^-]}$$

where $1.72 - 0.059 \, pH$ represents E'_{ClO^-} as a function of pH. The Pt electrode potential is dependent upon the pH, and this can be used to control the selectivity of the detector.

Pulsed Amperometry

Oxidation-reduction also is the main phenomenon with pulsed amperometry. One of the problems associated with nonpulsed amperometry is that, in addition to the reaction products often covering the electrode surface, a thin film of oxide coating is formed, and this reduces the sensitivity markedly. The solution is to apply the voltage only for a short time, then apply another pulse of the opposite charge to clean the electrode. Gold and platinum electrodes are used. This is particularly good for carbohydrates. Other systems that work well are I°, II°, and III° amines; alcohols; aldehydes; glycols; amino acids; mercaptans; thiols; and sulfides. Figure 24-12, p. 286, shows a pulsed detector.

UV/Vis

Very few ions are directly detectable by uv/vis absorption spectroscopy. However, if a postcolumn reaction is performed to chelate a cation with a chromophore (color producing group), to add a fluorescing agent, or to simply react the compound with another compound, such as using ninhydrin with amino acids, then uv/vis spectroscopy can be used. An example of such a system is the separation of lanthanides shown in the cation section.

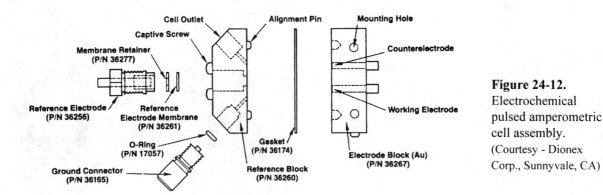

Figure 24-12. Electrochemical pulsed amperometric cell assembly.
(Courtesy - Dionex Corp., Sunnyvale, CA)

Another technique is *vacancy* spectroscopy. In this application, an eluent is used that absorbs uv or visible radiation. When the sample passes through the detector, less radiation is absorbed because of the *vacancy*. For example, the benzene ring in a mixture of 4 mM salicylic acid - (Tris) is measured at 254 nm. Alternately, the eluent for cations can be Ce^{+3}, and the decrease in the yellow color monitored in the visible region.

CATIONS

Figure 24-13 shows how alpha-hydroxyisobutyric acid can be used to separate 14 lanthanides. Post column derivatization with 4-(2-pyridylazo)resorcinol (PAR) permitted photometric detection at 520 nm.

ANIONS

One type of anion exchange column used for nonsuppressed applications is the ION-20 from Interaction Co. It is a quaternary ammonium group on a styrene-divinyl benzene polymer in the form of 9 μm particles with an exchange capacity of 100 μeq/g. Figure 24-14 is an example of a multiple anion separation.

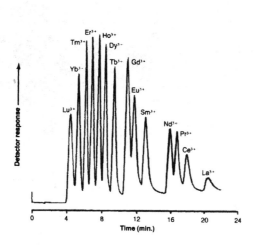

Figure 24-13. Separation of lanthanides.
(Courtesy: Heberling, S.S., Riviello, J.M., Shifen, M., and Ip, A.W., Dionex Corp., Sunnyvale, CA)

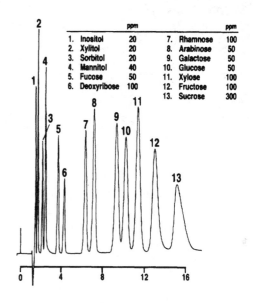

Figure 24-14. Separation of carbohydrates using pulsed amperometric detection on a gold electrode.
(Courtesy - Dionex Corp., Sunnyvale, CA)

GRADIENT ELUTION

Gradient elution can be used if some type of suppression is used, otherwise the background changes too much to give a good working range. Modern systems use computer controlled pumps that can handle up to six eluents. Figure 24-15 is an example of what can be done with a synthetic mixture.

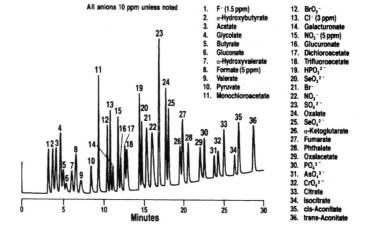

Figure 24-15. Gradient separation of a synthetic mixture of 36 anions in 24 minutes.
(Courtesy - Dionex Corp., Sunnyvale, CA)

SAMPLE CONCENTRATOR

One way to improve sensitivity is to use a precolumn concentrator. One such apparatus is shown in Figure 24-16. According to the Wescan advertising brochure, "Sample pre-concentration uses a cartridge packed with an appropriate ion exchanger in place of the sample loop in the standard 6-port rotary injector. Ions from large volumes of dilute samples are concentrated on the head of the cartridge while the depleted liquid flows through to the drain. The sample is then eluted onto the column."

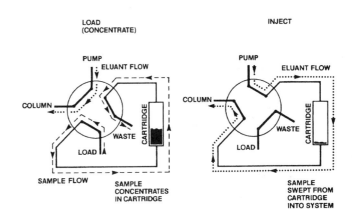

Figure 24-16. Sample preconcentrator
(Courtesy - Wescan Instruments Inc., Santa Clara, CA)

REVIEW QUESTIONS

1. What was the development that changed ion exchange chromatography into ion chromatography?
2. What types of compounds can ion chromatography separate?
3. What is a typical column efficiency, and what are normal detection limits?
4. Why is electrical conductivity a favorite method of detection?
5. Why is an OH⁻ column used as a suppressor if the eluting agent is an acid?
6. What is an advantage of carbonate as an eluting agent?
7. What is a disadvantage of the phenate ion as an eluent?

8. What is one way to eliminate the suppressor column?
9. What is the suppressor reaction product for potassium benzoate?
10. What is the possible problem if you have a severely fronted peak?
11. Why is it recommended that ordinary HPLC columns not be used for ion chromatography?
12. How do the capacities of columns for HPLC compare to those for ion chromatography?
13. What is a disadvantage of silica-based particles?
14. What is a disadvantage of resin-based particles?
15. On the anion membrane discussed, how are anions excluded?
16. What is the regenerate compound for cation membrane suppressors?
17. What type of power supply is needed for electrochemical regenerated ion suppression?
18. What is the anode reaction during electrochemical regeneration of the suppressor during an anion analysis?
19. What is the cathode reaction during electrochemical regeneration of the suppressor in a cation analysis?
20. What is equivalent conductance?
21. What is the SI unit for equivalent conductance?
22. What ions have the highest ionic equivalent conductances?
23. What is the preferred electrode combination for the d.c. amperometric detector?
24. What is the purpose of pulsing in an amperometric detector?
25. What is the principle of vacancy spectroscopy?
26. What is the basis for a sample preconcentrator?

IF IT EXISTS - CHEMISTRY IS INVOLVED
IF YOU CAN BUY IT - A CHEMIST WAS INVOLVED SOMEWHERE

25
ION RETARDATION, ION EXCLUSION, AND LIGAND EXCHANGE

ION RETARDATION

PRINCIPLES

Ion retardation is a form of separation that involves a resin with both cation and anion groups adjacent to each other and separates ions by "retarding" them rather than completely exchanging one for another. This technique can be used to separate electrolytes from nonelectrolytes, particularly large size nonelectrolytes. The interesting part is that both the cation and anion portions of a molecule are affected.

In ordinary ion exchange, you can expect the following:

$$-COOH + Na^+ \rightarrow -COONa + H^+$$

$$-R_3NOH + Cl^- \rightarrow -R_3NCl + OH^-$$

In ion retardation, salts are adsorbed without an exchange.

$$-COO^- \ldots\ldots {}^+NR_3- + NaCl \rightarrow -COO\ NaCl\ NR_3-$$

An aqueous solution of the sample to be desalted is added to a column of the resin and eluted with water, and the inorganic salts are retarded more than the organic compounds.

$$-COONa\ NaCl\ NR_3- \rightarrow -COO^- \ldots\ldots {}^+NR_3- + NaCl$$

Bio-Rad's explanation is, "A flow of water washes away the mobile ions, driving the reaction to the right. As the salts move down the column, they are repeatedly absorbed and desorbed, and thus retarded."

Compounds with common anions such as NH_4Cl, $ZnCl_2$, $ZnSO_4$, and $FeSO_4$ can be separated. This works as well for common cations, and NaCl can be separated from NaOH.

The hydrogen ion usually is held so tightly by the anion portion that it is not eluted with water. This makes this resin excellent for the rapid quantitative removal of acids. Some other applications are the continuous separations of glucose-salt mixtures, sucrose from molasses, and sodium dodecyl sulfate from proteins (albumin, ovalbumin, cytochrome c); salt from blood, milk, and whey; and kinins and amino acids from salts and buffers.

Figure 25-1, p. 290, is a detail of the resin composition, Bio-Rad AG 11A8, in which the anion portion is a quaternary amine and the cation portion is carboxylic acid. Figure 25-2, p. 290, illustrates this technique with the same resin. This type of resin was developed in 1957 (Hatch, M.J., Dillon, J.A., and Smith, H.D., *Ind. Eng. Chem.*, **49**, 1812). They used a resin that had very little cross-linking (~ 1%), called a *macroreticular resin*, which has very large openings and can act like a cage. They then polymerized *snakes* of both anion and cation resin inside of the cage, obtaining a *snake-cage* resin. The cage served to hold the linear strands of polymer near to each other, and the opposite charges caused the caged polymers to pair off across from each other.

The above resin was prepared by polymerizing an acrylic acid inside of a standard Dowex 1 resin. With both cationic and anionic exchange sites on the same resin, the charges effectively neutralize each other. This means that ions proceeding down the column are held only weakly when they are exchanged and are *retarded*. Excess water is often

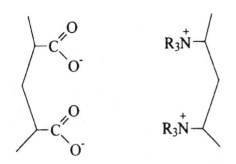

Figure 25-1. An ion retardation resin structure (AG 11A8).

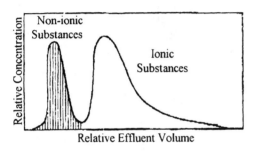

Figure 25-2. Desalting of biochemicals.
(Courtesy - Bio-Rad Co., Hercules, CA)

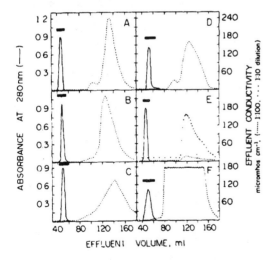

Figure 25-3. Separation of 10 mg of albumin from NaCl at pH 7.5-7.7. A, 1.0 M NaCl at 6 mL/hr, regenerated. B, Same, not regenerated. C, 33 mL/hr, not regenerated. D, 6.0 mL/hr, regenerated. E, 0.1 M NaCl at 6 mL/hr, regenerated. F, 15 mL, 5.0 M NaCl, 6 mL/hr, regenerated.
(Courtesy - Reis, M.L., Draghetta, W., and Greene, L., *Anal. Biochem.*, **81**, 346, 1977)

all that is necessary to strip the ions from the resin, but because of the various ionic charges of the sample's ions, a good separation usually is possible.

The above type resin loses its capacity after 6-8 regenerations because the *snakes* slowly wash out of the cage. An improved version (Bogoczek, R., *J. Chromatogr.*, **102**, 131, 1974) is to cross-link the *snakes* to each other within the cage, making a net out of the system. The result is a *net-cage* resin. The cross-linking reagents for basic resins are 2-33% divinyl benzene, 2-100% sorbic acid, and 2-100% muconic acid. For acid resins, the reagents are 2-100 % divinyl pyridine or 2-33% divinyl benzene. These resins have been regenerated 50 times with no measurable decay.

TECHNIQUES

One commercial resin is the AG 11A8 polymer made by Bio-Rad. These can be obtained prepacked with 10 mL bed volumes. The particle size varies from 180 to ~300 μm with a wet bed capacity of 0.7 meq/mL. The approximate m.w. exclusion is 1000. The resin can be used from pH 0 - 14. *This is not a fast method*, but 99+% of a salt can be removed from a protein. Figures 25-3 and 25-4, p. 291, show some typical separations. The sample volumes were 5 mL, unless indicated otherwise. The column was 0.9 x 200 cm, except for D, which was 2.0 x 40 cm. Notice that the simple ionic compounds separate better than the more complex compounds, but that a good separation is obtainable. The albumin was measured at 280 nm, and the salt by conductivity.

Resin Regeneration

While the resin can be used repeatedly without special regeneration, water elution simply being continued until the effluent is salt free, there comes a time when extra clean-up may be required. In some cases, the resin may retain an organic or salt component. Bio-Rad suggests the following for regeneration. "Regeneration may be accomplished using a concentrated salt solution followed by a water wash. Due to the high affinity of the carboxyl group for the hydrogen ion, a regeneration procedure using acid is even more effective than one using salt. However, the acid must be followed by a neutralization step to remove absorbed hydrogen ions. Likewise, if large quantities of an acidic solution (pH < 4) have been desalted, a neutralization step may eventually be necessary to restore the original high desalting capacity (start with step A2 of [Table 25-1], p. 291). Any salt or acid may be used, but the procedure given in Table [25-1] will return the resin to the form in which it was shipped."

Ion retardation, ion exclusion, and ligand exchange

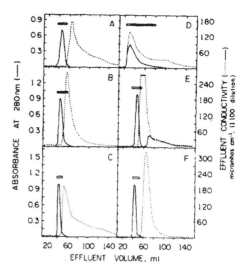

Figure 25-4. Separation of 10 mg of albumin from acid salts at pH 7.5-7.7. A, 0.33 M sodium citrate, 6 mL/hr, regenerated. B, Same, not regenerated. C, Same, 24 mL/hr, regenerated. D, Same, 6 mL/hr, regenerated. E, 5 mL 1 M sodium phosphate + 5 mL 0.1 M sodium phosphate, 6 mL/hr, regenerated. F, 1 M sodium acetate, 6 mL/hr, regenerated.
Courtesy - Reis, M.L., Draghetta, W., and Greene, L., *Anal. Biochem.*, **81**, 346, 1977)

Table 25-1. Regeneration procedure

Type of Regeneration	Step	Reagent	Quantity (Bed Volumes)	Linear Flow Rate (cm/min)
Salt	S1	1 M NH$_4$Cl	5*	2
	S2	H$_2$O	20**	2
Acid	A1	1 M HCl	2	4
	A2	1 M NH$_4$OH in 0.5 M NH$_4$Cl	4	4
	A3	1 M NH$_4$Cl	1	4
	A4	H$_2$O	20**	2

* or until adsorbed substances are eluted.
** or until the effluent is chloride free.

ION EXCLUSION

PRINCIPLES

Ion exclusion is a technique using ion exchange resins to separate ionic compounds as a class from nonionic compounds as a class. It is based on the differences in solubility of ionic compounds and nonionic compounds between the resin beads and the aqueous surroundings. The only requirement is that either the cation on the resin must be the same as the major cation in the sample or the anion on the resin must be the same anion as the sample. No ion exchange takes place, so the resin never needs to be regenerated.

If an aqueous mixture of ionic compounds (in the sodium form) and nonionic compounds is passed down a column (in the sodium form), the nonionic compounds are *more soluble* in the organic resin matrix than the ionic compounds, no exchange takes place, but a partitioning does take place. The nonionic compounds divide about equally between the resin and the aqueous surroundings, whereas the ionic compounds are more concentrated in the aqueous mobile phase. The net effect is that the ionic compounds are *excluded* from the resin beads by comparison to the nonionic compounds. If sample and water are added alternately to the top of the column, ionic compounds can be separated from nonionic compounds.. Figure 25-5, p. 292, is a diagram of a continuous operation for optimum conditions. Figure 25-6, p. 292, shows a simple separation done years ago, and Figure 25-7, p. 292 shows the separation of several nonionic components,

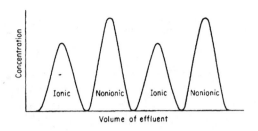

Figure 25-5. Diagram of two cycles of an ion exclusion separation.
(Courtesy - E.W. Berg, *Physical and Chemical Methods of Separation*, McGraw-Hill, New York, NY, 1963)

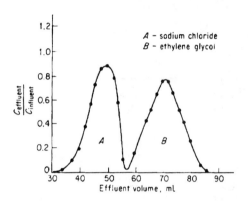

Figure 25-6. Separation of NaCl and ethylene glycol by ion exclusion.
(Courtesy - E.W. Berg, *Physical and Chemical Methods of Separation*, McGraw-Hill, New York, NY, 1963)

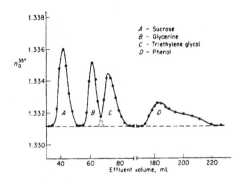

Figure 25-7. Separation of a multicomponent system. Dowex 50-8X, 50-100 mesh.

A nonionic can be any nondissociated compound. Organic acids can be made nonionic by adding a small amount of strong acid to the eluent. Furthermore, recent techniques permit the separation of nonionics from each other. This is shown in Figure 25-8, p. 293. Six organic acids were separated on a 100 x 7.5 mm highly sulfonated, polystyrene-divinyl benzene, ion exclusion column.

APPLICATIONS

The most common type of application is the separation of salt impurities from nonionic compounds. Some examples are the determination of sulfite in beer and seawater ions in seaweed roots; speciation of As(V) and As(III); sulfite in cellulosics; separating volatile and nonvolatile acids in sugar solutions; separating sucrose and non-sucrose materials in beet molasses; separation of hydroxycarboxylic acids from inorganic solids in alkaline pulping liquor; removing aromatic impurities in steamed hemicellulose hydrolyzate of birch wood; determining fatty acids produced by anaerobic bacteria in groundwater and SO_2 in grapes; separation of organic acids from carbohydrates in fruit juices; determining nitrite in cured meats or borate in soil, sediment, and water samples; separating ionic-nonionic surfactants; and detecting NO_2 in ambient air, organic acids in beer and wort, and the degradation products of ethylene glycol in chromium plating and electropolishing solutions. As you can see, this can be a versatile separation technique.

It is less expensive than ion exchange because there is no cost for regeneration of the resin. The practical maximums are about 8% for the ionic concentration and about 40% for nonionics. It is easily automated. The biggest disadvantage is that the *resin and the solution must have common cations or anions,* which usually limits the ionic species to a single contaminant. Ion exclusion works *backwards,* in that the major component of a mixture is removed when separating ionic compounds from nonionic compounds.

The first applications of size exclusion chromatography involved typical vertical gravity flow columns in which sample and solvent were added alternately as described above. In the late 1980's ion exclusion columns were placed in HPLC instruments and a one-pass-size exclusion separation was made. The advantage of this technique is that, in a complex mixture, the ionics are separated immediately from the nonionics, and then the nonionics can be separated in a regular manner.

One example is to separate mono- and dicarboxylic acids from alcohols and carbohydrates using a cation resin in the H^+ form with sulfuric acid as the eluate. Strong acids are highly ionized and elute at the void volume of the column, whereas the nonionic alcohols and carbohydrates, which are completely protonated, are neutral and enter the resin. They generally emerge from the resin in the order of the smallest molecule first.

The use of the *guard column* to do the ion exclusion is a novel approach to permit you to use a regular HPLC system, yet modify it temporarily for ion exclusion. As an example, carbohydrates interfere with the determination of organic acids in fruit juices. An anion exchange guard column in place of the sample loop permits selective retention of the acids and nonretention of glucose and fructose during sample loading. Elution with sulfuric acid then is performed.

THEORY

Terms:

V_s = volume occupied by the solid portion of the resin.

V_i = solution volume between the resin beads, ~30% of the total.

V_o = solution volume inside of the beads (occluded), ~40 % of the total.

D = distribution ratio of nonionics = $C_{resin}/C_{solution}$

V_r = retention volume = $V_i + DV_o$.

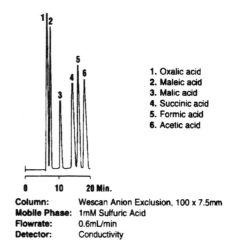

Figure 25-8. Organic acid standards by ion exclusion.
(Courtesy - Alltech Associates, Inc., Deerfield, IL)

Refer to the description of the Donnen membrane theory presented in Chapter 23, p. 273. Use the resin in the K⁺ form. Let C_o be the concentration of salt in the sample outside of the resin, and C_2 be that inside. Let X be the amount of salt diffusing into the resin.

$$[K^+]_{solution} [A^-]_{solution} = [K^+]_{resin} [A^-]_{resin} \tag{25-1}$$

therefore

$$[C_o - X][C_o - X] = [C_i + X]X \tag{25-2}$$

The proportion of salt initially present that has diffused into the resin = X/C_o. This is

$$\frac{X}{C_o} = \frac{C_o}{C_2 + 2 C_o} \tag{25-3}$$

As the exchange capacity of the resin increases, the proportion of the salt the diffuses into the resin decreases, with the result that the ionic component is excluded from the resin to a large extent. D is usually less than 1 for ionic components and from 1-2 for nonionics, with values over 50 in some special cases. Generally, it is close to 1, unless some additional interaction between the resin groups and the sample compound is taking place. D for the nonionic components depends upon the size of the molecule and increases with chain length and solution concentration. D for the ionic components varies with the type of resin, the amount of cross-linking, and the inherent capacity of the resin.

TECHNIQUE
COLUMN PACKINGS

Ordinary ion exchange resins are used with the vertical gravity flow columns. HPLC columns are pellicular, 5-10 μm diameter beads usually, with either strong acid or strong base functional groups.

LIGAND EXCHANGE

PRINCIPLES

A ligand is a group, ion, or molecule coordinated to a central atom. *Ligand exchange is a way to use conventional ion exchange resins to separate nonionic compounds.* Although the idea was mentioned in 1954 (Stokes, R.H., and Walton, H.F., *J. Am. Chem. Soc.*, **76**, 3327) and some preliminary separations were made by Thomas, C.L., in 1958 (U.S. Pat 2,865,970) and Giesen, J., and Mueller, F., (U.S.Pat. 2,916,525) in 1959, it was developed to a workable system by F. Helfferich in 1961. The explanation below is abstracted from his early papers. The mathematical treatment can be found in *J. Am. Chem. Soc.*, **84**, 3237 (1962), and the first practical testing can be found in the same issue on p. 3242.

Refer back to what takes place in a regular ion exchange process. Consider a resin in the H^+ form such as:

$$\begin{cases} -SO_3^-, H^+ \\ -SO_3^-, H^+ \end{cases}$$

If a solution containing Ca^{+2} ions is passed through the column, an *ion exchange* takes place:

$$\begin{cases} -SO_3^- \\ -SO_3^- \end{cases} Ca^{+2} \quad + \quad 2\,H^+$$

The original resin can be regenerated by adding a high concentration of H^+:

$$Xs\,H^+ \;+\; \begin{cases} -SO_3^- \\ -SO_3^- \end{cases} Ca^{+2} \;\longrightarrow\; \begin{cases} -SO_3^-\;H^+ \\ -SO_3^-\;H^+ \end{cases} + \; Ca^{+2}$$

The same process can be achieved by ligand exchange, if ligands of different coordinating valences are used and the concentrations are changed as needed.

An ion exchange resin containing a complexing metal ion, such as Cu^{+2}, Ni^{+2}, Ag^+, Zn^{+2}, Co^{+2}, or Fe^{+3}, as the counterion forms what is called the *solid sorbent*. $Ni(H_2O)_6^{+2}$ will be used as an example. The resin, such as a carboxylic acid, chelates with the nickel and replaces two of the water ligands.

$$\begin{cases} -COO^- \\ -COO^- \end{cases} \begin{matrix} H_2O \\ \cdots \cdots OH_2 \\ Ni^{+2} \\ \cdots \cdots OH_2 \\ H_2O \end{matrix}$$

The four water molecules are *ligands* and are held by the nonbonding electrons of the oxygen molecule. If other ligands now are poured down the column, the water ligands can be exchanged by the other ligands. The extent depends upon the strength of the cation-nonbonding electron bonds.

$$4\,NH_3 \;+\; \begin{cases} -COO^- \\ -COO^- \end{cases} \begin{matrix} H_2O \\ \cdots OH_2 \\ Ni^{+2} \\ \cdots OH_2 \\ H_2O \end{matrix} \;\longrightarrow\; \begin{cases} -COO^- \\ -COO^- \end{cases} \begin{matrix} NH_3 \\ \cdots NH_3 \\ Ni^{+2} \\ \cdots NH_3 \\ NH_3 \end{matrix} \;+\; 4\,H_2O$$

According to Helfferich, "Separations of ligands by ligand exchange can be carried out by the usual chromatographic procedures (elution development, gradient elution, displacement development, frontal analysis). Except for the advantage of high selectivities, attained by virtue of the high specificity of complex formation, there is no essential difference between ligand exchange chromatography and other chromatographic techniques. For separating compounds having different coordinative valences, however, ligand exchange offers the additional advantage that, within wide limits, the selectivity can be adjusted at will by varying the concentration of the external solution. Dilution of the solution increases the preference of the resin for the ligand having the higher coordinative valence and vice versa. If a suitable metal ion is chosen, the selectivity can even be reversed so that one ligand is preferentially taken up from dilute mixtures, and the other from concentrated mixtures.

"A suitable ligand exchanger prefers the ligands of the higher valence when the solution is dilute, and those of lower valence when the solution is concentrated. The ligand or ligands with the higher valence can thus be selectively removed from *dilute* mixtures in exchange for a ligand of lower valence. The ligand or ligands with the higher valence can then be recovered in high concentration, and the bed regenerated, by displacement with a *concentrated* solution of the ligand with the lower valence. No additional regeneration step, except for washing with solvent, is required to restore the bed for the next cycle."

Helfferich used the separation and concentration of 1,3-diaminopropanol-2 (DAP) from water to test the concept.

$$0.005\ M\ DAP\ \text{in water} + \left[\begin{array}{c}-COO^-\cdots NH_3\\ \quad\quad Ni^{+2}\\ -COO^-\cdots NH_3\end{array}\ \substack{NH_3\\ \\ NH_3}\right] \longrightarrow \left[\begin{array}{c}-COO^-\cdots NH_2R\\ \quad\quad Ni^{+2}\\ -COO^-\cdots NH_2R\end{array}\ \substack{NH_2R\\ \\ NH_2R}\right] + 4\ NH_3$$

which can be regenerated by the addition of concentrated NH_4OH and the DAP recovered.

$$Xs\ NH_3 \longrightarrow \left[\begin{array}{c}-COO^-\cdots NH_3\\ \quad\quad Ni^{+2}\\ -COO^-\cdots NH_3\end{array}\ \substack{NH_3\\ \\ NH_3}\right] + 4\ NH_2R$$

REVIEW QUESTIONS

1. What is the composition of an ion retardation resin?
2. What is the primary use for an ion retardation resin?
3. What are the *snakes* in a snake-cage resin?
4. What causes the caged polymers to pair off across from each other?
5. What is a *net-cage* resin?
6. What are some crosslinking compounds for basic net-cage resins?
7. What is the wet bed capacity of Bio-Rad's AG11A* resin?
8. How do net-cage resins compare to snake-cage resins with regard to regeneration stability?
9. What is the basis for separation in ion exclusion?
10. What is the major requirement?
11. What actually decides how the separation takes place?
12. Where are the ionic compounds more concentrated?
13. What is a nonionic compound as related to ion exclusion?
14. How can organic acids be made nonionic for ion exclusion purposes?
15. Why would ion exclusion be preferred to regular ion exchange?
16. What are the practical maximums for both ionic and nonionic samples?
17. What is the advantage of an ion exclusion column placed in an HPLC?
18. What is the advantage of using a guard column to do ion exclusion?
19. What does D for nonionic components depend upon?
20. What is the nature of an HPLC ion exclusion column?
21. How does ligand exchange differ from ion exchange?
22. What acts as the solid sorbent in ligand exchange?
23. What produces the high selectivity of ligand exchange?
24. If a dilute solution of ligands is being separated, what is the preference?
25. A dilute solution of H_2NNH_2, H_2NNHCH_3, and $H_2NN(CH_3)_2$ is passed through an Ni^{+2} column. What would you predict would be the order of separation?

IF IT EXISTS - CHEMISTRY IS INVOLVED
IF YOU CAN BUY IT - A CHEMIST WAS INVOLVED SOMEWHERE

SEPARATIONS INVOLVING ELECTRIC FIELDS

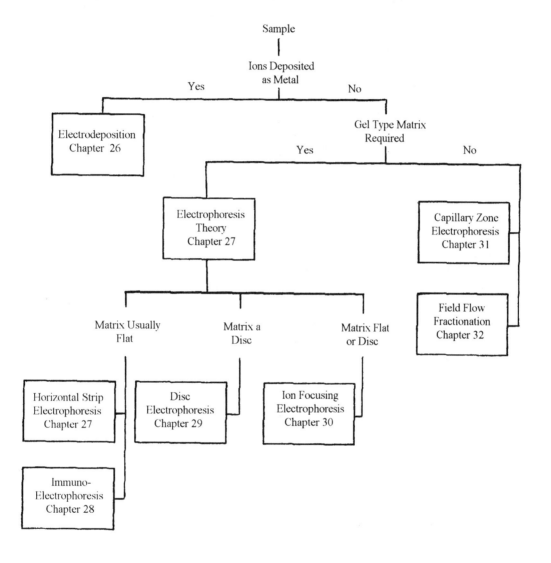

The application of a voltage to a conducting solution has resulted in some very valuable separation techniques. The basic concept is to place at least two electrodes into a conducting solution, one positive and one negative. What happens is what is expected; negative ions move to the positive electrode and vice-versa. If the potential is high enough, the ions can be reduced to a metal and deposited on the cathode. Potential controls which ions are deposited, and current controls how fast. By properly controlling the voltage and current, it is possible to make separations to 1 part in 5000. This is called electrodeposition, and the factors involved in this are discussed in Chapter 26, along with some safety tips. Commercially, this is the method of choice to electroplate metals on almost any surface desired.

If there are differences in size and charge, then a separation can take place in the solution. This original concept is known as electrophoresis and is discussed in Chapter 27. While good separations can take place as described above, it was hard to remove the separated ions from the solution without remixing them. Subsequently, many modifications have been developed to improve the separation and permit easier detection of the compounds. The introduction of materials that would turn the solution into a gel, and placing this gelled solution on a strip of paper or plastic has been of real value. It reduces diffusion, can improve the separation by a sieving effect, makes handling much easier, and permits an increased ease of detection. The first of these gel techniques was known as *horizontal electrophoresis,* discussed in Chapter 27. However, modifications of this technique include vertical placement as well as draping the strip over a ridge pole.

Immunoelectrophoresis, Chapter 28, is a special modification involving detection. It relies on antibody-antigen interactions to produce a precipitate with the desired compounds. It can be very sensitive and does not destroy all of the separated compound, which can then be removed for further examination.

The use of polyacrylamide as a gel matrix and placing it in a vertical cylinder is known as *disc electrophoresis.* It gets its name from the very sharp discs of compounds that are formed when made visible by staining (Chapter 29). It can be much better than horizontal electrophoresis in that in one case reported it separated over 30 components from a mixture that horizontal methods had separated only 7. The addition of a pH gradient between the electrodes can produce increased resolution, because the ions will move to where they are electrically neutral, their *isoelectric point,* and stay there as long as the voltage is on. This can be done either on a flat strip or in a disc, and is known as *isoelectric focusing*, Chapter 30. Another portion of that chapter describes the use of sodium dodecylsulfonate (SDS) to aid in the movement of neutral compounds.

Capillary zone electrophoresis (Chapter 31) at the present time provides the ultimate in separation capability and uses the smallest samples. Sample sizes of nanoliters are usual in a capillary 50 µm in diameter and 30 cm long.

An entirely different concept is discussed in Chapter 32: *field flow fractionation* (FFF). As an illustration, consider a stream bed with gravel on the bottom. The water rushing over the bed will move the smallest rocks farthest downstream, and eventually sand bars are formed. So it is with FFF. Two plates, a few cm wide, and spaced a few mm apart, have a voltage applied across them. When a solution containing a mixture of ions of different size is passed between the plates, the ions are driven to the plate of opposite charge, just like the gravel in the stream bed. The smaller the ion, the farther downstream it will move, and separation takes place. While the initial concept involved an electrical field to provide the driving force between the plates, crossed flow solutions, magnetic fields, and centrifugal force are some of the additional ways that have been developed.

Overall, electrophoresis is to ions what chromatography is to neutral molecules. There is one report of a two-dimensional separation involving ion focusing and SDS to resolve over 1000 compounds on a single sheet. That is amazing.

26
ELECTRODEPOSITION

PRINCIPLES

Electrodeposition involves removing of a metal ion from solution and depositing it on a conducting surface by reducing it to the metal by the transfer of electrons. The conducting surface is called an electrode. This principle is of vast importance in industry; the 11th *Collective Index of Chemical Abstracts* lists over 2500 citations. Its major use in industry is in electroplating, that is, coating a desirable metal over an undesirable one. Examples would be covering an iron car bumper with chromium or an iron spoon with silver. Other common uses are corrosion control, electromachining, electropolishing, and the removal of metals from waste waters. Although these are perfectly valid uses as a separation technique, with more control, it is possible to separate many metals from solution in a quantitative manner. In laboratories, this is called *electrodeposition* and is more likely to be done by analytical chemists.

The terms *potential* and *voltage* are often used interchangeably, but there is a difference. Potential is a measure of charge separation and the unit of measurement is the volt. It is correct to say, "The electrode potential is 1.5 volts."

Potential determines what metals are deposited, and current determines how fast.

Typically, the potential is adjusted to deposit the metal desired, and then the current is adjusted to deposit it as fast as possible and still get a good quality deposit. Regardless of the application, all are based on the law developed by Michael Faraday in 1833. He was the first to use the word electrode and used it for either electrode. Now the negative electrode is called the *cathode* (cation attracting), and the positive electrode is called the *anode* (anion attracting).

The number of gram-equivalent weights of a substance deposited, liberated, dissolved, or reacted at an electrode is equal to the number of Faraday's of electricity transported through the electrolyte.

$$g = \frac{i \times t \times a.w.}{n \times F} \qquad (26\text{-}1)$$

where: g = grams of metal deposited. i = current in amperes. t = time in seconds. a.w. = atomic weight of the metal. n = number of electrons in the reduction/atom. F = Faraday's constant: 96,485 coulombs. (1 coulomb = 1 ampere/second)

The first known quantitative separations (copper) were done in 1864 by W. Gibbs and in 1865 by C. Luckow.

EXAMPLE CALCULATION
How long must a car bumper be immersed in a Cr^{+3} solution to deposit 0.500 gram on it, if the current is 10.0 amperes?

ANSWER

$$0.500 = \frac{10.0 \times t \times 52.0}{3 \times 96,485}; \quad or \quad t = \frac{0.500 \times 3 \times 96,485}{10.0 \times 52.0}; \quad t = 278 \text{ sec } or \text{ 4.64 min}$$

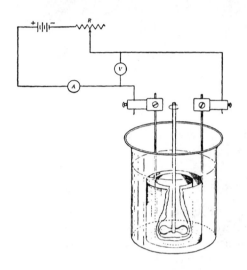

Figure 26-1. Diagram of an electrodeposition apparatus.
(Courtesy - H. Diehl, *Quantitative Analysis*, J. Wiley & Sons, New York, NY, 1952)

Figure 26-2. Electrolytic analyzer, Sargent-Slomin
(Courtesy - Eberbach Corp., Ann Arbor, MI)

In an analytical laboratory, electrodeposition is used to determine how much material is in a solution by selectively depositing it on an electrode. The electrode is weighed before and after, and the percentage determined based on the sample weight. You could determine the concentration of chromium in the above solution by taking a small aliquot and depositing the chromium onto a weighed platinum electrode.

Figure 26-1 is a diagram of the basic laboratory apparatus required, and Figure 26-2 shows one type of commercial laboratory apparatus for quantitative separations.

An accuracy of 1 part in 5000 can be attained with this technique, and it is quite useful to determine those metals that are major constituents in an alloy. The experiment explained later involves copper and nickel. However, other elements that have been separated include Sb, As, Bi, Cd, Co, Au, In, Fe, Pb, Hg, Mo, Pd, Pt, Rh, Ag, Te, Tl, Sn, and Zn. Actually, if you just want to separate metals as a group from a solution, such as trace contaminants, then nearly every metal and metalloid can be separated.

EARLY HISTORY

The study of electricity probably began with amber, a pale yellow to reddish brown fossil resin from the extinct pine tree *Pinites succinifera*. It is found along the Baltic coast and is mined in what was east Prussia. As early as 600 B.C., Thales of Miletus attracted feathers to amber that had been rubbed with a cloth.

1600   William Gilbert, an English physician, showed that amber attraction was different from lodestone attraction (iron only).

1646   Sir Thomas Browne, an English physician, used the word *electricity* for the first time. *Elektron* (Gr. "amber").

1700's   Stephen Gray, an Englishman, showed that some substances conduct electricity and others do not.

1752   Benjamin Franklin showed that lightning is electricity. He also proposed the use of positive and negative for the opposite charges.

1785   Charles Agustin de Coulomb, a French chemist, worked out the laws of attraction and repulsion between charged bodies. *The force of repulsion of two like charged particles is proportional to the product of their charges and inversely proportional to their distance of separation.*

1786   Luigi Galvani, an Italian anatomy professor, connected a frog leg to a copper wire hook and hung the hook over an iron rail. The leg twitched when the copper touched the railing. He concluded (wrongly) that the leg contained electricity that was released when the leg touched the metal.

1800 Alessandro Volta, Italian physicist, discovered why the frog's leg twitched. He built the first battery (voltaic pile) of stacked Ag-Zn plates separated by paper or cloth soaked in saltwater.

1819 Hans Christian Oerstad, a Danish scientist, discovered that a magnetic needle was deflected by an electric current - electromagnetism.

1826 Thomas John Seebeck, a German physicist, discovered that if two dissimilar metals were placed together and heated at one end, a potential was generated - thermal electricity.

1826 Andre Marie Ampere, a French mathematical physicist, published his classical work "Theory of Electrodynamic Phenomena"

1829 Georg Simon Ohm, a German physicist, discovered a relationship among voltage, current, and resistance- *Voltage equals the product of the current and the resistance* - (Ohm's law).

1831 Michael Faraday, an English physicist, believed that if electricity could produce magnetism then magnetism could produce electricity. He then showed that a moving magnet could produce an electrical current in a coil of wire. This is the basis of electric generators and transformers. He developed the concept of lines of force.

1833 Faraday developed what is now known as Faraday's Law.

1881 H. L. F von Helmholtz, a German physicist, recognized that something must carry the electrical current and suggested the existence of electrons.

1897 Joseph J. Thompson, an English physicist, is credited with discovering the electron and that electrons carry electrical current. He won the Nobel Prize in 1906 for this work.

1889 Walther Nernst, a German physical chemist, developed the relationship to relate the voltage obtained from a battery to the concentration of the chemicals present.

1913 Robert Andrews Millikan, an American physicist, determined the amount of charge on an electron. He was awarded the Nobel Prize in 1923 for this work.

We now know that an electron is about 5.6×10^{-13} cm in diameter, that it has an estimated rest mass of 9.1×10^{-28} g and a charge of 1.6×10^{-19} coulombs, that electrical charge is due to an increase or decrease in the number of electrons on a surface, and that an electrical current is due to the electrons moving through a conductor.

ELECTROCHEMICAL CELLS

An electrochemical cell (voltaic cell) is a device in which an oxidation-reduction reaction occurs and electrochemical energy (electricity) is produced. The *oxidation* reaction occurs at the *anode*, and the *reduction* reaction occurs at the *cathode*. Electrons are produced at the anode (a half cell) and are transported through the external circuit (connecting wires and measuring meters) to the cathode (a half cell). The internal circuit is completed by a *salt bridge*, the salt solution being held in place by an agar gel or by porous plugs. Figure 26-3 shows one combination of two half cells to form a *battery*.

Half reactions:

$Cu^{+2} + 2e^- = Cu \quad E^o = +0.34$ v

$Zn^{+2} + 2e^- = Zn \quad E^o = -0.76$ v

To determine the spontaneous direction of a voltaic battery using reduction potential half cells, write the most positive half cell reaction on top (as above) and the reaction will proceed in a clockwise manner.

Cathode reaction: (reduction)

$Cu^{+2} + 2e^- = Cu$

Anode reaction: (oxidation)

$Zn = Zn^{+2} + 2e^-$

Overall reaction:

$Zn + Cu^{+2} = Zn^{+2} + Cu$

$E_{battery} = E_{cathode} - E_{anode}$ (26-2)

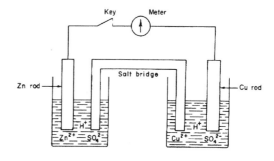

Figure 26-3. Diagram of a voltaic cell. (Courtesy - Pietrzyk, D.J. and Frank, C.W., *Analytical Chemistry*, Academic Press, New York, NY, 1974.)

EXAMPLE CALCULATION

Refer to Table 26-1. (A) What spontaneous reaction takes place if you place Ag_2O and Cu in a basic solution? (B) What is the potential of such a system?

ANSWER

A. Write the half cell reactions as reduction potentials with the most + on top.

$Ag_2O + H_2O + 2e^- = 2 Ag + 2 OH^-$ $E^o = +0.344$
$Cu_2O + H_2O + 2e^- = 2 Cu + 2 OH^-$ $E^o = -0.361$

The reaction will go spontaneously in a clockwise direction. Ag_2O will go to Ag metal, and Cu metal will go to Cu_2O.

B. $E_{battery} = E_{cathode} - E_{anode}$: $+0.344$ v $- (-0.361$ v$) = +0.705$ v.

The values in Table 26-1 are called *thermodynamic potentials*, that is, they are ideal values. Several factors, such as *overpotential, activation energy, complexation,* and *pH*, can change these values. These will be discussed later.

Table 26-1. Reduction potentials

Half Reaction		E^o, volts
The more + the potential, the better the oxidizing agent.		
$H_2O_2 + 2 H^+ + 2e^-$	$= 2 H_2O$	+1.77
$Ce^{+4} + e^-$	$= Ce^{+3}$	1.61
$MnO_4^- + 8 H^+ + 5e^-$	$= Mn^{+2} + 4 H_2$	1.52
$PbO_2 + 4 H^+ + 2e^-$	$= Pb^{+2} + 2 H_2$	1.46
$Au^{+3} + 3e^-$	$= Au$	1.42
$Cr_2O_7^{-2} + 14 H^+ + 6e^-$	$= 2 Cr^{+3} + 7 H_2$	1.36
$Tl^{+3} + 2e^-$	$= Tl^+$	1.21
$Br_2 + 2e^-$	$= 2 Br^-$	1.085
$Hg^{+2} + 2e^-$	$= Hg$	0.854
$Ag^+ + e^-$	$= Ag$	0.799
$Fe^{+3} + e^-$	$= Fe^{+2}$	0.77
$Sb_2O_5 + 6 H^+ + 4e^-$	$= 2 SbO + 3 H_2O$	0.64
$VO^{+2} + 2 H^+ + e^-$	$= V^{+3} + H_2O$	0.361
$Fe(CN)_6^{-3} + e^-$	$= Fe(CN)_6^{-4}$	0.36
$Ag_2O + H_2O + 2e^-$	$= 2 Ag + 2 OH^-$	0.344
$Cu^{+2} + 2e^-$	$= Cu$	0.344
$Hg_2Cl_2 + 2e^-$	$= 2 Hg + 2 Cl^-$ (satd)	0.244
$Sn^{+4} + 2e^-$	$= Sn^{+2}$	0.14
$2 H^+ + 2e^-$	$= H_2$	0.00
$Fe^{+3} + 3e^-$	$= Fe$	-0.036
$Pb^{+2} + 2e^-$	$= Pb$	-0.126
$Sn^{+2} + 2e^-$	$= Sn$	-0.14
$Cu_2O + H_2O + 2e^-$	$= 2 Cu + 2 OH^-$	-0.361
$Fe^{+2} + 2e^-$	$= Fe$	-0.426
$S + 2e^-$	$= S^=$	-0.508
$Cr^{+3} + 3e^-$	$= Cr$	-0.71
$Zn^{+2} + 2e^-$	$= Zn$	-0.761
$ZnO_2^{-2} + 2 H_2O + 2e^-$	$= Zn + OH^-$	-1.216
$Th^{+4} + 4e^-$	$= Th$	-2.06
$Mg^{+2} + 2e^-$	$= Mg$	-2.34
$Na^+ + e^-$	$= Na$	-2.714
$Ba^{+2} + 2e^-$	$= Ba$	-2.90
$Li^+ + e^-$	$= Li$	-3.024

The potential that is applied determines which ions are reduced. The ions that have the most positive reduction potential will react first.

> **EXAMPLE**
> Ag^{+1}, Au^{+1}, and Rh^{+2} are in a solution. In what order would they plate out?
>
> **ANSWER:**
> 1st $\quad Au^{+1} + e^- = Au \quad E^o = +1.68$ v
> 2nd $\quad Ag^{+1} + e^- = Ag \quad E^o = +0.80$ v
> 3rd $\quad Rh^{+2} + 2e^- = Rh \quad E^o = +0.60$ v

ELECTROLYTIC CELLS

If you dip a metal rod into a water solution of its ions, the atoms may lose electrons, form ions, and go into the solution, or ions in the solution may gain electrons from the metal rod, form more metal, and deposit on the rod. Each type of atom or ion has different amounts of positive and negative charges around it, and it only seems reasonable that they should either attract electrons or donate electrons to electrodes under different conditions. You want to take advantage of this difference in attraction to electrodes to separate various systems. For example, sodium, will immediately give up an electron and go into solution as a sodium ion. Gold will not do this very readily at all. This is why you can pan for gold in a river, but you don't pan for sodium.

This tendency of a metal to give up electrons and go into solution is called its *electromotive force* (EMF). It is measured in *volts* and is given the symbol E. If E is determined with unit activities for all of the components of the reaction, then the potential obtained is called the *standard potential*, E^o. Table 26-1, p. 302, lists several standard reduction potentials.

Let us see what would happen in a solution if we had a mixture of the bromides of Sn^{+2} ($E^o = -0.140$ v), Zn^{+2} ($E^o = -0.762$ v), and Mn^{+2} ($E^o = -1.05$ v) in a cell like that shown in Figure 26-1, p. 300. Notice in this case that a battery has been added as a part of the external circuit. The purpose of the external battery is to force reactions to take place in the cell that would not do so spontaneously by applying a potential larger and of opposite sign than the potential of the voltaic cell. This is called an *electrolytic cell*.

Overpotential is all of the extra potentials required to actually electrodeposit a metal and will be discussed later on. *We will ignore overpotential and the decomposition of water in order to simplify the explanation and cover the basic concepts.*

Let us consider only the cathode (negative electrode in electrolytic cells) at this time. What happens if we make the cathode slightly negative? Electrons are forced into the cathode by the external battery. The negative charge attracts the positive ions (Mn^{+2}, Zn^{+2}, Sn^{+2}) in the solution. If the metal ions pick up these electrons, they will be reduced to the metallic state and *plate out on the electrode*. We find that the metal ions behave quite differently as different potentials are applied. If we apply only -0.11 v to the electrode, the Sn^{+2} ions ($E^o = -0.104$ v) will be reduced to metallic Sn, but the Zn^{+2} and Mn^{+2} ions, having higher reduction potentials, will stay in solution. If we keep the electrode at this potential for a long period of time, nearly all of the Sn^{+2} ions will be removed from the solution. If the electrode is now made more negative, to -0.76 v, then the Zn^{+2} ions ($E^o = -0.76$ v) are able to pick up electrons from the electrode and are reduced to metallic zinc. At this potential, the Mn^{+2} ions remain in solution, while the zinc is plated out. If the electrode potential then is made -1.05 v, the Mn^{+2} ions ($E^o = -1.05$ v) will be reduced. How fast the ions will be removed from solution depends upon the current applied. The greater the current, the faster the ions will be removed.

While the metal ions plate out at the cathode, what reactions take place at the anode? The solution contains a mixture of bromides in water.

$$Br_2 + 2e^- = 2 Br^- \qquad\qquad E^o = +1.085 \text{ v}$$

$$O_2 + 4 H^+ + 4e^- = 2 H_2O \qquad\qquad E^o = +1.23 \text{ v}$$

Those bromide ions give up electrons to the anode (become oxidized) and form bromine more readily than does water. The electrons return from the anode to the battery to complete the circuit. Table 26-2, p. 304, lists the reduction potentials of several common anions.

Notice that the sulfide ion, S^{-2}, is oxidized very easily to elemental sulfur. This often causes problems, because sulfur adsorbs on the anode and *poisons* it (makes the process much less efficient or even impossible to continue the process). Generally, sulfur-containing compounds are avoided if possible in electrochemical work.

Table 26-2. Reduction potentials of some anions

Half Reaction			E^o, volts
$S + 2e^-$	=	S^{-2}	-0.50
$SO_4^{-2} + 4 H^+ + 2e^-$	=	$H_2SO_3 + H_2O$	+0.20
$I_2 + 2e^-$	=	$2 I^-$	+0.54
$CN^- + 2e^-$	=	$CO_3^{-2} + N_2$	+0.66
$NO_3^- + 3 H^+ + 2e^-$	=	$HNO_2 + H_2O$	+0.94
$ClO_4^- + 2 H^+ + 2e^-$	=	$ClO_3^- + H_2O$	+1.00
$Br_2 + 2e^-$	=	$2 Br^-$	+1.08
$O_2 + 4 H^+ + 4e^-$	=	$2 H_2O$	+1.23
$Cl_2 + 2e^-$	=	$2 Cl^-$	+1.36
$F_2 + 2e^-$	=	$2 F^-$	+2.85

POTENTIALS INVOLVED IN AN ELECTROLYTIC CELL

Five major potentials are involved in an electrodeposition: the single electrode potentials at each electrode, the polarization potential at each electrode, and the IR drop through the solution. Figure 26-4 shows a plot of the relative magnitude of each for a copper electrodeposition.

Referring to Table 26-1, you see that reducing Fe^{+2} to Fe will require -0.44 volts. This usually is not the case in actual practice. The value can be altered by stirring the solution, by the type of electrode, by the pH, whether a complexing ion is present, and what reaction is taking place at the anode.

Reversible Back Electromotive Force

Consider the plating of copper from a water solution with no other factors considered. Electrons must flow in a circuit. To get an electron flow, you must consider the reaction taking place at the anode. From Table 26-1:

At the cathode $Cu^{+2} + 2 e^- \longrightarrow Cu$ $E^o = -0.34$ v

At the anode $2 H_2O \longrightarrow O_2 + 4 H^+ + 4 e^-$ $E^o = +1.23$ v

$$E_{back} = E^o{}_{anode} - E^o{}_{cathode} = 1.23 \text{ v} - 0.34 \text{ v} = +0.89 \text{ v}$$

Therefore, in order to get the reaction to go, you must apply at least 0.89 v. This will change, depending on the anion associated with the copper.

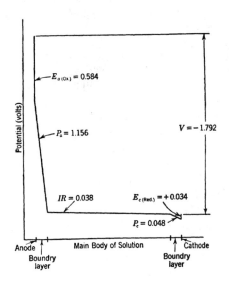

Figure 26-4. Various potentials in an electrodeposition of copper from a nitric acid solution.
(Courtesy - H. Diehl and Smith, G.F., *Quantitative Analysis*, J. Wiley & Sons, New York, 1952)

Concentration Overpotential or Polarization

Consider an unstirred solution of a metal ion. As the metal ion is plated from the solution, its concentration decreases next to the electrode compared to the bulk concentration. This is because the metal ion must diffuse through the solvent to get to the electrode. This rate of diffusion depends on the concentration of the ion, the ionic strength of the solution, and indirectly on the current applied. The more current, the faster the metal ions are removed near the electrode surface. A concentration profile is shown in Figure 26-5.

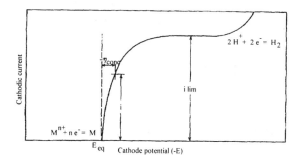

Figure 26-5. Cathodic polarization curve of a Pt electrode in an acidic solution.
(Courtesy - Laitinen, H.A., *Chemical Analysis*, McGraw-Hill, New York, 1960)

The change in the thermodynamic value can be approximated by using the *Nernst equation*.

$$E = E^o - \frac{2.3\,R\,T}{n\,F} \log \frac{[Products]}{[Reactants]} \qquad (26\text{-}3)$$

where: $R = 8.314$ Joules/deg/mol. T = temperature in K. F = Faraday's constant, use 96,500. The $2.3\,R\,T/F = 0.059$. n = number of electrons involved in that electrode reaction. [Products] and [Reactants] are based on the equation as written as a reduction potential.

At the cathode; $\quad E = E^o - \dfrac{0.059}{n} \log \dfrac{[Cu]}{[Cu^{+2}]} \quad$ As Cu^{+2} decreases, E^o becomes more negative which increases the overpotential.

At the anode; $\quad E = E^o - \dfrac{0.059}{n} \log \dfrac{[O_2]\,[H^+]^4}{[H_2O]^2} \quad$ As O_2 increases, E^o becomes more positive which increases the overpotential.

The concentration change can amount to a 1000-fold difference and about 0.18 v in unstirred single electron change solutions. This is called *concentration overpotential* or *polarization*. The technique to reduce this effect is to stir the solution and to use as low a current as is reasonable to complete the separation.

Activation Overpotential

When an ion is on the surface of the electrode, electrons must transfer from the electrode to the ion. How easily this takes place depends on the electrode metal and the nature of its surface. You must build up a charge first until the potential difference is such that a transfer will take place. This is called *activation overpotential*. Figure 26-6 shows how this varies as the current density changes.

Two electrodes are of importance, platinized Pt and Hg. Electrons transfer easiest from a platinized Pt electrode, so this type is preferred for cathodes. Hg resists the transfer of electrons and is preferred for metals more negative than H^+, because it reduces the hydrogen evolution so reductions can take place. This electrode is preferred for many electrochemical reactions. To platinize a Pt electrode, simply deposit a thin film of Pt on it, usually from a dilute chloroplatinic acid solution.

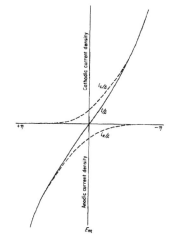

Figure 26-6. Activation over potential.
(Courtesy - Laitinen, H.A., *Chemical Analysis*, McGraw-Hill, New York, 1960)

Hydrogen Overpotential

The potential at which water decomposes at an electrode plays a major role in limiting which metals actually can be deposited.

$$2 H_2O \longrightarrow OH^- + H_3O^+$$

At the anode, oxygen can be formed from any OH^- present.

$$2 OH^- + 4e^- \longrightarrow O_2 + 2 H^+ \quad E^o = +1.23 \text{ v at 1 M } H^+$$

At the cathode, hydrogen can be formed from any H_3O^+ present.

$$2 H^+ + 2e^- \longrightarrow H_2 \quad E^o = 0.00 \text{ v at 1 M H}$$

Because H_2 forms so easily, it can form before the desired metal deposits and prevent the metal from depositing. If however, an electrode has a significant hydrogen overpotential, then many more metals can be deposited. Refer to Table 26-1, p. 302; all those metals more positive than H_2 can be plated out. Those more negative require an electrode with a high hydrogen overpotential to be deposited. Therefore, the Hg electrode is a favorite for many applications.

If high currents are used, hydrogen may form in small amounts in the metal deposit. Even a few hundredths of a percent make the deposit spongy, and particles are broken off easily. In addition, for electroplating, the deposit is very brittle and must be baked to remove the hydrogen.

In a mixture of metal ions, the one with the most positive reduction potential will be reduced first. The metals which are positive to hydrogen can be deposited in acid solutions, because the metal is reduced more easily than the hydrogen ion.

Overpotential shifts the potential at which hydrogen is discharged to more negative values, usually much more so than the metals, so even metals as negative as Zn^{+2} ($E^o = -0.76$ v) can be deposited.

In alkaline solutions, the H^+ concentration is very low, so the hydrogen discharge potential goes more negative (-0.59 v at pH 10).

Figure 26-7 shows the effect of hydrogen overvoltage for several metals. Note that platinized Pt has the least effect.

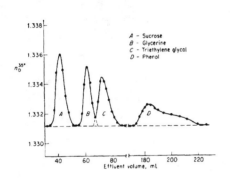

Figure 26-7. Hydrogen overpotential for several metals.
(Courtesy - Diehl, H., *Quantitative Analysis*, John Wiley & Sons, New York, NY, 1952)

COMPLEXATION AND CHELATION

The net effect of either complexation or chelation of a metal ion is to reduce the concentration of the metal ion at the electrode. This changes the potential and can be determined from the Nernst equation. Usually, a half cell potential such as

$$Fe^{+3} + e^- \longrightarrow Fe^{+2} \quad E^o = +0.77 \text{ v}$$

was determined with either perchlorate or nitrate as the anion, because neither forms effective complexes with metal ions, particularly the former. However, if other anions are involved, then the potential can be altered considerably.

$$Fe(CN)_6^{-3} + e^- \longrightarrow Fe(CN)_6^{-4} \quad E^o = +0.416 \text{ v}$$

$$Fe(ophen)_3^{+3} + e^- \longrightarrow Fe(ophen)_3^{+2} \quad E^o = +1.06 \text{ v}$$

$$Au^{+3} + 3e^- \longrightarrow Au \quad E^o = +1.50 \text{ v}$$

$$Au(OH)_3 + 3 H^+ + 3e^- \longrightarrow Au + 3 H_2O \quad E^o = +1.45 \text{ v}$$

$$AuCl_4^- + 3e^- \longrightarrow Au + 4 Cl^- \quad E^o = +1.00 \text{ v}$$

$$AuBr_4^- + 3e^- \longrightarrow Au + 4 Br^- \quad E^o = +0.87 \text{ v}$$

$$Au(CN)_2^- + e^- \longrightarrow Au + 2 CN^- \quad E^o = -0.60 \text{ v}$$

The following example of the silver cyanide complex (from the late Professor H. Diehl) illustrates how the calculation is made.

$$Ag^+ + e^- \rightarrow Ag \qquad E^\circ = +0.799 \text{ v}$$

$$Ag(CN)_2^- \rightarrow Ag^+ + 2 CN^- \qquad K_{dis} = \frac{[Ag^+][CN^-]^2}{[Ag(CN)_2^-]} = 3.8 \times 10^{-19}$$

Combining this with the Nernst equation:

$$E = E^\circ - 0.059 \log K_{dis} - 0.059 \log \frac{[Ag(CN)_2^-]}{[CN^-]^2}$$

For a solution 1 M in Ag and 1 M in CN⁻

$$Ag(CN)_2^- + e^- \rightarrow Ag + 2 CN^- \qquad E^\circ = -0.286 \text{ v.} \quad \text{This is a shift of 1.085 v.}$$

Advantage can be taken of complexing in a way you might not have considered. Brass is a mixture of copper and zinc.

$$Cu^{+2} + 2e^- \rightarrow Cu \qquad E^\circ = +0.344 \text{ v}$$
$$Zn^{+2} + 2e^- \rightarrow Zn \qquad E^\circ = -0.76 \text{ v}$$

If you try to obtain a brass plating on a base metal, you can see that you cannot plate out both metals simultaneously this way. However, if cyanide complexes are made, the following E°'s occur.

$$Cu(CN)_3^{-2} + e^- \rightarrow Cu + 3 CN^- \qquad E^\circ = -1.09 \text{ v}$$
$$Zn(CN)_4^{-2} + 2e^- \rightarrow Zn + 4 CN^- \qquad E^\circ = -1.26 \text{ v}$$

They are now close enough together that they plate out simultaneously.

pH

The effect of altering the pH can be seen by examining the Nernst equation (26-3) and those half cell reactions that involve acid or base. *As the solution becomes more acidic the E° value becomes more positive.*

Consider the following reaction:

$$SbO^+ + 2 H^+ + 3e^- \rightarrow Sb + H_2O \qquad E^\circ = +0.212 \text{ v}$$

Substituting into the Nernst equation at pH = 7:

$$E = 0.212 - \frac{0.059}{3} \log \frac{[Sb][H_2O]}{[SbO^+][H^+]^2}$$

Assume [SbO⁺] = 0.01 M; [Sb] = 1 (solid); [H₂O] = 1 (solvent):

$$E = 0.212 - 0.002 \log \frac{[1][1]}{[0.01][10^{-7}]^2} = 0.209 \text{ v}$$

With three electrons involved, the change is not much. At pH = 1: $E = 0.212$ v
Under basic conditions, a different compound is formed:

$$SbO_2^- + 2 H_2O + 3e^- \rightarrow Sb + 4 OH^- \qquad E^\circ = -0.66 \text{ v}$$

This changes the E^o 0.87 v in the beginning, so you must be careful to know what compound you have if pH is involved.

CONSTANT CURRENT ELECTRODEPOSITIONS

This is the process normally used for electroplating, because either only one metal is present or the conditions can be controlled so other metals do not interfere. One way to control the conditions is to use a potential buffer. *A potential buffer consists of a mixture of the reduced form and the oxidized form of the metal, both as ions.* As an example:

$$U^{+4} + e^- \longrightarrow U^{+3} \qquad E^o = -0.61 \text{ v}$$

U^{+4} is reduced (cathode) to U^{+3}, rather than to metallic uranium, and the U^{+3} is oxidized to U^{+4} at the anode. Because neither plates out, the system is stable. Suppose you wished to reclaim zinc from galvanized iron.

$$Zn^{+2} + 2e^- \longrightarrow Zn \qquad E^o = -0.76 \text{ v}$$

$$Fe^{+2} + 2e^- \longrightarrow Fe \qquad E^o = -0.44 \text{ v}$$

If a uranium ion buffer is added, the zinc can be deposited without the iron interfering.

CONSTANT VOLTAGE ELECTRODEPOSITIONS

As a rule of thumb, *to get a good separation, a 0.2 v difference must exist between the metals to be separated.* To be able to maintain this during an entire deposition is very difficult and usually cannot be done unless a special effort is made to maintain a constant potential. Such an apparatus is called a *potentiostat*. It requires a third electrode to monitor the potential, and then a way to adjust the potential as needed. A diagram of the minimum requirements is shown in Figure 26-8.

Although the main advantage of a controlled current separation is speed, the *main advantage of a controlled potential separation is selectivity*.

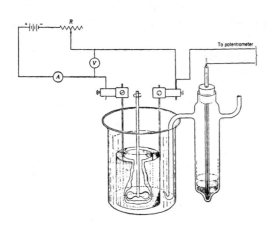

Figure 26-8. Minimum components for a controlled potential electrodeposition.
(Courtesy - Diehl, H., *Quantitative Analysis*, J. Wiley & Sons, New York, NY, 1952)

Three electrodes are required. Two have already been discussed; the working electrode where the deposition occurs, and the auxiliary electrode needed to complete the circuit. The potential of the working electrode must be kept as near to the desired voltage as possible. A reference electrode is placed as close to the working electrode as possible without touching it. This is to minimize any bulk solution effects. This electrode is commonly a calomel electrode, and the system is adjusted to maintain the voltage on the working electrode in reference to it. One arrangement devised to do this automatically is shown in Figure 26-9. Professor Hawley's description of the operation of the potentiostat is, "Most electrochemical cells consist of three electrodes: the working (W) and auxiliary (A) electrodes, which conduct all cell current, and the reference (R) electrode, which monitors the potential of the working electrode. The potential of the working electrode may be controlled at a potential that is either positive or negative with respect to that of the reference electrode. The potential difference between the working and reference electrodes is maintained by an electronic device called a *potentiostat*. For example, if the potential of the working electrode is such that oxidation occurs at its surface, then the working electrode functions as the anode. Since all current that flows through the working electrode must also flow through the auxiliary

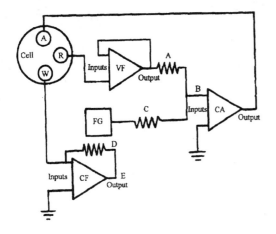

Figure 26-9. A universal potentiostat.
(Courtesy - Hawley, M.D., Kansas State University, Department of Chemistry, Manhattan, KS)

electrode, the auxiliary electrode in this example will function as the cathode and be negative in potential with respect to the working electrode. Because the reference electrode carries no cell current, there is no iR drop between the working and reference electrodes.

"The potentiostat circuit (Figure 26-9) consists of three operational amplifiers and are depicted in the schematic as triangles: the *current follower* (CF), through which all cell current passes; the *voltage follower* (VF), which is an impedance matching device that draws essentially no current from the reference electrode, and the *control amplifier* (CA), which causes the auxiliary electrode to assume whatever potential is necessary so that the potential between the reference and working electrodes is properly maintained. All cell current also passes through the control amplifier. To understand the operation of a potentiostat, it is important to recognize three major properties of (ideal) operational amplifiers. First, no current flows into or out of either of the two inputs. By convention, the two inputs are depicted in drawings as the connections to a face of the triangle. Second, the two inputs reside at the same potential. As a corollary, the voltage gain of the operational amplifier can be infinitely large. And third, the output, which is depicted as the connection to the apex of the triangle, can sink or source an infinite amount of current.

"In operation, the working electrode (W) resides at virtual ground or zero potential. Although the working electrode is not connected directly to ground, virtual ground for the working electrode is the consequence of connecting the second input of the current follower to ground *(vide supra:* property number two above). Since no current flows into or out of the inputs (property number one), the cell current that flows through the auxiliary and working electrodes can flow only through the current measuring resistor, D, and, subsequently through the output of the current follower. This causes a voltage to be developed at the output (E) of the current follower that, according to Ohm's Law ($V=iR$), is proportional to the cell voltage.

"The potential of the reference electrode is determined by the output voltage of the function generator (FG) and the relative magnitudes of the resistors A and C. If A and C are equal, then the output of the voltage follower, VF, must be equal to and opposite in sign to the voltage of the function generator (note that the potential at point B is at or near zero). Since the output of the voltage follower is fed back to the (inverting) input of the voltage follower, the reference electrode, which is connected to the remaining (non-inverting) input of the voltage follower, must also reside at the same potential as the output of the voltage follower. The voltage follower, therefore, performs the important function of not requiring the reference electrode to furnish the current that is used to determine the cell potential. This is important, since many reference electrodes will change their potentials if currents are drawn from them.

"Finally, the auxiliary electrode resides at the same potential as the output of the control amplifier, CA. If the potential of the reference electrode should deviate slightly from the control value, the potential at point B would rise slightly (microvolts) above or below zero. Since the voltage gain of the control amplifier is large (for example, one million), any small deviation at point B from zero volts causes the output voltage of the control amplifier, and thus the potential of the auxiliary electrode, to be adjusted in the opposite direction so that the desired potential of the reference electrode is always maintained. For example, if the potential of the reference electrode were to become slightly positive of the control potential, point B would also become slightly positive. This would cause the output of the control amplifier to move in the negative direction, thereby causing the auxiliary electrode, and subsequently, the reference electrode, to move to more negative potentials in order to regain potential control. This is an example of *negative feedback* and is used in all operational amplifier circuits where stable operation is required."

This potentiostat configuration is universal in that it can operate with any type function from the function generator (e.g., ramp, square wave, sine wave).

TECHNIQUES

GENERAL SAFETY PRECAUTIONS WHEN WORKING WITH ELECTRICAL DEVICES

Electrical current in a vacuum flows with the speed of light and nearly with the speed of light in electrical conductors. Individual electrons move about 0.02 cm/sec. How can this be? We will use a water hose filled with water as an analogy. If the hose is full of water, as soon as you turn on the faucet, water immediately comes from the nozzle. This cannot be the water that just entered the hose at the other end! What happened is that the water molecules entering the hose pushed the water molecules already there out of the way. Those, in turn, pushed those next to them out of the way, and so on. The net effect is a very rapid motion along the hose, yet each water molecule moved only a very short distance.

The following was taken from an article concerning electrical deaths (Wright, R.K., and Davis, J.H., *J. of Foren. Sci.,* **25** (3), 514, 1980).

"The amount of current flow, or amperes, is the most important single factor in human electrocution. By direct measurement and extrapolation from animal studies the following approximations as to the various effects of current flow in humans for 60 cycle alternating current are generally accepted:

"0.001 ampere   Barely perceptible tingle.
0.016 ampere   'Let go' current.
0.020 ampere   Muscular paralysis.
0.100 ampere   Ventricular fibrillation.
2.000 amperes   Ventricular standstill - death.
20.000 amperes Common household fuse blows

"The human body has a minimum internal resistance of less than 500 ohms. Hands and feet have minimal values of 1,000 ohms. Dry skin easily reaches resistances of 100,000 ohms. The resistance of human skin usually protects us from electrocution, as does the high resistance of the structures in which humans are ordinarily found. In any electrocution, there must always be a source of electrons under sufficient force to overcome the resistance of the body and a low resistance pathway to ground.

"Applying the dry skin resistance for hand-to-hand contact with 120 volts:

"Amperes = 120 volts/100,000 ohms = 0.001 (tingle sensation).

"However, for water or sweat soaked skin the resistance drops to 1,000 ohms:

"Amperes = 120 volts/1,000 ohms = 0.120 (ventricular fibrillation)."

Therefore, in keeping with the above facts, the following precautions are taken by scientists when they are working around electrical devices.

1. Remove all rings and bracelets when working around electrical devices.

2. Be sure all instruments are grounded. If you touch an instrument and feel a tingle, then check to see if it is grounded.

3. Do not poke around on the inside of an instrument with a screwdriver. Even if it is turned off, many instruments have capacitors that retain several hundred volts for hours.

4. When examining the electronic portion of an instrument, place one hand in a pocket or behind your back to reduce the possibility of getting a shock by shorting yourself out.

5. If you don't have any idea of what is going on, forget your pride, and get some knowledgeable help.

DEFINITIONS

Current: The rate at which charge is transferred across a section of a conductor; q/t - and given the symbol i or I. Current is measured with an *ammeter*. An ammeter that can measure microamperes is known as a *galvanometer*. Devices that can measure currents to 10^{-16} amperes are known as *electrometers*.

Ampere: The unit of measurement of current. 1 ampere = 1 coulomb/second. This is a fundamental SI unit from which many other electrical units are developed. If two infinitely long wires are placed parallel to each other, 1 meter apart, and an electrical current is passed through them such that 1 meter of length contains a magnetic force of 2×10^{-7} Newtons between the two wires, then 1 ampere of current is flowing in each wire.

Ohm: If 1 ampere of current passing through a conductor causes 1 watt of power to be dissipated, then the resistance of the conductor is 1 ohm. The symbol is either omega (Ω) or R. A variable resistor is called a *rheostat*.

Volt: The potential at a point in an electrostatic field is 1 volt, if 1 Joule of work per coulomb is done against electrical forces when a charge is brought from infinity to a point. A more usable definition is that *1 volt = 1 ampere flowing through a resistance of 1 ohm. $E = I \times R$.* Voltages are measured with voltmeters.

Coulomb: A quantity of charge that in 1 second crosses a section of a conductor in which there is a constant current of 1 ampere.

1 coulomb = 1 ampere/second.   (26-4)

Faraday: 96,487 coulombs.
Cathode: Where reduction takes place.
Anode: Where oxidation takes place.

In discussions of voltaic and electrolytic cells, much confusion often exists about what is the cathode and anode and what direction the current flows. Figure 26-10, p. 311, illustrates these differences.

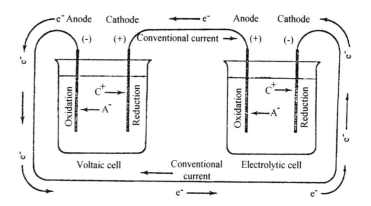

Figure 26-10. Conventions of nomenclature for a voltaic and an electrolytic Cell.
(Courtesy, Sisler, H., Vanderwerf, C., and Davidson, A., *College Chemistry*, 3rd ed., 1967, Macmillan Co., New York, NY)

ELECTRODES

The electrodes usually are made of platinum for analytical applications and of carbon for commercial electroplating. Figure 26-11 shows several designs.

For the best results, use a Pt gauze cathode with a rotating Pt gauze anode inside of it. A glass stirring rod in the center can be used quite successfully and is much less expensive, if a simple coiled wire anode is used. However, the best results are obtained if the solution passes through the cathode from side to side, rather than up and down. This means that the stirrer should be like the anode shown in the figure. It is corrugated, so that it directs solution outward when it turns. The stirring anode is 45 mesh Pt gauze, 110 mm tall, 22 mm at the top, and 17 mm at the bottom and weighs 13 g. The cylindrical cathode (52 mesh) is 52 mm tall and 32 mm in diameter. It weighs 11 g. The tapered cathode can handle higher current densities. The corrugated design is to provide greater surface area for a given diameter.

Caution: These electrodes are very expensive, several hundred dollars each, so know who has them at all times.

SAMPLE CONTAINER

A tall form beaker without a lip is preferred. These are called *electrolytic beakers*. One is shown in Figure 26-12, p. 312. The beaker often is covered with split watch glasses to prevent spattering during the electrodeposition.

COMMERCIAL ELECTROPLATING TECHNIQUE

Electroplating is the practical use of an electrolytic cell. The purpose is to deposit a metal of your choice onto the surface of an object. The steps necessary to get a uniform plate deposited on a conducting surface such as iron, steel, brass, or bronze are shown in Table 26-3, p. 312.

An article to be plated must be free from oil, dirt, or oxide films or the plate will not adhere and will not be smooth. More time often is spent on cleaning than on doing the actual plating. Inorganic oxides and other surface

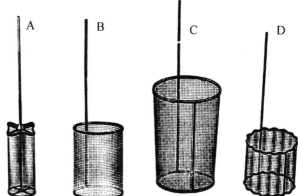

Figure 26-11. Platinum electrodes. A. Corrugated anode, B, C, and D, cathodes: cylindrical, tapered, corrugated.
(Courtesy - Eberbach Corp., Ann Arbor, MI)

Figure 26-12. Electrolytic beaker with a split watch glass.
(Courtesy - Eberbach Corp., Ann Arbor, MI)

Table 26-3. Steps in electroplating

For Good Adhesion	For Decorative Purposes
1. Solvent clean	1. Solvent clean
2. Blast (sand-alumina)	2. Polish, buffing wheels
3. Vapor degrease	3. Vapor degrease
4. Acid or alkaline etch	4. Acid or alkaline etch
5. Water spray rinse	5. Water spray rinse
6. Strike plate - high current	6. Strike plate - high current
7. Regular plate - low current	7. Regular plate - low current
8. Postplate bake out	8. Coloring plate
9. Final polish, if needed	9. Seal
	10. Final polish

compounds are removed by dipping the article in acid or *pickling* it for a few minutes. The bath usually contains solutions of sulfuric, nitric, hydrochloric, hydrofluoric, or chromic acid.

An article also can be cleaned electrically by making it the anode. The surface will be etched in about 1 minute in H_2SO_4 (700 mL/L.) dip at 27-37°C at a voltage of 3 to 12 v and a current density of 10 to 150 amperes/ft^2.

Degreasing

Degreasing is done by washing the article in hot trichloroethylene or by a vapor degreasing. In vapor degreasing, the article is suspended over the hot cleaning solvent. The vapors condense on the article, dissolve the grease and oil, and drain back into the solvent bath. The solvent is revaporized repeatedly to be used again. Thus, fresh solvent is always attacking the grease.

The last cleaning step is always an alkaline soak or electrochemical etch. This is necessary to remove organic soil from the surface. The article is either soaked for several minutes or made the anode and etched for a few minutes.

Polishing

The initial polishing is done to get better adhesion. Decorative articles are polished with a buffing wheel with an abrasive compound such as jeweler's rouge (Fe_2O_3) on it. For industrial applications where good adhesion is needed, the surface must be roughened much more. This is done by using a compressed air blast of sand or aluminum oxide on iron and steel or ground walnut shells on brass and bronze.

Final polishing can be done with a buffing wheel, but is more likely done by making the article the anode for a few minutes in a polishing solution.

Strike Plating

The purpose of strike plating is to cover the base metal with a more active metal or one that will adhere better than the final plating metal. It also is done to eliminate *immersion deposits* (spontaneous electrodepositing of unwanted metals). The strike plating solution has a low metal concentration and a high salt content (usually cyanide) so it will pass a high current. The high current increases the *throwing power* (the ability to cover irregularly shaped surfaces quickly). The strike plating is done only for a short time - 2 to 5 minutes. If it is done for a longer period, the plate becomes porous and no longer adheres to the surface. In addition, a large amount of hydrogen may be evolved, which embrittles the metal.

Plating Solution

A much lower current is used for plating than for striking, and the metal concentration in the solution is much higher. The plating is continued until the desired coverage is obtained. Times range from 15 minutes to 5 to 6 hours. The voltage should be adjusted to minimize hydrogen evolution. After plating, the article is heated to 175 °C for about 2 hours to bake out the hydrogen. The hydrogen is only normally present in ppm quantities, but it embrittles the metal. It can reduce the tensile strength of a piece of steel from 300,000 p.s.i. to 20,000 p.s.i. in a very short time.

REVIEW QUESTIONS

1. What does current control in an electrodeposition?
2. If a U.S. nickel that weighs 5.000 g contains 75% copper, how long will it take to plate it out if a current of 1.5 amperes is used? What would you recommend?
3. Why do metal ions plate out at different voltages?
4. Who coined the word *electricity*?
5. Who proposed the use of positive and negative for electrical charge?
6. Who suggested the existence of electrons?
7. Who discovered electrons?
8. What would happen if you mixed a solution containing Ce^{+4} with a solution containing Sn^{+2}? Refer to Table 26-1, p. 302.
9. What would be the potential of a battery prepared as in question 8?
10. A waste stream contains traces of Zn^{+2}, Ag^+, Fe^{+3} and Pb^{+2}. In what order would they deposit on an electrode?
11. Refer to Table 26-2, p. 304. If you had an aqueous NaCl solution and applied a voltage to it, what would happen at the anode?
12. What are the major potentials involved in an electrolytic cell?
13. Why are sulfur-containing compounds avoided if possible in electrodepositions?
14. If you had a water solution of Cr^{+3} ions, how much potential would you need to apply to deposit Cr, if only the back electromotive force is considered?
15. Why is the solution stirred during an electrodeposition?
16. If current controls how fast ions are deposited, why are low currents preferred for quality separations?
17. What causes activation overpotential?
18. What is the problem with hydrogen evolution during an electrolysis?
19. What is the problem if hydrogen gets into a deposited metal?
20. As a general rule, what metals can be plated out in acid solutions?
21. What is the effect of complexation or chelation on an electrodeposition?
22. What is a preferred anion to reduce complexation?
23. What is the effect of pH on an electrode potential whose half cell reaction involves H^+?
24. What is the E^o value for 0.1 M $KMnO_4$ at pH 2? Use Table 26-1, p. 302, and equation 26-3, p. 305. Assume Mn^{+2} = 10^{-5} M
25. What must take place for an system to be a potential buffer?
26. As a general rule, what must be the difference in potentials between two ions before they can be separated quantitatively?
27. A constant potential electrodeposition requires that three electrodes be present. What are they for?
28. What is the purpose of a potentiostat?
29. At what amperage does muscular paralysis occur when working with 60 cycle a.c. current?
30. Why is it suggested that you place one hand behind your back when you are making the first tests of the electrical components of an instrument?
31. What is a coulomb?
32. At what electrode does oxidation take place?
33. How can you clean an electrode electrically?
34. How can you polish an electroplated article electrically?
35. What is the purpose of "strike plating"?

IF IT EXISTS - CHEMISTRY IS INVOLVED
IF YOU CAN BUY IT - A CHEMIST WAS INVOLVED SOMEWHERE

27
ELECTROPHORESIS - GENERAL

PRINCIPLES

Electrophoresis is a technique in which *ions in solution are separated based on their differences in size and charge when a high voltage is applied to the solution. Positive ions migrate to the negative electrode, and negative ions migrate to the positive electrode.* The higher the charge and/or the smaller the ion, the faster the migration is. This is illustrated in Figure 27-1, p 316. Electrophoresis was developed into a workable system by Arne Tiselius of Sweden in the 1930's for which he received the Nobel Prize in chemistry in 1948. It was called the *moving boundary method* at that time. Figure 27-2, p. 316, shows the results from a current apparatus and techniques. The samples were streaked on a gel bed.

THEORY

The rate of movement is called the *mobility, U*. In order for a particle to migrate in an electrical field, it must possess a net electrostatic charge. This charge is an integral multiple of 4.8×10^{-10} esu, the charge on an electron. Many compounds that normally do not have a charge can be separated electrophoretically, provided some charging process is used. Charging processes that have been found to be successful are reactions with acids and bases, dissociation into ions by polar solvents, hydrogen bonding, chemical reactions, polarization, and ion pair formation.

The widest application of electrophoresis is in separations for the isolation and purification of small amounts of compounds. The best method of referring to such separations and of comparing one system to another has been the *mobility* of the compounds involved. Therefore, correct mobilities are important, and anyone using electrophoresis should be familiar with the various factors influencing the mobility and know what corrections should be made.

Consider a particle that is placed in a container of liquid, saturated with a buffer, and has a potential applied to it. The force, F, exerted on the particle is equal to the charge of the particle, Q, times the field strength, E.

$$F = QE \tag{27-1}$$

At first glance, it would appear that the particle would move toward one end of the container with an increasing velocity. However, it turns out that the velocity is constant. The reason for this is that as the particle moves through the buffer, it meets a retarding force caused by the viscosity of the solvent. This retarding force increases linearly with the particle acceleration, thus maintaining the velocity of the particle essentially constant. G. G. Stokes (~1845) showed that this opposing force for a sphere moving in a viscous medium can be expressed as

$$F_s = 6 \pi r \eta \upsilon \tag{27-2}$$

where F_s is the viscous retarding force, r is the radius of the particle in cm, η is the viscosity of the medium in poises, and υ is the electrophoretic velocity in cm/sec. When these forces are equal

$$F_s = QE = 6 \pi r \eta \upsilon \tag{27-3}$$

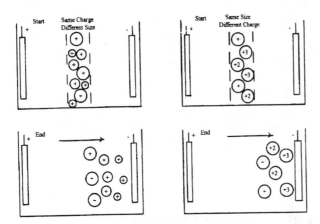

Figure 27-1. Diagram of ion motion in an electric field.

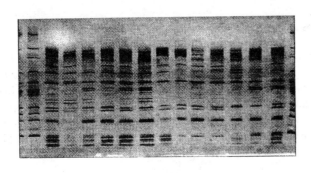

Figure 27-2. A 16-channel electropherogram to identify potato varieties.
(Courtesy - Pharmacia Biotech Inc., Piscataway, NJ)

The mobility of a particle, U, is defined as

$$U = v/E \tag{27-4}$$

where U is in cm^2 volt^{-1} sec^{-1} and E is in volt cm^{-1}. Substituting equation 27-3 into equation 27-4, we get

$$U = \frac{Q}{6\pi\eta r} \tag{27-5}$$

EXAMPLE CALCULATION
Calculate the mobility of the Ba^{+2} ion in a 0.0625 M solution of BaCl$_2$. The radius of Ba^{+2} in solution is 2.78 Å and the viscosity of this solution is 10.310 millipoises.

ANSWER
$Q = 2 \times 4.8 \times 10^{-10}$ esu; $r = 2.78 \times 10^{-8}$ cm; $\eta = 1.03 \times 10^{-2}$ poises
Because mobility has the units cm^{2}volt^{-1}sec^{-1}, the factor 300 is needed to convert from practical volts to esu's.

$$U = \frac{2 \times 4.8 \times 10^{-10}}{6 \times 3.14 \times 2.78 \times 10^{-8} \times 1.03 \times 10^{-2} \times 300} = 5.9 \times 10^{-4} \text{ cm}^2\text{volt}^{-1}\text{sec}^{-1}$$

The mobility calculated in this way with equation 27-5 cannot be verified experimentally, because the negative chloride ions moving in the opposite direction from the Ba^{+2} ions have an additional retarding effect on the mobility.

This tendency to decrease the mobility of the Ba^{+2} is related directly to the ionic strength of the solution. By incorporating the Debye-Huckel equation, the following correction term is obtained:

$$\frac{1}{1 + r A \sqrt{u}} \qquad (27\text{-}6)$$

where u is the ionic strength, A is a constant ($0.233 \times 10^{+8}$ for water at 25 °C), and r is the radius of the particle in cm. The ionic strength is

$$u = 1/2 \sum c\, z^2 \qquad (27\text{-}7)$$

where c is the concentration in mol/L, and Z is the valence of the ion.

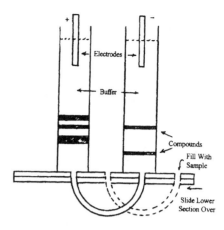

Figure 27-3. Diagram of an original moving-boundary-type electrophoresis apparatus.

EXAMPLE CALCULATION

Referring to the previous example, what is the mobility of the Ba^{+2} ion when the other ions are considered?

ANSWER

$$U = \frac{Q}{6\pi \upsilon \eta r} \times \frac{1}{1 + r A \sqrt{u}}$$

$$\sqrt{u} = [1/2(0.0625 \times 2^2) + (2 \times 0.0625 \times 1^2)]^{1/2}$$

$$\sqrt{u} = 0.433$$

The correction term is then

$$\frac{1}{1 + 2.78 \times 10^{-8} \times 0.233 \times 10^8 \times 0.433} = 0.781$$

Multiplying the mobility obtained in the previous example by 0.781 gives

$$U = 5.9 \times 10^{-4} \times 0.781 = 4.62 \times 10^{-4}\ cm^2 volt^{-1} sec^{-1}$$

The preceding equations were developed from theory; however, in practice, a relationship for calculating mobilities that involves more readily obtainable measurements is desirable. Recalling that the units involved are volts/cm, the equation can be rewritten as

$$U = \frac{P_m\, l_m}{V\, t} \qquad (27\text{-}8)$$

where P_m is the distance the ion travels in cm; l_m is the distance between the electrodes in cm; V is the voltage in volts (not to be confused with E, which is volts/cm); and t is the time in seconds. For some electrophoretic cells, the distance l_m is hard to obtain accurately, so volts/cm is determined from

$$E = \frac{i}{q\, k} \qquad (27\text{-}9)$$

where E is the field strength in volts/cm, i is the current in amperes, q is cross sectional area of the cell in cm^2, and k is the specific conductivity of the solution in ohm^{-1}cm^{-1}. The value of k is determined from

$$k = C/R \qquad (27\text{-}10)$$

where R is the resistance of the solution in the cell in ohms, and C is the cell constant in cm^{-1}. The cell constant is determined by the classical method of J. Jones, and B.C. Bradshaw (*J. Chem. Soc.* **55**, 1780, 1933) in which they found that if 7.457 g of dry KCl was added to 1 liter of water at 20 °C, the resulting conductivity was 0.007138 ohm^{-1}cm^{-1} at 0 °C.

EXAMPLE CALCULATION

A 3% solution of egg albumin buffered at a pH of 7.2 was placed in an electrophoresis cell having a diameter of 1.2 cm. A current of 20 mA was used, and the egg albumin took 208 minutes to move 5.87 cm toward the + electrode. If the cell constant was 18.3 cm^{-1} and the resistance of the system was 7690 ohms at 0 °C, calculate the mobility of the material.

ANSWER

Step (1) Calculate the specific conductivity using equation 27-10.

$$k = \frac{18.3 \ cm^{-1}}{7690 \ ohms} = 0.00238 \ ohm^{-1} \ cm^{-1}$$

Step (2) Calculate E using equation 27-9.

$$E = \frac{0.020}{(3.14 \times 0.6^2)(0.00238)} = 7.43 \ volts/cm$$

Step (3) Calculate U using equation 27-8. Remember that the l_m has been incorporated into the E value, because $E = V/l_m$. The sign is negative because the ion was negative, as indicated by motion toward the positive electrode.

$$U = \frac{-5.87}{(208 \times 60)(7.43)} = -6.3 \times 10^{-5} cm^2 volt^{-1} sec^{-1}$$

Heat Production

A problem that often arises in determining mobilities is heat production. The current passing through a cell develops heat, and this not only causes convection currents, which change the mobilities, but also increases the rate of evaporation of the buffer solution when strip electrophoresis is employed. This loss of buffer causes a change in mobility because of changes in the potential and current; it also causes a capillary action in the paper or gel, because it will take up solution from the buffer tank. The usual maximum amount of heat that can be tolerated is 0.15 watts/cm^3. The heat produced can be calculated from the equation:

$$H = \frac{i^2}{q^2 \ k} \qquad (27\text{-}11)$$

where H is the heat developed in W/cm^3, and i, q, and k are as previously defined.

EXAMPLE CALCULATION
Determine the heat developed in the solution described in the previous example.

ANSWER

$$H = \frac{(0.020)^2}{(1.13)^2 \ (0.00238)} = 0.132 \ W/cm^3$$

"Smiling"
Small white, translucent, spots appear on strips when the strips get too hot and the buffer evaporates. On slabs, in which a series of strips are run, uneven heating across the slab produces what is called smiling. This is shown in Figure 27-4 for a series of DNA profiles.

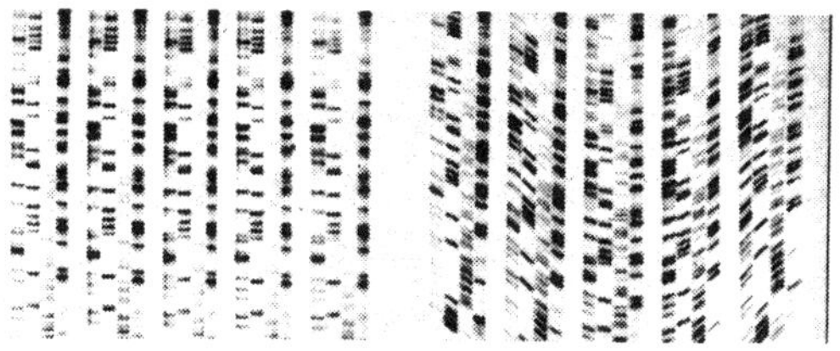

Correction for Electro-Osmosis
If water is placed in a capillary tube, and positive and negative electrodes are placed in contact with the water, the water will migrate toward the negative electrode, indicating that the water is positively charged. This movement of the solution

Figure 27-4. "Smiling" caused by uneven heating (right) vs. even heating (left).
(Courtesy - Pharmacia Biotech Inc., Uppsala, Sweden)

in an electric field is called *electro-osmosis*. It is described in much more detail later on, particularly in Chapter 31, p. 359. This migration can be altered by placing other ions in the system and, in fact, can be reversed.

Consequently, the mobilities must be corrected further, because the compound migrating in the electrophoretic system can be held back or speeded up by the movement of the buffer solution. This is similar to the action of a tail wind or head wind on an airplane.

To determine the electro-osmosis correction, a material with no inherent mobility is selected. Therefore, if it moves, it does so because of the solution movement carrying it along. This movement can be either added or subtracted from the mobility of the compound under investigation to correct for the electro-osmosis. Dextran, which has a mobility of only $-0.16 \ x \ 10^{-5}$, is used commonly to determine the correction factor for proteins. The corrected mobility is given by:

$$U_{cpd} = \frac{P_{cpd} \pm P_{dex}}{E \ t} \qquad (27\text{-}12)$$

where: P_{cpd} = is the distance the compound moved. P_{dex} = is the distance the dextran moved. If the dextran moves opposite to the compound, then P_{dex} is added (+) to P_{cpd}; if it moves in the same direction, then P_{dex} is subtracted (-).

MODIFICATIONS AND IMPROVEMENTS
The moving boundary method had disadvantages. (1) It was difficult to remove the individual components without remixing them and (2) even a small amount of heating caused convection currents that tended to remix the compounds. This limited the separation to large molecules with less mobility. Improvements came rather quickly. The first was *horizontal strip* paper electrophoresis developed by D. von Klobusitzki and P. Konig in 1939 (*Arch. Exp. Path. Pharmakol.* **192**, 271), who placed the sample in a buffer-soaked paper strip placed horizontally between two electrodes. Modifications of this were the acetate strip, agarose and starch gels (Chapter 27, p. 320), and *immunoelectrophoresis* (Chapter 28, p. 339). Subsequently, *disc electrophoresis* (Chapter 29, p. 345), *polyacrylamide gel, isoelectric focusing, SDS, two-dimensional* (Chapter 30, p. 351), were developed and more recently, *pulsed field electrophoresis, capillary zone* electrophoresis (Chapter 31, p. 359), and *field flow fractionation* for large molecules, (Chapter 32, p. 371). The specifics are discussed in each chapter.

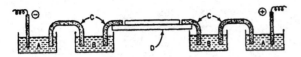

Figure 27-5. Diagram of a horizontal strip electrophoresis apparatus.

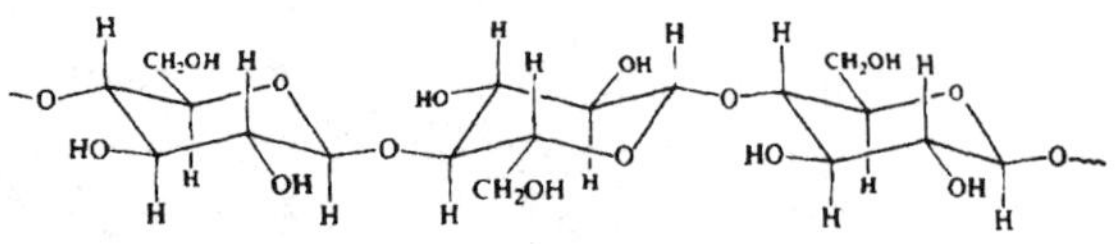

Figure 27-6. Cellulose.

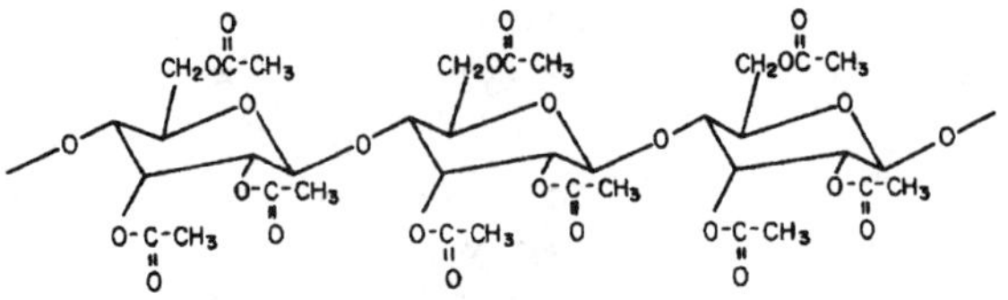

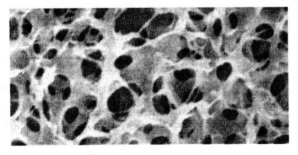

Figure 27-7. Cellulose acetate.
(Courtesy - M. Melvin, *Electrophoresis*, John Wiley & Sons, New York, NY, 1987)

Figure 27-8. Magnified cellulose acetate.
(Courtesy - H.P. Chin, *Cellulose Acetate Electrophoresis, Techniques and Applications*, Ann Arbor-Humphrey Science Publ., Ann Arbor, MI, 1970)

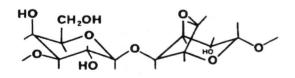

Figure 27-9. An agarose segment.

TECHNIQUES

The following are some general techniques as well as descriptions for a few areas for which no experiments are presented.

SUPPORTS

Paper Strips

The first improvement was to suspend a strip of filter paper saturated with buffer horizontally between the electrodes (Figure 27-5). Although this was suggested by P. Konig in 1937, probably the first actual use was by T. Weiland and E. Fisher in 1948 (*Naturwisenschaften*, **35**, 29). The sample was spotted in the middle, and *spots* migrated to each electrode depending upon their charge. The paper was removed, dried, and then sprayed with a reagent that would react with the separated compounds to reveal *spots*. This was a vast improvement in separating the compounds. However, it was difficult to measure the intensity of the developed spots through the paper. Paper used for electrophoresis must be at least 96% cellulose in order to absorb sufficient water (Figure 27-6). Paper also has a large electro-osmotic effect.

Cellulose Acetate

The next improvement was to use a strip of cellulose acetate, which is transparent (Figures 27-7 and 27-8). This was proposed by J. Kohn in 1957 (*Biochem J.*, **65**, 9). It is more homogeneous than paper and gives sharper resolution, and few stains stick to it, so the background is reduced. In addition, much shorter running times are needed. It is brittle when dry, but quite pliable and stronger than paper when wet.

When cellulose acetate support strips are prepared they should be laid carefully on the surface of the buffer and not dipped into the buffer. Air bubbles often stick to the surface if strips are dipped, which causes an uneven penetration of the buffer. The support is best left to soak overnight, so it is wetted from the bottom and the air is forced out the top. The excess buffer is then blotted off with cloth or paper towels. Low currents of 0.4 mA/cm width of support are recommended.

Cellulose acetate can be made nearly transparent by immersion in cottonseed oil or decalin. It can be dissolved in acetone, glacial acetic acid, methylene chloride, or phenol, and the compounds recovered.

Agar and Agarose Gels

The use of a gel to hold the buffer permits higher currents to be used, but the most significant advantage is the sieving action of the pores in the gel polymer, which adds to the separation capability. This can be demonstrated by changing the percent of gelling material in the mix. Agar is a polysaccharide complex of chains having alternating 1,3-linked β-D-galacto-pyranose and 1,4-linked 3,6-anhydro α-L-galacto-pyranose. A 1% solution forms a stiff jelly on cooling. It is extracted from the agarocytes of algae found in the Pacific and Indian Oceans and the Sea of Japan. Agarose (Figure 27-9) is separated from it as a neutral gelling fraction. As gels go, it allows the most diffusion, which is usually a drawback. However, with immuno-electrophoresis, this is a desired trait. A 2% solution in a buffer usually is chosen. Because most of the gel is water, this gives protein mobilities close to those of the moving boundary method. Other advantages of agarose are its negligible adsorption effects and lower content of charged groups, which reduces endo-osmosis. However, it has minimal sieving effects.

Starch Gels

In 1955, O. Smithies (*Biochem. J.*, **61**, 629) used starch gels to separate serum proteins. Typically, 10 to 20 g of starch are added to 100 mL of buffer, heated to about 90 °C, and stirred until it dissolves. The air bubbles are removed by applying a vacuum, and the viscous mixture is poured in a tray (over fill 1-2 mm) to provide a slab about 6 mm thick and let cool. Before pouring, a *comb* is placed vertically in the tray to provide wells for samples to be added later on. The extra 1 to 2 mm is sliced off. The gel then is placed on the flat portion of the apparatus with the ends dipping into the buffer, and sample (sometimes mixed with 2% starch) is added to each well. A sheet of cellophane is placed over the gel to retard evaporation. A voltage of 15 V/cm for 4 hr is common. After a run, the gel can be sliced horizontally to give three 1-2-mm-thick layers. The gels are placed in hot (70 to 80 °C) glycerol for 20 to 30 seconds to clear and then dried for scanning.

Polyacrylamide Gels (PAGE)

In 1959, S. Raymond and L. Weintraub (*Science*, **130**, 711) described the use of a polymer of acrylamide cross-linked with N,N'-methylene-bis-acrylamide to form a gel slab. In 1961, B.J. Davis and L. Orstein (*Disc Electrophoresis*, Distillation Products Division, Eastman Kodak Co., Rochester, NY) described the use of this polymer in tubes. See Chapter 29, p. 345.

Polyacrylamide (Figure 27-10) is more flexible, stronger, and easier to handle than the other gels. It is transparent to visible and uv radiation. Electro-osmosis and diffusion are negligible. A wide range of buffers can be used, the polymer is essentially nonionic, it is quite inert chemically, and it is thermostable. All of these characteristics permit shorter running times. The pore size can be varied from about 2 nm for 30 g of acrylamide/100 mL to 5 nm for 7.5 g/100 mL. Figure 27-11 shows a 10% gel. The dimension bar at the top is 100 nm. Notice that the porosity changes, getting larger as you approach the casting surface. The catalysts are ammonium or potassium persulfate or N,N,N',N'-tetramethyl-ethylenediamine (TEMED).

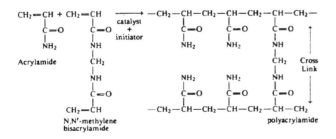

Figure 27-10. The polyacrylamide polymer.
(Courtesy - M. Melvin, *Electrophoresis*, J. Wiley & Sons, New York, NY, 1987)

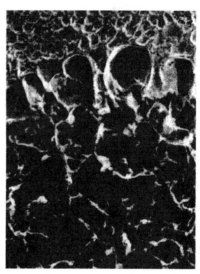

Figure 27-11. Photomicrograph of polyacrylamide gel.
(Courtesy - R.C. Allen, C.A. Saravis, H.R. Maurer, *Gel Electrophoresis and Isoelectric Focusing of Proteins*, de Gruyter, Berlin, Ger., 1984)

Disadvantages include the slow removal of the staining dyes and the toxicity of the monomer. It is recommended that prepared gels be purchased if possible to minimize the toxicity problem. Acrylamide has a neurotoxic effect by inhalation and skin contact. Wash any exposed area with soap and water immediately. The effects are drowsiness, fatigue, tingling in the extremities, and weakness in the legs. See Appendix A, Experiment No. 41 for the preparation techniques.

BUFFERS

Consider the following reactions of glycine, an amino acid. In acid solution, it is positively charged, and in basic solution, it is negatively charged. Somewhere in between, it is neutral - its *isoelectric point*, pI. It will move to either

$$\overset{+}{H_3N}-CH_2-C\underset{OH}{\overset{O}{\diagdown}} \underset{}{\overset{H^+}{\rightleftharpoons}} H_2N-CH_2-C\underset{OH}{\overset{O}{\diagdown}} \underset{}{\overset{OH^-}{\rightleftharpoons}} H_2N-CH_2-C\underset{O^-}{\overset{O}{\diagdown}}$$

$$pKa_1\ \ 2.34 \hspace{3cm} pKa_2\ \ 9.60$$

the cathode or anode depending upon the pH of the solution it is in. To get reproducible results, it is necessary to maintain the pH during the entire electrophoresis; therefore, a buffer is needed. Base is generated at one electrode (cathode) and acid at the other (anode). This will change the pH of the solution on the strip unless they are neutralized to the desired pH. This is why large buffer tanks are used. Large molecules, such as proteins, consist of many different amino acids, some on the surface and others buried. This means that both positive and negative charges are likely to exist on the same molecule at the same time, and the motion of the particle will depend upon the pH of the solution it is in. Advantage is taken of this in a process called *isoelectric focusing* described in Chapter 30, p. 352, to separate compounds with great resolution. The best buffer is found by experiment. In addition to pH, ionic strength is important.

The higher the ionic strength, the sharper the resolution is; the lower the ionic strength is, the more rapid the migration of the ionic species. A value of about 0.1 M is a reasonable compromise. Table 27-1 shows the composition of two buffers.

Table 27-1. Composition of selected buffers

pH 4.5	β-Alanine	31.2 g		pH 8.8	Tris(hydroxymethyl)amino methane	109 g
	Acetic acid	8.0 mL			Disodium ethylenediaminetetra-acetic acid	5.87 g
	H$_2$O to 1 L				Boric acid	2.3 g
					H$_2$O to 1 L.	

SAMPLE ADDITION

Although most samples in chromatographic separations were *spotted*, in electrophoresis they are usually *streaked*. Adding sample uniformly into a set gel is done either by cutting with a razor blade section (Experiment 38) or more commonly, by molding wells into the gel as it sets and then filling the wells. The apparatus for making the wells is called a *comb*, and it can be purchased with any number of teeth desired. Figure 27-12 shows the preparation of a gel with two sets of combs in place. The comb is removed, and the sample (1 to 50 µL) is added to the well by pipetting.

DETECTION

Most of the compounds separated are colorless, and those that are colored are in such low concentrations that they are almost always appear colorless. The most common methods of detection are *staining, antibody reactions*, and *autoradiography*.

Staining

The main process to detect compounds is to apply a stain to the strip or slab, let it react, then wash away the unreacted stain. Table 27-2, p. 323, lists several common stains. None will stain all of the compounds, but many will stain several. The desire is to stain only those of interest. *Although hundreds of compounds may be present on a strip, only those that react with the stain will show up.* Therefore, you must realize that the spot you have may be

Figure 27-12. Two sets of six-tooth combs.
(Courtesy - EDVOTEK, West Bethesda, MD)

contaminated heavily with other compounds. As a general rule, Coomassie blue R-250 is 100 times more sensitive than Amido black; silver staining is 100 times more sensitive than Coomassie blue R-250, and autoradioography is 150 times more sensitive than silver staining. There is no free lunch. The more sensitive the detection, the longer the development time is.

Table 27-2. Some stains for selected functional groups

Stain	Compound
Amido black 10 B	Proteins
Coomassie blue	Proteins
Silver nitrate	Proteins
Ethidium bromide	DNA and RNA
Sudan black	Lipids and lipoprotein
Schiff-periodic acid	Carbohydrate and glucoprotein
Ninhydrin	Amino acids

Coomassie Brilliant Blue

Coomassie brilliant blue R-250 (Fazekas de St. Groth, S., Webster, R.G., and Datyner, A., *Biochem. Biophys. Acta.*, **71**, 377, 1963) requires an acidic medium and then is believed to be electrostatically attracted to the NH_3^+ groups of the protein. It is three times more sensitive than Fast green. In a 0.01 M citrate buffer at pH of 3, it has a λ_{max} at 555 nm and at 549 nm for the protein dye complex.

Ethidium Bromide

This is the stain of choice for DNA detection. Ethidium bromide has a λ_{max} at 506 nm and intercalates double-stranded DNA and RNA.

Silver Nitrate

Developed in 1979 (Switzer, R.C., Merril, C.R., and Shifrin, S., *Anal. Biochem.*, **98**, 231), it can detect as little as 0.1 ng of protein. Many modifications of this procedure have been made. The one outlined below is a combination of several and is preferred for the most sensitive work. J.S. De Olmos (*Brain Behav. Evol.*, **2**, 213, 1969), found that the system was more sensitive if copper acetate was added first. The belief is that the Cu^{+2} forms a N-peptide complex, which then acts like metallic Cu electrochemically and will reduce Ag^+ in solution to metallic Ag on the N-peptide. This acts as a nucleation center for more Ag^+ to react. The gel is soaked in a solution containing Cl^-, then citrate ion, then $AgNO_3$. The system then is irradiated with white light for 4 minutes. The AgCl formed produces increased sensitivity and nondiffusion because of its solubility. Acetate and citrate enhance the sensitivity. The membrane then is placed in a solution containing hydroquinone and formaldehyde. Ionic Ag^+ is reduced to Ag, formaldehyde is oxidized to formic acid, and the hydroquinone is oxidized to quinone. Thiosulfate is added to remove the AgCl and lighten the background. The thiosulfate complex is removed by a water wash.

Fluorescence

Both pre- and post-electrophoresis staining can be done. However, post-staining usually is preferred, because it does not alter the separation. Anilinonaphthalene sulfonate can detect 20 µg of protein. Dansyl chloride reacts with amines, amino acids, proteins, and phenols. Exposure of protein for 1 to 2 minutes at 100 °C to dansyl chloride produces a fluorescence capable of detecting 8 to 10 ng. Similarly, exposing primary amine-containing compounds to fluorescamine at room temperature and an alkaline pH permits the detection of 6 ng of myoglobin. A newly popular compound is 2-methoxy-2,4-diphenyl-3(2H)-furanone (MDPF). It can detect 1 ng of protein, and it is linear from 1 to 500 ng.

Autoradiography

Autoradiography involves incorporating a radioactive component into the compound to be detected, usually with ^{3}H (amino acids), ^{32}P (phosphate), ^{35}S (methionine), ^{14}C (amino acids), or ^{125}I (iodoacetamide). The components are separated by any of the electrophoretic techniques, then placed on a sheet of x-ray film to expose the film, sometimes for days if the components are minimal. Components show up as black areas. If properly done, this technique is 150 times more sensitive than staining, but it can take considerable time at low levels. ^{3}H (0.018 MeV), ^{35}S (0.167 MeV), and ^{32}P (1.71 MeV) are β-emitters. ^{125}I emits an x-ray at 0.035 MeV. The weak β-particles from ^{3}H do not penetrate the gel, so a fluorographic method is used.

The fixed wet gel is impregnated with 2,5-diphenyloxazole (PPO). The β-particles excite these molecules to emit photons in the blue region, so blue-sensitive x-ray film is used. Development at low temperatures (-70 to -80 °C) increases detection by 10-to-12 fold for ^{3}H and by 9-fold for ^{14}C and ^{35}S.

Intensification screens often are employed for the higher energy β-particles and γ-rays because these emissions can pass through the film. Calcium tungstate is one compound used in an intensification screen. Particles emitted from radioactively labeled samples in the gel pass through the film and strike the intensifying screen. Inorganic phosphors absorb the energy, converting it to ultraviolet or blue radiation which then expose the film. This permits shorter exposure times and lower detection levels. This is illustrated in Figure 27-13.

DRYING

After the strip or slab has been stained and destained, the support must be dried so it is easier to handle. This can be done by placing the support in an oven, but vacuum drying is preferred, because large amounts of heat are not needed. Figure 27-14, p. 325, shows one type. The advantages are that even drying can be obtained, which reduces cracking, and it is reproducible. It has a heated aluminum-bottom platen to counter the cooling effect of evaporation and uses an oil-free vacuum pump. The model shown can be temperature adjusted from 40 to 80 °C. This is hot enough to rapidly evaporate the water without burning the gel. A transparent silicone rubber sheet covers the gel, so the gel can be viewed. *The sheet will feel cool until the gel is dry.* To reduce cracking and produce even drying of >10% polyacrylamide gels, *a porous polyethylene sheet is placed on top of the gel.*

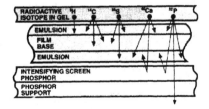

Figure 27-13. Diagram of the operation of intensifying screens.
(Courtesy - Fisher Scientific, Pittsburgh, PA)

2,5-Diphenyloxazole

FILM REMOVAL OR BINDING

Sometimes, it is desired to remove the gel from its backing support. For example, if blotting is to be done. Some backings can be dissolved, leaving the dried gel film. Another technique is to spray the backing with something like a 2% solution of dimethyldichlorosilane in octamethylcyclotetrasiloxane, a low toxicity, environmentally safe solvent (Repel-Silane ES). This prevents both polyacrylamide and agarose gels from sticking to glass during the casting process. A way to remove polyacrylamide and agarose films from plastic supports is to use a spring-loaded stainless steel wire (Figure 27-15). It will do so without distorting the gel. The separated gel can be removed with support film, filter paper, or (if it is to be blotted) a blotting membrane soaked in transfer buffer. It can remove gel thicknesses from 0.2 mm thick upward and sizes up to 200 x 245 mm in seconds.

Figure 27-14. A vacuum gel dryer, Hoefer model SE 1175.
(Courtesy - Pharmacia Biotech Inc., Uppsala, Sweden)

The reverse process, *binding*, which is sometimes necessary when polyacrylamide gels are formed on glass, can be done by spraying the glass with a solution of γ-methacryloxy-propyl-trimethoxysilane (Bind-Silane).

BLOTTING

After the components have been separated, it is often critical that their relative positions and concentrations do not change, such as in DNA typing or disease identification. Diffusion of the components always occurs within the wet gel, but this is usually small compared to the movement of the components during electrophoresis. However, once the electrical field is shut off, random diffusion is a problem for high quality work. In addition, dry gels are fragile. The solution to this problem is *blotting*. This process probably was first used by F. Sanger, G.C.. Brownlee, and B.G Barrell (*J. Mol. Biol.* **13**, 373, 1965) to determine nucleic acids. The basic process is to transfer the separated compounds from the fragile gel onto a more hardy membrane, which not only fixes the compounds in place, but selectively removes the compounds of interest, leaving the others behind.

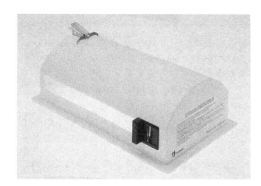

Figure 27-15. Film remover.
(Courtesy - Pharmacia Biotech Inc., Uppsala, Sweden)

Blotting Membranes

Many types of blotting membranes are available. Listed below are a few of the most popular.

Nitrocellulose

This is the favorite right now. It is a negatively charged medium and has a selective absorptivity for proteins and nucleic acids. It usually is treated for 5 minutes with 2% Tween 60 in a 25 mM veronal/2.5 mM citrate/1 mM oxalate buffer at pH 8.6 before use. Pore sizes are 0.3 to 0.5 μm, with 0.45 μm being usual. Its capacity for proteins is 80 to100 μg/cm^2. It appears to bind proteins by a hydrophobic interaction. Nonionic detergents can remove large amounts of bound protein.

Nylon 66

This has a higher binding capacity than nitrocellulose and is more rugged. It is quite resistant to chemical attack. Positively charged nylon membranes (PCM) have a high protein binding capacity: > 500 μg/cm^2.

Glass Fiber Filter Paper

The glass fiber is modified by a silane reaction to add primary amines or quaternary ammonium groups to better retain proteins.

Polyvinylidene difluoride (PVDF)

This is quite inert. It is good for chemiluminescent detection with luminol and 1,2-dioxetane substrates.

Blocking

Using proteins as an example, the purpose of blocking is to cover sites on the membrane that do not contain protein so you get nonspecific binding. Proteins adhere to nitrocellulose. If you block the other areas to prevent staining other compounds then you do not need to remove these other compounds. Some good blocking agents for proteins are nonfat dry milk, bovine serum albumin (BSA), and polyvinylpyrrolidone (PVP).

Southern Blotting

Southern blotting was developed by E.M. Southern (*J. Mol. Biol.*, **98**, 503, 1975) to transfer DNA fragments from an agarose gel to a nitrocellulose membrane. The gel strip is placed on a paper towel saturated with 0.15 M NaCl and 0.015 M sodium citrate. A strip of nitrocellulose, soaked in the same solution, and is placed on top of the gel; a piece of dry paper toweling is put on top of that. The compounds diffuse upward carried by the capillary action of the liquid. The DNA selectively binds to the surface of the nitrocellulose, and diffusion stops. The membrane is quite flexible and easy to handle. Figure 27-16 shows the sequence of steps. Southern blotting usually is done with agarose gels and horizontal electrophoresis. *Southern blotting* has become very popular since the use of DNA patterns for the identification of people.

Western, Northern, and Southwestern Blotting

Western, northern, and southwestern blotting are likely spoofs on the term Southern blotting. All follow the same basic technique. The initial step is the separation of a class of compounds, such as DNA, RNA, or proteins from the mixture. Electrophoresis is performed on the class to separate the individual compounds, which then are blotted. The blotting membranes can be treated to retain different compounds.

Western blotting (Towbin, H., Staehelin, T., and Gordon, J., *Proc. Nat. Acad. Sci.*, U.S.A., **76**, 4350, 1970) is designed to detect proteins. The gel is a thin layer of polyacrylamide and vertical electrophoresis is used. The membranes are nitrocellulose or polyvinylidene difluoride (PVDF). Usually an antigen-antibody reaction (immuno-blotting) is used for detection.

Northern blotting (Altwine, J.C., Kemp, D.J., and Stark, G.R., *Proc. Natl. Acad. Sci., USA*, **74**, 5350, 1977) uses an agarose gel and retains RNA fragments.

Southwestern blotting is used to retain DNA bound to proteins.

Electrical Blotting

Electrical blotting greatly speeds up the blotting process. It was pioneered by Towbin et al. (see reference above). This technique employs electrophoresis at 90° to the original gel direction (Figure 27-17, p. 327). The gel is placed against an immobilizing membrane, and the combination placed between a pair of electrodes. The anode, in a plastic

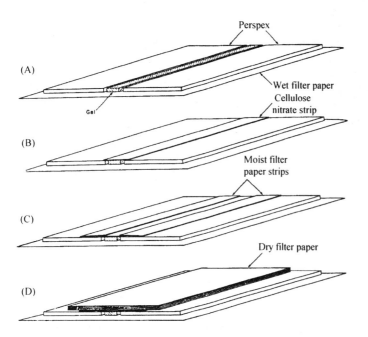

Figure 27-16. Steps for Southern blotting.
(Courtesy - Southern, E.M., *J. Mol. Biol.* **98**, 506, 1975)

sheet, contains 11 ft of palladium wire at 1.25 cm intervals, and the cathode is porous passivated stainless steel. The buffer tank holds up to 6 L of solution. The power supply (EC-420) voltmeter has two ranges, 0 to 25 and 0 to 250, and the ammeter has two ranges, 0 to 250 mA and 0 to 1A.

Constant voltage is preferred. However, a problem is that the current increases as the operation proceeds. This is caused by joule heating, which reduces the solution resistance, buffer breakdown into more conducting ions, and electrolyte elution from the gel into the transfer buffer. The net result is a steady increase in current. Currents of 300 to 400 mA are acceptable, but if the currents get into the 400 to 500 mA range, then cooling should be done.

Vacuum Blotting

The apparatus shown in Figure 27-18 is one type of a commercial design for vacuum blotting. The manufacturer says it is "Designed to handle all types of nucleic acid separations, the precision controlled vacuum pump maintains the exact vacuum level needed to blot all lengths of nucleic acids, from depurinated RNA or DNA fragments to chromosome size. No wicks, cassettes or stacks of filter paper are needed. It can do up to 10 transfers a day with close to 100% DNA recovery."

APPARATUS
Horizontal Strip

Figure 27-19 shows a commercial apparatus. The sample is applied as a narrow strip or small spot on top of the buffer-saturated paper. The electrodes are usually platinum strips and are immersed in buffer tanks. Notice that the buffer tank is separated into two compartments with the electrode in the one furthest away from the paper strip. The purpose of the buffer is to maintain the pH at the desired level during the entire separation. This is a problem, because the high voltage applied tends to generate acid at one electrode and base at the other. These can migrate into the strip, change the ionic nature of the compounds, and ruin the separation. Therefore, rather large buffer tanks are needed and the divider is to keep as much acid and base from the strip as possible yet permit electrical contact. After the separation, the positions of the compounds are visualized the same as with paper chromatography. This was a vast improvement over the moving boundary technique, because it allowed the separated components to be removed and more easily visualized.

Covering

The apparatus is covered for two main reasons: to prevent the evaporation of water by the heat generated, and as a safety precaution. With 3 to 8 milliamperes/strip and with several strips in the tank, this is a potentially hazardous situation. All commercial systems are arranged so that you cannot make contact with the buffer or strip when the power is on. The cover has a multiple-pronged plug that will connect only to the power supply when the cover is on.

The buffer on each side of the apparatus must be at the same height. If it is not, then siphoning can occur when the strips are put in place, which changes the mobility and can make

Figure 27-17. An Electroblot transfer system. EC model 4313-H75.
(Courtesy - A.H. Thomas Scientific, Swedesboro, NJ)

Figure 27-18. A vacuum blotting system.
(Courtesy - Pharmacia Biotech, Inc., Uppsala, Sweden)

Figure 27-19. A flat bed electrophoresis apparatus.
(Courtesy - Gelman Instruments, Inc., Ann Arbor, MI)

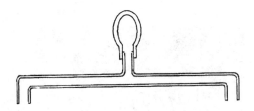

Figure 27-20. Buffer leveling device.

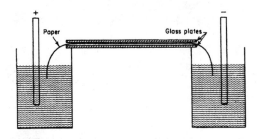

Figure 27-21. A sandwich modification.

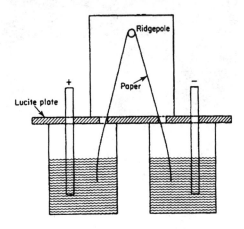

Figure 27-22. Ridgepole type apparatus.

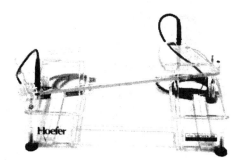

Figure 27-23. S submarine gel electrophoresis apparatus.
(Courtesy - Pharmacia Biotech Inc., Piscataway, NJ)

identification difficult. Most commercial apparatus have leveling marks on the buffer tanks, but a homemade apparatus requires a different technique. Figure 27-20, p. 328, shows a simple method to level the buffers.

The pipet bulb is squeezed, and one end is placed in each tank. The pipet bulb is released, drawing buffer up into the tube. In a few seconds, the buffer siphons from the high tank to the lower tank.

Blow a gentle stream of helium into the closed apparatus. Helium, with its high atomic velocity, is excellent at carrying away heat from the strips.

Use the minimum amperage necessary to get a separation. Heat is a problem, not only because it evaporates water from the strip but because it causes convection currents in the strip and makes separation more difficult. Cool it, man, cool it.

Marker Dye

Place a spot of marker dye alongside of where the sample was spotted to help you decide when the separation is finished. Use a dye that is a little faster than the fastest component you want to measure. The dye will indicate how far the separation has progressed.

Sandwich Type

The sandwich type (Figure 27-21) is preferred when high voltages are used. The plates are cooled to dissipate the heat generated by the current flowing through the strip. This type uses two glass or plastic plates, one on top and one on the bottom of the strips. The purpose is to prevent the buffer from evaporating.

Ridgepole Type

Figure 27-22 shows a very economical ridgepole type system because it can be made from a plastic, single-loaf bread box.

Solvent Immersion (Submarine) Type

In the submarine type, the strip containing buffer is immersed in a liquid immiscible with water, such as chlorobenzene. The purpose is to dissipate heat. Chlorobenzene often is used if paper strips are used, because its density is similar to that of buffer-saturated paper so the paper will stay submerged and won't float. Figure 27-23 shows a commercial apparatus, and Figure 27-24, p. 329, is a diagram of this type of apparatus.

Vertical Slab Apparatus

When many strips are to be run or if two-dimensional separations are desired, than a *slab* is preferred. This can be either horizontal or vertical. The vertical often is preferred, because the effect of gravity can help pull the compounds apart and cooling is easier. Figure 27-25, p. 329, shows a commercial apparatus, and Figure 27-26, p. 329, illustrates how a slab is loaded.

Continuous Electrophoresis

Figure 27-27, p. 330, illustrates a continuous electrophoresis setup developed by E.L. Durrum in 1951.

Preparative Electrophoresis

Although a pattern of components may be all that is desired to identify a disease, the small amounts of material separated often are not enough for further characterization. In those instances, a *prep scale* separation is required. The main types at the moment are batch, continuous flow, and recycle. These are illustrated in Figures 27-28 to 27-30, p. 330 and 331. Figure 27-28, p. 330, illustrates the method developed by B.J. Radola (*Ann. N.Y. Acad. Sci.,* **209,** 127, 1973). This involves ion focusing, which will be described later. According to Radola, "A slurry consisting of gel filtration media (Sephadex G-200), re-swollen in water and 2% carrier ampholytes of the desired pH interval, is poured in a tray (1) and spread in a uniform layer. The excess water is then evaporated until a consistent gel layer is obtained (2). After a short prefocusing step, a trench is dug into the gel layer (3) and the Sephadex beads intimately mixed with the protein sample to be purified. The slurry is then placed back in the trench (4), and isoelectric focusing separation continued until steady-state conditions are reached (5). Finally, the focused protein pattern is revealed by a paper print technique (6) which consists of removing a thin layer of liquid (with focused sample) on a filter paper strip followed by a quick staining step. Once the protein of interest is localized, it is scooped up from the granulated bed and eluted in saline solution.

"As much as 5-10 mg/mL of gel suspension can be loaded. In a 40 x 20 cm trough, 1 cm gel thickness, as much as 10 g of pronase E has been loaded; protein precipitation was in general not a problem, since the bed is horizontal and the precipitate is held in its zone by the Sephadex slurry."

Figure 27-29, p. 331, shows an apparatus similar to that shown in Figure 27-27, except that it is several layers thick. It is described by A. Strickler (*Sepn. Sci.,* **2,** 335, 1967) and K. Hanning (*Electrophoresis: Theory, Methods and Applications,* ed. M. Bier, Academic Press, New York, NY, 1967). This thickness causes some problems in that the flow through the system is parabolic, and the sample emerges in crescent-shaped peaks. "A free film of buffer flows between two planar glass plates, in general 0.5 mm apart, which form a narrow electrophoresis cell. Perpendicular to the laminar buffer flow, an electric field is applied. Near to the top of the cell a continuous, thin sample stream is introduced into the vertically flowing electrophoresis buffer. The sample components migrate laterally at different velocities, depending on their surface charge densities. This apparatus has been used to separate intact cells, organelles and membranes, bacteria and viruses and even parasites. In its modern version, the separation chamber is composed of two parallel glass plates (18 x 4 cm) held 0.3 mm apart. The chamber can separate between 3 and 10^7 cells/h at such a rate that they remain in the electric field for only 30 s."

Figure 27-30, 331, shows a free-flow recycling apparatus developed by M. Bier and Egan, N.B. in 1979 (*Electrofocus/78,* Haglund, H., Westerfall, J.C. and Ball, J.T. Elsevier, Amsterdam). "The basic idea behind recycling in electrophoresis is that solutes do not have to be completely separated in a single pass

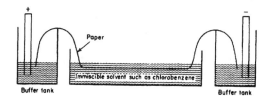

Figure 27-24. Solvent immersion type electrophoresis apparatus.

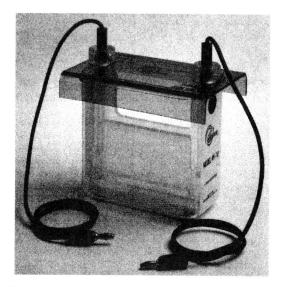

Figure 27-25. A vertical electrophoresis apparatus, EDVOTEK Model MV10.
(Courtesy - EDVOTEK, West Bethesda, MD)

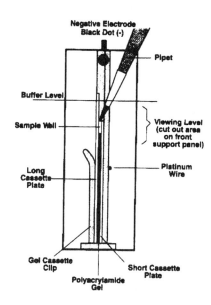

Figure 27-26. Loading instructions for the MV10.
(Courtesy - EDVOTEK, West Bethesda, MD)

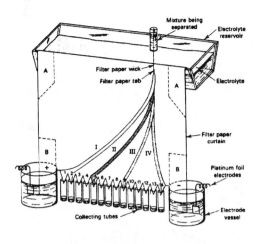

Figure 27-27. Principle of fractionation by continuous electrophoresis.
(Courtesy - Durrum, E.L., *J. Am. Chem. Soc.*, 73, 4875, 1951)

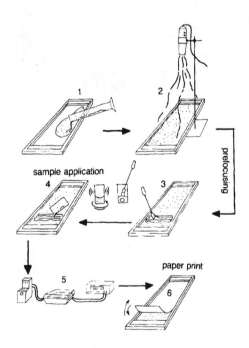

Figure 27-28. Sequence of steps for performing preparative separations by isoelectric focusing in granulated gel layers.
(Courtesy - *Advances in Electrophoresis*, Volume 5, Chrambach, A., Dunn, M.J., and Radola, B.J. Ed., VCH Publishers, New York, NY, 1992)

through the apparatus, but they can be repeatedly cycled through the device until separation is complete. There are two obvious advantages to this approach: (1) the quality of the separation can be carefully controlled and (2) the electric field strength can be reduced, since solutes do not have to be separated in a single pass. The heart of the apparatus is an adiabatic, multichannel slit partitioned into compartments by closely spaced, fine porosity nylon screens which damp inter-compartmental convection, but freely transmit macroions as they migrate in the electric field. During a run, each compartment develops a characteristic pH which is highest at the cathode and lowest at the anode, and which changes sharply at the nylon mesh screens. Once the pH gradient is established, the carrier ampholytes (i.e. the ampholyte buffers needed to create and stabilize the pH gradient) and the macromolecules are circulated continuously though the chamber until all components have migrated to their isoelectric points. The electric field is then switched off and the apparatus and reservoirs drained of products. Heat generated by electrical transport is removed by a multichannel heat exchanger located downstream. The various channel streams are monitored for temperature, pH, uv absorbance and conductivity."

POWER SUPPLIES

Figure 27-31, p. 331, shows one of the more sophisticated power supplies for electrophoresis. This unit can supply up to 3,500 v, and 150 ma at 100 watt operation. It has nine programs, each with nine phases. It can control time (hr, min); current (1 A to 150 mA); volts (volt hr); and break times. According to the manufacturer," The volt-hour mode automatically compensates for differences in field strength caused by variations in the gel thickness or ionic strength. The mA-hr mode can be used to compensate for differences in field strength caused by temperature variations." Much less expensive supplies are available.

DENSITOMETERS

A densitometer is simply a uv/vis spectrophotometer with the optics vertical rather than horizontal and narrow slits to improve resolution. The purpose of a densitometer is to measure the concentration of the separated components. Usually the separated components are stained with a dye; the more concentrated the component is the more intense the color developed will be. The strip or sheet is dried; possibly stripped from the backing; and then placed in the densitometer, which scans across it. The model shown in Figure 27-32, p. 332, is top of the line, and has a resolution of > 40 μm and an absorbance range from 0 to 4 absorbance units and can do this better than the human eye, thus providing more useful information.

PULSED FIELD ELECTROPHORESIS

The pulsed field is the most dramatic development in electrophoresis in recent years. It permits the separation of molecules of molecular weight at least two orders of magnitude

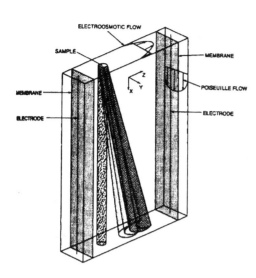

Figure 27-29. Continuous free-flow preparative electrophoresis apparatus.
(Courtesy - Chrambach, A., Dunn, M. J., and Radola, B. J., eds.*Advances in Electrophoresis*, Vol. 5, VCH Publishers, New York, NY, 1992)

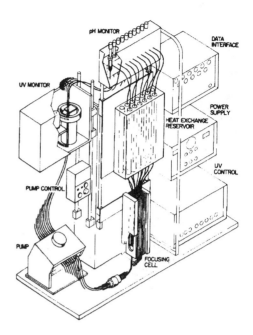

Figure 27-30. Preparative free-flow recycling electrophoresis apparatus.
(Courtesy - Chrambach, A., Dunn, M. J., and Radola, B. J., eds. *Advances in Electrophoresis*, Vol 5, VCH Publishers, New York, NY, 1992)

larger than normal electrophoresis. DNA molecules in excess of 5 Mbp (bp = base pair) can be separated, and separation of molecules even larger is predicted. This is the result of some imaginative thinking on just what might happen to a DNA molecule in an electric field. DNA is normally coiled, but when in an electric field, it can elongate and move through the pores easier. However, if it became snagged, then motion would stop. If the field were reversed for a short time, the molecule might back up and free itself, ready to start all over again. If the field were stopped, the molecule might recoil. Figure 27-33, p. 332, shows what happens if the field is pulsed, first in one direction, then in the opposite direction. If one direction is longer, then there is a net migration to one end. Notice that the time interval is critical and that the higher molecular weight molecules are the most affected. Suppose instead of just going forwards or backwards, the direction was changed, so that the recoiled or freed molecule would go in another direction off to one side and would be less likely to get snagged again in the same pore. The above was the early theory of what was happening. A newer theory, *racheting*, supported by experimental evidence is presented later in the crossed field section.

Pulsed Field Gradient Gel Electrophoresis

D.C. Swartz and C.R. Cantor (*Cell*, **37**, 67, 1984) were the first to propose what they called pulsed field gradient gel electrophoresis (PFGGE). Their electrode arrangement is shown in Figure 27-34, p. 333. "The electric field geometry consists of two orthogonal arrays (X and Y) of point electrodes distributed around the perimeter of the device. The Y array consists of a set of negative electrodes and a single positive point electrode and is activated for the set pulse time. The X array contains a full set

Figure 27-31. An electrophoresis power supply, Pharmacia model EPS 3500 XL.
(Courtesy - Pharmacia Biotech Inc., Piscataway, NJ)

Figure 27-32. A densitometer: the Pharmacia Image Master. (Courtesy - Pharmacia Biotech Inc., Piscataway, NJ)

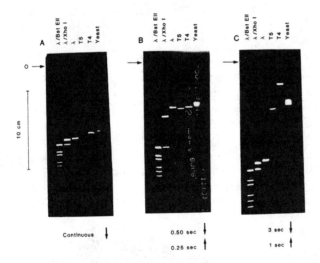

Figure 27-33. Field Inversion Gel Electrophoresis of DNA Molecules from 15 To 200 kbp. (Courtesy - Chrambach, A., Dunn, M.J., Radola, B.J., Eds. *Advances in Electrophoresis* Vol 1, VCH Press, Weinheim, Germany, 1987)

of electrodes and thus generates a uniform electric field when activated for the set pulse time. The electrodes are diode isolated to prevent current flow in the inactive arrays. B shows a typical separation of large DNA molecule." Notice how they curve to one corner.

Transverse Alternating Field Electrophoresis

K. Gardiner, W. Laas, and C.L. Smith (*Somat. Cell Molec. Genet.*, **12**, 185, 1986) solved the problem of nonuniform migration paths by an electrode arrangement shown in Figure 27-35, p. 333. "The transverse alternating field electrophoresis (TAFE) system consists of a vertically mounted gel submerged and in contact with buffer on all sides. The electrode pairs are mounted parallel to the gel plane so that the negative electrode is toward the top and offset

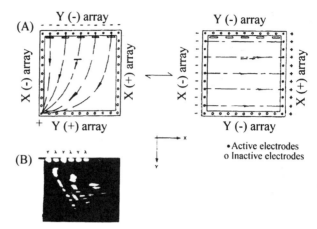

Figure 27-34. Pulsed field gradient gel electrophoresis.
(Courtesy - Chrambach, A., Dunn, M. J., Radola, B.J., Eds. *Advances in Electrophoresis* Vol 1, VCH Press, Weinheim, Germany, 1994)

Figure 27-35.
Transverse alternating field electrophoresis.
(Courtesy - Chrambach, A., Dunn, M.J., Radola, B.J., Eds. *Advances in Electrophoresis* Vol 1, VCH Press, Weinheim, Germany, 1987)

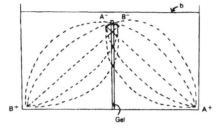
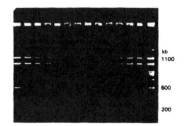

to one side and the positive electrode is at the bottom and offset to the opposite side. The electric field thus passes through the plane of the gel. The second electrode pair is mounted in a similar fashion, but on opposite sides of the gel. As the TAFE electrode pairs are alternately activated, the DNA molecules zig-zag through the thickness of the gel. The migration path is straight down the gel and shows little or no distortion in any lane (Figure 27-35 right side). The electrodes in the TAFE system are positioned to achieve a nominal interaction angle of 115° at the beginning of a separation. The electrical field strength and interaction angle progressively change as molecules migrate, similar to PFGGE, except that these changes occur only in one direction, vertically down the gel."

Crossed Field Pulsed Electrophoresis

The crossed field problem on nonuniform fields was solved by mechanical means (Figure 27-36, p. 334) (Southern, E.M., Anand, R., Brown, W.R.A., and Fletcher, D.S., *Nucleic Acids Res.*, **15**, 5925, 1987). An early problem was that inactive, but still conducting electrodes in the buffer chamber produced a non-uniform electric field in most pulsed field apparatus. By using only two electrodes, but rotating the gel between two positions separated by 110°, a nearly uniform field could be produced and has the advantage that the field angle can be altered at will. Two disadvantages include the impossibility of rotating the platform at rapid speeds (< 5 seconds) and the need for a mechanical method to rotate the platform position.

Figure 27-37, p. 334, is a diagram of what is believed to take place based on experimental data. It is a racheting mechanism. The shapes of the molecules have been observed by fluorescence microscopy. According to Southern and Elder, "This simple theory starts from the assumption that DNA molecules must adopt a stretched confirmation as they pass through gels in which the pore size is much smaller than the contour length of the molecules. At the end of each pulse a molecule will be left stretched in the orientation of the field. It is postulated that if a new field is applied at an angle oblique to the orientation of the previous field, the molecule is more likely to take off by movement which is led by what was its back end. The cumulative effect of this ratcheting motion is to subtract from the molecule's forward motion at each step, an amount which is proportional to its length."

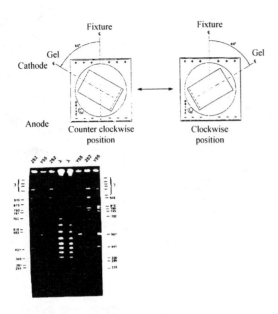

Figure 27-36. Crossed field pulsed electrophoresis. (Courtesy -, Chrambach, A., Dunn, M.J., Radola, B.J., Eds. *Advances in Electrophoresis* Vol 1 VCH Press, Weinheim, Germany, 1987)

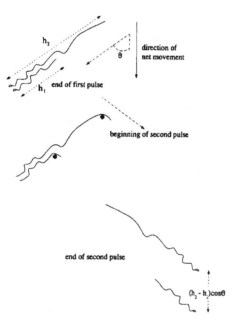

Figure 27-37. Movement of a DNA chain through an agarose gel under the influence of a pulsed field. (Courtesy - Monaco, A.P., Ed., *Pulsed Field Gel Electrophoresis* IRL Press, Oxford, NY, 1994)

Racheting Theory of Migration

"Originally, it was assumed that molecules were led through the gels by a physical end; it now seems that they may also be led by loops. This makes the racheting mechanism even more likely, as the loop at the front end will be snagged when the field is changed, so that movement of the molecule must be led by the trailing, free end. Looping may provide an explanation that all molecules above a certain size do penetrate the gel at a limit mobility. All DNA molecules above a certain size pass through a gel by conventional electrophoresis at approximately the same rate. If they are extended their back ends will trail behind their front ends at distances which depend on the lengths of the molecules. If the direction of the field is then changed in such a way as to push the DNA molecule sideways, it can only begin to move by leading off with one of its ends, or by forming a new 'end' by kinking. Movement led from within the molecule requires more work and probably occurs less often than movement led by an end. If the new field direction is at an acute angle to the old, movement will be led by the front end. But if it is greater than 90° it will be led by the old back end for two reasons; first, because leading with the old front end would turn the molecule through a sharper angle, which would require more work; and second, because the old front end is likely to be looped, as explained above. As the back ends of the molecules were left at different positions in the previous pulse, the starting point for the new movement is different for molecules of different length.

"This model predicts that if the angle between the fields is less than 90°, DNA molecules will behave as they do in a gel that is run conventionally, i.e. with no turning, whereas if the angle is greater than 90°, they will be held back in the gel at each turn by a distance that is proportional to their length; this prediction was confirmed by experiments which showed that the separation between molecules is roughly proportional to the difference in their lengths. A further prediction of the model is that the separation will be proportional to the total number of pulses and, up to a certain limit, independent of pulse length.

"The model may be expressed more precisely as follows. The distance moved in a single pulse by the front end of any molecule along the direction of the field is:

$$d = v\,t \qquad (27\text{-}13)$$

where v is the velocity of DNA in conventional gel electrophoresis under the same conditions and t is the duration of the pulse. The distance moved by the front end along the axis is:

$$d \cos \theta \qquad (27\text{-}14)$$

where θ is the angle of the field to the axis of the gel. After one pulse, the distance along the axis of the gel between the starting position of the front end of the molecule and the final position of the back end of the molecule is:

$$m = (d - h) \cos \theta \qquad (27\text{-}15)$$

where h, the molecules apparent end-to-end length, is less than or equal to the contour length. After one pulse, the distance separating the back ends of two molecules i and j will be:

$$m_i - m_j = -(h_i - h_j) \cos \theta \qquad (27\text{-}16)$$

After N pulses, the net distance moved by a molecule i will be

$$M_i = Nm_i = N(d - h_i) \cos \theta = (D - Nh_i) \cos \theta \qquad (27\text{-}17)$$

where D is the distance moved by DNA in conventional electrophoresis under the same conditions and for the same total time. The separation between molecules i and j will be:

$$\boldsymbol{M_i - M_j = N(m_i - m_j) = -N(h_i - h_j) \cos \theta} \qquad (27\text{-}18)$$

Thus, as observed from experiments, separation is proportional to the number of pulses N. Furthermore, for a given run time T, $N = T/t$. Hence, *separation is inversely proportional to pulse length*, since there will be more short pulses, and each one leads to the same amount of separation.

"During a pulse, the distance moved by the back end of a molecule is $d - h$. If h is greater than d, the back end will not progress past the position occupied by the front end on the previous pulse. This separation will extend up to molecules whose apparent length is equal to d. Since d is proportional to the pulse length, *the size limit for separation increases in proportion to the pulse length.*"

Contour Clamped Homogeneous Electric Fields

The contour clamped modification was introduced by Chu, G., Vollrath, D., and Davies, R.W. (*Science*, **234**, 1582, 1986). Figure 27-38 shows a diagram of the electrode arrangement and the results, and Figure 27-39, p. 336, shows a commercial apparatus.

According to Gemmill (*Advances in Electrophoresis*), "The CHEF modification used the principles of electrostatics to generate uniform electric fields across the gel. Electrodes are distributed around the perimeter of a hexagonal gel chamber and set (clamped) to a predetermined electrical potential. This is achieved by placing specific resistances between electrodes that were distributed around the closed contour. The potentials are chosen so that the electric field is uniform across the gel. This field mimics one that would result from a pair of infinite line electrodes. The CHEF design is one possible design that leads to completely uniform electric fields. This advance established beyond doubt that gradient electric fields were not required to achieve separation of large DNA. In CHEF, θ is fixed at 120° and the excellent separations achieved are straight and largely undistorted."

One advantage of the CHEF array is that electronic switching of fields can be achieved very rapidly for the separation of molecules smaller than 100 kbp. However, CHEF has the limitation that θ cannot be modified.

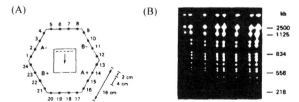

Figure 27-38. Contour clamped homogeneous electrical fields (CHEF).
(Courtesy - Chrambach, A., Dunn, M.J., Radola, B.J., Eds. *Advances in Electrophoresis* Vol 4, VCH Press, Weinheim, Germany, 1994)

Figure 27-39. Hexagonal electrode kit for pulsed electrophoresis.
(Courtesy - Pharmacia Biotech Inc., Uppsala, Sweden)

IMMUNO ELECTROPHORESIS is discussed in Chapter 28, p. 339, and **ISOELECTRIC FOCUSING, SDS, AND TWO-DIMENSIONAL TECHNIQUES** are discussed in Chapter 30, p. 351.

REVIEW QUESTIONS

1. If you had three ions of the same charge, but of different size, which would migrate the fastest?
2. What is meant by "mobility"?
3. What are some "charging" processes?
4. Why doesn't an ion accelerate as it moves toward an electrode?
5. Calculate the mobility of the Ba^{+2} ion in a 0.0200 M solution of $BaCl_2$. The radius of Ba^{+2} in solution is 2.78 Å, and the viscosity of the solution is 9.300 millipoises.
6. What is the ionic strength for the above solution and the correction factor associated with it?
7. What is the difference between V and E?
8. What is k for a cell having a cell constant of 17.5 cm^{-1} and a resistance of 8150 ohms?
9. What is E for a cell that has a current of 18 milliamperes and is 1.00 cm in diameter?
10. If a solution of an albumin was placed in the above cell and took 196 minutes to move 5.20 cm toward the + electrode, what is the mobility?
11. What is the heat developed in the solution described above (18 ma, 1.00 cm dia., and $k = 0.00215$ ohm^{-1} cm^{-1})?
12. What is electro-osmosis?
13. What direction will water move if placed in an electrical field, and what does this say about the nature of water?
14. Why is dextran usually used to determine the correction for electro-osmosis?
15. A compound has an uncorrected mobility of 5.20 cm in a system with an E of 10.00 volts/cm in 190 minutes and dextran moves 0.3 cm away from the compound. What is its corrected mobility?
16. It is suggested that acetate strips not be dipped in buffer, but laid on top of it. Why is this?
17. Where does agar come from?
18. What is the purpose of a comb?
19. What are some advantages of polyacrylamide gels?
20. What is the purpose of a buffer?
21. Rank Coomassie blue R-250, silver nitrate, and autoradiography in regard to sensitivity.
22. What is MDPF as a fluorescence agent and how good is it?
23. What are the major isotopes used for autoradiography?
24. When would you prefer to remove a gel from its backing support?
25. What is the purpose of blotting?
26. What is the mechanism of blotting?
27. What is the primary purpose of Southern blotting?
28. What is the purpose of blocking?
29. What is the purpose of a marker dye?
30. Why is an electrophoresis apparatus covered?
31. Why should the buffer be the same height on each tank?
32. What is the purpose of pulsed field electrophoresis?
33. What is the ratcheting theory of migration?
34. What is the advantage of contour clamped homogeneous electrical fields?
35. Egg albumin was found to have a net charge of -11.4, a radius of 27.8 Å, and a mobility of -2.0×10^{-4} cm^2 $volt^{-1}$ sec^{-1}. What is the viscosity of the solution?
36. What is the solution radius of the acetate ion if a 0.125 M solution of sodium acetate showed a mobility of -2.49×10^{-4} cm^2 $volt^{-1}$ sec^{-1}. Viscosity is 9.633 millipoises.

37. A 0.07 M solution of $La(NO_3)_3$ was placed in an electrophoresis cell. The viscosity was found to be 8.282 millipoises. What is the radius of lanthanum ion, if it had a mobility of $+6.09 \times 10^{-4}$ cm^2 $volt^{-1}$ sec^{-1}?

38. At pH 4.03, a serum albumin in a sodium acetate-acetic acid buffer had a mobility of $+7.3 \times 10^{-5}$. At pH = 7.68, its mobility was -1.46×10^{-4}. What was the net charge change in changing pH? The radius of 14.4 Å and the viscosity of 8.056 millipoises remain constant.

39. 10 mm^3 of a 0.1 M solution of $Cd(NO_3)_2$ required 2 hours and a half to move 88.5 mm when a potential of 100 volts and a current of 10 ma were applied. Calculate the heat generated and the mobility of the Cd^{+2} ion. The cell used was 1.1 cm in diameter, and its cell constant was 24.7 cm^{-1}. A solution of the electrolyte placed in the cell had a resistance of 10,500 ohms.

40. One and one fourth hours after a 0.1 M $Cu(NO_3)_2$ solution had been spotted in an electrophoresis apparatus, the Cu^{+2} ion had moved 62 mm. A current of 10 ma and a potential of 100 volts had been applied. Calculate the amount of heat evolved and the mobility of the Cu^{+2} ion. A 1 cm cell with a constant of 12.6 cm^{-1} had a resistance of 4150 ohms.

41. The electrophoretic fractionation of a pig embryo-plasma produced a spot 58 mm away from the origin 13 hours and 36 minutes after a current of 9 ma had been applied. The cell constant is 15.2 cm^{-1} for the 8 mm cell, and a resistance of 2500 ohms was obtained when a 0.02 M sodium phosphate - 0.15 M NaCl buffer was used. Calculate the mobility and the heat evolved.

42. The red blood cells of a guinea pig were found to migrate 38 mm in 86.5 minutes when placed in a 0.15 phosphate buffer. A current of 12 ma was used. Calculate the mobility and the heat evolved. The cell was 1.0 cm in diameter and had a constant of 18.9 cm^{-1} and resistance of 8140 ohms when the buffer was in the cell.

IF IT EXISTS - CHEMISTRY IS INVOLVED
IF YOU CAN BUY IT - A CHEMIST WAS INVOLVED SOMEWHERE

28
IMMUNOELECTROPHORESIS

PRINCIPLES

Immunoelectrophoresis is a separation technique based on forming a precipitate when an antigen (sample) reacts with an antibody. Most known antigens are proteins; some are polysaccharides or lipid-carbohydrate-protein complexes. Antibodies are formed in mammals to combat the invasion of antigens. This technique is of primary importance for the detection of biological proteins. Figure 28-1, p. 340, is a photograph of what such a reaction produces. Figure 28-2, p. 340, is a diagram of how it is obtained. The sample (antigen) is placed in a well in the center of the gel, and normal electrophoresis performed. Troughs are cut lengthwise along the sides of the gel, and an antibody (antisera) added. Both begin to diffuse. When the antigen contacts the antibody, a precipitate can form, which then is observed. The precipitates form arcs because the sample spot is round and diffuses circularly, whereas the antibody diffuses linearly out of the trough. This can be a very sensitive procedure and can detect sample components that do not react with stains or other detection schemes. This original method (Graber, P., and Williams, C.A., *Biochem. Biophys. Acta*, **10**, 193, 1953) is not quantitative, but it can be used to visualize the location of antigenic compounds from animal and plant tissues, body fluids, cell material, and microorganisms.

ANTIBODIES

An antibody is prepared by injecting a small amount of a foreign compound, *antigen*, into an animal, such as a horse, rabbit, mouse, chicken, donkey, goat, or sheep. The animal usually will form *antibodies* to combat the invasion. If these antibodies then are separated from the animal's body fluids, usually blood, they will react selectively with the antigen in another system. Although research chemists and biochemists can make their own for a specific use, dozens of antibodies are commercially available. There are monoclonal (one compound) and polyclonal (many compounds) antibodies. The first is used to isolate a specific compound, whereas the latter is used to isolate a class of compounds.

It is important that the right concentration ratio of antigen-antibody compounds is attained. The correct ratio will produce a precipitate called *precipitin*, whereas other ratios are more soluble or totally soluble. If the precipitin is slight, it can be visualized by staining.

IMMUNOGLOBULINS

From Pfizer Diagnostics, "Immunoglobulins are protein molecules that function as antibodies. Each molecule of all five classes of immunoglobulins, IgG, IgA, IgM, IgD, and IgE [Table 28-1, p. 341], is composed of a pair of long amino acid heavy chains and a pair of short amino acid light chains linked by disulfide bonds. The heavy chains of each immunoglobulin class differ chemically. A Greek letter identifies the heavy chains of each class - γ for IgG, α for IgA, μ for IgM, δ for IgD and ϵ for IgE. The light chains are of two types, kappa (K) and lambda (λ). Both light chains are common to all immunoglobulin classes, one type per molecule. Each immunoglobulin molecule, therefore, consists

Figure 28-1. Immunoelectrophoresis of human serum with horse polyvalent anti-human serum. (Courtesy - Simpson, C., and Whittaker, M., Eds., *Electrophoretic Techniques*, Academic Press, London, 1983)

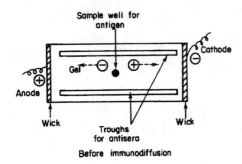

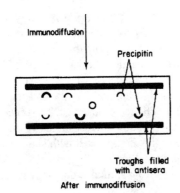

Figure 28-2. Classical immunoelectrophoresis. (Courtesy - *Electrophoresis*, Melvin, M., J. Wiley & Sons, New York, NY, 1987)

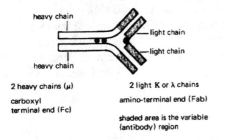

Figure 28-3. Diagram of the IgG tetrachain monomer. (Courtesy - Pfizer Diagnostics Div., New York, NY)

of a pair of heavy chains specific for its class and a pair of either kappa or lambda light chains." See Figure 28-3 and Figures 28-4 and 28-5 on p. 341.

"All immunoglobulin molecules have a constant carboxyl terminal end and a variable amino-terminal end. The amino-terminal end is the 'antibody active site' or the 'antigen combining site', with an amino acid sequence that varies to correspond with the configuration of challenging antigens.

"IgG protects body fluids, IgA protects body surfaces, IgM protects the bloodstream, IgD has no known antibody function and IgE is found in allergy and hypersensitive states."

TECHNIQUES

Many modifications have been made since the original technique was described, several of them permit quantitative evaluation. The most common of these methods are *rocket electrophoresis, crossed electrophoresis, fused rocket electrophoresis, intermediate gel electrophoresis, and tandem electrophoresis*. The basic arrangement for each is described below.

A double constriction pipet (Figure 28-6, p. 341) is preferred for adding the sample to the well. "The double constriction pipet has the advantage over the single constriction pipet of not being fully emptied. This prevents air being blown into the filled wells, causing overflow and contamination of the surrounding gel. The double constriction pipet is filled to the upper constriction and emptied by blowing gently until the meniscus reaches the lower constriction. The tip of the pipet must touch the bottom of the sample well in order to deliver precise and reproducible volumes. These pipets are available in various sizes from 1 µL to several mL" (Deyl, Z. *Electrophoresis, Part A*, Elsevier Scientific Publishing Co., Amsterdam, The Netherlands, 1979).

TECHNIQUE MODIFICATIONS
Electroimmunoelectrophoresis (Rocket Electrophoresis)

Refer to Figure 28-7, p. 342. The antibodies are distributed evenly throughout the gel, and the sample (antigen) is placed in

Table 28-1. Characteristics of immunoglobulins
(Courtesy - Pfizer Diagnostic Div., New York, NY)

	IgG	IgA	IgM	IgD	IgE
Heavy chain	γ gamma	α alpha	μ mu	δ delta	ε epsilon
Light chain	κ or λ	κ or λ	κ or λ	κ or λ	κ or λ
Molecular weight	160,000	170,000	900,000	180,000	196,000
% of total gamma globulin	80	14-20	5	0.2	0.002
g% in serum (range)	0.8-1.6	.14-.42	0.05-0.19	0.0005-0.04	0.00001-0.0007
Half-life (days)	23-30	6-7	5-10	2-8	2-3
Rate of synthesis/day mg/kg of weight	33 mg/kg	24 mg/kg	6.7 mg/kg	0.4 mg/kg	0.02 mg/kg
Placental transfer	+	0	0	0	0
Complement fixing	+	0	+	0	0

a small well at one end. In this case, the antigen moves toward the + electrode. As it does, it diffuses outward and reacts with the antibody, forming a precipitate. As the antigen moves forward, it meets more antibody, but because it is now less concentrated, it will diffuse slower so the line of precipitate is not as far from the center of migration. Eventually, all of the antigen will have moved forward and reacted. The resulting shape, Figure 28-7, p. 342, right side, looks like a rocket (Laurell, C.B., *Protides Biol. Fluids,* **14,** 499, 1967). Either the length or the area is measured and correlated with standards to quantitate the amount.

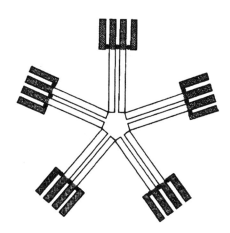

Figure 28-4. Diagram of the IgM tetrachain pentamer.
(Courtesy - Pfizer Diagnostics Div., New York, NY)

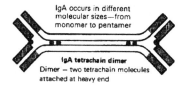

Figure 28-5. Diagram of the IgA tetrachain dimer.
(Courtesy - Pfizer Diagnostics Div., New York, NY)

Figure 28-6. A double constriction pipet.
(Courtesy - Deyl, Z., Ed. *Electrophoresis,* Part A. Elsevier Scientific Publishing Co., Amsterdam, The Netherlands, 1979)

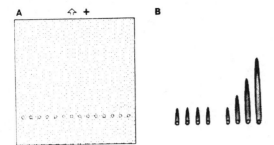

Figure 28-7. Template for rocket electrophoresis and the results.
(Courtesy - Deyl, Z., Ed., *Electrophoresis,* Part A, Elsevier Scientific Publishing Co., Amsterdam, The Netherlands, 1979)

Fused Rocket Electrophoresis

Suppose you have collected several fractions from a gel filtration, ion exchange, chromatographic, or any other similar separation and want to know in which fraction/s the compounds of interest are. Figure 28-8 shows the spotting template and the results. Because rockets from several fractions may diffuse together, this is known as *fused rocket electrophoresis*. The black dotted sample wells in Figure 28-8 are fractions that have been pooled before the run.

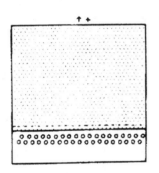 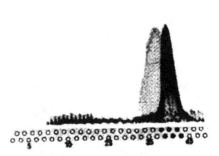

Figure 28-8. Template for fused rocket electrophoresis and a set of results.
(Courtesy - Deyl, Z., Ed., *Electrophoresis,* Part A, Elsevier Scientific Publishing Co., Amsterdam, The Netherlands, 1979)

Crossed Immunoelectrophoresis

Developed in 1965 (Laurell, C. B., *Anal. Biochem.*, **10**, 358, 1965), this is a two-dimensional separation. Figure 28-9, p. 343, shows the template arrangement. According to P. J. Svendsen, "A 1.5 mm thick agarose gel film is placed on a 10 x 10 cm glass plate and 5 wells punched into the set gel. Four wells are usually used for sample and the 5th is used for a marker dye, such as albumin stained with bromophenol blue. Normal electrophoresis is run at about 15 V/cm for about an hour at 15 °C or until the blue marker has moved about 5 - 5.5 cm. The agarose gel is then cut into five slabs as shown in [28-9 A], the marker slab being discarded. The first dimension gel slabs are then transferred, with a long razor blade, to the second-dimension glass plates [Figure 28-9B] which have been coated with agarose solution and dried. The glass plates are placed on a level surface and 12 mL of antibody containing solution (11 mL of 1% agarose + 1 mL of anti-serum) are poured onto the upper part of the glass plate, corresponding to the shaded area of [28-9B]. After 10 minutes the gel is ready and the plate is placed in the electrophoresis apparatus and connected to the electrode buffer by means of filter paper wicks. Immunoelectrophoresis in the second-dimension is performed at 3 V/cm for 18-20 hr."

Crossed Immunoelectrophoresis with an Intermediate Gel

Crossed immunoelectrophoresis with an intermediate gel is a technique to identify a specific component in a mixture, such as that shown in Figure 28-10, p. 343. The technique (Svendsen, P.J., and Axelsen, N.H., *J. Immunol. Methods*, **1**, 169, 1972) is to add an intermediate gel that contains an antibody specific for the compound of interest and remove it from the pattern. The example described below is to identify human transferrin. Figure 28-10 shows the same second dimension spotting template as in Figure 28-9, with the addition of the intermediate gel application.

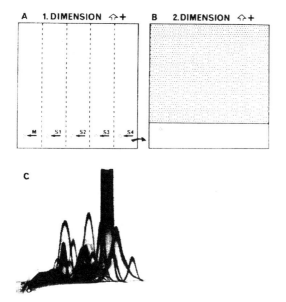
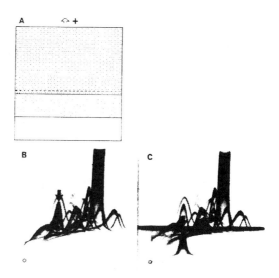

Figure 28-9. Spotting templates for crossed immuno-electrophoresis.
(Courtesy - Deyl, Z., Ed., *Electrophoresis,* Part A., Elsevier Scientific Publishing Co., Amsterdam, The Netherlands, 1979)

Figure 28-10. Spotting template for intermediate gel crossed immunoelectrophoresis and the results for one set of samples.
(Courtesy - Deyl, Z., Ed., *Electrophoresis,* Part A., Elsevier Scientific Publishing Co., Amsterdam, The Netherlands, 1979)

According to J.P. Svendsen and Axelsen, "The first dimension agarose gel electrophoresis and transfer of the agarose gel slabs to the coated second dimension glass plates is the same as for crossed immuno-electrophoresis. A 1 mm thick glass plate is placed 2.2 cm from the first dimension gel slab, along the dotted line. A 3.2 mL volume of agarose solution mixed with 75 µL of anti-human transferrin is poured into the open space between the gel and the glass plate. A reference plate is made in parallel, replacing the anti-human transferrin with 0.1 M sodium chloride solution. When the gel is solidified, the glass plate is cut free, the gel is trimmed along the solid line below the dotted line, and the thin gel strip is discarded. An 8.2 mL volume of agarose solution is mixed with 800 µL of anti-human serum and poured onto the remaining part of the glass plate and the same is done to the reference plate. After immunoelectrophoresis, rinsing, pressing, drying and staining the plates are ready [28-10B and C]. On the plate in [Figure 28-10B] all antigens have passed freely through the intermediate gel. The skewed baseline of the immunoprecipitates is due to decreasing mobilities of the antigens and electro-osmotic transport of slowly migrating antibody molecules. The transferrin is marked with a heavy arrow, but you would not know that is it until the part in [Figure 28-10C] is examined. On the plate all antigens except transferrin have passed through the intermediate gel. A simple comparison of the two plates easily indicates which of the peaks is transferrin. The horizontal line in [Figure 28-10C] is caused by an excess of antigen used to absorb the antibody preparation to make it monospecific.

"The slab is enlarged in an enlarger (9X), and with a light spot planimeter the curves are integrated for quantitation. Peak height is not as accurate because many peaks are un-symmetrical."

Tandem-Crossed Immunoelectrophoresis

The tandem crossed method (Kroll, J., *Scand. J. Clin. Lab. Invest.*, **22**, 79, 1979) is another way to identify a specific component in a mixture. Figure 28-11, p. 344, shows the spotting template and the results. "It is essentially the same as crossed immuno-electrophoresis except that two samples are applied in the first dimension gel, the sample wells having a centre to centre distance of 10 mm." Again using transferrin as the example, "Human serum, 2 µL, is applied in the rear sample well and 4 µL of the pool containing the pure human transferrin is applied in the other well in the

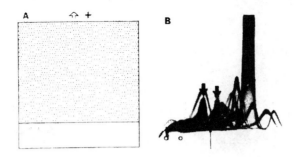

Figure 28-11. Spotting template for tandem crossed immunoelectrophoresis and a set of results identifying human transferrin.
(Courtesy - Deyl, Z., Ed., *Electrophoresis,* Part A., Elsevier Scientific Publishing Co., Amsterdam, The Netherlands, 1979)

front. Before electrophoresis in the first dimension, the plate is left for diffusion for 60 minutes on the cooled surface of the electrophoresis apparatus at 15 °C. The diffusion causes the contents of the sample wells to mix partially in the gel. If identical antigens are present in both wells, the corresponding peaks will show reaction of identity, i.e., fuse into a double peak, as shown in [Figure 28-11B]."

AFFINITY CHROMATOGRAPHY
Afinity chromatography is the same as immunoelectrophoresis in all respects, except that the antigen-antibody reaction is not used, but something else that will form a precipitate with a specific sample component is used. An example is the use of lectin to precipitate sugars, as in glycoproteins.

REVIEW QUESTIONS

1. What is an antigen in general terms?
2. What is an antibody?
3. What are monoclonal antibodies used for?
4. What are immunoglobulins?
5. What does IgM protect?
6. What is the half-life of IgA?
7. What is the advantage of using a double constricted pipet?
8. What causes "rocket formation"?
9. What is the major use of fused rocket electrophoresis?
10. What is the basis for crossed immunoelectrophoresis?
11. What is the purpose of adding an intermediate gel in crossed immunoelectrophoresis?

IF IT EXISTS - CHEMISTRY IS INVOLVED
IF YOU CAN BUY IT - A CHEMIST WAS INVOLVED SOMEWHERE

29
POLYACRYLAMIDE GEL DISC ELECTROPHORESIS

PRINCIPLES

One of the main requirements for high resolution in a multistage separation is to have as small of a starting region as possible. This was difficult to do with electrophoretic techniques until the concept of *disc electrophoresis* was developed by Leonard Ornstein and Baruch J. Davis in 1959 (reprinted in *Annals NY Acad. Sci.*, **121**, 404, 1964). In this case, the sample is placed in a polyacrylamide gel in a vertical cell between stacked buffer chambers. Gravity helps, but the sample is first focused electrically between a layer of fast moving ions and a layer of slow moving ions, then one layer is removed, and the sample separates by a combination of normal electrophoresis and the sieving action of the polymer. This can concentrate a sample to a layer 10 micrometers thick, starting from 1 centimeter. Figure 29-1, p. 346, is a diagram of the basic operations.

Some examples of applications from the medical area are: the detections of multiplemyloma, cancer, diabetic tendency, leukemia onset and remission, tuberculosis, schizophrenia, pneumonia, milk allergy, cystic fibrosis of the pancreas, and the iron binding capacity of various proteins. Bacterial and viral materials such as tuberculin bacteria are being prepared in greater purity so that they can be easy to study. The transfer of proteins from the mother to the fetus, radiation damage, drug reactive patients, brain proteins in allergic encephalomyelitis, and mental retardation. Pregnancy is indicated and miscarriage may be forecast by the level of HCG in urine precipitates.

Figure 29-2, p. 346, shows a commercial disc electrophoresis apparatus.

THEORY

The following is taken directly from Orstein and Davis's papers, abstracted to present the bare essentials.

"As early as 1897, Kohlrausch observed that, under conditions set by his 'regulating function' if two solutions of ions were layered one over the other, such that a solution with ions of substance γ of high mobility were placed below a solution of a slow ion, α, of the same charge, then the boundary between the two ionic species would be sharply maintained as they migrated down-wards in an applied electric field because the two species (on each side of the boundary) would have been 'arranged' to move at the same speed (provided that the lower solution was the denser of the two, and that the potential was applied with such polarity that the ions of α and γ moved downwards).

"If concentrations different from those specified by this regulating function existed when the potential was applied, it was shown that the concentrations at the boundary would, in time, 'regulate' automatically to those required by the regulating function. Large departures from 'regulating' concentrations would, however, lead rapidly to convective instabilities, preventing the ready formation of the moving boundary. On the other hand, in gels and porous media (which prevent convection), given sufficient time, a sharp moving boundary will form and be maintained independent of the densities of the starting solutions and their initial concentrations, provided that the faster ions precede the slower.

$$\frac{(A)}{(\Gamma)} = \frac{x_\gamma\, c_\alpha}{(x_\alpha\, c_\gamma)} = \frac{m_\alpha\, z_\gamma\, (m_\gamma - m_\beta)}{m_\gamma\, z_\alpha\, (m_\alpha - m_\beta)} \tag{29-1}$$

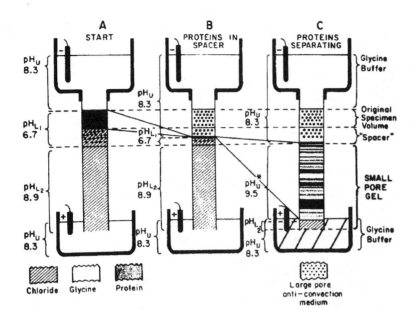

Figure 29-1. Sequences in a disc electrophoresis separation. (Courtesy - Distillation Products Industries, Eastman Kodak Co., NY)

Figure 29-2. A commercial disc electrophoresis apparatus. (Courtesy - Buchler Instruments, Inc., Ft. Lee, NJ)

where x is the fraction of dissociation; c is the concentration, m is the mobility; and α, γ, and β are the ions (with β of opposite charge to α and γ and common to both solutions) provides us with the ratio of the total concentrations of α substance (A) to γ substance (Γ) for initiating and maintaining a stable moving boundary.

"For the purpose of illustrating how [equation 29-1] can be used to produce starting zones, α will be the glycinate ion, β the potassium ion, and γ the chloride ion. Then,

$$m_\alpha = -1; \quad z_\gamma = -1; \quad m_\alpha = -15 \text{ mobility units}; \quad m_\beta = +29 \text{ mobility units}; \quad m_\gamma = -29 \text{ mobility units}$$

and therefore, (glycine)/(chloride) = $(A)/(\Gamma)$ = 0.58 (essentially independent of pH or either solution from pH 4 to 10, within which range neither the hydrogen nor hydroxyl ions will appreciably contribute to conductivity, provided that the chloride concentration is greater than 10^{-8} M).

"Above pH 8.0, most serum proteins have free mobilities in the range from -0.6 to -7.5 units. If the effective mobility of glycine, $m_\alpha x_\alpha$, were less than -0.6, the mobilities of the serum proteins would fall between that of the glycine and that of chloride. This requirement is satisfied when $x_\alpha = 1/30$ since the glycinate ion has a mobility of -15 units. The pH at which this degree of dissociation of glycine occurs can be calculated. From [Equation 29-2],

$$pH = pK_a - \log[(1/x_\alpha) - 1] \tag{29-2}$$

Therefore the pH of the glycine solution must be 8.3 if $x_\alpha = 1/30$.

"If a protein molecule with a mobility of -1.0 units is placed in the glycine solution (pH 8.3) one centimeter above the glycine/chloride boundary, by the time the boundary (or a glycine molecule) has migrated one centimeter, the protein will have migrated two centimeters and will then be located at the boundary. It would continue to run at the boundary because the mobility of the chloride is greater than that of the protein.

"If, a very large number of albumin molecules (mobility approximately -6.0 units) are placed in the glycine solution (pH 8.3), they will concentrate at the boundary between the chloride and the glycine at a concentration satisfying [equation 29-1] where α is now the serum albumin, β, the potassium ion, and γ, the chloride ion.

"Then,

$z_\alpha = -30$ (approximate charge of albumin molecule at pH 8.3); $z_\gamma = -1$
$m_\alpha = -6.0$ mobility units; $m_\beta = +29$ mobility units; $m_\gamma = -29$ mobility units.

and therefore (albumin)/(chloride) = 9.3 x 10^{-3}. If (chloride) = 0.06 M, then (albumin) = 0.00056 M. That is, *the albumin (m.w. 68,000) will automatically concentrate to about 3.8% behind the chloride* and would then stay at constant concentration. If the initial concentration of albumin had been 0.01% in the glycine buffer, and if 1 mL of this mixture is placed on top of the chloride solution in a cylinder of one square centimeter cross section, then after the chloride boundary has moved about one millimeter, all of the albumin will have concentrated into a disc (right behind the chloride), which would be about 25 *microns* thick. (In practice, this might be done by using a porous anti-convection medium all through the volume occupied by the chloride and through the one centimeter height of the glycine column). In this manner, a 380-fold increase in concentration can be achieved in a few minutes and the protein is reduced to a very thin lamina or disc. If the original concentration had been 0.0001%, *the same final concentration* would result, but the total change would now be 38,000-fold. If the column were 100 cm in length, *the same amount* of protein would have been concentrated.

"If, instead of a single protein, a 1 mL mixture with mobilities ranging from -1.0 to -6.0 units is placed over the chloride solution, by the time the boundary has migrated 1 cm in the applied electric field, *all of the proteins will have concentrated into very thin discs, one stacked on top of the other in order of decreasing mobility, with the last followed immediately by glycine.*

"If the boundary (and the following stack of discs) is permitted to pass into a region of higher pH, e.g., a pH of 9.8 (the pK_a of glycine), at which x_{gly} equals 1/2 and therefore $m_{gly}x_{gly}$ equals -7.5 mobility units, the glycine will now overrun all the protein discs and run directly behind the chloride, and the proteins will now be in a uniform linear voltage gradient, each effectively in an extremely thin starting zone, and will migrate as in 'free electrophoresis'. The required stationary pH boundary is quite easily established subject to degradation only by diffusion."

Combining Pore Size Control and Thin Starting Zones

"We will now combine the mechanisms of size control and a thin starting zone. An electrophoretic matrix can be prepared into which discontinuities in pH and gel pore sizes as well as Kohlrausch conditions are incorporated.

"The protein is placed as shown in A of [Figure 29-1] and, as a voltage is applied, instead of the proteins running ahead of the glycine to catch up with the chloride (as described above), the chloride overtakes the proteins, which then `sort' out according to their mobilities into highly concentrated discs of proteins, stacked exactly as described above [Figure 29-1B].

"A `large pore' (approximately 3% acrylamide) is usually used as the porous anti-convection medium in this region as well as in the 'spacer' region. The protein sample is mixed into the 'spacer mixture' and is usually gelled in place on top of the spacer. As the glycine following the stack of proteins moves through these regions, the pH is maintained at 8.3.

"At some time after stacking is complete (this time depending on the thickness of the spacer gel), the proteins reach the small pore gel where changes in their mobilities occur. Because of the special viscous properties of the gel, proteins of *equal free mobility* but of appreciably *different molecular weight (*different diffusion constant) will migrate with markedly different mobilities and will easily be separated [Figure 29-1C]. A 7½ % acrylamide gel has proven to effect useful separations of human serum proteins. The fastest pre-albumin has a mobility less than -5.0 units *in such a gel*. We, therefore, arrange for a 'running pH' of about 9.5 where the effective mobility of glycine ($m_{gly}x_{gly}$) is about -5.0 units in the gel.

"Given the concentration of chloride, (Γ), and the pH of the upper buffer as 8.3, it is possible to compute the concentration of glycine, (A), and the proper pH, pH_{L1} for the large-pore gel (3%) and pH_{L2} for the small pore gel (7.5%). In calculating $(B)_{L2}$ and pH_{L2}, a pH^*_U of 9.5, the 'running pH', and $x^* = 1/3$ [the value of $x\alpha$ at pH_U rather than 8.3 and 1/30 are used in equations [29-16 and 29-17, p. 350]."

The Regulating Function for Weak Electrolytes

"The electrical conductivity, λ, of a solution of ions is a function of the concentration of the *ith* ion, c_i, its mobility, m_i, and its elementary charge, z_i, such that,

$$\lambda = E \sum c_i m_i z_i \tag{29-3}$$

where E is the charge of the electron. If we consider acids and bases at pH's near the pK_a or $(14 - pK_b)$, only part of the population of molecules will be charged at any one time. If x_i is the fraction of dissociation (i.e., the ratio of charged molecules to the sum of the charged and uncharged forms), then each molecule can be viewed as being charged x_i of the time and uncharged the rest of the time. The average velocity of migration, s_i of the molecule in the voltage gradient, V, will be

$$S_i = V m_i x_i \tag{29-4}$$

"If two solutions, L (the lower) containing substances γ, and U (the upper) containing substance α (ions of α and γ of like charge with $m_\gamma x_\gamma$ greater than $m_\alpha x_\alpha$), are layered U over L in a cylinder, and a potential is placed across the cylinder (from end to end), the concentration of substance in these solutions for the velocity, α, to equal γ, may be derived as follows:

"Let

$$s_\alpha = V_U m_\alpha x_\alpha = s_\gamma = V_L m_\gamma x_\gamma \tag{29-5}$$

"Since the current, Y, through both solutions (which are electrically in series) is the same, it follows from Ohm's law that, $V_L = Y/\lambda_L S$, and $V = Y/\lambda_U S$; where S is the cross sectional area of the cylinder, from [29-1] and [29-3],

$$\frac{m_\alpha x_\alpha}{\sum c_{iU} m_{iU} z_{iU}} = \frac{m_\gamma x_\gamma}{\sum c_{iL} m_{iL} z_{iL}} \tag{29-6}$$

"Equation [29-6] is a modified form of the Kohlrausch regulating function. When the conditions specified in the equation are satisfied, then substances α and γ will migrate down the cylinder with equal velocity and the boundary between them will be maintained. If a molecule of α were to find itself in the bulk of solution L (where, from equation [29-5]) V_L is less than V_U, it will migrate more slowly than the molecules of γ (and therefore, more slowly than the boundary) and will be overtaken by the boundary. Conversely, a molecule of γ in the bulk of U will move faster than the boundary and will overtake it, thereafter migrating at the same velocity as the boundary. Let us now consider two solutions with one common ion, β, and two ions α and γ with charge opposite to. From [29-4],

$$\frac{m_\alpha x_\alpha}{c_\alpha m_\alpha m_\alpha + c_{\beta_U} m_\beta z_\beta} = \frac{m_\gamma x_\gamma}{c_\gamma m_\gamma m_\gamma = c_{\beta_U} m_\beta z_\beta} \tag{29-7}$$

The condition of net macroscopic electrical neutrality in each solution requires that,

$$c_\alpha z_\alpha = -c_{\beta_U} z_\beta \qquad c_\gamma z_\gamma = -c_{\beta_L} z_\beta \tag{29-8}$$

Therefore, from [29-7] and [29-8],

$$\frac{m_\alpha z_\alpha}{c_\alpha z_\alpha (m_\alpha - m_\beta)} = \frac{m_\gamma x_\gamma}{c_\gamma z_\gamma (m_\gamma - m_\beta)} \qquad (29\text{-}9)$$

Let the *total concentration* of molecular species i, be $(I) = c_i/x_i$ (29-10)
then,

$$\frac{(A)}{(\Gamma)} = \frac{x_\gamma c_\alpha}{(x_\alpha c_\gamma)} = \frac{m_\alpha z_\gamma (m_\gamma - m_\beta)}{m_\gamma z_\alpha (m_\alpha - m_\beta)} \qquad \text{The same as equation [29-1]}$$

"The relationship in this equation is relatively insensitive to temperature. Mobilities change with temperature mainly as a result of the sensitivity of the frictional resistance of the medium to temperature (the temperature coefficient of viscosity). Proportional changes in all mobilities cancel out in equation [29-1].

"From the Henderson-Hasselbalch equation for pH, where $pH = pK_a + \log c_i/(iH)$, where iH is an acid, and $pH = (14 - pK_b) + \log (i)/c_i$ where i is a base, and from [29-10],

$$x_\alpha = c_\alpha/(A) = c_\alpha/[c_\alpha + (\alpha H)] = 1/(1 + 10^{(pK_a - pH)}) \qquad (29\text{-}11a)$$

$$x_\beta = c_\beta/(B) = c_\beta/[c_\beta + (\beta)] = 1/(1 + 10^{-[(14 - pK_b) - pH]}) \qquad (29\text{-}11b)$$

Moving and Stationary pH Boundaries

If the pH of the lower solution is set so that the concentration of uncharged base is equal in both upper and lower solutions, then as the anionic boundary sweeps towards the anode (leaving uncharged base behind), the pH right behind the moving boundary will remain fixed and equal to the pH of the upper buffer (upper solution). This is achieved as follows:

Let $(\beta)_U = (\beta)_L$, then from the Henderson-Hasselbalch equation,

$$pH_U - (14 - pK_b) + \log c_{\beta_U} = pH_L = (14 - pK_b) + \log c_{\beta_L} \qquad (29\text{-}12)$$

Therefore,

$$pH_L = pH_U + \log c_{\beta_U}/c_{\beta_L} \qquad (29\text{-}13)$$

and from equations [29-8] and [29-10]

$$c_{\beta_U} = -x_\alpha(A)z_\alpha/z_\beta \qquad (29\text{-}14U)$$

and

$$c_{\beta_L} = -x_\gamma(\Gamma)z_\gamma/z_\beta \qquad (29\text{-}14L)$$

Therefore,

$$pH_L = pH_U + \log (A)x_\alpha z_\alpha/(\Gamma)x_\gamma z_\gamma \qquad (29\text{-}15)$$

From equations [29-15] and [29-11a], pH_L can be computed. From equations [29-11b] and [29-14L]:

$$(B)_L = c_{\beta_L} + (\beta)_L = [-x_\gamma(\Gamma) \, z_\gamma/z_\beta](1 = 10^{[(14 - pK_b) - pH_L]}) \qquad (29\text{-}16)$$

From equations [29-11b] and [29-14U]:

$$(B)_U = [-x_\alpha(A)z_\alpha/z_\beta](1 + 10^{-[(14 - pK_b) - pH_L]}) \qquad (29\text{-}17)$$

If, in addition, we wished to set up a stationary pH boundary, with pH_U above and pH^*_U below, which will remain at the original boundary between L and U, then pH^*_U, and x_α^* (x_α at pH^*_U) rather than pH_U and x_α are used in calculating pH_L, and therefore $(B)_L$. When the potential is applied, the pH of the solution above the original boundary will remain equal to pH_U, but the pH between that boundary and the moving boundary will equal pH^*_U (see Figure 29-1,C).

TECHNIQUES

MARKER DYE
A dye, faster moving than any of the sample components, is used to determine when to stop the separation. The top buffer contains 0.001% bromophenol blue. When a voltage is applied to the sample a small portion of this dye concentrates and moves down the tube. It is usually easy to see, and the separation should be stopped before it gets to the end of the tube.

POLYMERIZATION
To ensure a flat junction between polymer layers, add one drop of water to the top of the unpolymerized polyacrylamide before polymerization. The boundary between the water and the polyacrylamide will be flat, and the meniscus will be on the top of the water. The water is removed after polymerization by tipping the tube over and dabbing the remaining water out of the tube with a small section of a paper towel.

RIMMING
A fine syringe needle attached to a small syringe filled with water is used to scrape the polymerized acrylamide from the walls of the tube. The tube containing the gel is placed in a large beaker or bucket of ice water to cause the gel to contract. The syringe is filled with cold water, the needle is inserted carefully along the edge of the tube a short distance and the tube is turned carefully, so the needle cuts the gel away from the glass. A small amount of cold water then is forced into the space to lubricate the area. This is continued until the entire gel is cut loose. A small pipet bulb filled with water then is used to force the gel out of the tube into a 10 cm long test tube. The sample and spacer gel may fall off at this time.

REVIEW QUESTIONS

1. What is the arrangement of the buffer tanks in disc electrophoresis?
2. What are the three basic gel layers in a tube?
3. What is the basis of Kohlrausch's regulating function?
4. What are two provisions required to allow the regulating function to operate?
5. For a protein separation, chloride and glycine are placed in the spacer gel and sample in the gel above it. How does the sample get between the chloride and the glycine and then what happens to it?
6. What is the purpose of bromophenol blue in the top buffer?
7. How do you ensure a flat junction between each of the polymerized layers?
8. What is used to hasten the polymerization of the polyacrylamide gel?
9. If the polymer does not form within 15 minutes, what is likely to be the cause?

30
ION FOCUSING, SDS, AND TWO-DIMENSIONAL ELECTROPHORESIS

ION FOCUSING

PRINCIPLES

In Chapter 27, p. 322, you learned how a molecule can be neutral or negatively or positively charged, depending upon the pH of the solution it is in.

$$\overset{+}{H_3N}-CH_2-C\overset{O}{\underset{OH}{\nwarrow}} \underset{}{\overset{H^+}{\rightleftharpoons}} H_2N-CH_2-C\overset{O}{\underset{OH}{\nwarrow}} \underset{}{\overset{OH^-}{\rightleftharpoons}} H_2N-CH_2-C\overset{O}{\underset{O^-}{\nwarrow}}$$

$$pKa_1 \quad 2.34 \qquad\qquad pKa_2 \quad 9.60$$

Suppose such a molecule was placed on a gel strip, and the pH of the gel was acidic at one end (anode) and basic at the other (cathode) with a smooth gradient in between. It would take on the charge forced upon it by the pH of the surrounding solution. If an electric field was applied, it would move according to that charge. As it moved, it might change its net charge due to the surrounding liquid, become neutral, and cease to move. If it diffused in one direction, it might become positively charged and move toward the negative electrode and become neutral again. If it diffused in the opposite direction, just the reverse would occur. The molecule would be focused in a narrow band in the gel. This is now called *ion focusing* and was first proposed by H. Svensson (*Acta Chem. Scand.*, **15**, 325, 1961), who called it *isoelectric fractionation*. Figure 30-1, p. 352, is a diagram of what happens.

The pH at which the molecule is focused is called the *isoelectric point*, pI. This point varies from molecule to molecule more than one might expect. A very large molecule can contain many acid or base sites, and some can be buried within the molecule and not affect the net migration significantly. This type of molecule is called a *zwitterion*, which means that ionization of the + and - groups is equal. The molecule at this point has minimum conductivity, osmotic pressure, and viscosity. Proteins show the least swelling, they coagulate best, and contain the least amount of inorganic matter. Advantage is taken of this to separate molecules. Differences of as little as ± 0.01 pH unit can be resolved in favorable cases. Table 30-1, p. 352, lists pI's of several compounds.

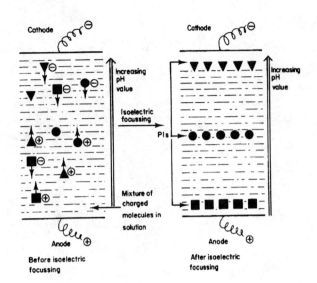

Figure 30-1. Diagram explaining isoelectric focusing.
(Courtesy - Melvin, M., *Electrophoresis*, J. Wiley & Sons, New York, NY, 1987)

Table 30-1. The pI of several compounds

Compound	pI	Compound	pI
Alanine	6.0	Methionine	5.74
Arginine	10.76	Phenylalanine	5.48
Asparagine	5.41	Proline	6.38
Aspartic acid	2.77	Serine	5.68
Cysteine	5.07	Threonine	5.60
Cystine	4.60	Tryptophan	5.89
Glutamine	5.65	Tyrosine	5.66
Glutamic acid	3.22	Valine	5.96
Glycine	5.65	Glutinin	4.5
Histidine	7.59	Gelatin	4.7
Hydroxyproline	5.83	Egg albumin	4.8
Isoleucine	6.02	Serum albumin	5.4
Leucine	5.98	Oxyhemoglobin	6.8
Lysine	9.74	Gliadin	9.2

TECHNIQUES

Developing the pH Gradient

There are two main techniques for establishing a pH gradient along a strip: nongel and gel. Both require compounds that have a variable charge but will become neutral at a certain pH. These are called *ampholytes*.

The nongel technique (Wrigley, C., *Sci. Tools*, **15** (2), 17, 1968) involves placing a sucrose density gradient in a vertical column to obtain a convection-free medium. The sample can be layered or placed throughout. The pH gradient is formed by the carrier ampholytes when a voltage (200 to 1200 V, dc, yielding a current up to 10 mA) is applied across the mixture. The proteins migrate to the point where they are electrically neutral. To reduce convection, the column is thermostated during operation. The time required for electrofocusing ranges from 24 to 72 hours. At the end of the run, the contents of the column are drained through a capillary column to a fraction collector.

The agarose or polyacrylamide gel technique was developed to reduce the convection and speed up the process, requiring only from 1 to 3 hours in many cases. The buffer in each tank is usually different, for example, acetic acid on the low pH side and tris on the high pH side for the wide range gradients. The liquids on the strips or slabs are very fluid; and if the ends of either are dipped into the buffer tanks as in other techniques, the liquids tend to drain away. The technique is to place some strips of paper toweling soaked in buffer under the edge of the strip and let the paper toweling hang into the buffer tank.

Ampholytes

When you studied the Periodic Table, you learned about amphoteric elements, those that would act like a base in some cases and like an acid in others. When a molecule has that ability and is used in an electrophoresis separation, it is called an *ampholyte*. Ampholytes are used as pH markers and to establish the pH gradient. There are two basic systems, natural and immobilized.

Natural Ampholytes

These are mixtures of aliphatic polyamino-polycarboxylic acids with molecular weights between 300 and 700. Each is synthesized to have a specific pI and usually cover the range from pH 3.5 to 10 (Figure 30-2). The gel is prepared as usual; however, about 2 to 3% of the ampholyte mixture is added just before polymerization. After polymerization, the pH of the gel is the same at all points. Once electrophoresis starts, each ampholyte migrates to its position and establishes the pH in the gel at its pI. The series thus produces a pH gradient. A sample placed in the gel moves to its appropriate position and stops. The sample also can be mixed in at the beginning. Because its molecules are usually larger and thus slower moving, the gradient is established first. Sample sizes from 10 to 50 µg up to 20 mg for prep scale are typical.

Immobilized Ampholytes

Available from Pharmacia Biotech as *Immobiline,* these are "derivatives of acrylamide with different dissociation constants. For acidic Immobiline (with pK 3.6 or 4.6), carboxylic acid is chosen as a functional group; for the bases, tertiary amino groups are selected. When different pK values of Immobiline are mixed in appropriate proportions, together with acrylamide, bisacrylamide, TEMED, and ammonium persulfate, the buffering groups responsible for creating the pH gradient become covalently attached to the polyacrylamide base.

"The result is a gel with an immobilized pH gradient. Because the pH gradient is completely immobilized, you can generate stable, reproducible linear gradients with a slope as low as 0.01 pH units/cm and separate proteins with pI differences of 0.001 pH units. This system provides the means to form any narrow or wide pH gradient in the 3.8 to 10 range. " The ampholytes are stained by the normal stains and must be removed before staining. One way to do this is to soak the strip in 5 % trichloroacetic acid and then stain with Coomassie brilliant blue."

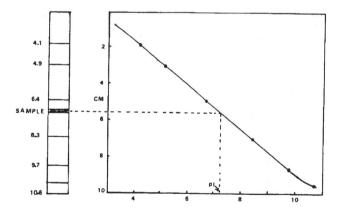

Figure 30-2. Example of the use of marker proteins. Acetylated cytochrome c at pI 4.1, 4.9, 6.4, 8.3, 9.7, cytochrome c at pI 10.6.
(Courtesy - Pomeranz, Y., and Meloan, C.E., *Food Analysis - Theory and Practice,* Chapman-Hall, New York, NY, 1994)

Determining the pH

The pH of the gel can be determined by inserting a microelectrode at points along the gel, or a small section can be cut out and dissolved, and the pH measured with a regular set of electrodes. A common set of microelectrodes includes antimony wire for the indicator electrode and a micro calomel electrode for the reference. If you use this pair, you must correct the pH meter reading based upon the half cell reaction.

$$Sb_2O_3 + 6H^+ + 6e^- \rightarrow 2Sb + 3H_2O \quad E^\circ = +0.152 \ V$$

$$E = 0.152 + \frac{0.059}{6} \log \frac{[Sb]^2[H_2O]^3}{[Sb_2O_3][H^+]^6}$$

$$E = 0.152 + 0.059 \log [H^+] \quad\quad E = 0.152 - 0.059 \ pH$$

Heat Production and Control

If you are not concerned about denaturing proteins or conformational changes in other types of compounds, then heat production is limited to ensuring that the strips or slab do not dry out. However, if denaturation is not wanted, then control of the temperature in the system is necessary. One joule of heat raises the temperature of 1 mL of water about 0.25 °C or 0.1 mL about 2.5 °C. For slab gels, it is recommended to keep the power at less than 0.5 watts.

EXAMPLE CALCULATION
If a 30 cm² slab of gel is heated for 1000 seconds, and it is desired to keep the power at less than 0.5 watts, what temperature would the gel attain? The thickness of the gel is 1 mm, so each cm² contains 0.1 mL of water.

ANSWER
0.5 watts/30 cm² = 15 mW/cm² or 15 x 10⁻³ J/s/cm². Therefore, 15 x 10⁻³ x 1000 x 2.5 = 30 °C with no heat dissipation. This is in 15 minutes, so in a half an hour the protein has been "cooked" unless cooling is provided.

Stone or glass cooling slabs usually are used (metal is seldom used because it conducts); and if the heat capacity of stone is the same as water, 1 kg would protect 3 mL of gel by lowering the heat rise about 300-fold with good contact. Tubes should be submerged in cold buffer.

The voltage needs to be increased as the separation progresses, because when the compounds reach their isoelectric point and become neutral, they also become nonconducting, so the voltage must be increased to maintain the same current.

SODIUM DODECYLSULFONATE ELECTROPHORESIS (SDS)

PRINCIPLES

SDS is a technique that can be used to separate neutral compounds or high molecular weight soluble compounds such as proteins. There are differing explanations on how SDS functions, depending upon the types of molecules involved and the degree of polymerization of the electrophoretic medium. For systems of low porosity or just liquids, a dilute aqueous solution of a polar detergent such as the anionic sodium dodecylsulfonate $H_3C(CH_2)_{10}CH_2OSO_3^-Na^+$ exists as single molecules. However, when the concentration is increased, there is a point where the nonpolar ends associate, being repelled by the polar water, with the polar ends pointing outward. These aggregates are known as *micelles*. Neutral molecules tend to penetrate to the inside of the micelle and be loosely held there. This causes the molecules to dissociate from each other and become easier to separate. The micelle, being negatively charged, is attracted to the anode and carries the neutral particle with it. Now the proteins are all anionic, and although they differ in size, they carry about the same charge and have the same electrophoretic mobility in free solution. However, if they are placed in a gel whose pore size is small, then they separate based on their molecular weights.

For proteins the explanation of J. Svasti and B. Panijpan (*J. Chem. Ed.*, **54** (9) 560, 1977) appears to be reasonable. Refer to Figure 30-3. "Proteins and polypeptides are made up of amino acids covalently linked by peptide bonds, regularly arranged on the backbone; the larger the molecular weight, the larger the number of peptide bonds. In their native state, the peptide chains of globular proteins are tightly folded to form a compact three-dimensional structure. Treatment with SDS in the presence of a reducing agent causes the polypeptide chains to unfold and assume a rod like structure in which the polypeptide core is coated by SDS molecules. Proteins of higher molecular weight bind more SDS than those of lower molecular weight. Since the SDS molecule has one net negative charge at neutral pH, larger SDS-poly-peptides will be more negatively charged. Refer to [Figure 30-3]. This shows the denaturation of a globular protein by 1% SDS in the presence of β-mercaptoethanol. The tightly folded polypeptide chain unfolds to form an extended rod, with the polypeptide at the core being coated with negatively charged SDS molecules.

"If the larger molecules have more charge, then why do the smaller molecules move faster? This is due to the sieving effect of the gel. There are two limits. For a 10% polyacrylamide gel, the normal linear separation range is from 15,000 to 70,000. The larger molecules are too big to get through the pores and do not migrate, while the small molecules are unretarded by the pores and all migrate together. A 15% gel can separate down to about 6,000. Higher molecular weights, up to 200,000 can be resolved, but the curve is sigmoid. If standards are employed, the molecular weight of an unknown can be estimated in favorable conditions to ±2%.

"A cationic detergent is needed for acidic proteins and tetramethyl ammonium bromide, $(CH_3)_4N^+,Br^-$, *N*-hexadecyl-pyridinium chloride, and cetyltrimethyl ammonium bromide are used.

"For molecular weight studies denaturation of the protein is preferred whereas for other studies, the original random coil structure is preferred. The addition of 6 M urea not only helps preserve the structure, but makes it easier to remove the SDS later. After SDS treatment, the SDS can be removed by dialysis, but it takes a long time, and the protein does not all regain its original conformation.

"In SDS - polyacrylamide gels, the mobility of reduced or carboxy methylated proteins is *inversely proportional to the logarithm of the molecular weight*. [Figure 30-4] shows a log plot of 37 different polypeptide chains from 11,000 to 70,000 vs. mobility.

"Why is the plot semi-logarithmic? This can be explained by thinking of the sieve pores as perfect circles of various diameters. The molecules could be imagined as spheres occupying the same effective space as that described by their constant tumbling. The ease with which a molecule goes through the pores is then dependent on the cross sectional area, which is directly proportional to the square of its diameter. The molecular weight is proportional to volume and hence to the cube of the diameter of the equivalent sphere. Taking logarithms of the molecular weight should thus help linearize its relationship to the electrophoretic mobility."

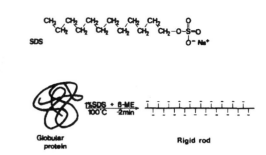

Figure 30-3. Showing the binding of SDS to globular proteins.
(Courtesy - Svasti, J., and Panijpan, B., *J. Chem. Ed.*, **54** (9), 560, 1977).

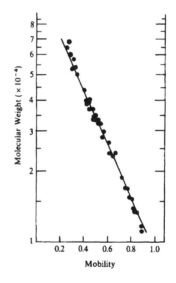

Figure 30-4. A plot of log of molecular weight vs. mobility.
(Courtesy - Weber, K., and Osborn, M., *J. Biol. Chem.*, **244**, 4406, 1969).

TECHNIQUES

Protein molecules are held in their conformation in large part by -S-S- bonds. If a reducing agent such as 1% 2-mercaptoethanol is added, these bonds are broken and -SH groups are formed. This results in the secondary, tertiary, and quaternary structures being disrupted to produce polypeptide chains in a random coil. When these protein molecules are treated with a detergent, it binds to the hydrophobic portion of the protein. So many SDS molecules bind that they mask the charge on the protein, and all protein molecules have essentially the same negative charge because of the anionic detergent, SDS. With proteins, the SDS appears to bind with -SH groups. These are formed from the -S-S- bonds in protein by heating the protein for 3 hours at 37 °C with 1% 2-mercaptoethanol (a reducing agent) and 1% SDS in a 0.1 M sodium phosphate buffer at pH 7.2. In the presence of a reducing agent, soluble proteins bind to SDS in the ratio of 1.4 g SDS/1.0 g protein. A preliminary 2 minute heating at 100 °C usually deactivates enzymes, which often interfere with the results.

TWO-DIMENSIONAL ELECTROPHORESIS

PRINCIPLES

Two-dimensional electrophoresis (2-DE) is based upon the same concept as two-dimensional chromatography, in that the compounds at the end of the first dimension should be in as narrow a band or spot as possible before starting the second dimension. The most common technique for this in two-dimensional electrophoresis is to use isoelectric focusing (IEF) in the first dimension and SDS in the second dimension. This was first proposed by P.H. O'Farrell (*J. Biol. Chem.*, **250**, 4007, 1975), who was able to resolve 1100 different components from *E. coli*. Proteins differing by only a single charge can be separated, and amounts from 10^{-4} to 10^{-5} of the total protein content can be detected. Figure 30-5 shows the overall scheme.

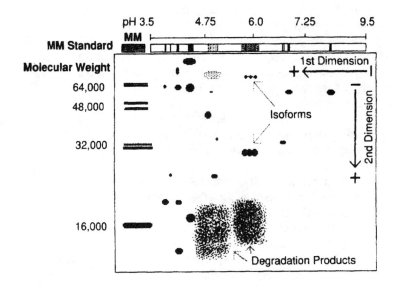

Figure 30-5. A two-dimensional electrophoresis separation with isoelectric focusing in the first dimension and SDS in the second dimension.
(Courtesy - Helena Laboratories, Beaumont, TX)

REVIEW QUESTIONS

1. What is the basis for isoelectric focusing?
2. What is the isoelectric point?
3. What is a zwitterion?
4. How is a pH gradient obtained?
5. How long does it take to establish a pH gradient?
6. What are natural ampholytes?
7. If the sample is mixed in with the ampholytes at the beginning of the electrophoresis, how can it be separated by isoelectric focusing?
8. What are immobilized ampholytes, and what advantage do they have?
9. What is a common electrode pair used to determine the pH of a gel at any point?
10. If a 40 cm^2 slab of gel is heated for 900 seconds and it is desired to keep the power at less than 0.45 watts, what temperature would the gel attain if its thickness is 1 mm?
11. Why are metal slabs usually not used for cooling purposes?
12. Why does the voltage need to be increased continuously as the separation progresses?
13. What is a micelle?
14. What is the purpose of 2-mercaptoethanol as related to globular protein separations?
15. When would you consider using SDS to aid in a separation?
16. If larger molecules have more charge because of SDS, why do the smaller molecules move faster through the gel?
17. Why is the plot of mobility vs. molecular weight linear if plotted semilogarithmetically?
18. Why is isoelectric focusing often preferred as the first dimension in two-dimensional electrophoresis?
19. What is a typical resolution and detection limit for 2-DE?

IF IT EXISTS - CHEMISTRY IS INVOLVED
IF YOU CAN BUY IT - A CHEMIST WAS INVOLVED SOMEWHERE

31
CAPILLARY ZONE ELECTROPHORESIS

PRINCIPLES

The general approach was first described by S. Hjerten (*Chromatogr. Rev.* **9**, 122, 1967) using 3 mm bore columns. R.Virtanen (*Acta Polytech. Scand. Chem. Incl. Metall. Ser.* **123**, 1, 1974) and F.E.P. Mikkers (Mikkers, F.E.P., Everaerts, F.M., and Verheggen, Th. P.E.M., *J. Chromatogr.* **169**, 111, 1979) made early advances. However, it was the work of J.W. Jorgenson and K.D. Lukacs (*Anal. Chem.* **53**, 1298, 1981) with 25 µm bore columns and 30 kV applied potential that made it popular. It is a microelectrophoresis system in which the separation takes place in a 10 to 100 µm internal diameter, fused quartz, hollow, capillary tube from 30 to 100 cm long, with each end immersed in a buffer (Figure 31-1, p. 360). The same buffer is used as the electrolyte and up to 300 V/cm d. c. are applied through graphite electrodes; uv and fluorescence are common means of detection. Usually two forces act to cause the separation, normal *electrophoresis* of the charged particles and *electro-osmosis*, the flow of the buffer in a charged field.

The zone broadening difficulties associated with normal electrophoresis are overcome by the use of a capillary. These problems are due to (1) density differences between the sample and the buffer; (2) heat caused by the passage of electrical current through the buffer; (3) variable sample ion concentration, which causes changes in the local conductivity; (4) molecular diffusion; (5) parabolic flow of liquids in a broad front; and (6) eddy migration, if gels are used. These disadvantages are all overcome or greatly minimized by using a capillary column.

It is believed that the radial position of a particle remains the same as it migrates down a tube. If it starts on the outside, it stays on the outside. This magnifies any differences between the center of the tube and the wall and causes zone broadening. With capillary tubes, the diameter is so small that even though the particle doesn't move very far radially, it can move back and forth many times, which tends to average any center-wall effects, and all particles tend to move with the same net velocity, thus providing for a narrower zone of particles. Figure 31-2, p. 360, shows what can be done in an ideal situation. This type of plot is called an *electropherogram*.

The temperature difference between the center of a column and the wall varies as the square of the radius. Therefore, a small radius reduces the temperature gradient significantly.

ELECTRO-OSMOSIS

The internal surface of the quartz capillary contains -OH groups as shown in Figure 31-3A, p. 361. These ions plus cations from the buffer are attracted to this charged layer and form a "double layer". The buffer ions, surrounded by water, act as if they had a positive surface; and when a potential is applied, the buffer migrates in bulk toward the cathode. This bulk movement of liquid toward the cathode is called *electro-osmosis* and carries all types of sample molecules along with it. In a capillary, with a high potential

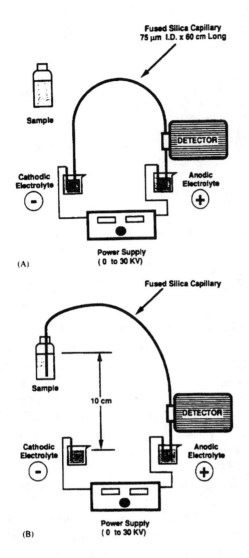

Figure 31-1. Diagram of a capillary zone electrophoresis apparatus. A. Normal operation. B. Hydrostatic filling position.
(Courtesy: Jones, W.R, and P. Jandik, *Am. Lab.*, June, 56, 1990).

applied, this can be a significant rate, moving particles along faster than they would by normal electrophoresis. This effect is so strong that everything is carried to the negative electrode, even triple-charged anions.

The electro-osmotic flow can be quite beneficial. Because the sample is added at one end, without this flow, only cations or anions could be analyzed during any one run, and neutrals could not be done. With the electro-osmotic flow, the order of elution is anions, neutrals, and cations toward the anode. In addition, the flow speeds up the movement of ions, so the very large or weakly charged ions can be separated with a much shorter column.

ELECTROPHORESIS

The other factor is the *electrophoretic separation* of charged particles. The rate of this migration is dependent upon the viscosity of the solution; the size and charge of the particles; and most importantly, the potential applied. In normal electrophoresis, the potential is a compromise between the need for speed and the need to reduce heat. With capillaries, high potentials can be applied, as much as 300 V/cm to date.

Therefore, all particles move; even the neutral ones are carried along by the electro-osmotic flow. Because most of the charge is at the surface and the diameter is so small, the normal parabolic flow profile of a liquid flowing through a pipe is changed to a flat front, and "zones" of ions can be formed.

In addition, the cations and anions are being separated by the normal electrophoretic process. Because of the high potential, this separation can be quite fast. The resolution is excellent, because the small diameter of the tube minimizes concentration broadening, diffusion broadening, and heat broadening. The absence of a gel or particles eliminates eddy migration.

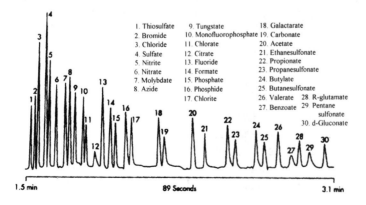

Figure 31-2. 30 anions in 89 seconds; 30 kV using a Quanta 4000 CE system, 50 µm x 60 cm capillary.
(Courtesy - W.R. Jones, P. Jandik, and R. Pfiefer,. *Am. Lab.*, May, 40, 1991)

A. *Quartz surface.*

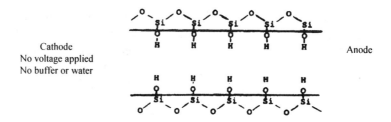

B. *Surface in the presence of buffer or just water.* When a neutral or alkaline buffer is added, the H⁺ ions dissociate, and the surface becomes negatively charged.

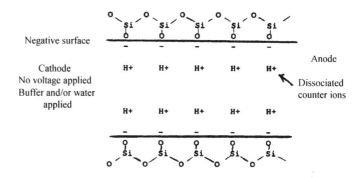

C. *Potential applied.*

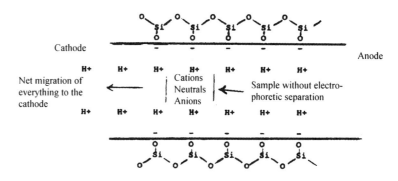

Figure 31-3. The electro-osmotic effect.

If the sample concentration is small compared to the buffer concentration, then the current is carried mostly by the buffer, so minimal broadening occurs from conductivity changes. Because of the small volume, little convective diffusion occurs from heating. The system is fast, usually requiring only a few minutes; the separation is excellent, up to 400,000 plates/meter; and only a few nL of sample are required. Compounds with mobility differences of 10^{-6} cm^2/cm have been separated. One picoliter (pL) can be detected in favorable cases. This is truly a micromethod. In addition, it is easy to automate. The column can be purged in a few seconds, new buffer added, and another sample injected automatically.

THEORY

The following equations and discussion are abstracts of those presented by J.W. Jorgenson and K.D. Lukacs (*Anal. Chem.* **53**, 1298, 1981, and *Science*, **222**, 266υ, 1983). "Consider an electrophoresis system consisting of a tube filled with a buffering medium across which a potential is applied. Charged species introduced at one end of the tube migrate under the influence of the electric field to the far end of the tube. The migration velocity of a particular species is given by

$$\upsilon = \mu E = \mu V/L \tag{31-1}$$

where υ is the velocity, μ is the electrophoretic mobility, E is the electric field gradient, V is the total applied potential, and L is the length of the tube. The time, t, required for a zone to migrate the entire length of the tube is

$$t = L/\upsilon = L^2/\mu V \tag{31-2}$$

This equation predicts that the analysis time is proportional to the square of the tube length and inversely proportional to the applied potential. "If molecular diffusion alone is responsible for zone broadening, the spatial variance, σ_L^2, of the zone after a time, t, is given by the Einstein equation

$$\sigma_L^2 = 2Dt \tag{31-3}$$

where D is the molecular diffusion coefficient of the solute in the zone. Substituting the expression for time from equation [31-2] into this expression yields

$$\sigma_L^2 = 2DL^2/\mu V \tag{31-4}$$

"The concept of separation efficiency expressed in terms of theoretical plates may be borrowed from chromatography. The number of theoretical plates, N, is defined as

$$N = L^2/\sigma_L^2 \tag{31-5}$$

Substituting [31-4] into this expression results in

$$N = \mu V/2D \tag{31-6}$$

N is directly proportional to applied potential and column length and analysis time have no effect. The sum of this is that for the best separations, you should use high potentials and short columns. However, the heat generated, which is not directly apart of these equations, determines how much potential can be applied and how short the column can be and still dissipate the heat.

"The above equations neglected the electro-osmotic effect. In an unobstructed capillary, the shape of the electro-osmotic flow profile is piston like. The flow velocity is constant over most of the tube cross section and drops to zero only near the tube walls. This is fortunate, as the flat flow profile of electro-osmosis will add the same velocity component to all solutes, regardless of their radial position, and thus will not cause any significant dispersion of the zone. Electro-osmotic flow causes the above equations to be modified. The migration time becomes

$$t = \frac{L^2}{(\mu + \mu_{osm})V} \tag{31-7}$$

and the separation efficiency is now

$$N = \frac{(\mu + \mu_{osm})V}{2D} \tag{31-8}$$

Equation [31-8] is based on theory. A more practical method is needed for day to day use and equation [31-9] based on common HETP measurements for chromatography can be applied.

$$N = 5.54\left(\frac{t}{w}\right)^2 \tag{31-9}$$

where w is the full peak width at the half-maximum points. "In the presence of electro-osmosis, equation [31-7] predicts an inverse relationship between applied potential and analysis time. Equation [31-8] suggests a misleading approach to improved separation efficiency. This is to promote very large values of μ_{osm}, electro-osmotic flow, in the same direction as the electrophoretic mobility. J.C.Giddings [*Sep. Sci.*, **4**, 181, 1969] derived an expression for resolution in electrophoresis as

$$R_s = \frac{N^{1/2}\Delta v}{4\ \bar{v}} \tag{31-10}$$

where R_s is the resolution and $\Delta v/\bar{v}$ is the relative velocity difference of the two zones being separated. This ratio is equal to

$$\frac{\Delta v}{\bar{v}} = \frac{\mu_1 - \mu_2}{\bar{\mu}} \tag{31-11}$$

where μ_1 and μ_2 are the mobilities of the two zones, and $\bar{\mu}$ is their average mobility. However, in the presence of electro-osmosis this becomes

$$\frac{\Delta v}{\bar{v}} = \frac{\mu_1 - \mu_2}{\bar{\mu} + \mu_{osm}} \tag{31-12}$$

It is readily apparent that a large value of μ_{osm} will decrease the relative velocity difference of the two zones. By substituting the expressions for the relative velocity difference [31-12] and the number of theoretical plates [31-8] into [31-10], we obtain

$$R_s = \frac{1}{4}\left[\frac{(\mu + \mu_{osm})V}{2D}\right]\left[\frac{\mu_1 - \mu_2}{\bar{\mu} + \mu_{osm}}\right] \tag{31-13}$$

and by rearranging

$$R_s = 0.177(\mu_1 - \mu_2)\left[\frac{V}{D(\bar{\mu} + U_{osm})}\right]^{1/2} \quad (31\text{-}14)$$

This equation indicates that resolution is affected by the applied potential, the molecular diffusion co-efficient, and the differences between the mobilities of the ions under both electrophoretic and electro-osmotic conditions.

"Now it is clear that a large component of electro-osmotic flow in the same direction as the electrophoretic migration will decrease the actual resolution of the two zones. In fact, it may be seen that the best resolution will be obtained when the electro-osmotic flow just balances the electrophoretic migration or

$$\mu_{osm} = -\mu \quad (31\text{-}15)$$

at which point substances with extremely small differences in mobility may be resolved. This resolution will be obtained, however, at a large expense in time, as may be seen by referring to equation [31-7] and imagining μ and μ_{osm} being nearly equal, but opposite. This is shown in Figure [31-4.]"

Figure 31-5 is a photograph of one model of a capillary zone electrophoresis apparatus.

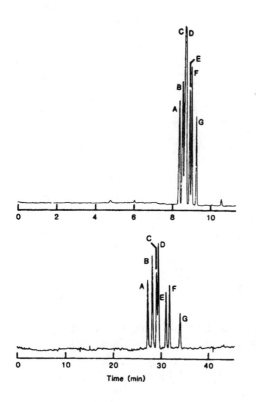

Figure 31-4. Effect of electro-osmosis on resolution and analysis time. Top, Untreated. Bottom, Prereated with 10% trimethyl-chlorosilane in dichloromethane for 20 minutes. (A) asparagine, (B) isoleucine, (C) threonine, (D) methionine, (E) serine, (F) alanine, (G) glycine.
(Courtesy - Jorgenson, J.W., and Lukacs, K.D., *Anal. Chem.*, **53**, 1301, 1981)

Figure 31-5. Isco model 3850 capillary zone electrophoresis apparatus with an integrator.
(Courtesy - Isco Co., Lincoln, NE)

TECHNIQUES

COLUMNS

The capillaries are made either of borosilicate glass or fused quartz. Pyrex glass has a cutoff at about 280 nm, so fused silica is often used. The capillary tube is coated with a polyimide to enhance flexibility by preventing the hydration of strained siloxy bonds. The columns vary in internal dia-meter from 10 to 100 μm and in length from 30 to 100 cm, with 50 to 100 cm being common. The capillary side wall is usually thick by comparison, 300 to 600 μm, to act as a heat sink. A small portion (1 to 2 mm) of the coating is burned away if a photometric detector is used. Figure 31-6 shows typical dimensions.

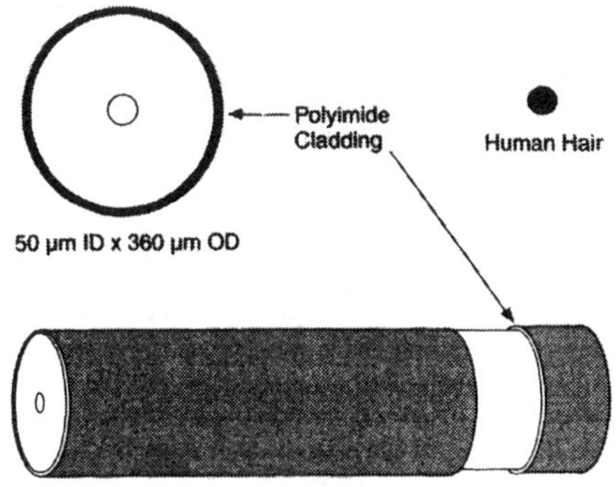

If the electro-osmotic flow is not wanted, then either the surface of the side wall is coated or a modifier is added to the buffer. Trimethylchlorosilane is used commonly to neutralize the surface and eliminate the electro-osmotic flow. Other silanes are aminopropyl and octadecyl. One of the best methods to coat the surface is to use glycol-containing groups such as polyethylene glycol (>5000 m.w.)

The most effective means found thus far to chemically coat the column is to first silate the column with (gamma-methacryloxypropyl)trimethoxysilane followed by cross-linking the surface-bound methacryl groups with poly-acrylamide.

Figure 31-7 shows the current method to add or exchange columns. They are placed in a cassette and simply snapped into place.

Figure 31-6. Typical capillary dimensions. (Courtesy - Bio Rad Laboratories, Hercules, CA)

Figure 31-7. Capillary column cassettes. (Courtesy - Hewlett-Packard, Wilmington, DE)

SAMPLE

The sample is small, on the order of 5 to 50 nL. The total volume of a 50 μm tube, 50 cm long would be about 975 nL. The sample is added by placing one end of the capillary into a sample container raised to a set height (10 cm) for a measured time (30 seconds) and letting a siphoning action take place. An alternate method is to place one end into the sample and apply a potential for a short time. This method tends to cause some preliminary separation of the sample, favoring the low m.w. ions, which can be an advantage if low m.w. compounds are desired. Usually this separation is considered a disadvantage, but it can be used as a preconcentration step and significantly lower the detection limits.

The sample volume required is only about 1/3000 to 1/5000 that used with ion chromatography. For maximum resolution, the quantity injected is determined by

$$Q = \frac{(\mu + \mu_{osm})VACt_{inj}}{L} \tag{31-16}$$

where Q is the quantity injected (in moles, not L), A = the cross-sectional area of the capillary, C = the sample concentration, and t_{inj} = the duration of the injection.

Table 31-1, p. 366, shows a comparison of the *hydrostatic* and *electromigrative* sample injection methods. This indicates that anion analysis can be performed at sub ppm levels using hydrostatic injection and at low ppb levels with electromigrative enrichment.

Table 31-1. Detection limits in ppb ($S/N = 3$)
(Courtesy - Jones, W.R., Jandik, P., and Pfiefer, R., *Am. Lab.*, May, 44, 1991)

Anion	Hydrostatic 10 cm for 30 sec	Electromigrative Enrichment 5 kV for 45 sec
Bromide	383	1.1
Chloride	147	0.5
Sulfate	173	0.8
Nitrite	331	1.2
Nitrate	347	1.5
Fluoride	99	0.4
Phosphate	380	1.7

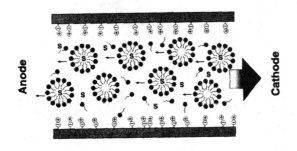

Figure 31-8. Schematic of MEKC separations. (S) Sample.
(Courtesy - Bio Rad Laboratories, Hercules, CA)

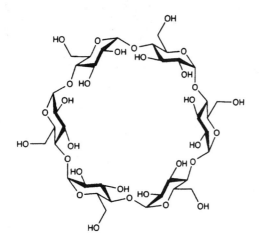

Figure 31-9. Alpha cyclodextrin.
(Courtesy - *Merck Index*, Merck and Co., Rahway, NJ, 1989)

SEPARATION OF NEUTRAL MOLECULES

Not all "neutral" molecules are neutral, in the sense that they usually have some polarity and an end or section that is more hydrophobic or hydrophilic than the rest of the molecule. Other "neutral" molecules in a mixture may be slightly different and advantage is taken of this. If a low concentration of a surfactant, such as sodium dodecyl sulfonate (one end hydrophobic, one end hydrophilic) is added to the system, these surfactant molecules at first stay separated. As their concentration increases, the hydrophobic ends tend to associate and they form small clusters called *micelles*, which have a hydrophilic outside (net negative charge) and a hydrophobic inside.

The more hydrophobic neutral molecules tend to penetrate the micelle to the hydrophobic center and associate for an equilibrium amount of time. As a result, they are carried along for a short time with the micelle toward the anode. This large molecule lags further behind the main bulk of the system. Different neutral molecules penetrate and associate for various lengths of time. This means that neutral molecules can become separated. This technique has been called *micellar electrokinetic capillary chromatography* (MECC), Figure 31-8.

SEPARATION OF OPTICAL ISOMERS

Several methods can be used for separation. A current favorite is to use a series of compounds called *cyclodextrins*, which have been found to associate preferentially with one form of optical isomer more than the other. Figure 31-9 shows a diagram of alpha cyclodextrin. They are obtained by the action of *Bacillus macerans* amylase on starch to form homogeneous cyclic alpha- 1,4-D-glucopyranose units. Alpha is the hexa ring, beta is the hepta ring, and gamma is the octa ring. They have hydrophobic cavities and form inclusion compounds with many compounds. They have a net negative charge and are carried to the anode. Compounds that are hydrophobic tend to associate in the cavity, and optical isomers associate sufficiently differently that one form is carried along more than the other; in many cases, the net result is a separation.

POTENTIAL, CURRENT, AND TEMPERATURE

The speed of an electrophoretic separation depends upon the applied potential. The volume of liquid inside of a capillary is small compared to the surface area and side wall thickness, so heat caused by the passage of current can be dissipated rapidly. This means that a high potential on the order of 300 V/cm can be applied.

The effect of temperature is to increase the mobilities about 2%/°C. Because the column of liquid cools most at the side wall, the ions closer to the center move faster, thus providing a parabolic front and a corresponding zone broadening. If the capillary diameter is small, this center effect is minimal, and a flat zone front occurs, which permits much higher resolution.

BUFFERS

The buffer is used to carry the majority of the current, and it should be selected so that its components are significantly different from the analytes, and detection by difference is easier. Osmotic flow modifiers (OSM) are electrolyte additives that control the flow of buffer through the capillary. 0.5 mM concentrations are common. Some buffer systems that have been used are 10 mM mannitol, 0.05 M pH 7 phosphate, and a borate/gluconate mixture.

Cations are left at the injection end. To analyze for cations, a positive power supply is needed and cationic electrolyte additives are required. 6.5 mM hydroxybutyric acid at pH 4.4 has been used effectively.

Nonaqueous electrolytes have been used. 0.05 M tetraethylammonium perchlorate and 0.01 M HCl in acetonitrile were used to separate isoquinoline from quinoline.

MODIFIERS

These are compounds that can be added to the electrolyte to alter the direction and rate of the electro-osmotic flow. *Flow rate varies inversely with ionic strength and is independent of column diameter.* The effect of pH between 2 and 12 and applied potential increase the electro-osmotic flow rate linearly. The flow can be reversed by adding quaternary amines, such as cetyltrimethylammonium bromide or tetradecyl-trimethylammonium bromide. Zero flow occurs when s-benzylthiouronium chloride is added to the buffer. Organic molecules such as methanol and putrescine reduce the flow, whereas acetonitrile increases the flow. Covalently bonded polyethylene glycol reduces the flow.

RELATIVE APPARENT MOBILITIES

H. Carchon and E. Eggermont (*Electrophoresis*, 3, 263, 1982) proposed *apparent mobilities* and *relative apparent mobilities* as a way to determine the type of buffer to use. Figure 31-10 shows how they can be obtained. As defined and explained by them, the apparent mobility is the reciprocal value of the time, t, an ion needs to migrate along the capillary under a constant potential. These apparent mobilities (AM) can be used to define the relative apparent mobilities (RAM). RAM is the percentage value of the AM of a given ion to that of a fully dissociated reference ion. Because RAM values are independent of the assay conditions, they can help in selecting the optimal conditions for separation.

The following criteria are proposed to select the proper pH: (1) All RAM values must be higher than a preset lower limit, (2) the difference between two successive RAM values must be higher than a preset lower limit, (3) the pH dependency of the different ionic species must be minimal, (4) the variation of the pH dependency for the various compounds must be minimal, (5) the mean difference between successive RAM values must be

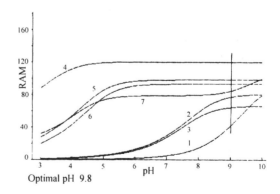

Figure 31-10. How to estimate relative apparent mobility of several anions as a function of pH.
(1) Glycine, (2) glycyl-glycine, (3) glycyl-glycyl-glycine, (4) formic acid, (5) acetic acid, (6) propionic acid, and (7) glutamic acid.
(Courtesy - Carchon, H., and Eggermont, E., *Am. Lab.* 69, January 1992)

maximal, (6) the variation of the differences between successive RAM values must be minimal, and (7) the selected pH must be as close to pH 7 as possible.

Table 31-2 is a summary of the results of the above plot. Using the above criteria, a pH of 9.0 was selected based on the data shown below.

Table 31-2. pH optimization of anionic compounds obtained from the relation between pH and relative apparent mobility (Courtesy - Carchon, H., and Eggermont, E., Am. Lab., 69, January 1992)

Minimum RAM value (upper limit 40): 40
Difference in RAM value (upper limit 4): 4
Optimal pH: 9.0

Compound	Estimated RAM
Formic acid	119.9
Acetic acid	98.1
Propionic acid	92.8
Glutamic acid	84.2
Glycyl-glycine	76.6
Glycyl-glycyl-glycine	64.0
Glycine	41.5

The order of appearance can be calculated from the estimated RAM values.

DETECTION

Detection is a continuing problem because the sample volume passing through the detector at any one time is so small, a few picoliters (pL). A sample concentration of 10^{-6} M is the normal limit, but one method has been demonstrated to detect 10^{-15} M materials; uv spectroscopy, fluorescence, Raman, and electroconductivity are used for detection.

uv Spectroscopy

The common source for a uv detector is a Hg discharge lamp or a xenon discharge lamp with suitable filters to isolate the desired wavelengths. Compounds with benzene rings are detected readily using the Hg source, but other compounds are not. A way to improve this is by *vacancy enhancement*. What is done is to place a compound in the buffer that is readily detectable. When the sample passes the detector, the sample has created a "vacancy" in the liquid, and the detector response diminishes, indicating the presence of a compound. Chromate is a favorite at the moment as a high mobility ion. Phthalate and *p*-hydroxybenzoate are lower mobility ions.

Fluorescence

Not many compounds fluoresce naturally. However, some compounds, when added to another compound, cause that compound to fluoresce. These compounds are called *fluorophores*. Dansyl chloride and fluorescamine are two that are used to react with primary amines. *o*-Phthalaldehyde is used as a postcolumn reactant to produce fluorescent compounds. Other compounds are naphthalenedialdehyde (NDA), fluorescein isothiocyanate (FITC), and phenylthiohydantoin (PTH). See Chapter 19, p. 203, for more detail.

Xe arc and Hg discharge lamps with filters to isolate the desired wavelengths are common sources. You can expect a 10- to 1000-fold increase in sensitivity, if you have a compound that fluoresces either naturally or when a fluorophore is added to it. The reason for this increase is that you are measuring at right angles to the source, and the high background is eliminated. Remarkable sensitivity has been achieved recently by using a coiled 30 m column and shining a laser down its length! The detector is placed at the end of the column. The capillary acts like a light pipe, and the laser beam bends around the coil. As the separation progresses, every compound, even if it has been separated, is in the beam, and the detector response is constant. However, when a compound emerges and drips off of the end, the signal now changes, and this can be recorded. Results indicate that a vacancy technique allows detection of 6 attograms of K^+ from a single red blood cell. HeCd, HeNe, and Ar^+ lasers have been used.

Conductivity and Amperometry

Conductivity is difficult to do with high sensitivity, because the buffer usually conducts so much current that detecting the sample in this background is difficult. Metal ions and organic acids have been done down to 10^{-6} M. See Chapter 19, p. 206, for more detail.

Amperometric detection is useful for those compounds that are oxidizable or reducible. The difficulty is the high potential applied to the column. R.A. Wallingford and A.G. Ewing (*Anal. Chem.* **59**, 1762, 1987) used a porous glass joint to couple two pieces of capillary together and electrically isolate the detector from the column. Carbon fiber electrodes were used, and detection limits of 200 to 400 attomoles of catecholamines were reported.

Mass Spectrometry

An electrospray ionization interface has been used with detection limits in the femtomole range. Reproducibility is a problem at the moment. See Chapter 19, p. 207, for more detail.

Radioisotopes

A beta detector consisting of a small semiconducting wafer of CdTe at the outlet of the capillary will respond to labeled beta emitters. 10^{-9} M solutions of compounds containing ^{32}P have been reported.

COLLECTING FRACTIONS

A problem in handling such small samples is to collect them after separation for further analysis. The experimental membrane fraction collector made by Millipore (Figure 31-11) appears to work well. According to H.J. Goldner (*R & D Magazine*, August, 1992), "The assembly comprises a polyvinyldifluoride (PVDF) membrane cover, a buffer reservoir consisting of two layers of 3 MM chromatography filter paper, and a stainless steel plate. The plate completes the electrical circuit for electrophoretic separation by acting as the ground electrode. The entire collection assembly is rotated by a stepping motor at 2.2 revolutions per hour while electrophoretic separation takes place. Analytes are deposited onto the membrane at discreet locations as they emerge. The collected fractions can be either stained in place or removed from the membrane for further treatment."

GENERAL OBSERVATIONS

1. An unsymmetrical peak shape indicates that the sample size is too large.
2. Adsorption to the column surface results in tailing.
3. Adsorption to the surface modifies the surface and changes the electro-osmotic flow, thereby changing the migration patterns.
4. Proteins usually elute in the order of their iso-electric points.
5. Electro-osmotic flow is minimal in gel filled capillaries, because the gel is attached to the wall.
6. As a general rule, strongly acidic anions migrate faster than weakly acidic anions and inorganic anions usually migrate much faster than organic anions.

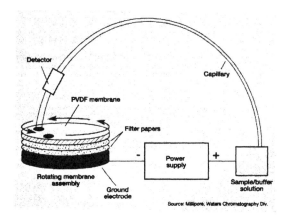

Figure 31-11. A membrane system fraction collector. (Courtesy - Goldner, H.J., *R & D Magazine*, August 1992)

REVIEW QUESTIONS

1. What are a typical diameter and length of a capillary in CZE?
2. What are considered to be the two main forces involved in CZE separations?
3. How does the temperature difference between the wall and the center of the capillary vary?
4. What happens to the surface of a quartz capillary in the presence of buffer or just water?
5. What direction does the net migration proceed when a potential is applied to the capillary?

6. What causes electro-osmosis?
7. What is the expected order of elution in a mixture of neutral molecules along with cations and anions?
8. What is the most important factor in controlling the rate of mobility in the electrophoretic portion?
9. What two factors would you apply to get the best separations?
10. What would you try to do if you wanted to resolve substances with extremely small differences in mobility?
11. What can be done to eliminate the electro-osmotic flow if it is not wanted?
12. What is the sample size range, and how is it added to the capillary?
13. What is an advantage of the electromigrative method of adding sample?
14. What is one way to separate optical isomers using CZE?
15. What causes a parabolic front to form in a liquid flowing in a tube, and why is this minimal in a capillary?
16. What is the main purpose of a buffer in CZE?
17. How would you analyze for cations?
18. What is the purpose of a modifier, and what is one compound that will produce zero flow?
19. What is the purpose of determining a relative apparent mobility?
20. What is the experimental parameter involved with determining a relative apparent mobility?
21. What is meant by vacancy enhancement?
22. How much can detection levels be lowered, if you can make a compound fluoresce?
23. What fluorescence technique has been used to detect compounds down to 10^{-15} M?
24. How can fractions be collected?
25. What can cause an unsymmetrical peak to emerge?
26. As a general rule, what moves faster; strongly acidic anions or weakly acidic anions?

IF IT EXISTS - CHEMISTRY IS INVOLVED
IF YOU CAN BUY IT - A CHEMIST WAS INVOLVED SOMEWHERE

32
FIELD FLOW FRACTIONATION

PRINCIPLES

Field flow fractionation (FFF) *is a separation technique based on forming a concentration gradient in a narrow channel and then applying a parabolic flow of liquid through it.* The original concept was developed by J.C. Giddings (*Sep. Sci.*, **1**, 123, 1966). The channel is from 75 to 300 μm thick, 2 to 5 cm wide and 30 to 100 cm long. Figure 32-1, p. 372, illustrates the concept.

Assume that the sample contains a mixture of charged particles. When these are in an electrical field in a narrow channel, they tend to stack vertically. A liquid flowing through the channel has a parabolic flow profile. The molecules near the center of the channel are pushed by a faster flow than those near the edge, and this results in a separation. Any field that stack molecules or particles can be used, and the particles need not be charged. This includes heat and centrifugal force.

This is a superb technique for separating very large molecules or particles. The verified operating range for particles is from a few thousand daltons to 10^{15} daltons, and for cells and microorganisms, 12,000 plates per foot are possible theoretically. Figure 32-2, p. 372, illustrates the mass ranges of the subtechniques.

THEORY

A complete mathematical treatment of the theory for retention was developed by Giddings in 1968 (*J. Chem. Phys.*, **49**, 81). A simplified *FFF as a random walk* theory by Giddings is presented below.

"The universal failure to realize infinitely sharp peaks and thus infinite resolution in separation techniques is most often due to the nonidentical activities of identical molecules. Driven by random entropic (or thermal) influences, they each take an individual and unique pathway in the system and as a consequence emerge separately - part of a smeared zone.

"Recall that the mean altitude of a solute zone in FFF is l. By normal fluctuations, any given solute molecule will occasionally be found at an altitude beyond the mean molecule height, l and occasionally inside l. When beyond l, the molecule is exposed to a high flow velocity and thus travels at a speed above average. When less than l, the speed along the column is below average. A solute peak would be literally smeared over the entire column if each molecule remained at its original altitude. Instead, the molecules diffuse randomly back and forth, sampling first one velocity, then another. Their actual velocity is the average of the various velocities sampled. Since each molecule samples velocities in the same velocity domain, they travel along at about the same speed. However individual molecular fluctuations which involve an undue occupancy time in high or low velocity regions cause departure from the mean. This broadens the solute zone. It is these fluctuations which can be approximated by a simple random walk.

"Suppose the molecule could sample but two velocities, one above and one below normal. The single high velocity state represents the collection of above average velocities, and vice versa. For the sake of constructing an approximate model, suppose the high velocity state is at an altitude $2l$ and is traveling at velocity $2V$, twice the mean zone velocity V. The low velocity state is at an altitude 0 and has zero downstream velocity. The average length of time a molecule persists in one of these states is the time required to diffuse the distance $2l$ between them - at time expressed approximately by

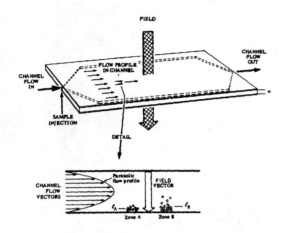

Figure 32-1. Principle of field flow fractionation.
(Courtesy - Giddings, J.C., Graff, K.A., Myers, M.N., and Caldwell, K.D., *Sep. Sci. Technol.*, **15**, 615, 1980)

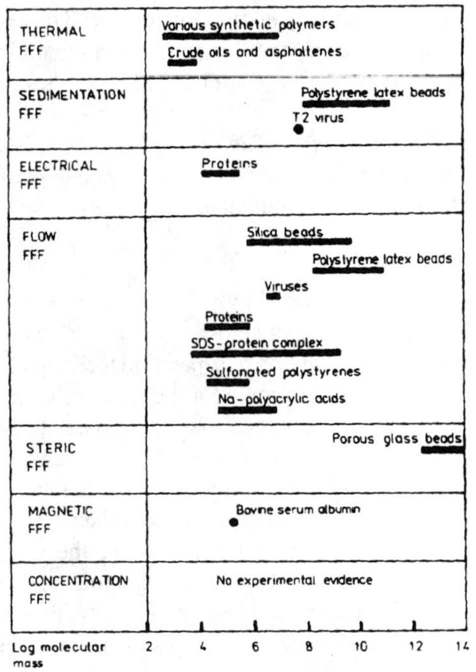

$$t_D = \frac{(diffusion\ distance)^2}{2D} = \frac{(2l)^2}{2D} = \frac{2l^2}{D} \qquad (32\text{-}1)$$

where D is the diffusion coefficient. (Time $t_D = 2l^2/D$ can also be shown to equal the time needed to cover distance $2l$ by field induced drift. The shuttle of molecules back and forth between states occurs by a combination of diffusion and drift.)

"A molecule in the high velocity state (altitude $2l$) is traveling twice as fast as the zone as a whole, and in time t it gains on the zone by a distance

$$\varsigma = (2V - V)t_D = Vt_D \qquad (32\text{-}2)$$

A molecule in the low velocity state falls behind by the same amount. Distance ς may be regarded as the approximate length of step in a random walk process - the distance moved forward or backward with respect to the zone position before some random event (diffusion) reshuffles altitudes and thus velocity states.

"A large number of random steps creates a Gaussian zone having a variance

$$\sigma^2 = \varsigma^2 n \qquad (32\text{-}3)$$

where n is the number of random steps. Quantity n is simply the total migration time, L/V needed to reach the end of the column at distance L, divided by the time, t_D, required for one step.

$$n = L/Vt_D \qquad (32\text{-}4)$$

The substitution of equations [32-2] and [32-4] into [32-3] leads to

$$\sigma^2 = LVt_D \qquad (32\text{-}5)$$

Column plate height, as in chromatography, is defined as $H = \sigma^2/L$, and is, therefore, equal to

$$H = \frac{\sigma^2}{L} = Vt_D \qquad (32\text{-}6)$$

Figure 32-2. Survey of practical applications of field flow fractionation techniques.
(Courtesy - Deyl, Z., Ed., *Separation Methods*, Elsevier, Amsterdam, The Netherlands, 1984)

When t_D from equation [32-1] is substituted into this, H becomes

$$H = \frac{2l^2 V}{D} \quad (random\text{-}walk\ model) \tag{32-7}$$

The rigorous theory of zone dispersion yields the equation

$$H = \frac{4l^2 V}{D} \quad (rigorous\ nonequilibrium\ theory) \tag{32-8}$$

which is identical in functional form to equation [32-7], but is a factor of two larger.

"Longitudinal diffusion also causes a degree of zone spreading; its contribution to variance and plate height is expressed in a form analogous to that for chromatography, $\sigma^2 = 2Dt$ and $H = 2D/V$, respectively. The overall plate height is the sum of its various constituents

$$H = \frac{2D}{V} + \frac{4l^2 V}{D} + \sum H_i \tag{32-9}$$

where $\sum H_i$ includes all extraneous contributions to H, including those from instrumental dead volumes, solute relaxation times, and so on."

TECHNIQUES

Many modifications of the original technique are now used. These involve different ways to establish the gradient and changes in the shape of the flow profile. The main subdivisions are thermal, sedimentation, flow, electrical, steric, magnetic and concentration. These are discussed below.

THERMAL FIELD FLOW FRACTIONATION (TFFF)

The cross-flow gradient in the TFFF arrangement is formed by placing the channel between a hot surface and a cold surface. Typical temperature gradients are from 20 to 75 °C, with the lower range for low m.w. particles. Two basic arrangements are used, the coiled capillary system shown in Figure 32-3 and the rectilinear channel system shown in Figure 32-4, p. 374.

H. Thompson, et al. state, "The coiled capillary system is composed of two Al cylinders with coiled capillary tubing sandwiched between them. The top cylinder is solid (except for the center hole) and is heated with a 110 V

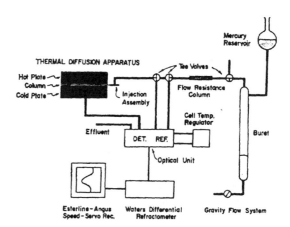

Figure 32-3. Schematic diagram of the coiled capillary thermal gradient system.
(Courtesy - Thompson, H., Myers, M.N., and Giddings, J.C., *Anal. Chem.*, **41** (10), 1219, 1969)

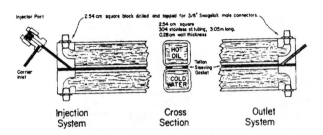

Figure 32-4. Diagram of the rectilinear channel thermal system.
(Courtesy - Thompson, H., Myers, M.N., and Giddings, J.C., *Anal. Chem.*, **41** (10), 1219, 1969)

electric stove element; the bottom cylinder is milled out to permit the flow of coolant (usually water) and the channel enclosed by means of a gasketed steel plate. Different sizes can be made. The largest size has a diameter of 35.6 cm and is 5.1 cm tall. The best results for the large size coiled system have been with Teflon capillaries 54.6 m long, 0.56 mm i.d., wall thickness 0.25 mm.

"The rectilinear channel system [Figure 32-4]. Two 304 SS tubes, 2.54 cm square outside (wall thickness 0.27 cm) and 3.05 m long are closed at both ends by 304 SS blocks, drilled and taped for 3/8 in. Swagelok male connections. An oil bath provides the heat source for the top tube. The oil is heated to the desired temperature by means of immersion heaters and the oil pumped through by means of a centrifugal pump. Cooling water is led through the bottom tube and run countercurrent to the hot oil. The rest of the apparatus is the same as for the coiled capillary system."

It was found necessary to *reverse program* the temperature if the m.w. range was wide. Figure 32-5 shows the results of separating a mixture from 51,000 m.w. to 860,000 m.w.. Normal temperature programming involves increasing the temperature with time, and reverse programming involves reducing the temperature with time. The high m.w. compounds do not move when the temperature gradient is large. After the lower m.w. compounds have been separated the temperature is lowered, so the higher m.w. compounds will move. The scale above Figure 32-5 shows the range for that separation. Notice that these are very slow separations, taking a few days.

SEDIMENTATION FIELD FLOW FRACTIONATION (SFFF)

In the SFFF modification, the stacking is produced by a centrifugal force. Figure 32-6 illustrates the concept.

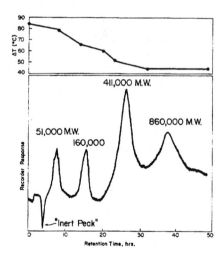

Figure 32-5. Fractogram of a mixture of 20, 30, 100, and 100 µL of 51,000, 160,000, 411,000, and 860,000 m.w. polystyrene solutions. Flow = 2.4 cc/hr. Reverse temperature programming with T_c = 26 °C.
(Courtesy - Thompson, H., Myers, M.N., and Giddings, J.C., *Anal. Chem.*, **41** (10), 1219, 1969)

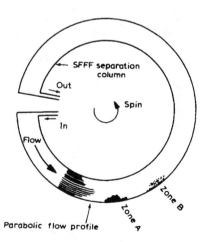

Figure 32-6. Principle of sedimentation flow fractionation.
(Courtesy - Giddings, J.C., Yang, F.S., Myers, M.N., *Anal. Chem.*, **46** (13), 1917, 1974).

Figure 32-7. A model S101 sedimentation field flow fractionator. (Courtesy - FFFractionation, Inc., Salt Lake City, UT)

The heavier molecules are pushed outward against the side wall stronger than lighter molecules and, thus, are retained longer. Figure 32-7 shows a commercial instrument. It has the following characteristics: channel dimensions: length, 89 cm, thickness 0.0254 cm, breadth 2.0 cm, void volume 4.5 mL; field strength: 2500 rpm (~1000 g); flow rate: 0.1 to 10 mL/min.; and radius of rotation: 15.1 cm. It has a built-in rpm and flow control for programmed separations. Particles from 0.05 to 100 μm can be separated, and particles differing by 10 % are resolved.

It is particularly good for colloids, emulsions and particles. Samples separated thus far include lattices, latex aggregates, silica sols, alkyd resins, perfluorocarbon emulsions, milk, carbon particles, hematite, clay, water-borne particles, liposomes, subcellular particles, viruses, and polymerized proteins.

Height Equivalent to a Theoretical Plate (HETP)

The HETP is determined by equation 32-10. This equation is important, because it illustrates what variables a chemist has to minimize the HETP and provide for a better separation.

$$H = \frac{432}{\pi} \frac{kT\eta v}{d^5 G^2 (\Delta\rho)^2} \quad (32\text{-}10)$$

where d = particle diameter, angstroms. v = zone velocity in the column, cm/sec. η = viscosity, poises. T = temperature, °K G = centrifugal field, g. $\Delta\rho$ = solute density - solvent density. k = Boltzman constant. Use 1.38×10^{-16} ergs.

For a given size particle, lowering the temperature, viscosity and flow rate will decrease the HETP. The larger the particle and the higher the centrifuge speed, the lower the HETP will be. Small differences in density between the solvent and solute are magnified as the square of the difference. Figure 32-8 is a plot of plate height vs. mean flow velocity, and Figure 32-9 is a plot of retention parameter, λ, (l/w) vs. particle diameter. Figure 32-10, p. 376, shows a typical sedimentation profile for a latex paint.

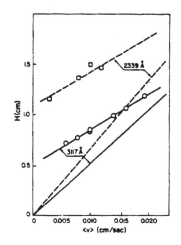

Figure 32-8. Plate height vs. mean flow velocity for polystyrene beads.
(Courtesy - Giddings, J.C., Yang, F.J.F., and Myers, M.N., *Anal. Chem.*, **46** (13), 1917, 1974)

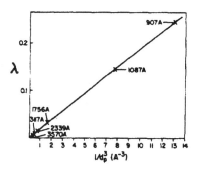

Figure 32-9. Variation of retention parameter with reciprocal field strength for various particle diameters.
(Courtesy - Giddings, J.C., Yang, F.J.F., and Myers, M.N., *Anal. Chem.*, **46** (13), 1917, 1974)

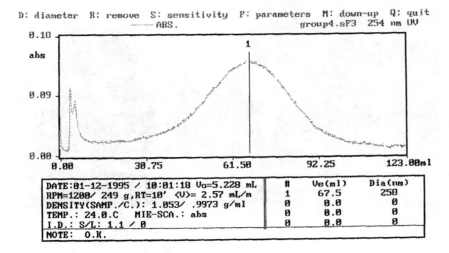

Figure 32-10. An example of the data from an SFFF instrument.

CALCULATIONS

The following equations are involved in sedimentation FFF. The retention ratio, R:

$$R = \frac{t^o}{t_r} = \frac{V^o}{V_e} = 6\lambda\left[\coth\frac{1}{2\lambda} - 2\lambda\right]; \quad R \approx 6\lambda \pm 5\% \tag{32-11}$$

where V^o is the void volume, V_e is the mean elution volume necessary to wash the given particle through the channel, and λ is the thickness of the sample equilibrium distribution.

The force (F) and field strength (G) are expressed as follows:

$$F = V_p \Delta\rho\, G = \frac{1}{6}\pi d^3 \Delta\rho\, G; \quad G = \omega^2 r_o = 4\pi^2 \left[\frac{rpm}{60}\right]^2 r_o \tag{32-12}$$

where F is the force acting on a particle in the field, and G is the gravitational force. The dimensionless thickness of the sample equilibrium distribution, λ, is given as:

$$\lambda = \frac{6kT}{\pi d^3 \Delta\rho\, \lambda G w}; \quad d = \left[\frac{6kT}{\pi \Delta\rho\, \lambda G w}\right]^{1/3} \tag{32-13}$$

where k is the Boltzman constant, 1.38×10^{-16} ergs/K; T is the temperature in K; d is the particle diameter in cm; ρ is the density of the particle; and w is the channel thickness. After the external field is applied, a stop flow period is required for the sample to reach equilibrium (relaxation). The appropriate relaxation time varies with the sample and the applied external field.

To determine the amount adsorbed on a surface:

$$m_2 = \left[\frac{kT}{Gw\left(1 - \frac{\rho_o}{\rho_2}\right)}\right]\left(\frac{1}{\lambda_2} - \frac{1}{\lambda_1}\right) \tag{32-14}$$

where λ_1 and λ_2 are the retention parameters recorded at a fixed field strength G for the bare and coated particles, respectively, and ρ_o and ρ_2 are the densities for the carrier fluid and coating material, respectively. Because the retention volume for the particles gives their size, d, if the density is known, the surface, A, of a particle is:

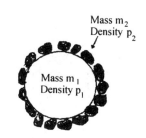

$$A = \pi d^2 = \pi \sqrt[2/3]{\frac{6 k T}{\lambda_1 \pi \Delta\rho G w}} \qquad (32\text{-}15)$$

The surface concentration, illustrated in Figure 32-11, is:

$$\Gamma = \frac{m_2}{A} \qquad (32\text{-}16)$$

Figure 32-11. Particle surface adsorption

The instrument constant, K^*, is determined as follows using two reference densities.

$$\rho_1 - \rho_2 = K^* (T_1^2 - T_2^2) \qquad (32\text{-}17)$$

where ρ_1, ρ_2 = densities. K^* = the instrument constant. T_1, T_2 = instrument parameters (readings). For the gravity, G, calculation (This is divided by 980.6 cm sec^{-2} to get x g):

$$G = \omega^2 r_o = 4 \pi^2 (rpm/60)^2 r_o \qquad (32\text{-}18)$$

EXAMPLE CALCULATION

Refer to Experiment 44, in Appendix A, for the details of the instrument and experimental conditions. Given the following data: air (T_1 = 1.697780, ρ_1 = 1.185 x 10^{-3} g/cm^3; water (T_2 = 2.082762, ρ_2 = 0.997 g/cm^3); imidazole (T = 2.082895), calculate the instrument constant K^* and the density of the imidazole.

ANSWER

For the densitometer that was used, apply equation 32-17. Use air and DI water as the two references to get K^*.

$$1.185 \times 10^{-3} - 0.997 = K^* [(1.697780)^2 - (2.082762)^2]; \quad K^* = 0.68420$$

Now use K^* to determine ρ for imidazole.

$$0.997 - \rho = 0.68420 [(2.082762)^2 - (2.082895)^2]; \quad \rho = 0.9974 \text{ g/cm}^3$$

EXAMPLE CALCULATION

Determine the particle size for the PS 252 standard (V^o = 5.23 mL Ve = 67.5 mL), R = 0.077452, λ = 0.0133), and when it is coated with casein (Ve = 80.1 mL, R = 0.065268, λ = 0.0111). rpm = 1200 and r_o = 15.5 cm. T = 24 °C, w = 0.0254 cm, ρ standard PS252 = 1.053 g/cm^3, ρ bare = 0.997 g/cm^3, λ coated = 0.99738 g/cm^3.

ANSWER

Use equation 32-18 to determine G. The acceleration due to gravity = 980.6 cm s^{-2}

$$G = 4 \times 3.14^2 (1200/60)^2 \times 15.5\,cm = 244{,}518\,cm\,sec^{-2};\ 244{,}518\,cm\,sec^{-2}/980.6\,cm\,sec^{-2} = \sim 250 \times g$$

Use equation 32-11 to determine λ for each material.
PS 252; $R = 5.23\,mL/67.5\,mL = 0.07745$; β-casein $= 5.23\,mL/80.1\,mL = 0.06529$.
$\lambda = R/6 = 0.07745/6 = 0.0129$ and $\lambda = 0.06529/6 = 0.0109$

Use equation 32-13 right to determine the diameter of the bare particles..

$$d = \sqrt[3]{\frac{6 \times 1.38 \times 10^{-16} g\,cm^2/sec^2/K \times 297\,K}{3.14 \times 0.056\,g/cm^3 \times 0.0129 \times 244{,}518\,cm/sec^2 \times 0.0254\,cm}} = 2.58 \times 10^{-7} cm\ or\ 258\ nm$$

Using the same process, d for the coated particles is 273 nm.

EXAMPLE CALCULATION
Calculate the area of a bare particle.

ANSWER
Use equation 32-15.

$$A = 3.14 \times (258 \times 10^{-9} m)^2 = 2.09 \times 10^{-13} m^2$$

EXAMPLE CALCULATION
Using the above data, calculate the surface concentration for the β-casein complex if $\rho_2 = 1.37\,g/cm^3$ and $\rho_o = 0.99738\,g/cm^3$ (imidazole).

ANSWER
Use equation 32-14.

$$m_2 = \left[\frac{1.38 \times 10^{-16} g\,cm^2/sec^2 K \times 297\,K}{244{,}518\,cm\,sec^{-2} \times 0.0254\,cm\left(1 - \frac{0.99738\,g/cm^3}{1.37\,g/cm^3}\right)}\right]\left(\frac{1}{0.0109} - \frac{1}{0.0129}\right) = 3.45 \times 10^{-16} g$$

EXAMPLE CALCULATION
Calculate the surface concentration of the coating.

ANSWER
Use equation 32-16.

$$\Gamma = \frac{3.45 \times 10^{-13} mg}{2.09 \times 10^{-13} m^2} = 1.72\ mg/m^2$$

FLOW FIELD FLOW FRACTIONATION (FFFF)
As seen in Figure 32-12, p. 379, a solution passing through two ceramic frits serves to provide the cross-flow field. The particles are retained on a semipermeable membrane. The field is a pressure field formed by the liquid flowing between the frits. According to J.C. Giddings (*Science*, **260**, 1456, 1993), "The sample particles are carried to the channel accumulation zone with the cross flow. Each species then diffuses into an equilibrium based zone of a specific

thickness as determined by its molecular weight or size. Axial channel flow separates the different species, as it displaces the zones along the accumulation wall. Elution time is governed by the sample diffusion coefficient, which can be translated to a molecular weight or particle size. This is a simpler system than SFFF and can not only separate much smaller particles, it can separate all types of particles, even neutrally buoyant ones, independently of density. Separation is based on size alone, with t_r approximately proportional to diameter. This first power dependence ($S_d = 1$) means that t_r is less sensitive to diameter than that in SFFF ($S_d = 3$). Nonetheless, selectivity is still relatively high, well exceeding that of size exclusion chromatography ($S_d = 0.2$). Because S_d is lower for flow FFF than sedimentation FFF, the t_r range is more condensed and programmed field operation is usually unnecessary. [Figure 32-13] shows the component parts of a flow field flow fractionator." Commercial apparatuses have an effective analysis range from 10 nm (2000 daltons) to 50 μm. The sample size is up to 50 μL. The channel dimensions are 2 cm x 26 cm x 0.0254 cm; void volume = 1.3 mL. Maximum field cross flow flux = 0.20 mL/min/cm^2."

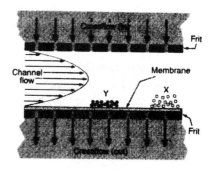

Figure 32-12. The principle of flow field flow fractionation.
(Courtesy - Giddings, J.C., *Science*, **260**, 1456, 1993)

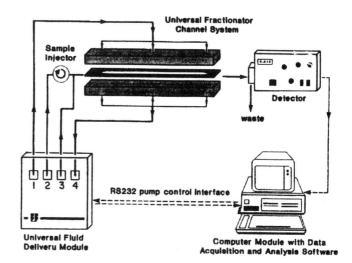

Figure 32-13. Component parts arrangement for flow field flow fractionation.
(Courtesy - FFFractionation, Inc., Salt Lake City, UT)

ELECTRICAL FIELD FLOW FRACTIONATION (EFFF)

EFFF, Figure 32-14, was developed in 1972 (Caldwell, K.D., Kesner, L.F., Myers, M.N., and Giddings, J.C., *Science*, **176**, 296) to complement electrophoretic methods where differential zone velocities arise because of mobility differences alone. "It provides an electrically induced separation different from electrophoresis, requires relatively low potentials, avoids adverse thermal effects, it is a one phase method and thus avoids support interactions and is an elu-

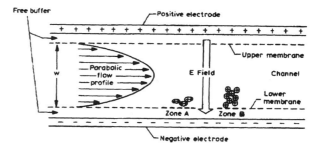

Figure 32-14. Illustration of electrical field flow fractionation.
(Courtesy - Kesner, L.F., Caldwell, K.D., Myers, M.N., and Giddings, J.C., *Anal. Chem.*, **48** (13) 1834, 1976)

tion technique." The membranes (dialyzer tubing) are stretched to form the sides of a ~43 x 2.5 cm channel approximately 0.0356 cm thick. The side walls are Plexiglas. Buffer of pH 4.5 and ionic strength 0.02 M are used for the acid systems.

STERIC FIELD FLOW FRACTIONATION (STERIC FFF)

Steric FFF was proposed in 1978 (Giddings, J.C. and Myers, M.N., *Sep. Sci. Technol.*, **13** (8), 637) as a high field limit to normal FFF for the larger particles (~0.5 to 200 µm) such that diffusion is suppressed and no longer plays a major role in the retention. Refer to Figure 32-15.

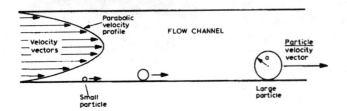

Figure 32-15. Principle of steric field flow fractionaton.
(Courtesy - Giddings, J.C. and Myers, M.N., *Sep. Sci. Technol.*, **13** (8) 637, 1978)

"Consider a sphere of diameter d (for example 10 µm) driven toward the accumulation wall by field forces. Motion is halted by the physical (steric) barrier of the wall when the particle center approaches distance $x = d/2$ above the wall. Thus the equilibrium elevation of the particle depends upon its size. Larger particles, more highly elevated than their smaller counterparts, are swept downstream more rapidly. Thus large particles elute before small particles, a trend opposite to that found for the normal mode FFF.

"In theory, any kind of external field (electrical, sedimentation, etc.) can be utilized for steric FFF. However, gravity provides a practical field for most particles in the 1 to 100 µm diameter range unless they are in a neutrally buoyant medium. Surfaces that are inert and flat should be employed. Furthermore, the flow velocity should be sufficiently high that the viscous forces dragging and rolling the particles along should exceed the gravitational forces pulling the particle against the surface. In this way the particle would tend to be pulled immediately free of any ensnaring influence. Therefore, high flow velocities appear to be advantageous for steric FFF separations."

MAGNETIC FIELD FLOW FRACTIONATION (MFFF)

Proposed in 1980 (Vickery, T.M., and Garcia-Ramirez, J.A., *Sep. Sci. Technol.*, **15** (6), 1297), with a full mathematical treatment of the theory. MFFF is a method to separate paramagnetic from diamagnetic molecules. In addition, diamagnetic molecules can be tagged by paramagnetic ions (Ni^{+2}, Fe^{+2}), so they can be separated. Figure 32-16 shows the design, which is fairly standard for MFFF. A field strength of 4000 Gauss was used. Figure 32-17, p. 381, shows the component parts. The system was first tested with 5×10^{-5} M Ni^{+2} bovine serum albumin in a 5.6 cm coil of 304 cm length at 0.1 mL/min.

The authors put forth the following assessment: "Aside from the obvious purification and separation of paramagnetic metal-containing species, the possible utility of this technique as a characterization tool should be mentioned. There are many biochemical problems which involve the observation of a paramagnetic species during the course of an enzymatic reaction while the non-functioning enzyme is diamagnetic. The fraction containing paramagnetic centers is removed from the non-reacting (and therefore diamagnetic) counterpart.

"Another possible application is the 'tagging' of diamagnetic large molecules with a paramagnetic center (formation of a metal complex, for example) which can be removed after separation."

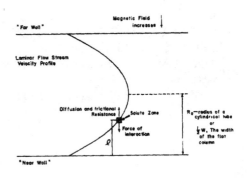

Figure 32-16. Diagram of the lateral shift induced by the interaction of a magnetic field with a plug of paramagnetic material.
(Courtesy - Vickery, T.M., and Garcia-Ramirez, J.A., *Sep. Sci. Technol.*, **15** (6), 1297, 1980)

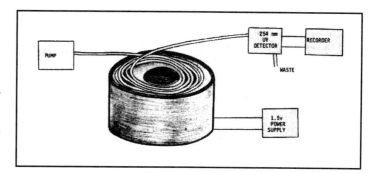

Figure 32-17. Schematic diagram of a MFFF apparatus.
(Courtesy - Vickery, T.M., and Garcia-Ramirez, J.A., *Sep. Sci. Technol.*, **15** (6), 1297, 1980)

REVIEW QUESTIONS

1. What is the basic principle of how large molecules are separated in field flow fractionation?
2. What is the basis for Gidding's theory of FFF separations?
3. What is D in the overall plate height equation for FFF?
4. What is the temperature gradient for low m.w. compounds in thermal FFF?
5. What is meant by reverse temperature programming?
6. What produces the force in sedimentation FFF?
7. What is the range of particle sizes that can be separated with SFFF?
8. What is the effect on HETP of lowering the viscosity of the mobile phase?
9. What keeps the particles from washing through the side wall in flow FFF?
10. How does the selectivity of flow FFF compare to size exclusion chromatography?
11. How do you explain the fact that larger particles move faster than smaller ones in steric FFF?

IF IT EXISTS - CHEMISTRY IS INVOLVED
IF YOU CAN BUY IT - A CHEMIST WAS INVOLVED SOMEWHERE

SEPARATIONS INVOLVING FLOTATION

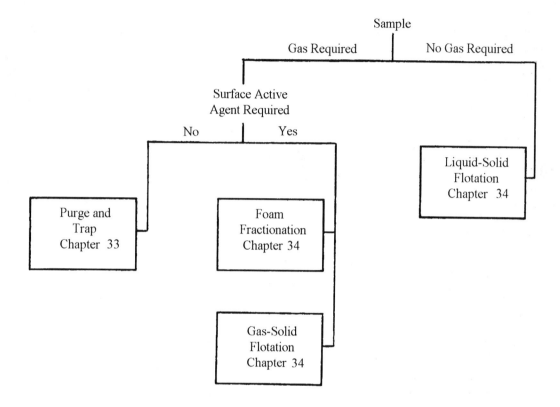

Flotation, not floatation, is the term used to describe the use of a gas to cause selected solid particles to rise to the surface from a mixture of particles suspended in a liquid. A newer term, *adsubble*, has been proposed, and may eventually win out. Parts of the two chapters in this section are not flotation in the strictest sense in that one technique does not separate solid particles, and the other does not use a gas, yet all four techniques are closely related.

Purge and trap (PAT), Chapter 33, involves bubbling a gas through a liquid which contains volatile or semi-volatile compounds. These compounds transfer into the gas bubble based on the second law of thermodynamics; *systems tend to maximum disorder, or compounds go from high concentration to low concentration*. In this case, the inside of the gas bubble is at zero concentration so volatile compounds in the liquid transfer into the bubble in the form of a vapor. The bubbles of gas rise to the surface, pass through an adsorption tube, and the contents trapped. This is not only a separation technique, but it is valuable in concentrating trace materials. The most common use of this technique is to separate EPA's priority pollutants from water supplies.

A way to separate many nonvolatile compounds is to form a foam and have the desired material attach itself to the surface of the gas bubble. The compounds attached are not sufficiently large to be classified as a solid. An example of this is the removal of detergents from waste water by bubbling a gas through the water and collecting the detergent in the foam that is formed. This is known as *foam fractionation*, Chapter 34.

The true *flotation* technique is described in Chapter 34. This technique is of tremendous importance industrially. Two examples are the removal of industrial grade diamonds from the surrounding granite and the concentration of taconite iron ore to an economically high percentage of iron. This can become quite involved. The surface of the solid particle must be made attractive to the gas bubble. This may require an activator on the surface to hold the collector that will then attract the gas bubble. Much chemistry is involved.

A flotation without involving a gas (Chapter 34) is to use a liquid that will selectively coat the solid particles. In some cases, the desired particles rise to the top, and in others, the undesirables rise to the top. The experiments described are the official methods for determining fly eggs and maggot fragments in fruits and vegetables; mites, aphids, and thrips on vegetables; and insect fragments in corn meal and flour. These are described in the experimental section. Bon appetit!

33
PURGE AND TRAP AND DYNAMIC HEADSPACE ANALYSIS

PRINCIPLES

Chapters 33 and 34 involve the application of bubbles of gas to effect a separation. In Chapter 33 diffusion of a volatile compound into a gas bubble is the operating principle (purge and trap). Chapter 34 discusses foam fractionation in which the gas bubbles are coated with a surfactant. Dissolved compounds adhere to the surface and are removed as a foam. Later on in Chapter 34 flotation, in which small solid particles are coated such that gas bubbles adhere to them and they are lifted out of the solution, is discussed.

Purge and trap (PAT), Figure 33-1, p. 386, is the removal and collection of volatile compounds from a liquid by diffusion of these materials into a stream of gas bubbles passing through it and trapping the expelled components. The purpose for purging is the removal of one or more components. The purpose of trapping is to concentrate the purged components for further identification and quantitation. Dynamic head space applies to other matrices, such as solid foods, polymers, or arson debris. The most common example of PAT is the detection of industrial chemicals in water, called *priority pollutants* by the EPA. Air, nitrogen, or helium is bubbled through a water sample; and in this case, the volatile industrial chemicals (plus those from nature) transfer to the gas, are expelled from the water, and are trapped on a combination of charcoal, silica gel, and Tenax. The trapped materials then are driven off into a gas chromatograph with a mass spectrometer detector. The concentration factor can be thousands, depending upon the original concentrations.

An example of a gas being purged by a gas is the removal of oxygen from a water solution by passing pure nitrogen through the solution or the removal of air in an HPLC solvent by passing helium through it. These are purges only because the purged material usually is not trapped. An example of the removal of a liquid from a solution is the removal of mercury vapor from a fish digest to determine the mercury. An example of the removal of a liquid from a solid is the determination of ethylene dibromide from bakery goods. An example of removal of a solid from a solid is the removal of volatile pesticide residues from food extracts.

The basis for all of these separations is one interpretation of the Second Law of Thermodynamics; that molecules, if left to themselves, would spontaneously tend toward maximum disorder. In the Chapters on Osmosis and Dialysis, this is used to explain the movement of molecules of solvents or solutes from a high concentration (HC) region to a low concentration (LC) region. The same is true here, only a gas bubble is the region of low concentration.

Consider a glass of champagne. When a bottle is opened, there is a sudden release of CO_2 bubbles. Each bubble will rise to the top, expanding as it rises because of the reduced pressure surrounding it. It contains 0% ethanol when it is first formed. This can be considered low concentration (LC) ethanol. If the beverage contains dissolved ethanol, it is at a higher concentration than 0%. This is now high concentration ethanol (HC) by comparison, and the ethanol in the beverage will transfer to the CO_2 bubble until equilibrium is established. When the gas bubble reaches the top of the liquid, it bursts and the ethanol escapes into the air. This is *purging* (to get rid of). If a gas is forced through the liquid, then the process is called *sparging*. In purge and trap, the ethanol vapors would be adsorbed onto a solid such as silica gel and later desorbed by heating.

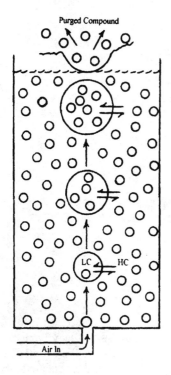

Figure 33-1. Diagram of the purge principle.

When a bubble of nitrogen enters a water solution from the bottom, as by a gas dispersion tube, it will rise to the top, expanding as it rises because of the reduced pressure surrounding it. It contains 0% oxygen when it first enters the solution. This can be considered low concentration (LC) oxygen. If the water solution contains dissolved air, then it contains oxygen and at a higher concentration than 0%. This is now high concentration oxygen (HC) by comparison, and the oxygen in the solution will transfer to the nitrogen bubble until equilibrium is established. When the gas bubble reaches the top of the liquid, the oxygen can be trapped and measured or allowed to escape. This is a very efficient process. Oxygen can be removed from water in just a few minutes, if the bubbles are small and many.

In each of the above cases, the initial concentration of any of the components in the gas bubble will be 0%, and any component in the sample with a vapor pressure will be at a higher concentration and tend to move into the gas bubble. Theoretically, every component in the mixture will move some molecules into the gas bubble. However, only those with the higher vapor pressures will be concentrated to a significant extent and those desired can be trapped selectively. The gas chromatograph will further separate the trapped components, and the mass spectrometer can be set to detect only the desired components. Such an operation is done many times daily for the priority pollutants and requires about 3 hours from the start to the final calculations.

As a general rule, any organic compound that boils below 180 °C is considered a volatile compound (VOC).

PURGE EFFICIENCY

Vapor pressure, solubility, temperature, sample size, method of purge, and purge volume all affect purge efficiency (from Tekmar Co.).

Vapor pressure. The higher the vapor pressure, the higher purge efficiency is.

Solubility. The greater the solubility in the sample matrix, the harder it is to remove the compound from that matrix. Methanol is very soluble in water and is only 10 % removed.

Temperature. An increase in temperature always increases purge efficiency. The amount of water adsorbed onto the trap doubles for each 10°C rise in temperature. Table 33-1 shows how purge efficiency changes with temperature for one compound.

Table 33-1. Efficiency of purging 1,1,2,2-tetrachloroethane in water vs. temperature

Temperature °C	% Efficiency
25	54
40	78
60	88

Sample size. An increase in sample size requires an increase in purge volume. If the purge volume is not increased, you can expect only a 70 to 80% increase in sensitivity for a 100% increase in sample size.

Purge volume. An increase in the purge volume will improve purge efficiency; 30 to 50 mL/min are common.

Purge method. Given the same purge volume, the dispersion of fine bubbles will be more efficient than that of large bubbles. A fritted sparger is more efficient than a needle sparger.

TECHNIQUES

The complete apparatus consists of several components: a sparger, a trap, a desorber, a liquid N_2 trap, a gas chromatograph, usually a mass spectrometer as a detector, a data file, and now a computer to do most of the work. The sparger is used to remove compounds from the sample. However, it may take up to half an hour, and the compounds are so dilute that they could not be detected. They are concentrated by adsorbing them on solids such as silica gel or charcoal. They are then removed (desorbed) by heating at several hundred degrees for just a few minutes. This is still too slow for plug injection into a gas chromatograph, so the vapors are trapped in a U tube cooled with liquid N_2. The liquid N_2 is removed, the U tube is heated, and the plug of compounds is injected into a chromatograph. Depending on the concentration level, either the regular GLC detectors or a mass spectrometer can be used.

Heating is not always necessary. For example; when ethylenedibromide is to be detected in bakery goods, it is adsorbed onto a small plug of Tenax and later eluted with 1 mL of hexane.

Spargers

Figure 33-2 shows a needle type sparger for slurries, soils, solid, foaming liquid samples, and viscous samples. This style comes in either 5 or 25 mL sizes. The sample is added to the tube, and a needle is inserted to the bottom of the sample to introduce the gas. The bulge near the top is to reduce any foaming.

Figure 33-2. A needle sparger.
(Courtesy - Corning Glass Co., Corning NY)

Figure 33-3 shows two types for liquids. The side tube is a reservoir for the sample when gas is not flowing. The sample is added to the tube on the left and connected to the apparatus. Gas then is admitted into the top of the right tube. If a solution is the sample, then a frit is used to disperse the gas. If a soil or solid is the sample, then the frit is removed.

Traps

The Tekmar Corporation lists the following as requirements for a good trapping material: (1) retain analytes of interest, (2) allow O_2, CO_2, and H_2O to pass, (3) release the analytes quickly and easily upon heating, (4) remain stable and not contribute volatiles, (5) operate without causing catalytic reactions, and (6) be of reasonable price. At the present time, the following are the most commonly used: Tenax, silica gel, coconut charcoal, 3% OV-1 on Chromosorb WHP, molecular sieves, and carbo traps. Tenax adsorbs weakly polar compounds and no water. Silica gel adsorbs water so tightly that it usually is retained during desorption. Charcoal adsorbs most organic compounds, but they can be desorbed with moderate heating. OV-1 is a lot like Tenax and molecular sieves trap O_2, CO_2, and H_2O effectively.

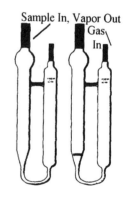

Figure 33-3. Purge and trap spargers.
(Courtesy - Corning Glass Co., Corning NY)

The trap for industrial chemicals in water is a borosilicate glass tube 5 mm in diameter and 25 cm long. (Figure 33-4, p. 388). This was proposed by T.A. Bellar and J.J. Lichtenberg in 1974 (*J. Am. Water Works Assoc.*, **66**, 739). After trapping, the compounds are desorbed for 4 minutes at 180 °C.

Figure 33-5, p. 388, shows the preferred arrangement for adsorbents in the trap and the direction of gas flow during purging and desorption.

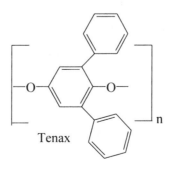

Tenax

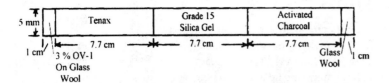

Figure 33-4. A trap for halocarbons in water.

Apparatus

Figure 33-6 shows a single unit PAT apparatus. Multiple units involving automatic operation are now common, as shown in Figure 33-9, p. 389.

Figure 33-7, p. 389, shows a diagram of the other components needed to do a PAT separation. In this case, the detectors on the gas chromatograph are used rather than connecting the column to a mass spectrometer.

Figure 33-8, p. 389, shows a static dilution bottle used to prepare standards of known concentration. The top has a gas tight valve for a syringe. The bulb volume is known, and a known amount of gas can be added with a syringe. The valve is open when green, and closed when red.

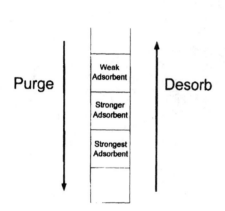

Figure 33-5. Trap composition and gas flow.
(Courtesy - Tekmar Corp., Cincinnati, OH)

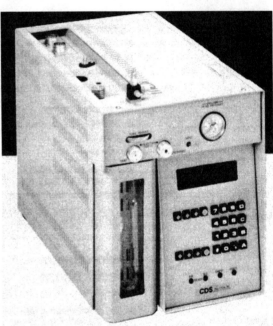

Figure 33-6. A CDS Analytical Inc., model 6000 single unit purge and trap apparatus.
(Courtesy - CDS Analytical Inc., Oxford, PA)

EXAMPLE CALCULATION

The bulb shown in Figure 33-8 is 2.00 L. How many µL of benzene must be added to produce a 10 ppm standard? Is this practical?   Air, $d = 1.23$ g/L, benzene 0.88 g/mL.

ANSWER

$$\frac{1.23 \ g}{L} \times 2.00 \ L \times \frac{0.010 \ g}{1,000 \ g} \times \frac{1.00 \ mL}{0.88 \ g} \times \frac{1,000 \ \mu L}{1.00 \ mL} = 0.028 \mu L$$

No. Dilute a larger, weighable, quantity of benzene in acetone and inject the combination

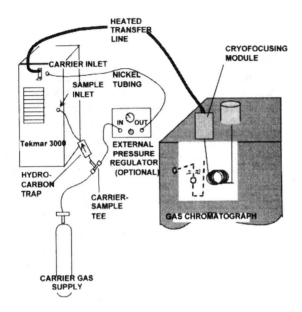

Figure 33-7. Component arrangement for a purge and trap separation.
(Courtesy - Tekmar Corp., Cincinnati, OH)

Figure 33-8. A static dilution bottle.
(Courtesy - Fisher Scientific Co., Pittsburgh, PA)

Figure 33-9 shows a bank of traps and heaters. Figure 33-10 shows the arrangement for a split capillary interface to a gas chromatograph and Figure 33-11, p. 390, shows a splitless interface with cryofocusing. Cryofocusing is concentrating the sample in a small volume by freezing the components as they are trapped.

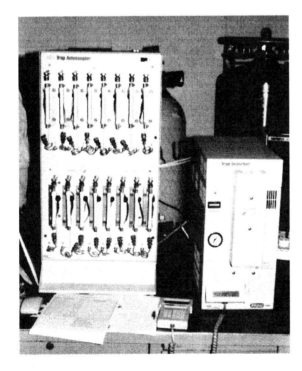

Figure 33-9. A bank of traps and heaters.
(Courtesy - Tekmar Corp., Cincinnati, OH)

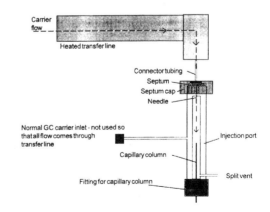

Figure 33-10. A split capillary interface.
(Courtesy - CDS Analytical Inc., Oxford PA)

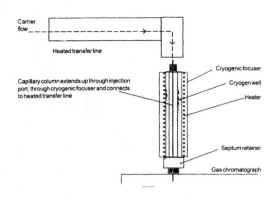

Figure 33-11. A splitless interface with cryogenic focusing.
(Courtesy - CDS Analytical Inc., Oxford, PA)

Refer to Chapter 19, p. 207, and Chapter 20, p. 238, the sections on mass spectrometers.

PRIORITY POLLUTANTS

Perhaps the greatest use of PAT at this time is to detect what are known as the *priority pollutants*. These originally included 128 industrial compounds that the Environmental Protection Agency considered as their first priority to monitor and regulate in our nations streams, lakes, and surface water. This list has since been expanded slightly. The tolerances set are very low, a safety factor of 1000 in most cases, and are usually not detectable by ordinary chemical means. However, the compounds can be concentrated by PAT and then are readily detectable. With computer assistance and a databank of standards for a mass spectrometer detector, the entire 128 compounds can be screened, detected, and quantitated at the ppb level and the report written in about 45 minutes. The technique is to choose a mass for each compound and set the mass spectrometer to monitor that mass. This eliminates most of the fragments and speeds up the identification and calculations. Table 33-2 is a list of the priority pollutants. Take a look at this list. If in later years you handle any of these compounds, be extra careful how you dispose of them.

Table 33-2. Priority pollutants
(Courtesy - Continental Analytical Services, Salina, KS)

Analyte	Water reporting limits, µg/L	Soil reporting limits, µg/kg
Volatiles (28)		
Acrolein	25.	25.
Acrylonitrile	25.	25.
Benzene	5.0	5.0
Bromodichloromethane	5.0	5.0
Bromoform	5.0	5.0
Bromomethane	5.0	5.0
Carbon tetrachloride	5.0	5.0
Chlorobenzene	5.0	5.0
Chloroethane	5.0	5.0
2-Chloroethylvinyl ether	5.0	5.0
Chloroform	5.0	5.0
Dibromochloromethane	5.0	5.0
1,1-Dichloroethane	5.0	5.0
1,2-Dichloroethane	5.0	5.0
1,1-Dichloroethylene	5.0	5.0
1,2-Dichloroethylene (Total)	5.0	5.0
1,2-Dichloropropane	5.0	5.0
1,3-Dichloropropene (Total)	5.0	5.0
Ethylbenzene	5.0	5.0
Methylene chloride	5.0	5.0
1,1,2,2-Tetrachloroethane	5.0	5.0
Tetrachloroethylene	5.0	5.0
Toluene	5.0	5.0
1,1,1-Trichloroethane	5.0	5.0
1,1,2-Trichloroethene	5.0	5.0
Trichloroethylene	5.0	5.0
Vinylchloride	5.0	5.0

Table 33-2. Continued

Extractables (57)

Acenaphthene	5.0	0.5
Acenaphthylene	5.0	0.5
Anthracene	5.0	0.5
Benzidine	50.	5.0
Benzo (a) anthracene	5.0	0.5
Benzo (a) pyrene	5.0	0.5
Benzo (b) fluoranthene	5.0	0.5
Benzo (ghi) perylene	5.0	0.5
Benzo (k) fluoranthene	5.0	0.5
Bis (2-chloroethoxy) methane	5.0	0.5
Bis (2-chloroethyl) ether	5.0	0.5
Bis (2-chlotoisopropyl) ehter	5.0	0.5
Bis (2-ethylhexyl) phthalate	5.0	0.5
4-Bromophenyl phenyl ether	5.0	0.5
Butyl benzyl phthalate	5.0	0.5
p-Chloro-*m*-cresol	5.0	0.5
2-Chloronaphthalene	5.0	0.5
2-Chlorophenol	5.0	0.5
4-Chlorophenyl phenyl ether	5.0	0.5
Chrysene	5.0	0.5
Dibenz (a,h) anthracene	5.0	0.5
1,2-Dichlorobenzene	5.0	0.5
1,3-Dichlorobenzene	5.0	0.5
1,4-Dichlorobenzene	5.0	0.5
3,3"-Dichlorobenzidine	50.	5.0
2,4-Dichlorophenol	5.0	0.5
Diethyl phthalate	5.0	0.5
Dimethylphthalate	5.0	0.5
Di-*n*-butylphthalate	5.0	0.5
2,4-Dimethylphenol	5.0	0.5
4,6-Dinitro-o-cresol	50.	5.0
2,4-Dinitrophenol	50.	5.0
2,4-Dinitrotoluene	5.0	0.5
2,6-Dinitrotoluene	5.0	0.5
Di-*n*-octylphthalate	5.0	0.5
1,2-Diphenylhydrazine	5.0	0.5
Fluoranthene	5.0	0.5
Fluorene	5.0	0.5
Hexachlorobenzene	5.0	0.5
Hexachlorobutadiene	5.0	0.5
Hexachlorocyclopentadiene	5.0	0.5
Hexachloroethane	5.0	0.5
Indeno (1,2,3-cd)pyrene	5.0	0.5
Isophorone	5.0	0.5
Naphthalene	5.0	0.5
Nitrobenzene	5.0	0.5
2-nitrophenol	5.0	0.5
4-Nitrophenol	5.0	0.5
n-Nitrosodimethylamine	5.0	0.5
n-Nitrosodi-n-propylamine	5.0	0.5
n-Nitrosodiphenylamine	5.0	0.5
Pentachlorophenol	5.0	0.5
Phenanthrene	5.0	0.5
Phenol	5.0	0.5
Pyrene	5.0	0.5
1,2,4-Trichlorobenzene	5.0	0.5
2,4,6-Trichlorophenol		

Table 33-2. Continued

Organochlorine Pesticides/polychlorinated Biphenyls (26)

Aldrin	0.1	0.01
Dieldrin	0.1	0.01
Chlordane	0.1	0.01
4,4'-DDT	0.1	0.01
4,4'-DDE	0.1	0.01
4,4-DDD	0.1	0.01
A-Endosulfan	0.1	0.01
B-Endosulfan	0.1	0.01
Endosulfan sulfate	0.1	0.01
Endrin	0.1	0.01
Endrin aldehyde	0.1	0.01
Heptachlor	0.1	0.01
Heptachlor epoxide	0.1	0.01
A-BHC	0.1	0.01
B-BHC	0.1	0.01
G-BHC	0.1	0.01
D-BHC	0.1	0.01
PCB-1233	0.5	0.05
PCB-1254	0.5	0.05
PCB-1221	0.5	0.05
PCB-1232	0.5	0.05
PCB-1248	0.5	0.05
PCB-1260	0.5	0.05
PCB-1016	0.5	0.05
Toxaphene	5.0	0.5
Methoxychlor	1.0	0.1

Metals (13)   (Reporting limit)

Antimony	0.006 mg/L	5. mg/kg
Arsenic	0.01	1.
Beryllium	0.004	0.5
Cadmium	0.005	0.5
Chromium	0.01	1.
Copper	0.02	2.
Lead	0.1	10.
Mercury	0.0002	0.1
Nickel	0.04	4.
Selenium	0.005	0.5
Silver	0.01	1.
Thallium	0.002	1.
Zinc	0.02	2.

Cyanide/phenolic Compounds (4)

Cyanide	0.01 mg/L	0.3 mg/kg
Phenolic compounds	0.005	0.3

REVIEW QUESTIONS

1. What is meant by "purging"?
2. In polarographic analysis, the dissolved oxygen forms two waves that overwhelm the signal from the trace metals present in the solution. How is the oxygen purged?
3. What scientific law is the driving force behind purging?
4. What types of compounds are considered "volatile organic compounds (VOC's)"?
5. What are some of the factors that control purging efficiency?
6. Why are organohalogens easy to remove from water by purging?
7. What are some common purge volume rates?

8. What is a normal sequence of events for a purge and trap analysis?
9. For what types of samples is a needle type sparger preferred?
10. At this time, what are the most common materials used in traps?
11. What types of compounds does Tenax adsorb?
12. What is a common size for a trap?
13. A 20 ppm standard of dimethyldisulfide, f.w. = 94.16, d = 1.06 g/mL, is to be prepared in a 2.00 L static dilution bottle. 100 μL have been dissolved in 10 mL of ethanol, d = 0.789 g/mL. How many μL are required?
14. As a group, what are the priority pollutants?

IF IT EXISTS - CHEMISTRY IS INVOLVED
IF YOU CAN BUY IT - A CHEMIST WAS INVOLVED SOMEWHERE

34
FOAM FRACTIONATION, GAS-, AND LIQUID-ASSISTED FLOTATION

FOAM FRACTIONATION

PRINCIPLES

Foam fractionation is a *separation based upon transfering one or more components in a liquid to the surface of gas bubbles passing through it and temporarily collecting the separated components in a foam at the top of the liquid.* The foam is skimmed away and broken, and the concentrated components are collected. The compounds collected are called the *colligend*. Surfactants are used as the foaming agents. The term *adsubble methods* is now used to include all adsorptive bubble separation methods. Figure 34-1, p. 396 is a diagram that illustrates the surface arrangement of the foam. Figure 34-2, p. 396 shows the components needed.

Air is dispersed through a liquid that contains a small amount of surfactant (Figure 34-3, p. 396). The surfactant collects at the gas-liquid interface with the nonpolar ends oriented into the gas bubble. This makes the surface of the gas bubble polar, its charge depending upon the type of surfactant used. This charge attracts molecules of the opposite charge to the surface. The surfactant is designed to form sufficiently strong bubbles such that a foam is produced when the bubbles emerge from the liquid. The foam, containing the desired components on each bubble's surface, is skimmed off and then defoamed, The desired components then can be collected for further treatment.

The *enrichment factor*, the ratio of the concentration in the collapsed foam to the concentration in the feed, varies considerably as expected, but can be from 10 to several hundred.

The applications of foam fractionation are many. Its use is primarily to concentrate materials present in low concentration (as low as 10^{-10} M/L) in liquids where other processes are not economical. On a commercial scale, this process is used to treat waste water to remove organic waste material and to remove ions from process streams such as the oxyanions of Re(VIII), Mo(VI), Cr(VI), W(VI) and V(V) and the cyanide complex anions or the chloride complexes of Zn(II), Cd(II), Hg(II), and Au(III). In the laboratory, it is used to concentrate proteins such as bovine albumin, cytochrome-c, barley malt, alpha amalyase, and beta casein, as well as proteins and enzymes from various plant and animal systems.

Compare a glass of champagne to a glass of beer. When a bottle of each is opened, there is a sudden release of CO_2 bubbles. Each bubble will rise to the top, expanding as it rises because the reduced pressure surrounding it. It contains 0 % ethanol when it is first formed. This can be considered low concentration (LC) ethanol. If the beverage contains dissolved ethanol, it is at a higher concentration than 0%. This is now high concentration (HC) ethanol by comparison, and the ethanol in the beverage will transfer to the CO_2 bubble until equilibrium is established. When the gas bubble reaches the top of the champagne, it bursts, and the ethanol escapes into the air. However, beer contains proteins, which act as a foaming agent, and the CO_2 bubbles containing the ethanol are trapped in the foam. The foam now contains a significant quantity of ethanol, as any beer drinker will confirm. If this foam were skimmed away and broken by adding a defoamer, the ethanol could be collected.

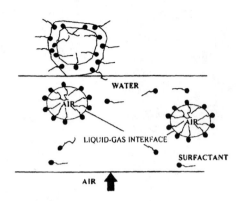

Figure 34-1. Micelles in foam fractionation. (Courtesy - Tharapiwattananon, N., Scamehorn, J., Osuwan, S., Harwell, J., and Haller, K., *Sepn. Sci. Technol.*, **31**(9), 1233, 1996)

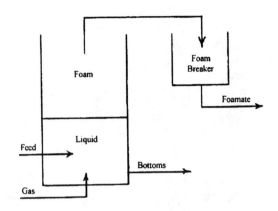

Figure 34-2. The foam fractionation process.

This is a modified foam fractionation in that not all of the adsorbed compound is on the surface, but the desired compound is trapped in the foam. This is a very efficient process, if the bubbles are small and many. In purge and trap, the ethanol vapors would be adsorbed onto a solid such as silica gel and later desorbed by heating.

Foams

A foam is an irregular dispersion of a gas in a liquid with the largest volume usually being the gas. The surface of the gas bubbles is coated with a thin layer of liquid called the *lamella*. The thickness of this layer varies from 30,000 Å down to about 50 Å, at which point the layer breaks apart, and the bubble collapses. Layers of 200 to 2,000 Å are typical. Once the gas bubbles collect into a foam and press against each other, they assume a dodecahedral shape, the liquid between them starts to drain away.

A dodecahedron is a 12 sided figure with 12 congruent, regular, pentagonal faces. Gas and vapors diffuse through the lamellae, and because the pressure inside of a small bubble is greater than that in a large bubble the small bubbles tend to get smaller and the large bubbles get larger. The gas that dissolves into the lamellae then diffuses through them to a region of lower pressure (any adjacent large bubble) After a short time, only 10% of the number of bubbles may be present in the same size foam, yet the foam has not collapsed.

Figure 34-4, p. 397, shows a capillary between bubbles. Liquid is in a layer between the bubbles and in the vacancies where the bubbles join, called capillaries.

Drainage under the influence of surface tension occurs because the pressure within the liquid at point O is less than elsewhere due to the negative curvature and the liquid is drawn from the lamellae into the capillary. Gravity then causes it to drain from the foam in general. The net effect is that the liquid surrounding the bubble gets thinner with time and the bubble will usually break within a few seconds unless some stabilizing agent is added.

Theory

(Abstracted with permission from R. Lemlich, "Principles of Foam Fractionation," E.S. Perry Ed., in *Progress in Separation and Purification*, Vol. 1, Wiley-Interscience, New York, NY, 1968.)

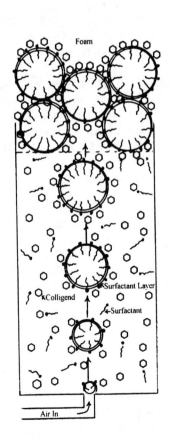

Figure 34-3. Diagram illustrating the principle of foam fractionation.

Equation [34-1] shows the equilibrium adsorption of dissolved material at a gas-liquid interface.

$$d\gamma = -RT\sum \Gamma_i \, d \ln a_i \qquad (34\text{-}1)$$

where: γ = surface tension. a_i activity of the ith component. R = the gas constant. T= °K. Γ = surface concentration, g-moles/cm².

This is a difficult equation to evaluate because of the activities. However, if the surfactant concentration is low, below the critical micelle concentration (cmc) where activities are constant, then:

$$\Gamma = -\frac{1}{RT}\frac{d\gamma}{d \ln C} \qquad (34\text{-}2)$$

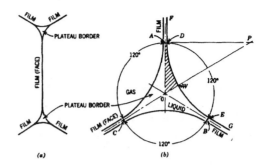

Figure 34-4. Plateau borders in cross section. (a) two capillaries with a film between them, (b) magnified view of one capillary showing the six fold symmetry. (Courtesy - Lemlich, R. ,*Principles of Foam Fractionation*, Wiley-Interscience, New York, NY, 1968)

where C = the bulk concentration.

If a surfactant is in low concentration in water, it is dispersed uniformly throughout the water. If a gas bubble passes through the water, the surfactant can collect on the bubble's surface, forming a film and stabilizing the bubble. This surface can then adsorb compounds, the type depending upon the charge on the surfactant. If the surfactant concentration increases, there will be a concentration where the surfactant molecules gather together to form a microball called a *micelle*. Its surface will look like the gas bubble in that it will have the same charge and it also can adsorb molecules to its surface, however, at a different concentration because the surface tension is different. Figure 34-5 illustrates the change in surface tension with surfactant concentration. Experiments indicate that the same constant Γ applies above the cmc, and the addition of further surfactant goes primarily to the formation of micelles. Γ is in the range of 3×10^{-10} g-moles/cm², which is monolayer coverage.

Figure 34-6 shows the effect of concentration in the liquid on concentration on the bubble surface. The portion of the curve on the left is linear. This can be represented by a Langmiur isotherm to determine Γ on the surface of a bubble without micelle interference.

$$\Gamma = K C / (1 + K^1 C) \qquad (34\text{-}3)$$

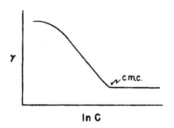

Figure 34-5. Surface tension vs. ln concentration of surfactant. (Courtesy - Lemlich, R., *Principles of Foam Fractionation*, Wiley-Interscience, New York, NY, 1968)

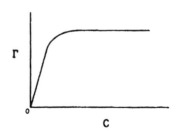

Figure 34-6. Effect of concentration in the liquid on concentration on the surface. (Courtesy - Lemlich, R., *Principles of Foam Fractionation*, Wiley-Interscience, New York, NY, 1968)

where K and K^1 are constants.

Micelles compete for the colligend molecules and since they stay in solution, they reduce the effectiveness of the separation. Assume the monolayer is E times as effective in adsorbing the colligend as is the surface of the micelle.

Let the concentration of unadsorbed colligend counterions be C_1 in the absence of micelles, but at sufficient surfactant concentration to assure a surfactant monolayer. Let C_2 be the concentration of unadsorbed colligend counter-ions in the presence of micelles. Then:

$$\Gamma_1 = K C_1 \quad \text{and} \quad \Gamma_2 = K C_2 \tag{34-4}$$

In the latter case the ratio of colligend adsorbed at the surface to surfactant adsorbed at the surface is KC_2/Γ_s, where Γ_s is the concentration of surfactant at the surface. Assuming all surfactant added above the cmc goes to the formation of micelles, the quantity of colliged counter-ion adsorbed by the micelles per unit volume of solution is then $(C_s - C_{sc})/KC_2/\Gamma_s E$ where C_s is the surfactant concentration in the solution, and C_{sc} is the cmc. A mass balance for the colligend counter-ion in a unit volume of pool liquid yields:

$$\frac{(C_s - C_{sc})KC_2}{\Gamma_s E} + C_2 = C_1 \tag{34-5}$$

This neglects the adsorption at the surface of the bubbles in the liquid pool since the holdup at the surface is generally small compared to the material in the pool. Combining equations 34-4 and 34-5:

$$\Gamma_2 = \frac{\Gamma_1}{1 + (C_s - C_{sc})K/(\Gamma_s E)} \tag{34-6}$$

This equation gives the effect of the micelles on Γ for the colligend counter-ion.

For a surfactant, or for a colligend adsorbed on the surfactant layer, Γ can be found by foam fractionating in the simple mode. In the foam, the component in question is partly in the interstitial liquid and partly adsorbed at the bubble surfaces. For spherical bubbles of various diameters d_i, each present in proportion to the number n_i, the rate of surface overflow in the foam is $6G/d$, where G is the volumetric gas flow rate and d is determined according to equation 34-7.

$$d = \frac{\sum n_i d_i^3}{\sum n_i d_i^2} \tag{34-7}$$

Thus, the rate of overflow of the component in the adsorbed state is $6G\Gamma/d$.

If there is no bubble coalescence in the foam, the concentration in the interstitial liquid will be essentially that in the liquid pool, C_w. Thus:

$$C_Q Q = C_w Q + 6G \Gamma/d \tag{34-8}$$

where Q is the volumetric rate of foam overflow on a gas-free (collapsed) basis, and C_Q is the concentration in the collapsed foam. Upon rearrangement:

$$\Gamma = \frac{(C_Q - C_w)Qd}{6G} \tag{34-9}$$

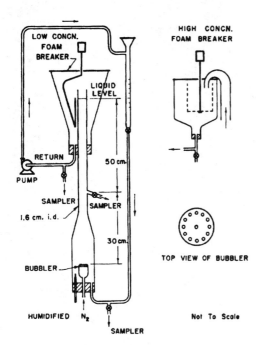

Figure 34-7. Foam fractionator to evaluate reaction parameters.
(Courtesy - Brunner, C.A., and Lemlich, R., *Ind. Eng. Chem. Fundamen.*, **2** (4), 297, 1963)

Thus, Γ can be found from measurements of Q, d, G, C_Q, and C_w. Figure 34-7 shows an apparatus developed by Brunner and Lemlich for making such measurements. The collapsed foam is returned to the bottom in order to maintain steady-state conditions. Prehumidified nitrogen is used in order to reduce any evaporative effects. A low height of foam is employed to reduce foam residence time and minimize coalescence within the foam. Bubble diameters are measured either optically or photographically through the glass walls of the column.

The following examples will help illustrate the magnitude of the variables commonly encountered and how to determine Γ.

EXAMPLE CALCULATION

Estimate the surface concentration of the surfactant and of the colligend for the following conditions if coalescence is negligible.

A continuous foam fractionation column is operating in the simple mode with 4 cm^3/sec of prehumidified nitrogen. Collapsed foam, which is collected overhead at the rate of 0.2 cm^3/sec, contains 5×10^{-7} g-mole/cm^3 of surfactant and 5×10^{-12} g-mole/cm^3 of a colligend counter-ion. The average bubble size is 0.1 cm. Liquid withdrawn from the pool contains 2×10^{-7} g-mole/cm^3 of surfactant and 2×10^{-13} g-mole/cm^3 of the colligend.

ANSWER

Part 1: Applying equation 34-9 to the surfactant,

$$\Gamma = \frac{[(5 \times 10^{-7}) - (2 \times 10^{-7})] \times 0.2 \times 0.1}{6 \times 4} = 2.5 \times 10^{-10} \ g\text{-}mole/cm^2$$

Part 2: Applying equation 34-9 to the colligend,

$$\Gamma = \frac{[(5 \times 10^{-12}) - (2 \times 10^{-13})] \times 0.2 \times 0.1}{6 \times 4} = 4 \times 10^{-15} \ g\text{-}mole/cm^2$$

EXAMPLE CALCULATION

For the liquid of the previous problem with the same concentration of colligend, if the critical micelle concentration of the surfactant is 4×10^{-7} g-moles/cm^3, estimate the equilibrium surface concentration of adsorbed colligend when the concentration of the surfactant in the liquid is 6×10^{-7} g-mole/cm^3. Assume that the surfactant molecules constituting the micelles are equally effective in adsorbing colligend as the surfactant molecules that constitute the surface monolayer.

ANSWER

Step 1: For this system, $E = 1$. $K = \Gamma_1/C_1 = (4 \times 10^{-15})/(2 \times 10^{-13}) = 0.02$ cm.

Step 2: Substituting into equation 34-6:

$$\Gamma_2 = \frac{4 \times 10^{-15}}{1 + \frac{[(6 \times 10^{-7}) - (4 \times 10^{-7})] \times 0.02}{2.5 \times 10^{-10}}} = 2.35 \times 10^{-16} \ g\text{-}mole/cm^2$$

TECHNIQUES

Foamers

Use a material of opposite charge to the sample to make a good foam. Use about 10^{-4} M.

Ethoquad	+ charged	
Duponol	− charged	A mixture of the sodium salts of sulfated fatty alcohols made by reducing the mixed fatty acids of coconut oil or of cottonseed oil or of fish oil.
Tergitol	0 charged	Polyoxyethylene glycol fatty alcohol ethers, R(OCH$_2$CH$_2$)$_n$OH, where R is a long chain alkyl group or mixture of alkyl groups.

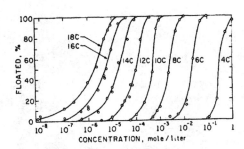

Figure 34-8. The effect of hydrocarbon chain length on the flotation of quartz in alkylammonium acetate solutions.
(Courtesy - Somasundaran, P., *Sep. Sci.*, Technol., **10** (1), 93, 1975)

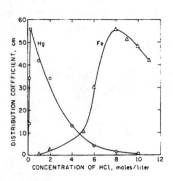

Figure 34-9. Distribution coefficients for Hg and Fe in the presence of a cationic surfactant.
(Courtesy - Karger, P.L., Poncha, R.P., and Miller, M.W., *Anal. Lett.*, **1**, 437, 1968)

Defoamers

Quaternary amines, benzene, and silicones are all popular. Only a small amount is required.

Chain Length of the Surfactant

As the chain length of the nonpolar end of the surfactant increases, its adsorption at the gas-liquid interface and, therefore, its separation increases. Figure 34-8 shows this effect. Figure 34-8 is actually a flotation system, but the results are the same.

Surfactant Concentration

Figure 34-8 also shows that the concentration of the surfactant is important and *up to a limit, the separation increases as the concentration of the surfactant increases.* The concentration of the surfactant also changes the bubble size, with the size getting smaller as the surfactant increases. This makes for a creamier foam.

Effect of pH

pH is quite important because it can alter the ionic form of the species present. Figure 34-9 shows how the distribution coefficients for Hg and Fe changes with pH.

Sintered-Glass Pore Size

Figure 34-10 shows that pore size affects the enrichment ratio for the removal of surfactants from a water system. Figure 34-11, p. 401, shows an apparatus for small scale laboratory separations.

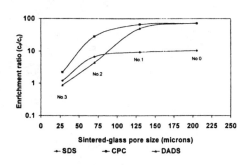

Figure 34-10. Effect of sintered-glass porosity on enrichment ratio. SDS, sodium docecyl sulfate; CPC, cetylpyridinium chloride; DADS, sodium n-hexyl diphenyloxide disulfonate.
(Courtesy - Tharapiwattananon, N., Scamehorn, J.F., Osuwan, S., Harwell, J.H., and Haller, K.J., *Sepn. Sci. & Tech.*, **31** (9), 1233, 1996).

GAS-ASSISTED FLOTATION
PRINCIPLES

The separation of desired particles from a heterogeneous mixture of particles dispersed in a liquid is called *flotation*, not floatation. A gas is bubbled through the mixture and collects selectively on the desired particles, thereby lowering their density and floating them to the top where they can be skimmed away. This is known as gas assisted flotation, froth flotation,

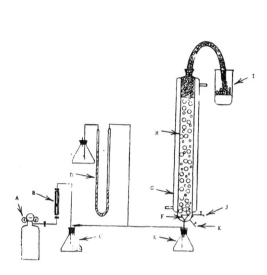

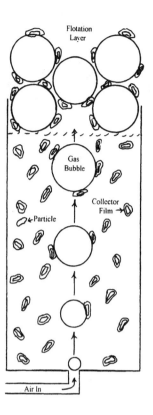

Figure 34-11. Schematic diagram of the foam fractionation apparatus. (A) air cylinder; (B) gas rotameter; (C) gas saturator (H_2O); (D) Hg U tube; (E) liquid trap; (F) fritted glass gas dispenser; (G) constant temperature water jacket; (H) foam fractionating column; (I) foam receiver; (J) tap for withdrawing foamed solution; (K) tap for draining foam fractionating column. (Courtesy - Skomoroski, R.M., *J. Chem. Ed.*, **40** (9), 470, 1963)

Figure 34-12. Diagram of the basic flotation process.

benefication, or simply flotation. In the next chapter, the use of liquids to alter the density of particles will be discussed.

Although this is a process that can be used in the laboratory, since the 1920's, it has become of extreme importance in large-scale industrial applications, mainly to concentrate ores and the recovery of industrial grade diamonds. For example, the diamonds in a ton of ore can be separated in 10 minutes. Figure 34-12, shows a diagram of the basic process and Figure 34-13, shows a photomicrograph of the separation of galena (PbS) particles. This is not as easy as it appears. Chemists must determine how to get the gas bubbles to stick to the desired surface.

Overview For Minerals
(Hercules Inc., Wilmington, DE)

"Most metal-bearing minerals do not occur in massive formations, but are found for the most part as small particles embedded in a matrix of more or less worthless rock, clay, sand, or other matter. The whole mass is known as *ore*, and with most ores it is necessary to separate at least a part of the worthless material, or 'gangue,' from the desired mineral before smelting. This process is known as *ore dressing* or *concentration* and can be done in several ways, one of which is the flotation process. This method for concentrating ores has been intensely

Figure 34-13. Photomicrograph of the bubble-particle attachment in flotation. (Courtesy - H. Rush Spedden, in Berg, E.W., *Physical and Chemical Methods of Separation* by E.W. Berg, McGraw-Hill, New York, NY, 1963)

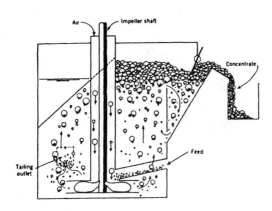

Figure 34-14. A froth flotation cell.
(Courtesy - R.E. Kirk and D.F. Othmer, eds., *Encyclopedia of Chemical Technology*, J. Wiley & Sons, New York, NY, 1980)

developed to the point where the processing costs of many ores have been lowered considerably.

"To concentrate by flotation, it is necessary first to reduce the whole ore to a size sufficiently fine so that the individual particles of mineral are detached substantially from the gangue and are small enough to be floated by attached air bubbles. Since grinding is expensive, as coarse a grind as possible is used. The crushed ore is fed to a flotation *cell*, where it is mixed with appropriate flotation agents in an aqueous phase. This mixture is referred to as the *pulp*. A flotation cell (3 to 30 m^3) is a tank fitted with a stirrer and an air-blowing device [Figure 34-14] that can keep the pulp vigorously stirred and can form a supernatant foam or froth of air bubbles. Within a few minutes after starting the cell, a mineral-laden foam is formed and is skimmed off. Usually, the desired mineral is taken off in the foam; the gangue is left behind with the *tailings* in the cell.

"A mineral particle will attach itself to an air bubble and float when its surface is hydrophobic, or water-repellent; and conversely, a particle with a hydrophilic, or water-attractive, surface will not become attached to an air bubble. Practically all minerals as they occur are hydrophilic. It is the function of one class of flotation agents, called promoters or collectors, to convert the surface of the mineral to be floated from a hydrophilic to a hydrophobic state. At the same time, the collector must not alter the hydrophilic surface of the gangue, which is to stay behind. One theory as to how this is done is that the collector is selectively adsorbed and oriented on the mineral surface to form a hydrophobic film.

"The action of a collector may be altered by the presence of secondary additives known as modifiers or conditioners. One that promotes collection is called an *activator*, and one that inhibits it is called a *depressor* or depressant. The pH of the pulp can be an important factor, and acids or alkalies are often added to bring the pH to the best level for the particular collector being used. Collectors and conditioners together can be regarded as one class of flotation agents whose function is to modify a mineral surface so that it will attach itself to an air bubble.

"The second class of flotation reagents is the *frothing agents*. To float minerals, foam must have the correct characteristics. It must be stable enough to resist disruption by the circulating mineral particles and yet brittle enough to break down after removal from the flotation cell. It should not be too copious, and the bubbles should be of the right size to match both the chemical nature and the particle size of the mineral being floated. The most commonly used substances are pine oil, certain aliphatic alcohols, polypropylene glycol ethers, and cresylic acids. These frothers are effective in extremely low concentrations that are well within their solubility range. A typical pulp may contain 2 tons of water, 1 ton of ore, and 0.1 lb of frothing agent.

"Some of the minerals concentrated by flotation are: hematite (Fe_2O_3), phosphate rock ($Ca_3(PO_4)_2$), barite ($BaSO_4$), fluorspar (CaF_2), kyanite (Al_2O_3), ilmenite (TiO_2), scheelite ($CaWO_4$), limestone ($CaCO_3$) and spodumene (Li_2O-Al_2O_3-$4SiO_2$)."

Theory

According to E.W. Berg (*Physical and Chemical Methods of Separation*, McGraw-Hill, New York, NY, 1963) "It is possible to show through basic energy requirements that a gas bubble will spontaneously adhere to a solid if its contact angle is finite [Figure 34-15, p. 403]. Consider, for example, a gas bubble and a solid suspended separately in a liquid medium [Figure 34-16, p. 403]. If the gas bubble spontaneously adheres to the solid, then there must be a loss in energy accompanying the process. The two-dimensional interfaces that previously existed between the gas-liquid and liquid-solid phases have been replaced by a one-dimensional line of contact where the three phases coalesce. The energy change can be evaluated from the change in interfacial surface area and the three interfacial energies. If the measured contact angle is introduced, the unknown solid-liquid and solid-gas interfacial energies are removed. The energy change then can be evaluated in terms of the contact angle, the surface energy or tension of the liquid, and the area of interface involved. For the special case where the bubble is small and the solid particle is large, the surface of the bubble can be considered planar, since the interface is confined to a surface of the particle.

"The decrease in free energy that occurred when the bubble and the particle coalesced is the sum of the two interfacial energies less the sum of the terminal surface energies. The adherence of the bubble with radius R to the solid

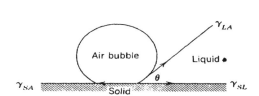

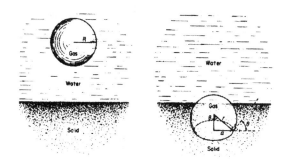

Figure 34-15. Contact angle formed by an air bubble on a solid submerged in water.
(Courtesy - R.E. Kirk and D.F. Othmer, eds., *Encyclopedia of Chemical Technology*, J. Wiley & Sons, New York, NY, 1980)

Figure 34-16. Illustrating interface terms.
(Courtesy - Berg, E.W., *Physical and Chemical Methods of Separation*, McGraw-Hill, New York, NY, 1963)

surface is shown in [Figure 34-15]. Let a be the radius of the one-dimensional three-phase interface inscribed on the surface of the solid and r equal the radius of the adhering bubble. The height and area of the spherical portion remaining in contact with the liquid phase are $r(1 + \cos \theta)$ and $4 \pi r(1 + \cos \theta)/2$, respectively. The change in free energy or work of adhesion will be equal to:

$$W = (A\gamma_{SL} + 4\pi R^2 \gamma_L) - \left[(A - \pi a^2)\gamma_{SL} + \pi a^2 \gamma_S + 4\pi R^2 \frac{1 + \cos \theta}{2} \gamma_L \right] \tag{34-10}$$

where A is the total area of the solid surface and γ_{SL}, γ_L and γ_S are the surface energies (tensions) of the solid-liquid, liquid-gas, and solid-gas interfaces, respectively. The terms in A cancel, and the equation can be reduced to:

$$W = 4\pi R^2 \gamma_L + (\pi a^2 \gamma_{SL} - \pi a^2 \gamma_S) - 4\pi r^2 \frac{1 + \cos \theta}{2} \gamma_L \tag{34-11}$$

$$W = \pi a^2 (\gamma_{SL} - \gamma_S) + \left(4\pi R^2 - 4\pi r^2 \frac{1 + \cos \theta}{2} \right) \gamma_L \tag{34-12}$$

By substituting in $\gamma_{SL} - \gamma_S = -\gamma_L \cos \theta$ and $a = r \sin \gamma$, the work of adhesion becomes:

$$W = \pi r^2 \sin^2 \theta \cos \theta \gamma_L + \left(4\pi R^2 - 4\pi r^2 \frac{1 + \cos \theta}{2} \right) \gamma_L \tag{34-13}$$

Equation 34-13 can be simplified by letting $r = fR$. Then,

$$W = -\pi f^2 R^2 (1 - \cos^2 \theta) \cos \theta \gamma_L + \left[4\pi R^2 \gamma_L - 4\pi f^2 R^2 \frac{(1 + \cos \theta)}{2} \gamma_L \right] \tag{34-14}$$

$$W = 4\pi R^2 \gamma_L \left[\frac{-f^2 (1 - \cos^2 \theta) \cos \theta}{4} + 1 - \frac{f^2 (1 + \cos \theta)}{2} \right] \tag{34-15}$$

$$W = 4\pi R^2 \gamma_L \left[\frac{f^2(-\cos\theta + \cos^2\theta - 2 - 2\cos\theta)}{4} + 1 \right] \tag{34-16}$$

$$W = 4\pi R^2 \gamma_L \left[1 - f^2 \frac{2 + 3\cos\theta - \cos^3\theta}{4} \right] \tag{34-17}$$

"If it is assumed that there is no volume change in the gas bubble when it adheres to the solid, then the parameter f can be evaluated. The volume of a spherical segment with height $h = r(1 + \cos\theta)$ and radius r is $1/3\, \pi h^2(3r - h)$. From this it follows that the volume of the independent bubble is equal to:

$$4/3\,\pi R^3 = 1/3\,\pi R^2(1 + \cos\theta)^2[3r - r(1 + \cos\theta)] \tag{34-18}$$

"The parameter f is evaluated by substituting fR for r in equation [34-18] and solving for f.

$$4/3\,\pi R^3 = 1/3\,\pi f^2 R^2(1 + \cos\theta)^2[3fR - fR(1 + \cos\theta)] \tag{34-19}$$

$$4/3\,\pi R^3 = (\pi f^3 R^3 - 1/3\,\pi f^3 R^3 - 1/3\,\pi f^3 R^3 \cos\theta)(1 + 2\cos\theta + \cos^2\theta) \tag{34-20}$$

$$4/3\,\pi R^3 = \pi f^3 R^3 (1 - 1/3 - 1/3\cos\theta)(1 + 2\cos\theta + \cos^2\theta) \tag{34-21}$$

$$f^3 = \frac{4\pi R^3}{3\pi R^3 (2/3 - 1/3\cos\theta)(1 + 2\cos\theta + \cos^2\theta)} \tag{34-22}$$

$$f^3 = \frac{4}{2 + 3\cos\theta - \cos^3\theta} \tag{34-23}$$

$$f = \left(\frac{4}{2 + 3\cos\theta - \cos^3\theta} \right)^{1/3} \tag{34-24}$$

Substitute f into equation [34-17] and let $F = 1/f$:

$$W = 4\pi R^2 \gamma_L \left(1 - f^2 \frac{1}{f^2} \right); \quad = 4\pi R^2 \gamma_L (1 - F) \tag{34-25}$$

"The work of adhesion can thus be shown to be equal to the product of the surface of the original bubble, the specific surface energy of the liquid, and the parameter $(1 - F)$, which is a function of the contact angle θ only. The parameter F is positive and less than unity for all values of θ between 0 and 180°, so the energy change is in a direction favoring the spontaneous adhesion of the bubble to the particle. In other words, a particle can be stabilized in a gas-liquid interface if the contact angle is finite. In actual practice, adherence of bubble and particle is favored by angles not too near to zero and by sharp particle edges.

"For a large bubble and a small particle, the surface of the bubble can be considered planar. It can then be shown that the work of adhesion equals:

$$W = S\gamma_L(1 - \cos\theta) \tag{34-26}$$

where S equals the area of the interface involved and γ_L is the surface free energy of the liquid. Again, one can

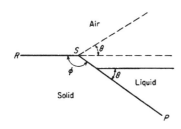

Figure 34-17. Contact angle changes at an edge.
(Courtesy - Berg, E.W., *Physical and Chemical Methods of Separation*, McGraw-Hill, New York, NY, 1963)

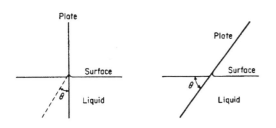

Figure 34-18. The rotating plate method for measuring contact angles.
(Courtesy - Berg, E.W., *Physical and Chemical Methods of Separation*, McGraw-Hill, New York, NY, 1963)

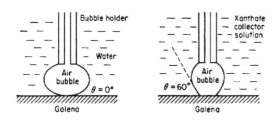

Figure 34-19. The captive-bubble method for measuring contact angle.
(Courtesy - Berg, E.W., *Physical and Chemical Methods of Separation*, McGraw-Hill, New York, NY, 1963)

see how dependent the stabilization of a particle in a gas-liquid interface is on the contact angle. The work of adhesion is real for all finite values of the contact angle, but is dependent also on the size of the particle.

"When a solid particle is stabilized in an air-liquid interface, there is a decrease in the potential energy of the system as the two independent interfaces are replaced by a single interface. Therefore, a large bubble will tend to extend spontaneously and cover the entire plane solid surface that is available. It will extend to a point of sharp change in the orientation of the solid-liquid interface. Perhaps this can be more easily seen if we consider the displacement of a gas by a liquid on a solid surface. In [Figure 34-17] the liquid is spreading over the solid surface PS with its characteristic constant angle θ. When the liquid front reaches the edge of the solid, then it will not proceed down the surface RS until its characteristic contact angle has been established. This requires that the liquid front be rotated through an angle 180° - Φ. Obviously energy is required to swing the liquid front through this angle. As a result, the interface has a tendency to become stable or static at the edge or sharp corner of the solid surface. This is precisely why a tumbler can be filled above its rim without overflowing. Overflow does not occur until the hydrostatic head builds up sufficiently to overcome the interfacial energy at the edge. This edge effect is also a contributing factor in floating a steel needle or razor blade or dense particles with small contact angles."

Contact Angle Measurement

A measurement of the contact angle will provide an indication of how successful a given system for flotation might be. *The greater the contact angle, the better the chance is for bubble adhesion.* The rotating plate method (Adam, N.K., and Jessop, G., *J. Chem. Soc.*, **127**, 1865, 1925), is shown in Figure 34-18. A clean plate of the solid is immersed in the liquid and then tilted until there is no deformity where the liquid contacts the plate. The contact angle then equals the angle of tilt.

An alternate method (Taggart, A.F., Taylor, T.C., and Ince, C.R., *Trans. AIMME*, **87**, 285, 1930) is to place a capillary just above a clean solid surface and gently produce a gas bubble at the end of the capillary. The contact angle is measured by observing the edge of the bubble with a microscope (Figure 34-19).

TECHNIQUES

Variables (Hercules Inc., Wilmington, DE)

Particle Size

The best particle size range has been found to be from 45 to 65 mesh, particularly for the sulfide ores, and up to 20 mesh for the nonmetallic minerals. If particles are too coarse, they stay in the tailings, if they are too fine, they *slime*. *The best adhesion occurs if the particles have sharp edges.*

Conditioning
This is the process of mixing the reagents before the flotation step. The crushed ore is further ground as the pulp to remove surface stains and then deslimed, usually by adding acids or alkalis. The solids content is adjusted to about 25-30%. The addition of the reagents is usually (1) pH regulators - lime, soda ash, acid; (2) activating or depressing reagents; and (3) collectors and frothers.

Modifiers
These are used if necessary to get the collector to work correctly. They include pH regulators, depressants, activators, sulfidizers, dispersants, and deflocculators.

pH Regulator
These are usually lime or soda ash on the alkaline side and sulfuric acid on the acid side.

Depressants
These are used to prevent flotation of one material while floating another. Lime and NaCN are examples. Both of these depress pyrite. Dichromate is a depressant for galena. HF is used to depress quartz in the flotation of feldspar with amine collectors. Sodium silicate is also a quartz depressant. Quebracho depresses calcite and dolomite in the flotation of fluorite with fatty acids.

Activators
If a mineral is not normally floatable, an activator can be used. $CuSO_4$ is the standard activator for sphalerite. NaS_x is the sulfidizing agent for oxidized ores.

Dispersants and Deflocculators
Premature flocculation can be an interference. Sodium silicate is used as a dispersant. Starch, casein, and glue are used to disperse both gangue and carbonaceous matter. They also serve as protective colloids to prevent reflocculation, such as in the flotation of gold.

Collectors
Few minerals will float naturally by dispersing a gas through finely divided particles, because polyfunctional surface ions make their surfaces hydrophilic. Collectors are compounds that adhere to the particle's surface and make it hydrophobic. They are selected to be charged opposite to the particle's surface charge. They must adsorb strongly enough to displace water from the surface of the particle. Chemisorption is the preferred mechanism, because it is thought to provide more selectivity. In some cases, an otherwise good collector will not adhere, so an activator is added first to condition the particle's surface so the collector will adhere.

Anionic Collectors
Collectors are used to enhance the attachment of the particle to be floated to the gas bubble. These are usually the unsaturated 18-carbon fatty acids, oleic and linoleic acids; obtained from the distillation of crude tall oil and used in a mixture of about equal amounts. Hercules Trade Names for combinations of these two are Pamak and Pamolyn.

$$H_3C(CH_2)_7CH=CH(CH_2)_7C\overset{O}{\underset{OH}{\diagdown}}$$
Oleic Acid

$$H_3C(CH_2)_5CH=CH-CH=CH(CH_2)_7C\overset{O}{\underset{OH}{\diagdown}}$$
Linoleic Acid

The carboxyl group has the negative charge and is the polar group. These acids are used for metallic and nonmetallic minerals usually in neutral or alkaline systems.

Cationic Collectors
Hercules cationic collectors include a rosin-derived primary amine and its acetic acid salt. Dehydroabietylamine is the primary amine, with the $-NH_2$ group being the polar group, and is positively charged in solution with acetic acid, the addition of which makes it water soluble. They are soluble in kerosene, fuel oil, and other frothers.

Dehydroabietylamine

Frothers

Frothers are heteropoly organic compounds having one or more water-repellent, nonpolar, hydrocarbon groups attached to a hydrophilic, or water attracting, polar group. They collect at the air-water surface and lower the surface tension of the water phase of the pulp. The most widely used frothers are pine oil, aliphatic alcohols, polypropylene glycol, alkyl ethers of polypropylene glycol, and cresylic acids. The preferred alcohols contain 5 to 8 C atoms, such as mixed amyl alcohols, methylisobutylcarbinol, and certain heptanols and octanols.

Modern industry is serious about flotation, with over 400,000,000 tons of ore processed yearly in nearly 300 plants. Figure 34-20 shows a flow sheet to separate copper and zinc from an ore.

Alpha-Terpineol

Quantities

Table 34-1 shows typical quantities of flotation agents applied per ton of ore.

Table 34-1. Pounds of reagent per ton of ore
(Courtesy - Hercules Inc., Wilmington, DE)

Reagent	lbs/Ton
Frothers	0.01 to 0.050
Froth modifiers	
(hydrocarbon oils)	0.50 to 5.00
Collectors	0.02 to 2.00
Modifiers	
pH regulators	0.02 to 5.00
Depressants	0.02 to 1.00
Activators	0.50 to 4.00

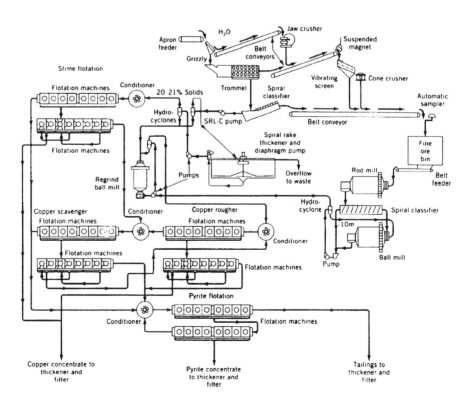

Figure 34-20. Complex copper ore concentration flow sheet.
(Courtesy - Denver Equipment Co., Denver, CO)

Figure 34-21. A Kilborn flask setup.

Applications

Some applications listed in the *Encyclopedia of Chemical Technology* by Kirk-Othmer in addition to minerals include removal of inks, pigments, and coatings from paper; the flotation of sewage and industrial waste; purification of water; removal of oil from refining waste; concentration of naphthalene from coke oven gas cooling tower waste; flotation of ions and precipitates; separation of bitumen from tar sands; flotation removal of seeds, pods, and other impurities from peas; the separation of the oilrich fraction of seeds; removal of wax from wool; flotation of dyes; flotation of microorganisms; and separation of textile fibers from process streams.

LIQUID-ASSISTED FLOTATION

PRINCIPLES

Liquid-assisted flotation is a liquid-liquid process by which solids are selectively coated with one of the liquids and are separated from the other components by becoming dispersed in that liquid. It includes two basic processes: (1) floating the desired material and (2) floating the undesired material.

An example of the first technique is to detect insect fragments in cornmeal or wheat flour by coating the fragments with mineral oil and floating them on top of a 6 M HCl solution. The same process can be used to separate mites, aphids, and thrips from fruits and vegetables. An example of the second type is to coat tomato pulp with petroleum ether and float it away from fly eggs and maggot segments to determine insect contamination. All three of these examples will be described in Appendix A, Experiments 48 to 50 using methods of the Association of Official Analytical Chemists, International. It is recommended that you at least read through those experiments to see how the FDA examines the food you eat for filth.

TECHNIQUES

Only two pieces of special apparatus are required: a Kilborn flask and a Wildman trap.

Kilborn Flask

A Kilborn flask is a cylinder of glass about 20 cm long and 7 to 8 cm in diameter drawn down to a single opening about 8 mm in diameter and 5 cm long at the bottom end. Figure 34-21 shows a flask and how it is arranged. The nature of the sample is the reason for this design. Small insects and particularly insect fragments are floated in hot mineral oil. These fragments must be filtered from the oil so they can be examined under a microscope. This means the oil layer must be separated from the aqueous layer, which usually contains 45 grams of grain particles. A large flask with a large opening at the bottom is called for. It has to act like a separatory funnel without a typical stopcock, whose bore is too small and becomes plugged.

Figure 34-22. Kilborn flask modified by the author.

A problem with the Kilborn flask is that the opening at the bottom is small and often plugs. This can be remedied by squeezing on the rubber tube. However, when the tube opens, rapid drainage from the flask occurs that is hard to stop quickly, and some of the oil can be lost. A technique developed by the author for the FDA laboratories in Lenexa, KS is shown in Figure 34-22. The top of a 5 pint acid bottle has been cut off and turned upside down. A rubber stopper fitted with a long glass rod is used as the plug. This is easy to remove, and the outflow rate is equally easy to control.

Wildman Trap Flask

Figure 34-23, p. 409, shows a Wildman trap. Figure 34-24, p. 409, shows it used as a stirrer in the initial part of the separation (left) and in place to separate the oil layer (right). The trap is a rubber disk 5 cm in diameter and 2 mm thick placed on the end of a 3 mm diameter brass rod about 30 cm long. It is needed to separate the mineral oil containing insects from fruits and vegetables. The density of the fruits and vegetables is very close to that of the oil, and the chopped up foods are too bulky to pass through a Kilborn flask. The trap is pushed through the neck of a 1 L Erlenmeyer flask, and then the chopped food, 6 M HCl, and mineral oil are added. The mixture is brought to a rolling

Figure 34-23. A Wildman trap.

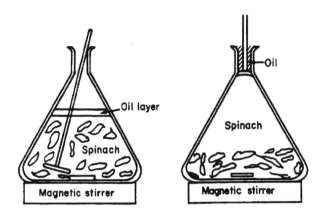

Figure 34-24. Left, a Wildman trap as a stirrer. Right, locked in place.

boil, then cooled by adding more of the HCl solution until the bottom of the oil is just to the bottom of the neck of the flask. The trap is now raised with a gentle swirling action to be just under the oil layer and on top of the food particles. It then is pulled up just into the bottom of the neck of the flask so it sticks there when the flask is tilted. If it is pulled too hard, the sample is lost. If it is not pulled enough, the oil goes back into the flask. It takes a bit of practice, which should be done on a blank system.

REVIEW QUESTIONS

1. What does the term "adsubble methods" mean?
2. What is the enrichment factor?
3. For what types of applications is foam fractionation most useful?
4. What is a foam?
5. What is the purpose of the surfactant?
6. What causes small bubbles to disappear and large bubbles to form in a foam?
7. What is the shape of most bubbles in a foam?
8. Why do you not want micelle formation in foam fractionation?
9. Estimate the surface concentration for the surfactant and for the colligend for the following conditions if coalescence is negligible. A continuous foam fractionation column is operating in the simple mode with 4 cm^3/sec of prehumidified nitrogen. Collapsed foam, which is collected overhead at the rate of 0.22 cm^3/sec, contains 4.7 $\times$ 10^{-7} g-mole/cm_3 of surfactant and 4.7 $\times$ 10^{-12} g-mole/cm^3 of a colligend counter-ion. The average bubble size is 0.12 cm. Liquid withdrawn from the pool contains 1.8 $\times$ 10^{-7} g-mole/cm^3 of surfactant and 1.8 $\times$ 10^{-13} g-mole/cm^3 of the colligend.
10. For the liquid of the previous problem with the same concentration of colligend, if the critical micelle concentration of the surfactant is 3.5 $\times$ 10^{-7} g-moles/cm^3, estimate the equilibrium surface concentration of adsorbed colligend when the concentration of the surfactant in the liquid is 5 $\times$ 10^{-7} g-mole/cm^3. Assume that the surfactant molecules constituting the micelles are equally effective in adsorbing colligend as the surfactant molecules that constitute the surface monolayer.
11. What is a recommended concentration of foamer for foam fractionation?
12. Of what class of compounds is Tergitol a member?
13. What are some classes of compounds that make good defoamers?
14. What is likely to happen to the efficiency of a foam fractionation separation as the concentration of surfactant is increased?
15. What does the term "flotation" mean?
16. What is required for a mineral particle to attach itself to an air bubble?
17. What is the purpose of promoters or collectors?
18. What is an activator?

19. What are some typical frothing agents?
20. In practice, what conditions favor the adherence of a gas bubble to a particle?
21. Why will a gas bubble stop expanding along a surface when it reaches an edge of the particle?
22. How does the rotating plate method for determining contact angles work?
23. About how many lbs of a collector are used per ton of ore?
24. What is the best particle size range?
25. How is a particle deslimed?
26. What is a common pH regulator on the acid side?
27. What is the purpose of a dispersant?
28. What is the composition of Pamak, an anion collector manufactured by Hercules?
29. What is the purpose of adding acetic acid to dehydroabiettylamine?
30. What is the basic composition of a frother?
31. Why is the Kilborn flask designed as it is?
32. What is a Wildman trap?
33. What is the purpose of a Wildman trap?

IF IT EXISTS - CHEMISTRY IS INVOLVED
IF YOU CAN BUY IT - A CHEMIST WAS INVOLVED SOMEWHERE

SEPARATIONS INVOLVING MEMBRANES

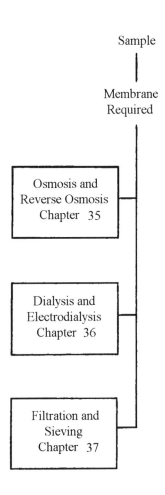

Almost everyone has filtered a mixture, even if it is only sifting sand in a sandbox. The piece of screen wire is not what is normally considered to be a membrane, but it serves to convey the concept. Membranes are usually considered to be thin, pliable, semipermeable materials. They may be of either natural or synthetic origin, and the pores can be so small that only water molecules or small ions can pass through. Four very important separation techniques involving membranes are *osmosis* and *reverse osmosis*, discussed in Chapter 35, and *dialysis* and *electrodialysis*, discussed in Chapter 36. Osmosis, concerned with solvents, and dialysis, concerned with solutes, have as their main driving force the Second Law of Thermodynamics. Molecules or ions at higher concentration move across the membrane to a region of lower concentration. This rate can be increased by the application of an electrical field, *electrodialysis*. If an external force of sufficient strength is applied on the low concentration side, then *reverse osmosis* can occur, a system used to desalt water.

The development of synthetic membranes that can withstand high pressures has permitted the rapid filtration of even bacteria from a solution. Filtration is usually not considered to employ the Second Law, but gravity or higher pressures. These membranes are discussed in Chapter 37. However, once the topic of filtration is described, then related uses of it in addition to membrane filtration seem reasonable to mention. Therefore, not only regular filtering techniques but sieving techniques are described in this chapter.

35
OSMOSIS AND REVERSE OSMOSIS

PRINCIPLES

In 1748 Abbe Nollet (*Lecons de Physique-Experimentale,* Hippolyte-Louis Guerin, Paris) did an experiment that was supposed to answer if digested food passed through a membrane. He placed red wine in a bladder, placed the bladder in a cylinder filled with water, and waited for the red dyes to come through. Instead, the bladder swelled and in some cases even burst, and there was no red dye in the outside water. Today we explain this by the *Second Law of Thermodynamics* - systems tend toward maximum disorder. Another way to state this with solutions is that *high concentration (HC) goes to low concentration (LC)*. Consider the water concentration on each side of the membrane. The water is 100% water or high concentration water. Wine is 12.5% ethanol by volume (plus the other wine components) or 87.5% water. This is then low concentration water by comparison, and water from the high concentration will pass through the membrane until the concentration is equal on each side or until it is stopped by *osmotic pressure*. The high concentration water has no impurities so it is more ordered. The water molecules in the wine are mixed with ethanol molecules plus all of the other wine components and, therefore, are disordered. To make the overall system as disordered as possible, water must move into the wine. This process, when it involves the *solvent*, is now called *osmosis* (Gr. *osmos* - "push"). When *solute* molecules are involved, it is called *dialysis*.

The important point to remember when examining an osmotic process is, *high concentration goes to low concentration*. Some examples of the effects of osmosis are as follows. The skin on ripe tomatoes on the vine will sometimes split after a rain. The rain water (HC) passes through the membrane to the inside (LC), and the cells get so large that they burst. A swollen ankle is soaked in a bucket of Epsom's salt ($MgSO_4$) to reduce the swelling. Water from the swollen cells in the ankle (HC water) pass out into the $MgSO_4$ water (LC water) to reduce the swelling. Meat shrinks during cooking when salt is added. The salt dissolves in the surface water and forms LC water. The water inside of the meat is HC water by comparison, so it is drawn out of the meat, making the meat tougher. This is why salt should not be added until the end of the cooking. Meat, particularly hams, are salted for storage. If any bacteria land on the surface, they are dehydrated and die. Their cells contain HC water in comparison to the thin film of LC water on the surface of the ham, so water passes out of the bacteria. Butter has salt added for the same reason. The same process takes place with candy. Did you ever see mold grow on a piece of hard candy? Plants will not grow in salty soil. The HC water in the roots passes out into the LC water in the soil, and the plant dies. Freshwater fish use their kidneys to pump out excess salt whereas saltwater fish *gulp water* and excrete excess salts. See if you can use osmosis to explain why a fresh glazed doughnut will become sticky on the outside yet be dry on the inside when you eat it? Can you now explain how water gets to the top of trees, and why trees are not 1000 ft tall?

THEORY

The first quantitative experiments are believed to have been done by E. Pfeffer and J. DeVries (*Z. Physik. Chem.* **3**, 103, 1889) and J.H. van't Hoff (*Z. Phys. Chem.*, **1**, 481, 1887). They did work with water-sucrose solutions and observed that a film of Cu(II) ferrocyanide on earthen pots was permeable to water, but not sucrose. The following relationship was developed (Suess, M. J., *J.Chem. Ed.*, **48** (3), 190, 1971).

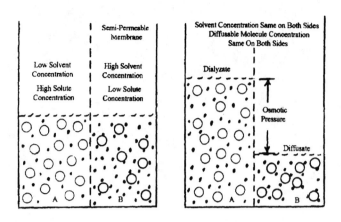

Figure 35-1. Diagram of the principle of osmosis. Left: before osmosis; right: after osmosis.

Figure 35-1 illustrates the process of osmosis. The process of osmosis takes place "due to a difference in the thermodynamics potential - the Gibbs free energy of the pure solvent being higher than that of the solution. Because of the tendency of any system to reach equilibrium, a state at which the free energy of both the free solvent and the solution are equal, the initial difference in the free energy of the system can be considered as the driving force of the above described process called osmosis.

"The volume of the solvent, which is transferred through the membrane, is capable of performing work against a pressure, π, denoted the osmotic pressure. If the Gibbs free energy is defined as

$$G = H - TS \tag{35-1}$$

and the enthalpy is defined as

$$H = E + PV \tag{35-2}$$

then

$$G = E + PV - TS \tag{35-3}$$

For the change in free energy, equation [35-3] becomes

$$\Delta G = \Delta E + P\Delta V + V\Delta P - T\Delta S - S\Delta T \tag{35-4}$$

"For a reversible process, and osmosis is one, $\Delta q = T\Delta S$. If all of the work done is expansion, then $\Delta w = P\Delta V$. From the first law of thermodynamics it is recalled that $E = q - w$ and $\Delta E - \Delta q + \Delta w = 0$. Thus equation [35-4] can be rewritten in the form:

$$\Delta G = V\Delta P - S\Delta T \tag{35-5}$$

For a process taking place in isothermal conditions $S\Delta T = 0$, and the final equation of reversible and isothermal work becomes simply $\Delta G = V\Delta P$, or for small incremental changes

$$\Delta G = V\Delta P \tag{35-6}$$

"At equilibrium, the difference in pressure ΔP is the pressure difference between solution (P_2) and solvent (P_1). Thus

$$\Delta P_{eq} = P_2 - P_1 = \pi \tag{35-7}$$

"Remembering that the equation describing the free energy change of a dilute solution is $\Delta G = RT \ln(X_A)$, X_A being the mole fraction of the solvent in the solution, then for an equilibrium state to be established, the increase in free energy of the solution (change of concentration of solute) must be balanced by the increase in pressure; and

$$\pi V + RT \ln(X_A) = 0 \tag{35-8}$$

"If n_B is the number of moles of solute in solution and V_s is the volume of the solution:

$$\pi V_s = RT n_B \tag{35-9}$$

"This has the same form as the ideal gas law and R, determined experimentally to be 0.083, is the same as the gas law R of 0.0821. However, osmotic pressure is not believed to be due to random collisions with a surface as is gas pressure."

Concentrations in osmotic pressure calculations, like all other colligative property calculations, are molal, not molar.

If 1.0 mole of a component is dissolved in 22.4 L of solvent it would exert an osmotic pressure, π, of 1.0 atmosphere (760 mm Hg) at 0 °C. To determine R so P is in cm of water:

$$R = 76 \text{ cm} \times 13.6 \text{ g/cm}^3 \times 22.4 \text{ L/mol}/273 \text{ K} = 84.8 \text{ cm/mol/deg}$$

At 25 °C, 1.0 mol/L of solution has an osmotic pressure of 84.8 cm/mol/deg $\times$ 1.0 L/mole $\times$ 298 K = 2.5×10^4 cm of water. What this means is that a 1.0 M solution exerts about 25 atmospheres of pressure! Fortunately, few systems have a 1 M solution on one side of a membrane and 0.0 M on the other. However, you know from experience that osmotic pressure can be quite large. How many times have you seen a concrete sidewalk or driveway raised up and cracked by tree roots? In the days of wooden ships if cargoes of dried beans got wet, they were thrown overboard at once. The wet beans would swell so much and the force would be so great that it could push apart the hull of the ship.

EXAMPLE CALCULATION
What would be the osmotic pressure of a solution containing 1.50 g of a polymer with a molecular weight of 18,000 dissolved in 100 mL of solution at 25 °C? Could this pressure be measured realistically with an Hg manometer?

ANSWER

$$\pi = 82.1 \times 298 \times \frac{1.50}{18,000 \times 100} = 0.0203 \text{ atm or } 15.5 \text{ mm Hg}$$

This is easily read with a Hg manometer.

REVERSE OSMOSIS

Refer to Figure 35-1, p. 414. Suppose you apply a pressure on the left side to force the solvent back across the membrane to the other side. This is called *reverse osmosis* and, as a separation technique, has many commercial applications. The most notable is the desalinization of sea water. The removal of dissolved salts in hard water to make it soft water is done extensively. Many chemistry laboratories use reverse osmosis to obtain "distilled water". It is used to treat industrial waste water and to concentrate fruit juices.

About 3,500,000 people in the United States have water with > 1,000 ppm salts. Good water is usually considered to have from 200 to 300 ppm, with 500 ppm being the upper limit. Calcium and magnesium salts in the water tend to hold cells together, and beans cooked in hard water can be too hard to eat. A city in Texas had water with salts between 1700 and 2100 ppm, and people had to haul in water for nearly all uses. A reverse osmosis plant was set up using 16 bundles of membranes, 14 in. in diameter, and 7 ft long (85,000 ft²/bundle). The unit produced 100,000 gallons per day of 300 to 400 ppm water, 46,000 gallons of 5,500 ppm water, plus 1500 lb of $MgSO_4$, 300 lb of $NaHCO_3$, 450 lb of $CaSO_4$ and 150 lb of NaCl. The town was able to survive.

Referring to the equations above, "the reversible isothermic work for the reverse osmosis process can be described by

$$-W = (P_2 - P_1)V_s = RTn_B \tag{35-10}$$

The negative sign refers to work done on the system to cause the process to take place, and to transfer a unit volume of the solvent from the solution to the pure solvent. For such a process, $(P_2 - P_1)$ is greater than π. If work overcomes only the osmotic pressure of the solution, $(P_2 - P_1)$ would equal π, and

$$-W = \pi V_s = RTn_B \tag{35-11}$$

This is the work required to keep the system continuously at its initial volumetric balance against the osmotic forces.

"However, to achieve the transfer of solvent from the solution to the pure solvent, that is, a reverse-osmosis process

$$|-W'| > |-W| \tag{35-12}$$

and

where $\Delta P' \neq \Delta P$, and stands for any pressure in addition to π. The removal of the solvent from the solution will change the concentration of the solution, but not the number of moles of solute, n_B, in solution. However, the volume, V_s, of the solution will change. To fulfill the condition of reversibility, the solvent removed must at all times be in equilibrium with the solution, thus calling for a differential treatment of equation 35-13.

$$-\Delta W' = \pi'\Delta V_s + V_s\Delta\pi' \tag{35-14}$$

If the pressure on the system is kept constant, namely, $V_s\Delta\pi' = 0$, then the final equation for the change of required work with change of volume becomes

$$-\Delta W' = \pi'\Delta V_s \tag{35-15}$$

and the total work required to transfer a volume of solvent equals

$$-W^1 = \pi^1 \int_{V_{s_1}}^{V_{s_2}} \Delta V_s \tag{35-16}$$

X_A and X_B are defined as the mole fractions of the solvent and solute, respectively, both being smaller than unity; n_A and n_B are defined as the moles of the solvent and the solute, respectively. Also, by definition $X_A = 1 - X_B$ and $X_B = n_B/(n_A + n_B)$. For very dilute solutions $X_B \ll X_A$, and $n_B \ll n_A$. Thus, $X_B \approx n_B/n_A$ and $\ln(X_A) = \ln(1 - X_B) \approx -X_B$. Therefore, the osmotic pressure equation (35-8) can be rewritten

$$\pi V \approx n_B/n_A\, RT \tag{35-17}$$

Furthermore, if V is the volume of 1 mole of *solvent* in a solution containing n_B moles of solute, and V_s is the volume of the *solution* similarly containing n_B moles of solute, then $Vn_A = V_s$ and

$$\pi = n_B/V_s\, RT \tag{35-18}$$

Also, if w is the percent concentration of the solute, Wt and GMW are respectively, the weight in solution and the gram molecular weight of the solvent, and n_B = Wt/GMV, then Wt = $w/(100 - w) V_s$ and the osmotic pressure becomes equal to

$$\pi = \frac{w}{(100 - w)} \frac{RT}{(GMW)} \tag{35-19}$$

This is also the value of the pressure which has to be applied to the solution against the osmotic pressure in order to keep the system in its initial state (i.e., no transfer of solvent in any direction).

"To calculate the additional pressure, ΔP, needed to transfer additional volume, ΔV, of solvent from the solution through the membrane to the pure solvent side, a system can be considered in which the initial volume of the solution is smaller by ΔV, or $V_s' = V_s - \Delta V$, and in which the total amount of solute does not change, namely, $n_B' = n_B$ and $W' = W$. As already explained, this new state will result in a higher 'osmotic potential', hence a higher osmotic pressure, $\pi' > \pi$. Consequently, the applied pressure needed for keeping the system in a constant state will equal the new osmotic pressure. Thus

$$\pi' = \pi + \Delta P = \frac{n_B'}{V_s}RT = \frac{n_B}{V_s - \Delta V}RT \tag{35-20}$$

Inserting the above given expressions for n_B and Wt, one obtains

$$\pi' = \frac{V_s}{(V_s - \Delta V)} \frac{w}{(100 - w)} \frac{RT}{(GMW)} \qquad (35\text{-}21)$$

If an infinite volume is given, for which $\Delta V \ll V_s$, $V_s \bar{\rho}$, then $\lim V_s/(V_s - \Delta V) = 1$, and $\lim \pi' = \pi$, $V_s \bar{\rho}$ and $\bar{\rho} V/V_s \bar{\rho} 0$."

If there is solute on both sides of the membrane, then equation 35-21 is modified as follows.

$$\pi \frac{RT}{V_s}(n_{B1} - n_{B2}) = \frac{RT}{V_s}\Delta n_B \qquad (35\text{-}22)$$

As

$$\Delta n_B = \frac{V_s}{(GMW)}\left(\frac{w_1}{100 - w_1} - \frac{w_2}{100 - w_2}\right) \qquad (35\text{-}23)$$

It follows that

$$\pi = \frac{RT}{(GMW)}\left(\frac{w_1}{100 - w_1} - \frac{w_2}{100 - w_2}\right) \qquad (35\text{-}24)$$

and $\pi' = \pi + \Delta P$."

EXAMPLE CALCULATION
Calculate the reversible energy required to transfer 4 L of water from an infinite volume of a 0.100% (weight) NaCl solution at 25.0 °C to an (A) infinite volume of pure water, (B) finite volume of 1000 L of pure water, (C) infinite volume of 0.01% (weight) NaCl solution, and (D) a finite volume of 1000 L of 0.010% (weight) NaCl solution, all at the same temperature.
Assume 1 mL of water = 1 cm^3. $R = 82.05$ atm cm^3/deg-mole; $T = 298.15$; m.w. NaCl = 58.44; RT/GMW = 418.604 atm cm^3/g mole. Also 1 atm cm^3 = 0.10133 joule = 0.010332 kg m.

ANSWER
(A) Use equation 35-19.

$$\pi = \frac{0.100}{(100 - 0.100)} \times 418.604 = 0.419 \; atm$$

Thus, the pressure that has to be applied to the NaCl solution to transfer 4,000 mL to the pure solvent side is
$E = 0.419 \times 4000$ mL = 1676 atm cm^3 or 169.8 joules or 17.3 kg m.

(B) Use equation 35-21.

$$\pi' = \frac{1000}{(1000 - 4)} \frac{0.100}{(100 - 0.100)} \times 418.604 = 0.4206 \; atm$$

$E = 0.4206 \times 4000 = 1682$ atm cm^3 or 170.5 joules or 17.38 kg m.

$$\pi' = 418.604\left(\frac{0.1}{99.9} - \frac{0.01}{99.99}\right) = 0.3772 \; atm$$

(C) Use equation 35-24.
$E = 0.3772 \times 4000 = 1{,}499$ atm cm^3

(D) Given $V_s = 1{,}000$ L and $V = 4$ L

$$\pi^\prime = \frac{1000}{1000 - 4} \times 0.3772 = 0.3785 \; atm$$

$E = 0.3785 \times 4000 = 1514$ atm cm^3

Figure 35-2. A commercial reverse osmosis unit.
(Courtesy - Osmonics, Inc., Hopkins, MN)

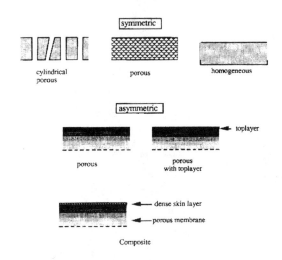

Figure 35-3. Schematic representation of membrane cross sections.
(Courtesy - *Basic Principles of Membrane Technology*, Mulder, M., Kluwer Academic Publishers., Dordrecht, The Netherlands, 1991)

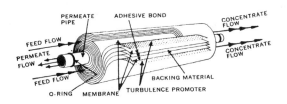

Figure 35-4. Diagram of a membrane module.
(Courtesy - Kunz, A.L., *Bull. Parenteral Drug Assoc.*, **27** (6), 266, 1973)

TECHNIQUES

Figure 35-2 shows one type of reverse osmosis unit. This one can produce 43,000 gallons of water per day.

MEMBRANES

A piece of wire screen can be considered as a very large-scale membrane. Depending on its use, it can let small grains of sand pass through, but not larger stones. It can let air molecules pass through, but not flies. Because it lets some things pass through, but not others, it is called a semipermeable membrane. All living materials depend upon semipermeable membranes to transport molecules to the desired locations. They are very effective, but they are not sufficiently rugged for commercial purposes. Over the years, chemists have prepared many types of membranes.

Commercial membranes are cellophane, cellulose acetate, cellulose triacetate, polysulfone, polyimine, or almost any polymer. Figure 35-3 shows one way to classify membranes. The symmetric membranes are from 10 to 200 µm thick, and the thicker the film, the slower the transfer rate is. S. Loeb and S. Sourirajan (*Adv. Chem. Ser.*, **38**, 117, 1962) developed asymmetric membranes. These consist of a dense, but thin, top layer (0.1 to 0.5 µm) on top of a porous, but strong, sublayer 50 to 150 µm thick. The top layer provides the selectivity and the bottom layer the strength for high pressures. Pressures from 600 to 1000 p.s.i. can be obtained without rupturing the membrane. At these pressures, flow rates of 0.1 to 0.15 gallon/ft^2/day are common. This is low; however, a typical cylinder contains over 80,000 ft^2. Figure 35-4 shows a cross-section of one type of a membrane module and Figure 35-5 shows a photomicrograph of a membrane cross-section. As you can see, thousands of square feet of membrane can be rolled into a small volume and the surface area is quite large.

A.L. Kunz (*Bull. Parenteral Drug Assoc.*, **27** (6), 266, 1973) explains the difference between salt and organic molecule rejection at a membrane. "Salt rejection occurs because of the repulsion of the salt ions from the surface of the membrane and the adsorption of water to the membrane surface (Figure 35-6, p. 419). Due to the physical and chemical nature of the cellulose acetate membrane, a pure water layer of about two water molecules thickness (i.e. 10 Angstroms) develops at the surface of the membrane. Salts are repelled from the surface of the membrane with the

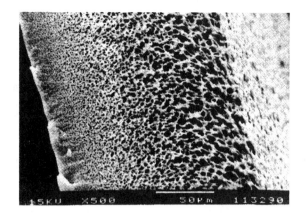

Figure 35-5. Cross section of an asymmetric polysulfone membrane.
(Courtesy - *Basic Principles of Membrane Technology*, Mulder, M., Kluwer Academic Publ., Dordrecht, The Netherlands, 1991)

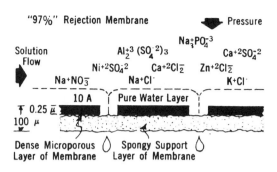

Figure 35-6. Salt rejection mechanism.
(Courtesy - Kunz, A.L., *Bull. Parenteral Drug Assoc.*, **27** (6), 266, 1973)

higher valent ions being repelled to a greater distance. If salts of two different valences, e.g. Ca with a valence of +2 and 2 Cl each with a valence of -1, are in solution, the Ca^{+2} tends to hold the Cl$^-$ with it, and both ions are repelled from the membrane as if both had a valence of two. The highest valence ion of a salt is usually the dictating species in predicting membrane rejection performance.

"Organic rejection [Figure 35-7] is based on a sieve mechanism related entirely to the size and shape of the organic molecule. Organic molecules are not repelled from the surface of the membrane since they have no net charge. They tend to lower the interfacial tension between the solution and the membrane allowing low molecular weight organics (< 100) to be enriched at the membrane surface. The pores in the 97% NaCl rejection membrane are 20 Angstroms or 0.002 μm in diameter. The small pore size will reject substantially all of the organics over 200 m.w. and will reject a percentage of those between 100 and 200 depending on their shape." Formaldehyde (m.w. = 30) passes through easily and is used to sterilize the equipment by a formaldehyde flush.

Figure 35-8 shows one type of an apparatus for measuring membrane characteristics. The membrane is stretched between two flanges, greased, and held together with spring or screw type clamps. Pressure is obtained by nitrogen gas. A conductivity cell is used to measure solute concentration.

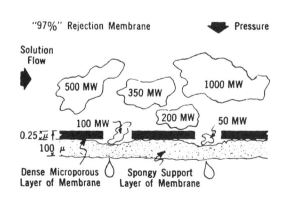

Figure 35-7. Organic rejection mechanism.
(Courtesy - Kunz, A.L., *Bull. Parenteral Drug Assoc.*, **27** (6), 266, 1973)

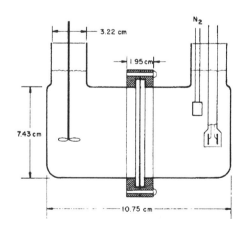

Figure 35-8. Diagram of a cell for osmosis studies.
(Courtesy - Garbarini, G.R., Eaton, R.F., Kwei, T.K., and Tobolsky, A.V., *J. Chem. Ed.*, **48** (4), 226, 1971)

Membrane Parameters

Some of the parameters used to evaluate membranes are the diffusion coefficient, D; the permeability coefficient, P; the solubility constant, S; the filtration coefficient, L_p; the solute permeability coefficient, ω; and the reflection coefficient, σ. The *diffusion coefficient* can be obtained by the time lag method illustrated in Figure 35-9.

$$D = l^2/6\lambda \quad (35\text{-}25)$$

where l is the membrane thickness in cm and λ is the time lag and the intercept on the time axis.

The *permeability coefficient* is expressed as gram (or mole) of diffusant/second/cm² of area/cm of thickness when the concentration of the external solution is 1 g (or mole)/ cm³. This is a way to normalize all membranes to the same concentration gradients. It can be obtained from the slope of the curve in Figure 35-9 at steady state.

$$P = \frac{(Slope)(l)}{(A)(\Delta C)} \quad (35\text{-}26)$$

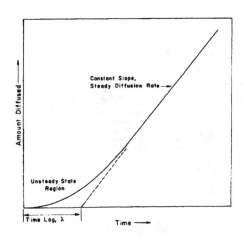

Figure 35-9. Illustration of the time lag method for diffusion through a polymer membrane. (Courtesy - Garbarini, G.R., Eaton, R.F., Kwei, T.K., and Tobolsky, A.V., *J. Chem. Ed.*, **48** (4), 226, 1971)

The *solubility constant* can be determined from the surface concentration and the external concentration by

$$C = SC^{ext} \quad (35\text{-}27)$$

and

$$P = DS \quad (35\text{-}28)$$

The *filtration coefficient* relates the water flux through a membrane to the applied pressure.

$$L_p = \frac{-J_w}{\Delta P} \quad (35\text{-}29)$$

J_w is the rate of transfer per unit area of the section and ΔP is the applied pressure. L_p has units g/cm²/atm/sec.

REVIEW QUESTIONS

1. Why do solvent molecules move through a semipermeable membrane from the high solvent concentration side to the low solvent concentration side?
2. What might happen if only a dilute solution of $MgSO_4$ was used to try to reduce swelling?
3. What is the First Law of Thermodynamics?
4. What is the difference between a 1 molal (m) sucrose solution (m.w. = 342) and a 1 molar (M) solution?
5. What would be the osmotic pressure of a solution containing 3.5 g of resin with a m.w. of 8400 dissolved in 250 mL of water at 25 °C?
6. What is meant by reverse osmosis?
7. Calculate the reversible energy in Joules required to transfer 50 L of water from an infinite volume of a 0.200% NaCl solution at 25 °C to an infinite volume of pure water. 1 atm cm³ = 0.10133 joule.
8. Use the same data as in question 7, but transfer the 50 L to a finite volume of 1,000 L of pure water.
9. What is meant by a semipermeable membrane?
10. What is the basic difference between a symmetric membrane and an asymmetric membrane?
11. What are typical flow rates of water through a commercial membrane?

12. If a cylinder contained 85,000 ft² of membrane and it had a rate of 0.12 gallons/ft²/day at 750 p.s.i., how many gallons of purified water could be obtained in 8 hours?
13. Why are salts repelled at a cellulose acetate membrane surface?
14. What is organic rejection of a membrane based on?
15. What is a good compound to sterilize osmosis and reverse osmosis equipment?
16. What are some membrane parameters used to evaluate membranes?
17. The following data on carrot sections soaked in sucrose solutions at 22 °C were obtained by two of the author's grandchildren, Dane (3rd grade) and Julie (7th grade). Follow the example in Experiment 51 and determine the osmotic pressure in carrots.

	Weight of carrot sections	
Molarity	Initial Weight (g)	Final Weight (g)
0.0	5.07	5.98
0.1	5.14	5.74
0.2	4.57	5.02
0.3	4.73	5.14
0.4	4.58	4.91
0.5	4.70	4.90
0.6	3.96	3.89
0.7	5.44	5.07

IF IT EXISTS - CHEMISTRY IS INVOLVED
IF YOU CAN BUY IT - A CHEMIST WAS INVOLVED SOMEWHERE

36
DIALYSIS AND ELECTRODIALYSIS

PRINCIPLES

DIALYSIS

Dialysis is *the removal of low molecular weight solute molecules or ions from a solution by their passage through a semipermeable membrane driven by a concentration gradient.* In the previous chapter, p. 413, the principle of osmosis was discussed. The same process can take place with solute molecules or ions and it is called *dialysis*. The apparatus for doing a dialysis is called a *dialyzer*. Dialysis has many applications. A person with failing kidneys has to be put on a dialysis machine every 3 days to remove the salts collecting in the blood stream. Its biggest use in the biomedical field is to remove salts from biological preparations. High salt concentrations tend to denature proteins and change cell volumes. This phenomenon was discovered in 1861 when Thomas Graham (*Phil. Trans. Roy. Soc.*, **151**, 183) was studying the separation of various substances through a membrane. He observed that those compounds that permeated the membrane would form crystals when the solvent was evaporated, so he named them *crystalloids*. The compounds that would not go through were gummy when the solvent was evaporated, and he called these *colloids*.

TECHNIQUES

Dialysis is usually a slow process, the speed depending on the concentration gradient between the sides of the membrane. To hasten the transfer, fresh water usually is passed over one side. As an example, suppose a protein extract needed to have the trace metal ions removed. They are there in a low concentration even at the start. The protein solution or suspension is placed in a dialysis tube, closed off, and suspended in pure water. As the metal ions transfer, they will reach equilibrium, and the transfer will stop. If however, the water is continuously changed, the transfer will continue. It is best to change a small volume of water often rather than dialyze in one large volume.

The fact that dialysis is slow does not mean that it cannot be used efficiently or quickly. For example, the Autoanalyzer contains a dialyzer unit for biological materials. If you pass the sample through the dialyzer coils, only a few percent of the ions or molecules transfer before the sample passes through. However, this is a very reproducible few percent, and it is easy to quantitate these materials.

TUBING AND MEMBRANES

Many commercial membranes are available. Visking (sausage casing), collodion, cellophane, and cellulose are common. Spectra/Por uses cellulose esters or regenerated cellulose. These are prepared to have a wide range of cutoff values from 1,000 to 300,000. Commercial dialysis tubing is not always pure and should be cleaned for exacting work. Common contaminants are glycerol added as a humectant, sulfides used to plasticize the tubing, and traces of heavy metals. Figure 36-1, p. 424, shows dialyzer tubing filled with solution.

A cleaning procedure recommended by R.F. Boyer (*Modern Experimental Biochemistry*, Addison-Wesley Publishers, 1986, Reading, MA) is, "Soak the tubing in 1% acetic acid for 1 hour then transfer to glass-distilled water. To remove metal ions, the tubing is then boiled in basic EDTA solution (1% Na_2CO_3, 10^{-3} M EDTA) for 1 hour. Repeat with fresh EDTA. Rinse in hot distilled water five times. Store in distilled water at 4 °C with either a few drops of $CHCl_3$ or 0.1% NaN_3 as a preservative."

Tubing usually comes as a flattened tube 10 m long and 10 to 120 mm wide. *Caution*: This is not the size of the opened tube. A 10 mm flat will produce a 6.4 mm diameter tube and a 16 mm flat will produce a 10 mm tube. Flat sheets are available, 200 x 200 mm.

Cellulose esters should not be used with acetone, methyl ethyl ketone, or dioxane.

Leave about 10% air in the top of the bag so the bags will float above the stirring bar.

Dialyze at refrigerator temperatures, if possible.

Remember: the solvent is moving into the bag, so the sample is becoming more dilute.

Figure 36-2 shows a commercial arrangement that can handle up to 47 tubes (20 2.5 mL and 24 5.0 mL). A reversible motor rotates the tubes 360° at 10 rpm. The tank is 4 L.

Figure 36-3 shows several microdialyzers, from 5 well to 96 well and for sample sizes from 20 µL to 60 µL. They are advertised for desalting macromolecules, changing buffer or pH of sample solutions, purifying viruses after a sucrose gradient, binding fatty acids to proteins, and concentrating antibodies. Because of the small volume, equilibrium is achieved in 1 hour.

Hollow-Fiber Filter Dialysis

Dialysis is usually quite slow, sometimes taking a few days to reach the desired level of purity. A newer technique is to use a hollow fiber-filter system, which can attain the same purity in just a few hours. Figure 36-4, p. 425, shows an electron micrograph of a fiber, and Figure 36-5, p. 425, shows an apparatus. The fiber is placed in the sample solution, and fresh water or buffer is passed through the center. The rate of dialysis is increased because of the very large surface area and the comparatively rapid flow of water.

ELECTRODIALYSIS

Although both neutral molecules and ions pass through a membrane based on concentration differences, it may be possible to hasten the process for ions by applying a voltage to the system, using a cathode on one side of the membrane and an anode on the other. In 1869 Maignot and Sabetes (Ger. Patent 50,473) used a three-compartment system with the sample in the center compartment and membranes serving to make the inner compartment. This was found to work fairly well and was and is used to desalt biological preparations. Figure 36-6, p. 426, shows a Mini-Perfusion system developed by World Precision Instruments to desalt the contents of just a few cells.

However, it was believed that if the membrane could be made ion selective, a more efficient process could be obtained. The current systems are based on the work of W. Juda and W.A. McRae (*J. Am. Chem. Soc.*, **72**, 1047, 1950). They used a membrane impregnated with ion exchange resin. Today membranes are reinforced with an inert woven fabric for increased strength, so the pressure of a fast flowing

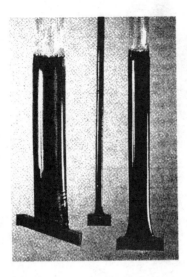

Figure 36-1. Dialzer tubing.
(Courtesy - A.H. Thomas Scientific Co., Swedesboro, NJ)

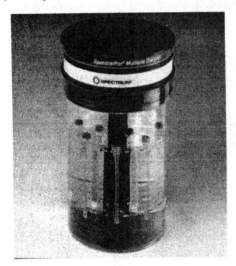

Figure 36-2. A MultiDialyzer.
(Courtesy - Fisher Scientific Co., Pittsburgh, PA)

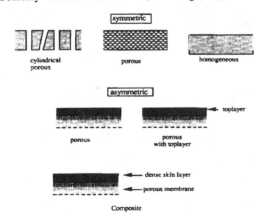

Figure 36-3. Micro dialyzers.
(Courtesy - Fisher Scientific Co., Pittsburgh, PA)

stream would not rupture them. The membranes are essentially water impermeable and very thin (0.5 mm) so diffusion and exchange are fast, as is electro-conductivity. Pore sizes vary from 10 to 100 Å with 10 to 20 Å being more common. They have a capacity of from 1.6 to 3.0 meq/dry g. There are four variations of electrodialysis: *electrolytic, concentrating-diluting, ion substitution,* and *reversal*. The diagrams and text that follow were abstracted from discussions and literature sent to the author by Ionics, Inc., Watertown, MA.

Electrolytic Dialysis

For electrodialysis, a membrane is placed between electrodes of an otherwise conventional electrolysis cell to carry out standard electrochemical processes. The membrane in this type of application serves to separate the products of the electrode reaction. This is shown in Figure 36-7, p. 426.

Concentrating-Diluting Dialysis

A number of chambers called diluting and concentrating cells are separated by alternating anion and cation membranes and are arranged between two electrodes. This is shown in Figure 36-8, p. 426.

Compartments 1 and 6. These compartments contain the metal electrodes (Pt plated on Ti). Chlorine gas, oxygen gas, and H^+ ions are produced at the anode. Hydrogen gas and OH^- ions are produced at the cathode.

Compartment 2. Cl^- ions pass through anion membrane (A) into compartment 3. Na^+ ions pass through cation membrane (C) into compartment 1.

Compartment 3. The Na^+ ions cannot pass through the anion membrane and remain in compartment 3. The Cl^- ions cannot pass through the cation membrane and remain in compartment 3.

Compartment 4. Cl^- ions pass through the anion membrane into compartment 5. Na^+ ions pass through the cation membrane into compartment 3.

Compartment 5. The Na^+ ions cannot pass through the anion membrane and remain in compartment 5. The Cl^- ions cannot pass through the cation membrane and remain in compartment 5.

The overall effect shows that compartments 2 and 4 have been depleted of ions and the ions have concentrated in compartments 3 and 5. When properly manifolded, this unit will yield two major and separate streams; demineralized and concentrate waters, and two minor streams from the electrode compartments. Figure 36-9, p. 427, shows how the manifolds are connected.

Ion Substitution Electrodialysis

For ion substitution electrodialysis, the arrangement is to effect double decomposition reactions and ion substitutions, which have the net effect of modifying the ionic composition. To accomplish the latter, the liquid to be modified is passed between two membranes of like electrical charge and a liquid containing the makeup ion (or ions) is passed through an adjacent compartment.

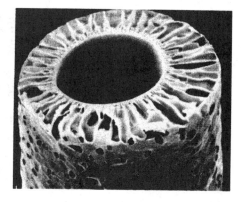

Figure 36-4. Electronmicrograph of a hollow fiber-filter dialysis tube.
(Courtesy - M. Billadeau in R.F. Boyer, ed., *Modern Experimental Biochemistry*: Addison-Wesley Publishers, Reading, MA, 1986)

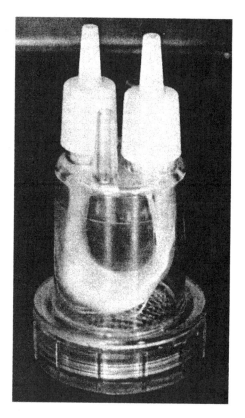

Figure 36-5. Apparatus for hollow fiber filter dialysis.
(Courtesy - R.F. Boyer, ed., *Modern Experimental Biochemistry*, Addison-Wesley Publishers, Reading, MA, 1986)

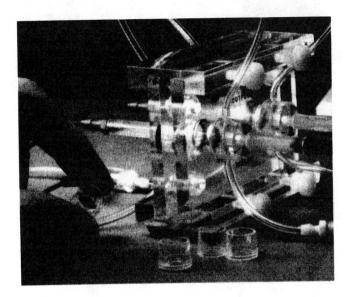

Figure 36-6. A Mini-Perfusion system.
(Courtesy - World Precision Instruments, Sarasota, FL)

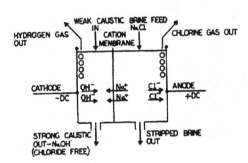

Figure 36-7. An electrolytic cell with a cation membrane.
(Courtesy - Ionics Inc., Watertown, MA)

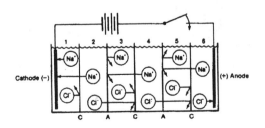

Figure 36-8. Arrangement for concentrating-diluting electrodialysis.
(Courtesy - Ionics, Inc., Watertown, MA)

Figure 36-10, p. 427, shows the arrangement for cationic substitution. Notice that this is ACCACCACCA rather than ACACACAC. In this particular case, a caustic product feed is acidified by replacing cations M^+ with hydrogen ions (H^+) from an acidic makeup stream. Other applications of this principle are replacing M^+ with ammonium ions (NH_4^+) or X^- with hydroxide (OH^-) or acetate (OAc^-) ions.

Electrodialysis Reversal

Electrodialysis reversal is the same arrangement as the concentrating-depleting apparatus, except that provisions are made to reverse the polarity periodically. This is done to prevent membrane scaling and fouling. This innovation has permitted the large-scale commercial application of electrodialysis for either the purification of sea and brine waters or the ultrapurification of ordinary waters. For example, the world's largest electrodialysis reversal system was put into production in January 1995 to produce high quality potable water. This 12.0-million-gallons-per-day plant's feed water source is a 24,000 acre well field and ecological reserve with 11 production wells. Typical feed waters have total dissolved solids of 1,300 mg/L, over 2 ppm H_2S, saturated with $CaSO_4$ and CaF_2 and some wells have a high pH (Ionics, Inc., Watertown, MA).

Calculations

Although many calculations are required for an individual unit, a few basic calculations are common. These are based on Faraday's law, Ohm's law, and current efficiency.

Faraday's Law
Faraday's Law as related to electrodialysis (ED) states that the passage of 96,500 amperes of electrical current for one second will transfer one gram equivalent of salt. This is equal to 26.8 amperes of current passing for 1 hour.

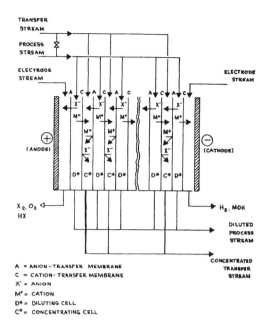

Figure 36-9. Demineralization in a multicell pair stack of concentrating-diluting cells.
(Courtesy - S.R. Jain, *Am. Lab.*, Oct., 1979)

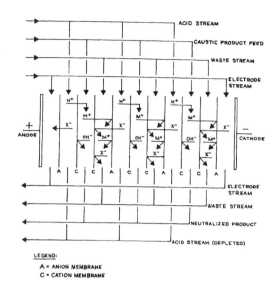

Figure 36-10. Ion substitution in an electrodialysis stack, ion-modifying cells.
(Courtesy - S.R. Jain, *Am. Lab.*, Oct., 1979)

The amount of sodium chloride transferred by one Faraday will be 23 g of sodium to the cathode and 35.5 g of chloride ions to the anode at 100 % current efficiency. Faraday's law therefore is the basis for calculating the amount of electric current needed in an ED system for transferring a specific quantity of salts. When put in terms for use in ED calculations, Faraday's law is:

$$I = \frac{F^* \times F_d \times \Delta N}{e \times N^*} \tag{36-1}$$

where I = current in amperes. F^* = Faraday's constant, 96,500 ampere seconds/equivalent. e = the current efficiency. ΔN = change in normality of the demineralized stream between the inlet and outlet of the membrane stack. N^* = the number of cell pairs. F_d = the flow rate of the demineralized stream through the membrane stack (L/sec).

EXAMPLE CALCULATION

What is the current in amperes required for a 500-pair stack operating at 87.2% current efficiency, if the initial water flowing at 969.3 gpm is 680 ppm, the final water is 142 ppm, and the average equivalent weight is 80.49?

ANSWER

680 ppm/80,490 ppm/N = 0.008368 N; 142 ppm/80,490 ppm/N = 0.001764 N. ΔN = 0.00668 N.
969.3 gal/min x 3.7 L/gal x 1.0 min/60 sec = 59.8 L/sec.

$$I = \frac{96,500 \times 59.8 \times 0.00668}{87.2 \times 500} = 0.88 \; amperes$$

Ohm's Law

To determine the voltage requirements for a given system, the current is determined from Faraday's law, and the resistance is determined by the components of the membrane stack and the solution under treatment.

Ohm's law is:

$$E = I \times R \qquad (36\text{-}2)$$

where E = the voltage in volts. I = the current in amperes. R = the resistance in ohms. The components which contribute to the resistance of a membrane are expressed in the following equation: All units are ohm-cm².

$$R_{cp} = R_{cm} + R_{am} + R_c + R_d \qquad (36\text{-}3)$$

where R_{cp} = resistance/unit area of one cell pair. R_{cm} = resistance/unit area of the cation membrane. R_{am} = resistance/unit area of the anion membrane. R_c = resistance/unit area of the concentrate stream. R_d = resistance/unit area of the demineralized stream.

Factors which affect resistance include temperature, ionic species and solution concentration. The higher the temperature and solution concentration, the lower the solution resistance and hence stack resistance. Typically, for every 0.55 °C change in temperature, there is a 1.1% change in stack resistance. Concerning the ionic species, the resistance decreases as the % Na^+ and Cl^- (based on the total cation and anion epm) increase.

EXAMPLE CALCULATION
A 500 cell pair unit required 333 volts at 5.83 amperes to operate. What is the total resistance?

ANSWER
$333 = 5.83 \times$ ohms. Ohms = 57.1

Current Efficiency

If 100 electrons are transferred and all do what you want them to do, then the current efficiency is 100%. If 90 do what you want and the other 10 react in another way, then the current efficiency is 90%. In general, the lower the current, the higher the current efficiency is, but the slower the process. Factors that reduce current efficiency in ED are cross-leakage and back diffusion. The current efficiency (e) is calculated as:

$$e = \frac{F^* \times F_d \times \Delta N \times 100}{I \times N^*} \qquad (36\text{-}4)$$

EXAMPLE CALCULATION
Calculate the current efficiency for the following system: feed water, 0.0480 N (2805 ppm); product water, 0.0078 N (456 ppm); dilute flow (F_d), 9765 L/hr (43 gal/min); amps, 25.5; number of cells, 450; Faraday's constant, 26.8 amp-hr.

ANSWER
$\Delta N = (0.0480 - 0.0078) = 0.0402$. Therefore:

$$e = \frac{26.8 \times 9765 \times 0.0402 \times 100}{25.5 \times 450} = 91.7\%$$

CELL PAIRS AND STACKS

The actual construction of a cell pack involves more than is implied by the previous diagrams. The membranes should not warp; therefore, a spacer is required. If large surface areas are involved, then pressure must be applied to force a reasonable flow, so the membranes must be rugged and supported. Figure 36-11, p. 429, shows one cell pair. It consists of a cation transfer membrane, a demineralized water flow spacer, an anion transfer membrane, and a concentrate water flow spacer.

Membranes have essentially the same configuration as the spacer shown in Figure 36-12, p. 429. Membranes must have the following properties: (1) low electrical resistance, (2) insoluble in aqueous solutions, (3) semirigid for ease of handling during stack assembly, (4) resistant to pH changes from 1 to 10, (5) operate at temperatures in excess of 46 °C, the softening point of the spacer, (6) resistant to osmotic swelling when being cycled between 220 ppm and

30,000 ppm salt solutions, (7) long life expectancy, (8) resistant to fouling, and (9) impermeable to water under pressure.

A cation membrane is of the sulfonic acid type, which repels anions and permits cations to pass. The anion membranes are tetramethylammonium ions, which repel cations and permit anions to pass. Both are water impermeable and usually 0.5 mm thick.

Spacers form the concentrating and demineralizing flow paths within the membrane stack. Ionics Inc. spacers are made of low density polyethylene sheets with die-cut flow channels. In the manufacture of a spacer, two sheets of plastic are die-cut and then glued together to form a turbulence producing "tortuous path". The stream entering the spacer (A of Figure 36-12) is split into two mirror-image streams (B,C), which are each divided into three or four parallel flow paths that wind their way through the spacer until they exit at the outlet manifold (D). However, they are only one ply thick and thus will force water to flow in an over-and-under manner (Figure 36-13). This type of flow causes turbulence in the stream, allowing higher electric current flow per unit area and, therefore, more efficient use of the membrane area in the stack. Turbulence promotes mixing, which aids the transfer of ions to the surface of the membrane and breaks up boundary layers and slimes at the membrane surface. Remember - the water flow is not through the membranes, but across them. Figure 36-14, p. 430, shows a stack of membranes. Stacks vary from 300 to 500 cells/stack.

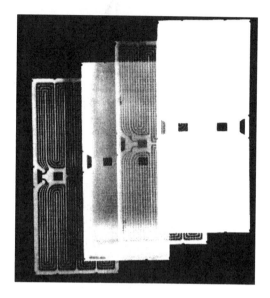

Figure 36-11. Components of a cell pair.
(Courtesy - Ionics, Inc., Watertown MA)

Figure 36-12. A spacer.
(Courtesy - Ionics, Inc., Watertown, MA)

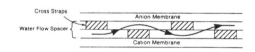

Figure 36-13. Cross-section of a spacer illustrating the flow around the cross straps.
(Courtesy - Ionics, Inc., Watertown, MA)

REVIEW QUESTIONS

1. How did the name *colloid* originate?
2. What is an important factor in determining the rate of dialysis?
3. What are some common contaminants in commercial dialysis tubing?
4. Why is a little air left in a dialysis bag if the bag is not held in place?
5. What is an advantage of a hollow-fiber dialysis system?
6. What is the purpose of electrodialysis compared to dialysis?
7. What is the anion-cation membrane arrangement in a concentrating-diluting electrodialysis system?
8. What happens in compartment 3 of Figure 36-8, p. 426?
9. What is the anion-cation membrane arrangement in an ion substitution electrodialysis system?
10. What is the purpose of ion substitution electrodialysis?

Figure 36-14. An electrodialysis stack.
(Courtesy - Ionics Inc., Watertown, MA)

11. What is the purpose of electrodialysis reversal?
12. Where do you believe the term "potable water" arose?
13. What is the current in amperes required for a 500-pair stack operating at 93.1% current efficiency, if the initial water flowing at 998.1 gpm is 1,297 ppm, the final water is 214 ppm and the average equivalent weight is 61.47?
14. A 500-cell pair unit required 476 volts at 15.32 amperes to operate. What is the total resistance?
15. Calculate the current efficiency for the following system: feed water, 0.052 N (2600 ppm); product water, 0.0081 N (405 ppm); dilute flow (F_d), 9235 L/hr (41.6 gal/min); amps, 24.6; number of cells, 450; Faraday's constant, 26.8 amp-hrs, avg equivalent weight, 50.0
16. What is the purpose of a spacer?
17. If a membrane is impermeable to water, how does it conduct current?
18. Why do you want turbulent flow around the membranes?

IF IT EXISTS - CHEMISTRY IS INVOLVED
IF YOU CAN BUY IT - A CHEMIST WAS INVOLVED SOMEWHERE

37
FILTERING AND SIEVING

PRINCIPLES

Filtering and sieving have been used as a means of separation since before recorded history. Their main use then was to prepare *potable water*. Muddy water was passed through layers of sand until it was clear and then was put in pots for drinking - potable water. *To filter or sieve is to selectively remove a portion of a mixture by passing it through a semi-porous barrier.* If the barrier is a mat of fibers or a porous polymer with small pore sizes, it is called *filtering*. If the barrier is of interwoven metal wire (screen); natural cloth (burlap, net); or polymer fibers in a regular pattern and having regular, rather large, pore sizes, it is called *sieving*. The barrier may be as coarse as a piece of fishing net or it may be as fine as a biological membrane. Today, many uses exist for filtering and sieving and the principle is used in a variety of ways. Sand can be filtered from rocks, soot particles from smokestacks, bacteria from water, and fine particles from extracted samples before injection into an high performance liquid chromatograph (HPLC). Particulate matter in air is sampled by filtering a measured volume of air. Cheese curd is filtered from the whey by cloth, and precipitates can be filtered from solutions. Flies can be sieved from houses by screen wire. It may not be chemical, but it is a rewarding separation. Particle size analysis is extremely important in many industrial applications such as paint manufacturing. The principle of filtering and sieving is easy to understand. Figure 37-1, p. 432, lists several relative sizes of particles. The purpose of this chapter is to acquaint you with some of the apparatus and techniques available to you.

TECHNIQUES

AIR

Filters, Pumps, Flow Meters

Air is filtered for many types of particles, such as dust, molds, pollen, acid rain droplets, aerosols, stack gas particles, and bacteria. Figure 37-2, p. 432, shows an air filter using a sheet of paper, glass fiber, or plastic barrier.

The air is drawn through the barrier at a calibrated rate. By measuring the time, you can determine the volume of air. For example, the Topeka Kansas Allergy Center draws air in for 10 minutes at a rate of 14.5 L/min to determine pollen. Figure 37-3, p. 432, is a photograph of a typical pump. This provides either a vacuum (to 27" Hg) or a pressure (to 25 psi). Figures E53-1 and E53-2 of Experiment 53 in Appendix A, are photographs of several airborne particles. Figure E53-3, also in Experiment 53, shows a commercial setup to determine pollen and molds in air. Figure 37-4, p. 433, is a *high efficiency particulate air* (HEPA) filter. This is becoming the standard for clean rooms and analytical laboratories where trace analysis is done. These filters, made of an acrylic copolymer supported on a nonwoven nylon backing, will filter particles down to 0.3 µm at 99.96% efficiency. Figure 37-5, p. 433, shows one type of a gas flow meter. These meters are calibrated, and the flow rate is measured by how high a glass or metal float ball rises in the glass tube. The one shown has a magnifying window and leveling screws to ensure vertical alignment.

432 Techniques

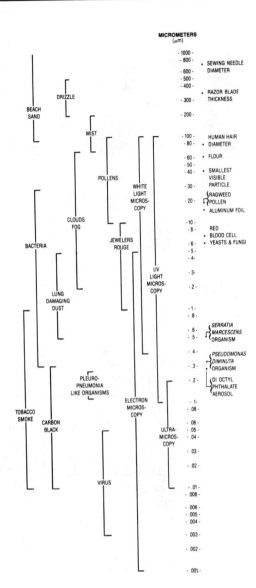

Figure 37-1. Relative sizes of small particles.
(Courtesy - Gelman Sciences, Ann Arbor, MI)

Figure 37-2. An air filter assembly.
(Courtesy - Millipore Corp., Bedford, MA)

Figure 37-3. A pump for either drawing in or pushing out air.
(Courtesy - Millipore Corp., Bedford, MA)

Figure 37-4. An HEPA filter cartridge.
(Courtesy - Gelman Sciences Inc., Ann Arbor, MI)

Figure 37-5. A gas flow meter.
(Courtesy - Fisher Scientific Co., Pittsburgh, PA)

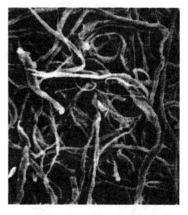

Figure 37-6. A depth filter. fiberglass bonded with an acrylic ($2,600x$)
(Courtesy - Millipore Corp., Bedford, MA)

Calculations

Many commercial units, like the Samplair shown in Figure E53-3, in Experiment 53 in Appendix A, come with the volume calibrated. However, an additional calculation usually is needed. The number desired when doing particulate air sampling is the number of particles per cubic meter per 8 hours or 24 hours of exposure. Because the air is sampled for only a few minutes and only a small volume of air actually is sampled, a calculation is needed to convert to the desired value. In addition, the sampling often is done several times daily, so an average value per sample time is used. Equation 37-1 shows how each factor is obtained and how the count/m³/day is calculated. The first factor corrects for the small volume, and the second factor corrects for the short time of sampling. The 1440 min would be changed to 370 min for an 8-hour exposure.

$$Particles \times \frac{1000\ L/m^3}{Total\ Liters/Count} \times \frac{1440\ min/24\ hr}{Total\ Time(min)} = Particles/m^3/24\ hr \tag{37-1}$$

EXAMPLE CALCULATION

14.5 L/min of air are drawn into an air sampling apparatus for 10 minutes every 3 hours during the day. The oak tree pollen counts were 7, 6, 3, 8, 10, 4, 7, and 5. What is the oak tree pollen count per cubic meter per day? First, average the particles/count.

ANSWER

$$6.2\ (Avg\ of\ 8) \times \frac{1,000\ L}{10 \times 14.5\ L} \times \frac{1,440\ min}{8 \times 10\ min} = 770\ oak\ tree\ pollen\ grains/m^3/24\ hrs$$

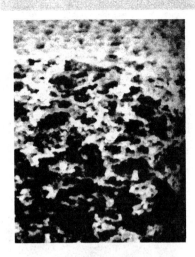

Figure 37-7. A screen filter.
Courtesy - Millipore Corp., Bedford, MA)

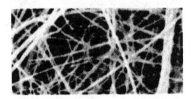

Figure 37-8. Fiberglas sheet (2,600x).
(Courtesy - Millipore Corp., Bedford, MA)

Figure 37-9. 0.45 µm mixed cellulose ester (1,000x) showing bacteria trapped on the surface.
(Courtesy - Millipore Corp., Bedford, MA)

Barriers

Paper is the most convenient barrier for most filtering, but porous polymers with small and fairly regular size pores are now common and needed for the smaller particles. Pore sizes down to 0.2 µm are obtainable. Glass fiber mats are available. They are resistant to organic solvents and high temperature, but break if bent; therefore, they must be used flat. Barriers ruled with crosshatching (grid style) are preferred if large particles are to be counted, so you can keep track of where you are.

There are two main types of barriers: *depth* and *screen*. A depth filter retains particles both on its surface and within its matrix. Figure 37-6, p. 433, shows a depth filter made of fiberglass bonded with acrylic resin to prevent pieces from getting into the filtrate. Depth filters have a random matrix of fibers. The pore structure is irregular so they retain a variety of particle sizes. These filters are given a nominal rating, and they will retain 98% of all particles larger than that size.

Screen filters retain particles on their surface (Figure 37-7). The pore size can be regulated during manufacture so the particle size retained can be known fairly well. They tend to clog rapidly but can be very thin and retain little liquid.

Figure 37-8 shows a Fiberglas sheet (2600x). Notice the irregular pore size. Figure 37-9 is a mixed cellulose ester membrane (1000x) with 0.45 µm pore size showing bacteria trapped on the surface. This is a popular size for filtering HPLC solutions.

"Most synthetic polymers are hydrophobic. However, the surface can be modified through the addition of covalently bonded chemical moities. This process is used to produce the Durapore hydrophilic membrane [Figure 37-10, p. 435]. The base polymer is polyvinylidene fluoride; the surface is the -OH adduct" (Millipore).

Materials used for membranes are cellulose acetate, cellulose triacetate, polytetrafluoroethylene, polyvinylchloride, nitrocellulose, polypropylene, nylon, Fiberglas, and paper.

LIQUIDS

Gravity Filtering

Funnels

The common funnel for analytical separations has a 70 mm wide top, a 60° taper, and a 150-mm-long stem (Figure 37-11, left, p. 435). This was once the easiest to make. However, it had very little surface area for drainage, which required a special folding technique (see Figure 37-12, p. 435). To offset poor training, newer funnels have a 58° angle so the paper can be folded without an offset. Table 37-1 shows the effects of flutes on filtering speed.

Another design to increase flow rate is to put small drainage canals, called *flutes*, in the top (Figure 37-11 center, p. 435). The paper need not be offset when it is folded, although it usually is done as a matter of better technique.

Table 37-1 Effect of flutes on filtering speed
(Courtesy - Fulton, B. and Meloan, C.E., *J. Chem. Ed.*, **61** (6), 554, 1984)

Type of funnel	Time (200 mL of 0.02 M Fe(OH)$_3$)
60° regular	12 min 44 sec
60° fluted	10 min 33 sec
Helicoid	9 min 12 sec

Figure 37-11 right shows the newest design, the helical coil. This uses 60° folding and is the fastest drainage yet. A comparison of the filtering speeds under similar conditions gave the following results (averages of five trials).

Papers

The filtering membrane for gravity filtering in an analytical laboratory is a high quality paper that is usually free of any sizing or filling agents. This is known as *filter paper*. It is made from selected fibers of wood or cotton and by the wet-laying process under various conditions to alter the thickness, flow rate, strength, and retention. Papers for analytical work come in circles from 2 cm to 50 cm in diameter. Table 37-2, p. 436, shows the characteristics of several filter papers.

Paper Drainage Classifications. The most common way to classify how fast a paper will drain is, slow, medium and fast. This is usually accomplished by changing the pore size or increasing the thickness of the paper. The goal is to remove all of the particles desired in the shortest time. There are two standards for determining flow rate: the Association for Standard Testing Methods (ASTM) and the Herzberg method.

ASTM: the time for 100 mL of prefiltered water to pass through 15 cm quadrant-folded paper.

Herzberg: the time for 100 mL to pass through 10 cm² of filter paper with a constant 10 cm head pressure of water.

Ashless. This type of filter paper has been treated with a dilute solution of HF to volatilize any silica particles. The residue upon ashing is less than 0.1 mg for an 11 cm circle, which is less than the detection limit of a common analytical balance.

Phase Separation Paper. This paper (Whatman 1PS) has been treated with a silicone to make it hydrophobic yet let organic materials pass through. This is a convenient way to separate water from organic mixtures, particularly emulsions. The organic droplets or layer pass through while the water droplets or layer stays behind. It is folded and used just like any other paper, the only difference being that water will not penetrate it.

Table 37-2, p. 436, shows several different types of filter paper and their characteristics.

Folding. Figure 37-12 shows how to fold a circle of filter paper for placing in a 60° funnel. When a piece of paper is folded into quarters, it will form a 60° angle when opened. This causes it to make contact with the entire side wall of a 60° funnel and retards the transfer of liquid. To correct for this, the paper is offset about 2 to 3 mm on the second fold so when it is opened the angle will be somewhat less than 60°, and much more area is available for draining.

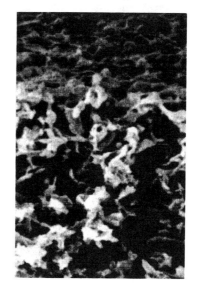

Figure 37-10. Polyvinylidene fluoride hydrophilic screen membrane (5000x).
(Courtesy - Millipore Corp., Bedford, MA)

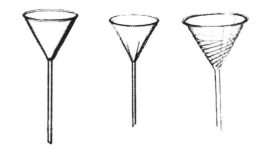

Figure 37-11. Gravity filtering funnels. Left: 60° smooth. Center: 60° fluted. Right: helicoid.
Courtesy - Fisher Scientific Co., Pittsburgh, PA

Figure 37-12. Steps for folding filter paper.

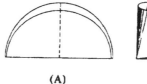   

(A)   (B)   (C)   (D)

Table 37-2. Qualitative and quantitative filter papers
(Courtesy - Whatman Inc., Clifton, NJ)

Whatman Grade	Retention μm	Porosity	Rate, sec. ASTM	Rate, sec. Herzberg	Applications
Qualitative					
P8	>20	Coarse		3.0	Gelatinous precipitates, clarifying solutions, oils, sugar analysis
P5	5 - 10	Medium		9.5	Clinical and spot testing, clarifying pharmaceuticals
P4	2 - 5	Medium-fine		25	General purpose
P2	1 - 3	Fine		90	Boiler water testing, pigment studies
Quantitative-ashless					
40	>8	Medium	75	340	Medium crystalline precipitates
41	> 20 - 25	Coarse	12	54	Gelatinous precipitates; air pollution monitoring
42	>2.5	Fine	240	1870	For critical gravimetric analysis
43	>16	Medium	40	155	Foodstuff and soil analysis, air pollution monitoring
44	>3	Fine	175	995	Thinner than 42
Quantitative - hardened low ash					
50	>2.7	Fine	250	2685	Retains fine crystalline precipitates
48	>7	Medium	55	235	General purpose filtering
54	>20 - 25	Coarse	10	39	Gelatinous and coarse particles
Quantitative - hardened ashless					
540	>8	Medium	55	200	Gravimetric analysis of metals in acid/base solutions
541	>20 - 25	Coarse	12	34	Coarse gelatinous precipitates; useful for analysis under strongly acidic or basic conditions.
542	>2,7	Fine	250	2510	Retains fine particles under demanding conditions

Proper Filtering Technique

The time-honored good technique for gravity filtering through paper is to do the following: (1) use the proper grade of filter paper, (2) decant, (3) use a long-stem funnel, (4) use a narrow-diameter stem rather than a wide-diameter stem, (5) use a fluted funnel if possible, (6) fold the paper with a 1/8- to 1/4- inch offset on the second fold, (7) tear the paper at the top of the fold to prevent air intake, (8) keep the long stem full of solution, and (9) touch the end of the funnel stem to the side of the beaker. This is shown in Figure 37-13, p. 437. In 1981 (Meloan, C.E., *J. Chem. Ed.* **58** (1), 73) the author decided to determine if this was true and to measure each step. The precipitate chosen was iron hydroxide, a gelatinous precipitate that is hard to filter. Good technique was compared to poor technique, which was essentially doing the opposite of the good technique steps. Table 37-3 shows the effect of paper porosity. One variable was changed at a time, and the filtering rates were timed. The tables below show averages of three trials.

Table 37-3. Effect of Paper Porosity

Paper	Good Technique	Poor Technique
Whatman No 41 (fast)	4 min 28 sec	8 min 19 sec
Whatman No 40 (medium)	5 min 51 sec	24 min 25 sec
Whatman No. 42 (slow)	15 min 19 sec	48 min 46 sec

The use of good technique clearly saves considerable time. The proper choice of filter paper is very important. *Never use a paper more porous than necessary to retain the precipitate.* Two additional factors are related to good technique that are not apparent from the time measurement alone. The filtered precipitate was considerably wetter in the poor case than in the good case. The combined weights of the precipitate and wet paper for each of the three trials (Whatman 40) were 29.78 g for the poor case and only 22.65 g for the good case. This means that even more time will be saved later on because less water must be driven off of the precipitate before the paper can be burned.

The second benefit was the nature of the placement of the precipitate on the filter paper. In the good technique case, all of the precipitate was 6 to 8 mm below the top of the paper, and the paper could be removed, folded over, and placed immediately into a crucible. In the poor case, the precipitate was 1 to 2 mm from the top of the paper, and a further washing would be necessary before the paper could be folded and transferred.

Table 37-4 shows the effect of decanting. The argument is that if you stir the solution, the precipitate will clog the pores early on, making it more difficult for the following solution to pass through. The good technique parameters were used, except that the solution was decanted in one set and stirred before each addition in the other set.

Table 37-4. Effect of decanting (Whatman No. 40 Paper)

Decanted	Not Decanted
3 min 58 sec	5 min 58 sec

The solutions had been standing for 1 extra day, and apparently the precipitate had digested sufficiently more to increase the overall rate of the filtration when compared to Table 37-3.

Table 37-5 illustrates the effect of stem length and column of water. The argument is that the weight of a column of water will help pull the water through the paper faster. The 4 to 5-mm long stem was not filled with water, the 4 to 5 mm being the few drops at the end of the stem during filtering.

Note: The funnels were placed in hot HNO_3 for 30 minutes and rinsed with water, ether, acetone, and then distilled water to clean the stems. It is almost impossible to maintain a full stem of solution if the stems are dirty and will not wet.

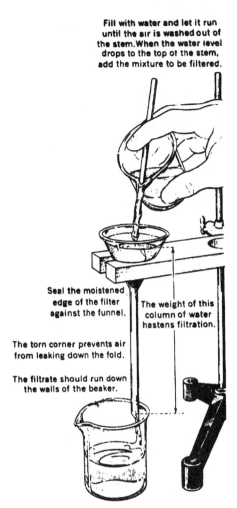

Figure 37-13. Proper gravity filtering technique. (Courtesy - Frantz, W., and Malm, L.E. - *Chemical Principles in the Laboratory*, 2nd ed., W.H. Freeman Co., San Francisco, CA, 1977)

Table 37-5. The Effect of stem length and column of water

142 mm Stem (Filled)	53 mm Stem (Filled)	4-5 mm Stem (Empty)
3 min 58 sec	5 min 18 sec	10 min 19 sec
22.15 g	24.19 g	27.99 g

Clearly, the length of the water column is very important not only in affecting the speed of filtration, but also for the weight of water remaining in the flocculent precipitate. The longer column of water draws out more water from around the precipitate, thus making the drying steps much quicker.

Use a long narrow stem and keep it full of solution. A narrow stem is much easier to keep full of solution.

This experiment also shows the importance of tearing off the corners of the folds, particularly the outside fold, because this is where air most easily enters the system.

Air bubbles in the stem markedly decrease the rate of filtration. Even with a full column of water, if one air bubble occurs at the joint between the stem and the base of the funnel, a measurable slowing down of the filtration occurs. *Care must be taken that the outside fold does not overlap a flute, or air will readily enter the stem.*

Recommended Procedure to Obtain a Full Column of Water

Place the properly folded filter paper in the funnel and thoroughly wet it. Press it gently into place and apply pressure to the folds to flatten them out. Place a forefinger at the lower end of the paper to shut off the flow of water, and add water to the filter paper until the paper is nearly full. When the column is full of water (you may have to remove the forefinger briefly, then add more water to the funnel), wait a few seconds until the air bubbles rise 10 to 12 mm above the stem. Place the forefinger of the other hand at the bottom of the paper and with gentle upward strokes, squeeze out the air bubbles. Gently press against the folds to squeeze air away from them. Fill the funnel to within 2 to 3 mm of the top with water. Have the beaker containing your precipitate readily available. Remove your finger from the end of the stem and let the water drain (it goes fast). A quick bit of pressure on the outside fold will generally seal off an air leak. If the column remains solid, immediately begin filtering. Do not wait for the water level to drain to the level of the flutes, or air is likely to enter the system. During the decantation process, keep the liquid level 2 to 6 mm from the top of the paper. When the precipitate enters and begins to close the pores of the paper, the liquid level can be lowered, because air is less likely to enter the system. Table 37-6 shows the effect of flutes and the offset fold.

Table 37-6. The effect of flutes and the offset second fold
(Long-stem funnels, Whatman No. 40 paper)

Smooth inside, 58°, paper 60°	6 min 2 sec
Fluted inside, 60° paper 60°	5 min 9 sec
Fluted inside, 60° paper offset 2 mm	4 min 58 sec

Touching the stem tip to the side of the beaker prevents splashing. The breaking of surface tension by touching the tip to the sidewall to allow the water to flow faster out of the stem was not measured.

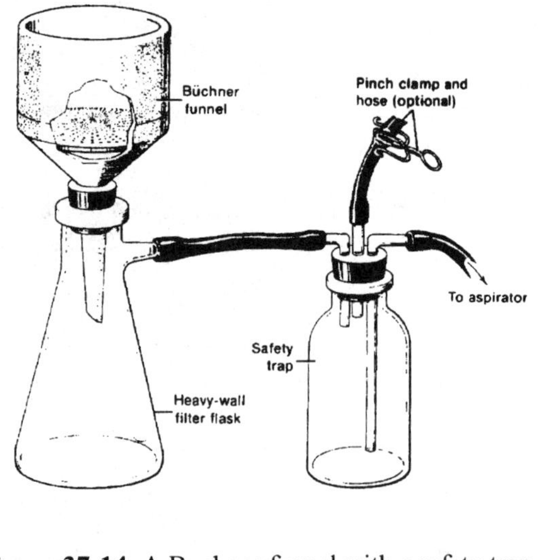

Vacuum Filtering

To hasten the rate of filtering of either a gelatinous or large volume precipitate, a vacuum can be applied. The vacuum is low level, and a water aspirator is sufficient. For large volumes of precipitate, a Buchner funnel is used. Figure 37-14 shows the most common arrangement with a safety trap attached. Circles of filter paper are placed in the bottom of the funnel to cover the holes. A small layer of charcoal is sometimes added on top of the paper, if the solution is to be clarified.

Figure 37-14. A Buchner funnel with a safety trap.
(Courtesy - Frantz, W., and Malm, L.E., - *Chemical Principles in the Laboratory*, 2nd ed., W.H. Freeman Co., San Francisco, CA, 1977)

Dental Dam

One of the advantages of using a vacuum filtration is that the air being drawn through the filter cake will rapidly dry the crystals. However, a more likely case is shown in Figure 37-15, p. 439, in which large cracks have appeared in the filter cake. Any air drawn through this mass will go only through the cracks, and the filter cake will remain wet.

A technique used to prevent the cracks from forming and to speed the filtration is to use a *dental dam* as shown in Figure 37-15, bottom. A dental dam is a piece of rubber, 0.0004 cm thick and 15 cm square. Dentists cut a small hole in it and place it around a tooth to keep the tooth dry while it is being filled. As a filtering aid, it is placed on top of the filter cake and is immediately drawn down tightly on top of the crystals by the vacuum. This applies a uniform pressure to the filter cake and quickly squeezes the solvent out, so any cracks are closed immediately by the pressure of the dam. The vacuum is released, the dam is removed and scraped clean, and then suction is continued to draw air through the filter cake to finish the drying. Very little cracking then occurs. If the precipitate volume is small, then a Hirsch funnel (Figure 37-16) is preferred. This is the same as a Buchner funnel, except that it is tapered to force the precipitate into a smaller volume.

For smaller volume precipitates, such as those collected in a crucible when the precipitate is to be weighed, then porcelain or glass crucibles with sintered glass bottoms of various porosities are used. Some are shown in Figure 37-17. Porosities are ultrafine (UF): 0.9 to 1.4 µm; very fine (VF): 2 to 2.5 µm; fine (F): 4 to 4.5 µm; medium (M): 10 to 15 µm; and coarse (C): 40 to 60 µm. Several crucible holders are shown in Figure 37-18, p. 440.

Figure 37-19, p. 440, shows a vacuum filter used with liquids for the analysis of biological and particulate matter. This one is 47 mm in diameter with a stainless steel support screen.

Pressure Filtering

In some situations, fast filtering is needed and a vacuum is inappropriate, such as filtering a sample before injection into an HPLC or removing bacteria and particulate matter from water to be used for biological work. The simplest system is to fill a syringe with the sample solution, attach a filter of the desired porosity, and force the sample solution through the filter by pushing on the syringe plunger. This is fast, efficient, and economical. One such arrangement is shown in Figure 37-20, p. 440.

For the rapid filtering of large volumes of solution, higher pressures are needed. The standard apparatus at the moment is a disc filter like the one shown in Figure 37-21, p. 440. Figure 37-22, p. 440, shows exploded views of the 142-mm and a 293-mm holders. The 142-mm holder can filter up to 10 L at one time. The pressure is supplied either by a pump as shown in Figure 37-3, p. 432, or by a pressurized tank shown in Figure 37-23, p. 441. They are used for clarification of laboratory solvents, vitamins, makeup water, polymeric coatings, photoresists, cell counter solutions, tissue culture media, and blood fractions.

Figure 37-15. Top: Cracked filter cake. Bottom: Dental dam in place.

Figure 37-16. Hirsch funnels, porcelain and glass.
(Courtesy - Fisher Scientific Co., Pittsburgh, PA)

Figure 37-17. Filtering crucibles with sintered glass bottoms.
(Courtesy - Fisher Scientific Co., Pittsburgh, PA)

Figure 37-18. Types of crucible holders, (A) Inner Funnel. (B) Bailey. (C), Filtervac. (D) Walter's.

A   B   C   D

Figure 37-19. Filter holder.
(Courtesy - Millipore Corp., Bedford, MA)

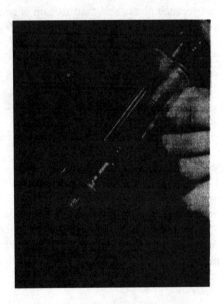

Figure 37-20. Filtering with a syringe.
(Courtesy - Gelman Sciences, Ann Arbor, MI)

Figure 37-21. A 293mm disc filter holder.
(Courtesy - Gelman Sciences, Ann Arbor, MI)

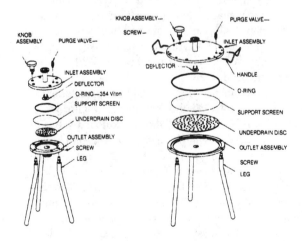

Figure 37-22. Exploded diagrams of the 142 mm and 293 mm disc filter holders.
(Courtesy - Gelman Sciences, Ann Arbor, MI)

Figure 37-23. Portable pressure vessel.
(Courtesy - Gelman Sciences, Ann Arbor, MI)

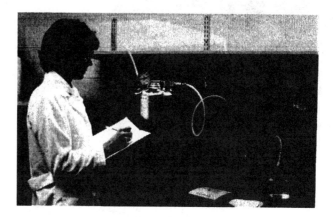

Figure 37-24. A high pressure filtering arrangement.
(Courtesy - Gelman Sciences, Ann Arbor, MI)

Pressure Vessel

Figure 37-23 shows one type of pressure vessel. It can process up to 19 L or 11 L (two sizes) of liquid for sterilization and clarification. It operates at pressures up to 7 kg/cm (100 psi) at a maximum temperature of 38 °C. Figure 37-24 shows a typical laboratory arrangement.

SOLIDS

The material on solids was abstracted in part from Gilson Co. literature by permission. Sieves for most laboratory work are round and vary from 3 to 12 inches in diameter. They are made of brass to be less static and magnetic. The screen itself is either woven wire, electroformed metal, woven cloth, or machined plastic. Each sieve is designated by its *mesh*. Mesh is *the number of holes per linear inch of screen*. A 10 mesh screen would contain 10 holes along 1 inch. *Caution:* A 10 mesh screen has 10 x 10 or 100 holes per in^2, not just 10 holes as you might suspect.

If a coarse separation is all that is needed, then shaking by hand usually is done. You need to do this only a few times, and you will try to purchase a shaker or find a "friend". The biggest mistake made is to try to force the particles through the sieve. *Do not push on the particles, let them fall of their own accord.* The screen wire is fragile, and even if not broken, the wires can be shifted, thus changing the mesh in parts of the sieve.

Figure 37-25 shows a stack of sieves on a shaker. The sieves are stacked with the smallest number on top and the highest on the bottom. The bottom pan is for collecting the very fine particles, and the top sieve has a cover. The crushed, pulverized, and/or ground sample is placed in the top sieve, the cover put in place, the clamp closed, and the shaker turned on. Each particle will fall until it can go no farther. After 10 to 15 minutes, the shaker is turned off, the sieves are separated, and each is examined. If you want the 60 to 80 fraction, as for most chromatography column packings, you collect those particles that passed through the 60 mesh screen, but not the 80 mesh screen.

The shaker shown has a vertical 2-way tapping action that automatically reverses 'up' and 'down' taps. 'Up' taps mix, reorient, and distributes

Figure 37-25. A Gilson model SS-8R sieve shaker and sieves.
(Courtesy - Gilson Co., Inc., Worthington, OH)

Figure 37-26. Left: Woven mesh. Right: Electroformed mesh.
(Courtesy - Gilson Co. Inc., Worthington, OH)

Figure 37-27. Polyester monofilament fabric.
(Courtesy - Gilson Co. Inc., Worthington, OH)

the sample so that new particles and new orientation are periodically presented to sieve surfaces. "Down" taps assist passing of near-size materials to the next sieve and clear the mesh for new particles. In addition, there is perfect circular rotation. Particles are forced to roll in all directions relative to the linear pattern of the woven sieve mesh.

Figure 37-26 shows closeups of woven mesh and electroformed mesh. The electroformed mesh is precisely formed from nickel using an electrodeposition process resulting in a planar mesh with very consistent openings. Sieves of mesh size 3 or larger holes are made of stainless steel wire cloth. Sieves of 2.8 mm and smaller have brass or phosphor-bronze wire cloth. Polyester (Figure 37-27) has good abrasion resistance, great resistance to most acids and alkalies up to pH 10, and very good wet stability. Figure 37-28 is a list of mesh sizes. The sieve number is the Tyler No. Sieves are easy to damage and should be stored properly. Figure 37-29, p. 443, shows a storage rack.

Figure 37-30, p. 443, shows an ultrasonic bath sieve cleaner. When ultrasonic energy waves are transmitted to a liquid, a pattern of microscopic bubbles forms and collapses immediately after generation. This rapid cavitation keeps particles in constant motion and frees lodged particles. It is most effective for fine mesh sizes up to about No. 30. Cleaning time is typically 2 to 5 minutes.

For accurate work, the sieves may need to be calibrated. Four organizations prepare *standard reference materials* (SRM's); the National Institute of Standards and Technology (NIST), Bethesda, MD; the Community Bureau of Reference (BCR), Brussels; Whitehouse Reference Standards; Brussels, and Duke Scientific, Palo Alto, CA. NIST uses glass beads and silicon nitride particles; BCR uses quartz beads; Whitehouse uses soda-lime glass; and Duke uses polystyrene-divinylbenzene or glass spheres. NIST and BCR standards are expensive. Figure 37-31, p. 443, shows a set of standards.

Sonic Sieving

For sonic sieving, illustrated in Figure 37-32, p. 443, a flexible wall at the bottom of the column of stacked sieves is compressed providing a vertically oscillating column of air that first lifts the particles,. upon release, then carries them back against mesh openings at 3600 pulses per minute.

Added vertical or alternating vertical and horizontal tapping actions help clear binding of near-size particles or de-agglomerate samples with electrostatic, hygroscopic, or other adhesion problems. This technique is for small samples and fine particles. The screen size is 3 inches, the sample size is 10 g for particles smaller than 38 μm and 20 g for larger particles. The particle range is from No. 20 to 5 μm. Figure 37-33, p. 444, shows a commercial model.

Coarse Series				
Order as: Sieve Designation Std.¹	Alt.	Permissible Variation of Avg. Opening from Std.	Nominal Wire Dia.² mm	Tyler Screen Scale Equivalent
125mm	5 in.	±3.7mm	8.00	
106mm³	4.24in.³	±3.2mm	6.40	
100mm	4 in.	±3.0mm	6.30	
90mm³	3-1/2 in.³	±2.7mm	6.08	
75mm	3 in.	±2.2mm	5.80	
63mm	2-1/2 in.	±1.9mm	5.50	
53mm³	2.12 in.³	±1.6mm	5.15	
50mm	2 in.	±1.5mm	5.05	
45mm	1-3/4 in.	±1.4mm	4.85	
37.5mm	1-1/2 in.	±1.1mm	4.59	
31.5mm	1-1/4 in.	±1.0mm	4.23	—
26.5mm³	1.06 in.³	±0.8mm	3.90	1.050 in.
25.0mm	1 in.	±0.8mm	3.80	—
22.4mm	7/8 in.	±0.7mm	3.50	.883 in.
19.0mm	3/4 in.	±0.6mm	3.30	.742 in.
16.0mm	5/8 in.	±0.5mm	3.00	.624 in.
13.2mm³	0.530 in.³	±0.41mm	2.75	.525 in.
12.5mm	1/2 in.	±0.39mm	2.67	—
11.2mm	7/16 in.	±0.35mm	2.45	.441 in.
9.5mm	3/8 in.	±0.30mm	2.27	.371 in.
8.0mm	5/16 in.	±0.25mm	2.07	2-1/2mesh
6.7mm³	0.265 in.³	±0.21mm	1.87	3 mesh
6.3mm	1/4 in.	±0.20mm	1.82	—
5.6mm³	No.3-1/2³	±0.18mm	1.68	3-1/2mesh
4.75mm	No. 4	±0.15mm	1.54	4 mesh

Intermediate and Fine Series				
4.00mm	No. 5	±0.13mm	1.37	5 mesh
3.35mm	No. 6	±0.11mm	1.23	6 mesh
2.80mm	No. 7	±0.095mm	1.10	7 mesh
2.36mm	No. 8	±0.080mm	1.00	8 mesh
2.00mm	No. 10	±0.070mm	0.900	9 mesh
1.70mm	No. 12	±0.060mm	0.810	10 mesh
1.40mm	No. 14	±0.050mm	0.725	12 mesh
1.18mm	No. 16	±0.045mm	0.650	14 mesh
1.00mm	No. 18	±0.040mm	0.580	16 mesh
850μm	No. 20	±35μm	0.510	20 mesh
710μm	No. 25	±30μm	0.450	24 mesh
600μm	No. 30	±25μm	0.390	28 mesh
500μm	No. 35	±20μm	0.340	32 mesh
425μm	No. 40	±19μm	0.290	35 mesh
355μm	No. 45	±16μm	0.247	42 mesh
300μm	No. 50	±14μm	0.215	48 mesh
250μm	No. 60	±12μm	0.180	60 mesh
212μm	No. 70	±10μm	0.152	65 mesh
180μm	No. 80	±9μm	0.131	80 mesh
150μm	No. 100	±8μm	0.110	100 mesh
125μm	No. 120	±7μm	0.091	115 mesh
106μm	No. 140	±6μm	0.076	150 mesh
90μm	No. 170	±5μm	0.064	170 mesh
75μm	No. 200	±5μm	0.053	200 mesh
63μm	No. 230	±4μm	0.044	250 mesh
53μm	No. 270	±4μm	0.037	270 mesh
45μm	No. 325	±3μm	0.030	325 mesh
38μm	No. 400	±3μm	0.025	400 mesh
32μm	No. 450	±3μm	0.028	—
25μm	No. 500	±3μm	0.025	—
20μm	No. 635	±3μm	0.020	—

Figure 37-28. Sieve characteristics.
(Courtesy - Gilson Co. Inc., Worthington OH)

Filtering and sieving

Figure 37-29. A sieve storage rack.
(Courtesy - Gilson Co. Inc., Worthington, OH)

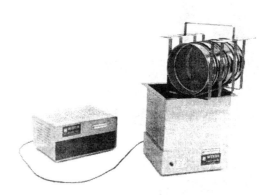

Figure 37-30. Ultrasonic bath sieve cleaner.
(Courtesy - Gilson Co. Inc., Worthington, OH)

The Autosiever is designed to automatically take full advantage of the unique capabilities of sonic sieving. The intensity of the sieving action may be varied by changing the amplitude of the sonic pulses. Test times vary with sample size, density, texture, and quantity. Five minutes is typical, but sieving time can be as brief as 30 seconds. Electronic controls give exact repeatability of programmed time and amplitude sequence. Amplitude may be set precisely with a digital LED 0-99 even increment scale.

Wet Sieving

Wet sieving is used to overcome particle agglomeration, pan caking, electrostatic attraction and other problems that inhibit proper sieving of many fine materials on a dry basis. There are two types, open air and vacuum. The sieves for open air have a higher wall to retain the liquid. While water is most often used, it has a high surface tension, resulting in low flow rates through small screens. Other polar liquids or hydrocarbons containing some dispersant may be used." Figure 37-34, p. 444, shows a sieve for hand swirling. Figure 37-35, p. 444, shows a sieve used under a water faucet. The nozzles shown below easily attach to any standard faucet.

For fine particles and to increase speed, a Wet-Vac can be used. This is shown in Figure 37-36, p. 444,. It adds a slight vacuum (2 inches of Hg) to pull the water through the sieve, then recycles it. Figure 37-37, p. 444, compares the results of dry sieving a coal sample to a Wet-Vac.

Typical Wet-Vac uses include coal and mining and mineral process design or control, investigations of soils or ocean and river sediments, control of slurries or non-soluble chemicals and powders.

Figure 37-31. Sieve reference standards.
(Courtesy - Gilson Co. Inc., Worthington, OH)

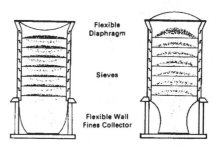

Figure 37-32. Diagram of the operating principle for sonic sieving.
(Courtesy - Gilson Co. Inc., Worthington, OH)

Figure 37-33. The Gilson model GA-6 Gilsonic Autosiever.
(Courtesy - Gilson Co. Inc., Worthington, OH)

Figure 37-34. Wet-washing sieve.
(Courtesy - Gilson Co. Inc., Worthington, OH)

Figure 37-35. Faucet spray wet sieving.
(Courtesy - Gilson Co. Inc., Worthington, OH)

Figure 37-36. Gilson Wet-Vac.
(Courtesy - Gilson Co. Inc., Worthington, OH)

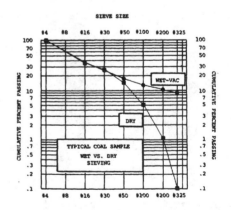

Figure 37-37. Comparison of dry and Wet-Vac sieving on a coal sample.
(Courtesy - Gilson Co. Inc., Worthington, OH)

REVIEW QUESTIONS

1. What is considered to be a sieve?
2. What is the particle size range of tobacco smoke?
3. What does HEPA stand for?
4. How good are HEPA filters?
5. What is the standard number required when determining particles in air?
6. If 13.5 L/min of air are drawn into an air sampling apparatus for 10 minutes every 2 hours during a day and ragweed pollen counts were 12, 15, 18, 16, 15, 11, 10, 13, 9, 9, 7, and 5, what is the ragweed pollen count per cubic meter per day?
7. What are the properties of a depth filter?
8. Why are most new gravity type analytical filtering funnels made at 58°?
9. What is the purpose of flutes on a gravity filtering funnel and are they effective?
10. How do they make ashless filter paper?
11. What is phase separation paper and how is it used?
12. What is the difference in ASTM flow
13. rates between Whatman No. 41, 40, and 42 filter papers?
14. How much faster is gravity filtering with medium-grade filter paper using good technique than poor technique?
15. What is the purpose of having a long stem on a gravity-type filtering funnel?
16. Why does just one air bubble in the funnel stem slow down the rate of filtering?
17. Why should the tip of the stem be placed against the sidewall of the beaker?
18. What is a dental dam, and what is it used for?
19. What does 50 mesh mean?
20. Is it acceptable to rub the sample over the screen wire mesh in order to do a final fine grinding?
21. If you wanted to get 150 µm particles, what size sieve would you use?
22. What is the recommended way to clean the above sieve?
23. What is an SRM?
24. What materials does NIST use for its reference sieve standards?
25. What is the purpose of wet sieving?
26. If you wanted to filter 200 mesh coal particles, what would you expect to happen if you used a Wet-Vac instead of a dry sieve?

IF IT EXISTS - CHEMISTRY IS INVOLVED
IF YOU CAN BUY IT - A CHEMIST WAS INVOLVED SOMEWHERE

SEPARATIONS INVOLVING MISCELLANEOUS TECHNIQUES

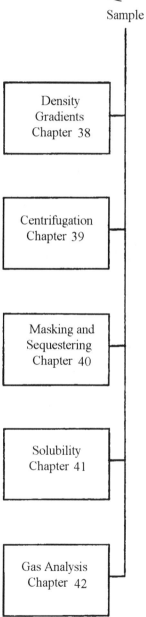

These last five chapters involve techniques that are important, but which do not lend themselves to be related to other subdivisions. Density is placed first because the use of *density gradients,* Chapter 38, is important in forensic science and in biochemical separations. Once the concept of density gradients has been established, then the principles of the *ultracentrifuge*, Chapter 39, become clear.

The use of *masking and sequestering agents*, Chapter 40, are probably the truest chemical separations there are. These reagents react with interfering materials in the solution and stay there, thus eliminating one or more subsequent steps.

It is not generally considered good technique to attempt to selectively dissolve one component from a mixture because it is seldom quantitative. However, selective *solubility* can be used to concentrate and recover large amounts of material. Even more basic is the problem of how to dissolve the materials to start with. Putting both inorganic and organic compounds into solution is discussed in Chapter 41.

Before the advent of the gas-liquid chromatograph, O_2, CO, CO_2, N_2, olefins, and several saturated gaseous organic compounds were separated and measured by selectively absorbing the gases and combusting the organic compounds. Chapter 42 illustrates the concepts. The volume of the reactants and the combustion products are then measured. This includes the volume of O_2 consumed, the CO_2 produced, and the contraction due to the formation of water. This in known as the Orsat apparatus. It is still in use, although very sparingly. It is included because it involves a true volumetric analysis, in that the measurement of volume is the only requirement for quantitation and the separation of gases by selective absorption is truly a chemical separation. The portable Orsat is quite convenient for measuring CO and CO_2 in buildings. The author used this to determine CO_2 at the bottom of a 100 ft tall dried-hay storage bin. Two workers were overcome and no one knew why.

38
DENSITY GRADIENTS

PRINCIPLES

Density is a physical property of matter that is specific to the sample being measured and can be used as a means of identification or comparison, whichever is required. The density (D) of a substance is defined as a mass (M) per unit volume (V) and given by the following equation:

$$D = \frac{M}{V} \tag{38-1}$$

Densities of gases usually are expressed in g/L, and those of liquids and solids as g/mL or cm^3. Engineers prefer lb/ft^3. Table 38-1 shows the densities for several liquids and solids.

Table 38 - 1. The densities of various liquids and solids

Liquids		Solids	
Compound	Density(g/mL)	Compound	Density(g/cm^3)
Gasoline	0.68	Cork	0.22 - 0.26
Rubbing alcohol	0.785	Bone	1.7 - 2.0
Kerosene	0.82	Glass, window	2.47 - 2.56
Olive oil	0.916	Flint	2.63
Water	1.0	Aluminum	2.70
Sea water	1.025	Iron	7.86
Milk	1.028 - 1.035	Brass, yellow	8.44 - 8.70
Chloroform	1.492	Lead	11.34
Mercury	13.6	Gold	19.3

Density measurements are extremely important in industry, because a large amount of information can be obtained about a material with a reasonably simple measurement. By measuring the density of liquids, we can determine how much electrical charge is in your car battery, at what temperature your radiator will freeze, the salt content of a pickling brine, the alcohol content of a brandy, and certain diseases from urine samples. The densities of solids can tell us the hydration degree and porosity of Portland cement, the cotton-polyester blend in a fabric, the velocity of detonation of an explosive, and whether or not a particular soil found on a suspect matches that on a murdered person's head scarf. Rotten eggs float, whereas good eggs sink in water. The flux used on welding rods is made so that after it has cleaned a metal's surface, it will float to the top of the molten weld and not weaken the bond. Sawdust added to pepper as an adulterant can be detected because wood floats. Density measurements or the application of density are involved to some extent with almost all commercial products. The purposes of this chapter's experiments are to show you some of the various techniques for determining density and how density can be used as a separation technique.

Most people know that a nail will sink if placed on water, but few realize that a nail will float on mercury. A log of mahogany or beechwood will float on water but will sink if placed in gasoline.

An object placed in a fluid will sink if its density is greater than that of the surrounding fluid, float if its density is less than that of the surrounding fluid, and stay suspended if the densities are the same.

Galileo, about 350 years ago, illustrated this concept by placing a layer of fresh water over a layer of very salty water and then adding a ball of tallow. The tallow sank through the fresh water layer but floated on the more dense salty water.

An object immersed in a fluid displaces a volume of fluid equal to its own volume.

This principle was discovered by Archimedes (287-212 B.C.) when he was given the problem of determining whether the crown on King Herion the II's head was pure gold or a cheaper alloy. 1.0 cm^3 of gold weighs 19.3 grams, but 19.3 grams of lead (d = 11.3 g/cm^3) will occupy 1.7 cm^3 of volume. By weighing the crown and then submersing it in water, Archimedes was able to determine its volume to see if it was made of pure gold. It was not.

SPECIFIC GRAVITY

The specific gravity of a liquid or a solid is the *ratio of the density of the material to the density of water.* Therefore, it is essentially independent of temperature. For gases, the comparison material is air.

EXAMPLE CALCULATION

The density of oil of wintergreen is 1.184 g/mL at 25 °C, whereas that of water is 0.9970 g/mL at 25 °C. What is the specific gravity of oil of wintergreen?

ANSWER

$$\frac{1.184 \text{ g/mL}}{0.997 \text{ g/mL}} = 1.188 \qquad \text{(Notice that specific gravity has no units)}$$

DENSITY OF SOLIDS

If a solid object has a regular shape such as a cube, rectangle, rod, or sphere, then its density may be obtained by weighing it to obtain its mass and then calculating its volume by geometry. If the object is irregular in shape, such as a piece of a belt buckle, a coat hanger, a rock, or a spider, then the object is weighed first in air and then submersed in water and weighed again to obtain its volume. It will lose 1.0 g for each cm^3 of water it displaces.

EXAMPLE CALCULATION

A pair of eyeglasses weighs 60.0 g in air and 26.0 g when weighed completely submersed in water. What is the volume occupied by the eyeglasses?

ANSWER

$$\text{Volume (cm}^3\text{)} = \text{weight in air - weight in water} \qquad (38\text{-}2)$$
$$= 60.0 \text{ g} - 26.0 \text{ g}$$
$$= 34 \text{ cm}^3$$

EXAMPLE CALCULATION

A Susan B. Anthony dollar weighs 8.027 g in air and 7.127 g when weighed suspended in water. What is its volume and what is its density?

ANSWER

$$\text{Volume (cm}^3\text{)} = \text{weight in air (g) - weight in water (g)}$$
$$= 8.027 \text{ g} - 7.127 \text{ g}$$
$$= 0.900 \text{ cm}^3$$

$$\text{Density (g/cm}^3\text{)} = \frac{\text{Weight air (g)}}{\text{Volume (cm}^3\text{)}} = \frac{8.027}{0.900} = 8.92 \text{ g/cm}^3$$

TECHNIQUES

DETERMINING THE DENSITY OF SMALL IRREGULARLY SHAPED OBJECTS

How can you determine the density of particles that are so small that you may have to use a magnifying glass to see them? One way, used in forensic science, is described below.

Recall that *an object placed in a fluid will sink if its density is greater than that of the surrounding liquid, will float if its density is less than the surrounding fluid, and will stay suspended if the densities are the same.*

The basic idea is to place a liquid more dense than the sample in the bottom of a small vessel. The sample then is added and will float on top of the liquid. A second, less dense liquid, then is added dropwise with stirring until the sample object just stays suspended. The density of the object is now the same as the liquid. The unknown then is added and if the densities are the same, it will behave the same as the reference object; otherwise, it will sink or float. This is a quick means to compare samples without needing to know the exact densities and is particularly valuable when comparison density measurements are needed only occasionally. The problems proposed below will serve to illustrate the practical application of this principle.

Animal hairs:	Heavy liquid: $ZnCl_2$; 5 g/ 10 mL	Light liquid: Water
Explosives:	Heavy liquid: $ZnCl_2$ 0.5 g/25 mL	Light liquid: Water
Cloth fibers:	Heavy liquid: $ZnCl_2$; 8 g/10 mL	Light liquid: Water
Watch glass:	Heavy liquid: Bromoform	Light liquid: Bromobenzene
Tire sidewall rubber:	Heavy liquid: $ZnCl_2$; 5 g/10 mL	Light liquid: Water
Real or plastic turquoise:	Heavy liquid: Bromoform	Light liquid: Bromobenzene

DENSITY GRADIENTS

In the discussion above, a solution was prepared that matched the density of the standard sample. What was that density? What was the density of the sample that either floated slightly or just sank? Do we have to repeat the experiment several times until we find out? There must be a better way. The better way is called a density gradient, and a diagram of a density gradient tube is shown in Figure 38-1, p. 452. The lowest density liquid is on the top, and the density gradually increases until the most dense material is reached at the bottom. An object dropped into the top of the tube will sink until it reaches liquid of its same density and then will remain suspended there. *Particle size makes no difference, except that big particles will reach their equilibrium density faster than smaller particles.*

Producing Density Gradients

There are three main ways to prepare a density gradient column. The first is to prepare several liquids of different densities and pipet a 2 cm layer of each, heaviest first, into a column about 30 cm x 1 cm, closed at the bottom. After the column stands for a day or so, a smooth gradient usually is obtained. The second method begins the same as the first, except that a small magnet is placed inside of the column and moved slowly vertically several times by using a second magnet on the outside of the column. The third method is more involved, but once the apparatus is available, it is easy to prepare large columns over any selected range of densities. Figure 38-2, p. 452, shows a commercial apparatus, and Figure 38-3, p. 453, shows a homemade apparatus.

The apparatus in Figure 38-2, p. 452, employing two vessels, is capable of producing gradients from 0.8 to 2.7 g/cm³, to 0.0001 g/cm³, requiring 2 to 4 hours. The jacket operates from below ambient to 85°C. It contains a mesh basket that can be raised or lowered at 2 cm/min to dispense and retrieve density beads or samples.

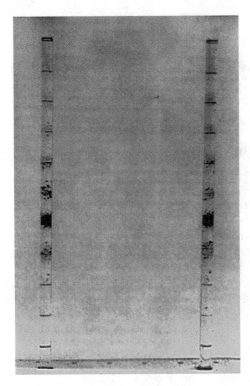

Figure 38-1. Soil density profile.
(Courtesy - New Jersey State Police)

Figure 38-2. A density gradient column preparer.
(Courtesy - VWR Scientific Products, McGraw Park, IL)

The apparatus in Figure 38-3, p. 453, is much the same, except that all controls are manual. Close all stopcocks. Pour 500 mL of the more dense liquid into flask A and 500 mL of the less dense liquid into flask B. Start the stirrer and open stopcock A to achieve pressure head equilibrium. Open stopcock B and regulate the flow so that the column will take a minimum of 1.5 hours to fill.

Determining the Density

The various densities along the column can be determined by adding objects of known density and noting their positions. The least expensive method is to add a few crystals of several different compounds that are insoluble in the density gradient liquid. Table 38-2 lists several crystals than can be used in organic liquids.

Table 38-2. Crystals of different densities

Compound	Density (g/cm^3)	Compound	Density (g/cm^3)
Potassium acetate	1.57	Potassium fluoride	2.48
Borax decahydrate	1.73	Sodium chlorate	2.49
Sodium tartrate dihydrate	1.82	Potassium perchlorate	2.52
Potassium ferricyanide	1.85	Sodium fluoride	2.56
Potassium nitrite	1.91	Magnesium sulfate	2.66
Potassium chloride	1.98	Potassium dichromate	2.65
Potassium nitrate	2.10	Potassium permanganate	2.70
Sodium chloride	2.16	Sodium chromate	2.72
Sodium nitrate	2.26	Silver perchlorate	2.80
Potassium chlorate	2.32		

Small balls, 1/8 inch in diameter, can be obtained commercially from VWR or Fisher Scientific Co. covering the range from 0.8 to 2.0 or 0.8 to 2.6 g/cm^3 with 2, 3, or 4 place accuracy. Liquids insoluble in water are used for aqueous columns, and crystals insoluble in organic solvents are used for the nonaqueous systems. Once the reagents are added, the tubes are let stand for 24 hours. The 2nd Law of Thermodynamics takes over, and a smooth gradient is obtained. A faster method, but which requires some care, is to use a small magnet on the inside of the tube and one on the outside to mix the liquids. Be careful *not* to make the solution homogeneous.

Calculating the Density of a Mixture

Equation 38-2 can be used to determine the amount of each solvent to use to obtain a certain density.

$$\text{Desired Density} = \frac{D_1 \times V_1 + D_2 (V - V_1)}{V} \quad (38\text{-}2)$$

where D_1 = the density of the lighter liquid. D_2 = the density of the heavier liquid. V_1 = the volume of the lighter liquid. V = the total volume.

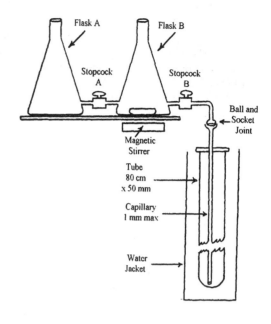

Figure 38-3. A simple density gradient forming apparatus.
(Courtesy - The Textile Institute, Manchester, England)

EXAMPLE CALCULATION

A mixture of hexane (d = 0.660 g/cm^3) and carbon tetrachloride (d = 1.589 g/cm^3) with a density of 1.28 is needed to dissolve the synthetic polymer Kodel IV. If you pipet 20 mL of hexane into a beaker, how many mL of carbon tetrachloride will have to be added?

ANSWER

$$1.28 \text{ g/cm}^3 = \frac{0.660 \text{ g/cm}^3 \times 20 \text{ mL} + 1.589 \text{ g/cm}^3 (V - 20 \text{ mL})}{V}$$

V = 60 mL; therefore, 60 mL - 20 mL = 40 mL of CCl$_4$ to be added.

An Application

Probably the best way to illustrate the practical applications of the use of a density gradient column is to use an actual case. About 1902, the German criminalist, Dr. Gerhard Popp was faced with the following problem. A young lady had been strangled with her head scarf, and the police had a prime suspect. He denied doing it or even being in the area. Dr. Popp noticed that there was soil on the head scarf, and a search of the suspect's room revealed a pair of pants with soil on the knees. Were they the same? Soil is made up of many components, and each has its own density. Would a density profile of the soil on the head scarf match that of the soil on the pants? How similar would a soil a few feet away be? Using a density gradient column, Dr. Popp was able to show that the soil on the head scarf, on the pants, and at the scene were the same, yet the control sample was quite different. Figure 38-1 is similar to what Dr. Popp obtained.

REVIEW QUESTIONS

1. What units are the densities of liquids and gases usually expressed in?
2. What is specific gravity, and what are its units?
3. The density of milk at 25 °C is 1.032 g/cm^3. What is its specific gravity, if water is 0.997 g/cm^3 at 25 °C?

4. I weigh 226 lb in air and 9 lbs when swimming. What is my volume, if water weighs 62.4 lb/ft^3?
5. What is meant by a "density gradient"?
6. Examine Figure 38-1 and notice that there appears to be concentrations of particles around where the gradient marks are. What causes this? Is this good or bad and can it be altered?
7. Soil samples are taken from three places, a head scarf, trousers, and an area in a park at the crime scene. A portion of each sample is dried and then ground to a fine powder. Did the grinding ruin the samples?
8. A 43/57 hexane ($d = 0.660$ g/cm^3) to CCl$_4$ ($d = 1.589$ g/cm^3) is needed to dissolve Darvan, a synthetic polymer. What is the density of this mixture?

THE SCIENTIFIC METHOD

1. CHANGE ONLY ONE VARIABLE AT A TIME.
2. RUN A BLANK OR A CONTROL.
3. THE FINAL EXPERIMENTS SHOULD BE BLIND OR DOUBLE BLIND.

The *results* of the scientific method are theory and law.

39
CENTRIFUGATION

PRINCIPLES

Centrifugation is a technique that *separates materials based upon their difference in densities, the rate of separation being amplified by applying a rotational force.* The force is called a *centrifugal force,* and the apparatus providing the rotational force is called a *centrifuge.*

Centrifuges commonly are divided into three groups based on their rotational speed and the gravitational force, g, generated; bench top (0 to 15,000 $x\ g$), high speed (15,000 to 75,000 $x\ g$), and the ultracentrifuges (75,000 to 390,000 $x\ g$ or higher). Bench top types require no cooling or special shielding from flying fragments in case of a malfunction. The high speed types require cooling, maybe a reduced pressure H_2 atmosphere, and special shielding, and the ultracentrifuges require a vacuum, cooling, and heavy shielding. It has been calculated that a fragment from a 78,500 rpm centrifuge will produce the same force as the projectile from an 8 inch gun! This is why they (you) are protected by heavy steel armor plate.

The most common separations involve separating a solid or mixture of solids from a liquid, such as a fine suspension of $BaSO_4$ precipitate. This can take hours if left to settle naturally, but can be separated in a few minutes in a centrifuge. Centrifugation often is used to break emulsions. Modern ultracentrifuges can be used to separate at cellular and molecular levels and to determine molecular weights.

The sample is placed in a cell, tube, or bottle and placed in a holder attached to a rotor. When the rotor turns, it spins the sample in a circular motion, and the force of gravity increases as the speed increases. This added force drives the most dense particles to the bottom of the tube; the less dense movie there as well, but slower. The liquid on top of the collected precipitate is decanted away. Under well-controlled conditions, it is possible to separate a mixture into discrete layers in the suspending liquid.

Most scientists will use a bench top centrifuge sometime in their career to separate a precipitate too small to filter, a few will use a high speed centrifuge, and only a very few will use an ultracentrifuge. Because the theory for centrifugation encompasses all types, the theories presented below are for centrifugation in general and the ultracentrifuge in particular.

THE ULTRACENTRIFUGE

Why would you want a very high speed centrifuge? If you have ever played crack the whip, you know that even at that speed, a heavy particle (person) has a large force pulling them away from the line. If you place a variety of particles in a tube and spin it, the particles will be forced to the bottom of the tube. You can easily see big particles do this, but what about particles at the molecular level? They are usually more dense than water molecules so they should eventually go to the bottom as well. However, they are so small that is difficult for them to overcome the frictional forces from the solvent, and because they are so small, they want to diffuse as well. If they are to be forced to the bottom of the tube, then a large force must be applied, and if this is to be done in a short time, then an even larger force is required. With a modern ultracentrifuge, it is possible to determine molecular weights, sedimentation coefficients that can be used to identify compounds, and diffusion coefficients in various solvents.

T. Svedberg pioneered the first high speed centrifuges in the mid-20's in an attempt to measure the molecular weights of proteins. The first centrifuges that used optics to determine the position of the sedimenting layer were called *ultracentrifuges*. Today, an ultracentrifuge is more related to very high speed, regardless of the detection system. Spinning a tube of sample in a circle at high speeds can cause many problems. The sample heats considerably from the friction of the air, so a vacuum must be used. The rotor bearings get hot, so a cooling gas such as hydrogen is sometimes used or the system is refrigerated. The construction material must be strong to prevent fragmentation, and if the progress of the sediment is to be monitored, then some means must be made to see through the tube as it rotates at a very high speed. If the system is out of balance by even a small amount, disaster can occur.

OPTICAL SYSTEMS

In order to determine the molecular weight of a substance or to separate compounds at the cellular level, the rate of settling needs to be known. Once the rotation begins, the compounds will begin to collect and move down the tube. The portion at the very top is now solvent and has a *meniscus*. Where the compounds begin to collect is called the *boundary*. This boundary is detected and followed for a few centimeters to determine the rate. There are several ways to do this; the oldest is discussed first.

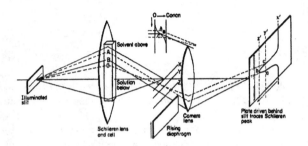

Figure 39-1. A scanning schlieren system.
(Courtesy - Bowen, T.J. and Rowe, A.J., *An Introduction to Ultracentrifugation*, Wiley-Interscience, New York, NY, 1970)

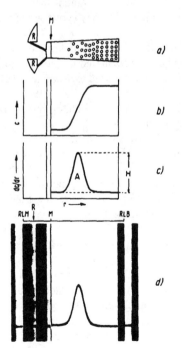

Figure 39-2. Concentration distribution; evaluation with a schlieren cell.
(Courtesy - Nicolau, C., *Experimental Methods in Biophysical Chemistry*, John Wiley & Sons, New York, NY, 1973).

Schlieren Optics.

This is the German name for streaks and not the name of a scientist. The optical diagram is shown in Figure 39-1.

Figure 39-2 shows how the boundary looks and is followed.

Abstracting from T.J. Bowen and A.J. Rowe (*An Introduction to Ultracentrifugation*, Wiley-Interscience, New York, NY, 1970), we have an illuminated slit, a lens which we now call the schlieren lens, and an image of the slit formed at X. The schlieren lens has no faults, but we have placed near it a cell in which there is an optical "fault" in that the cell contains solvent at the top forming a boundary with solution below. B marks the plane of maximum concentration (i.e. refractive index) gradient, while A and C represent two other planes where the concentration gradient is equal, but of course, less than at B. Rays suffer maximal deflection at B to give the image at Z, while rays from B and C meet at Y where, incidently, they can interfere so that the images below X actually show interference banding due to the unequal path lengths traversed. Let us move a horizontal diaphragm up in the focal plane, while simultaneously moving the plate sideways behind a slit. If the diaphragm is below Z, no light is intercepted. When the diaphragm reaches Z, a dark shadow appears on the plate corresponding to the position B in the cell. When the diaphragm reaches Y, the dark shadow on the plate corresponds to the region from A to B in the cell. When the diaphragm reaches X, no light at all can reach the plate. Since the plate moved continuously behind the slit, we have produced a black image (on a positive print) in the form of a peak corresponding to the gradient or refractive index in the cell.

Refer to Figure 39-2. (a) Cell with particles, R: reservoir connected to the sector with a capillary. (b) Concentration as a function of the

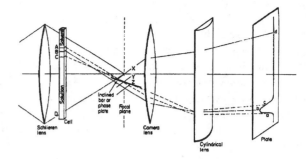

Figure 39-3. Philpot-Svensson cylindrical lens system.
(Courtesy - Bowen, T.J. and Rowe, A.J., *An Introduction To Ultracentrifugation*, Wiley-Interscience, New York, NY, 1970)

distance r to the rotor axis. (c) Concentration gradient as a function of the distance r to the rotor axis. (d) Schlieren pattern.

The Philpot-Svensson Optics

Figure 39-3 shows the cylindrical lens system as described by Bowen and Rowe. "The illuminated slit to the left has been omitted in the diagram. The cylindrical lens scans the inclined bar from side to side and can be considered to have replaced the necessity for the moving plate in the scanning arrangement. By using an inclined diaphragm to cut the rays X, Y, and Z, the necessity for a moving horizontal diaphragm is obviated. The same result will be produced in that a peak appears on the plate."

The Rayleigh Optics

Rayleigh optics is another modification of the basic Schlieren optics. Figure 39-4 is a diagram of such a system, and Figure 39-5 shows more detail of the fringe pattern.

An interferometer is an apparatus used to measure the refractive index of a gas or liquid. Light from a slit is rendered parallel by means of a collimating lens and then passed through two slits in front of a sample cell and a reference cell. A second lens focuses these two beams on a plate.

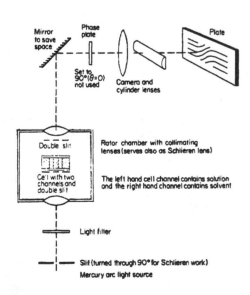

Figure 39-4. Arrangement for producing Rayleigh fringes modifying a schlieren system.
(Courtesy - Bowen, T.J. and Rowe, A.J., *An Introduction to Ultracentrifugation*, Wiley-Interscience, New York, NY, 1970)

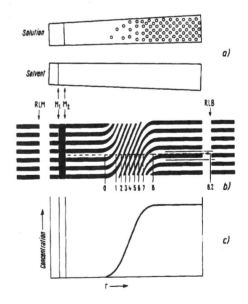

Figure 39-5. Concentration distribution in the double sector cell and its evaluation with the optical interference system.
(Couretsy - Nicolau, C. ed., *Experimental Methods In Biophysical Chemistry*, Wiley-Interscience, New York, NY, 1973)

If the refractive index is the same in both paths, then the two beams reinforce each other (constructive interference), and a bright line appears. If the paths differ, then destructive interference occurs and a dark line appears. Refractive index is a measure of concentration, so the change in concentration in the cell can be measured.

Philpot and Cook (*Research*, **1**, 234, 1948) showed that the introduction of a cylindrical lens to focus the cell contents in the final image made it possible to obtain a set of fringes, each of which gave a curve relating refractive index to the height of the cell. This is shown in Figure 39-4, p. 457.

Refer to Figure 39-5, p. 457: (a) A double sector cell. At the top, a solution with particles, below, solvent alone. (b) Scheme of an interference pattern. The number of fringes crossing the horizontal is approximately 8.2. (c) Concentration as a function of the distance r to the rotor axis.

Spectrophotometric Method

A beam of radiation, which contains one or more wavelengths that are absorbed by the compounds under investigation, is passed through the cell. The amount absorbed can be detected either by photographic means or by a phototube. A double-sector system usually is used, with one side containing solvent and the other side sample. The solvent is subtracted from the spectrum and the results are similar to that shown in Figure 39-6: (a) A single sector cell with half-sedimented particles. (b) Concentration as a function of the distance r to the rotor axis. (c) Camera picture. (d) Scan recording centrifugation in the double-sector cell. At the right are the calibrating steps, each 0.2 absorption units; below is the first derivative of the concentration distribution.

The calibration is done by generating several standard-size pulses corresponding to 0.2 absorption differences. If the system is scanned with phototubes, then the results are available immediately, and a computer calculates everything you want to know.

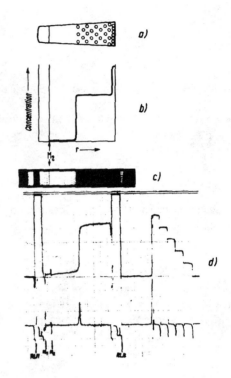

Figure 39-6. Evaluation by means of an optical absorption method.
Courtesy - Nicolau, C., ed., *Experimental Methods in Biophysical Chemistry*, Wiley-Interscience, New York, NY, 1973)

THEORY

Centrifugal force is the force pulling an object away from the center of a rotating system.

$$F = ma = m\omega^2 r \tag{39-1}$$

where F = the force on the particle in dynes. m = the mass of the particle in grams. ω = the angular velocity of rotation in radians per second. r = the radial distance of a particle from the axis of rotation in cm. This can be expressed in grams, if F is divided by g, the gravitational constant, 980.7 cm/sec².

$$F = m\omega^2 r/g \tag{39-2}$$

$$F = 1.118 \times 10^{-5} \times m \times r \times N^2 \tag{39-3}$$

where N = the speed of rotation in rpm.

A more useful concept is the *relative centrifugal force* (RCF), which is the force acting on a given particle in a centrifugal field in terms of multiples of its weight in the earth's gravitational field, and

$$RCF = 1.118 \times 10^{-5} \times r \times N^2 \tag{39-4}$$

Because the radial distance from the axis of rotation and the depth of the liquid in the centrifuged tube vary widely with different types of centrifuges, reporting results in terms of the speed of rotation in rpm is inadequate. Reporting data in terms of centrifugal force ($x\,g$) calculated from equation 39-4 is more meaningful.

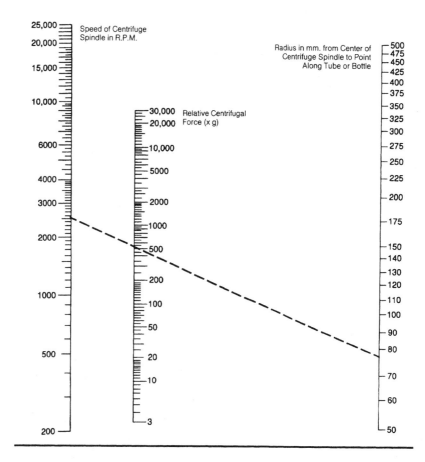

Figure 39-7. Nomogram for computing relative centrifugal forces.
(Courtesy - Corning Glass Co., Corning, NY)

EXAMPLE CALCULATION

A bench type centrifuge spins at a constant 1760 rpm, and the average distance from the center of the rotor to the tube is 6 inches. What is the RCF?

ANSWER

$$RCF = 1.118 \times 10^{-5} \times 6.0 \times 2.5 \times 1{,}760^2 = 520\,g$$

Figure 39-7 is a nomograph for determining RCF. Measure the radius in mm from the center of the spindle to the particular point on the tube. Draw a line from this radius value on the right hand column to the appropriate centrifuge speed on the left hand column. The RCF value is the point of intersection with the middle column.

The velocity at which a particle settles depends on the viscosity of the suspending liquid and also on the buoyancy effect of the liquid. Every particle is buoyed up by the volume of liquid it displaces and the density of that liquid. This is Archimedes' principle.

$$m_{eff} = m - V\rho \tag{39-5}$$

where V = particle volume in the solution. ρ = density of the solution.

$$F_{eff} = (m - V\rho)\omega^2 r \tag{39-6}$$

If the particle is a sphere

$$F_{eff} = \left(\Delta\rho \frac{\pi}{6} D^3\right)\omega^2 r \tag{39-7}$$

where D = the particle diameter. $\Delta\rho$ = the difference between the density of the particle and the suspension liquid. Just as in electrophoresis, there is a retarding force, Stokes law.

$$F = 3\pi\eta D V_s \tag{39-8}$$

where η = the viscosity of the liquid. V_s = the velocity of the particle moving through the liquid. Combining equations 39-7 and 39-8:

$$V_s = \frac{\Delta\rho D^2 \omega^2 r}{18\eta} \tag{39-9}$$

Note: The old unit for viscosity was the Poise = 1 dyne-sec/cm². The new SI unit is the Pascal-second. To convert, 1 Poise x 0.1 = 1 Pascal-second (Pa s)

EXAMPLE CALCULATION

What is the settling velocity of a 1.0 micron diameter BaSO₄ particle, density 4.50, placed in water at 25 °C, density 0.997, viscosity 893 µPa s, in a centrifuge tube at an average of 6.00 inches from the center of a rotor turning at 1760 rpm?

ANSWER

1.0 microns = 1 x 10⁻⁴ cm.

$$V_s = \frac{(4.50 \, g/cm^3 - 0.997 \, g/cm^3)(1 \times 10^{-4} \, cm)^2 \left(\frac{2 \times 3.14 \times 1760}{60 \, sec}\right)^2 \times 6.0 \, in \times 2.54 \, cm/in}{18 \times 0.00893 \, g/cm\text{-}sec} = 0.112 \, cm/sec$$

Sedimentation Coefficients

You can actually see a particle of precipitate being separated from its surrounding liquid. However, if the particle is a nuclear fragment or a large molecule, it may not be seen, and even if many fragments or molecules are compressed together, they may be seen only with special optical systems described in the detector section. One way to help identify compounds is to determine how fast they settle during a centrifugation. This requires a detector that will determine where they are in the tube, both early and later in the separation. If the distance traveled is determined as well as the time to travel that distance, then a *sedimentation coefficient*, s, can be calculated for that compound.

$$s = \frac{1}{r\omega^2} \frac{dr}{dt} = \frac{2.303}{60 \, \omega^2} \left(\frac{d \log x}{dt}\right) \tag{39-10}$$

A plot of the logarithm of the distance of a peak from the center of rotation (cm) *vs.* time (minutes) should give a straight line, and s can be determined from the slope. s for most cellular compounds varies from 10 to 200 $\times$ 10^{-13} seconds.

One Svedberg unit (1 S) is equivalent to $s = 10^{-13}$ seconds.

Figure 39-8 shows several sedimentation coefficients corrected to standard conditions.

The above sedimentation coefficient is valid for the conditions at which it was obtained. However, to make sedimentation coefficients of value for identification purposes, they must be converted to a set of standard conditions. The reference solution is water at 20 °C ($s_{20,w}$), and the variables requiring correction are the viscosity (η); the density (ρ); and the partial specific volume ($\bar{v}$), which is the volume change occurring when a particle is placed in an excess of solvent. Equation 39-11 below shows how it is corrected.

$$s_{20,w} = s_{obs} \left(\frac{\eta_t}{\eta_{20}} \right) \left(\frac{\eta_{sol}}{\eta_w} \right) \left(\frac{1 - \bar{v}\rho_{20,w}}{1 - \bar{v}\rho_{t,sol}} \right) \quad (39\text{-}11)$$

Viscosity is a measure of the resistance to flow. Viscosities are measured with a capillary viscometer (Figure 39-9) in a constant temperature bath. In actual practice, only the viscosity relative to water is needed. Therefore, the ratio of the time for the solvent to pass two points to that of water can be substituted directly into the equation. The viscometer shown requires 7 mL of liquid. The time it takes for the liquid to pass from mark 1 to mark 2 is measured.

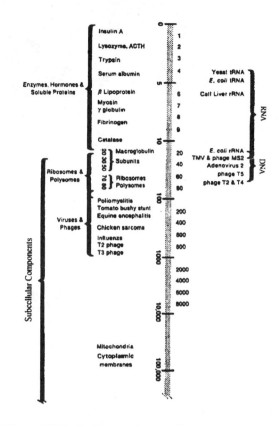

Figure 39-8. Sedimentation coefficients, $s_{20,w}$. (Courtesy - Beckman Instruments Co., Fullerton, CA)

$$\frac{\eta_1}{\eta_2} = \frac{\rho_1 t_1}{\rho_2 t_2} \quad (39\text{-}12)$$

Density is best measured with a *pycnometer* (Figure 39-10). The weight of the empty container is determined, then the container is filled with the sample liquid and weighed again. If the volume of the container is known, then the density can be calculated. If the volume is not known, then water can be determined first, and the sample and water values compared. This gives a specific gravity instead of density, but usually the difference is negligible for dilute aqueous solutions.

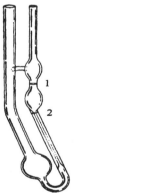

Figure 39-9. A Cannon-Fenske viscometer. (Courtesy - Ace Glass Co., Vineland, NJ)

Figure 39-10. A Weld pycnometer. (Courtesy - Ace Glass Co., Vineland, NJ)

EXAMPLE CALCULATION

What is the sedimentation coefficient in Svedberg units if at 16.0 minutes, x was 12.582 cm and at 24.0 minutes, x was 12.720 cm in a centrifuge rotating at 52,500 rpm?

ANSWER

Log 12.720 = 1.1045; log 12.582 = 1.0997; therefore, $d \log x = 0.0048$
$\omega = 2 \pi \text{ rpm}/60 = 0.1047 \text{ rpm}$.

$$s = \frac{2.303}{60 \, (0.1047 \times 52{,}500)^2} \frac{0.0048}{(24 - 16)} = 7.62 \times 10^{-13} \text{ seconds.} \quad s = 7.62 \, S$$

Determination of Diffusion Coefficients

The diffusion coefficient, D, is involved in many calculations because it must be overcome in order to completely sediment small particles and is necessary to determine molecular weights. The ultracentrifuge method is to form a sharp boundary in the cell using a high speed, then to lower the speed and allow the boundary to spread, but not sediment it. The most accurate method is to use the interference fringes as they change over time. The basic equation is 39-13, and the correction equation is 39-14.

$$D = \frac{\Delta r^2}{4 y^2 t} = \frac{\Delta r^2}{t} \frac{1}{3.64} \tag{39-13}$$

$$D_{20,w} = D_{obs} \left(\frac{293.2}{T} \right) \left(\frac{\eta_1}{\eta_{20}} \right) \left(\frac{\eta_{sol}}{\eta_w} \right) \tag{39-14}$$

The example calculation is courtesy of C.H. Chervenka, Beckman Instruments Inc., Fullerton, CA. For each exposure make the following measurements. Determine the total fringe shift across the boundary, J, and compute one-fourth and three-fourths of this value. Tabulate the results as in Table 39-1. Determine the radial position of the $J/4$th and $3J/4$th fringes. Because $J/4$ and $3J/4$ most likely contain fractional parts, determine these positions by a reversal of the procedure used to determine fractional fringe numbers. That is, align the horizontal cross hair of the comparator on the center of a light fringe in the solvent region of the cell image, then move the carriage vertically until the cross hair is higher on the pattern by an distance equal to the product of the fractional part of the fringe and the fringe spacing. Next move the carriage horizontally the number of whole fringes in $J/4$, and center the vertical cross hair on the last whole fringe counted. Record the scale reading $x_{1/4}$, and repeat the process for $x_{3/4}$. Compute Δx, and divide by the magnification factor (2.15) to obtain Δr. Make a plot of Δr^2 in cm^2 vs time in seconds. The slope of the plot divided by 3.64 is the apparent diffusion coefficient in cm^2/sec for the concentration of solute employed. (3.64 is equal to $4 y^2$, where y is a constant from probability tables for 1/4 and 3/4ths concentration levels of a Gaussian concentration distribution.)

Table 39-1. Data for calculating a diffusion coefficient

Frame number	Time (sec)	J (fringes)	$J/4$ (fringes)	$x_{1/4}$ (cm)	$3J/4$ (fringes)	$x_{3/4}$ (cm)	Δx (cm)	$\Delta r = \Delta x / 2.15$	Δr^2
1	430	16.92	4.23	14.512	12.69	14.599	0.087	0.0404	1.63×10^{-3}
2	910	16.90	4.23	14.500	12.68	14.621	0.121	0.0563	3.17×10^{-3}
3	1390	16.88	4.22	14.486	12.66	14.635	0.149	0.0694	4.81×10^{-3}
4	1870	16.87	4.22	14.478	12.66	14.648	0.170	0.0791	6.26×10^{-3}
5	2350	16.86	4.22	14.467	12.66	14.658	0.191	0.0887	7.86×10^{-3}
6	2830	16.87	4.22	14.460	12.66	14.668	0.208	0.0966	9.31×10^{-3}
7	3310	16.83	4.21	14.453	12.63	14.678	0.225	0.1045	10.92×10^{-3}
8	3790	16.84	4.21	14.447	12.63	14.687	0.240	0.1115	12.43×10^{-3}

EXAMPLE CALCULATION

Calculate the diffusion coefficient for the solute used to obtain the data in Table 39-1.

ANSWER

The slope of Δr^2 vs. $t = 3.175 \times 10^{-6}$ cm^2/sec.

$$D = \frac{3.175 \times 10^{-6} \, cm^2/sec}{3.64} = 8.72 \times 10^{-7} \, cm^2/sec$$

Fringe number		x scale reading (cm)	r (cm)	r^2	c (fringes)	$\log_{10} c$
0	(meniscus)	14.196	6.792	46.131	3.02	0.480
0.5		14.268	6.827	46.408	3.52	0.547
1.5		14.377	6.879	47.321	4.52	0.655
2.5		14.457	6.917	47.845	5.52	0.742
3.5		14.531	6.953	48.344	6.52	0.814
4.5		14.587	6.979	48.706	7.52	0.876
5.5		14.639	7.004	49.056	8.52	0.930
6.5		14.684	7.026	49.365	9.52	0.979
7.5		14.726	7.046	49.646	10.51	1.001
8.5		14.759	7.062	49.872	11.52	1.061
9.5		14.796	7.079	50.112	12.52	1.098
10.5		14.824	7.093	50.311	13.52	1.131
11.15	(bottom)	14.850	7.105	50.481	14.17	1.151

Last dark fringe is number 11.0   Fractional fringe = 0.15

Figure 39-11. Data for the calculation of molecular weight by conventional sedimentation equilibrium.

Molecular Weight Determination
(Conventional Sedimentation Equilibrium Method)

Although separating compounds is the most important use of a preparative ultracentrifuge, the determination of a molecular weight is the most important function of the analytical ultracentrifuge. Several equations are used to determine the molecular weight, but one that is convenient is:

$$M = \frac{2RT}{(1 - \bar{v}\rho)\omega^2} \frac{2.303(d \log c)}{d(r^2)} \quad (39\text{-}15)$$

A second equation is needed to obtain additional required data.

$$c_m = c_o - \frac{r_b^2(c_b - c_m) - \Delta j \sum_{r_m}^{r_b} r^2}{r_b^2 - r_m^2} \quad (39\text{-}16)$$

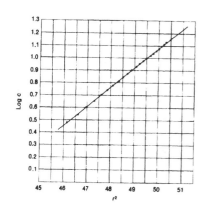

Figure 39-12. A log c vs. r^2 plot.

The data and explanation for this example calculation are courtesy of Dr. C.H. Chervenka of Beckman Instruments, Fullerton, CA. For more details on this and other ultracentrifuge methods, see *A Manual of Methods for the Ultracentrifuge* by C.H. Chervenka.

Determine the total fringe shift, J, for each of the synthetic boundary exposures and make a plot of J vs run time. Extrapolate the plot to zero time, and record the value of the number of fringes obtained as c_o. Select the best equilibrium exposures and determine the radial distance to the upper sample meniscus. Next, determine the radial distance to the center of each fringe: start at the meniscus on a dark fringe, and make the first reading on the next bright fringe ($\Delta I = 0.5$). Continue to read the position of each bright fringe (one fringe intervals) to the bottom of the cell. Read the position of the last dark fringe and the position of the bottom meniscus (oil-sample). Determine the fraction of a fringe remaining between the last dark fringe and the bottom meniscus. Make up a table as shown in Figure 39-11, and compute r^2 for each measured position.

The requirement for the summation term in equation 39-16 is estimated (where $\Delta j = 1.0$) by summing the values for r^2 for each bright fringe, plus a contribution for the fractional fringe. This contribution is determined by averaging the r^2 values for the last whole fringe and the bottom of the cell and multiplying by the fractional fringe number. (Alternately, on graph paper make a plot of r^2 on the y-axis vs fringe number on the x-axis. The summation is the area

EXAMPLE CALCULATION

Using the data in Figure 39-11 and the following information, calculate the molecular weight of the compound in question. $(1 - \upsilon\rho) = 0.277$; rotor speed = 15,000 rpm; temperature = 12.4 °C; initial concentration 7.23 fringes.

ANSWER

$$\Delta j = \sum_{r_m}^{r_b} r^2 = 544.545$$

Using equation 39-16

$$c_m = 7.23 \, \frac{50.481(11.15 - 0) - 544.545}{50.481 - 46.131} = 3.02$$

From Figure 39-12 ($\log c$ vs. r^2 plot),

$$\frac{d \log c}{d(r^2)} = 0.156$$

Using equation 39-15,

$$\frac{RT}{(1 - \nu\rho)\omega^2} = \frac{(8.313 \times 10^7)(285.6)}{(0.277)(2.467 \times 10^6)} = 3.474 \times 10^4$$

$$M = 2(3.474 \times 10^4) \times 2.303 \times 0.156 = 2.504 \times 10^4$$

Figure 39-13. IEC model HR-SII bench top centrifuge.
(Courtesy - A.H. Thomas Scientific, Swedesboro, NJ)

Figure 39-14. Marathon model 21K/R refrigerated centrifuge.
(Courtesy - Fisher Scientific, Pittsburgh, PA)

under the curve between $r = r_m$ and $r = r_b$). Determine the concentration at the meniscus, cm, from equation 39-16, using the total fringe shift as $c_o - c_m$. The concentration of fringes at various radial positions in the cell is determined by adding the fringe shift at each radial position to the concentration at the meniscus. A plot is made of $\log_{10} c$ vs r^2 (Figure 39-12, p. 463) and the slope of this plot, which is $d \log c/dr^2$, is substituted into equation 39-15 to calculate the molecular weight. Determine v and ρ as described earlier.

TECHNIQUES

APPARATUS

Dozens of different models of centrifuges are on the market. If there is any sustained use for a particular separation, then a centrifuge is made for it. This includes the size of the tubes, their shape, the speed, and refrigerated or not. The most recent models are for microsize samples at relatively high speed. Figure 39-13, p. 464, shows a modern bench top centrifuge, and Figure 39-14, p. 464, shows a high speed floor model. The bench top model has maximum speeds of 6400 rpm with a horizontal rotor and 8650 rpm with an angle rotor. It contains a tachometer to measure speeds to 9000 rpm and an automatic 60-minute timer. It can accept 20 different types of rotors. Cost: $2,100.

The high speed model reaches speeds of 13,300 rpm (20,350 $x\ g$) and can handle sample volumes up to 1160 mL (4 x 290 mL). The chamber temperature can be set between -20 to + 40 °C. The speed is variable and reads to ±100 rpm. It has a braking system to slow the rotor after a run, and an automatic shutoff if overheating or an imbalance occurs. Cost $6,500.

Rotors

The rotor is the part that turns. There are two major types, *swinging bucket* and *fixed angle*. Figure 39-15 shows an example of each type. Fixed angle types usually are used for sample sizes less than 39 mL and on the bench top models. The angle is commonly 45 °.

Safety Shields

Several stainless steel types are shown in Figure 39-16. The sample tubes may break when rotated at high speeds, so a safety shield or liner is placed in the

Figure 39-15. Rotors. *Top*, swinging bucket. *Bottom*, fixed angle.
(Courtesy - A.H. Thomas Scientific Co., Swedesboro, NJ)

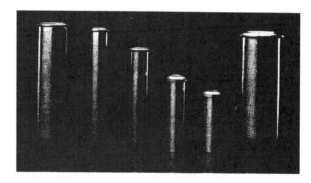

Figure 39-16. Stainless steel safety shields.
(Courtesy - A.H. Thomas Scientific Co., Swedesboro, NJ)

Figure 39-17. Adapters.
(Courtesy - Fisher Scientific Co., Pittsburgh, PA)

rotor and the tube is placed inside of it. If the tube breaks, the parts are confined, and the sample also is contained and easier to clean up.

Adapters

Rather than have to buy several rotors, it is more economical to buy a large size rotor and then to buy adapters for use with either smaller tubes or speciality tubes. Figure 39-17, p. 465, shows several types of adapters.

Tubes

There are many types of centrifuge tubes, round bottom to tapered, 0.1 mL to 300 mL, plain to graduated, open top to stoppered top, and plastic to glass to metal. Several of these are shown in Figure 39-18. In addition, there are several *specialty tubes,* such as those shown in Figure 39-19.

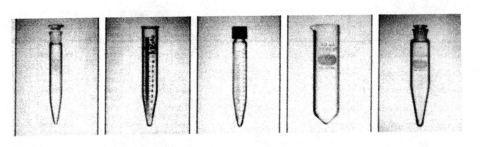

Figure 39-18. Several types of glass centrifuge tubes.
(Courtesy - Fisher Scientific Co., Pittsburgh, PA)

These are for the direct determination of fat in various products. They are all modifications of the Babcock method for fat in milk. They are designed so that if a measured volume of sample is placed in the tube and then centrifuged, the percent fat can be read directly on the stem. These are from left to right, skim milk, (0.5% in 0.01 increments), milk (8 % in 0.1 increments), ice cream (20% in 0.2 increments), cream (30 or 50% in 0.5 increments), and cheese (20% in 0.2 or 50% in 0.5 increments). Unsaturated hydrocarbons in gasoline can be handled in a similar manner, as can essential oils in extracts.

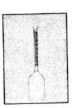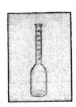

Figure 39-19. Special Tubes For Fat. Left to right; skim milk, milk, ice cream, cream, and cheese.
(Courtesy - Fisher Scientific Co., Pittsburgh, PA)

Types

Types refer to processes using the ultracentrafuge such as differential, zonal, and isopycnic.

Differential Centrifugation

This is a separation done in a tube whose liquid portion is the same density from top to bottom. Figure 39-20 is a series of diagrams to illustrate the concept. It consists of a series of centrifugations at increasing, but set, speeds.

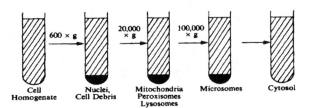

Figure 39-20. Differential centrifugation of a cell homogenate.
(Courtesy - Boyer, R.E., *Modern Experimental Biochemistry*, Addison-Wesley Publ. Co., Reading, MA, 1986)

A regular high speed centrifuge can be used for the lower speeds, but then an ultracentrifuge is required for the later stages.

Density Gradient Centrifugation

The density gradient centrifugation process involves a separation in which the supporting liquid varies in density from the top to the bottom of the tube. Two types are common: *zonal* and *isopycnic*. Figure 39-21 compares differential, zonal, and isopycnic.

In the zonal type, a density gradient is established in the tube prior to adding the sample. A common method is to prepare several sucrose solutions of various concentrations and layer them in the tube. Allow the tube to set overnight, and a smooth gradient is obtained by diffusion. The sample then is added to the top of the tube in a thin layer and when the centrifuge is turned on each component will settle to its density or the vicinity thereof. If sedimentation coefficients are desired, then the system is stopped before the sediment reaches the bottom of the tube. A solution of 60% sucrose has a density of 1.28 g/cm^3. Use this to prepare less dense and much less sticky solutions. Glycerol also is used.

In the isopycnic type, the gradient is formed during centrifugation. Several compounds can be used to develop the gradient. Originally, a solution of cesium chloride was prepared, the sample mixed in, and the mixture added to the centrifuge tube. During the centrifugation, the cesium chloride establishes a gradient, and the sample particles find their level, going up or down as necessary. Cesium salts are used because they are dense and very water soluble, and the chloride can produce a density of 1.8 g/cm^3. Cesium sulfate also can be used.

Recently, improved synthetic preparations have become available. Ficoll, a highly branched, hydrophilic polymer of sucrose can produce densities up to 1.2 g/mL and has better osmotic properties than sucrose. It can be autoclaved at neutral pH at 110 °C for 30 minutes without degradation but is unstable to hydrolysis below pH 3.

Percoll consists of silica particles (15 to 30 nm diameter) coated with nondialyzable polyvinyl-pyrrolidone (PVP). It is nontoxic and almost chemically inert and does not adhere to membranes. Percoll can form densities of 1.0 to 1.3 and is iso-osmotic throughout.

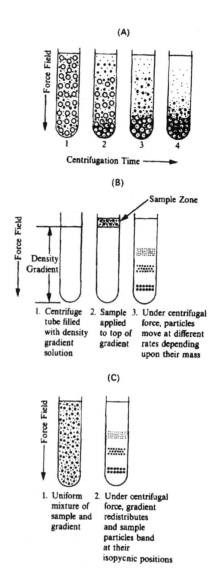

Figure 39-21. (A) Differential centrifugation. (B) Zonal centrifugation. (C) Isopycnic centrifugation. (Courtesy - Boyer, R.E., *Modern Experimental Biochemistry*, Addison-Wesley Publishing Co., Reading, MA, 1986)

Figure 39-22. Density changes and density marker beads. (Courtesy - Pharmacia Biotech, Uppsala, Sweden)

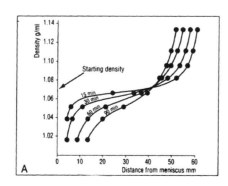
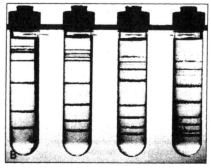

Standards

The density changes continuously during the time of centrifugation, and it is often useful to know how this is happening and what the density is. Figure 39-22, p. 467, shows how the density changes and how it is measured with calibration beads for the Percoll system. The beads of different densities are color coded. The system was run at 20,000 $x\,g$ in 0.15 M NaCl.

REVIEW QUESTIONS

1. What is the common force range generated by a bench top centrifuge?
2. What forces must be overcome by an ultracentrifuge?
3. Why is a vacuum applied when using an ultracentrifuge?
4. In schlieren optics, what is the direction of motion of the horizontal diaphragm?
5. What does step C in Figure 39-2, p. 456, show?
6. What improvement over the schlieren optics is provided by the Philpot-Svensson optics?
7. What is the basis for the Rayliegh optics?
8. What does part B of Figure 39-5, p. 457, show?
9. What is the main advantage of using a double-beam spectrophotometric detection method when phototubes and a computer are used?
10. A bench type of centrifuge spins at 3,000 rpm and the rotor distance is 5 inches. What is the RCF?
11. Use the nomograph in Figure 39-7, p. 459, to determine the g forces on a particle in a centrifuge spinning at 10,000 rpm at an r of 900 mm.
12. What is the settling velocity of a 2.5 micron particle of AgCl, density 5.56 g/cm^3, placed in a buffer at 20 °C, density 1.062 g/cm^3, viscosity 948 µPa s, in a centrifuge tube at an average of 3.5 inches from the center of a rotor turning at 1500 rpm?
13. What is a Svedberg unit?
14. What is a pycnometer?
15. What is the sedimentation coefficient in Svedberg units if at 24.3 minutes x was at 12.840 cm and at 32.3 minutes x was at 12.993 cm in a centrifuge rotating at 39,750 rpm?
16. Calculate the diffusion coefficient for a solute that after 2,339 seconds had a Δr^2 of 7.68 x 10^{-3} cm^2?
17. What is the purpose of a liner in a rotor?
18. What is a Babcock tube?
19. What is the technique for differential centrifugation?
20. What is meant by zonal density gradient centrifugation?
21. What are some systems used to form gradients by the isopycnic type of centrifugation?
22. How can the density be measured and followed during a centrifugation?

IF IT EXISTS - CHEMISTRY IS INVOLVED
IF YOU CAN BUY IT - A CHEMIST WAS INVOLVED SOMEWHERE

40
MASKING AND SEQUESTERING AGENTS

PRINCIPLES

A *masking agent* forms a complex ion (no rings) with a metal ion to prevent it from interfering with an analysis. It separates the ion from the measurement reaction by altering its activity. Examples are $Cu(NH_3)_4^{+2}$ and FeF_6^{-3}. A *sequestering agent* serves the same purpose as a masking agent, only a chelate (ring compound) is formed. Masking agents, because of their charge, usually are left in the solution, whereas a sequestering agent, which is usually a neutral molecule, can be extracted if necessary. Ethylenediaminetetraacetic acid (EDTA) is unique in that it can chelate an ion yet still be charged and remain soluble in the aqueous layer.

COMPLEXOMETRIC TITRATIONS

A complexometric titration is a technique of titrimetric analysis in which the formation of a chelate or complex is used to determine the concentration of a metal and to indicate the endpoint. It is also known as "chelometry". A Swiss chemist, Gerold Schwarzenbach, Figure 40-1, p. 470, was the first to propose the use of complexing agents as titrants in 1951. He was working with uramildiacetic acid, and when washing up the reaction vessels in hard tap water observed the change of color produced by calcium ion. Continued investigation of this phenomenon enabled him to recognize the analogy to acid-base indicators and led to the idea of metal indicators.

Ethylenediaminetetraacetic Acid (EDTA) (H_4Y)

In the mid 1930's, the I.G. Farbenindustrie in Germany manufactured a compound with the trade name of Trilon B. After the War, Gerold Schwarzenbach, at the University of Zurich, investigated this compound and introduced it as a titrant in 1951.

Ethylenediaminetetraacetic Acid
(EDTA or H_4Y)

2,4-Pentanedione

The pK_a's are 2.0, 2.76, 6.16, and 10.3. Two protons are readily lost. The acid is quite insoluble in water (0.2 g/L) so the disodium salt dihydrate usually is used to prepare a standard solution (Na_2H_2Y).

Walter Biedermann and Schwarzenbach published the first titrimetric determination for the hardness of water using Erio-T as an indicator in January of 1948. In the next 10 years, over 1200 papers would be published on the uses of EDTA.

Figure 40-1. Gerold Schwarzenbach (Courtesy - Diehl, H., *Quantitative Analysis*, J. Wiley & Sons, New York, NY, 1972)

EDTA is a *hexadentate*, that is, it has six positions that can coordinate with a metal ion, forming an octahedral arrangement as shown below left. The lines are NOT bonds, but are used to show the octahedral configuration. EDTA, having six coordinating centers, can fill all of these positions, but it seldom does. Usually one or two acid groups do not coordinate, which leaves the chelate charged and quite water soluble. EDTA forms stable chelates with nearly all of the positive ions, regardless of charge, except the alkali metals. The diagram below shows how EDTA can "surround" the Zn^{+2} ion. Table 40-1 lists the formation constants of several EDTA metal chelates.

Table 40-1. Formation constants of metal-EDTA chelates

Metal Ion	Log K_f	Metal Ion	Log K_f
Sodium	1.66	Titanium	17.3
Lithium	2.79	Plutonium (4)	17.66
Silver	7.32	Dysprosium	17.75
Barium	7.76	Holmium	18.04
Strontium	8.63	Plutonium (2)	18.12
Magnesium	8.69	Americium	18.16
Calcium	10.70	Lead	18.3
Bismuth	10.95	Curium	18.45
Manganese	13.6	Palladium	18.5
Ferrous	14.3	Nickelous	18.6
Lanthanum	15.40	Cupric	18.8
Cerous	15.8	Erbium	18.98
Neodymium	16.05	Lutetium	19.07
Aluminum	16.16	Californium	19.09
Cobaltous	16.20	Gallium	20.26
Zinc	16.3	Mercuric	21.8
Plutonium (6)	16.39	Thorium	23.2
Praseodymium	16.5	Indium	24.95
Cadmium	16.6	Ferric	25.1
Gadolinium	17.2	Cobaltic	36.

TECHNIQUES

To determine metals, the organic components must be removed. There are two ways to do this: (1) wet ashing, described in Chapter 41, p. 475, and (2) dry ashing, described below. Dry ashing is best done by slowly burning them away in a *muffle furnace*. A muffle furnace is a small, very high temperature, heavily insulated oven capable of temperatures of 1125 °C, with some going to 1,700 °C. Figure 40-2, p. 471, shows the Fisher Model 160, which is a mid-size furnace (2.5 L inside chamber). Modern furnaces have either a needle or LED display and a programmable temperature control.

Porcelain items only should be placed in a muffle furnace. No glass items should be placed in a muffle furnace. The glass will melt and cause a real mess. If such an accident happens, the furnace must be shut off, and cooled completely, and the glass carefully chipped out. You will likely never get it all out, so a ceramic plate must be placed on the bottom to keep other items from sticking.

Other items necessary when working around a muffle furnace are gloves, long-handled tongs, and safety glasses. Figures 40-3, 40-4, and 40-5 show these. The gloves used to be made of asbestos which worked very well. However, these are being phased out and replaced with other materials. ZETEX plus is a material that can withstand temperatures up to 2,000 °F. It conducts heat radially along the fabric, not through it. Just like the government - one size fits all.

Normal safety glasses do not protect against the intense ultraviolet and infrared radiation emitted when the door of a hot furnace is opened. Welders glasses are usually too dark, but suitable glasses are available. A pair of glasses that can reduce uv (220 to 380 nm) 99.9%, blue (430 to 440 nm) 90% and infrared (770 to 1700 nm) 81% costs less than $10.00. The visible region is generally from about 400 to 700 nm, so you are still able to see what you are doing.

The tongs are just like crucible or beaker tongs, except that they have longer handles, 53 to 56 cm, instead of the usual 33 cm, and the better ones are nickel plated.

REVIEW QUESTIONS

1. What is the difference between a masking agent and a sequestering agent?
2. How did the idea of a complexometric titration originate?
3. What is meant by "hexadentate"?
4. EDTA is used extensively in industry to chelate with many metals. What is one property that makes EDTA different from most other chelating agents?
5. Refer to Table 40-1, p. 470. Can you explain why CaEDTA is given to help cure chronic lead poisoning?
6. If you needed to heat an object for a long time at a high temperature, what apparatus might be used?
7. Why should glass items never be placed in a muffle furnace?
8. Why should you use long-handled tongs and insulated gloves when transferring materials in and out of a muffle furnace?
9. If you are going to work around a muffle furnace for more than one transfer, why is it recommended that you wear glasses that protect against uv and infrared radiation?

Figure 40-2. A bench-top muffle furnace.
(Courtesy - Fisher Scientific Co., Pittsburgh, PA)

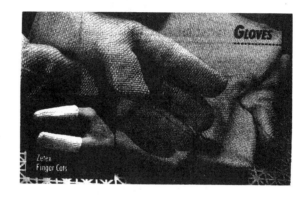

Figure 40-3. ZETEX gloves, mittens and finger cots.
(Courtesy - Lab Safety Supply, Janesville, WI)

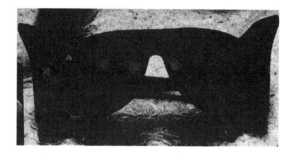

Figure 40-4. uv and infrared safety glasses.
(Courtesy - Lab Safety Supply, Janesville, WI)

Figure 40-5. Muffle furnace style tongs.

IF IT EXISTS -CHEMISTRY IS INVOLVED
IF YOU CAN BUY IT - A CHEMIST WAS INVOLVED SOMEWHERE

41
SOLUBILITY

PRINCIPLES

Solubility is *the amount of a substance that will dissolve in another substance.* The solubility of inorganic compounds is based on how many grams will dissolve in 100 mL of water, usually at 25 °C. A compound that dissolves less than 1 g/100 mL is generally considered insoluble.

Trying to selectively dissolve one component from a mixture is usually not a quantitative operation, but it can be used to remove a major portion of an interfering substance.

Solubility is important because the sample, or some part of the sample, must be dissolved before a chemical separation can take place.

SAFETY

Please be careful. Compounds used to dissolve other compounds or materials are in general hazardous to handle unless common sense safety precautions are taken. Most organic solvents are flammable and acids and bases are usually highly corrosive. Use as little as necessary to do the job. Remember, you are your nation's most valuable resource.

INORGANIC COMPOUNDS
General Principles
1. Try the simplest system first.
2. Use the minimum amount of acid or flux so as to reduce the addition of foreign ions.
3. Use volatile acids, bases, or organic solvents first. HCl, b.p. of water azeotrope = 108.6 °C; HNO_3, b.p. of water azeotrope = 120 °C; $HClO_4$, b.p. = 203 °C; H_3PO_4, b,p, = 250 °C; H_2SO_4, b.p. = 260 °C
4. Metals with more positive (negative) reduction potentials to H_2 need an oxidizing acid such as HNO_3; HNO_3 + HCl; or hot $HClO_4$ > 50%. Examples are: Sb, As, Bi, Cu, Ag, Hg, Pt, Au, and Rh.
5. Metals with more negative (positive) reduction potentials to H_2 need a nonoxidizing acid. H_2SO_4; HCl; H_3PO_4; or cold $HClO_4$ < 50%. Examples are: Pb, Sn, Ni, Co, Cd, Fe, Cr, Zn, Mn, Al, and Mg.

Water
Solubility Rules for Inorganic Compounds

Over the years, many trends have been observed regarding the solubilities of inorganic compounds in water. These have led to the *solubility rules* shown below as tabulated by Professor Allan Clifford, Virginia Polytechnic Institute, Blacksburg, VA.

1. All salts of the alkali metals and ammonium ion are soluble. Potassium perchlorate is only slightly soluble.

2. All nitrates, perchlorates, and acetates are soluble. Silver acetate is only slightly soluble.

3. All chlorides of the metals are soluble. Those of AgCl, Hg_2Cl_2, and $PbCl_2$ are slightly soluble in cold water and moderately soluble in hot water.

4. All sulfates are soluble, except those of barium, strontium, and lead. Those of calcium, mercury (II), and silver are only slightly soluble.

5. All fluorides are insoluble, except those of the alkali metals, silver, and thallium (I).

6. All carbonates (where they exist) and phosphates are insoluble, except those of the alkali metals.

7. All sulfides and hydroxides of the metals are insoluble, except those of the alkali and alkaline earth metals. The hydroxides of the alkaline earth metals are only slightly soluble.

8. All silver salts are insoluble, except the nitrates, perchlorates, and fluorides. Silver acetate, sulfate, and chlorate are slightly soluble.

Some further observations are made with regard to solubilities, many of which also apply to organic crystals.

1. Small crystals are usually more soluble than large ones. Ions at corners and edges contact more solvent and dissolve faster.

2. If a salt is derived from either a weak base or a weak acid, its solubility may be increased by formation of the unionized base or acid. Thus, ammonium vanadate, NH_4VO_3, is much more soluble in base because of the formation of the unionized NH_4OH.

$$NH_4VO_3 = NH_4^+ + VO_3^-$$
$$\quad\quad\quad\quad\quad | OH^-$$
$$\quad\quad\quad\quad\quad |_____ NH_4OH$$

Commercial HCl (37%, $d = 1.19$ g/mL; 12.0 M)

$2\ Cl^- = Cl_2 + 2e^-\quad E^o = -1.36\ v$; therefore, it is a poor oxidizing agent.

It is the lowest boiling of the common acids. It dissolves metals with more negative reduction potentials than H_2 and dissolves carbonates and oxide minerals or ores. In high concentrations, it is a mild reducing agent and advantage is taken of this to dissolve the higher oxides of Pb and Mn. Pyrolusite (MnO_2) is insoluble in dilute HCl or HNO_3 but will dissolve in hot concentrated HCl.

$$MnO_2 + 4\ HCl = MnCl_4 + 2\ H_2O + Cl_2$$

Excess HCl is removed by boiling or adding H_2SO_4 and heating. $HgCl_2$ (b.p. = 302 °C), CrO_2Cl_2 (b.p. = 117 °C), $AsCl_3$ (b.p. = 110 °C), $GeCl_2$ (decomposes), and $SbCl_3$ (b.p. = 160 °C) are volatile.

Commercial HNO₃ (72%, $d = 1.42$ g/mL, 16.2 M)

$NO_3^- + 3\ H^+ + 2e^- = HNO_2 + H_2O\quad\quad E^o = +0.94\ v$
$NO_3^- + 4\ H^+ + 3e^- = NO + 2\ H_2O\quad\quad E^o = +0.96\ v$
$2\ NO_3^- + 4\ H^+ + 2e^- = N_2O_4 + 2\ H_2O\quad\quad E^o = +0.81\ v$

It forms an azeotrope (68% HNO_3 + 32% H_2O), which boils at 120.5 °C. It is a strong acid and a strong oxidizing agent. It dissolves metals with more positive reduction potentials than H_2 such as Hg, Ag, Cu, and Sb. A problem can occur if just HNO_3 is used to dissolve Sb, Sn, or W; a white precipitate will form.

$Sb + HNO_3 = Sb_2O_5 \downarrow$
$Sn + HNO_3 = SnO_2 \cdot 2H_2O \downarrow\ (H_4SnO_2)$ meta stannic acid.
$W + HNO_3 = WO_3 \downarrow$

It is used for attacking alloys and for very insoluble sulfides such as CuS (8.5×10^{-45}), PbS (3.4×10^{-32}) and FeS_2. The excess is removed by displacement with H_2SO_4 or destruction with excess HCl (reduces it).

Commercial H₂SO₄ (96%, $d = 1.84$ g/mL, 18.0 M)

$2\ SO_4^= + 4\ H^+ + 2e^- = S_2O_6^= + H_2O\quad E^o = -0.20\ v$
$SO_4^= + 4\ H^+ + 2e^- = H_2SO_3 + H_2O\quad E^o = +0.20\ v$

It is considered a nonoxidizing acid when cold or dilute and a mild oxidizing agent when hot. It is also an excellent dehydrating agent. It is used to dissolve the ores of Al and Ti. It is not used much, because removing the excess is hard. The white fumes are $(SO_3)_2$. The concentrated acid has an E^o of +0.34 v.

Commercial H₃PO₄ (85%, $d = 1.69$ g/mL, 17.7 M)

$H_3PO_4 + 2\ H^+ + 2e^- = H_3PO_3 + H_2O\quad E^o = -0.276\ v$

It is a nonoxidizing acid but will dissolve porcelain and granite ware when hot.

Commercial HF (48%, $d = 1.15$ g/mL, 27.6 M)

$2\ F^- = F_2 + 2e^-\quad E^o = -2.87\ v$

Its primary use is to dissolve silicates, forming the volatile SiF_4.

$$SiO_2 + 4\,HF = 2\,H_2O + SiF_4 \uparrow$$
$$MSiO_3 + 6\,HF = 3\,H_2O + MF_2 + SiF_4 \uparrow$$

An excellent solvent for Sn-Sb alloys is a solution containing 2 to 3% HF and 2 to 3% HNO_3. Be careful when handling HF. It can produce some of the most painful burns you can imagine. Wear gloves at all times.

Commercial HI (57%, d = 1.70 g/mL. 7.6 M)

$$2\,I^- = I_2 + 2e^- \quad E^o = -0.54\,v$$

Its primary use is to dissolve HgS ($H_2S\uparrow$), SnO_2 ($SnI_4 \uparrow$), AgX, CaF_2, $BaSO_4$, $SrSO_4$, and $PbSO_4$. HI is colorless if pure, but is usually red because I_2 is present. Distill with H_2PO_2 (hypophosphorous acid) to remove the iodine, if necessary.

Commercial $HClO_4$ (70-72%, d = 1.66 g/mL, 11.6 M)

$$ClO_4^- + 8\,H^+ + 7e^- = 1/2\,Cl_2 + 4\,H_2O \qquad E^o = +1.34\,v$$

When cold and dilute, it is just another strong acid. It is not an oxidizing agent until it is 50% and 130 °C. However, hot and concentrated to 85%, it has an E^o of >2.0 v. Then it is as good as O_3 or F_2. It dissolves stainless steel and ferro alloys and particularly chrome steels. Almost all perchlorates are soluble, and this solubility increases about tenfold as the temperature is raised from 0 °C to 100 °C.

It is the only acid that slowly releases its O and in proportion to its concentration (46.4% O). Advantage is taken of this in that NH_4ClO_4 is used as part of solid rocket fuels. Its oxidation potential changes with concentration. Table 41-1 shows how the potential varies with concentration, and Table 41-2 shows the percent $HClO_4$ in the various solution compositions.

Table 41-1. % $HClO_4$ and oxidation potential

50%	0.80 - 0.85 v
57	1.3 - 1.4
60	1.5 - 1.6
85	approx 2.1

50 to 60%: varies linearly.

> 60%: rises rapidly.

Table 41-2. % $HClO_4$ and composition

61%	$2\,HClO_4 \cdot 7\,H_2O$
65	$HClO_4 \cdot 3\,H_2O$
69	$2\,HClO_4 \cdot 5\,H_2O$
72	$HClO_4 \cdot 2\,H_2O$
85	$HClO_4 \cdot H_2O$

The 72% acid is perfectly stable when cold and cannot be made to detonate. The 85% acid is a white solid, m.p. 50 °C, and is so stable it can be shipped by common carrier. However, the 100 % (anhydrous) is not stable. It is a heavy oily liquid and eventually will detonate. It should not come in contact with organic matter or easily reduced inorganic compounds. It must be kept cool and should not be made with the intentions of storing it. Perchloric acid hoods are available for using this acid when fumes might escape. These hoods are equipped with a wash down system to prevent any formation of anhydrous acid in the exhaust duct. *Remember*: cold and dilute, it is just another strong acid. Individual flask air condensers are available from the G.F. Smith Chemical Co., Columbus, OH, that can be used if only a few samples are to be dissolved and conventional hoods are used. Figure 41-1 shows such an arrangement.

Perchloric Acid and Organic Compounds

Obtain the free booklet from the G.F. Smith Chemical Co, Columbus, OH, titled the *Handling and Properties of Perchloric Acid* if you are going to use perchloric acid.

Although perchloric acid does not dissolve organic compounds, it digests them. This section is added here to help overcome the fear of using perchloric acid digestions of organic compounds. If handled properly, perchloric acid is the most efficient method available to destroy organic compounds. If not, trouble can occur.

Perchloric acid and its compounds are widely used worldwide. The 11th *Collective Chemical Abstracts Index* lists 38 pages of new references!

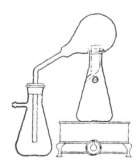

Figure 41-1. Oxidation reactor and fume eradicator assembly.
(Courtesy - G. F. Smith Chemical Co., Columbus, OH)

Nationwide, 1.5 million lb of $HClO_4$ are produced annually. Therefore, it seems appropriate that you become familiar with this acid and how to handle it with organic compounds.

Most troubles occur with the formation of ClO_4^- esters. Ethyl perchlorate is shock sensitive and a violent explosive. Therefore, use a mixture of HNO_3 and $HClO_4$ if alcohols, carbohydrates, and dehydrating conditions are present. The HNO_3 destroys the alcohol groups, preventing the formation of the esters before the $HClO_4$ becomes sufficiently concentrated and hot enough to be an oxidizing agent.

$$ROH \; + \; \underset{HO}{\overset{O}{\underset{\|}{\overset{\|}{Cl}}}}\!\!\!\!\!\overset{O}{\underset{O}{}} \longrightarrow \underset{RO}{\overset{O}{\underset{\|}{\overset{\|}{Cl}}}}\!\!\!\!\!\overset{O}{\underset{O}{}} \; + \; H_2O$$

Alcohols should not be heated with a mixture of $HClO_4$ and H_2SO_4, unless H_2O and HNO_3 attack has been made first.

With an unknown system, start with 50 to 100 mg of sample and see how it behaves.

You can filter cold $HClO_4$ solutions, but wash the paper thoroughly or it may explode if placed in a drying oven. Wash with NH_4ClO_4, if possible. Wash down the fumes. Use an air condenser like that in Figure 41-1, p. 475 (inner part) as a minimum, if a perchloric acid hood is not used. Place lead shot in the flask to keep it from tipping and cover the shot with NaOH solution. Use a gentle suction.

Most difficulties come from adding too little $HClO_4$. The reason is that $HClO_4$ is not reactive until it is hot and concentrated. If only a few mL are used, as a mixture with HNO_3, when the $HClO_4$ becomes active, much heat and little acid are present, so it can go anhydrous quickly and cause a problem. *If you become concerned, add water immediately to stop the reaction.*

Catalysts are helpful; ammonium vanadate is the most common. 3 to 4 mg of $K_2Cr_2O_7$ often is added as an indicator. The solution is green if C is present and then yellow orange when it is gone. If $K_2Cr_2O_7$ has been added, another 15 to 20 minutes may be needed to completely oxidize the formic acid so the color will leave. CrO_3 is soluble in hot 70% $HClO_4$, but precipitates quantitatively when it is cold. Table 41-3 shows which catalysts are preferred for a given concentration range, and Figure 41-4 shows the effect of a catalyst.

Table 41-3. Recommended catalysts

% $HClO_4$	Catalyst
60 - 68	NH_4VO_3
68 - 72	NH_4VO_3 or $K_2Cr_2O_7$
> 72	$MoO_4^=$

Table 41-4. 5 g oxalic acid + 20 mL 72 % $HClO_4$ + 10 mg $K_2Cr_2O_7$ indicator

VO_3 mg	0	0.15	0.30	0.45	0.60	0.75
Time; min	9.0	5.75	4.25	3.25	1.75	1.15

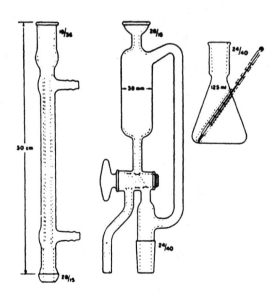

Figure 41-2. The Bethge digestion apparatus.
(Courtesy - G.F. Smith Chemical Co., Columbus, OH)

Table 41-5 shows the effect of concentration vs. time to digest 0.8 g of filter paper + 20 mL of $HClO_4$ + 10 mg NH_4VO_3, and Table 41-6 shows how the $HClO_4$ composition changes with temperature.

Table 41-5. Effect of $HClO_4$ concentration on digestion time

% $HClO_4$	Time (min)
50.0	>375
57.7	60
60.1	52
62.5	45
64.0	17
66.0	9

Table 41-6. $HClO_4$ composition vs. temperature

% $HClO_4$	Temp. (°C)	% $HClO_4$	Temp. (°C)
56.3	140	64.5	175
57.7	154	66.0	182
58.8	157	67.5	190
59.2	158	68.4	195
60.1	160	72.5	203
62.6	170		

Use the Bethge apparatus (Figure 41-2, p. 476) to control the composition. The sample and acid are placed in the flask. The stopcock is turned to drain away the condensed water until the desired composition is reached, then turned so the condensed water is returned to the flask to maintain a constant composition. If any fat is in the sample, be sure to clean the apparatus thoroughly between digestions or an explosion may occur (speaking from experience, while digesting a hot dog).

The author prefers to use a burner rather than a hot plate to heat the mixture. If the flask should crack, the $HClO_4$ just spills onto the bench top. If a hot plate is used, it can become anhydrous and be a much more serious problem.

The Liquid Fire Reaction

The liquid fire reaction is the most common method to do a perchloric acid digestion and is the safest. The initial reaction mixture is 3 parts concentrated HNO_3 and 1 part $HClO_4$ with 1 mg/mL NH_4VO_3.

1. Heat slowly at first: alcohols and aldehydes react below 120 °C. Reaction is indicated by NO and NO_2 fumes forming and frothing.
2. 120.5 °C (azeotrope; HNO_3 = 68%, H_2O = 32%): HNO_3 is an avid oxygen doner at this concentration and temperature. The reaction usually intensifies here.
3. 120 to 130 °C: HNO_3 is removed, the solution clears up somewhat and the reaction slows down.
4. 150 °C: $HClO_4$ begins to lose oxygen and some frothing usually occurs. The reaction picks up.
5. 160 °C: The reaction continues normally. Usually - just at the end - vigorous frothing occurs for a few seconds (don't panic), and it is all over. The sample is digested.

TECHNIQUES

Figures 41-3 and 41-4, p. 478, show the simplest type of apparatus that can be used to obtain satisfactory solubility measurements of inorganic compounds.

The water bath could be a regulated constant-temperature bath, if K_{sp} values for very insoluble compounds are to be determined. S.A. Butter's description *(J. Chem. Ed.*, **51**, 70, 1974) is, "a weighed amount of solute is placed in the tube on top of a medium glass frit. A measured volume of solvent is added, the condenser put in place, and a gentle stream of air or nitrogen is passed through the mixture. The temperature is measured and increased as necessary by a hot plate if higher temperatures are desired. After equilibrium is attained (need to repeat the process to determine the time), the condenser is removed and air or nitrogen is forced in the opposite direction to filter the undissolved solute away from the solvent. The solvent is collected, a measured volume placed in a weighed weighing dish, evaporated to dryness and constant weight. The solubility and/or solubility product is then calculated. A simple mercury pressure-relief bubbler may be inserted into the gas line to avoid pressure buildup during filtration."

DISSOLVING ORGANIC SAMPLES

The first dry cleaner was Jean-Baptiste Jolly in Paris in 1845. He used benzene. "The ideal solvent is a volatile, mobile, safe liquid that is chemically unreactive, readily available and inexpensive. Only about 250 compounds meet these requirements and only about 50 are made in quantity" (John Emsley).

Some very good old standbys are being phased out because of excessive safety requirements.

Carbon tetrachloride - liver damage
Diethyl ether - flammable
Benzene - suspect carcinogen
Dichloromethane - causes cancer in rats at prolonged exposure to 2000 ppm. No known human cancers.
Tetrachloroethane - liver and kidney damage
Ethylene glycol - still OK, but forms oxalic acid in the body. Used in water-soluble paints.
Diethylene glycol - $HOCH_2CH_2OCH_2CH_2OH$. Same as ethylene glycol.

Those without a hydrogen can damage the ozone layer. To replace all Cl's with F's is not the answer, because CH_2F, CF_4, and CF_3CF_3 are gases.

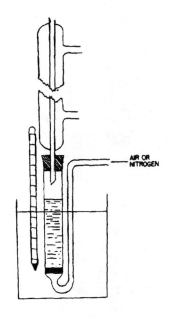

Figure 41-3. Equilibration apparatus to determine solubilities.
(Courtesy - Butter, S.A., *J. Chem. Ed.*, **51**, 70, 1974).

Perfluorodecalin

Oxygen dissolves so well in prefluorocarbons that a mouse enclosed in a plastic bag can be submerged in the liquid. These compounds are the bases of artificial blood, Fluosol-DA is an example.

The next generation:

PER	Perchloroethylene	$CCl_2=CCl_2$
TRI	Trichloroethylene	$CHCl=CCl_2$
	FC113 1,1,2-trichlorofluoroethane	$CF_2Cl\text{-}CFCl_2$
	1,1,1 Trichloroethane	$CH_3\text{-}CCl_3$
Propylene glycol		$HOCH_2CH(CH_3)OH$
Glyme	Ethylene glycol methylether	$CH_3OCH_2CH_2OH$

Solvent for paints and printing inks. Can cause malfunctions in the reproductive system.

Diglyme   Diethylene glycol dimethyl ether
          $CH_3OCH_2CH_2OCH_2CH_2OCH_3$
Dissolves metal compounds.

DMF   Dimethyl formamide $HC(=O)N(CH_3)_2$
DMSO Dimethyl sulfoxide $O=S(CH_3)_2$

The now generation:
Ethylene glycol n-propyl ether $H_3CCH_2CH_2OCH_2CH_2OH$
The propyl group is not metabolized by the body, and the extra methyl group has no effect on the solvent properties.

Propylene glycol methyl ether $H_3COCH_2CH(CH_3)OH$
The toxicity safety increases from 30 to 3,000 ppm (no effect level).

Diester solvents: Mixtures of the methyl esters of succinic, glutaric and adipic acids. Paint will smell nice.

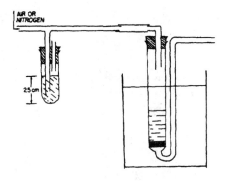

Figure 41-4. Filtration arrangement.
(Courtesy - Butter, S.A., *J. Chem. Ed.*, **51**, 70, 1974).

The solvents below can dissolve a wide range of both polar and nonpolar solutes. They are highly polar, but do not have a hydroxyl group to cause unwanted interactions.

Sulfolane   N-methyl-2-pyrrolidine   Hexamethylphosphoramide   Dimethylpropylene urea

Hexamethylphosphoramide (HMPA) causes cancer in animals but is not known to in humans. It is an excellent solvent.

Dimethylpropylene urea (DMPU) causes no damage to chromosomes.

The most basic generalization is like dissolves like - or polar dissolves polar and nonpolar dissolves nonpolar.

SOLVENT PREDICTORS

Chemists have searched for centuries for that one equation that would allow them to predict which solvents would dissolve a given solvent and to what extent. One of the difficulties with all solvent predictors is that the polarity changes depending upon the surroundings. Several of the best attempts are discussed briefly below.

The Reichert Solvent Polarity System, E_T

The Reichert solvent polarity system uses one dye that dissolves in all solvents and changes color depending upon the polarity. The compound is pyridinium-*N*-phenoxide (shown to the right). It has been used on 250 liquids. Values range from 30 for nonpolar liquids such as paraffins to 63 for water. It does not work for acidic solvents.

The Kamlet, Taft, Doherty, Abraham Equation

$$P_{sol} = P + s(\pi^* + d\delta) + a\alpha + b\beta + h\delta_H + e\varepsilon \qquad (41\text{-}1)$$

J. Org. Chem., **48**, 2877 (1983) gives π^*, α, and β values for several hundred solvents.

The Swaim and Swaim Equation

$$P = Po + aA + bB \qquad (41\text{-}2)$$

where A = acity = anion solvating ability. B = basity = cation solvating ability

Hildebrand's Solubility Parameter

This is the most commonly used equation, although usually only the last terms are used.

$$2.3\ RT \log\left(\frac{A_2}{X_2}\right) = V_2\ \theta_1^2\ (\delta_2 - \delta_1) \qquad (41\text{-}3)$$

where A_2 = activity of the solute. X_2 = mole fraction of the solute. V_2 = solute's liquid molal volume. θ_1, = volume fraction of the solvent. $\delta_2 - \delta_1$ = solubility parameters of the solute and solvent.

If like dissolves like, then if $\delta_2-\delta_1$ is small, the solubility should increase.

Table 41-7 compares various parameters for I_2 in several solvents.

Table 41-7. I_2 in various solvents at 25 °C

Solvent	Dipole μ (Debyes)	V	$100X_2$	A_2/X_2	δ_1	δ_2
n-heptane	0.0	227.	0.0185	1400.	5.7	14.2
trans- $C_2H_2Cl_2$	0.0	77.4	1.477	18.2	9.0	14.5
cis-$C_2H_2Cl_2$	1.89	75.8	1.441	17.1	9.1	14.5
1,1-$C_2H_4Cl_2$	2.09	84.7	1.531	16.9	9.1	14.4
1,2-$C_2H_4Cl_2$	1.18	79.5	2.20	11.7	9.8	14.9
1,2-$C_2H_4Br_2$	0.81	86.6	7.82	3.30	10.4	14.1

Compare $100X_2$ with $\delta_2-\delta_1$ 0.018   14.2 - 5.7 = 8.5
     7.8   14.1 - 10.4 = 3.7

Note that no correlation exists between μ and X_2 in this case
What is δ?

$$\delta = \left(\frac{-\Delta H_{vap}}{V}\right)^{\frac{1}{2}} \quad (41\text{-}4)$$

Therefore, it varies with T. V is the molal volume (the volume occupied by 1 mole of the compound as a solid, liquid or gas). The ΔH_{vap} is used because it can be thought of as the energy necessary to break the bonds of the solvent to provide a space for the solute molecule to fit into.

$$\Delta H_{298} \text{ cal/mole} = -2900 + 23.7T_b + 0.020T_b^2 \quad (41\text{-}5)$$

EXAMPLE CALCULATION
What is δ for acetone, m.w. = 58.08, if it boils at 56.5 °C and has a density of 0.788 g/mL?

ANSWER
ΔH_{298} cal/mole $= -2900 + 23.7(273 + 56.5) + 0.020(273 + 56.5)^2$
$= 7,080$ cal/mol (7,143 literature)

$V = 58.08$ g/mol ÷ 0.788 g/mL = 73.70 mL/mol

$\delta = (7,080/73.70)^{1/2}$; δ = 9.84

Another way to determine δ.

$$\delta = D\left(\frac{\Delta H_{vap} - RT}{M}\right)^{\frac{1}{2}} \quad (41\text{-}6)$$

where D = density M = m.w. of a single compound or the repeat unit of polymer. R = gas constant; 1.98 cal/mol.

EXAMPLE CALCULATION
Repeat the above example using equation 41-6.

ANSWER
$\delta = 0.788[(7,080 - \{1.98 \times 329.5\})/58.08]^{1/2}$
$= 9.34$

Table 41-8 lists the densities of various polymers.

Table 41-8. Densities of various polymers

Compound	Density	Compound	Density
Butadiene rubber	0.9		
Butyl rubber	0.93	-1500	1.15 - 1.21
Cellulose	1.5	-4000	1.20 - 1.21
Epoxies	1.2	-6000	1.21
Melamine-formaldehyde	1.5	Polyethylene glycol-	
Natural rubber	0.92	p-isooctylphenyl ether	1.489
if vulcanized up to	1.3	Polyethylene terephthalate	1.38
Nylon 4,7	0.97-1.15	Polymethylmethacrylate	1.2
Polyacrylonitrile	1.2	Polypropylene	0.90 - 0.92
Polycarbonate	1.25	Polystyrene	1.04-1.06
Polychloroprene (Neoprene)	1.2	Polytetrafluoroethylene (Teflon)	2.25
Polyethylene	0.92 - 0.94	Polyurethanes	1.1-1.3
High density	0.95 - 0.97	Polyvinyl chloride	1.67-1.85
PEG-400	1.11 - 1.14	Silicones	0.96-1.5
-600	1.19		

Still another way to estimate δ;

$$\delta = D \sum \frac{E}{M} \tag{41-7}$$

where E = the molar attraction constant. See Table 41-9.

EXAMPLE CALCULATION
Calculate δ for acetone using the above equation, previous information, and the data from Table 41-7, p. 479.

ANSWER
Calculate $\sum E$ for acetone, $H_3C-C(=O)-CH_3$.
$2 \times -CH_3 = 2 \times 148 = 296$
$1 \times C=O = 1 \times 275 = 275$
$ \overline{}$
$ 571$
$\delta = (0.788 \times 576)/58.08$
$ = 7.74$

Table 41-9. Molar attraction constants at 298 °K, $(Cal/cm^3)^{1/2}$/mole

Group		E	Group		E
-CH$_3$		148.	-O-	epoxide	176.
-CH$_2$-		131.5	-COO-		326.5
>CH-		86.	>C=O		263.
>C<		32.	-CHO		293.
CH$_2$=		126.5	(CO)$_2$O		576.
-CH=		121.5	-OH-	H bond	226.
>C=		84.5	-OH	aromatic	171.
-CH=	aromatic	117.	-H	acidic dimer	-50.5
-C=	aromatic	98.	-NH$_2$		226.5
CH=C-		285.	-NH-		180.
-C=C-		222.	-N-		61.
-O-	ether, acetal	115.	-C=N		354.5
			-NCO		358.5

Group	E	Group	E
-S-	209.5	H variable	80-100.
-F	161.	CO ketones	275.
-Cl primary	205.	COO esters	310.
>Cl₂ twinned	260.	Cis	-7.
-CCl₃ triple	250.	Trans	-13.
-Br	340.	Ortho	9.
-I	425.	Meta	6.
-CF₂	150.	Para	40.
-CF₃	274.	-S- sulfides	225.
Phenyl	735.	-SH	315.
Phenylene (o,m,p)	658.	-ONO₂ nitrates	~440.
Naphthyl	1146.	-NO₂ nitro	~440.
Ring, 5 membered	105-115	PO₄ organic phosphate	~500.
Ring, 6 membered	95-105	Si silicones	~-38.
Conjugation	23.		

Table 41-10 provides several solubility parameters for nonpolar solvents. Table 41-11 for moderately polar solvents and Table 41-12 for solvents than have hydrogen bonding capabilities.

Table 41-10. Solubility parameter values. Nonpolar solvents Table compiled by R. B. Seymour.

Group	δ (H).	Group	δ (H)
Acetic acid nitrite (acetonitrile)	11.9	Ethane, nitro	11.1
Anthracene	9.9	Ethane, pentachloro	9.4
Benzene	9.2	Ethane, 1,1,2,2-tetrachloro	9.7
Benzene, chloro	9.5	Ethanethiol (ethyl mercaptan)	9.2
Benzene, 1,2-dichloro	10.0	Ethane, 1,1.2-trichloro	9.6
Benzene,ethyl	8.8	Ethane trichlorotrifluoro (Freon 113)	7.3
Benzene, isopropyl (cumene)	8.5	Ethene, (ethylene)	6.1
Benzene, 1-isopropyl-4-methyl (p-cymene)	8.2	Ethene, tetrachloro (perchloroethylene)	9.3
		Ethene, trichloro	9.2
Benzene, nitro	10.0	Heptane	7.4
Benzene,propyl	8.6	Heptane, perfluoro	5.8
Benzene, 1,3,5-tri methyl (mesitylene)	8.8	Hexane	7.3
Benzoic acid nitrite (benzonitrile)	8.4	Hexene	17.4
Biphenyl, perchloro	8.8	Malonic acid dinitrile (malononitrile)	15.1
1,3-Butadiene	7.1	Methane	5.4
1,3-Butadiene, 2-methyl (isoprene)	7.4	Methane, bromo	9.6
Butane	6.8	Methane, dichloro (methylene chloride)	9.7
Butanoicacid nitrite	10.5	Methane, dicliloro-difluoro (Freon 12)	5.5
Carbon disulfide	10.0	Methane, dichloromonofluoro (Freon 21)	8.3
Carbon tetrachloride	8.6	Methane, nitro	12.7
Chloroform	9.3	Methane, tetachlorodifluoro (Freon 112)	7.8
Cyclohexane	8.2	Methane, trichloromonofluoro (Freon 11)	7.6
Cyclohexane, methyl	7.8	Naphthalene	9.9
Cyclohexane, perfluoro	6.0	Nonane	7.8
Cyclopentane	8.7	Octane	7.6
Decalin	8.8	Pentane	7.0
Decane	8.0	Pentane, 1-bromo	7.6
Dimethyl sulfide	9.4	Pentane, 1-chloro	8.3
Ethane	6.0	Pcntanoic acid, nitrite (valeronitrile)	9.6
Ethane, bromo (ethyl bromide)	9.6	Pentene	16.9
Ethane, chloro (ethyl chloride)	9.2	Phenanthrene	9.8
Ethane, 1,2-dibromo	10.4	Propane	6.4
Ethane, 1,1-dichloro (ethylidene chloride)	8.9	Propane, 1-bromo	8.9
Ethane, difluorotetrachloro (Freon 112)	7.8	Propane, 2,2-dimethyl (neopentane)	6.3

Table 41-10 Continued

Group	δ (H).
Propane, 1-nitro1	6.3
Propane-2-nitro	9.9
Propene (propylene)	6.5
Propene, 2-methyl (isobutylene)	6.7
Propenoic acid nitrile (acrylonitrile)	10.5
Propionic acid nitrite	10.8
Styrene	9.3
Terphenyl, hydrogenated	9.0
Tetralin	9.5
Toluene	8.9
Xylene, m-	8.8

Table 41-11. Solubility parameter values and moderately polar solvents

Name	δ (H).
Acetic acid, butyl ester	8.5
Acetic acid, ethyl ester	9.1
Acetic acid, methyl ester	9.6
Acetic acid, pentyl ester	8.0
Acetic acid, propyl ester	8.8
Acetic acid amide. N,N-diethyl	9.9
Acetic acid amide, N,N-dimethyl	10.8
Acrylic acid, butyl ester	8.4
Acrylic acid, ethyl ester	8.6
Acrylic acid, methyl ester	8.9
Adipic acid, dioctyl ester	8.7
Aniline, N,N-dimethyl	9.7
Benzene, 1-methoxy-4-propenyl	8.4
Benzoic acid, ethyl ester	8.2
Benzoic acid, methyl ester	10.5
Butanal	9.0
Butane, 1-iodo	8.6
Butanoic acid, 4-hydroxylactone (butyrolactone)	12.6
2-Butanone	9.3
Carbonic acid. diethyl ester	8.8
Carbonic acid, dimethyl ester	9.9
Cyclohexanone	9.9
Cyclopentanone	10.4
2-Decanone	7.8
Diethylene glycol, monobutyl ether (butyl carbitol)	9.5
Diethylene glycol, monoethyl ether (ethyl carbitol)	10.2
Dimethyl sulfoxide	12.0
1,4-Dioxane	10.0
Ethene, chloro (vinyl chloride)	7.8
Ether, 1,1-dichloroethyl	10.0
Ether, diethyl	7.4
Ether, dimethyl	8.8
Ether, dipropyl	7.8
Ethylene glycol, monobutyl ether (butyl Cellosolve)	9.5
Ethylene glycol, monoethyl ether (ethyl Cellosolve)	10.5
Ethylene glycol, monomethyl ether (methyl Cellosolve)	11.4
Formic acid amide, N,N-diethyl	10.6
Formic acid amide, N,N-dimethyl	12.1
Formic acid, ethyl ester	9.4
Formic acid, methyl ester	10.2
Formic acid, 2-methylbutyl ester	8.0
Formic acid, propyl ester	9.2
Furan	9.4
Furan, tetrahydro	9.1
Furfural	11.2
2-Heptanone	8.5
Hexanoic acid, 6-aminolactam	12.7
Hexanoic acid, 6-hydroxylactone	10.1
Isophorone	9.1
Lactic acid, butyl ester	9.4
Lactic acid, ethyl ester	10.0
Methacryfic acid, butyl ester	8.3
Methacryllc acid, ethyl ester	8.5
Methacrylic acid, methyl ester	8.8
Oxalic acid, diethyl ester	12.6
Oxalic acid, dimethyl ester	11.0
Oxirane (ethylene oxide)	11.1
Pentane, 1-iodo	8.4
2-Pentanone	8.7
Pentanone-2,4-hydroxy-4-methyl	9.2
Pentanone-2,4-dimethyl (mesityl oxide)	9.0
Phosphoric acid, triphenyl ester	8.6
Phosphoric acid, tri-2-tolyl ester	8.4
Phthalic acid, dibutyl ester	9.3
Phthalic acid, diethyl ester	10.0
Phthalic acid, dibexyl ester	8.9
Phtatic acid, eimethyl ester	10.7
Phthalic acid, di-2-methylnonyl ester	7.2
Phthalic acid dioctyl ester	7.9
Phthalic acid, dipentyl ester	9.1
Phthalic acid, dipropyl ester	9.7
Propane, 1,2-epoxy (propylene oxide)	9.2
Propionic acid, ethyl ester	8.4
Propionic acid, methyl ester	8.9
4-Pyrone	13.4
2-Pyrrolidone, 1-methyl	11.3
Sebacic acid, dibutyl ester	9.2
Sebacic acid, dioctyl ester	8.6
Stearic acid, butyl ester	7.5
Sulfone, diethyl	12.4
Sulfone, dimethyl	14.5
Sulfone, dipropyl	11.3

Table 41-12. Solubility parameter values and hydrogen-bonded solvents

Name	δ (H).	Name	δ (H).
Acetic acid	10.1	1-Hexanol	10.7
Acetic acid amide, N-ethyl	12.3	1-Hexanot-2-ethyl	9.5
Acetic acid, dichloro	11.0	Maleic acid anhydride	13.6
Acetic acid, anhydride	10.3	Methacrylic acid.	11.2
Acrylic acid	12.0	Methacrylic acid amide, N-methyl	14.6
Amine, diethyl	8.0	Methanol	14.5
Amine, ethyl	10.0	Methanol, 2-furil (furfaryl alcohol)	12.5
Antine, methyl	11.2	1-Nonanol	8.4
Ammonia	16.3	Pentane, 1-amino	8.7
Aniline	10.3	1,3-Pentanediol, 2-methyl	10.3
1,3-Butanediol	10.9	1-Pentanol	11.6
1,4-Butanediol	10.0	2-Pentanol	12.1
2,3-Butanediol	8.7	Piperidine	11.1
1-Butanol	13.0	2-Piperidone	11.4
2-Butanol	12.6	1,2-Propanediol	10.8
1-Butanol, 2-ethyl	11.9	1-Propanol	10.5
1-Butanol, 2-methyl	11.5	2-Prpanol	10.0
Butyric acid	10.5	1-Propanol, 2-methyl	10.5
Cyclohexanol	10.6	2-Propanol, 2-methyl	11.4
Diethylene glycol	11.8	2-Propenol (allyl alcohol)	12.1
1-Dodecanol	9.9	Propionic acid	8.1
Ethanol	10.0	Propionic acid anhydride	12.7
Ethanol, 2-chloro (ethylene chlorohydrin)	12.6	1,2-Propanediol	12.2
		Pyridine	14.6
Ethylene glycol	10.7	2-Pyrrolidone	12.1
Formic acid	14.7	Quinoline	13.9
Formic acid amide, N-ethyl	10.8	Succinic acid anhydride	16.1
Formic acid amide, N-methyl	15.4	Tetraethylene glycol	16.5
Glycerol	9.9	Toluene, 3-hydroxy (meta cresol)	10.3
2,3-Hexanediol	0.2	Water	9.4
1,3-Hexanediol-2-ethyl	23.4		

Figure 41-5, p. 485, shows a scheme developed by R.L. Shriner and R.C. Fuson (*The Systematic Identification of Organic Compounds*, 3rd ed., J. Wiley & Sons, New York, NY, 1948), to narrow the range of possibilities when determining an unknown organic compound based on its solubility in reaction solvents. The Class descriptions follow.

"Class S_1. Compounds soluble in water and ether, usually of low molecular weight. Exceptions are low-molecular weight hydrocarbons and their halogen derivatives, which fall in Class I. Low molecular weight compounds that have two or more functional groups usually belong in Class S_2.

"Class S_2. Compounds soluble in water and insoluble in ether. Water soluble salts of all kinds, most of the low-molecular weight bifunctional compounds, and many polyfunctional compounds.

"Class A_1. Compounds insoluble in water, but soluble in dilute sodium hydroxide solution and in dilute sodium bicarbonate solution. Acids and a few negatively substituted phenols, such as picric acid and *s*-tribromophenol.

"Class A_2. Compounds insoluble in water and in dilute sodium bicarbonate, but soluble in dilute sodium hydroxide solution. Weakly acidic compounds belong in this class. Weakly acidic properties are usually exhibited by oximes, imides, amino acids, sulfonamides of primary amines, primary and secondary nitro compounds, enols and phenols. Certain mercaptans also are weak acids.

"Class B. Compounds insoluble in water and in alkali, which react with dilute HCl to yield soluble products. Amines are in this class. Diaryl- and triarylamines are exceptions, being nearly neutral compounds. Amphoteric compounds are classed as A_1(B) or A_2(B). Water-insoluble acids such as calcium oxalate also are in Class B. Likewise, certain acetals, which are readily hydrolyzed by dilute acids, may fall in this class.

"Class M. Neutral compounds, insoluble in water, which contain elements other than C, H, O, and the halogens. Nitro compounds, amides, negatively substituted amines, nitriles, azo compounds, hydrazo compounds, sulfones, sulfonamides derived from secondary amines, mercaptans, thio ethers, and many less common types of compounds are classified in the miscellaneous group.

"Class N_1. Neutral compounds insoluble in water and soluble in concentrated sulfuric acid and in syrupy phosphoric acid. Low-molecular weight alcohols, aldehydes, cyclanones, methyl ketones, and esters make up this class. In most of these series, the upper limit of solubility in phosphoric acid is in the neighborhood of the members containing nine carbon atoms.

"Class N_2. Neutral compounds insoluble in water and in syrupy phosphoric acid and soluble in concentrated sulfuric acid.

In addition to alcohols, aldehydes, cyclanones, ketones, and esters that may have more than nine carbon atoms, this class contains many quinones, ethers, and unsaturated hydrocarbons. Anhydrides, lactones, and acetals may be found here as well as in Classes S_1 and N_1.

"Class I. Compounds, insoluble in water, that dissolve in none of the reaction solvents. Saturated aliphatic hydrocarbons, aromatic hydrocarbons, and halogen derivatives of these hydrocarbons constitute this class."

POLYMERS

Some polymers just aren't soluble. They will swell a bit but will not go into solution. Solubility is defined as taking place without degradation. It is perfectly fair to work above the melting point of the polymer, but degradation must be carefully avoided.

Most polymers, especially those that have been heat-treated, are composed of a crystalline phase and an amorphous phase. The ability of solvent molecules to penetrate the crystalline phase may be quite limited, in turn, limiting solubility. For this reason, many polymers are soluble only near their crystalline m.p..

Many polymers have two melting points, that for the crystalline portion and that for the amorphous portion. If solubility is to occur, the crystalline segments must be broken apart. This occurs at the crystalline m.p.

Solubility can often be achieved at temperatures significantly below the m.p. This is the solvent depresses the m.p. and as the m.p. is approached, crystallinity decreases. Thus linear polyethylene, is soluble in many liquids at temperatures above 100 °C, even though its crystalline melting point, T_m = 135 °C, and polytetrafluoroethylene, T_m 325 °C, is soluble in some of the few liquids that exist above 300 °C.

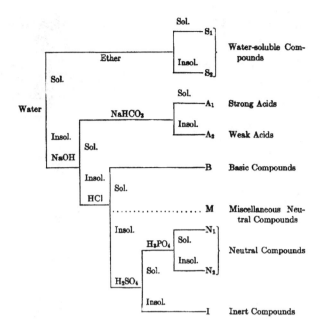

Figure 41-5. Division of organic compounds into solubility classes.
(Courtesy - R.L. Shriner and R.C. Fuson, *The Systematic Identification of Organic Compounds*, 3rd Ed., J. Wiley & Sons, New York, NY, 1948)

Cross-linked - polymers can't dissolve but go through two stages:
1. Gel formation - swelling.
2. Disintegration - may take days for high m.w. polymers.

The free energy of mixing must be negative for solubility to occur.
$$\Delta G = \Delta H - T\Delta S. \tag{41-8}$$

Because the entropy of mixing is nearly always positive, then the ΔH term is of overriding importance in determining solubility. Using Hildebrand's solubility parameter, solubility can be expected if the difference between δ_1 and δ is less than 2 in the absence of H-bonding.

$$\Delta H = V_1 V_2 (\delta_1 - \delta_2)^2 \tag{41-9}$$

where V_1 = volume fraction of the solvent. V_2 = volume fraction of the solute.

If H-bonding is evident with the particular polymer, then some salt must be added to break it up. For example, linear chains - LiCl or $ZnCl_2$ in methanol. Use the metal to break the H bonds between chains.

$$\delta_{mix} = \frac{X_1 V_1 \delta_1 + X_2 V_2 \delta_2}{X_1 V_1 + X_2 V_2} \tag{41-10}$$

where X = mole fraction.

EXAMPLE CALCULATION
What is the δ for an equimolar mixture of chloroform ($CHCl_3$, δ = 9.3 d =1.484, m.w. = 119.4) and hexane (C_6H_{14}, δ = 7.4, d = 0.660, m.w. = 86.2)?

ANSWER

$X_1 = 0.5$; $X_2 = 0.5$, $V_1 = (119.4 \text{ g} \times \text{mL}/1.484 \text{ g} = 80.45 \text{ mL})$; $V_2 = (86.2 \text{ g} \times \text{mL}/0.660 \text{ g} = 130.6 \text{ mL})$

$$\delta = \frac{(0.5000 \times 80.45 \times 9.3) + (0.5000 \times 130.60 \times 7.4)}{(0.5000 \times 80.45) + (0.5000 \times 130.60)} \qquad \delta = 8.12$$

EXAMPLE CALCULATION

A medium m.w. for polyvinyl acetate [-(CH_2=CH-OCOCH$_3$)-] pellets is 167,000 and the density is 1.01 g/cm^3. What range of δ value solvents might dissolve this polymer?

ANSWER

Apply equation 41-7 and the values from Table 41-9, p. 481.

Group	E	m.w.
-CH$_2$-	131.5	14
>CH-	86	13
-O-	176	16
-C=O	263	28
-CH$_3$	148	15
	804.5	86

$\delta = 1.01(804.5/86) = 9.45$ Any solvent with a δ of 9.45 ± 1.8 should work.

Code Numbers

In an attempt to make recycling of plastics easier, a code system has been developed to allow a person to separate various plastic items into the same class. The code numbers, what they represent, and a few examples are shown in Table 41-13.

Table 41-13. Plastic container code system

Number	Plastic	Examples
1	Polyethylene terephthalate (PET)	Large soda bottles
2	High density polyethylene (HDPE)	Milk containers
3	Vinyl/polyvinyl chloride (PVC)	Containers
4	Low density polyethylene (LDPE)	Glad wrap
5	Polypropylene (PP)	Some clothing
6	Polystyrene (PS)	Cups
7	All other resins and layered materials	

The chart shown in Figure 41-6, p. 487, by H. Clotier and J.E. Prud'homme can be used to identify 29 thermoplastic polymers based on solubility and is the basis for Experiment No. 57.

REVIEW QUESTIONS

1. Iron is soluble in HCl and copper is not. Can you use HCl to dissolve all of the iron out of an Fe-Cu alloy?
2. What are some acids that might be used to dissolve Co, Cr, and Mn?
3. What inorganic fluorides are soluble in water?
4. What is one reason why small particles are more soluble than large ones of the same compound?
5. Commercial HNO_3 is said to have a density of 1.42 g/mL and to be 72% acid. Can you calculate its molarity and see how close you come to 16.2 M, the accepted value?
6. Why is H_2SO_4 usually used as a last resort when dissolving inorganic compounds?

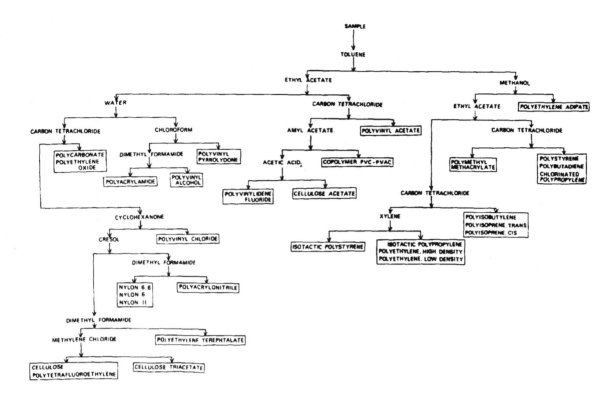

Figure 41-6. Solubility chart for several thermoplastic polymers. Soluble to the right, insoluble to the left. (Courtesy - H. Clotier and J. E. Prud'homme, *J. Chem. Ed.*, **62** (9), 815, 1985)

7. A person comes up to you and says that you cannot dissolve a porcelain crucible. What acid would you use to prove him wrong?
8. Under what conditions does $HClO_4$ become an oxidizing agent?
9. What is the difference between a perchloric acid hood and a regular hood and why?
10. What types of perchlorate compounds are the most dangerous?
11. What is a common catalyst for $HClO_4$ acid digestions?
12. Table 41-5, p. 477, shows how the digestion rate is related to the % $HClO_4$. How can you control the % composition?
13. What acids make up the solvent for the *liquid fire* reaction, and what is the purpose of HNO_3?
14. What types of compounds have been found to dissolve large quantities of oxygen?
15. What is diglyme?
16. What is implied by - like dissolves like?
17. What is the basis for determining polarity in the Reichert system?
18. Why is ΔH_{vap} of the solvent used when determining Hildebrand's solubility parameter?
19. Using equation 41-4, what is the δ for sulfolane, m.w. = 120.16, if it boils at 285 °C and has a density of 1.2606 g/mL?
20. Using equation 41-6, determine δ for sulfolane with the data in question 19.
21. Calculate δ for propylene glycol methyl ether, $H_3COCH_2CH(CH_3)OH$, using equation 41-7. d = 1.04 g/mL.
22. Refer to Figure 41-5, p. 485. What type of compound would be insoluble in water, insoluble in NaOH, insoluble in HCl, but soluble in H_2SO_4 and in H_3PO_4?
23. You have a compound you suspect to be an oxime. What Shriner and Fuson class would you expect to find it in?
24. A drinking cup for hot coffee has the number 6 stamped on its bottom. What does this mean?
25. You have a polymer that melts when it is warmed. It is soluble in toluene, insoluble in ethyl acetate, soluble in CCl_4 and amyl acetate, and insoluble in acetic acid. What is it likely to be?
26. Many polymers contain two phases, amorphous and crystalline. Which is the hardest one to dissolve, and what can be done to help solubilize this type of polymer?
27. What can be done to help solubilize a polymer that has heavy hydrogen bonding?
28. If you pick up an empty 2-liter plastic soft drink bottle and look at its underside, you will see molded in the plastic a set of arrows in a triangle arrangement with a 1 inside. What does the 1 mean?

IF IT EXISTS - CHEMISTRY IS INVOLVED
IF YOU CAN BUY IT - A CHEMIST WAS INVOLVED SOMEWHERE

42
GAS ANALYSIS BY PORTABLE ORSAT

PRINCIPLES

Portable Orsat is a separation technique that involves the selective absorption and quantitative measurement of one gas from a mixture of gases. Such a mixture might be the exhaust gas from an engine, the gas in a mine shaft, the gas in a sewer, or the gas in a silo. This is a true volumetric procedure. All quantitative values depend on volumes measured. Few, if any, of you have ever really handled a gas. You may have prepared H_2, CO_2, or O_2 as a freshman chemistry laboratory experiment, but you have never had to collect and separate a mixture of gases. This chapter will show you several reagents that are used to absorb various gases, how to collect a gas sample, and how to separate it into several of its components.

Why not just use a gas chromatograph? If you have a low cost, light weight, battery operated, portable unit, then use the chromatograph. You might also get a sample in a gas-tight syringe and take it back to the lab for chromatographic analysis. Before the advent of gas chromatographs, the Orsat apparatus was the method of choice, and oil companies had rooms full of very large, elaborate units. Its use has declined dramatically in recent years, almost to nonexistence. However, if the amounts are large, and you want something reliable to take to the field, then the Orsat apparatus is a good choice. The author still gets a call every year or so from someone who wants to borrow his. When the author called the manufacturer, their representative said they still sell several in the United States and even more overseas. The main purposes here are to show you what has been done and what can be done and to describe the principles and techniques involved so you will have another option in your bag of tricks for the future.

The apparatus to do these separations is called the Orsat apparatus. The sole supplier is Burrell Scientific Inc., 2223 5th Ave., Pittsburgh, PA 15219 (412-471-2527). They have everything needed, including replacement parts, reagents, and directions.

METHODS OF GAS ANALYSIS

Gas absorption

For gas absorption a specific absorbent is used, and the difference in volume before and after is determined. Oxygen is absorbed from air by an alkaline solution of pyrogallol.

Gas volumetric

For gas volumetric the volume of a gas liberated from a solid or liquid is determined, such as CO_2 from $CaCO_3$.

Combustion analysis

For combustion analysis a measured amount of the gas is combusted with a known amount of air or O_2. This is described later. An example is:

$$CH_4 + O_2 \rightarrow CO_2 + H_2O$$

Gravimetric

For gravimetric methods the gas is absorbed and weighed, such as CO_2 being absorbed on Ascarite.

Titrimetric

For titrimetric methods the gas is allowed to react with a solution, and the excess solution is titrated.

$$Cl_2 + K_3AsO_3 \text{ and } CO_2 + Ba(OH)_2$$

Instrumental

Instrumental methods are fast and accurate, but usually expensive.

ABSORBENTS

Carbon dioxide

With a 30% solution of KOH, the capacity is enormous. Use KOH rather than NaOH, because K_2CO_3 is soluble in KOH, but Na_2CO_3 is not soluble in NaOH, and the precipitate clogs the lines.

Ethylenediamine, H_2N-CH_2-CH_2-NH_2, has been tried with some success.

Unsaturated hydrocarbons

Bromine water (containing 5% KBr) is preferred. The Br_2 adds across the double bond. If bromine water is used, then the gas must be passed through KOH to remove the bromine fumes.

Fuming H_2SO_4 (20%) is a reagent to handle with care. The 20% means that an extra 20% of SO_3 is present. The fumes are SO_3 dimer, not H_2SO_4 vapor.

Oxygen

Alkaline pyrogallol in equal volumes of 15% pyrogallic acid and 30% KOH. The capacity is 1:8, but it takes 3 minutes to completely absorb the O_2. CO is evolved after a while. Pyrogallol is not specific, but absorbs all acidic gases because of the KOH present. Because of the low capacity a tag usually is attached to the absorption container, and the number of mL of O_2 absorbed is recorded, so it can be replaced when necessary.

Pyrogallol

Alkaline Hydrosulfite

$$2\ Na_2S_2O_4 + O_2 + 4\ NaOH \rightarrow 4\ Na_2SO_3 + 2\ H_2O$$

Add anthraquinone-β-sulfonate as a catalyst. The capacity is 1:128 mL. It absorbs all acidic gases.

Yellow phosphorus (also known as white phosphorus) is specific for O_2. It is used as sticks, stored under water. If O_2 is present in more than 50%, an explosive mixture is formed. The reaction product is soluble in water; therefore, the capacity is quite large. Unsaturated hydrocarbons (0.01% C_2H_4) and H_2S poison it. K_2SO_4 solution can be used to remove red phosphorus, which is a more stable form and will form upon standing. It ignites at about 30 °C in moist air. Handle with forceps and keep under water. The fumes and the element itself are poisonous.

$$4\ P + 3\ O_2 + 6\ H_2O \rightarrow 4\ H_3PO_3$$

Chromous chloride, $CrCl_2$ (acidic solution), is sold commercially as OXSORBENT. It is specific for O_2. It is very rapid and quite accurate. However, the reagent is hard to prepare. The chromic salt must be passed through a Jones reductor (amalgamated Hg) to reduce the chromium. A Jones reductor is easy to prepare. Dissolve 1 to 2 g of $HgCl_2$ in a 125 mL Erlenmeyer flask. Add granulated Zn and swirl the flask for a few seconds. The Zn will become silvery and expand. Discard the solution, then rinse the amalgamated Hg with water and place it in a buret or ion exchange column. Pass the chromium solution through the column, and the chromium emerging will be Cr^{+2}.

Carbon monoxide

All CO absorbents employ cuprous salts. O_2 and unsaturated hydrocarbons interfere with all three of these absorbents.

Cuprous chloride (acid) should be stored under a H_2 atmosphere and over Cu wire.

$$Cu_2Cl_2 + HCl$$

Cuprous chloride (basic):

$$Cu_2Cl_2 + NH_4OH + NH_4Cl$$

In weak reagents, CO may be liberated:

$$Cu_2Cl_2 + 4\ H_2O + 2\ CO \longrightarrow Cu_2Cl_2 : 2CO : 4H_2O \text{ (unstable)}$$

Cuprous sulfate + β-naphthol is sold commercially as COSORBENT. Its capacity is 1:18 mL. Add Zn on a porcelain plate to generate H_2 and you can store it longer.

$$CuSO_4 + 2\ CO \longrightarrow Cu_2SO_4 : 2CO \text{ (stable)}$$

Nitric Oxide

Ferrous sulfate (slightly acidic):

$$2\ FeSO_4 + 2\ NO \longrightarrow (Fe(NO)_2(SO_4)_2$$

Alkaline $(KSO_3)_2NO$:

$$(KSO_3)_2NO \text{ (blue violet)} + NO \longrightarrow (KSO_3)_2(NO)_2 \text{ (colorless)}$$

Ozone

Neutral or Alkaline KI:

$$O_3 + 2\ KI + H_2O \longrightarrow O_2 + 2\ KOH + I_2$$

Sodium arsenite:

$$Na_3AsO_3 + O_3 \longrightarrow Na_3AsO_4 + O_2$$

MIXTURES

Generally, the order in which gases are absorbed is as follows. Carbon dioxide is removed first. Next are the unsaturated gases, such as ethylene, acetylene, butyne, or anything that has a double or triple bond in it. Then oxygen is absorbed, followed by carbon monoxide. Hydrogen and the other hydrocarbons are removed by combustion. Finally, what's left is considered as nitrogen.

Usual mixtures of combustion gases contain CO_2, CO, O_2, and N_2.

1. CO_2 (1st), O_2 (2nd), N_2 by difference.

2. CO_2 (1st), O_2 (2nd), CO last, but pass into KOH to remove HCl.

FRACTIONAL COMBUSTION

After O_2, CO_2, and the unsaturated hydrocarbons are removed, CO and H_2 can be determined by fractional combustion in the presence of saturated hydrocarbons.

$$CuO + H_2 \rightarrow Cu + H_2O \quad \text{At 275 °C}$$
$$CuO + CO \rightarrow Cu + CO_2 \quad \text{At 275 °C}$$
$$CuO + CH_4 \rightarrow \text{N.R., only above 400 °C}.$$

After the combustion, reoxidize the Cu so it will not interfere in subsequent work.

SLOW COMBUSTION

Saturated hydrocarbons are inert to being absorbed; therefore, they are determined by combustion with O_2. Figure 42-1 shows an uncovered slow combustion pipet, which also is shown in Figure 42-3, p. 495, on the left side. Mercury must be used as the retaining liquid. A measured volume of the gas or mixture of gases is combusted in a measured volume of O_2. The volume contraction, the amount of O_2 consumed, and the amount of CO_2 produced are measured. From these three pieces of data, you can calculate three unknown quantities. You can do this:

1. Determine CO, H_2, and one other combustible gas.
2. Determine three hydrocarbons, provided no more than two are in the same group (C_nH_{2n+2}). H_2 is considered a member of the hydrogen group.
3. Determine the average n for a mixture of gases all in the same group.
4. Determine the volume of each of two groups of gases present.
5. Identify a pure gas, C_nH_m.

Figure 42-1. A slow combustion cell. (Courtesy - Burrell Scientific Inc., Pittsburgh, PA)

EXAMPLE CALCULATION

Develop equations to determine the relative amounts of CO, H_2, and CH_4 in a mixture of these three.

ANSWER

Step 1. Write the combustion equations.
$$2\,CO + O_2 \rightarrow 2\,CO_2$$
$$2\,H_2 + O_2 \rightarrow 2\,H_2O$$
$$CH_4 + 2\,O_2 \rightarrow CO_2 + 2\,H_2O$$

Step 2. Develop equations for the contraction (C), the CO_2 formed, and the O_2 used.

Contraction: This is the volume contraction that occurred during the combustion because each component acting as if it was the only one present. The water formed is assumed to have 0 volume. *Compare* the original volume of the gas to the final volume. For CO, there were originally 3 volumes of gas (2 CO + O_2) compared to 2 volumes after combustion (2 CO_2); therefore, the contraction is 1 volume or 1/2 the amount of CO present.

Repeat this process for H_2 and CH_4. For H_2, there are 3 volumes initially (2 H_2 + O_2) and 0 finally (H_2O = 0); therefore, the contraction is 3/2 of the H_2 present. Yes it is. Think about it. For CH_4, there are 3 volumes initially (CH_4 + 2 O_2) and 1 volume finally (CO_2) for a contraction of 2 volumes or 2 times the CH_4 present. The total contraction equation is:

(1) $C = 1/2\, CO + 3/2\, H_2 + 2\, CH_4$

Similar equations for the CO_2 produced (2) and the O_2 consumed (3) follow. Remember: *always compare the volume difference to the volume of the original gas.*

(2) $CO_2 = CO + CH_4$

(3) $O_2 = 1/2\, CO + 1/2\, H_2 + 2\, CH_4$

You now have three equations for three unknowns. Solve any combination of them to obtain an equation for each of the three gases. In some mixtures, you can have more than one answer, depending upon which equations you prefer to work with.

To determine H_2 in the mixture.

Equation (1) - equation (3)

(4) $C - O_2 = H_2$ or $\mathbf{H_2 = C - O_2}$

To determine CH_4 in the mixture, combine equations to eliminate H_2, a three-step process.

2 x equation (1) - equation (2)

(5) $2\,C - CO_2 = 3\,H_2 + 3\,CH_4$

2 x equation (3) - equation (2)

(6) $2\,O_2 - CO_2 = H_2 + 3\,CH_4$

Equation (5) - 3 x equation (6)

(7) $2\,C - 6\,O_2 - 2\,CO_2 = -6\,CH_4$ or $\mathbf{CH_4 = 1/3(3\,O_2 + CO_2 - C)}$

For CO, do the same thing, only eliminate CH_4.

2 x equation (2) - equation (1)

(8) $C - 2\,CO_2 = -3/2\,CO + 3/2\,H_2$

2 x equation (2) - equation (3)

(9) $2\,CO_2 - O_2 = 3/2\,CO - 1/2\,H_2$

3 x equation (9) - equation (8)

(10) $4\,CO_2 - 3\,O_2 + C = 3\,CO$ or $\mathbf{CO = 1/3\,C - O_2 + 4/3\,CO_2}$

TECHNIQUES

TUTWEILER APPARATUS

This is a gas-volumetric-titrimetric combination used for determining sulfur, S, in iron and steel, mineral waters, sewers, and other materials. Figure 42-2, p. 494, shows a drawing of the apparatus.

In acid:

$$H_2S + I_2 + H^+ \longrightarrow 2\,HI + S$$

In base:

$$S^{-2} + 4\,I_2 + 8\,OH^- \longrightarrow SO_4^{-2} + 8\,I^- + 4\,H_2O$$

The main bulb is filled with starch solution. 100 mL of sample is added and confined. The leveling bulb then is lowered to drain out an additional 10 mL and create a vacuum inside of the main bulb. Iodine is added to the 10 mL buret and stoppered. The I_2 is added slowly to the main bulb to titrate the H_2S. At the endpoint the starch turns blue-violet.

Starch solution: 4 g/L.

Iodine solution: 1.7076 g/L;

1 mL = 1.7076 mg I_2 = 0.229 mg H_2S/mL)

Engineers prefer grains/100 ft^3

1 grain = 64.7 mg

1 mL = 100 grains H_2S/100 ft^3

The S formed turns the solution white. The system is sensitive to 10 grains of H_2S.

Figure 42-2. Tutwiler apparatus.
(Courtesy - Corning Glass Inc., Corning, NY)

EXAMPLE CALCULATION

A sewer, found to have a broken line, has backed up into the basement of an abandoned building. The workers nearby are complaining of a rotten egg odor and want this sewage removed. The foreman of the cleanup crew wants to know if it is safe to have his men enter and work in the basement without wearing breathing apparatus. You collect a 100 mL sample of the air and do a Tutwiler analysis on the scene. It takes 0.87 mL of I_2 to obtain a permanent blue color. What is the concentration of H_2S in this air, and is it safe for normal exposure? The LC_{50} (1 hr) for H_2S is 673 ppm. 1.00 L of air weighs 1.23 g.

ANSWER:

100 mL of air weighs 0.123 g; therefore, 1 ppm weighs 1.23×10^{-4} mg

$$0.87 \; mL \times \frac{0.229 \; mg \; H_2S}{mL} \times \frac{1 \; ppm}{1.23 \times 10^{-4} \; mg} = 1620 \; ppm$$

This far exceeds the limit of 673 ppm, so the workers must be provided with supplemental air.

THE PORTABLE ORSAT APPARATUS

Figure 42-3, p.495, shows a portable Orsat apparatus. Its use in step-by-step detail is fully described in the accompanying experiment. In general, all absorption bottles are filled to their zero mark, the gas buret (extreme right) is zeroed, and the sample collection bottle is connected by a rubber tube at point A. 100 mL of sample is drawn into the gas buret and then passed successively into each absorption bulb.

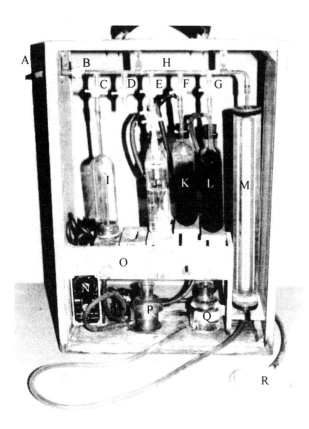

Figure 42-3. A portable Orsat apparatus.
(Courtesy - Burrell Scientific, Inc., Pittsburgh, PA)

Figure 42-4. A gas buret.
Courtesy - Burrell Scientific Inc., Pittsburgh, PA)

The loss in volume is measured by the gas buret. Figure 42-4 shows a gas buret, Figure 42-5, p. 496, shows a closeup of the scales on the gas buret, Figure 42-6, p. 496, shows two types of absorption bulbs, and Figure 42-7, p. 497, shows a manifold and a gas sample collecting bottle.

Gas Buret

The gas buret can contain slightly over 100 mL of a gas. It is water jacketed to maintain everything at constant temperatures. The heat capacities of the gases are low, so a water jacket around the tube will bring them to temperature equilibrium very quickly. The buret has two mL scales printed on it. One scale goes from 0 to 100 starting at the top, and the other goes from 0 to 100 starting at the bottom, as shown more clearly in Figure 42-5, p. 496. It is filled with water that is slightly acidic, so it will not absorb acid gases. A few drops of H_2SO_4 are preferred, because both HCl and HNO_3 produce small amounts of gas in the headspace, but H_2SO_4 does not. A few drops of methyl red indicator is added to indicate that the water is acidic. If the water color turns yellow, then it is not acidic. The buret is filled by using a *leveling bulb*, Q in Figure 42-3. The bulb is removed from its holder, so it can be raised and lowered. Sufficient water is placed in the bulb to fill the gas buret and have a small excess. Gas is drawn into the buret by lowering the bulb, and the volume is measured by placing the bulb along side of the buret and adjusting the level of the water, so it is the same in the buret and in the bulb. This corrects for atmospheric pressure.

Figure 42-5. Lower end of a gas buret.
(Courtesy - Burrell Scientific Inc., Pittsburgh, PA)

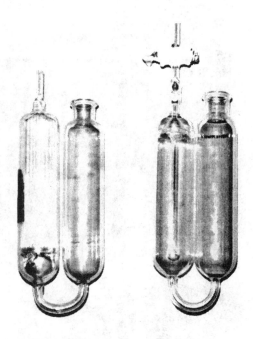

Figure 42-6. Gas absorption bottles. Left: filled with glass tubes. Right: single tube with holes in the bottom.
(Courtesy - Burrell Scientific Inc., Pittsburgh, PA)

Gas Sampling Bottle

Usually, a 5 cm piece of rubber tubing is attached to one end and an 8 to 10 cm piece on the other end of a gas sampling bottle. To obtain a sample of gas not contaminated by air, the bottle is filled with slightly acidic water, and the stopcocks are closed. The bottle is taken to where a sample is to be collected; the long piece of tubing is placed in the container, such as the exhaust pipe of a car; and both stopcocks are opened to let the water drain out. As the water drains out, it draws in a sample uncontaminated by air - if you're good. The stopcocks are closed and you have a sample saturated with water vapor so it will be compatible with the apparatus.

Absorption Bulbs

Absorption bulbs are modifications of the original Hemple pipets, and some people still use that name. The principle is to provide as much surface area as possible for the gas to come in contact with. One design is shown on the left of Figure 42-6. Several glass tubes are placed in one side of the bulb. This is the side connected to the manifold. The other side is to collect the displaced absorption solution when the gas enters. A small rubber bag is attached to the top of this side to hold the air being pushed out when the liquid fills this chamber. A bag is provided for each absorption bulb and one is shown as R in Figure 42-3. The bag allows liquid to transfer and absorption to take place, without contamination of the solution from the outside air.

The bulb on the right side of Figure 42-6 has a single tube inside of the bulb with several holes at the bottom. This is to disperse the gas, again providing for a large surface area. At the top of each bulb up near the manifold, on the manifold side, is a graduation mark. The liquid level is always brought to this mark before and after each measurement.

Manifold

This is easy to break. All of the lines are capillary tubing to keep the volume small. When you connect the gas buret or the absorption bulbs to it, be sure to get as close to glass to glass contact as possible.

Figure 42-7. Top: manifold for an Orsat apparatus. Bottom: a gas sampling bottle. (Courtesy - Burrell Scientific Inc., Pittsburgh, PA)

REVIEW QUESTIONS

1. What do we mean by a volumetric analysis?
2. What is the basis for a gas absorption separation?
3. What is the basis for a titrimetric gas analysis?
4. Why should a tag be placed on a pyrogallol absorption pipet?
5. What is COSORBENT?
6. What is a Jones reductor?
7. What are two reagents for ozone?
8. Derive formulas for calculating each component in the following mixtures:
    a. $H_2 + CH_4$
    b. $CH_4 + C_2H_6$
    c. $H_2 + CO$
    d. $H_2 + CO + C_2H_6$
    e. $CH_4 + C_2H_6 + C_3H_8$
    f. $C_2H_2 + C_2H_6$
    g. $CO + CH_4 + C_2H_6$
9. What is the value of 1 grain in mg?
10. Refer to the example problem. How many mL of I_2 would it take to just indicate the safe level of H_2S?
11. How are the scales arranged on a gas buret?
12. Why is the water in the Orsat apparatus made slightly acidic?
13. Where should a leveling bulb be placed when making a volume measurement and why?
14. How do you get a sample of a gas without contaminating it with air?
15. Why is a gas bag added to the back portion of an absorption tube?

IF IT EXISTS - CHEMISTRY IS INVOLVED
IF YOU CAN BUY IT - A CHEMIST WAS INVOLVED SOMEWHERE

APPENDIX A

EXPERIMENTS BASED ON COMMERCIAL PRODUCTS AND NATURAL SAMPLES

There are far more experiments presented here than you are expected to do. Your instructor will select or suggest those that are best suited to the needs of the class or your particular area of interest. They involve commercial products or natural samples for the most part. A wide variety of samples are involved because the techniques of sample preparation are very important for satisfactory results. While most of the experiments can be completed in one laboratory period, several cannot. In those experiments, you can either spend extra time, your instructor can suggest portions to omit, or some portions can be done for you. These are real world samples. They sometimes smell, they do not always dissolve as you would like them to, and most contain interfering substances. Therefore, the author believes it is important that you learn how to handle various samples and recommends that do as much as you can in the time allotted.

It is not enough to complete a separation. You must verify the extent of the separation (percent recovery), and the last part of each experiment is devoted to verification. While a Soxhlet extractor is the same for every laboratory, such instruments as spectrophotometers and gas-liquid chromatographs are not. Therefore, whereas the directions for the Soxhlet extactor can be the same, the various instrument directions need to be localized. To provide some direction for your instructor as to the sequence and basic settings, the directions for a specific instrument are provided for a system known to work.

The format for each experiment is to provide first a brief summary of the principles involved. A problem is presented that requires a separation to be made. A list of all of the equipment needed for one complete setup is provided. A list of all of the chemicals and samples needed is provided with the amount for each student (S) listed at the end, such as 15 mL/S. In the author's laboratory, each of these experiments is on audio tape so the student can proceed at his/her own pace. It is imperative then that the apparatus be set up to follow the directions. That is one of the purposes of the apparatus diagram. The second reason is so that the student can locate where the experiment is set up without being embarrassed because they didn't know what it looked like.

A one paragraph summary of what is to be done gives the student a broad view of what is expected. It also provides the instructor with a quick review of the separation so if a question is asked, the instructor can catch up quickly and provide an answer or a suggestion.

The directions are detailed. It is often cookbook and intentionally so. This is not Chem I with beakers and test tubes where "discovery" is paramount and the apparatus is inexpensive. These separations involve a limited amount of expensive equipment, and good techniques are to be learned. In addition, when real world samples are used, the experiments can take considerable time, so directions are necessary. Unless students are shown what good technique is, they will seldom learn it on their own, and unless they are shown how to take care of an instrument, it will be a "machine" and treated as roughly.

An example calculation is provided at the end. The samples often smell, won't dissolve easily, are messy, and may have interfering substances present. However, that is the way it is after graduation, so let's get used to it now.

If you have an option, do not do experiments that you have familiarity with. If you have had a summer job, or an undergraduate research position that involved some of these techniques, then do something different. You are here to get an education, try to make the best of it.

EXPERIMENT 1. VOLATILIZATION

The determination of CO_2 in either limestone or baking powder.

YOUR PROBLEM

Your company is looking for a source of $CaCO_3$, preferably limestone, to be used to remove SO_2 from its power plant stack emissions.

$$2\ CaCO_3 + 2\ SO_2 + O_2 \rightarrow 2\ CaSO_4 + 2\ CO_2$$

This reaction is about 85% efficient for ordinary conditions, but in times of high power usage, it may drop to 65%. The limestone supply must contain at least 70% $CaCO_3$ to be economically acceptable. Your job is to locate the nearest supply that meets this requirement.

Alternate Problem

There are areas in this country in which the above problem is not reasonable, and there are laboratories that do not have the grinding machines necessary. In that case, an alternate problem, the determination of CO_2 in baking powder, is possible.

Baking powders are required by law to contain sufficient ingredients to produce 12% CO_2 by weight when they are used. Baking powders slowly deteriorate during storage, so most companies adjust the formulation to provide from 12.5 to 14.0% CO_2 initially. Your job as a quality control chemist is to test the product given to you to see if it meets specifications before it is shipped.

EQUIPMENT

1	Aspirator, water		1 pr	Pliers
1	Balance, analytical ± 0.0001g		1	Ring, iron, 10 cm
1	Bottle, weighing		1	Rod, metal 60 cm x 1 cm
1	Burner, Bunsen		2	Ring stands
6	Clamps, buret		1	Spatula, ordinary size
1	Condenser, West		1	Stopper, rubber, no. 3 one hole
1	Dish, weighing, aluminum		1	Stopper, rubber, no. 6, cut 3 holes
1	Flask, Erlenmeyer, 250 mL		1	Towel or sponge
1	Funnel, separatory, 60 mL		1	Tube, drying
1	Gauze, wire, 10 cm x 10 cm		3	Tubes, absorption, small
1 pr	Glasses, safety		8 ft	Tubing, condenser
1	Holder, clamp		5 ft	Tubing, vacuum
1 bx	Matches		1 ft	Wire, copper
1	Oven, drying (110 °C)		1	Z tube, 8 mm glass, 75 cm long

If a field-collected limestone sample is used, then the following or similar apparatus are needed.

1	Brush, paint, 7.5 cm dia.		1	Paper, brown wrapping, 1 m x 1 m
1	Bucket, waste		1	Pulverizer, disc type
1	Crusher, Chipmunk, jaw type		1	Spatula, 30 to 40 cm long
1	Funnel, separatory, 60 mL			

CHEMICALS   S = Student

Anhydrone ($Mg(ClO_4)_2$)
Ascarite (NaOH on asbestos)
Cupric sulfate, $CuSO_4$, anhydrous preferred
Drierite, indicating ($CaSO_4$)
Glass wool

Glycerine, $C_3H_8O_3$
Hydrochloric acid, HCl, conc., 40 mL/S
Stopcock grease
Sulfuric acid, H_2SO_4, conc.

APPARATUS SETUP

A diagram of the apparatus is shown in Figure 1-2, p. 8. It is known as a *Knorr alkalimeter*. The evolution portion is mounted on one ring stand, and the absorption train on the other. All rubber tubing should be wired in place. In addition to the water condenser (F), this includes the short pieces of vacuum tubing between H, I, J, M, and N. These pieces should be kept short, and as much glass contact as possible should be made. A small amount of glycerine on the glass helps make a gas-tight seal. An additional water trap can be placed between the water aspirator and tube M if desired, although no problems have resulted thus far from its absence. It is suggested that one spare absorption tube and two spare U tubes be kept on hand.

THE BIG PICTURE

In a limestone area, go to any hill where a road is cut through and get a 2½ gallon bucket full of what you consider to be a representative sample of the layers of limestone present. The size of the rocks should be no larger than 2.5 inches in diameter. These will be ground in a jaw crusher to 1/4 inches in diameter and then finally to 100 mesh in a disc pulverizer. Cone and quarter the sample to reduce it to about 15 grams, which you will place in a weighing bottle. Place 1.0 gram into a 250 mL flask. Weigh the two U tubes (remove all glycerine) and replace them in the train. Slowly add HCl to the sample until the evolution of CO_2 stops. Turn on the burner, heat the solution to boiling, and continue heating for 5 minutes. Close all stoppers, shut off the vacuum and reweigh the U tubes, again being sure to remove all of the glycerine. Calculate the % CO_2 and % $CaCO_3$ in the sample.

DIRECTIONS

Collecting and Preparing a Limestone Sample

If your instructor provides you with a 100 mesh sample of limestone or baking powder, then go directly to step 30.

1. Collect about a 2.5 gallon bucket full of limestone rocks (no bigger than 2.5 inches in diameter) from an area of interest to you. Try to obtain as representative a sample of this area as possible.

2. Take a weighing bottle, a 7 to 8 cm wide paint brush, a large spatula (30 to 40 cm long), and a 1 m x 1 m sheet of brown wrapping paper to the Chipmunk grinder (Figure E1-1). This grinder will reduce your sample to pieces 1 to 7 mm in diameter or less.

3. Remove the sample collection tray and clean it out.

4. Turn on the grinder and brush off the entire grinder including the area where the sample collecting tray was removed while the grinder is vibrating. CAREFUL.

5. Replace the sample collecting tray.

6. Place one medium-sized rock into the top of the grinder. When the rock has been crushed, discard the material collected in the tray. This includes impurities that were in the grinder.

7. Grind the remainder of the sample. If you have more than the collecting tray will hold, then empty the tray onto the brown wrapping paper and continue with the grinding.

8. Remove all of the sample from the collecting tray. Brush off the grinder and shut it off.

9. You will have too much sample, but this is necessary if a representative sample is to be obtained. It is reduced in size by a process called "coning and quartering" as shown in Figure E1-2. Place your entire sample in the center of a 1 m x 1 m piece of brown wrapping paper or something similar.

10. Take corner A and, holding it 3 to 4 inches from the table top, pull it over to corner B.

11. Repeat step 10 but take corner C to corner D. Then B to A and D to C. *Note*: do not raise the corners so high that the sample slides across the paper. Keep the corners low so that the sample tumbles into itself, thus mixing the fine material in with the coarser particles.

Figure E1-1. A Braun Chipmunk ore crusher, jaw type.
(Courtesy - Gilson Company, Inc., Worthington, OH)

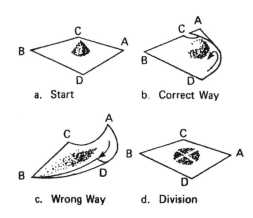

Figure E1-2. The steps in coning and quartering.

Figure E1-3. A Braun type disc pulverizer.
(Courtesy - Gilson Company, Inc., Worthington, OH)

12. Repeat steps 10 and 11 several times, then work the sample to the center of the paper (*coning*). Using a large spatula, divide the sample into 4 sections (*quartering*).
13. Discard opposite quarters.
14. Repeat steps 10 to 13 until you have reduced your sample to about two large handfuls.
15. Examine the disc pulverizer (Figure E1-3). This grinder consists of two plates, one fixed and one rotating. The separation between the discs is controlled by a screw adjustment. This grinder will reduce samples to approximately 100 mesh, sufficient for this experiment.
16. Remove the sample collecting pan and clean it out.
17. Turn on this grinder and brush it off, particularly the discs.
18. Have your instructor show you how to use this grinder.
19. Adjust the grinder until the discs just touch and then separate them about 4 to 5 mm.
20. Place the sample collecting tray in place.
21. Transfer your sample from the brown paper into a small box (shoe box) and place the paper under the sample tray to catch any sample that falls or is thrown out during the grinding.
22. Add the sample and reduce it in particle size.
23. Remove the sample and adjust the disc separation to about 3 to 4 mm.
24. Repeat steps 21 and 22.
25. Continue the process, each time reducing the space between the grinding discs until the discs just barely touch. When the discs just touch, pass the sample through one last time.
26. Place the sample on the brown paper and cone it.
27. Brush off the grinder and then shut it off.
28. Cone and quarter the sample until you have an estimated 15 grams or enough to fill a small weighing bottle about 1/3 full.
29. *Clean up this entire area before you leave.* Thank you.

The Knorr Alkalimeter

(Refer to Figure 1-2, p. 8)

30. Weigh (to ± 0.0002 g) a 0.8 to 1.0 gram portion of the undried sample and place it in a 250 mL Erlenmeyer flask. Record this weight.
31. Weigh an aluminum weighing dish or small watch glass and place 0.8 to 1.0 g (to ± 0.0002 g) of sample on it and record this weight.
32. Place this sample in a drying oven at 100 °C. At this time, go to the Knorr alkalimeter and be sure you fully understand what you are expected to do.
33. Check the Ascarite tube (B) and make sure it is still active. It should be a light brown over at least 2/3 of the tube length. If not, then recharge it. Ascarite is NaOH on asbestos and turns white (Na_2CO_3) when it is used up.
34. The concentrated H_2SO_4 in scrubber (F) need not be replaced.
35. Check scrubber (G). If more than 2/3 of the $CuSO_4$ has turned from blue to green, then recharge this scrubber with fresh $CuSO_4$.
36. Check tubes (H) and (I) and be sure that no more than 1/3 of the Ascarite sections (brown) have turned white. The white Anhydrone section on the right side of each tube is to retain the water released when CO_2 reacts with NaOH.

$$CO_2 + 2\,NaOH \longrightarrow Na_2CO_3 + H_2O$$

37. Check scrubber (J). The indicating Drierite (blue) should be replaced if more than 2/3 of it has turned pink.
38. Connect the rubber tube from scrubber (J) to a water aspirator.
39. Close the stopcocks and then remove the two center U tubes (H) and (I) from the volatilization apparatus. (CARE- these are made of soft glass to reduce static charge buildup and are easily broken)
40. Wipe the glycerine off of all of the connections on these tubes.
41. Weigh each tube (± 0.0002 g) and record the weight of each one.
42. Place a small amount of glycerine on the ends of each tube and connect the tubes back into the absorption train.
43. Wrap a small piece of copper wire around the rubber joint and tighten the wire on the joint with a pair of pliers.
44. Connect the 250 mL Erlenmeyer flask to the apparatus. Do this carefully, yet be sure of a tight fit.
45. Connect the condenser tubing to the water outlet (in at the bottom) and turn on the water. (NOTE- good technique requires that all hose connections be either wired or clamped).
46. Fill the separatory funnel, C, about ½ full with concentrated HCl.
47. Carefully connect the Z tube to the condenser and then to the H_2SO_4 scrubber.
48. Open all of the stopcocks along the line.
49. Slowly turn on the water aspirator. Use the H_2SO_4 scrubber as a gauge and adjust the vacuum until a slow stream of air bubbles is passing through the scrubber.
50. Carefully open the stopcock on the separatory funnel, C, and add the HCl dropwise. Watch the H_2SO_4 scrubber and add the acid at such a rate that the system is not overtaxed. Continue adding the HCl until the bubbling rate in the H_2SO_4 scrubber has subsided to about the rate it was before you started adding the HCl.
51. When it appears that most of the CO_2 has been evolved, light the Bunsen burner. Bring the HCl to a gentle boil and boil it for about 5 minutes.
52. Shut off the burner and close the stopcocks on U tubes (H) and (I).
53. Shut off the water aspirator.
54. Drain the remaining HCl from the funnel into the flask. Remove the flask from the apparatus and rinse it out.
55. Shut off the water to the condenser.
56. Remove the two U tubes from the train and using the same techniques as before, weigh them again. Record these weights.
57. Replace the U tubes back into the line, but don't bother to wire them in place.
58. Close all of the stopcocks on the absorption train lines.
59. Remove the sample from the drying oven if at least 1 hour has passed. If not, then wait awhile. When this sample has cooled, weigh it and record the weight.
60. Clean up the entire area and put everything back in its place. *Leave the area better than it was when you started.* Thank you.
61. Calculate the % CO_2 and % $CaCO_3$ in the sample two ways; on an "as received" basis and on an "oven-dried" basis. An example of how this is done is shown in Chapter 1.
62. If you are satisfied (or better still, if your instructor is satisfied) with your results, then you can throw away the rest of your sample and clean up the weighing bottle.
63. OPTIONAL: Repeat the experiment using a sample of known value such as those prepared by the Thorn Smith Co. or do a duplicate.

EXPERIMENT 2. VOLATILIZATION
The determination of mercury in hair by flameless atomic absorption spectroscopy
(*Association of Official Analytical Chemists*, International, Method 971.21, 16th ed., modified for hair)

BACKGROUND
Mercury forms strong covalent bonds with sulfur. If the sulfur happens to be in an amino acid (cystine, cysteine, methionine), in a protein, or an enzyme then poisoning can occur. One result of this interaction is that Hg accumulates in toe and finger nails as well as in hair. Almost everyone has some Hg in their nails and hair, but an excessive amount can indicate chronic heavy metal poisoning. A survey of Hg in the hair of 17 general population people indicated 6.0 ± 2.9 ppm. Dentists had 9.8 ± 4.7 ppm.

Hg in hair is determined by dissolving the hair in concentrated HNO_3 to form the Hg^{+2} ion. Stannous chloride is then added to reduce the Hg^{+2} to Hg. Air is passed through the solution and the Hg vapors transfer to the air bubbles by the 2nd Law of thermodynamics (high concentration goes to low concentration). The Hg vapor plus water vapor is passed through a drying tower to remove the water and the Hg vapor passed through the beam of an atomic absorption spectrophotometer and measured.

$$Hg^{+2} + Sn^{+2} \longrightarrow Hg + Sn^{+4}$$

YOUR PROBLEM
A person has the symptoms of chronic mercury poisoning, but there is no logical explanation of where the contamination is occurring. The problem is to determine if the person is in fact being slowly exposed to mercury vapors or a mercury compound. This will be done by testing for Hg in head hair. You are given a sample of hair to determine the Hg in it. However, you have never done this before, and you need a blank, in any case. You obtain some hair from a known healthy person and determine the Hg present. You run two samples; one as a blank and the other spiked with 10 ppb Hg to make sure you can detect it. You also run a set of five standards. This allows you to prepare a calibration curve and to determine the % recovery.

EQUIPMENT

1	Adapter, gas inlet, like Kontes K181000 24/40 joint	1	Funnel, Buchner, 5 cm
1	Atomic absorption spectrophotometer	1	Gas diffusion tube
1	Balance, analytical	1 pr	Gloves, plastic
1	Beaker, 400 mL	1 pr	Goggles, safety
2	Bottles, wash; for 1:9 HNO_3 and water	1	Hollow cathode, Hg
1	Bulb, pipet, 6 mL	1	Hot plate
1	Cell, 25 x 115 mm with quartz windows	1	Paper, filter, 5 cm
1	Clamp, pinch	1	Pipet, delivery, 1 mL
1	Cylinder, graduated, 100 mL	1	Pipet, Mohr, 1.0 mL
1	Cylinder, graduated, 10 mL	1	Pump, diaphragm, like Neptune Dyna-pump
1	Flask, digestion, 250 mL flat bottom, 24/40 joint	1	Recorder, 1 mv full scale.
3	Flasks, Erlenmeyer, 125 mL	1	Rod, glass, stirring, 6 x 150 mm
1	Flask, filtering, 125 mL	1	Stopper, rubber, 2 hole, no.. 5
5	Flasks, volumetric, 100 mL	1	Towel or sponge
2	Flasks, volumetric, 250 mL	1	Transformer, variable
1	Flask, volumetric, 1 L	4 ft	Tubing, Teflon, 16 gauge
		3 ft	Tubing, vacuum

CHEMICALS   S = student
Anhydrone, anhydrous magnesium perchlorate, $Mg(ClO_4)_2$, 10 to 15 g/S
Detergent, liquid, 1 mL/S
Diluting solution
    To a 1 L volumetric flask containing 300-500 mL of H_2O, add 58 mL HNO_3 and 67 mL H_2SO_4. Dilute to volume with H_2O.   800 mL/S
Hydroxylamine sulfate, $NH_2OH \cdot SO_4$, for the reducing solution, 7.5 g/S
Mercury standards

Stock solution - 1000 μg/mL (1000 ppm). Dissolve 0.1354 g $HgCl_2$ in 100.0 mL of H_2O
Working solution - 1 μg/mL. (1 ppm). Dilute 1 mL of the stock solution to 1,000 mL with 1 N H_2SO_4 (prepare fresh daily)
Daily standards: Requires five 100 mL volumetric flasks and one 200 mL volumetric flask

0.0	0.00 mL to 100 mL with the diluting solution	
2.5	0.25 mL of 1 ppm to 100 mL	= 2.5 ng/mL = 2.5 ppb
5.0	0.50 mL of 1 ppm to 100 mL	= 5.0
7.5	0.75 mL of 1 ppm to 100 mL	= 7.5
10	2.5 mL of 1 ppm to 250 mL	= 10
15	1.5 mL of 1 ppm to 100 mL	= 15

Nitric acid, HNO_3, conc, 90 mL/S, (1:9) 200 mL/S
Reducing solution;
    Mix 25 mL H_2SO_4 with ca 150 mL H_2O. Cool to room temperature. Add 7.5 g of NaCl, 7.5 g
    of hydroxylamine sulfate, $NH_2OH \cdot SO_4$ and 12.5 g of stannous chloride, $SnCl_2$. Dilute to 250 mL
Sample, hair. Can be yours or from a local barber or beauty shop. 5 g/S
Sodium chloride, NaCl, for the reducing solution, 7.5 g/S
Stannous chloride, $SnCl_2$, for the reducing solution, 12.5 g/S
Sulfuric acid, H_2SO_4, for the reducing solution, 25 mL/S

THE BIG PICTURE

A sample of hair is washed with soap and water to remove surface dirt, vacuum filtered, and air dried. Three samples of 1.0 g of hair are each placed in a flask. To one is added 5 mL of 1 μg/mL Hg. All are acid digested. While the digestion is taking place, a calibration curve is prepared by adding standards to a series of digestion flasks and evolving the Hg by adding the reducing solution. The reducing solution is added to the samples in turn, and the flask is attached immediately to the flameless apparatus. The amount of Hg in the hair is then determined two ways: by standard addition and by a calibration curve.

APPARATUS SETUP

The apparatus is set up as shown in Figure 1-3, p. 9, with the cell placed in the beam path of the atomic absorption spectrophotometer.

DIRECTIONS

NOTE: While the hair is digesting, become familiar with the atomic absorption instrument and obtain the values needed to prepare the calibration curve. *It is very important that the dead space volume in the apparatus is the same for both the standards and the samples. Be sure to make any final additions or dilutions to assure this.*

1. Put on a pair of disposable plastic gloves (hair may be dirty). Place 4 to 5 g of hair in a 400 mL beaker, add about 100 mL of water, and stir the hair until it is wet.

2. Add a few drops of liquid detergent and stir the mixture well for a few minutes. This is to remove dirt and oils from the hair.

3. Rinse the hair until you believe the detergent has been removed, then transfer the wet hair to a 5 cm Buchner funnel and vacuum filter it to remove the water. Let it run for 8 to 10 minutes to air dry the hair.

4. Weigh three 1.0 g samples of hair, putting each one in a 125 mL Erlenmeyer flask.

5. Pipet 1.0 mL of the 10 ng/mL (10 ppb) standard into one of the flasks and mark it SP for spike.

6. Add 10 mL of HNO_3 to each flask.

7. Set these on a hot plate in a hood. Turn the hot plate on to a medium temperature and let it react for 20 minutes. This will dissolve the hair, it will not digest it. Do not let it go to dryness. Go to step 8 now.

8. Rinse all of the flameless apparatus glassware with 1:9 HNO_3, then rinse with distilled-deionized water and set on a paper towel to drain. This includes the volumetric flasks for the standards, the evolution flask, and the vapor trap.

9. Remove the flasks from the hot plate and let them cool for 15 minutes. Go to step 10 now.

10. Place about 10 g of Anhydrone ($Mg(ClO_4)_2$) in the trap on the outlet between the sample flask and the AA to remove the water vapor.

11. Transfer each sample to a 50 mL volumetric flask and dilute to volume with distilled-deionized water.

    Do each of the five standards, the blank and the three samples one at a time from here on. The following order is suggested. The blank: the standards in order, lowest first; the two samples; and finally the spiked sample.

12. Check the atomic absorption instrument to be sure it is ready. Adjust the wavelength to 253.7 nm, the slit width to 160 μm, the lamp current to 3 ma, and the sensitivity scale to 2.5.

13. Adjust the flow rate through the pump to about 2 L/min by using a variable transformer. The instructor may give you a transformer setting to provide this rate. You can measure it by using a soap bubble flow meter and adjusting it to about 35 mL/min. (See Figure 20-34, p. 229).

14. Turn on the recorder and adjust it to zero.

15. Be sure the pinch clamp on the vapor trap is closed.

16. Add 50 mL of distilled-deionized water to the evolution flask. This will be the "sample" when you do the samples later on, but it is necessary to occupy space now.

17. Add 100 mL of diluting solution to the evolution flask.

18. Add 25 mL of the reducing solution to the evolution flask and close it immediately.

19. Turn on the pump and aerate the sample for about 30 seconds or until you are sure you have a blank signal. Refer to Figure 1-5, p. 11, for what to expect. Mark the strip chart - Blank.

20. Open the pinch clamp and flush the system clean. The recorder pen trace will go to zero.

21. Remove the evolution flask, empty it into a waste bottle, and rinse it with distilled-deionized water.

22. Repeat steps 14 to 21 for each standard EXCEPT:
    (A) Add 1.0 mL of a standard to obtain data for a calibration curve.
    (B) When doing the samples, add 50 mL of the sample rather than the distilled-deionized water.
    (C) DO NOT ADD any standards to the samples. One has already been spiked.

23. Plot absorbance vs. μg Hg and determine the best straight line between the points. You can do this with a calculator or be a real student and do it by hand.

24. If the μg Hg exceeds the curve, take a smaller aliquot and repeat the analysis.

25. Clean up all of the apparatus and the surrounding area. *Make it better than it was when you started.*

    Thank you.

EXAMPLE CALCULATION

A 1.9082 g sample of hair treated as above was found to contain 0.78 μg of Hg. A 1.9447 g sample spiked with 0.5 μg Hg was found to have a value of 1.19 μg Hg. What is the ppm of Hg in the sample of hair, what is the percent recovery, and what is the ppm corrected for recovery?

ANSWER

A.

$$ppm = \frac{0.78 \; \mu g}{1.9082 \; g} = 0.41 \; ppm$$

B. Determine the μg of Hg in the spiked sample from the hair (incurred residue).

$$\frac{1.9082 \; g}{1.9447 \; g} = \frac{0.78 \; \mu g}{X \; \mu g}; \quad X \; \mu g = 0.79 \; \mu g \; in \; the \; hair.$$

1.19 μg - 0.79 μg = 0.48 μg recovered from the spike.

$$\frac{0.48\ \mu g \times 100}{0.50\ \mu g} = 96\%\ \text{recovery}.$$

C.

$$\frac{0.41\ ppm}{X\ ppm} = \frac{0.96}{1.00} \qquad X = 0.43\ \text{ppm corrected for recovery}.$$

EXPERIMENT 3. ZONE REFINING
The separation of traces of methyl red from naphthalene

The only commercial small-scale zone refiners now being made are shown in Figures 2-10, p. 18, and 2-14, p. 19. However, there are some simple homemade zone refiners that work well. For very small-scale size purifications, the one shown in Figure 2-12, p. 18, is the simplest. The experiment that follows is based upon this apparatus.

BACKGROUND
Naphthalene is obtained from coal tar, being about 11 % of the dry tar. It is used as a starting material for making dyes and solvents. It has antiseptic qualities and was used for years as a moth repellent.

YOUR PROBLEM
A sample of naphthalene is suspected of being contaminated with anthracene, which may cause an unwanted side reaction in a synthesis involving naphthalene. Your task is to zone refine a few grams of naphthalene to either remove the anthracene or to ensure that the naphthalene is pure. The apparatus you have is a homemade one and you do not believe it will work. You decide to test it by adding a few mg of a dye to some naphthalene, so you can see if it actually works.

EQUIPMENT

1	Balance, top loading	1	Oven, drying, 110 °C
2	Beakers, 250 mL	2	Stands, ring
1	Board, 140 mm dia, 1/2" polystyrene as shown in Figure 2-12.	2	Stoppers, rubber, 1 hole, no. 4
		1	Stopper, rubber, 1 hole, no. 5
3	Clamps, buret	1	Transformer, variable
1	Container, plastic, 5 gallon, with stopcock at the bottom.	1 ft	Tubing, glass, 6 mm for the siphon.
		1 ft	Tubing, 10 mm closed bottom.
4	Clips, alligator, 2 to connect heaters and 2 to connect the transformer to the heaters.	3 ft	Tubing, rubber, to drain the overflow
		3 ft	Tubing, rubber, to fill the desiccator
		1 ft	Wire, to connect heaters
1	Cylinder, graduated, 100 mL	3 ft	Wire, double strand with male plug at one end and alligator clips at the other end.
1	Desiccator bottom, 160 mm with a 3/4" diameter hole near the top.*		
1	Funnel, 35 mm	5 ft	Wire, 25 gauge bare nichrome. 40 turns /loop
3	Holders, clamp		

*To drill the hole: The drill is a 3/4" diameter tube of brass or copper with eight 1/8" deep notches in one end. The other end is soldered to a 1/4" rod to fit into a drill press. (1) Build a dam with modeling clay. (2) Slurry 150 mesh SiC grit in water to the consistency of thin honey. (3) Place epoxy on the back side of the desiccator to reduce chipping. (4) Use modeling clay to help hold the desiccator in place. (5) Put the slurry in the dam and with a gentle pecking motion drill the hole. It is more of an abrasion than a true cut - James Hodgson - Senior Glass blower, Kansas State University, Manhattan, KS 66506. He will drill one for you for a small fee.

CHEMICALS   S = Student
- Glue, epoxy, 1 mL/apparatus
- Methyl red,   5 mg/S
- Naphthalene, $C_{10}H_8$, 30 g/S
- Plaster of Paris, calcium sulfate, $CaSO_4$, 25 g/apparatus
- Silicon carbide, SiC, 150 mesh grit, 5 g/apparatus

APPARATUS SETUP

The apparatus is shown in Figure 2-12, p. 18. It is set near a sink, so the siphoned water can drain away. The large reservoir container can rest on any shelf above the desiccator. Plaster of Paris was used to keep the asbestos shields from twisting, and two layers were needed to keep the melted zone small.

THE BIG PICTURE

5 mg of methyl red is added to about 30 g of naphthalene and placed in a 100 mL beaker. About 100 mL of water is added to a second beaker. This, a small funnel, and the glass sample tube are placed in an oven to melt the naphthalene and to heat the glassware. The molten contaminated naphthalene is poured into the warmed glass tube and set vertical to remove air bubbles. The cooled tube is placed in the apparatus, the heater wires connected, power applied, and the water adjusted to a steady drip, about as fast as you can count. The apparatus should be set in a large plastic container to catch any overflow water in case it does not siphon when it should, particularly if it is run overnight.

DIRECTIONS

Assemble the apparatus as shown in Figure 2-12 near a sink or other drain. It is recommended that the apparatus be set up and the temperature of the coils near what is desired before the sample is prepared. If the naphthalene is warmed too long in the oven, the fumes become obnoxious. Turn on the hood where ever it is to reduce the fumes as quickly as possible.

1. Turn on a drying oven, if it is not already on.
2. Adjust the lower heater to 2.5 cm above the rim and the top heater 11.5 cm above the rim.
3. Fill the reservoir container with tap water and place the tubing from its drain so it empties into the desiccator as far away from the heaters as possible.
4. Place the end of the siphon tube into the sink or drain and put its end as low as possible to ensure siphoning.
5. Place a thermometer inside of a rubber stopper and then inside of the top coil. Hold it in place with a buret clamp.
6. Connect the heating coils to the variable transformer and adjust it to about 15 volts (not the scale reading) for 5 to 6 minutes, then back it off to about 12 volts. It is easy to blow a fuse, so have a spare handy.
7. Adjust the heat to just above 82 °C, the m.p. of naphthalene.
8. Add a few mg of either methyl red or methyl green to about 30 g of naphthalene in a 250 mL beaker and place this in an oven *already heated* to 110 °C to melt the naphthalene.
9. Place a second beaker containing about 100 mL of water, a 100 mL graduated cylinder, a 10 mm x 30 cm length of glass tubing closed at one end, and a 35 mm diameter funnel in the oven. It takes about 30 minutes to get all of the items up to temperature.
10. Place a buret clamp on a ring stand, so it can hold a graduated cylinder.
11. Use tongs or a cloth towel and remove the graduated cylinder from the oven. Hold it in place on a ring stand with a buret clamp. Fill it to about 80 mL with the hot water.
12. Place the hot glass tube in the hot water and place the small funnel in the top.
13. Carefully pour the melted naphthalene (now red or blue) into the glass tube. Do this before it cools, or you get to start over.
14. Place a beaker on the lower shelf of the oven and place the glass tube containing the sample through the wire on the shelf and set its bottom in the beaker. Let it stand for about 10 minutes to remelt the naphthalene and to remove as many air bubble as possible. Let it stay in the oven until the temperature of the coils in the zone refiner is a little over 82 °C.

15. Take the sample tube to the zone refiner. Remove the thermometer. Place the sample tube through the loops so its bottom rests on the rubber stopper. Press it in tightly or the float will tip when it rises. Adjust the loops so the tube is vertical and can move freely.

16. Adjust the loops to be about 8 to 9 cm apart, so that when the top part of the sample reaches the bottom loop, it will be above the bottom loop at the beginning of the next pass.

17. Fill the desiccator with water to determine if the tube will rise smoothly, when it will siphon, and that it will siphon into the drain. Just see that it works.

18. Start the flow of water into the desiccator at the rate of about 20 drops every 15 seconds. You should see two melted bands about 1 cm wide, and the rest of the sample should be solid.

19. This separation is easy. You can watch it as it happens. Run for two cycles and the dye should be removed.

20. Shut off the power to the coils and stop the flow of water into the desiccator.

21. Careful: remove the sample tube and inspect it. You may keep this as a token of our friendship.

22. *Clean up the area.* Thank you.

EXPERIMENT 4. AZEOTROPIC DISTILLATION
The production of absolute ethanol from 95% ethanol-water

BACKGROUND

This is the commercial method for this preparation. Ethanol in various concentrations is produced either by fermenting sugar, starch, or cellulose or by the hydration of ethylene. The products then are concentrated to the 95% azeotrope by regular distillation. In 1982, 816×10^6 L of 190 proof ethanol were produced, 10% by fermentation, and the rest by synthesis. The synthetic method is less expensive ($0.50/L) in that 1 kg of ethanol requires 2 kg of sugar, 3.3 kg of corn, or 4.0 kg of molasses, whereas 1 kg of ethanol can be prepared from 0.6 kg of ethylene. 55% is used for solvents, and the rest for drugs, plastics, lacquers, polishes, plasticizers, perfumes, and cosmetics and now extensively in gasoline.

Absolute alcohol is required for the production of gasohol. Figure E4-1 shows a three-phase diagram of gasoline-ethanol-water. Notice how just a small amount of water can make the gasoline divide into two phases.

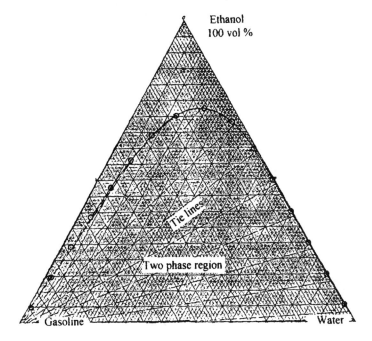

Figure E4-1. Phase diagram for ethanol-gasoline-water in volume percent at 76 °F (Courtesy - Unknown)

Noah is believed to have built for himself a vineyard in which he grew grapes that he fermented into a sort of alcoholic beverage.

HISTORICAL NOTE

In England, the *proof* was to pour some of the product over gunpowder. If the mixture exploded, which happened if the product was 11 parts ethanol to 10 parts water (about 50%), this was proof of the quality. In the United States, the proof is twice the alcohol content by volume; 95% ethanol = 190 proof.

The current synthetic process is as follows using a gas phase reaction over a phosphoric acid-impregnated Celite catalyst.

YOUR PROBLEM

An organic synthesis requires 50 mL of anhydrous ethanol as a solvent. Only 95 % ethanol is available. Removing the water by storing over molecular sieves or using a drying agent such as anhydrous sodium sulfate is not considered economical because too much water is present and time is short. An azeotropic distillation is decided upon. You get 150 mL of 95% ethanol and prepare the desired product.

$$CH_2=CH_2 + H_3O^+ \xrightarrow{Fast} H_2O + [\text{Pi complex}] \xrightarrow{\text{Rate determining step}}$$

$$CH_3CH_2OH + H_3O^+ \xleftarrow{+H_2O} CH_3CH_2OH_2^+ + \xleftarrow[Fast]{+H_2O} CH_3CH_2^-$$

EQUIPMENT

1	Adapter, 24/40 joints	1	Holder, heating mantle
1	Beaker, 5 mL	1 bx	Kimwipes, lens tissue or cotton
3	Clamps, buret	1	Pliers, wire cutters
1	Column, distilling, Vigreux 20 mm x 40 cm with 24/40 joints	1	Refractometer, Abbe'
1	Condenser, West or Liebig 30 cm, 24/40 joints	2	Rings, cork, 60 x 110 mm r.b. flask holders
1	Cylinder, graduated 100 mL	1	Ring, iron, 10 cm dia.
3	Droppers, medicine	3	Stands, ring
2	Flasks, r.b., 500 mL single neck, 24/40 joint	1	Thermometer, -10 to 150 °C 10/30 joint, 7.5 cm immersion
1 pr	Goggles, safety	1	Towel or sponge
1	Head, distilling, T type, 24/40 joints; 10/30 thermometer joint	1	Transformer, variable
		8 ft	Tubing, condenser
1	Heating mantle, 500 mL	1 ft	Wire, copper

CHEMICALS   S = student

Benzene, C_6H_6, 75 mL/S
Boiling chips, 4-5/S
Ethanol, H_3CCH_2OH, 95 %, 150 mL/S

Glycerine, $HO-CH_2CH(OH)CH_2-OH$, (in a dropping bottle), 1 mL/S
Joint grease, tube

GENERAL APPARATUS DESCRIPTION

A diagram of the apparatus is shown in Figure E4-2, p. 512. Although two ring stands are sufficient, three are recommended to reduce equipment breakage. The condenser tubing should be wired in place. Tape a spare fuse to the outside of the variable transformer and be sure that you do not add boiling chips to a hot solution.

THE BIG PICTURE

One hundred fifty mL of 95 % ethanol (b.p. = 78 °C) and 75 mL of benzene (b.p. = 80.1 °C) are added to a 500 mL r.b. flask. This forms a new azeotropic mixture that boils at 65 °C (74% benzene, 18.5% ethanol, 7.5% water). This removes the water from the alcohol, but a benzene-ethanol mixture remains. A new azeotrope (67% benzene, 33% ethanol), which boils at 68 °C, is now distilled off leaving anhydrous ethanol (b.p. = 78.5 °C).

Good technique is to add just enough benzene to remove the water, but not so much that excessive ethanol is lost when the excess benzene is removed.

CAUTION: You are advised not to sample the fruits of your labor. Traces of benzene provide an excellent hangover.

DIRECTIONS

NOTE: A refractometer is used for the quantitation step rather than either a gas-liquid chromatograph (GLC) or a high performance liquid chromatograph (HPLC) to provide a variety of exposure to various instruments.

1. Check the apparatus by comparing it to Figure E4-2, p. 512, and be sure all of the components are present.

2. Set a 500 mL r.b. flask (C) in a cork ring and add 150 mL of 95% ethanol.

3. Add 75 mL of benzene to the flask.

4. Add 4 to 5 porcelain boiling chips to the flask. (NOTE: Do not add boiling chips to a hot solution later on). Boiling chips produce a stream of air bubbles when they are heated, which reduces the possibility of the solution "bumping".

5. Place a small amount of grease on the lower joint of the Vigreux column (D) and then attach the r.b. flask to it.

6. Place the heating mantle (B) under the flask and adjust its holder so that the mantle is about 5 cm above the base of the ring stand.

7. Hold the Vigreux column in place with a buret clamp.

8. Place a small amount of grease on the two inner connections of the T distilling head (E) and connect the head to the top of the column. Twist the head slowly and with gentle pressure, until the grease spreads and makes a complete seal.

9. Place a small amount of grease on the standard taper section of the thermometer (F) and place it in the distilling head. Twist the thermometer slowly and with gentle pressure, until the grease spreads and makes a complete seal. The bulb of the thermometer should be opposite the T opening or just slightly below.

10. Attach the condenser (G) to the distilling head and hold it in place with a buret clamp.

11. Check the condenser tubing, particularly the ends that are attached to the condenser. If the tubing is old or cracks are beginning to appear, then either replace the tubing or cut off the bad ends.

12. Place a drop of glycerine on each of the tubing connections on the condenser and attach the condenser tubing.

13. Secure the tubing in place by wrapping a few turns of copper wire around the connection and then twisting it tight with a pair of pliers. (Tubing connections should always be either wired or clipped on. Tubing connections always seem to come apart when you turn your back or leave the room).

14. Connect the condenser tubing to the water inlet, being sure that the inlet water goes in at the bottom of the condenser.

15. Place the end of the outlet tubing into the sink. Be sure it goes well down into the sink. If not, tie a small weight to the end so it won't flop around when the water is turned on.

16. Turn on the water slowly and test the system. If you are satisfied that the connections will hold, then turn off the water.

17. Place grease on the lower connection of the condenser and attach the adapter (H) by applying gentle pressure and a twisting motion until a tight seal is obtained.

18. Depending upon the style of the adapter, it is held in place by one of two methods. If it is a smooth surface adapter, then use a buret clamp. If it has a vent or vacuum fitting, then you can hold it to the condenser with a piece of copper wire.

19. If the adapter has a standard taper outlet, apply a small amount of grease to it and then add a 500 mL r.b. flask (K).

20. Place a 10 cm diameter iron ring (M) under the flask to hold it in place.

21. Check the system and be sure that it has an opening to the outside. It should be either at point (I) or at point(J). Do not apply heat to a closed system.

22. Turn on the condenser water.

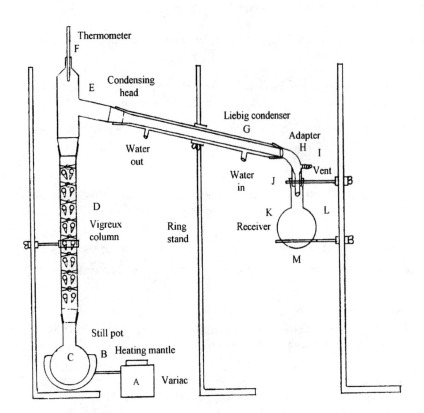

Figure E4-2. Apparatus for a laboratory-scale azeotropic distillation.

23. Turn on the variable transformer (Variac or Powerstat) and adjust it to about 1/3 of a full setting (80 on a 240 scale).
24. Carefully watch the reflux line proceed up the column and adjust the transformer as needed until the reflux line just covers the bulb of the thermometer and liquid begins to condense in the condenser.
25. Turn the transformer setting back about 1/4. This is usually sufficient to ensure a smooth distillation that is not too fast. About 1 drop a second is sufficient.
26. The temperature should be about 65 °C. It will be 65 °C at sea level and with a correct thermometer, so it likely will be a few degrees lower in your laboratory. Therefore, do not apply extra heat to try to make it reach 65 °C.
27. After about 5 minutes, the system should be distilling at the rate of 3 to 4 mL/min. This amounts to 17-18 drops every 15 seconds. Adjust the transformer so the distillation proceeds at this rate. If anything, distill at a slower rate. This distillation will take about an hour, so you have time to make minor adjustments.
28. Excess water is removed first, and then the benzene-ethanol-water azeotrope will distill. After the system has been distilling for 10 minutes or so and the temperature is in the 62 to 65 °C range, lower flask K and collect a few drops of distillate in a 5 mL beaker.
29. Replace flask K. Have your instructor show you how to use the Abbe refractometer and determine the refractive index of the azeotrope. A diagram of the light path of an Abbe refractometer is shown in Figure E4-3, p. 513, and a photograph of one of the more modern refractometers is shown in Figure E4-4, p. 513. Steps 31 to 48 give the details of using a refractometer if an instructor is not available at this time.

Directions for Using an Abbe Refractometer

The refractive index (n_D^{20}) of water at 20 °C relative to air (the sodium D line at 589 nm)
          1.33299 at 20 °C   1.33241 at 26 °C

31. Reach behind the refractometer and turn on the power switch.

32. With the forefinger of your left hand, push in on the button at the rear of the left side. Look into the eyepiece of the refractometer and you will see an illuminated scale.

33. With your right hand, reach to the right side rear of the refractometer to the double knob (large is coarse, small is fine) and adjust the instrument until the cross hairs on the scale read 1.33.

34. Remove your hand from the button on the left side and let the field become black again.

35. In the front of the instrument, there is a lever arm with a plastic knob on the left side of it. Push this down and out of the way for the moment. Locate the spring clip mechanism on the right front side that holds the two prisms together.

36. Place the forefinger of your left hand on top of the top prism and your left thumb under the spring clip. Use a counter-clockwise twist with your wrist and open the lenses.

37. With a medicine dropper, marked *water*, add 2 drops of distilled water to the top surface of the lower prism. Be careful not to touch the surface with the medicine dropper, because it can scratch the surface. Cover this liquid by moving the top prism back into place with a clockwise motion. Lock it in place.

38. Move the side arm back into place and turn the clear plastic knob until the black square on the top prism is illuminated.

39. Look into the eyepiece of the instrument and you will see a multicolored field, usually varying from black at the bottom to light or colored at the top. Adjust the arm until this is so.

40. What is needed is a sharp dividing line between the two fields, white on one side and black on the other. On the top front of the refractometer is the *compensator*. It is used to remove the colored portions between the white and black areas. Remove the cover from the compensator. It is held on magnetically. Place it over the name plate for temporary storage.

41 You should see a scale with numbers on it. Look into the eyepiece and then slowly turn the compensator until you remove as much of the color between the fields as possible to get the sharpest dividing line. Scratches on the prisms or loose prisms may make it impossible to remove all of the color. Do the best you can.

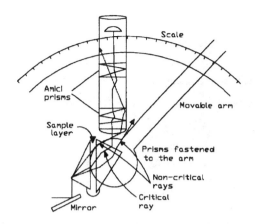

Figure E4-3. Optical diagram of an Abbe refractometer.
(Courtesy - Bausch & Lomb Inc. Analytical Systems Division, Rochester, NY)

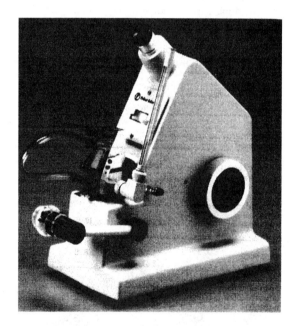

Figure E4-4. An Abbe refractometer.
(Courtesy - Bausch and Lomb Inc. Analytical Systems Division, Rochester, NY)

41. When the dividing line is as sharp as you can get it, replace the cover on the compensator.

42. Use the large knob on the right rear of the refractometer and adjust the dividing line to match the inter- section of the cross hairs in the field.

43. Use the small, fine adjust knob to make the last motions.

44. To determine the refractive index of the solution, use your left hand to push in on the button at the left rear of the instrument. Look into the eyepiece. The scale will appear, and you can determine the refractive index of the liquid. For water, it should be 1.3330 at 20 °C. If it is not, and it probably won't be because of differences in

temperature, determine what it is and make a corresponding correction for all subsequent readings on other liquids.

45. Open the lenses of the refractometer and use a piece of lens tissue to wipe the liquid off of the lenses.
46. Use a clean medicine dropper and add 1 to 2 drops of the benzene-cyclohexane mixture to the lower prism face. Obtain the refractive index of this solution and record the value. Correct it based on the water reading.
48. Open the prisms and clean the surfaces.

Continue the Distillation

49. Continue the distillation, and when the temperature rises to 68 °C (probably 65 to 67 if your laboratory is much above sea level), again collect a few drops of distillate.
50. Determine the refractive index of this benzene-ethanol azeotrope.
51. Continue the distillation until the temperature rises to 78.5 °C (probably 75 to 77 °C in your lab. It depends upon the atmospheric pressure at the time). The anhydrous product (ethanol) should now be distilling. Collect a few drops of this and measure its refractive index.
52. Collect another sample of the distillate and see if the refractive indices check each other. The n_D^{20} for absolute ethanol is 1.3611.
53. Measure the refractive index of the starting material, the original 95% ethanol solution.
54. If the refractive indices indicate that you are now distilling absolute ethanol, then shut off the transformer. If not, then distill for 5 to 10 minutes more and collect another sample. Repeat as necessary until you are satisfied with the result.
55. Turn off the transformer.
56. Lower the heating mantle from around flask C. When the distillation stops, turn off the condenser water.
57. Dump all of the organic material into the waste solvent can.
58. Put everything back in order. *Clean up the area until it is better than it was when you began the experiment.* Thank you.
59. Calculate the expected refractive index of the benzene-ethanol-water azeotrope and compare it with what you obtained. See the example calculation for how this is to be done.
60. Calculate the expected refractive index of the benzene-ethanol azeotrope and compare it with what you obtained. See the example calculation.
61. Calculate as nearly as possible the purity of your final product. See the example calculations.

EXAMPLE CALCULATION

Problem 1. The refractive index of a sample of 1,1,2-trichloroethane was found to be 1.4710 at 24 °C. The literature value for a pure sample is 1.4706 at 20 °C. How could you determine if your sample is pure?

ANSWER

Step 1. The prisms of the refractometer are at 24 °C. Place 1 or 2 drops of distilled water in the refractometer and wait a few minutes until the water comes to 24 °C. In this case, the refractive index was determined to be 1.3326. The refractive index of pure water at 20 °C is 1.3330.

Step 2. The temperature correction is
1.3326 - 1.3330 = -0.0004

Step 3. 1.4710 - 0.0004 = 1.4706. Therefore, the sample is pure.

NOTE: A refractometer that is used often will be connected to a constant temperature bath, and a correction such as this will not be necessary.

Problem 2. If we assume the solutions behave ideally, then the refractive index of a mixture is proportional to the mole fraction of the compounds involved. What is the calculated refractive index for a 3.1% water - 85% carbon tetrachloride - 11.9% t-butyl alcohol azeotrope that boils at 64.7 °C?

ANSWER
Physical data:
H₂O   m.w. = 18.02 n_D^{20} = 1.3330   b.p. = 100 °C
CCl₄   m.w. = 153.84 n_D^{20} = 1.4607   b.p. = 76.0
C₄H₁₀O m.w. = 74.12 n_D^{20} = 1.3847   b.p. = 82.5

Step 1. Assume a sample size. Any size will suffice, but to make it easy, assume a 100 g sample. Then:

$$H_2O \quad \frac{3.1 \text{ g}}{18.02 \text{ g/mole}} = 0.0172 \text{ moles}$$

$$CCl_4 \quad \frac{85.0 \text{ g}}{153.84 \text{ g/mole}} = 0.5525 \text{ moles}$$

$$C_4H_{10}O \quad \frac{11.9 \text{ g}}{74.12 \text{ g/mole}} = 0.1606 \text{ moles}$$

Total moles $\overline{0.8851}$

Step 2. Multiply the mole fraction by the refractive index.

$$H_2O \quad \frac{0.1720 \text{ moles} \times 1.3330}{0.8851 \text{ moles}} = 0.2591 \text{ contributed by water}$$

$$CCl_4 \quad \frac{0.5525 \text{ moles} \times 1.4607}{0.8851 \text{ moles}} = 0.9118 \text{ contributed by carbon tetrachloride}$$

$$C_4H_{10}O \quad \frac{0.1606 \text{ moles} \times 1.3847}{0.8851 \text{ moles}} = 0.2513 \text{ contributed by t-butanol}$$

The calculated refractive index of the azeotrope is   1.4222

Problem 3. During the purification of ethanol, a refractive index of 1.3618 was obtained. What is the purity of this compound, if we assume ideal solution behavior?
Physical data:
Ethanol   m.w. = 46 n_D^{20} = 1.3610
Benzene   m.w. = 78 n_D^{20} = 1.5011

Step 1. Obtain the total refractive index difference.
1.5011 - 1.3610 = 0.1401

Step 2. Obtain the impurity refractive index difference.
1.3618 - 1.3610 = 0.0008

```
1.5011 |<----------------- Total-------------------->| 1.3610
       -----------------------------------------------
                                     1.3618|<----->| 1.3610
                                              Impurity
```

Step 3. Determine the mole fraction of each component.

$$\frac{0.0008}{0.1401} = 0.0057 \text{ for benzene}$$

$$\frac{0.1393}{0.1401} = 0.9943 \text{ for ethanol}$$

Step 4. Determine the weights of each material present.
   Benzene   0.0057 moles $\times$ 78 g/mole = 0.44 g
   Ethanol   0.9943 moles $\times$ 46 g/mole = 45.74 g

   Total   46.18 g

Step 5. Determine the weight % of the ethanol.

$$\frac{45.74 \text{ g} \times 100}{46.18} = 99.05\% \text{ ethanol}$$

EXPERIMENT 5.   EXTRACTIVE DISTILLATION
Separation of cyclohexane from benzene using aniline

BACKGROUND
Over 1.2 billion lbs of cyclohexane are produced annually, mostly from the catalytic hydrogenation of benzene. 60% of this cyclohexane is used to make adipic acid and 30% to make caprolactam, both of which are used to make nylon apparel and carpets. Cyclohexane also is used as a solvent and in making derivatives of cyclohexanol and cyclohexanone, which are used in making dyes, pesticides, and other specialties.

Benzene + 3 H_2 $\xrightarrow{\text{Ni}, 200\ ^\circ C}$ Cyclohexane $\xrightarrow{\text{Air, Co salts}, 95\ ^\circ C,\ 150\ \text{p.s.i.}}$ Adipic acid

Cyclohexane made from benzene costs about $1.20/gal. It boils at 80.8 °C. Benzene boils at 80.1 °C, and the two form an azeotrope at 53% benzene-47% cyclohexane by weight. An extractive distillation is useful in removing the benzene from the cyclohexane by the process described in Chapter 4, p. 47.

YOUR PROBLEM
The freezing point depression constant for cyclohexane is 20.3, which is very high when compared to the 1.86 for water. You need 20-30 mL of pure material in order to make several measurements, and all you have available is some technical-grade material that was prepared by catalytic hydrogenation. You are to do an extractive distillation to remove the benzene impurity.

EQUIPMENT
   1   Adapter, 24/40 joints, vacuum type
   1   Beaker, 5 or 10 mL
   1   Bottle, wash, for acetone
   3   Clamps, buret

1 Column, distilling, Vigreux 37.5 cm long
1 Condenser, Liebig 300 mm, 24/40 joints
8 ft Condenser tubing
1 Cylinder, graduated, 100 mL
3 Droppers, medicine
2 Flasks, 500 mL r.b. single neck 24/40
1 Funnel, separatory, 250 mL, 24/40 lower joint
1 pr Goggles, safety
1 Head, distilling, Claisen,
 3 24/40 and 1 10/30 joints
1 Mantle, heating, 500 mL
1 pr Pliers, wire cutters
1 Refractometer, Abbe
1 Ring, cork, 60 x 110 mm
1 Ring, iron, 10 cm
3 Stands, ring
1 Thermometer, 7.5 cm immersion, -10 to 150 °C
1 Transformer, variable, Variac
1 ft Wire, copper

CHEMICALS   S = student

Aniline, C_6H_5-NH_2, 250 mL/S
Benzene, C_6H_6, 20 mL/S
Boiling chips, 4-5/S
Cotton or lense tissue, 1 box/apparatus
Cyclohexane, C_6H_{12}, 80 mL/S
Joint grease, 1 tube/apparatus

THE BIG PICTURE

80 mL of cyclohexane (b.p. 80.8 °C) and 20 mL of benzene (b.p. 80.1 °C) are mixed to form a technical-grade product. The refractive index is obtained on this mixture. 100 mL of aniline then is added (b.p. 184 °C), and all reagents are added to the 500 mL flask attached to the Vigreux column. Boiling chips are added. As soon as the solution starts to boil, aniline is added to the sidewalls at the top of the column at a steady dripping speed. After 10 to 15 minutes, 2 to 3 mL of the distillate is collected in a small beaker, and its refractive index obtained. Aniline is added to the column to continue to break up any cyclohexane-benzene azeotrope that may form up the column. The refractive indicies of the compounds involved are:

Cyclohexane 1.4266;   Benzene   1.5011;
Aniline   1.5863.

NOTE: The system as arranged usually will not produce pure cyclohexane, because the aniline added at the top is added too high up the column and some is carried over with the cyclohexane. This can be improved by adding a small section of bent glass to the separatory funnel tip and adjusting it so the addition is below the top of the condenser and directs the flow to the sidewall of the column. A small amount of aniline impurity will make a rather large change in the refractive index. If you do the distillation slowly, then a fair separation will take place.

DIRECTIONS

1. Check the apparatus and be sure you have all of the components shown in Figure E5-1. Be sure the tip of the separatory funnel is bent sufficiently so that any liquid from it will go down the side of the condenser and not drip directly down the center.

2. Set up the apparatus as shown but do not connect the 500 mL r.b. flask.

3. Add (100 mL graduated cylinder) 20 mL of benzene to the 500 mL r.b. flask.

4. Add (100 mL graduated cylinder) 80 mL of cyclohexane to the r.b. flask and mix the two liquids thoroughly.

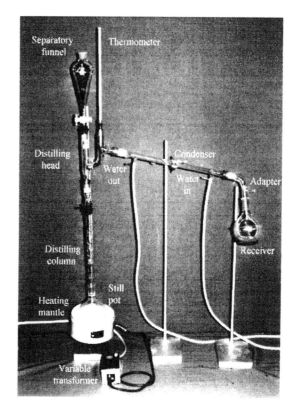

Figure E5-1. A simple apparatus for an extractive distillation.

5. Locate the Abbe refractometer and have the instructor show you how to use it. If you know how to use the refractometer, then go to step 25 after you measure this refractive index. If no instructor is available, then follow the detailed steps 7 to 24 to learn how to use the refractometer. Refer to Figures E4-3 and E4-4, p. 513, to refresh your memory.

6. Measure the refractive index of this mixture.

7. Use a clean medicine dropper and add 1 to 2 drops of the benzene-cyclohexane mixture to the lower prism face. Obtain the refractive index of this solution and record the value. Correct it based on the water reading.

7. Open the prisms and clean the surfaces.

8. Go to the distillation apparatus.

9. Add 100 mL of aniline to the cyclohexane-benzene mixture and mix it thoroughly.

10. Obtain the refractive index of this solution.

11. Add 4-5 boiling chips to the liquid in the r.b. flask to prevent bumping.

    NOTE: *Do not add boiling chips later on if the solution is hot*. That can be quite dangerous - hot liquid may gush from the flask and burn you. If you must add boiling chips later, let the liquid cool nearly to room temperature and then add them *carefully* one at a time.

12. Connect the 500 mL r.b. flask to the bottom of the fractionating column. Use a small amount of grease on the joint.

14. Turn on the condenser water and make sure the connections are secure.

15. Add about 150 mL of aniline to the 250 mL separatory funnel and attach it to the top of the distilling head. A small amount of grease may be used.

16. Turn on the variable transformer to about 75 to 80 volts (200 on a 280 scale).

17. After 10 to 15 minutes the liquid will begin to boil. Notice the reflux line and watch it rise up the column to just below the bottom of the thermometer. Turn the variable transformer back about 5 to -10 volts.

18. Adjust the flow of aniline into the top of the column to about 3 drops in 2 seconds. A fast drip is necessary. It is important that the aniline run down the side of the column and coat its surface. If the drops go down the middle of the column without touching the sides, then tilt the column slightly so the drops hit the column 4 to 5 cm below where the condenser outlet is.

19. The temperature usually will drop slightly at this time. Wait a few minutes and then readjust the variable transformer until the reflux line is just above the top of the thermometer bulb. Do this slowly as the liquids heat slowly.

20. Put the receiver in place and begin collecting the condensate.

21. After about 20 minutes, remove the receiver and collect a few drops coming from the end of the condenser (not from the receiver) in a small beaker.

22. Use a clean medicine dropper to transfer the liquid and obtain the refractive index of the distillate. Apply the temperature correction determined by the water measurement and record the value.

23. How clean is the product? This is a tough separation because of the possible formation of an azeotrope and because of the closeness of the boiling points of benzene and cyclohexane. A good fractionating column is required and extreme care is needed to ensure that no aniline gets into the condensate. The refractive index of aniline is high and just a small contamination will make the separation look worse than it is. A gas chromatographic separation would be better, but it is our intention to have you obtain experience with a refractometer. Commercially, 99+% purity is attained by this method routinely.

24. If you believe you can do better, have the time, and want to try, change the reflux rate and the rate at which the aniline is added.

25. Shut off the flow of aniline and pour the unused liquid back into the aniline container.

26. Turn off the variable transformer and lower the heating mantle a few centimeters.

27. Clean up the refractometer.

28. When the reflux line has receded into the r.b. flask, turn off the condenser water.

29. Pour the residue in the r.b. flask into the waste bottle and clean up the apparatus.

30. *Leave the area in better condition than you found it.* Thank you.

EXPERIMENT 6. STEAM DISTILLATION
The separation of oil of clove from the whole cloves

BACKGROUND
According to the Merck Index -- Clove oil, 82-85% eugenol, including about 10% acetyl eugenol, caryophyllene, small quantities of furfural, vanillin, and methyl ethyl ketone.

Eugenol   Acetyleugenol   Caryophyllene   Furfural   Vanillin

Oil of clove is a colorless to pale yellow liquid that becomes darker and thicker with a age. Its refractive index is 1.530, density 1.038 g/mL, and b.p. about 250 °C. It is used in confections, tooth powders, and as a local anesthetic for toothaches. Eugenol has an R.I. of 1.5410, a density of 1.05 g/mL, and a b.p. of 255 °C.

YOUR PROBLEM
Several of the major compounds present in oil of clove have been identified. It is desired to prepare a small amount of a sample of oil of clove for further separation by a gas chromatograph. You are given two small boxes of whole cloves and you are to steam distill them and obtain a water-free sample of the oil.

EQUIPMENT

5	Beakers, 50 mL	1 pr	Goggles, safety
1	Beaker, 600 mL for an ice bath	1 bx	Matches
1	Burner, Bunsen	1	Mortar and Pestle, large
5	Clamps, buret	1 pr	Pliers
1	Condenser, West, 30 cm 24/40 joints	1	Rack, funnel
		1	Refractometer, Abbe
1	Cylinder, graduated, 25 mL	2	Rings, iron, 10 cm dia.
1	Dropper, medicine	3	Stands, ring
1	File, triangular	4	Stoppers, rubber
1	Flask, r.b. 100 mL		no. 5 2 hole no. 4 2 hole
1	Flask, r.b. 500 mL		no. 2 2 hole no. 2 1 hole
1	Flask, r.b. 1000 mL	1	Towel, cloth
1	Funnel, 60° short stem	8 ft	Tubing, condenser
1	Funnel, separatory, 125 mL	4 ft	Tubing, glass, 8 mm o.d.
1	Gauze, wire, asbestos center	1	Vial, 1 dram
5	Glasses, watch, small	1 ft	Wire, copper

CHEMICALS   S = student

Boiling chips, 3-4/S
Glass wool
Glycerine in a dropper bottle
Hydrochloric acid, HCl, conc, 3/4 mL/S
Ice, 400 g/S

Potassium hydroxide, KOH, solid, 1 pellet/S
Sodium chloride, NaCl, solid, 0.5 g/S
Sodium sulfate, Na_2SO_4, 3-4 crystals/S
Whole cloves   2 bx/S

APPARATUS SETUP

The apparatus is set up as shown in Figure 5-1, p. 50, with the steam generator, trap, and sample flask on one ring stand. The glass tube used as a pressure regulator should be adjusted to within 2 cm of the bottom of the flask. A separatory funnel is used for the trap, which makes it easier to remove the condensed water. Glass wool is not used to break the emulsion, because a part of this experiment is to become familiar with different ways to break emulsions. Keep the connecting lines short to minimize the amount of condensate.

THE BIG PICTURE

The steam generator is filled half full with water and then heated to produce steam. 50 to 80 grams of whole cloves are ground in a mortar and pestle and then placed in the sample flask. The clove oil then is removed by steam distillation and collected in a receiving flask. The collected material usually will be a milky white or light yellow suspension and will eventually yield a *few small drops* of clove oil. Five different methods then will be tested to break this emulsion.

DIRECTIONS

1. Check over the apparatus as shown in Figure 5-1, p. 50, and make sure all of the components are present.
2. Check the rubber tubing on the condenser, and if the tubing has cracks or gives any appearance of being weak, replace it with new tubing. Place a drop of glycerine on the tubing connection and wire the tubing in place.
3. Fill flask B no more than ½ full with distilled water and add 3 to 4 boiling chips.
4. Connect the safety valve (C) and the safety trap (D) to the steam generator. In order for the system to function properly, the bottom of the glass tube that acts as the safety valve and pressure regulator must be below the water level in the steam generator. Usually a position 1 to 2 cm from the bottom of the flask is satisfactory.
5. Close the stopcock on the trap and begin heating the water in the steam generator. Do not connect the steam supply to the sample container (E) yet.
6. While the steam generator is heating, partially grind two boxes (80 g) of whole cloves. They are not to be ground to a powder, but just broken up a bit.
7. Place the partially ground cloves in the sample flask (E) and add 25 mL of distilled water to just dampen the cloves.
8. Get a beaker of ice for the ice bath (H).
9. Turn on the water to the condenser (F).
10. Open the stopcock on the trap and drain the condensate into a beaker and then discard it. Close this stopcock.
11. Wait until the steam generator is producing steam and then connect the entire system (CARE: steam can cause severe burns - wrap a towel around your hands when you connect the steam line to the sample flask.
12. Distill the cloves until you collect 70 to 80 mL of total distillate.
13. Turn off the Bunsen burner.
14. Drain the trap and disconnect the safety valve so the system is open.
15. Turn off the condenser water.
16. Gently swirl the receiver flask so any drops that have formed will settle to the bottom.
17. Place about 10 mL of the emulsified mixture (above the bottom few drops) into each of five, 50 mL beakers.
18. Add 0.5 g of NaCl to the first beaker and swirl it to dissolve the salt. Set the beaker aside and observe what happens while you continue with the other systems.
19. Add 3 to 4 drops of concentrated HCl to the second beaker and swirl it to mix the system. Set it aside and go to the next step.
20. Add one pellet of KOH to the third beaker and swirl the beaker until the pellet dissolves. Set it aside and go to the next step.
21. Filter the contents of the 4th beaker using a glass wool plug placed in the bottom of a small funnel.
22. Leave the 5th beaker alone for comparison purposes.

23. Cover the beakers with watch glasses and set them in a refrigerator, if one is available. If not, then place them in a drawer where they will be in the dark. Look at them 3 to 4 times during the day and record what happens.
24. Add 0.5 mL of concentrated HCl to the remainder of the distillate and let it set in a beaker of ice water until the oil of clove settles out.
25. When you estimate that about ½ of the oil of clove has settled out, pour off the excess water and transfer the oil of clove to a small vial.
26. With a medicine dropper, carefully remove the excess water from above the oil of clove in the vial.
27. Add 3 to 4 crystals of anhydrous Na_2SO_4 (gently) and let the crystals settle.
28. Measure the refractive index of the oil of clove. Check a *Merck Index* for the standard value.
29. *Clean up the entire area so that it is now in better condition than when you began the experiment.* Thank you.
30. Evaluate the results of the emulsion breaking tests.

EXPERIMENT 7.   IMMISCIBLE SOLVENTS DISTILLATION
The use of toluene to remove water from fruits, vegetables, and meats.

YOUR PROBLEM
The Kjeldahl determination for protein nitrogen requires a sulfuric acid digestion. This reaction is slowed considerably by the presence of water, so it is important that large amounts of water be removed before the digestion begins. Oven drying can be used in many cases, but some food samples are heat sensitive and oven drying often takes a long time. A toluene distillation is decided upon to remove the wate, and your job is to prepare one of the samples.

EQUIPMENT

1	Adapter, bushing, 34/45 to 24/40	1	Mantle, heating, 250 mL
1	Balance, analytical	1 pr	Pliers
1	Bottle, wash, 250 mL for toluene	1	Ring, cork, 7.5 cm dia.
1	Bottle, weighing	1	Ring, iron, 10 cm
2	Clamps, buret	1	Stand, ring
1	Condenser, West, 30 cm 24/40 at bottom only	1	Towel or sponge,
		1	Transformer, variable
1	Cylinder, graduated, 100 mL	1	Trap, Barrett, 20 mL, 24/40 joints
1	Flask, 250 mL with a 34/45 joint		
1	Gauze, wire, 10 *x* 10 cm	8 ft	Tubing, condenser
1	Glass, watch or Al weighing dish	1 ft	Wire, copper
1 pr	Goggles, safety		

CHEMICALS   S = student
Glycerine in a dropping bottle   Stopcock grease, 1 tube/apparatus
Toluene, C_6H_5-CH_3, 100 mL/S
A sample of your choice. A group of students might want to do a miniature research project. Each student would do either a different brand of the same food or different sizes of the same food such as large, medium, or small peas.

THE BIG PICTURE
You provide your own fruit, vegetables, or meat. 15 to 20 grams of a sample is weighed and placed it into the wide mouth flask. Toluene, in a wash bottle, is used to wash down the neck of the flask. 100 mL of toluene is added and the mixture is distilled. The water will settle to the bottom of the Barrett trap, and the toluene will recycle. The amount of water is measured, assuming 1 mL = 1 gram and the % water in the sample is calculated.

APPARATUS SETUP
Set up the apparatus as shown in Figure 5-4, p. 53. The flask must have a wide mouth or the larger pieces of fruit cannot be placed into the flask without squeezing out some of the water and causing an error. It is suggested that a spare wide-mouth flask be available because these seem to break easily.

Experiment 7

DIRECTIONS

Set this up in a hood, if possible.

1. You will need a sample of food that does not contain more than 20 g of water. Check Table 5-1, p. 54, for an estimate of the water content of your sample. Estimate the approximate size sample you will need to obtain 12 to 15 grams of water.
2. Get an aluminum weighing dish or a watch glass and weigh it to the nearest ± 0.001 g.
3. Place your sample on it and weigh it to the nearest ± 0.001 g.
4. Remove the bushing adapter from the flask and transfer your sample to the flask. Use a stream of toluene from a wash bottle to wash any water from the weighing dish into the flask.
5. Use a graduated cylinder and add 100 mL of toluene to the flask.
6. Use a wash bottle and spray a small amount of toluene on the outside of the bushing adapter and then place this adapter in the top of the flask.
7. Spray a small amount of toluene on the end of the Barrett trap and connect the flask to the trap.
8. Spray a small amount of toluene on the bottom end joint of the condenser and place the condenser in the trap.
9. Check the condenser tubing to be sure it is of good quality and connect the inlet water tube to the bottom connection of the condenser.
10. Place the outlet tube in the drain and, unless it goes well into the drain, tie a small weight to the end of the tubing.
11. Turn on the condenser water - slowly at first.
12. Be sure the stopcock on the Barrett trap works freely and then close it.
13. Be sure all toluene containers are closed and placed back well out of the way.
14. If the area appears safe, then turn on the heating mantle. Adjust it as necessary to produce a reflux line on the lower part of the condenser.
15. Distill the mixture. It is easy to see the water droplets form and settle to the bottom of the trap Eventually, the trap will fill with toluene and the excess will run back into the flask. If toluene droplets are trapped in the water, gently tap on the trap to free them so they float to the top.
16. Continue the distillation until you can no longer see water droplets coming over, then shut off the heating mantle.
17. Do not remove the sample flask at this time. Although toluene is not particularly toxic, it can cause a headache and dizziness if too much is inhaled. Wait until later when it cools to remove the flask.
18. Gently tap the trap to ensure that the toluene has all risen to the top, then let the system alone while you do the next step.
19. Obtain the weight of a weighing bottle and lid to the nearest ± 0.001 g.
20. Turn off the condenser water.
21. Carefully drain the water from the trap into the weighing bottle and then place the lid on the bottle.
22. Weigh this combination and determine the weight of the water obtained.
23. Remove the trap and pour the toluene into the waste solvent container.
24. Carefully pour the toluene out of the flask into the waste solvent container and recover your dry sample (careful: some break easily). Place the sample saturated with toluene on a watch glass or aluminum dish.
25. If you desire and your instructor is willing, you can place this sample in a drying oven at 60 to 80 °C for 10 to 15 minutes to evaporate the toluene. You now have a dry sample for further analysis. You can weigh it and determine if the weight lost equals the weight of the water recovered. You may have the sample as a token of our friendship.
26. Clean up the balance area and replace the cover on the balance.
27. *Clean up the equipment and the entire area so that it is better than when you started.* Thank you.
28. Calculate the % H_2O in the sample.

29. Assuming that you had 100% efficiency, how much toluene distilled in order to drive off the H_2O? Use equation 5-2, p. 50, and follow the example calculations if needed. Toluene b.p. 111 °C; water b.p. 100 °C; assume the mixture distills at 95 °C.

30. Plot the following data:

Vapor pressure (torr)	B. P. Toluene	B.P. Water	Vapor Pressure (torr)
760	111	100	760
400	89.5	90	526
100	51.9	80	355
40	31.8	50	93
10	6.4	20	18

Ask your instructor what a normal atmospheric pressure is for your area, measure it if you have a barometer in the laboratory, or get it from the local weather station, and then calculate the partial pressures of each component and the expected boiling point of the mixture.

EXPERIMENT 8. VACUUM DISTILLATION
A vacuum distillation to purify the solvent dimethylformamide

BACKGROUND
Dimethylformamide (DMF) has been called the universal organic solvent. It is a favorite in electrochemical investigations of organic compounds, because it is essentially nonconducting, andaprotic and has a rather high dielectric constant (36.7), leaving all of the electron transfer to the dissolved organic compound. Its Hildebrand solubility parameter is 12.1. World production is about 267,000 tons; the United States produces about 25,000 tons that is used mostly for reactions and crystallizing pharmaceuticals and as a solvent for acrylic polymers. It is used as a paint stripper and is probably the best solvent for carbonaceous deposits. It dissolves many salts, and it evaporates slowly.

It is seldom found naturally occurring, although it has been detected in sausages, cooked mushrooms, grapes, and wine. It is not carcinogenic in test animals. Its LD_{50} is 2800 mg/kg administered orally to rats and 4720 mg/kg absorbed through the skin of rabbits. However, the fumes should not be inhaled for prolonged periods. This is usually not a problem, because the dimethylamine impurity can be detected at > 1ppm.

It can react violently with strong reducing and oxidizing agents. Runaway reactions have been reported for HNO_3, MnO_4^{-1}, $Cr_2O_7^{-2}$, Br_2, Cl_2 gas, NaH, $NaBH_4$, and alkyl Al compounds, even though DMF is a good solvent for many of the above. The reasons for this are not yet understood.

YOUR PROBLEM
Your research director has a great idea on how to study the mechanism of an organic reaction that might be involved in a biological process. You are to do the experiment and find you will need about 50 to 60 mL of pure and dry DMF. You decide that a vacuum distillation is required.

Experiment 8

EQUIPMENT

This is a list of the component parts to build the vacuum rack shown in Figure 6-20, p. 68. The vacuum tubing is either wired on or attached with hose clamps; your choice. All joints are 24/40 standard taper with the exception of the thermometer, which is 10/30.

1	Adapter, 75°, two 24/40 inner with 10/30 outer	1 pr	Pliers
1	Adapter, straight, two 24/40 outer	1	Pump, vacuum
1	Adapter, vacuum takeoff, 24/40	1	Rack, vacuum, 4 ft long, 3 ft high ½ inch rods 3, 4 ft rods 4, 3 ft rods
1	Bar, stirring, 2.5 cm	1	Receiver, udder type
1	Beaker, 100 mL	1	Refractometer, Abbe
1	Bottle, aspirator, 250 mL	1	Ring, cork, 250 mL
3	Clamps, buret	4	Rings, cork, 50 mL
4	Clamps, 3-prong	2	Sacks, paper, to get dry ice in
4	Clamps, holder, right angle	1	Stopcock, straight, 2 mm
1	Column, distilling, Vigreux, 30 cm	1	Stopper, rubber, 1 hole, No. 3
1	Condenser, West or Liebig, 20 cm	1	Stopper, rubber, 1 hole, No. 7
16	Connectors for rack rods	1	Strip, terminal, 3 port minimum
1	Cylinder, gas drying, large	1	Support rack, stirring motor rack
1	Cylinder, graduated, 100 mL	1	10/30 Thermometer, 5 cm immersion, -10 to 150°C
1	Dropper, medicine		
2	Flasks, Dewar, 1000 mL	1	Towel to crush dry ice in
4	Flasks, 50 mL, r.b., 24/40 outer	2	Transformers, variable
1	Flask, 250 mL, r.b., 24/40 outer	2	Traps, vapor
1	Gauge, Mcleod, tilting type	3	Tubes, glass, connecting, T
1 pr	Gloves	10 ft.	Tubing, condenser
1 pr	Goggles, safety	15 ft	Tubing, vacuum, 1/4 x 3/16
1	Hammer	3 ft	Tubing, vacuum, 3/6 x 5/16
9	Holders, clamp	1	Valve, bleeding valve, Bunsen burner bottom or 3 way stopcock
1	Manometer, U-tube, 1 m		
1	Manostat, Lewis-Nester	15 ft	Wire, copper
1	Mantle, heating, 250 mL		
1	Motor, magnetic stirring		

CHEMICALS   S = student

Acetone or isopropanol, 1.5 L/S
Dimethylformamide, $HC(=O)N(NH_2)_2$
Dry ice, 700 g/S
Ice, 700 - 1000 g/S
Mercury, to fill manostat, manometer, and McLeod gauge as needed.

Aluminum foil, 0.4 m³
Drierite, $CaSO_4$, 400 g as needed
Glycerine in a dropping bottle
Labeling tape, 15 cm
Pump oil, as needed
Vacuum grease, tube

APPARATUS DESCRIPTION

If you have a vacuum rack on a cart, then use it. Or you can build something that will work at a minimum cost. The rack is made by clamping cross rods onto two tall ring stands and then using C-clamps to hold the ring stands to a table top. Figure 6-20, p. 68, shows the general arrangement of the various components, the directions below are based on that apparatus.

It is recommended that the entire set of directions be read before you begin the experiment. Remember- be deliberate around a vacuum line.

THE BIG PICTURE

150 mL of DMF is placed in a distillation flask, a forerun is obtained, and then two 25 mL fractions are obtained at about 40 torr and one small fraction at about 60 torr. The purity of each is determined by refractive index measurement.

DIRECTIONS

1. Check over the apparatus and become familiar with the names and locations of each component.

2. Check the oil in the vacuum pump (A). It should be clean and should reach the full mark. This is a precaution; the oil is usually satisfactory. *Do not attempt to change this oil unless your instructor agrees.*

3. Carefully remove the tubing from the side arm of the traps S and AA and pour out any liquid in them into a small beaker.

4. Replace these traps carefully; a *few drops* of glycerine is helpful.

5. Go to the dry ice chest, and use a hammer to break off several pieces. Place 600 to 700 g in a large beaker or paper sack.

6. Place a towel on the floor and pour the dry ice onto it.

7. Cover the dry ice with the rest of the towel and using a hammer, crush the dry ice into chunks no bigger than 2 cm in diameter.

8. Fill one Dewar (T) with isopropanol to within 5 cm of the top and fill the other Dewar (BB) with dry ice to the same height.

9. To Dewar T, add a small piece of dry ice. Record what happens.

10. Add a few mL of isopropanol to the Dewar (BB) with the dry ice in it. Compare what happens now to what happened in step 9.

12. Repeats steps 9 and 10 until the Dewar with the dry ice in it is filled to within 2.5 cm of the top with isopropanol and the Dewar that had the isopropanol in it initially is filled to within 2.5 cm of the top with dry ice. Which method is the fastest and least messy?

13. Check the drying tower (Q) and change the desiccant if you feel it is necessary. The blue particles ($CoCl_2 \cdot 2H_2O$) will be pink ($CoCl_2 \cdot 6H_2O$) if they are wet. Connect the tower to the vacuum line as shown in Figure 6-20, p. 68, if it is not already connected.

14. Connect the McLeod gauge (CC), the manostat (M), and the bleeding valve (P) into the vacuum line as shown in Figure 6-20, p. 68, if not already done.

15. Open the bleeding valve (P) several turns.

16. Tilt the McLeod gauge so the mercury is in the reservoir bulb (D of Figure 6-10, p. 64).

17. Tightly close the stopcock (Y) on the hose that will connect the distilling column to the trap. Do not connect the hose to the distilling apparatus.

18. Open valve N on the manostat.

19. Start the vacuum pump (A) and watch the U-tube manometer (R). When it settles down, what pressure is being registered?

20. Close the bleeding valve (P) a small amount and again read the pressure.

21. Close the bleeding valve completely and immediately read the pressure on the manometer.

22. Read the pressure by using the McLeod gauge. Refer to Figure 6-10, p. 64, to see how this is done.

23. Let the system pump out for 5 to 10 minutes and read the gauges again. Be sure to tilt the McLeod gauge back to its starting position before a new reading is made. Did pumping for a longer time have any effect?

24. Open stopcocks N and O on the manostat. This is not necessary on the Gilmont type.

25. Use the bleeding valve (P) and adjust the pressure in the manometer to about 60 torr.

26. Close stopcock (N), and when you get to 40 torr, close stopcock O. (Tilt the Gilmont type to set the pressure)

27. Close the bleeding valve a small amount and observe what happens to the mercury in the manometer. Did it change?

28. Open the bleeding valve a small amount and observe what happens to the mercury in the manometer. Did it change? About how closely does this manostat hold a pressure?

29. Turn stopcock O to open the manostat.

30. Open the bleeding valve and shut off the vacuum pump.
31. Check the condenser tubing and be sure it is of good quality. If not, replace it.
32. Set up the rest of the distillation apparatus as shown in Figure 6-20, p. 68. Ask your instructor to check it before you go any further.
33. Place a small piece of labeling tape on each of the 50 mL flasks.
34. Lower the stirring motor holder (D), the stirring motor (F), and the heating mantle (G). Catch the flask if it comes loose.
35. Remove the flask and set it in a cork ring.
36. Add 125 mL of N,N-dimethylformamide to the sample flask.
37. Add a magnetic stirring bar and then reassemble the flask, heating mantle, stirrer, and holder.
38. Cover the flask and the distilling head up to the thermometer bulb with aluminum foil (shiny side in). Leave a small opening at the front of the flask for observation.
39. Turn each variable transformer back to 0, then check the fuse (small round knob on the side) by twisting it a small amount to remove the cover and looking at it to see if it is good. Replace the fuse. It is recommended that a spare fuse be taped to the side of the transformer.
40. Make sure the heating mantle is plugged into the transformer and that both variable transformers (B and C) are plugged into the power strip.
41. Turn on each variable transformer.
42. Turn on the variable transformer controlling the heating mantle to 85 to 90 volts.
43. Connect the condenser tubing to the water supply and place the outlet tube in the drain. If the drain tubing does not go deeply into the drain, then tie a small weight onto the end of it.
44. Turn on the water slowly at first until you are sure the connections will hold.
45. Turn on the vacuum pump, close the bleeding valve, open stopcock Y, and pump the system down to a pressure of about 40 torr as before. The manostat should still be set.
46. Turn the udder so a 50 mL flask is in a collecting position. Mark this - forerun.
47. Collect the forerun. The pure material should boil at 76 °C at 39 torr.
48. When the pure material begins to come over, turn the udder and collect about 25 mL of the purified material. Mark this - 40 torr or what ever your pressure is.
49. Adjust the pressure to 60 torr by opening the bleeder valve slowly. Use the U-tube manometer.
50. Set the manostat by closing valve O or tipping the adjustable manostat.
51. Turn the variable transformer to 100%.
52. Close the bleeding valve and check the pressure with the manometer. If it is close to 60 torr, then go to the next step. If not, reset the manostat.
53. Turn the udder to another position and collect about 25 mL. Mark this 60 torr. What is the temperature of this fraction?
54. Shut off the variable transformer and lower the heating mantle.
55. Open stopcock O on the manostat.
56. Turn off the pump.
57. Slowly open the bleeding valve and bring the system to atmospheric pressure.
58. Turn off the condenser water.
59. Determine the refractive index of the following: a. the original solution, b. the forerun, c. the 40 torr fraction, d. the 60 torr fraction. See Experiment 4, p. 512, for the directions on the use of a refractometer, if you did not do this experiment.

The refractive index of DMF is reported in the literature as 1.4280 at 25 °C. Make your normal temperature corrections based upon the value of water as a reference standard.

60. Leave the isopropanol in the Dewar flasks.

60. Clean out the distillation apparatus. DMF is water soluble, and this small amount can be poured into the drain.

61. *Clean up all of the nonrack equipment and place it on the desk for the next student.* Thank you.

EXPERIMENT 9. MOLECULAR DISTILLATION

The determination of vitamin E (tocopherol) in margarine (*Ind. Eng. Chem. Anal. Ed.*, **19**, 707, 1946)

Vitamin E, or alpha tocopherol, occurs in large quantities in carrot and wheat germ oil as well as many other substances. Vitamin E was suspected in 1922, identified in 1925, and its structure was determined in 1938. It has eight forms, four tocopherols and four trienes. The most active form is alpha tocopherol, which has a m.p. of 2.5 to 3.5 °C and boils at 200 to 220 °C at 0.1 torr. The structures are shown in Table E9-1.

Table E9-1. The vitamin E contents of several materials

Material	Vitamin E (µg/g)	Material	Vitamin E (µg/g)
Margarine	544	Lard	27
Carrot oil	1620	Cottonseed oil	726
Egg lipid	130	Olive oil	69
Dried eggs	95	Almonds	274
Peanuts	114	Sunflower seeds, raw	495
Wheat germ	1330	Lettuce	4
Muskmelon	100		

Alpha tocopherol

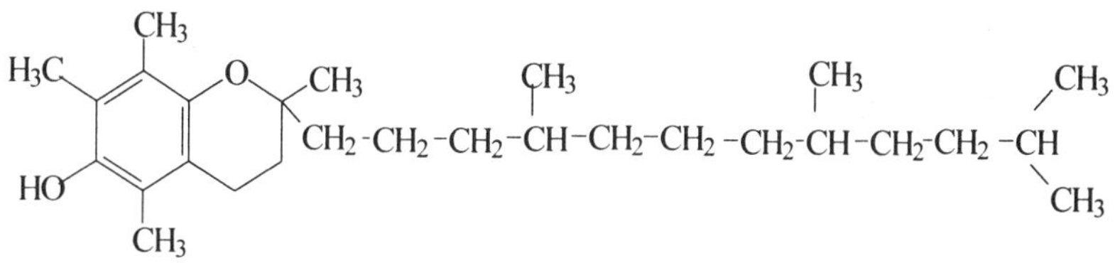

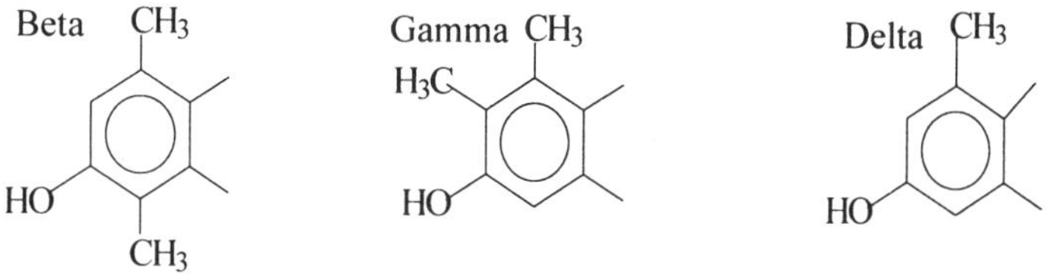

[Structure of alpha tocopherol: chromanol ring with CH_3 groups at positions 5, 7, 8, and 2; HO– at position 6; side chain CH_2-CH_2-CH_2-CH(CH_3)-CH_2-CH_2-CH_2-CH(CH_3)-CH_2-CH_2-CH(CH_3)$_2$]

Beta, Gamma, Delta [ring structures shown with different methyl substitution patterns]

The trieneols are the same, except that the side chain contains three double bonds. The alpha, beta, gamma, and delta forms correspond to the tocopherols.

These compounds often occur as acetate or succinate esters and, for a total analysis, these esters are cleaved by a KOH saponification. Vitamin E is obtained commercially by molecular distillation. This does not provide a pure product, but it does separate vitamin E from most of the other materials associated with it such as chlorophyll, xanthophylls, carotenes, ubiquinol, ubichromenol, steroids, and quinines, but not vitamin A and the beta carotenes.

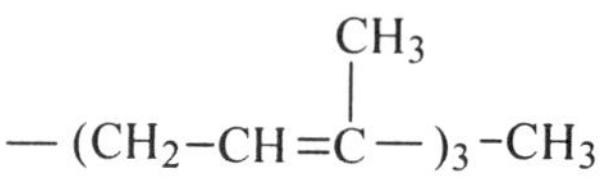

$$-(CH_2-CH=C(CH_3)-)_3-CH_3$$

YOUR PROBLEM

A commercial margarine states on the label that it has been fortified with Vitamin E. Your job is to determine if vitamin E is, in fact, present and if so, approximately how much. NOTE: Neither the apparatus you will use nor any other will make a clean separation of vitamin E from margarine. However, it will concentrate the vitamin E so a semiquantitative measurement can be obtained. A further extraction process is necessary to clean up the product for a more accurate quantitative measurement. In this experiment, you will not do an exhaustive distillation, but distill for only about 30 minutes to illustrate the principles involved.

EQUIPMENT

1	Adapter, socket 34/45 to 28/15		4	Holders, clamp
1	Balance, analytical		1	Motor, stirring
1	Bar, stirring, 5 cm		1	Pipet, delivery, 5 mL
1	Beaker, 50 mL		1	Pump, like Welch Duoseal
1	Bottle, wash, 250 mL		1	Pump, oil diffusion
1	Clamp, 28/15 ball and socket		1	Rack, vacuum
2	Clamps, buret		1	Sack, paper, size 8
2	Clamps, pinch, screw type		2	Spatulas
2	Clamps, versatile		1	Spectrophotometer, visible region
4	Clips, alligator		1	Still, molecular
2	Cords, extension, 2 meter, male one end, alligator clips other end.		1	Stopcock, 3 way
			1	Tape, electrical, plastic
1	Dish, crystallizing, 125 mm dia.		1	Thermometer -10 to 260 °C
1	Flask, Dewar, 600 mL		1	Towel, cloth
3	Flasks, volumetric, 25 mL		1	Trap, vapor
1	Flask, volumetric, 100 mL		2	Transformers, variable, Variac
1	Funnel, powder		5 ft	Tubing, vacuum
1 pr	Goggles, safety		2 ft	Wire, chromel, no. 20
1	Hammer			

CHEMICALS   S = Student

Acetone or isopropanol, 500 mL/S
Alpha tocopherol, 25 mL of 4 µg/mL, keep refrigerated, make fresh weekly.
Chloroform, $CHCl_3$, 50 mL/S
Dry ice, 1,500 g/S
Ferric chloride, $FeCl_3 \cdot 6H_2O$, 20 mL/S, See *J. Biol. Chem.* **156**: 653 and 661 (1944) for color developing instructions.
Glacial acetic acid, $H_3C\text{-}COOH$
Margarine, 1 g/S, Students bring their own sample. Label should say it is fortified with vitamin E.
Petroleum ether, 50 mL/S
Pump oil, diffusion, Myvoil, Eastman 9120 for example, as needed
Pump oil, fore pump, as needed
Silicone oil, Dow 550 or equivalent, 200 mL initially, then as needed.
Sulfuric acid, H_2SO_4
Sodium hydroxide, NaOH
Sodium sulfate, anhydrous, Na_2SO_4

APPARATUS SETUP

The apparatus is set up as shown in Figure E9-1, p. 529. An all-glass oil diffusion pump should be used so the student can observe how it operates. If either a Pirani or thermocouple gauge is available, it should be connected at I of Figure E9-1. The molecular still used here is a very simple design, is inexpensive, and can be made by almost anyone. Mineral oil *should not* be used for the oil bath. Its flash point is too close to the distillation temperature to be safe. (An oil bath is used rather than a circular heating mantle so the student can watch the entire operation). It is suggested that two to three spare fuses be taped to the top of the Variacs.

THE BIG PICTURE

The sample, 1 gram or less, is weighed and placed into the middle cup (Figure E9-2, p. 529). The oil bath is warmed to about 100 °C to melt the margarine. The apparatus is closed, and the system pumped down as low as possible to degas the sample. The vacuum is released. The cold finger is filled with dry ice and acetone, the pumps are turned back on, and the oil is heated to 210 to 220 °C for about 30 minutes. The system then is allowed to completely cool. The cold finger is removed carefully and the material washed off with petroleum ether. This solution (about 5 to 10 mL) is transferred to a 25 mL volumetric flask and diluted to volume with the tocopherol reagent. The absorbance is measured at 515 nm after about 15 minutes. A standard is prepared by adding 100 µg of pure tocopherol to a 25 mL volumetric flask and diluting to the mark with half petroleum ether and half tocopherol reagent.

DIRECTIONS

1. Weigh approximately a 1.0 g sample of margarine (a cube about 12 mm on a side) to the nearest 0.1 mg and place it in the sample holder (E).

2. Place the sample holder in the molecular still.

3. Place the crystallizing dish (D) under the molecular still so the oil is about 5 mm over the bottom of the still.

4. Turn on the stirring motor and adjust it so the stirring bar barely turns.

5. Check the electrical connections (C) to the crystallizing dish. Be sure they are covered with electrician's tape, so that they cannot short out with each other or against the support rack.

6. Turn on the variable transformer (A) and adjust the voltage to about 10 volts. The wires (C) will not glow, but they may smoke slightly as surface oils burn off. This is normal. After about 5 to 6 minutes, the oil temperature should be about 95 to 100 °C, and the margarine will melt.

7. Slowly turn up the Variac to 13 volts. *If you go over 15 volts, you will probably blow a fuse* in the variable transformer. Spares fuses should be taped to it.

8. Obtain 1000 to 1200 g of dry ice from a dry ice chest. Place it in a paper sack or large beaker and bring it back to the laboratory.

9. Place the dry ice on a cloth towel, fold the towel over to cover the dry ice, place this on the floor, and crush the dry ice to marble size by hitting it with a hammer.

10. Check the speed of the stirring bar in the oil. You may have to slow it down as the oil heats. Try to avoid getting an air bubble under the still. If that happens, stop the stirrer, CAREFULLY tilt the still slightly, and let the bubble escape.

11. Lower the Dewar flask (M) and check trap (L). If there is considerable liquid in the bottom, drain the trap.

12. Replace the Dewar and add as much dry ice to it as it will hold.

13. Very slowly pour either acetone or isopropanol into the Dewar until you fill it to about 2 cm from the top.

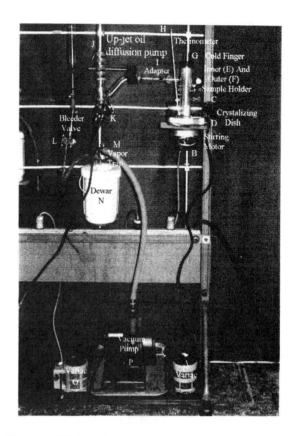

Figure E9-1. Apparatus used for the molecular distillation of vitamin E from margarine.

Figure E9-2. Closeup of the oil bath and molecular still.

14. The temperature of the oil bath usually will be up to about 140 to 150 °C at this time. Place the cold finger (G) in the molecular still. BE CAREFUL - Center it, push it in tightly and do not spill the hot oil.

    A REMINDER - Wear safety goggles, and long sleeved shirts or blouses, turn stopcocks slowly to avoid sudden pressure surges, and think about what will happen before you turn a stopcock.

15. The sample will now be degassed. Turn bleeding valve (K) to connect the diffusion pump with the trap. Check the bore on the stopcock to make sure this is correct.

16. Turn on the fore pump (O) and watch the sample. If it starts to spatter because the trapped air is leaving too fast, open the bleeding valve and, as the spattering subsides, slowly close it. In extreme cases, you may have to shut off pump (O) for a few minutes.

17. After the fore pump has been operating for 3 to 4 minutes, turn on the variable transformer (N). Apply about 42 to 43 volts.

18. The diffusion pump will not start to operate immediately. If will take 4 to 5 minutes to warm up. After the pump oil begins to reflux, pump on the system for about 15 minutes. If you have a low pressure vacuum gauge attached, measure the pressure of the system.

19. Shut off the fore pump.

20. The system is still under reduced pressure. Go to the bleeding valve (K) and turn it slowly clockwise until you hear a low hissing sound as air enters the system. Hold that stopcock position for about 30 seconds, then open the stopcock completely. Turn off the variable transformer (N).

21. Fill the cold finger (G) with dry ice.

22. Use a wash bottle filled with either acetone or isopropanol and carefully fill the cold finger. Be careful not to drip any solvent into the hot oil.

23. Close bleeding valve (K).

24. Turn on the variable transformer to the diffusion pump.

25. Turn on the fore pump.

26. Let the system pump down to whatever is its natural lowest pressure. If a Pirani or thermocouple gauge is attached to the system, measure this pressure every few minutes to determine how fast a system of this size can be pumped out. When the oil temperature reaches 200 °C, continue the distillation for another 30 minutes.

27. You should have a very thin film of tocopherol on the bottom of the cold finger plus considerable quantities of some light yellow low-boiling material that also distilled. Use what you have for the next step rather than trying to purify this further.

28. Shut off the variable transformer that controls the oil bath.

29. Shut off the diffusion pump heater.

30. Shut off the fore pump.

31. Slowly open the bleeding valve as before and let the system come to atmospheric pressure.

32. You will now use a spectrophotometer to measure the amount of tocopherol. Please have your instructor show you how to properly use the spectrophotometer in your laboratory.

33. Hold on to the molecular still with one hand and remove the cold finger and the top rubber stopper with the other hand. CAREFUL- The oil may still be quite hot.

34. Pour the dry ice - acetone mixture remaining in the cold finger in the sink.

35. Place 5 mL of petroleum ether in a 50 mL beaker.

36. Place the end of the cold finger into the petroleum ether and swirl it until all of the material has dis-solved from its surface.

37. Transfer the solution from the beaker to a 25 mL volumetric flask. Use a few mL of petroleum ether to aid in the transfer.

38. Pipet into this flask 10 mL of the tocopherol color reagent.

39. Dilute to the mark with petroleum ether and thoroughly mix the contents of the flask. Mark this flask " Sample".

40. Prepare a tocopherol calibration standard. Pipet 5.0 mL of the tocopherol standard into a 25 mL volumetric flask. Add 10 mL of the tocopherol color reagent and dilute to the mark with petroleum ether. Mix thoroughly. Mark this flask "Standard". The solutions in both flasks should be a faint red.

41. Prepare a blank as follows. Pipet 10 mL of tocopherol color reagent into a 25 mL volumetric flask and dilute to the mark with petroleum ether. Mark this flask - BLANK.

42. Take these three flasks to the spectrophotometer.

43. Adjust the instrument to 515 nm.

44. Place the blank in a cell in the reference side of the instrument. NOTE: Immediately wash off any material spilled on your hands. Glacial acetic acid does not burn your skin immediately but produces painful burns and large blisters after a short delay.

45. Zero the instrument.

46. Place the blank solution in a cell and place the cell in the sample side of the instrument.

47. Adjust the instrument to read 0 absorbance or 100% T.

48. Replace the blank solution in the sample cell with the standard and determine the absorbance of this solution. Record this value.

49. Repeat step 48 with the sample solution. Record the absorbance.

50. NOTE: If you do not adjust the 100% T or 0 absorbance with the blank solution, but with water or petroleum ether, then you must determine an absorbance value for the blank. Place the blank solution in the sample side and measure its absorbance.

51. Turn the instrument off, remove all of the cells, and clean up the area. Empty the cells, rinse them out, and thoroughly wash your hands.

52. Clean the sample from the outer tube of the molecular still. Ether and/or chloroform are good solvents.

53. Clean the sample tubes. You may leave the dry ice and acetone in the Dewar.

54. Make sure that everything is shut off and that the area is cleaner and in better order than when you began the experiment. Thank you.

55. Calculate the amount of vitamin E (tocopherol) in your sample, first as µg/g of sample, and then as %. Follow the example calculation shown below.

EXAMPLE CALCULATION

This example assumes that a blank was obtained. A 0.8634 g sample of margarine was distilled, and the vitamin E content measured. The following data were obtained when 1.00 cm cells and 25.0 mL volumetric flasks were used.

Blank absorbance   0.021
Standard absorbance 0.265
Sample absorbance   0.347

Calculate the µg of vitamin E per gram of margarine and the % vitamin E.

ANSWER

Use the Beer-Lambert Law; $A = abc$

where A = absorbance
a = absorptivity
b = cell thickness (1 cm unless specified otherwise)
c = concentration

Step 1. Calculate a. Remember - The standard is 100 µg/100 mL or 4.00 µg/mL.
$A = (0.265 - 0.021) = 0.244$
$0.244 = a \times 1 \text{ cm} \times 4.00 \text{ µg/mL}$
$a = 0.061 \text{ mL/cm-µg}$

Step 2. Calculate c.
$A = (0.347 - 0.021) = 0.326$

$0.326 = 0.061 \times 1 \times c$
$c = 5.34 \; \mu g/mL$

Step 3. Calculate the µg in this sample.
$5.34 \; \mu g/mL \times 25.0 \; mL = 155 \; ug$

Step 4. Calculate the % vitamin E in the sample.
$155 \; \mu g = 0.000155 \; g$

$$\% = \frac{0.000155 \; g \times 100}{0.8634 \; g} = 0.0179\%$$

EXPERIMENT 10   SUBLIMATION
The separation of benzoic acid from saccharin

BACKGROUND
2,3-dihydro-3-oxobenzisosulfonazole was discovered in 1879 by Ira Remsen. It was found to be about 500 times sweeter than sugar and was introduced commercially in the form of the water-soluble sodium salt known as *saccharin* in 1885. It is not changed by any body function and, therefore, has no caloric content. It was for this reason that it was banned in 1912. No calories were later considered to be an advantage as was its ability to provide sweetness in foods for diabetics, so it was reinstated. It has a bitter, metallic aftertaste which lowered it sales. When cyclamate was discovered (1942), it was found that about 10% added to saccharin removed the aftertaste effects for most people, and this mixture was used as a sweetener for many foods. One experiment showed cyclamate to be a possible carcinogen in animals, so it was banned. Saccharin is not considered to be a carcinogen itself, because it is not metabolized, but some suspect it of making other compounds carcinogens. The debate continues as to whether to ban it or not. One method of preparation is as follows:

The impurities in saccharin (m.p. = 229 °C) include *o*-sulfomyl benzoic acid (m.p. = 152 °C) and benzoic acid (m.p. = 122 °C).

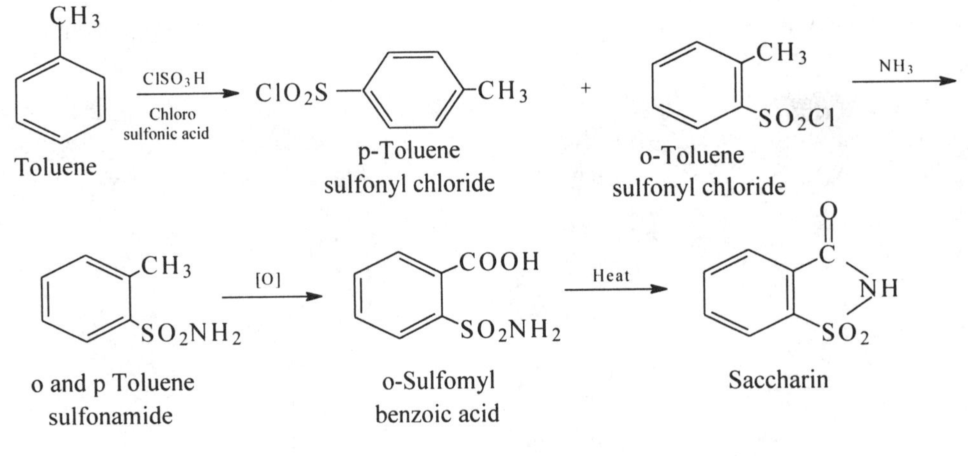

YOUR PROBLEM
Benzoic acid can be present in saccharin as an impurity. Because it also is used as a preservative, this is of no real concern as long as the levels are low. However, it is suspected that a company is adding an extra amount to dilute the saccharin and increase profits. Because it may be there anyway, no one will get suspicious. Saccharin costs 6.4 cents/g whereas benzoic acid costs only 1.4 cents/g. Your instructor has some of the suspected product, and your job is to determine the purity of the product by a sublimation separation.

EQUIPMENT
1 Apparatus, melting point (like Fisher-Johns)	1 Balance, triple beam
	3 Clamps, 3 pronged
1 Apparatus, sublimation	1 Dewar, 500 mL

1	Gauge, pressure, Mcleod	2	Stands, ring
1 pr	Goggles	1	Stopcock, 3-way or burner bottom
1	Hammer	1	Towel, cloth
3	Holders, clamp	1	Trap, vapor
1	Paper, weighing	1	Transformer, variable, Variac
1 pr	Pliers	8 ft	Tubing, rubber
1	Pump, vacuum, 1 torr	4 ft	Tubing, vacuum
2	Rings, iron, 7.5 cm dia.	1 ft	Wire, copper
1	Sack, paper, no. 8		
1	Spatula		

CHEMICALS   S = Student

Benzoic acid, C_6H_5-COOH, 1 g/S
Dry ice, 600 to 700 g/S
Glycerine in a dropper bottle

Isopropanol, H_3C-CH(OH)-CH_3, 500 mL/S
Saccharin, 98% purity, 5 g/S
Vacuum grease, tube

APPARATUS SETUP

The apparatus is set up as shown in Figure 7-19, p. 82. Any type sublimator can be used. The directions given are for an apparatus similar to the one shown in Figure 7-1, p. 72. The bleeding valve can be either a 3-way stopcock or the base of a Bunsen burner. The vacuum pump is an ordinary mechanical pump capable of pumping to 1 torr. A tilting Mcleod gauge normally is used. The sample can be a synthetic one prepared by the instructor by adulterating saccharin with benzoic acid, or it can be regular 98% purity saccharin (not food grade). Benzoic acid sublimes at 50 °C and 1 torr whereas it takes 150 °C is needed for saccharin to sublime. Figure E10-1, p. 534, is a temperature-composition diagram of mixtures of benzoic acid and saccharin to illustrate the m.p. depression.

DIRECTIONS

1. Check over the apparatus and compare it with Figure 7-19, p. 82, to be sure all of the components are present.
2. Set up the apparatus as shown in Figure 7-19, p. 82, if it is not already done.
3. Check the rubber tubing on the condenser of the sublimator (D) and be sure it is of good quality. If not, replace it. Use glycerine as a lubricant and wire the connections.
4. Place the outlet tubing of the condenser into the drain and if it doesn't go 10 to 12 cm into the drain, tie a small weight to the end so it will not come out when the water flows.
5. Connect the inlet water tubing to the water outlet and wire the connection.
6. Remove the Dewar flask (I) and check the trap (H). If there is liquid within 1 cm of the inlet tube, remove the trap and clean it out.
7. Replace the trap and Dewar.
8. Obtain about 600 to 700 g of dry ice from the dry ice chest. Carry it back to the lab in a paper sack or large beaker.
9. Place a cloth towel on the floor and lay the dry ice on top of it. Fold the towel over the dry ice and crush it to marble size with a hammer.
10. Place the dry ice in the Dewar and then slowly pour in isopropanol to within 2 cm of the top of the flask.
11. Check the oil level on the vacuum pump (J). It should be satisfactory. If not, *call the instructor* and add sufficient oil only if he/she agrees.
12. Open the bleeding valve (G) and tilt the Mcleod gauge (F) to the horizontal position.
13. Weigh 2 to 3 grams of sample on a triple beam balance and place it in the sample compartment of the sublimator. Save a small potion of the original material so a melting point can be obtained.
14. Place a small amount of vacuum grease on the joint and put the sublimator together.
15. Turn on the condenser water.
16. Turn on the variable transformer (Variac) (A) and apply about 15 volts (check with your instructor for the proper setting to obtain 50 °C).

17. Turn on the vacuum pump.
18. Slowly close the bleeding valve (G) and let the system pump to its lowest pressure.
19. After about 5 minutes, measure the pressure with the Mcleod gauge (F). If it is not close to 1 torr, then check the system for leaks. If leaks are present, try pushing the tubing tighter onto the glass tubing. At these pressures, this is usually sufficient.
20. Determine the m.p. of the impure sample by using a Fisher-Johns or similar m.p. apparatus.
21. Let the impurities sublime for about 1 hour. Measure the pressure occasionally.
22. Shut off the vacuum pump (J), open the bleeding valve (G), shut off the Variac (A), and the cooling water, and take apart the sublimator (D).
23. Remove two samples, one from the top of the remaining sublimand and one from the sublimate.
24. Determine the m.p. of each. Any type of melting point apparatus can be used. The ones shown in Figures E10-2 and E10-3 are quite easy to use.
25. Place a few crystals on the glass-covered heater under the magnifying glass.
26. Place a cover glass over the crystals (optional).
27. Turn on the power and turn on the light. Turn the heater transformer to about 50% and watch the crystals and the temperature on the thermometer. Raise the temperature rapidly (10 to 15 °C/min) until you get within a few degrees of the expected m.p., then slow down the rate to 1 to 2 °C/min. Benzoic acid starts to sublime at ~100 °C and melts at 122 °C. Saccharin melts at 228 °C, and the sodium salt melts much higher. Record the results.
28. Use Figure E10-1 and estimate the purity of both the sublimate and sublimand.
29. Clean out the sublimator.
30. *Leave the coolant in the Dewar flask and clean up the area so it is better than when you started.* Thank you.

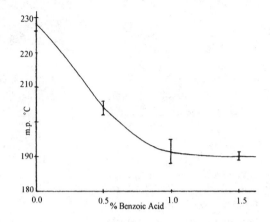

Figure E10-1. A temperature-composition diagram of mixtures of benzoic acid and saccharin.

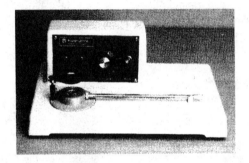

Figure E10-2. A Fisher-Johns melting point apparatus.
(Courtesy - Fisher Scientific Co., Pittsburgh, PA)

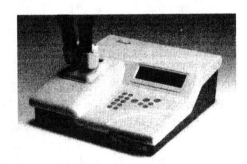

Figure E10-3. An Electrothermal Automelt automatic melting point apparatus.
(Courtesy - Fisher Scientific Co., Pittsburgh, PA)

EXPERIMENT 11.   ENTRAINER SUBLIMATION
The separation of caffeine from coffee by entrainer sublimation

Natives have for centuries brewed teas from plants that contain caffeine. They didn't know what was present, but they knew they received a stimulating effect from the brew. The active ingredient was found to be caffeine. It occurs in coffee (1.5%), tea (5%), and mate leaves, guanova paste, and cola nuts. It was isolated in pure form in 1820, although its structure was not known. It is an odorless, slightly bitter, solid that forms needle crystals. Small amounts stimulate the central nervous system, an effective dose being 150 to 200 mg, (about 1 to 2 cups of coffee or tea). Large amounts cause nervousness, loss of sleep, headaches, and digestive disturbances. The fatal dose is believed to be about 10 grams, but no deaths have been reported. It is one remedy for poisoning by alcohol, opium, and other drugs that suppress the CNS. 80 % is broken down to urea; the remainder is either excreted unchanged or as a methylated form.

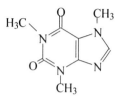

Caffeine
m.p. 238 °C
Sublimes at 178 °C
Sublimes rapidly at 160 °C and 1 torr. 1 g will dissolve
in 5.5 mL $CHCl_3$ or 5.5 mL of 80 °C water

YOUR PROBLEM
Your instructor has been given a new and simple entrainer sublimator that it is supposed to work quite well. You are to test it for him by separating caffeine from coffee. First you are to determine if anything can be sublimed from coffee. If there is any sublimate, you will try to identify it.

EQUIPMENT

1	Adapter, T type, 24/40	1	Hammer
2	Adapters, hose connection with stopcocks, 24/40 joints, one inner, one outer	1	Holder, thermometer
		1	Hot plate
		1	Manometer, T-type preferred
1	Beaker, 30 mL	2	Paper, chart
1	Beaker, plastic, 400 mL	1	Pump, vacuum, 1 torr
2	Bottles, wash, 250 mL	1	Rack, vacuum
1	Card, calibration, polystyrene	1	Sack, paper, no. 8
1	Cell, NaCl, sandwich	1	Spectrophotometer, infrared
1	Cell, NaCl, wedge	1	Sublimator, entrainer
5	Clamps, buret	1	Syringe, liquid, 1 mL
2	Columns, distilling, Hempel 25 cm, 24/40 joints	1	Tape, heating, 1/4" x 3'
		1	Thermometer, -10 to 260 °C
1	Desiccator, 170 mm	1	Towel, cloth
1	Flask, Dewar, 650 mL	1	Transformer, variable, Variac
1	Funnel, powder, 70 mm	1	Trap, vacuum pump
1 pr	Goggles	6 ft	Tubing, vacuum

CHEMICALS   S = student
Acetone or isopropanol   500 mL/S
Chloroform, $CHCl_3$, 10 mL/S
Coffee, 3-5 g, any non-decaffeinated brand.
Dry ice, 1 kg/S

Mercury, as needed for the manometer.
Pump oil as needed
Vacuum grease, tube

APPARATUS SETUP
The apparatus is set up as shown in Figure 7-20, p. 82. It consists of two Hempel distilling columns placed horizontally with a T adapter in the center to hold the sample. Two indentations can be added to the bottom of the T by heating it and then pressing it with a file. These ridges will help hold the coffee in place. Hose-connecting adapters with stopcocks serve to control the flow of air. The entrainer air is heated by wrapping one Hempel column and the sample chamber with heating tape. Any type manometer can be used. The T-type is recommended to provide the student with the experience of operating a different type. The trap and pump are standard items.

THE BIG PICTURE

About 2 to 3 g of coffee are placed in the sublimator, the pressure is reduced to 50 to 60 torr, and the temperature of a slow flow of entrainer air is raised to 170 to 180 °C. The entrainer gas is heated because, if cold gas passes over a hot sample, the caffeine will condense as a fog. Many of these particles then will be pumped through the system and not collected, thus causing low results. The caffeine will separate as white crystals in the cooled downwind side. These crystals are dissolved in chloroform and verified as being caffeine by infrared spectroscopy.

DIRECTIONS

1. Check the apparatus and compare it with Figure 7-20, p. 82. Ensure that all of the components are present.
2. We will assume that the last person to use this apparatus did not clean it out. This person may be found on the rack in the chemistry department dungeon. Carefully disassemble parts A to G of the sublimator.
3. Rinse them out carefully with chloroform (hood) and collect the waste chloroform in a beaker. Transfer this solvent to the waste solvent container.
4. Allow the parts to air dry for a few minutes.
5. Place a small amount of grease on all of the joints and reassemble the sublimator.
6. Carefully wrap the heating tape (C) around the heating tube (D) and the lower part of the sample holder (E).
7. Add 2 to 3 g of coffee to the sample holder. It need not be weighed. A small powder funnel may be of help.
8. Place the thermometer in the top of the T and adjust the height of the bulb so it extends into the stream of hot air. This may require gently shaking the sample tube to spread out the sample.
9. Obtain about 500 g of dry ice. A paper sack is a convenient way to carry it from the dry ice chest to the laboratory.
10. Lay a cloth towel on the floor. Pour the dry ice onto it, fold the towel over it, and crush the dry ice to marble size pieces by hitting it with a hammer.
11. Check trap (J) by lowering the Dewar flask (K). If there is liquid in the trap less than 1 cm from the bottom of the tube, disconnect the trap and empty it.
12. Replace the trap and Dewar flask.
13. Fill the Dewar to the top with dry ice.
14. Carefully add acetone or isopropanol to the Dewar until the liquid is within 2 cm of the top.
15. Open stopcock (H).
16. Open stopcock (A).
17. Open the stopcock on the manometer (I).
18. Turn on the transformer (M) and adjust it to the value stated by your instructor. Heating tapes vary considerably so this setting varies widely. Spare fuses are usually taped to the side of the transformer.
19. Turn on the vacuum pump.
20. Slowly turn stopcock (A) until the pressure in the manometer reads 50 to 60 torr (6.65 to 8.0 kPa).
21. Watch the collection tube and the thermometer. When the temperature reaches about 100 °C, a region of the collection tube will become cloudy as the caffeine begins to sublime. At 140 to 150 °C, the caffeine begins to collect fairly rapidly. Continue the sublimation for about 15 minutes after the temperature reaches 178 °C. NOTE: If the collected material begins to turn brown, lower the temperature, because other compounds are subliming.
22. The inside of the collection tube usually will be white from end to end. With some coffee a small amount of a yellow compound also will sublime and this will be close to the sample.
23. Shut off the transformer.
24. Shut off the vacuum pump and let the system come to atmospheric pressure. Let it cool sufficiently so that you can touch the collection tube without getting burned.
25. Remove the end plug (H) and let it hang.

26. Carefully remove the collection tube (G) with a twisting motion.

27. Take a wash bottle of chloroform, a 30 mL beaker, and the collection tube to a hood.

28. Hold the collection tube vertically over the 30 mL beaker. If you have some brown or yellow material, hold that end up. Put the tip of the wash bottle into the tube as far as it will go (past any brown material) and add one drop of $CHCl_3$. Let the drop run to the bottom, dissolving as much caffeine as possible along the way.

29. Repeat step 28 as often as necessary, turning the collection tube with each drop, but do not use more than 20 drops.

30. Take this beaker of sample and the desiccator containing the infrared cells and syringe to an infrared spectrophotometer.

31. Have your instructor show you how to use this particular instrument.

32. Place a piece of chart paper in the spectrophotometer.

33. Make sure the recorder pen contains ink and that the pen works.

34. Place a polystyrene calibration card into the sample side of the instrument. Adjust the %T to read somewhere between 90 to 95% and record only the first 4.9 μm of this spectrum. The bands at 3.5 μm serve to calibrate the chart paper, correcting for slippage and/or instrument problems.

35. Return the pen and the chart to the starting position.

36. Remove the polystyrene card.

37. Place an NaCl sandwich cell on a paper or cloth towel. CAREFUL: NaCl rapidly dissolves in water - do not breath on this cell or expose it to any water.

38. If your cells are the demountable type, remove the top plate of one of them, place 2 to 3 drops of the sample (syringe) on the lower crystal, and put the top plate back in place. If you have fixed thickness cells, remove the filling plugs. Use a syringe and add 2 to 3 drops of sample to one of the cells. Put the filling plugs back in place.

39. Place this cell in the sample side on the instrument.

40. Place chloroform in either a second sandwich cell or a wedge cell and place this cell in the reference beam of the instrument.

41. Look at Figures E11-1 (Caffeine) and E11-2, p. 538 ($CHCl_3$). Notice that the 8.2 and 13.2 μm bands of $CHCl_3$ will interfere with the 8.1 and 13.4 μm bands of the caffeine spectrum. If possible, the absorption due to $CHCl_3$ should be blanked out. A wedge cell, having a variable thickness, is excellent for this purpose. Set the chart at the short wavelength end of the scale (2.5 to 3.0 μm) and adjust the %T to about 80-85%.

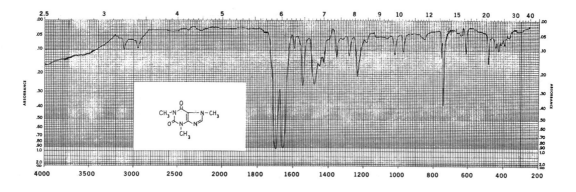

Figure E11-1. Infrared spectrum of caffeine.
(Courtesy - BIO-RAD Laboratories, Sadtler Division, Philadelphia, PA)

42. Set the chart at 15.0 μm (no caffeine band) and move the wedge cell so that the %T equals what it was at 2.5 to 3.0 μm. This effectively cancels out the absorption due to $CHCl_3$.

43. Record the entire spectrum of your sample, from 2.5 to 15 μm.

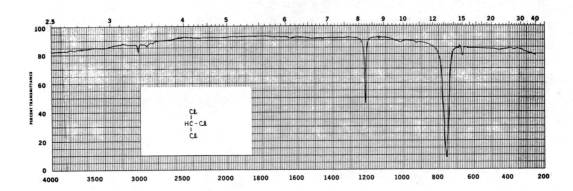

Figure E11-2. Infrared spectrum of chloroform.
(Courtesy - BIO-RAD Laboratories, Sadtler Division, Philadelphia, PA)

44. Remove the chart paper and replace it with a new one.
45. Remove the cells and flush out or wash off each one, first with $CHCl_3$ and then with acetone. When the plates are dry, reassemble the cells and place them in a desiccator.
46. Remove the chart pen and replace it in its storage container.
47. *Clean up this area, put the cover on the instrument*, and go back to the sublimation apparatus.
48. Remove your coffee sample from the apparatus and rinse out the tubes with $CHCl_3$.
49. Completely remove any caffeine from the collection tube.
50. Let the dry ice acetone mixture alone.
51. *Clean up the area.* Thank you.
52. Compare your spectrum of caffeine with the one shown in Figure E11-1, p. 537. Can you find any $CHCl_3$ solvent bands that were not blanked out?

EXPERIMENT 12. LYOPHILIZATION
Preparing freeze-dried beverages

BACKGROUND
Probably the most well known freeze-dried product is coffee. The commercial process is as follows: a coffee extract containing 20 to 25% solids is filtered, then frozen to concentrate the solids to 30 to 40%. This material then is frozen to -25 to -43 °C and crushed into small particles. These *particles* are freeze dried to remove the ice. A vacuum of about 0.2 torr (0.027 kPa) is used, and the final product contains 1 to 3% water. The entire process takes about 7 hours.

Water containing solids generally will freeze between -6 °C to -25 °C. Figure E12-1, p. 539, shows the DTA's for several fruit juices.

YOUR PROBLEM
You are to freeze dry a beverage of your choice, then rehydrate it, taste it, and see if the quality was preserved. NOTE: This may require an overnight operation depending upon how late in the day you start your experiment.

EQUIPMENT
This assumes a setup like that in Figure 8-9, p. 88.

1	Beaker, 2,000 mL	1	Cylinder, graduated, 50 mL
1	Bleeding valve, 3-way stopcock or Bunsen burner bottom	1	Flask, Dewar, 1,000 mL or larger
		1	Flask, r.b. single neck, 500 mL
3	Clamps, 3 prong	1 pr	Goggles, safety

1 Hammer
3 Holders, clamp
1 Manometer, U-tube
1 Pump, vacuum, mechanical and oil diffusion if possible
1 Ring, cork, 10 cm dia.
1 Sack, paper, no. 12
1 Spatula
1 Sponge

2 Stands, ring
1 Stopper, rubber, no. 6, 1 hole
1 Stopper, rubber, no. 8, 2 hole
1 Stopper, rubber, no. 8, 2 hole
2 Towels, cloth
1 Tube, test, 300 mm
3 ft Tubing, Glass, 10 mm
3 ft Tubing, vacuum

CHEMICALS   S = Student

Coffee, tea, milk, fruit juice, any liquid of your choice, 100 mL/S
Dry ice, 4 kg/S
Isopropanol, H_3C-CH(OH)-CH_3, 1,500 mL/S
Mercury, as needed for a U-Tube manometer

APPARATUS SETUP

A diagram of a simplified apparatus is shown in Figure 8-9, p. 88, and is used for this experiment. A 500 mL r.b. flask is used as the sample holder. A trap is made out of a 30 cm long test tube, which is cooled by a dry ice-isopropanol mixture held in a Dewar flask. An ordinary vacuum pump is used, and because this is not a high vacuum, the sample in some cases may have to be cooled partially by immersing it a short distance into the ice bath.

THE BIG PICTURE

The sample (40 to 50 mL) is placed in the round bottom flask and then frozen onto the side walls by using the technique described earlier. While the sample is freezing, the Dewar flask is filled with dry ice-isopropanol. After the sample is frozen, the rubber stoppers are put into place and the vacuum pump turned on. The system is allowed to freeze dry overnight, and the next day the product is examined, tasted, rehydrated, and tasted again.

DIRECTION

1. Check over the apparatus and compare it with Figure 8-9 to be sure all of the components are present.

2. Remove any isopropanol from the Dewar (J) and a 2 liter beaker (A) and return it to the isopropanol container.

3. Clean out the trap (I).

4. Place 50 mL of water in the trap and then place the rubber stopper (G) in place. Hold the trap vertically and then adjust glass tube (H) to ensure that it will be about 1 cm above the water level.

5. Remove the water from the trap.

6. Obtain 3 to 4 kg of dry ice. A paper sack is a convenient carrying device.

7. Place a cloth towel on the floor and place about 1 kg of dry ice on the towel. Fold the towel over the dry ice and crush the ice to marble size particles or smaller with a hammer.

8. Add this to the 2 liter beaker (A). Repeat steps 7 and 8 until the beaker is about 2/3 full.

9. Place the 30 cm test tube (I) in the Dewar, center it, d old it in place with a 3-pronged clamp.

10. Add crushed dry ice to the Dewar until it is filled to within 2 cm from the top. Put any remaining dry ice in the 2 liter beaker.

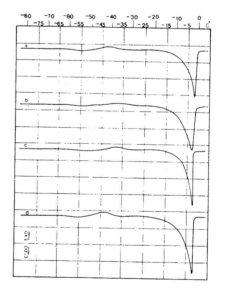

Figure E12-1. DTA (A) grape juice 18.5 °Bx. (B) raspberry juice 24.5° Bx. (C) grapefruit juice 20 °Bx. (D) apple juice 25 °Bx. Heating rate: 1 °C/min; Freezing rate: 6 °C/min. To -120 °C.
(Courtesy - Goldblith, S.A., Rey, L., and Rothmayr, W.W., ed's., *Freeze Drying and Advanced Food Technology* Academic Press, New York, NY, 1975)

11. Add isopropanol to both the 2 liter beaker and the Dewar until the liquid is about 2 cm from the top of each.

12. Connect the vacuum pump to the trap. BE CAREFUL: Check Figure 8-9, p. 88, to be sure the correct side is connected to the trap.

13. Obtain about 100 mL of a beverage of your choice such as coffee, tea, milk, or a fruit juice. Select something that you drink, so you have an idea of what it normally tastes like.

14. Place 40 to 50 mL of your sample into the 500 mL r.b. flask (B) and save 50 mL to compare with tomorrow. Store this reserve in a refrigerator if necessary. Record how many mL of sample you placed in the flask.

15. Place flask B into the dry ice-isopropanol mixture in beaker A. Tilt the flask 90° and slowly rotate it so that a thin film of sample will freeze all over the side walls. After you believe the sample is completely frozen, continue to freeze it for another 4 to 5 minutes. NOTE: A sharp cracking sound is not the flask cracking, but the ice inside.

16. Fasten the flask (B) in place with a 3-pronged clamp so that the bottom 1 cm of the flask is in the coolant.

17. Attach the stopper (C) to the sample.

18. Turn on the vacuum pump (L) and then go over the entire apparatus and tighten all of the hose and stopper connections, including the bleeding valve (F).

19. Every time you check this system, read the pressure and record both the time since the pump was turned on and the pressure. After the system has been under vacuum for about 5 minutes, the dry ice isopropanol mixture can be removed from beneath the r.b. flask, if the sample is still frozen. If the sample melts, then there may be a leak that is causing a poor vacuum. Check each connection and watch the U-tube manometer (E) as you do, because it will indicate if a leak has been closed. If everything is proper, the system should stay frozen without any additional coolant. If it melts, then touch only the bottom of the flask to the coolant.

20. It usually takes about 5 to 6 hours to lyophilize 50 mL of ice. It is suggested that you check the system in about 2 hours and then again before you go home for the day. If you are going to let it go over night, PLEASE add dry ice to the Dewar to ensure that it will last overnight.

21. Some time later: Before you shut off the pump, take one last pressure reading. Shut off the vacuum pump and slowly open the bleeding valve. As the system comes to atmospheric pressure, remove the vacuum tubing from one side of the trap to prevent pump oil from siphoning into the trap.

22. Examine the sample. It doesn't look very appealing does it? Remove a very small amount with a spatula and taste it. Record your impression.

23. Add the same amount of water as your original sample contained. Swirl the flask to rehydrate the sample and then either cool it or heat it as necessary to be as close as possible to the original. Taste it and compare this taste with the reserve sample you saved. Record your impressions and make any suggestions you care to.

24. Do you believe compounds other than water sublimed? In commercial processes, they often add flavor components to the dried products.

25. Wash out the flask and clean out the trap. Set the flask in a cork ring.

26. DO NOT place the solvent in the isopropanol container.

27. *Clean up the area so that it is in better condition than when you started.* Thank you.

EXPERIMENT 13   CONTINUOUS EXTRACTION-SOLVENT HEAVIER THAN WATER

The determination of caffeine in cola drinks using a continuous extractor; solvent heavier type.

BACKGROUND

Caffeine is found in tea, coffee, mate leaves, guarana paste, and cola nuts. See Experiment 11, p. 535, for additional information. The law on cola beverages is that they must not contain more than 0.02% caffeine (200 ppm), making no distinction between that added and that present naturally.

A partial cola nut analysis:

Starch	42.5%
Fibers	20.0
Water	13.65
Sugars and gums	10.67
Ash	3.2
Caffeine	2.21
Fat	1.52

A study published by Gilbert in 1981 estimated that worldwide, the average intake of caffeine by the 4.4 billion inhabitants was about 50 mg/day, of which 90% came from coffee and tea. Table E13-1 shows the average intake by age groups and source.

Table E13-1. Mean daily consumption of caffeine (mg/kg body weight) by source
(Courtesy - P.B. Dews, ed., *Caffeine*, Springer-Verlag, Berlin, 1984)

Age (yrs)	All sources	Coffee	Tea	Soft drinks	Chocolate
Under 1	0.18	0.009	0.13	0.02	0.02
1 - 5	1.20	0.11	0.57	0.34	0.16
6 - 11	0.85	0.10	0.41	0.21	0.13
12 - 17	0.74	0.16	0.34	0.16	0.08
18 and over	2.60	2.1	0.41	0.10	0.03

Table E13-2 lists the caffeine contents for several solid foods and beverages.

Table E13-2. Caffeine content of various beverages and solids (5 oz for liquids, unless otherwise 1 oz for solids indicated)
(Courtesy - P.B. Dews, Ed., *Caffeine*, Springer-Verlag, Berlin, 1984)

Sodas (mg/100 mL)		Coffees (solid)	
Diet Rite	12.2	*C. liberica*, roasted	2.19%
Metro Cola	11.0	*C. liberica*, raw	1.5
Simpson Spring	10.99	Java	1.48
Metro Cola	9.3	*C. arabica*	1.16
Pomeroy Sparkling	9.3	Mocha, raw	1.08
American Dry	8.4	Mocha, roasted	0.82
Country Club	7.9		
White Rock	7.8	Roasted and Ground	64 - 124 mg/5oz
Clicquot Club	6.6	Decaffeinated	2 - 5
Cott Low Cal	4.8	Instant	40 - 108
Bokay Sparkling	2.6	Decaffeinated	2 - 8
Clicquot	1.8	Drip grind	112
Big Y	1.75	Instant drip	29 - 176
Canada Dry Low Cal	1.7		
Red Fox	Neg	Teas (solid)	8 - 91
Mayfair	Neg	Bagged 5 oz	28 - 44
Ukon Club	Neg	Leaf 5 oz	30 - 48
Gala	Neg	Instant 5 oz	24 - 31
Surfine	Neg		
Gala Diet	Neg	Cocoa	2 - 7
Gala Low Cal	Neg	African	6
Cosco Diet	Neg	South American	42
Regular colas 6 oz	15 - 23 mg	Sweet chocolate	5 - 35
Decaf colas 6 oz	trace	Milk chocolate	1 - 15
Diet colas 6 oz	1 - 29	Chocolate bar 30 g	20
		Baking chocolate	18 - 118
		Chocolate milk	1 - 6

YOUR PROBLEM

You are to determine the amount of caffeine in a soft drink that is believed to contain caffeine.

EQUIPMENT

1 Aspirator, water
1 Balance, triple beam

1 Holder, clamp
1 Hot plate

542   Experiment 13

1 Beaker, Berzelius, 300 mL	1 Pipet, delivery, 1.0 mL
2 Beakers, 50 mL	1 Pipet, delivery, 2.0 mL
2 Bottles, wash, 250 mL	1 Pipet, delivery, 5.0 mL
1 Brush, test tube	2 Policeman
1 Bulb, pipet	1 Ring, cork, 250 mL size
3 Clamps, buret	1 Ring, iron, 10 cm dia.
1 Clamp, versatile	2 Rods, stirring, glass
1 Condenser, Allihn, 300 mm 24/40 lower	1 Spectrophotometer, visible range
1 Cylinder, graduated, 100 mL	1 Stand, ring
1 Extractor, solvent heavier	2 Stoppers, rubber, no. 3, 1 hole
1 Flask, filtering, 50 mL	2 Stoppers, rubber, no. 3, 1 hole
1 Flask, r.b., 250 mL, 24/40 single neck	1 Towel, cloth
2 Flasks, volumetric, 25 mL	1 Transformer, variable, (Variac)
2 Funnels, fritted glass, 15M	2 Tubes, test, side arm, 150 mm
1 pr Goggles, safety	8 ft Tubing, condenser
1 Heating mantle, 250 mL	4 ft Tubing, vacuum
	1 ft Wire, copper

CHEMICALS   S = Student

Acetone, CH_3COCH_3, 50 mL/S
Boiling chips, as needed
Caffeine standard, 1.0 mg/mL, 5.0 mL/S
Chloroform, $CHCl_3$, 120 mL/S
Hydrochloric acid, HCl, 1:9, 50 mL/S; 1:1, 2 mL/S
Sample: Bring your own sample of Coke, RC cola, Tab, Diet Pepsi, Pepsi, etc. -but not the Uncola; 100 mL is needed

Joint grease, tube
Phosphomolybdic acid, $20MoO_3 \cdot 2H_3PO_4$, $48H_2O$, 10 g/50 mL, 4 mL/S
Sodium hydroxide, NaOH, 10 N, 2 mL/S

APPARATUS SETUP

The apparatus shown in Figure 10-4, p. 109, is fastened to a ring stand. The filtering system is set up as shown in Figure E13-1. The test tube is placed in a flask filled with about 5 cm of water to act as a stand. The spectrophotometer can be any type that is capable of measuring at 440 nm.

THE BIG PICTURE

100 mL of $CHCl_3$ is placed in the pot. 20 mL of $CHCl_3$ is placed in the extractor chamber, and 100 mL of the soft drink added. The system is extracted for 1.5 to 2 hours. The soft drink is discarded, and the $CHCl_3$ in the pot evaporated. The caffeine is washed out of the pot with water, 1:1 HCl is added, and the caffeine precipitated with phosphomolybdic acid. The precipitate is digested about 20 minutes. The precipitate is filtered and then washed with 1:9 HCl. The precipitate is dissolved off of the fritted glass with acetone, and the solution diluted to 25 mL. The caffeine is determined spectrophotometrically at 440 nm vs. an acetone reference.

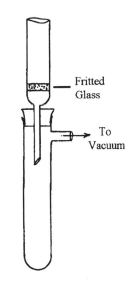

Figure E13-1. A vacuum filtering system

DIRECTIONS

1. Assemble the apparatus as shown in Figure 10-4, greasing all of the joints.
2. Add 100 mL of $CHCl_3$ to the distilling flask. Add 3 to 4 boiling chips.
3. Add 1.0 mL of 10 N NaOH.
4. Add 20 mL of $CHCl_3$ to the sample side of the extractor.
5. Add 100 mL of the cola drink to the sample side. Notice the flow of $CHCl_3$ from the bottom of the extractor into the distilling flask.
6. Turn on the Variac to 40 to 45 volts and then the condenser water.

7. Extract for at least 1.5 hours. Watch what happens to the CHCl₃ during the extraction, so you can under stand this extraction process.
8. Turn on the hot plate to a medium temperature about 10 minutes before you stop the extraction.
9. Shut off the transformer and let the receiver cool a few minutes.
10. Turn off the condenser water and remove the condenser.
11. Loosen the clamp holding the extractor, tilt the extractor, and drain the CHCl₃ from below the sample into the collecting flask, A few mL left is satisfactory.
12. Place the flask back in the heating mantle.
13. Pour the sample residue out of the extractor and place the extractor back in its original position.
14. Connect the condenser and turn on the condenser water. The apparatus should be back in its original form, except that no sample or CHCl₃ are in the receiver.
15. Adjust the transformer to about 50 volts and distill the CHCl₃ until the flask is *just dry*. The CHCl₃ collects where the original sample was placed. NOTE: You may not be able to see any caffeine. It is a thin film on the side of the flask. DON'T WORRY: It is there.
16. Shut off the transformer and lower the heating mantle immediately. Turn off the condenser water.
17. Remove the heating mantle and add 4 to 5 mL of water to the flask (wash bottle). Swirl this liquid around to rinse the sides of the flask and pour the contents into a 50 mL beaker. Use another 1 to 2 mL of water to rinse the flask. Add these washings to the beaker. Mark this beaker, S, for sample. Any brown residue is of no consequence.
18. Pipet 5.0 mL of the caffeine standard solution into another 50 mL beaker and label it C.
19. To each beaker, add 1.0 mL of 1:1 HCl.
20. To each beaker add 2.0 mL of phosphomolybdic acid and stir each solution to mix it thoroughly.
21. Place both beakers on the warm hot plate for about 15 to 20 minutes to allow the precipitate to digest and settle to the bottom of the beaker. While this is taking place, go immediately to step 22.
22. A yellow precipitate is indicative of the presence of caffeine. A clear yellow solution indicates that this sample was caffeine free. While the digestion is taking place, prepare the fritted glass bottom filtering funnels and practice connecting them to both the side arm test tube and the filtering flask (Figure E13-1, p. 542)
23. Sometimes the sample solution above the yellow precipitate is different in color than the pure caffeine standard. This is because of other materials that extracted along with the caffeine. These usually cause no problem. Place about 150 mL of water in a 250 mL beaker. Use this as a holder for the filtering system.
24. Filter (water aspirator assisted) each precipitate through the fritted glass filtering funnels. Use a policeman on a stirring rod and a small stream of 1:9 HCl to aid in the transfer.
25. Remove the filtering funnels and attach each to a side arm test tube.
26. Disconnect the vacuum tubing from the filtering flask first and then remove the filtering funnel. Place the filtering funnel in the side arm test tube.
27. Add 2 to 3 mL of acetone to the fritted glass funnels to dissolve the precipitate. Rinse down the sides of the funnel carefully. Wait about 5 to 10 minutes and rinse again. Do not use more than a total of 15 mL of acetone to dissolve the precipitate.
28. Transfer the dissolved material to a 25 mL volumetric flask. Rinse out the test tube and dilute to the mark with acetone.
29. Have your instructor show you how to use the visible range spectrophotometer, if you have not used it before.
30. Measure the absorbance of each solution at 440 nm against an acetone blank. Record your readings.
31. *Clean up the area around the spectrophotometer and the distillation area. Leave it in a cleaner condition than you found it.* Thank you.
32. Calculate the results as shown in the following example calculation.

EXAMPLE CALCULATION
Data obtained:
    Absorbance of 5.0 mg of caffeine standard   = 0.227
    Absorbance of caffeine from 100 mL of soda   = 0.452

ANSWER
Use Beer's Law, $A = abc$ and the standard solution to determine a.
    $0.227 = a \times 1$ cm cell thickness $\times 5$ mg/25 mL
    $a = 0.0454$
Now determine the unknown.
    $0.452 = 0.0454 \times 1 \times c$
    $c = 10.0$ mg/25 mL   Your entire sample was in this 25 mL; therefore, the concentration is 10 mg/100 mL or about 100 ppm.

EXPERIMENT 14.   CONTINUOUS EXTRACTION-SOLVENT LIGHTER THAN WATER
Separating the anti-amoebic alkaloids from Ipecac syrup

BACKGROUND
"Brazil root" (*Cephaelis ipecacuanha*), now known as *Ipecac,* was used by the natives of Brazil several hundred years ago as a cure for diarrhea. It was sold as a secret remedy to the French in 1658. It was not until 1912 that Vedder demonstrated that its effectiveness was due mainly to one of its components, *emetine,* that killed *Eschericia histolytica,* which we now know as one of the major causes of amoebic dysentery.

For some time, it was believed to be a specific cure for amoebic dysentery, but it is now known not to cure the disease, merely to put it into remission. However, it is one of the most effective drugs for rapidly controlling the symptoms of these intestinal infections. Its side effects are severe, being nausea, vomiting, and even mild diarrhea initially. It can be used two ways: (1) as an anti-amoebic by either deep subcutaneous injection of a maximum of 60 mg/day for 4-6 days, or by enteric coated pills, and (2) as an emetic or expectorant (produce vomiting) by giving it orally in the form of a syrup with doses for adults being 0.5 to 1.0 mL and doses for children being 5 drops plus 1 drop for each year of age.

It is a slow acting emetic, usually taking at least 30 minutes to produce an effect. As an anti-amoebic, it causes degeneration of the nucleus and reticulation of the cytoplasm of amoebae; it is thought to eradicate the parasites by interfering with the production of trophozoites (a protozoan during its vegetative state).

Dried roots contain from 2 to 3% alkaloids, of which emetine (60 to 75%) is the most effective. Cephaeline (methyl emetine) is next in abundance, but is not as effective as emetine. Other components are emetamine, psycotrine and methyl psycotrine. The structures of these are shown below.

YOUR PROBLEM
Several patients have complained to their doctors that the Ipecac prescriptions either do not have any effect or have minimal effect. It is suspected that low-strength drugs are being bottled and labeled to look like quality drugs. You are part of a regional survey to test the actual potency of Ipecac drugs and have been given one drug to test today.

EQUIPMENT

1	Adapter, bushing, 34/45 to 24/40
1	Adapter, bushing, 40/50 to 34/45
1	Balance, analytical
1	Beaker, 250 mL
1	Bottle, indicator
1	Bottle, wash, plastic, 250 mL
1	Bottle, weighing
1	Buret, 50 mL
2	Clamps, versatile
1	Condenser, Liebig, 300 mm 24/40 lower only
1	Cylinder, graduated, 10 mL
1	Cylinder, graduated, 100 mL
1	Extractor, solvent lighter 40/50 and 24/40 joints
1	Flask, r.b. one neck 24/40 500 mL
2	Flasks, Erlenmeyer, 125 mL
1	Flask, volumetric, 100 mL
1	Funnel, filter, 60°, long stem
1 pr	Goggles, safety
1	Holder, buret
2	Holders, clamp
1	Hot plate
1	Paper, filter, medium porosity
1	Pipet, delivery, 10 mL
1	Pipet, Mohr, 5 mL
1	Plate, titrating, 15 cm x 15 cm
1 pr	Pliers
1	Rack, funnel
2	Stands, ring
1	Spatula
1	Towel or sponge
1	Transformer, variable (Variac)
8 ft	Tubing, condenser
1 ft	Wire, copper

CHEMICALS   S = Student

Ammonium hydroxide, NH_4OH, conc, 1 mL/S
Boiling chips
Diethyl ether, $(H_3CCH_2)_2O$, 150 mL/S
Ethanol, absolute, H_3CCH_2OH, 20 mL/S
Glycerine in a dropping bottle for condenser tubing, as needed.
Ipecac - requires a prescription, 10 mL/S
Joint grease, tube
Methyl red, indicator, 0.02 g/100 mL 60% ethanol, a few drops as needed.
Potassium hydrogen phthalate, $C_6H_4COOHCOOK$, 2 g/TA to standardize the base.
Sodium hydroxide, NaOH, 0.02 N, standardized, 100 mL/S
Sulfuric acid, H_2SO_4, 0.10 N, standardized, 20 mL/S; and 1 N

TO BE DONE PRIOR TO CLASS

In order to get 100% recovery, the preliminary sample treatment described in the next paragraph must be done. If it is not done, then maximum recoveries are 90 to 95%.

This can be done by the instructor. Pipet 20 mL of Ipecac into a 100 mL volumetric flask, add ca. 5 mL of 1 N H_2SO_4, and with the aid of an air blast, evaporate on a steam bath on a low-temperature hot plate to about 10 mL. Then, while rotating the flask, add ca. 30 mL of water, cool to room temperature, and dilute to volume. Let stand overnight and filter through a dry filter, rejecting the first few mL of filtrate.

APPARATUS SETUP

The apparatus to be used is as shown in Figure 10-7, p. 111 (or very similar). Be careful when handling it so that the inner funnel does not fall out and get broken. If a burner is used instead of the heating mantle, then this experiment must be done in a hood. A hood is optional if a heating mantle is used, although *it should be done in a hood if possible*.

The only problem is adding too much sample. The weight of the ether will not be sufficient to force it out of the bottom of the funnel, and no extraction will take place.

THE BIG PICTURE

You will extract 10 mL of Ipecac with 125 mL of diethyl ether for about 2 hours. The ether will be evaporated from the sample, and the sample redissolved in ethanol. The alkaloids will be back titrated using both standard sulfuric acid and standard sodium hydroxide. Methyl red is the indicator.

DIRECTIONS

DO THIS IN A HOOD IF POSSIBLE

1. Pipet 10.0 mL of Ipecac syrup into a 125 mL Erlenmeyer flask.
2. Clean the pipet by rinsing it with water.
3. Add 60 mL of distilled water to the flask.
4. Add 8 drops (medicine dropper) of concentrated NH_4OH.
5. Mix the contents thoroughly and transfer them to the lower section of the extraction apparatus. Pour the solution in carefully, so it does not run into the side arm.
6. Tell everyone in the laboratory that you are working with diethyl ether and that there should be no open fires.
7. Add 150 mL of diethyl ether to the 250 mL round bottom flask, then add 3 to 4 boiling chips.
8. Connect the flask to the side arm of the extraction apparatus and place the heating mantle up under it.
9. Adjust the center collector so it is not tilted.
10. The extractor may require bushing adapters to accommodate the condenser. If this is so, then place a *very thin* film of grease on them and put them in place. Twist them slightly to ensure that they are firmly seated.
11. Connect the condenser to the top of the extractor and be sure the entire apparatus is vertical so the condensate will drip into the collection funnel. If grease is used, only a very thin film is necessary.
12. If not already done, connect the water lines to the condenser with the *inlet water at the bottom*. Be sure good tubing is used and that all connections are either wired in place or clips are placed on the joints. REMEMBER: Condenser hoses come off only at night when no one is around.
13. Turn on the condenser water, slowly at first, and check for leaks. Be sure the outlet hose is firmly placed in the drain.
14. Adjust the variable transformer (Variac or Powerstat) to 40 volts.
15. Watch the extractor to ensure that it is working properly. The ether should start to boil in a few minutes. Once it begins to condense in the condenser, be sure it drips into the collecting funnel. Adjust the tilt of the apparatus to make it work properly.
16. Adjust the variable transformer to provide a smoothly boiling solution.
17. Once you get a column of ether about 3 cm above the level of the sample, the ether will begin to bubble up through the sample. The first bubbles will be visible. However, the refractive index difference between these solutions is small, and you have to look closely to see the micro droplets of ether.
18. Extract for 2 hours after the ether begins to boil. About 15 minutes before the end of the extraction, turn on a hot plate (in a hood) to medium.

19. Shut off the variable transformer and lower the heating mantle 2 to 3 cm so that the heat is removed from the flask, but if the flask should come loose, it won't fall and break.
20. When the ether stops dripping in the funnel, raise the condenser about 10 cm above the extractor.
21. Loosen the clamps holding the extractor and the round bottom flask.
22. Remove the flask and the extractor and, holding one in each hand, slowly rotate the extractor and decant as much of the ether into the flask as possible.
23. Place the flask in a cork ring to hold it and place the extractor back in the ring stand.
24. Transfer the solution from the 250 r.b. flask into a 250 mL beaker, set it on a hot plate (in a hood) and evaporate the ether. Continue the evaporation until the sample is just moist. Do not allow the sample to char. Be sure the hood is turned on.
25. It will take 10 to 15 minutes for the ether to evaporate. Get ready to do the titrations.
26. Fill a clean 50 mL buret with 0.2 N NaOH and zero it.
27. While you are waiting for the ether to finish evaporating, determine a reagent blank. Pipet 10.0 mL of absolute ethanol and 25.0 mL of distilled water into a 125 mL Erlenmeyer flask.
28. Add 4 drops of methyl red indicator, and the solution should turn red.
29. Place a white titrating plate or a cloth towel under the flask and titrate this solution to the first permanent yellow color. This will serve as the blank. It will take only a few mL, so be careful.
30. Record the blank value, clean out the flask, and refill the buret.
30. The ether should be evaporated by now. To ensure that the ammonia is gone, add enough ethanol to cover the bottom of the beaker 3 to 4 mm and swirl to redissolve the residue. Any stopcock grease will not dissolve and will remain as a gummy mass. This is of no consequence.
32. Place the beaker on the hot plate and evaporate the solution to near dryness again.
33. Add another 10 mL of ethanol and then, with a aid of a wash bottle filled with water, transfer this solution to a 125 mL Erlenmeyer flask.
34. Pipet into the flask 10.0 mL of standardized 0.1N acid and add 4 drops of methyl red indicator. The solution should be red. If it is not, then add another 10.0 mL of acid.
35. Add water (wash bottle) until the volume is about the same as that of the blank. Swirl to ensure the system is mixed thoroughly.
36. Very carefully titrate this solution until it turns a permanent yellow. Record the volume.
37. Shut off the hot plate and turn off the condenser water.
38. Clean up the apparatus, being very careful not to break the center section.
39. Drain the base out of the buret, rinse the buret thoroughly, and loosen the stopcock so it will not freeze.
40. *Clean up the entire area and put it back in better order than you found it.* Thank you.

EXAMPLE CALCULATION
Given the following data, calculate the amount of Ipecac in the sample.
Acid = 0.1510 N and 5.00 mL were pipetted into the flask. Each mL = 24.4 mg of alkaloid calculated as emetine. Base = 0.0224 N. 1.0 mL of acid = 6.74 mL of base. Blank = 2.00 mL. Sample = 33.00 mL

ANSWER

$$\frac{33.00 \text{ mL} - 2.00 \text{ mL}}{6.74 \text{ mL}} = 4.50 \text{ mL of acid for the back titration.}$$

5.00 - 4.50 = 0.50 mL of acid used to neutralize the Ipecac alkaloids.
0.50 mL x 24.4 mg alkaloids/mL = 12.2 mg

> In this case, the bottle's label stated that 100 mL = 7 g of Ipecac, each g containing 2.0% active ingredient. This equals 140 mg of active ingredients/100 mL. We used a 10 mL sample or 14.0 mg of alkaloids was expected.
>
> 12.2/14.0 x 100 = 87% recovery. Expected, because no preliminary clean-up treatment was done.

EXPERIMENT 15. CONTINUOUS EXTRACTION - THE SOXHLET EXTRACTOR
Extracting oil from nutmeats
(*Association of Official Analytical Chemists*, International, Method 25.004)

BACKGROUND

Peanuts were native to South America, probably from southern Bolivia to northern Argentina. Excavations of Peruvian sites indicate they were used as early as 3000 B.C. They arrived in the United States in the 17th century. Pound for pound, peanuts have more protein, minerals, and vitamins than beef liver; more fat than heavy cream; and more calories than sugar. 1 kg of nut kernels provides the same energy as 5.1 kg of bread, 8.2 kg of steak, 27.1 kg of white potatoes, or 33.1 kg of oranges!

The world production of peanuts is about 17 million metric tons, of which about 10% is produced in the United States, 41% Georgia and 15% in Alabama. The average American consumes 4 kg/yr, 50% as peanut butter, 21% as salted nuts, and 16.5% in nut candy.

Peanut seeds contain about 50% oil and 25% protein and provide about 2600 cal/lb. The compounds found in peanuts are used in paints, varnishes, lubricating oils, leather dressings, furniture polish, insecticides, and nitroglycerine. Soaps are made from the saponified oil as well as several cosmetic bases. The protein fraction is used in the textile fibers Ardil and Sarelon. The shells are used in plastics, wallboard, abrasives, and as a fuel. The chemicals furfural, xylose, cellulose, and mucilage are obtained from peanuts. The tops are used for hay, and the press cake is used for animal feed and fertilizer.

Peanut oil is a nondrying edible oil. It accounts for about 1/6 of the world's vegetable oil. It can be obtained commercially by either of two methods. The first is by mechanically pressing the nut seeds to remove 1/2 to 2/3 of the oil, forming *press cake*. The residual oil is removed by an extraction with petroleum ether or hexane. The second method is an extraction with hexane at room temperature for 120 hr. This removes a bit over 80% of the oil. The residual hexane in the nut is removed by drying for 9 to 10 hrs.

YOUR PROBLEM

In order to control the homogeneity of the product and to ensure that the correct fat content is present, it is necessary to know the fat and oil contents of the nuts that are in the starting material. Some typical values for shelled nuts are shown in Table E15-1.

Table E15-1. Oil content of several species of nuts

Almonds	42.4	Filberts	62.4
Beechnuts	50.0	Hickory nuts	68.7
Black walnuts	59.3	Peanuts	48
Brazil nuts	66.9	Pecans	71.2
Cashews	45.7	Pistachios	53.7
English walnuts	64.0	Sunflower seeds	47.3

EQUIPMENT

- 1 Balance, analytical
- 1 Beaker, Phillips is preferred, 250 mL
- 1 Beaker, 250 mL
- 1 Beaker, 600 mL
- 2 Clamps, 3 prong
- 1 Cylinder, graduated, 100 mL
- 1 Cylinder, graduated, 100 mL
- 1 Dish, weighing, Al
- 1 Extractor, Soxhlet
    - Condenser
    - Flask, r.b., 500 mL
    - Thimble 1 pr
- Goggles, safety
- 2 Holders, clamp or heating mantle holder
- 1 Hot plate
- 1 Knife or razor blade
- 1 pr Pliers
- 1 Ring, iron 10 cm dia
- 1 Stand, ring
- 1 Towel, cloth
- 1 Transformer, variable (Variac)
- 8 ft Tubing, condenser
- 1 ft Wire, copper

CHEMICALS   S = Student

Boiling chips, 4-5 /S
Glass wool, 10-15 g

Diethyl ether, (H$_3$CCH$_2$)$_2$O, 300 mL/S
Sample - 5 g of any kind of shelled nuts

APPARATUS SETUP

The apparatus is set up in a hood as shown in Figure 10-10, p. 112. Either diethyl ether, petroleum ether, or hexane can be used as the extraction solvent. Diethyl ether is preferred because it can be removed quicker at the end of the extraction, thus reducing the time for this experiment.

THE BIG PICTURE

4 to 5 g of shelled nuts that have been cut into small pieces are placed in the extraction thimble and covered with a small amount of glass wool. No more than 300 mL of diethyl ether is placed in the collecting flask. The apparatus is assembled, the condenser water turned on and the extraction allowed to proceed for at least 4 hours, preferably overnight. The extract is transferred to a weighed Phillip's beaker (Figure E15-1) and the ether evaporated. The residue is weighed and brought to a constant weight. The percent oil, which is called *percent crude fat,* is determined.

DIRECTIONS
DO IN A HOOD - NO SMOKING - NO FIRES

1. Weigh an aluminum weighing dish or suitable substitute to the nearest ± 0.0001 g and record this weight.

2. Place 4 to 5 g of any kind of shelled nuts in the dish, weigh them, and record the weight.

3. If you have large nuts, use either a knife or a single edged razor blade and cut the nuts into smaller pieces.

4. Remove the extraction thimble from the apparatus and add the nut pieces to it. Cover the sample with a small plug of glass wool. The reason for the glass wool is to spread the solvent when it drips onto the sample, providing for a more uniform extraction.

5. Place the extraction thimble in the extraction chamber of the apparatus.

6. Add 300 mL of diethyl ether to the 500 mL r.b. collecting flask, but NO boiling chips. Attach this to the apparatus.

7. Place the heating mantle around the flask and adjust it to fit properly.

8. Place a sign by the apparatus that lets others know that you have an ether distillation proceeding.

9. Check the condenser tubing to be sure it is in good condition and that the tubing is either wired or clipped in place. If you are satisfied, attach the condenser to the apparatus and turn on the condenser water.

10. Turn on the variable transformer to about 45 volts. After the ether begins to boil, adjust it so the system cycles every 6 to 7 minutes.

11. The official method requires a 16-hour extraction. About 95% of the oil is removed in 4 hours. You be the judge as to how long you want to continue the extraction. You can let it go overnight if you want to, but PLEASE tell the instructor and be sure the hose connections are secure.

12. Just before you stop the extraction, place a hot plate in the hood and turn it on to a low setting.

13. Turn off the Variac and lower the heating mantle 2 to 3 cm from the flask so the liquid will cool rapidly.

14. Place 2 to 3 boiling chips in a Phillips beaker and weigh this combination to the nearest ± 0.0001 g. A Phillips beaker also is called a "fat flask". It is a tall flask with a narrow mouth. It is used because it has a high neck, which acts as an air condenser to slow the rate of evaporation, and a narrow top, which also reduces the rate of evaporation. A slower evaporation rate also reduces spattering and the loss of sample. If the ether evaporates faster than the hood can remove it, the excess fumes, being more dense than air, surround the hot plate and may ignite.

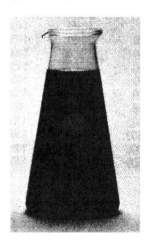

Figure E15-1. A Phillips beaker, "fat flask".
(Courtesy - Van Waters & Rogers Scientific Co., Greenbelt, MD)

15. Transfer a portion of the ether extract to the beaker to fill it about 1/2 full.

16. Place the beaker on the hot plate and evaporate the ether. Add more sample as the occasion demands. Evaporate the ether until you believe only the oils and fats are left. Do not scorch the sample. The oil will begin to turn brown as it oxidizes and it also will increase in weight.

17. Remove the beaker from the hot plate and weigh it and the contents as soon as it is cool. Record this weight.

18. Heat the beaker again for 5 minutes, cool, and reweigh.

19. Compare the readings. If they agree within 10 to 15 mg and the ether appears to be, use the last weighing for your calculation. If the difference is more than a 15 mg, heat again for 5 minutes.

20. Following the example shown below, calculate the "percent crude fat" in the sample and compare your results with those given in Table E15-1, p. 548.

21. *Clean up the area. Shut off the condenser water and the hot plate*. Thank you.

EXAMPLE CALCULATION
A sample of pecan nuts is suspected of having been partially stripped of oil. An analyst obtained the following data. Were the nuts partially stripped?
Sample weight 4.1872 g

First weighing		Second weighing	
Beaker and residue	92.2265 g	Beaker and residue	92.2249 g
Beaker empty	90.0876	Beaker empty	90.0876
Residue	2.1389 g	Residue	2.1373 g

ANSWER

$$\% \text{ Crude fat} = \frac{\text{Residue} \times 100}{\text{Sample wht.}} = \frac{2.1373 \times 100}{4.1872} = 51.0\%$$

Pecans should contain over 70 % oil. These do not, so a partial stripping has occurred.

EXPERIMENT 16. COUNTERCURRENT EXTRACTION

Separation of the phenolic flavor components, carvacrol and thymol, from oregano leaves.
(Courtesy - Dr. Walter Conway, Manish Sharma, and Lijing Chen, Dept. of Pharmaceutics, State University of New York, Buffalo)

(NOTE: The extraction and filtering takes about 45 minutes, followed by an overnight room-temperature evaporation of the extracting solvent. The experiment then is continued the following day).

BACKGROUND

Oregano is also called joy of the mountain and wild marjoram. It is a perennial plant of the mint family and grows 30 to 60 cm tall. It is native to the sunny Mediterranean slopes of Spain, Italy, and Greece. It was brought to North America by European colonists. The Mexican variety (*Origanum vulgare* L.) is stronger flavored. In the Middle Ages, it was used to cure spider and scorpion bites, improve eyesight, and cleanse the brain. It then was used to help preserve food because it has considerable antifungal activity. Now it adds a special flavor to tomato-based dishes, but is also used to flavor meat, sauces, and liquors and to scent soap and oral preparations. The United States imports about 2500 tons annually.

Spanish oregano oil contains up to 50% thymol and 7 to 8% α-pinene, cineole, linalyl acetate, linalool, dipentene, *p*-cymene, and β-caryophyllene. Other varieties contain carvacrol as the major component. Most contain carvacrol and thymol in varying amounts. Dried oregano will contain 0.2 to 0.4% of a yellowish volatile oil with thymol and carvacrol in varying amounts. The phenolic content may be 50 to 65%. *Carvacrol* is found in the oil of oregano, thyme, marjoram, and summer savory. It is practically insoluble in water. It is used as a disinfectant and in organic synthesis. Therapeutically, it is used as an anti-infective and anthelmintic.

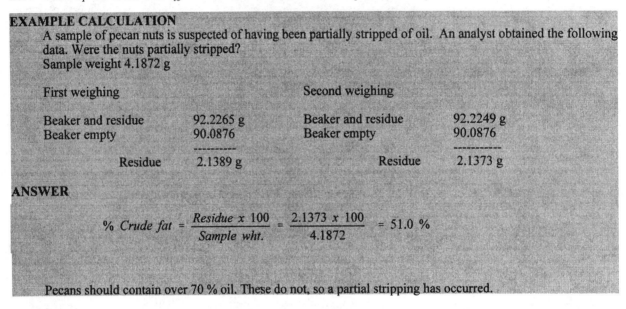
Carvacrol   Thymol

Thymol was isolated in 1719. In addition to a flavoring agent, it is used to prevent mold and preserve documents, art objects, and urine. It is also a stabilizer for trichloroethylene and halothane.

YOUR PROBLEM

A problem that occurs now and then is that a producer of herbs partially extracts them, separates the components, sells them separately, and then sells the partially extracted leaves as originals. This is a fraud, unless the label indicates a partial extraction. An activist group suspects that such an operation is going on involving the removal of carvacrol and thymol from oregano leaves and demands an immediate analysis. You are to extract some leaves they suspect and compare the amount recovered to a known quality product.

EQUIPMENT

1	Aspirator, water	1	Pencil, grease or magic marker
1	Balance, ± 0.01 g	1	Pipet, Mohr, 5 mL
2	Beakers, 250 mL	1	Pump, Milton Roy, mini-pump
1	Bottle, wash, 250 mL	1	Rack, test tube for small tubes
1	Brush, test tube, small	1	Ring, iron, 10 cm
1	Bulb, pipet	1	Spectrophotometer, uv/vis
2	Cells, uv, 1 cm	1	Sponge
1	Centrifuge, coil planet	1	Stand, ring
1	Coil, 12.2 mL, 1.07 mm i.d. PTFE tubing	2	Stopper, rubber, 1 hole, no. 6.5 & 7
1	Condenser, West. 30 cm	1	Syringe, 20 mL
1	Cylinder, graduated, 10 mL	2 ft	Tape, label
1	Cylinder, graduated, 100 mL	20	Tubes, test, 10 x 75 mm
1	Detector, uv, like Isco V-4	6 ft	Tubing, condenser
1	Filter, 0.45 μm, syringe type.	4 ft	Tubing, vacuum
1	Flask, Erlenmeyer, 250 mL	1	Valve, injection, 4-way, 0.20 and 0.50 mL injection loop
1	Flask, filtering, 500 mL	1	Valve, back pressure, Upchurch
1	Funnel, separatory, 1 L	1	Valve, back pressure, relief
1	Funnel, sintered glass, 15 M	1	Valve, flow switching
1 pr	Goggles, safety		
1	Hotplate		

CHEMICALS   S = Student

Carvacrol, 5-isopropyl-2-methylphenol, 4 mg/mL of mobile phase, 2 mL/S
Thymol, 2-isopropyl-5-methylphenol, 4 mg/mL of mobile phase, 2 mL/S
Sample: Oregano leaves, 10 g/S
t-Butyl methyl ether, $(H_3C)_3C\text{-}O\text{-}CH_3$, 100 mL/S
Separation phases: (1:1 heptane : 55% methanol-water); *CAREFUL: this warms considerably, vent often*).
   Mix in a separatory funnel held under a stream of cold water, use the top layer for the mobile phase and the bottom layer for the stationary phase
   Heptane, $H_3C(CH_2)_5CH_3$, 300 mL
   Methanol, $H_3C\text{-}OH$, 165 mL
   Water, 135 mL

THE BIG PICTURE

Refer to Figure E16-1, p. 552. Ten grams of oregano leaves is reflux extracted with *t*-butyl methyl ether for 20 minutes, and the solvent evaporated at room temperature overnight. The CCC column is filled with stationary phase, and then mobile phase is passed into it. Standards of both thymol and carvacrol are run, and 20 1-minute fractions are collected. The content of each fraction is determined at 276 nm, and their partition ratios are determined. The sample residue is taken up in 5 mL of mobile phase and 0.50 mL is added. The peak area is compared to that shown in Figure E16-2, p. 552, which was obtained under the same conditions.

APPARATUS SETUP

For the best results, the apparatus is set up as shown in Figure E16-1, p. 552. In the directions that follow, the in-

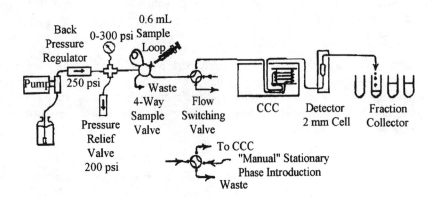

Figure E16-1. Flow diagram for a CCC system

line detector is not used, and any uv spectrophotometer can be used to measure the contents of each collected fraction. A Milton Roy minipump is used. The analytical CCC coil is fitted with a Valco HPLC injection valve and a 0.65 mL PTFE sample loop. An Upchurch (model no. P788) back pressure valve, set at 250 p.s.i., is connected in the pump effluent line prior to the injection valve to provide sufficient back pressure for reliable pump operation. A second back-pressure valve, set lower at 200 p.s.i., is connected to the flow line via a cross just prior to the injection valve to protect the PTFE column in the event of an inadvertent outlet blockage. A 0 to 300 p.s.i. glycerol-filled pressure gauge (Alltech no. 9226) is connected to the same tee. The flow switching valve is convenient for loading the small column with stationary phase manually by means of a 20 mL syringe. The mobile phase stream is bypassed to waste (or shut down) during this operation.

DIRECTIONS

1. In a hood, turn on a hot plate to medium.
2. Place 10 g of oregano leaves in a 250 mL Erlenmeyer flask. Record the weight.
3. Add 75 mL of *t*-butyl methyl ether to the flask.
4. Attach a condenser to the flask, connect the water lines (in at the bottom), set the flask on the hot plate, and reflux for 20 minutes.
5. Cool to room temperature and using a water aspirator, filter the contents through a medium-porosity sintered glass filter.
6. Rinse with 10 mL of *t*-butyl methyl ether. (Clean out the filter.)
7. Transfer the green solution to a weighed 250 mL beaker with a few mL of *t*-butyl methyl ether. Set this in a hood and allow it evaporate overnight. If a rotovap is used to hasten the evaporation, then do not let the residue go to dryness or some of the desired phenolic components may be lost.
8. After the evaporation, weigh the beaker and residue and determine the approximate amount of residue obtained (400 to 500 mg is normal).
9. Check the apparatus and compare it to Figure E16-2. The directions that follow involve a 12.2 mL column and collecting fractions manually. If you have a fraction collector then please use it. Otherwise, place a multiple unit test tube rack filled with 20 10 x 75 mm test tubes at the outlet.
10. Place a small piece of label tape at the top of each test tube and number them 1 to 20. Fill one test tube with 1.0 mL of mobile phase and mark its level with a grease pencil or magic marker. Then fill it to 5.0 mL and mark it. Mark each test tube based on this.

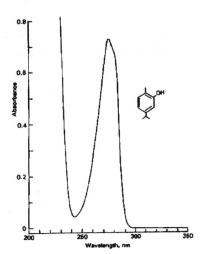

Figure E16-2. uv absorption spectrum of carvacrol, 50 µg/mL, 1 cm cell.

11. Fill the mobile phase reservoir bottle by the pump and insert the filter end inlet line into it.
12. Adjust the 4-way sample valve so that it is closed to mobile phase flow.
13. Fill a 20 mL syringe with stationary phase, attach it to the flow switching valve, and inject it into the head of the 12.2 mL coil. Collect the effluent in a small beaker and return it to its storage container.
14. Start the CCC and adjust its speed to 1,200 rpm.
15. Start the pump and adjust it to 1.0 mL/minute.
16. Open the switching valve and the 4-way sample valve so the mobile phase can flow into the CCC.
17. Put a 10 mL graduated cylinder in a 250 mL beaker and place this at the tail of the CCC 12.2 mL coil.
18. Collect the expelled stationary phase until the mobile phase front just emerges and measure the volume collected. Record this value and label it V_{co}. This is also equal to the mobile phase volume, V_m.
19. Fill a 0.2 mL sample loop with 4 mg/mL thymol and inject it into the column.
20. Collect 20 1 mL fractions, being sure to keep them in order.
21. Add 4.0 mL of mobile phase to each test tube and swirl to mix.
22. Use mobile phase in the reference cell, 276 nm as the wavelength, and determine the absorbance value for thymol in each test tube.
23. At home tonight, or anywhere else, plot A vs. volume and determine the retention volume for thymol.
24. Pour out the cell and test tube contents into a waste flask. Use the 500 mL filtering flask. Clean out the test tubes and place them in order back in the rack.
25. Repeat steps 21 to 26 with carvacrol.
26. Slurry the residue with 5 mL of the mobile phase. The residue is not completely soluble, so use the supernatant liquid for the sample.
27. Remove the 0.20 mL sample loop and replace it with a 0.50 mL loop.
28. Fill the 0.5 mL loop with the supernatant of the oregano extract that is equivalent to 1 g of sample.
29. Repeat steps 21 to 26 with the sample. Why do you only get one peak, although you know fairly well that both thymol and carvacrol are present? Refer to Figure E16-3.
30. Clean up time. Shut off the pump. Turn the flow switching valve to waste (and collect it). Shut off the CCC.
31. Let the TA do this step. Flush the coil with methanol or 2-propanol (which is miscible with heptane) and then blow out the contents with air or nitrogen.
32. Clean out the test tubes and invert them in the rack to drain and dry. Clean out the beakers and the glass frit (good luck here). *Leave this area cleaner and better organized than when you arrived.* Thank you.
33. The material in the waste flask is small and can be disposed of down the drain. Flush it with sufficient water so that heptane fumes do not collect in the bottom of the sink.
34. Compare either the peak area or the peak height that you obtained to that "officially" obtained in Figure 16-3 and make your conclusions. Be careful about this. Remember, you diluted your fractions fivefold so a 1 cm cell would equal a 2 mm cell.

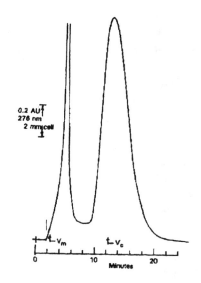

Figure E16-3. CC chromatogram of oregano extract on an analytical column. 1.07 mm i.d. 12.2 mL. Residue equivalent to 1 g oregano in 0.2 mL mobile phase; 1:1 heptane/55% methanol-water; 1200 rpm.

EXPERIMENT 17. SOLID PHASE EXTRACTION
Atrazine, Simazine, and Propazine in ponds, lakes, and rivers (Supelco Applications)

NOTE: This experiment requires an HPLC for final separation and detection. This sample can be saved for the HPLC experiment later on in place of the other experiments..

BACKGROUND

A herbicide is a chemical that kills plants. The desire is to find one that will kill weeds and leaves other plants alone and, if not used, will decompose so no carryover into the next season will occur. In 1896, Bonnet, a French grape grower noticed that the leaves of a common weed died when treated with Bordeaux mixture (copper sulfate + hydrated lime) when he applied the chemical to his grape vines to prevent downy mildew. Ammonium sulfate, ferrous sulfate and lead arsenate also were discovered to be weed killers. The first organic herbicide, 2-methyl-4,6-dinitrophenol, was recognized in 1932. Since that time, thousands of compounds have been made and tested to selectively control weeds. Weeds rob the crop plants of moisture, sunlight, and nutrients and make harvesting difficult, and in some cases impossible. If harvest is possible then the crop must be cleaned of the weed seeds and chaf, a time-and-energy consuming process. The cost of cultivating large fields is expensive, so the application of a herbicide at the time of planting is quite advantageous.

Some herbicides are sprayed on the leaves and others are applied to the soil and are taken up by the roots (systemics). Some herbicides are growth promoters and kill the plant by causing it to grow too fast. Others function by closing the stomata and "suffocating" the plant. The triazine herbicides in this experiment appear to stop photosynthesis by reacting with the enzymes catalyase and peroxidase. The herbicides have very low human toxicity.

Atrazine, a selective herbicide with an LD_{50} of 1900 mg/kg, was developed in 1959 by CIBA-Geigy and is used preemergence and early postemergence, applied at 1/4 to 4 lb/acre. Resistant crops, such as corn, contain the compound 2,4-dihydroxy-3-keto-7-methoxy-1,4-benzooxazine, which occurs naturally in their roots and removes the Cl, thus detoxifying it. It kills barnyard grass, mustards, chickweed, cocklebur, crabgrass, downy brome, foxtail, jimsonweed, lambquarters, nutgrass, quackgrass, purslane, ragweed, velvetleaf, wild oats, and many other weeds. It can be used for corn, guava, grasses grown for seed, macadamia orchards, rangeland, pastures, pineapples, sugarcane, sorghum, turf grass, sod, and Christmas trees.

Simazine, a preemergence herbicide with an LD_{50} of 5000 mg/kg, was developed in 1956 by CIBA-Geigy and applied at 1-4 lbs/acre. It is not as strong as Atrazine, but persists longer. It can be used for corn, orchards, cranberries, pineapples, sugarcane, Christmas trees, asparagus, alfalfa, and particularly for aquatics to control algae in ponds and swimming pools. Do not apply around tomatoes, tobacco, oats, spinach, curcubits, onions, cloves, carrots, rice, beets, soybeans, crucifers, and lettuce.

Propazine, a selective preemergent giving residual control, was developed in 1959 by CIBA-Geigy. Its LD_{50} is 7,700 mg/kg and is applied at 1-3.2 lbs/acre.. It can be used only for sorghum. Do not plant any other crop for at least 18 months except cotton, soybeans, or corn and then only after 12 months. It kills morning glory, carpetweed, lambsquarters, pigweed, ragweed, foxtail, smartweed, velvetleaf, and many other weeds.

YOUR PROBLEM

A farmer has a stock watering pond at the collection point of several fields on which herbicides have been used regularly. Under recent proposals by the EPA's Clean Water Act, they want to regulate private ponds. The farmer has no idea how much herbicide, if any, is in the water and has taken a sample for you to examine.

APPARATUS SETUP

The apparatus is set up as shown in Figure 12-8, p. 134.

THE BIG PICTURE

Two samples are run; one is 20 mL of pond water, and the other is 20 mL of spiked standard. 1 mL of methanol is added to each sample before passing it through the SPE cartridge. The cartridge is eluted with methanol, diluted to 2.0 mL, and injected into an HPLC either now or later on.

EQUIPMENT

- 1 Adapter, cartridge
- 1 Balance, micro preferred
- 2 Beakers, 100 mL
- 1 Bulb, pipet
- 1 Cartridge, SPE, 3 mL, LC-18
- 1 Chromatograph, HPLC with uv detector.
- 1 Column, HPLC, Supelcosil LC-8-DB 15 cm x 4.6 mm (5 μm packing)
- 1 Column, guard, 2 cm *x* 4.6 mm (5 μm packing)
- 1 Cylinder, graduated, 10 ml,
- 1 Cylinder, graduated, 100 mL
- 1 Flask, volumetric, 2 mL
- 1 Flask, volumetric, 1 L.
- 3 Flasks, volumetric, 100 mL
- 1 pr Goggles, safety
- 1 Holder, buret
- 1 Pipet, Mohr, 10 mL
- 1 Pipet, 20 mL
- 1 Stand, ring
- 1 Syringe, for 25 μL
- 1 Syringe, 50 mL

CHEMICALS   S = Student

Acetonitrile, H_3CCN, -15 mL/sample

Eluting solution, water: acetonitrile: methanol; 4:3:1, 1 mL/sample

Methanol, H_3C-OH, 4 mL/sample

Sample, pond or lake water, 20 mL/sample

Standards: Obtain 100 mg vials from Supelco Inc., Supelco Park, Bellefonte, PA, 16823-0048. Prepare 10 ppb, 100 ppb, and 1 ppm working standards from a 10 ppm stock standard. Prepare a 10 ppm standard by dissolving 10 mg of each herbicide in 1 L of water. A few mL of methanol will be necessary to get the Atrazine in solution. A microbalance is preferred. Dilute 10 mL of the stock to 100 mL to prepare the 1 ppm standard. Dilute 10 mL of the 1 ppm standard to 100 mL to prepare the 100 ppb standard and dilute 10 mL of this to 100 mL to prepare the 10 ppb standard.

DIRECTIONS

1. Either collect a water sample from a pond or small lake or use one the instructor has obtained. Avoid river water because it is too fast moving and dilutes the residues.

2. Place 50 mL in a 100 mL beaker, add 2.5 mL of methanol, and swirl to mix. The methanol reduces the interaction between the herbicides and the tube packing, promoting easy removal of the herbicides in 1 mL of methanol later.

3. Arrange the apparatus as shown in Figure 12-8, p. 134, except place a 100 mL beaker under the cartridge.

4. Condition the cartridge by pipetting 2 mL of methanol onto the cartridge, allow it to drain, then pipet 2 mL of distilled water onto the column and let it drain.

5. Pipet 20 mL of the pond water sample containing the methanol onto the cartridge and collect the drainings in the 100 mL beaker.

6. Discard this waste into the drain

7. Place a 2 mL volumetric under the cartridge.

8. Elute the herbicides with 1 mL of the eluting solution.

9. Dilute to volume with distilled water.

10. This sample requires an HPLC to detect the residues. HPLC has not been covered yet, so the sample can be saved for later separation, or if the analyst has previous experience with HPLC, it can be done now. The conditions are:

    Column: Supelcosil LC-8-DB, 15 cm x 4.6 mm, 5 μm particles.
    Guard column, Same, 2 cm x 4.6 mm, 5 μm particles.
    Mobile phase: acetonitrile: water; 45:55.
    Flow rate: 1.5 mL/min.
    Column Temp.:room.
    Injection volume: 25 μL
    Detection: uv (254 nm, 0.005, 0.010, 0.040 absorbance units full scale (AUFS) for 10 ppb, 100 ppb, and 1 ppm)

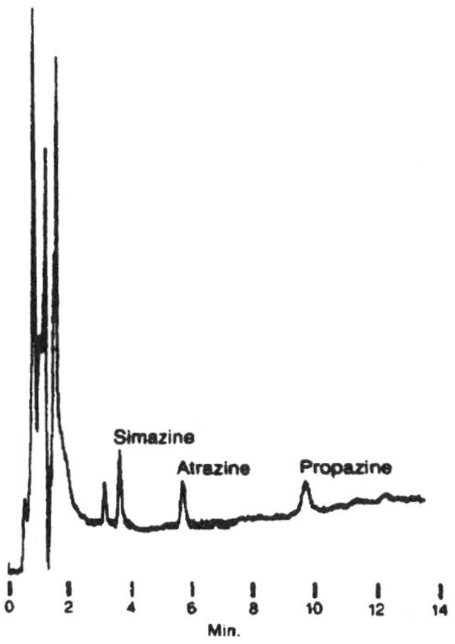

Figure E17-1. HPLC of Triazine herbicides at 10 ppb.
(Courtesy - Supelco Inc., Bellefonte, PA)

11. Figure E17-1, p. 555, shows a typical chromatogram at 10 ppb for each herbicide.
12. NOW - do it all again with the spiked sample.

EXPERIMENT 18 SUPERCRITICAL FLUID EXTRACTION
The separation of oil from pecans
(Courtesy - David Heikes and Marvin Hopper, FDA Total Diet Research Center, Lenexa, KS.)

BACKGROUND

Pecans (*Carya illioensis*), either Pih-KAHN or PEE-can, is the most valuable nut crop in the United States ranging from $60.5 million in 1970 to $309 million in 1991. The largest crop was 340,000,000 lbs in shell in 1981. An average year is 250,000,000 in shell lbs, of which 4/5th are marketed as shelled nuts. Texas is the largest producer of native pecans, but Georgia is the overall leader when improved varieties are added.

Pecans have a distinct flavor and texture, the highest fat content of any vegetable product, and a caloric content/g close to that of butter. The trees require 205 to 233 frost-free days for nuts to reach maturity and, therefore, grow mainly in the south. New trees start producing after 5 years and slowly increase production with age. They take about 20 years to be economically worthwhile, so short-lived fruit trees or truck crops are planted between the trees in the meantime. The nuts usually are harvested when they fall to the ground and vary in size from 30 to 90 nuts/lb. The very best trees produce about 500 lb/year. Native pecans average 46.6 cents/in shell lb, and the improved varieties average 72.4 cents/in shell lb. Pecan trees grow upright in closely spaced orchards but spread out as an ornamental, growing to 160 to 180 ft with a 4 to 6 ft trunk. Table E18-1 lists the composition of several nut varieties.

Table E18-1. Composition of several nut varieties
(Courtesy - Watt, B.K. and Merrill, A.L.,*Composition of Foods, Handbook No. 8*, USDA)

Nut	Water	Fat	Protein	Carbohydrate	Cal/100 g
Almonds	4.7	54.2	18.6	19.5	598
Brazil nuts	4.6	66.9	14.3	10.9	654
Butter nuts	3.8	61.2	23.7	8.4	629
Cashews	5.2	45.7	17.2	29.3	561
Hazel nuts	5.8	62.4	12.6	16.7	634
Hickory nuts	3.3	68.7	13.2	12.8	673
Macadamia nuts	3.0	71.6	7.8	15.9	691
Peanuts, raw	5.4	48.4	26.3	17.6	568
Pecans	3.4	71.2	9.2	14.6	687
Pili nuts	6.3	71.1	11.4	8.4	669
Pignolias	5.6	47.4	31.1	11.6	552
Pinon	3.1	60.5	13.0	20.5	635
Pistachio nuts	5.3	53.7	19.3	19.0	594
Walnuts, black	3.1	59.3	20.5	14.8	628
Walnuts, English	3.5	64.0	14.8	15.8	651

YOUR PROBLEM

Your administration has just purchased the SFE apparatus you have been wanting, so you are going to extract the oil from a sample to prove to them that this was a wise purchase. Many varieties of pecans are now available with new varieties being tested each year to improve the yield, get a larger nut meat, a thinner shell, and be insect resistant. A new variety has just produced its first fruit, and you are to determine the oil content.

EQUIPMENT

- 1 Balance, analytical
- 1 Beaker, 400 mL
- 1 Bulb, pipet, 6 mL
- 1 Cartridge, 10 mL
- 1 Filter, metal (HPLC type)
- 1 pr Forceps or tweezers
- 1 Funnel, powder, 60 mm
- 1 pr Goggles, safety
- 1 Hair dryer
- 1 Heater, restrictor
- 1 bx Kimwipes
- 1 Mortar and pestle, 70 mm
- 1 Pipet, 1 mL
- 2 Pumps, syringe, ISCO Model 260D or 100DX
- 1 Spatula
- 1 Supercritical Fluid Extractor (like ISCO SFX-2-10
- 1 Timer

Appendix A   557

CHEMICALS   S = Student
Celite 545, 5 g/S
Chloroform, $CHCl_3$ 1 mL/S
CO_2, liquid, cylinder
Glass wool, 1 g/S
Methanol, pesticide grade, cylinder
Sample, pecans preferred, 1/S

APPARATUS SETUP

The apparatus used for this experiment is the ISCO Model SFX-2-10 with two Model 260D or two 100 DX syringe pumps and a restriction temperature controller. Figure 13-9, p. 142, is a photo of this apparatus. Figure E18-1 is a photo of the restrictor heater and two collection vials and their holder.

THE BIG PICTURE

A single pecan nut meat is weighed, ground with Celite, and placed in an extraction cartridge. 1 mL of methanol is added. The void volume is filled with Celite. The oil is extracted with CO_2 and methanol, both static and dynamic. The collected oil is weighed in a previously weighed collection vial.

DIRECTIONS

It is highly recommended that before you start, you either have a brief demonstration of the apparatus by your instructor or you spend some time reading the instrument manuals. All limits have been set.

1. Refer to Figure 13-9, 142. Place a 400 mL beaker (H) near the right pump (B). A metal filter (D) on a Teflon tube from the inlet of pump (B) is placed in the beaker.

2. Go to the pump controller unit (J) and turn on the power. (1 of Figure E18-2).

Figure E18-1. ISCO restrictor heater and sample collection vials.
(Courtesy - ISCO Inc., Lincoln, NB)

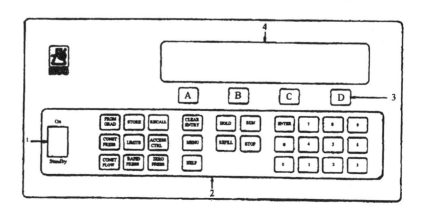

Figure E18-2. Pump controller (1) on/stby switch, (2) programming keypad, (3) softkeys to select menu items, (4) display.
(Courtesy - ISCO Inc., Lincoln, NB)

Figure E18-3. Pushing glass wool into a collection vial.

3. Push softkey D to select the pumps.
4. Select the modifier pump by pushing softkey B.
5. Open the inlet valve (I) counterclockwise on this pump.
6. Close (clockwise) the outlet valve (G).
7. Push "Run". It will shut itself off after pushing any liquid in it out into the collection beaker.
8. Zero the pump by pushing ZERO on the keypad and softkey B.
9. Place the metal filter and Teflon tube into a 400 mL beaker containing about 200 mL of methanol.
10. Select "refill". Choose softkey B. It will shut off when it is full.
11. This pump must now be purged of air. Select RUN and watch the filter in the methanol. When the air stops, then select STOP and softkey B and close the inlet.
12. Push softkey D and select the extractor pump A by pushing softkey A.
13. Open the outlet valve (E).
14. Close the inlet valve (C).
15. Remove the front cartridge port (P) and set it off of the extractor.
16. Open vent valve (N) and supply valve (O).
17. Push RUN. It will shut itself off after expelling any gas.
18. Zero the pump by pushing ZERO on the keypad and softkey A.
19. Fill the solvent pump (A). Close (clockwise) the pump outlet valve (E).
20. Open (counterclockwise) the pump inlet valve (C).
21. Open the main CO_2 cylinder valve.
22. Select "refill". Choose softkey A. It will shut off when it is full.
23. Close both the pump inlet valve and the CO_2 cylinder valve.
24. Press the orange MENU key and press softkey A -MORE.
25. Select 4. MULTI-PUMP.
26. Select MODIFIER by pressing the 3 on the numeric keypad.
27. Press softkey D previous twice to return to the main menu.
28. Push softkey C and enter 15% for the modifier and save by pressing ENTER.
29. Weigh about 1 g of glass wool and place it in a collection vial. Tamp it to the bottom with a rod. It should look like Figure E18-3.
30. Wipe off any fingerprints, weigh the vial and the glass wool to ± 0.001 g, and record the weight.
31. Place the vial under the front heater of the fixed restrictor (R of Figure 13-9, 142) and *carefully* push the vial up until the fixed restrictor rod is down into the glass wool. If you twist the vial, the rod will go through the glass wool.
32. Set the extractor temperature control to 80 °C as follows.
33. Refer to Figure E18-4, p. 559. Turn on the power (L). Go to the temperature con-troller (M) and press SV. Use the ∧ and ∨ keys to set 80 °C on the display. Press "Enter". The restrictor heater will be turned on later.
34. With tweezers, pick up a single half of a pecan nut meat, weigh it to ± 0.001 g, and record the weight.
35. Place the pecan in a small mortar and add 1 g of Celite 545 for every 100 mg of sample.
36. Crush the two together until they appear homogeneous.
37. Arrange a 10 mL cartridge for filling (Figure 13-12, Chapter 13). Be sure the S side on the filter is down and close the end finger tight.

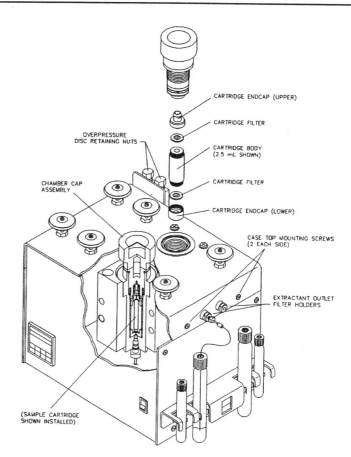

Figure E18-4. Exploded view of the SFX 2-10 extractor.
(Courtesy - ISCO Inc., Lincoln, NE)

38. With a small powder funnel and a spatula, transfer the sample to the cartridge. Tap the cartridge to settle the sample.
39. Add a small amount of Celite to the mortar and grind it with the pestle to remove the residual oil from the side walls.
40. Transfer this to the cartridge.
41. Wipe the sides of the mortar and the end of the pestle with a small amount of glass wool (use tweezers so your finger oil will not contaminate the sample). Add this to the cartridge.
42. Pipet 1 mL of methanol into the cartridge.
43. Fill the void volume with Celite 545 and tap the cartridge to settle it.
44. Close the top finger tight.
45. Place the cartridge in the front compartment of the extractor (C) and screw the cover on finger tight *carefully* - the compartment may be quite hot by now.
46. Set the restrictor heater to 165 °C. If you do not have a fixed heater restrictor, then you must use a stainless steel or fused silica restrictor and a hair dryer.
47. Turn the power on to Heater no. 1 and set the restrictor temperature to 165 °C with the + and - buttons. .
48. Use the toggle switches (C) and set the temperature to 165 °C.

This extraction will involve both a static and a dynamic process. The static portion will involve filling the cartridge with CO_2, adding it to the methanol already present, and leaving it there for 5 minutes. Then, methanol will be added to the CO_2 through valve F, and the combination will be pumped through the sample for 15 minutes. It is essential that the pressures in both pumps be the same initially and that there is no air in the modifier pump.

49. Close the outlet of the extractor ((Q).
50. Make sure vent (N) is closed (both front and back ports).
51. Open the supply port (O).
52. Open the outlet of the CO_2 pump (E).
53. Push softkey D and select extractor pump A by pushing softkey A.
54. Select CONST PRESS and enter the pressure. Save by pressing ENTER.
55. Press softkey B under MODIFIER. Word OFF will change to ON above modifier.
56. Press RUN. Note the time and continue for 5 minutes. Record the piston position. This step extracts some or all of the oil and keeps it in the cartridge.
57. Hold at pressure 5 minutes.
58. HOLD. Press RUN appears on the display.
59. Now make a dynamic run. Open outlet valve (G) of pump B. NOTE: If you do not have a heated restrictor, then you must apply heat to the restrictor with a hair dryer.
60. Open outlet valve (Q) of the extractor.
61. Message EQUILIBRATE will appear on the display.
62. Press RUN.
63. Once the message RUNNING is displayed, modifier will be delivered at the selected rate.
64. Record the pump stroke volume of pump A from the display.
65. Extract for about 15 minutes.
66. Press STOP on the keypad and softkey A.
67. Close the outlet valves on both pumps. Record the piston position.
68. Vent the extraction cartridge (N).
69. Shut off the CO_2 cylinder tank valve.
70. Remove the collection vial, wipe off fingerprints, and weigh it to ± 0.001 g.
71. Place it in a vacuum oven and heat the contents to remove any dissolved CO_2 and methanol. This is a measurable amount and is a necessary step for good work. Weigh to *constant weight*.
72. Remove the cartridge *carefully* - it may be still hot.
73. Remove the contents. Wash the cartridge with hot soapy water, drain, and let air dry. The frit is usually left alone but can be cleaned by a sonic cleaner, if deemed necessary.

EXAMPLE CALCULATION
A half of a pecan nut meat weighing 1.305 g was treated as above and placed in a collection vial containing a glass wool plug, the total weight being 25.782 g. After extraction, the weighings were 26.843, 26.768, and 26.765 g. What is the percent "fat" in this hybrid pecan?

ANSWER
26.765 g - 25.782 g = 0.983 g.   0.983 g/ 1.305 g x 100 = 60% fat.

EXAMPLE CALCULATION
The piston readings on the CO_2 pump were 203.38 mL and 173.38 mL and on the methanol pump were 96.56 mL and 91.97 mL. What was the percent modifier actually used?

ANSWER
CO_2: 203.38 mL - 173.38 mL = 30.00 mL.   Methanol: 96.56 mL - 91.97 mL = 4.59 mL.
4.59 mL/30.0 mL = 15.3%

EXPERIMENT 19. DISPLACEMENT COLUMN CHROMATOGRAPHY
The separation of *cis-trans*-azobenzene

BACKGROUND
An *actinometer* is a device to measure the amount of radiation emitted from a light source. It consists of either a single compound or a mixture of compounds that react with radiation and are converted to other compounds, the amount of the conversion being proportional to the amount of incident radiation. The best actinometers use both initial and final compounds that are easy to measure.

Diphenyldiazine, more commonly known as azobenzene, exists in two isomers, the *cis* and the *trans*. The more stable *trans* isomer can be converted to the *cis* form by the application of ultraviolet radiation. Reagent grade azobenzene normally has about 1% of the *cis* isomer present. However, if a solution of this compound is exposed to uv radiation, the concentration of the *cis* isomer can be increased. Both compounds are highly colored and easy to measure with a visible region spectrophotometer. The *trans* isomer is quite nonpolar and will not be retained on a polar alumina column. The *cis* isomer is quite polar and will be retained at the top of the column.

Trans — Opposite sides — Dipoles cancel each other — m.p. = 68 °C

$hv \rightarrow$

Cis — Same side — Dipoles add — m.p. = 61 °C

Therefore, separating the two compounds is rather easy, but it requires considerable time to remove the *cis* isomer unless a second, more polar, mobile phase is used as a second eluent.

Toxicity - This material has produced liver damage in rats. Chemically, it is used as an intermediate in the synthesis of many dyes.

YOUR PROBLEM
A research group needs to determine the power output of a new N_2 laser and have decided to use an azobenzene actinometer. Your job is to prepare several crystals of both *cis-* and *trans*-azobenzene, so the actinometer can be prepared and calibrated.

EQUIPMENT

1	Apparatus, melting point, Fisher-Johns or similar	1pr	Goggles, safety
1	Balance, triple beam	1	Hotplate
2	Beakers, 400 mL	1	Lamp, uv, short wavelength, powerful if possible
1	Brush, column	1	Pipet, delivery, 25 mL
1	Bulb, pipet	1	Rod, glass, stirring, 8 mm x 45 cm
1	Clamp, buret	1	Sponge
1	Column, glass, chromatographic, 2 cm x 30 cm	1	Spatula
1	Cylinder, graduated, 50 mL	1	Stand, ring
1	Dish, evaporating, 50 mL	2 ft	Tubing, rubber, condenser

CHEMICALS   S = Student
Alumina, Al_2O_3, Brockman activity 1, 80-200 mesh, 35 g/column
Azobenzene, C_6H_5-N=N-C_6H_5, 0.25 g/25 mL petroleum ether, 25 mL/S
Glass wool, 2 g/S
Petroleum ether, 200 mL/S
Petroleum ether + 1% methanol, 150 mL/S
Stopcock grease, 1 tube/lab

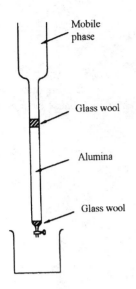

Figure E19-1. Apparatus setup.

APPARATUS SETUP
The apparatus is set up as shown in Figure E19-1.

THE BIG PICTURE
Thirty five grams of alumina is made into a slurry with petroleum ether and transferred to a chromatographic column. Twenty five mL of azobenzene solution is placed in an evaporating dish and irradiated for 30 minutes with uv radiation (the more intense the better). The sample then is added to the column. The *trans* isomer is eluted with petroleum ether, and the *cis* isomer is eluted with petroleum ether + 1% methanol. The solutions are collected, the solvent is evaporated, and the melting points are obtained to establish purity.

DIRECTIONS
1. Turn on a hotplate (in a hood) to the high position, so it will be ready for later use.
2. Place 25 mL of the azobenzene solution into an evaporating dish (flat bottomed preferred) and set it in the hood.
3. Place a high intensity uv radiation lamp close to the top of the azobenzene solution and turn on the lamp. The uv radiation provides energy to convert the *trans* form into the *cis* form. Note the time when this is begun. Irradiation for at least 30 minutes with a high intensity lamp is required. Check with your instructor for the time required with your lamp.
4. If possible, place a dark glass or plastic shield on a ring stand and set it between the lamp and other people in the lab to protect their eyes from the intense radiation.
5. Obtain a glass column with a large-bore stopcock, a glass frit-bottom plate, and if possible a solvent reservoir at the top. Avoid using a buret for column chromatography. The stopcock bore is too small and flow rates are slow. The bore can easily plug with viscous mobile phases and the solvent reservoir capacity is too small
6. If your column does not have a glass frit, add a small amount of glass wool and push it to the bottom with a glass rod.
7. Add 35 g of alumina to a 400 mL beaker.
8. Use a wash bottle filled with petroleum ether and make a slurry out of the alumina. This is called "wet packing" a column.
9. The object is to add the alumina to the column without trapping any air bubbles. The way to do this is to add a few mL of petroleum ether to the column first. Add 25 to 30 mL of petroleum ether to the column.
10. Use a wash bottle to spray a stream of petroleum ether onto the alumina slurry and wash it into the column.
11. When all of the alumina has been transferred to the column, let it set for a few minutes until the solids settle.
12. Wash down the sides of the column with a stream of petroleum ether, so that all of the alumina is together. If the column becomes full of petroleum ether, open the stopcock and drain the excess into a small beaker. Do not let the liquid level go below the top of the column packing.
13. If there are any air bubbles in the column, try to remove then by gentle stirring with a long glass rod. If this is unsuccessful, as it often is, then pour out the alumina and start over.
14. Place a small piece of glass wool on top of the packing and tamp it gently in place. This holds the packing in place and prevents the top of it from being stirred when the sample and mobile phase are added.
15. Drain the petroleum ether from the column until it is 2 to 3 mm above the packing.
16. When the azobenzene has been irradiated for about 30 minutes, shut off the lamp. NOTE: the time will vary on this depending on the lamp intensity. Thirty minutes is based on a Hanovia high intensity lamp with a visible region filter. *Be careful - the evaporating dish will be hot.* Use a towel when you pick it up.
17. Some of the azobenzene has probably crept up the sides of the container and the volume has decreased somewhat. This is of no consequence.

18. Pour all of the contents that will come out easily onto the top of the column packing. Do not try to rinse out the evaporating dish. You should have about 20 cm of alumina in the column and 5 to 7 cm of azobenzene solution on top of it.
19. Clean a 400 mL beaker with petroleum ether - no water.
20. Place this beaker under the column outlet to collect the eluent.
21. Slowly open the stopcock and let the azobenzene layer adsorb onto the top of the column. Stop when the liquid is about 1 mm above the packing. Do not let air get to the column.
22. Discard the washings into the solvent waste container.
23. Carefully - so as not to disturb the column packing - add petroleum ether until the reservoir is about half full.
24. Place a clean beaker under the column and open the stopcock until the flow rate is a fast drip.

    What will happen is that a yellow band (*trans*-azobenzene) will go down the column, and you will believe that the *cis* is not staying behind. This is not so. The *cis* is just present in a very low concentration and is hard to see at this time. It will become visible as time goes by. It will become a band about 2 to 3 cm wide and 3 to 4 cm down the column, and the trans isomer will be eluting off the column.

25. Add petroleum ether as needed until the *trans* compound has been eluted from the column, then lower the level of mobile phase to 1 to 2 mm above the column packing.
26. You have separated the azobenzenes. Set the collection beaker on the hot plate (in the hood) and begin evaporating the mobile phase. Check the evaporation periodically. You must remove the solvent, but you do not want to bake it dry. When it becomes syrupy (melted azobenzene), stop the heating. While this is taking place, begin to remove the *cis* compound from the column.
27. Run tap water over the outside of the beaker to cool the contents and cause the azobenzene to crystallize.
28. While the solvent on the *trans* compound is evaporating, elute the *cis* compound with petroleum ether containing 1 % methanol. Follow the same procedure as you did with the *trans* compound.
29. Evaporate this solution the same as the *trans* compound. Be very careful at the end, because excess heat will cause some of the *cis* isomer to revert to the *trans* isomer.
30. Shut off the hot plate.
31. Repeat step 27 with the *cis* isomer. There will usually be very few *cis* crystals, so tilt the beaker to collect them in one place.
32. Obtain the m.p. of each compound. Use a Fisher-Johns m.p. or similar apparatus. If you have not used this apparatus, then follow the directions below. If you have used it before, then go to step 41 after the m.p.'s have been obtained.

Operating a Fisher-John's Melting Point Apparatus
Refer to Figure E10-2, p. 534.

33. The 4 cm round disc at the left front is the actual hot stage of the apparatus. The magnifying glass on the holder immediately to the left of the hot stage is used to magnify the crystals, so you can more easily note small crystalline changes.

    Off to the right is a very good thermometer. The openings on the shield should be pointed away from you, so you cannot see the scale if you are sitting directly in front of the apparatus. This is done so you must first decide if the crystals are melting, then read the temperature. It keeps you honest. The bulb of the thermometer is imbedded in silver wool for better heat transfer.

    On a back raised section is the switch for the light and the control for the hot stage. In the center of the panel is a rheostat to control the rate of heating. It also controls the maximum temperature. To the right of the temperature control is an on-off switch to control the light with which to view the crystals. In a slight depression on the top of the apparatus are several thin glass discs about 1 mm x 1.5 cm. These are where the crystals to be examined are placed. They may be cleaned and used several times until they break.

34. Place 1 to 2 of your crystals on a clean glass disc and put it on the hot stage.
35. Turn on the light and adjust the magnifying glass so that the crystals are in focus.

36. Turn the heater on and set it at about 30. This produces a fairly fast heating rate - 3 to 4 °C/min. For the best results, a rate of 1 to 2 °C/min is desired. The usual technique is to do a fast heating until the crystals melt to determine the general region, allow the stage to cool 8 to 10 °C so the compound crystallizes, then heat the crystals again at 1 to 2 °C/min to obtain an accurate m.p.

37. Obtain a m.p. for the *trans* isomer and record the value.

38. Shut off the heater and allow the stage to cool 10 to 15 °C.

39. Use another glass disc and repeat the above steps for the *cis* isomer.

40. Shut off the heater and the light. Clean the apparatus and the glass discs.

41. Unpack the column. This can be done easily as follows. Take the column to a water faucet and attach a 1 to 2 ft length of condenser tubing to the water outlet and to the bottom of the column. Place a beaker in the sink. Open the stopcock and direct the top of the column into the beaker. Slowly turn on the water, which will force the packing out of the column. The glass wool plug acts as a scrubber, and the column is quite clean in one pass. Rinse it with distilled water.

42. Pour the alumina into a collection can, so it can be reactivated at a later date.

43. Use either methanol or petroleum ether and clean the evaporating dish containing the azobenzene residue.

44. *Put back all of the reagents and clean up the area.* Thank you.

EXPERIMENT 20. MULTIPLE COLUMN PARTITION CHROMATOGRAPHY

The separation of codeine, chlorpheniramine, and phenylephrine in a cough syrup
(Courtesy - Association of Official Analytical Chemists, International method 32.076; *J. Assoc of Off. Anal. Chem.* **48**, 594, 1965)

BACKGROUND

Coughing is a common physiological method to clear the respiratory passages. If it becomes uncontrolled, then medication often can help. Such medications are called *antitussives*. The cough control center has been located in the medulla of the brain by electrical stimulation techniques. Normally, coughing is caused by chemical or mechanical irritation of the respiratory tract. Therefore, to stop a cough, the source of the respiratory tract irritation should be eliminated and the signal to the medulla interrupted.

An effective cough medicine usually contains a mixture of components. A representative example is the product "Pediacof", which contains the following ingredients for a 5 mL dose:

Codiene phosphate	5.0 mg	Potassium iodide	75.0 mg
Phenylepherine HCl	2.5 mg	Sodium benzoate	10.0 mg
Chlorpheniramine maleate	0.75 mg	Ethanol	250 mg
Raspberry flavoring			

Opiates, of which codiene is one, are the most effective compounds to stop coughing. Codeine depresses the cough reflex signal to the medulla by a mechanism as yet not known. A dose of 15 mg orally is usually sufficient. It is a pain killer, depresses respiration, and stimulates the vomiting center. Phenylephrine is a nasal vasorestrictor to halt the production of material that irritates the throat and lungs. It reacts at the postsynaptic alpha receptors. Chlorpheniramine is an *analgesic*, which numbs the throat, thus numbing it and making it more insensitive to irritation. It is also a mild sedative to induce sleep lost because of excessive coughing. Salts usually are used because they are water soluble and reduce odor by reducing the vapor pressure of the organic compounds.

Codeine hydrochloride
MLD orally in guinea pigs, 120 mg/kg.
Medical use: Analgesic, antitussive.
Side effects: nausea, vomiting, dizziness.
Converted to morphine in the body.

Chlorpheniramine maleate
Medical use: Antihistaminic.
Side effects: drowsiness, headache, xerostomia, anorexia.

Phenylephrine hydrochloride
LD_{50} in mice, 1000 mg/kg.
Medical use: Sympathomimetic. Topically as a vasoconstrictor, mydriatic, and nasal decongestant.

Potassium iodide is an expectorant. It liquifies the mucous, so coughing is more productive when it happens. The sodium benzoate is a preservative. A syrup is used as a base to slow the passage of the analgesic through the respiratory tract, thus giving it more time to react. It also acts as a demulcent for coughs originating above the larnyx. Red, orange, and purple colors are used because patients, particularly children, associate them with cherry, strawberry, orange, and grape flavors that they recognize and like. Children often decide by the look of the medicine if they will even taste it.

YOUR PROBLEM

A sample of a commercial cough syrup is suspected of not containing the listed amounts of the drugs codeine, phenylephrine, and chlorpheniramine. Your job is to separate the three drugs from the syrup and flavoring agents and determine the actual concentration of at least one of the drugs, doing as many of the others as time permits.

EQUIPMENT

1	Balance, triple beam, ± 0.01 g		1	Hotplate
7	Beakers, 100 mL		1	Paper, filter, Whatman no. 41, 11 cm
4	Beakers, 250 mL		1	Pipet, 2 mL
1	Bottle, wash, small		1	Pipet, 10 mL
1	Brush, test tube		1	Pipet, 20 mL
1	Bulb, pipet		1	Pipet, Mohr, 1 mL
2	Cells, quartz, 1 cm		1	Pipet, Mohr, 10 mL
3	Clamps, buret		1	Rack, funnel
3	Columns, 2.5 cm x 20 cm		1	Rod, tamping
1	Cylinder, graduated, 10 mL		1	Scoop, Celite, 3 g size
1	Cylinder, graduated, 100 mL		1	Spatula
3	Flasks, volumetric, 25 mL		1	Spectrophotometer, uv capability
1	Flask, volumetric, 250 mL, bottom only for adding eluting solvent		1	Sponge
1	Funnel, 60°, long stem		1	Stand, ring, tall
1 pr	Goggles, safety			

CHEMICALS   S = Student

Acetic anhydride (Ac_2O).
Acetic anhydride (Ac_2O) - $CHCl_3$ 2:5, 20 mL/S
Celite 545, 10 g/S
Chloroform, $CHCl_3$ - water saturated, 150 mL/S
Chlorpheniramine maleate standard - 16 mg/L of 0.1 N HCl, 70 mL/S
Codeine standard - 84 mg/L of 0.1 N HCl, 70 mL/S
Glass wool, 3 g/S
Hydrochloric acid, HCl, concentrated, 5 mL/S
Hydrochloric acid, HCl, 0.1 N, 200 mL/S
Nitric acid, HNO_3 - 1 N, 5 mL/S
Phenylephrine standard - 40 mg/L of 0.1 N HCl, 70 mL/S
Potassium hydroxide-ethanol, KOH - ETOH, Satd Soln 5 mL/S
Sodium hydroxide, NaOH - 1 N, 10 mL/S
Sulfuric acid, H_2SO_4 - 1 N, 5 mL/S
Triethylamine - $CHCl_3$ 2:5 25 mL/S
Triethylamine - $CHCl_3$ 1:99 300 mL/S

NOTE TO TA'S

Triethylamine purification - if a blank absorbance is >0.01, reflux 100 mL ET_3N with 20 mL H_2O and 2 g Na hydrosulfite at least 8 hours. Wash with H_2O, dry by distilling into a Dean-Stark trap, then distill, collecting the first 75 mL. Store over anhydrous Na_2CO_3 or K_2CO_3.

APPARATUS SETUP

Three columns are packed with Celite and arranged vertically as shown in Figure 15-11, p. 163. The bottom of the lowest column is high enough so that a 250 mL beaker will fit under it.

Experiment 20

THE BIG PICTURE

Three Celite columns are prepared, one containing 2.0 mL of cough syrup and 0.5 mL of 1.0 N nitric acid, the second containing 2 mL of 1.0 N NaOH, and the third containing 2.0 mL of 1 N sulfuric acid. These are placed vertically on a ring stand, and the components eluted with chloroform. The drugs are retained on the columns, and the syrup, food dye, and flavoring agents are washed away. Each drug then can be eluted from its column, and the amount determined spectrophotometrically by comparison with standards.

DIRECTIONS

Generally, the sample is added in as narrow a band as possible at the top of the column. In this case, this is not necessary since the separation is so easy.

1. Pack each of the columns as shown in Figure 15-11, p. 163, using the technique previously discussed.
2. Arrange the columns so that the effluent from I flows into II and then into III.
3. Pass 150 mL $CHCl_3$ through the columns. Discard the $CHCl_3$, into a recylce bottle and do each column separately from here on. Do as many columns as time permits.

COLUMN I - Phenylephrine

4. Add 7 mL acetic anhydride (Ac_2O).
5. Wash the column with 95 mL Ac_2O + $CHCl_3$ (1 + 99)
6. Evaporate to dryness on a hot plate in a hood.
7. Add 5 mL alcoholic KOH. Heat 15 minutes on a *low temperature* hot plate.
8. Acidify with HCl. Filter, if the mixture is cloudy. Transfer to a 25 mL volumetric flask and dilute to the mark with 0.1 N HCl.
9. Prepare calibration curve standards as follows (use pipets):

    a. Standard original solution - 40 µg/mL
    b. 30 mL Std + 10 mL H_2O - 30 µg/mL
    c. 20 mL Std + 20 mL H_2O - 20 µg/mL

10. Measure unknown and standards at 272 nm with a uv spectrophotometer.
11. Plot absorbance (y axis) vs. concentration (x axis) and determine the concentration in the unknown. Does it match what the sample label indicated?

COLUMN II - Chloropheniramine

12. Add 7 mL triethylamine (ET_3N) + $CHCl_3$ (2 + 5). Collect in a 250 mL beaker.
13. Wash with 95 mL ET_3N + $CHCl_3$ (1 + 99).
14. Evaporate to dryness on a hot plate in a hood.
15. Dissolve in 0.1 N HCl, transfer it to a 25 mL volumetric flask, and dilute it to the mark with 0.1 N HCl.
16. Prepare calibration curve standards as follows (use pipets):

    a. Standard original solution - 16 µg/mL
    b. 30 mL of Std + 10 mL H_2O - 12 µg/mL
    c. 20 mL of Std + 20 mL H_2O - 8 µg/mL

17. Measure unknown and standards at 265 nm.
18. Plot absorbance (y axis) vs. concentration (x axis) and determine the concentration in the unknown. Does it match what the sample label indicates?

COLUMN III - Codeine

19. Add 7 mL ET_3N + $CHCl_3$ (2 + 5). Collect in a 250 mL beaker.
20. Wash with 95 mL of ET_3N + $CHCl_3$ (1 + 99).
21. Evaporate to dryness on a hot plate in a hood.
22. Dissolve in 0.1 M HCl, transfer it to a 25 mL volumetric flask, and dilute it to the mark with 0.1 N HCl.

23. Prepare calibration curve standards as follows (use pipets):

   a. Standard, Original solution - 84 µg/mL
   b. 30 mL of Std + 10 mL H$_2$O - 63 µg/mL
   c. 20 mL of Std + 20 mL H$_2$O - 42 µg/mL
   d. 10 mL of Std + 30 mL H$_2$O - 21 µg/mL

24. Measure the unknown and the standards at 285 nm.

25. Plot absorbance (y axis) vs. concentration (x axis) and determine the concentration of the unknown. Compare it with the label on the product.

To pass FDA regulations, the product should contain between 95% to 105% of what is declared on the label. Refer to the example calculation at the end of Chapter 15, p. 163, and calculate your results.

EXPERIMENT 21. AFFINITY CHROMATOGRAPHY
Separation of a lectin from soybeans
(Based on Allen, A.K. and Neuberger, A., *FEBS Lett.*, **50**, 362, 1975)

NOTE: This is neither a quick nor an easy experiment if done in its entirety, but it illustrates all of the steps required. It is done as an experiment by the Kansas State University Biochemistry Department and the directions are those they use.

Part A - Sample preparation.
Part B - Preparation of the affinity absorbent.
Part C - Separation of the lectin.
Part D - Elution of the lectin.
Part E - Determining the agglutination activity.
Part F - Determining protein content and degree of purification.

It is recommended that students do the last part of B, C, D, and E with the instructor doing Part A and the first part of B. (Biochemists may want to do all parts).

BACKGROUND

CAUTION: Purified lectins can be hazardous if handled improperly because of their ability to cause coagulation of many biological systems.

Soybean agglutinin (SBA) is a lectin (m.w. 120,000) believed-to consist of four equal subunits. It has been called *glycine max* which is the Latin name for soybeans. *Agglutinins* are compounds that bind cells together causing coagulation. *Lectins* are plant and animal proteins capable of binding selectively to certain carbohydrate groups on the cell surface. Examples are concanavalin A from jack beans and ricin from castor beans. During the past decade, when lectins have been obtainable in increasingly purer forms, they have become important for the study of cell surface reactions involving glycoproteins and the process of cell-cell recognition.

The agglutinating property of SBA is specifically inhibited by N-acetyl-O-galactosamine and, to a lesser extent, by D-galactose. Advantage is taken of this to prepare an affinity chromatography method of separation. Galactosamine is coupled to CH-Sepharose 4B by a carbodiimide reaction in aqueous solution with the production of a caproylamido linkage to C-2 of galactosamine. The adsorbed lectin can be eluted with D-galactose.

YOUR PROBLEM

You have begun to work for a research biochemist whose area of interest is how cells recognize other cells. Currently, he is using a soybean lectin that is separated from defatted raw soybeans by an affinity chromatography method. He needs large amounts of this material and believes it is less expensive to prepare it himself than to purchase it. In addition, the preparation can serve as a training aid for future work in this technique. Your first job is to learn how to prepare and use affinity chromatography gels. Last week, he hired a student who was able to defat several grams of raw soybeans before quitting to become a psychology major. You are to take over the project, prepare the adsorbent, and use it to obtain a purified soybean lectin.

EQUIPMENT

1	Balance, triple beam	1	Paper, filter, 7.5 cm
4	Beaker, 100 mL	1	Pipet, Mohr, 1 or 2 mL
1	Bottle, wash, 250 mL.	1	Pipet, Mohr, 10 mL
1	Brush, test tube, small	1	Plate, glass, spot
1	Centrifuge, 9000 x g, 50 mL tubes	1	Rack, funnel
1	Cheesecloth, 30 x 60 cm	1	Rack, test tube
2	Clamps, buret	1	Refrigerator, (4 °C)
1	Column glass, 1 cm x 7 cm w/ stopcock, and bottom frit	1	Rod, stirring, glass, 10 cm long
		1	Spatula
1	Column, Pharmacia K 9/15	1	Spectrophotometer, u.v. - (280 nm) or Visible (465 and 595 nm)
18	Corks, size 1		
1	Cylinder, graduated, 100 mL	1	Stand, ring
1	Cylinder, graduated, 10 mL	1	Towel, cloth
6	Droppers, medicine	2	Tubes, centrifuge, 50 mL
1	Flask, filter, 125 mL	18	Tubes, test, 10 x 75 mm
1	Funnel, Buchner, 7.5 cm	4 ft	Tubing, vacuum
1	Funnel, powder, 10 cm	1	Watch, w/second hand
1 pr	Goggles		
1	Meter, pH		

CHEMICALS   S = Student

Petroleum ether, 60 mL/S
CH-Sepharose 4B 1.5g/S
Phosphate buffered saline (PBS) 300 mL./S
    Sodium chloride, NaCl, 7.2 g
    Disodium hydrogen phosphate, 1.5 g   } dilute to 1 L, final pH is 7.2
    Potassium dihydrogenphosphate 0.43 g
PBS + 0.2 g sodium azide
D-gallactose, 0.5 M in PBS, 1.0 M in PBS
D-galactosamine HCl, 100 mg/S
1-Ethyl-3-(3-dimethylaminopropyl) carbodiimide HCl, (EDAC), 50 mg/S
Soybean meal (must be raw), 12.5 g/S
Glycine max, standard, 1 mg/mL of PBS, Sigma No. L 8004, 1 mg/S
Sephadex G-25, 1g/S
Rabbit red blood cells (Sigma Chem. Co., R 1629) 1.5% in PBS, 5 mL/S
Sodium chloride, NaCl, 0.5 M (25 mL/S), 1.0 M (25 mL/S), 2.0 M (25 mL/S)
Buffer, pH 6.8, 1 M in NaCl and 0.1 M in Tris-HCl, 25 mL/S
Buffer, pH 3.0, 1 M in NaCl and 0.05 M in sodium acetate, 25 mL/S

Sodium hydroxide, NaOH, 0.1 M, 5 mL/S
Coomassie Brilliant Blue G-250, 106 mg
Ethanol, H_3CCH_2OH, 95 %, 50 mL, dilute to 1 L
Phosphoric acid, H_3PO_4, 85 %, 100 mL

APPARATUS SETUP

The apparatus consists of a glass column mounted vertically on a ring stand. The column has a glass frit to retain the gel and a stopcock to adjust the flow rate. The agglutinating activity is determined by reacting rabbit red blood cells with the agglutinin and noting the amount of precipitate formed on a spot plate. Protein content is obtained by spectrophotometric measurement.

THE BIG PICTURE

Raw soybean meal is extracted with petroleum ether to remove the fats. The agglutinin is extracted from the soybean meal with a phosphate buffered saline solution. Sepharose is swollen and then reacted first with a carbodiimide, and then with D-galactosamine to form the affinity gel. The gel is purified, and placed in a column, and the soybean extract is passed through. The agglutinin is retained while the other materials pass through with about a 100-fold purification being obtained. The agglutinin is washed from the gel with a D-galactose solution, and fractions are collected. The excess galactose, which reduces the activity of the lectin, is removed by gel filtration. The activity of the separated agglutinin is determined by the precipitate it forms with rabbit red blood cells. The total protein present can be obtained by measurement with either a uv spectrophotometer at 280 nm or a protein-dye mixture can be measured at 595 nm.

DIRECTIONS

Directions for the complete separation are given. Begin where your instructor advises and go as far as you desire.

Part A - Defatting of the Soybean Meal

This is usually done in a cold room at 4 °C. The directions have been modified to use a refrigerator.

1. Place 12.5 g of raw soybean meal into a 100 mL beaker and add 60 mL of petroleum ether.

2. Stir the mixture until the solid is thoroughly wetted, cover with a watch glass, and place in a refrigerator to retard evaporation. Leave it overnight, stirring when convenient.

3. Suction filter.

4. Let the meal air dry. Discard the extracted fat.

5. Place 5 g of the meal in a 100 mL beaker and add 50 mL of PBS solution. Stir for a few minutes and place in a refrigerator overnight, stirring when convenient. This extracts the lectin.

6. Fold a 30 x 60 cm piece of cheesecloth repeatedly until it is 15 cm on a side (8 thicknesses).

7. Place this in a powder funnel and filter the meal, collecting the filtrate in two, 50 mL, centrifuge tubes.

8. Centrifuge for 15 minutes at 9000 x g.

9. Carefully pour off the supernatant into a graduated cylinder avoiding the loose pellet. Note the volume and save 2 mL for protein and agglutination assays.

Part B - Preparation of the Affinity Gel

This gel can be obtained commercially from Sigma Chemical Co., A 6764, at $197 for 25 mL. 1 mL will bind approximately 20 mg of *Glycine max*.

10. Place 1.5 g of CH-Sepharose 4B in a 100 mL beaker, add 15 mL of 0.5 M NaCl, stir gently occasionally, and allow the beads to swell for 1 hour.

11. Transfer the gel to a centrifuge tube and wash three times with 0.5 M NaCl by centrifugation.

12. Remove the gel and wash once with water, then make a slurry with 25 mL of water. Do this at room temperature.

13. Add 100 mg of galactosamine and adjust the pH to 5.0 with 0.1 M NaOH.

14. Over a period of about 5 minutes, add 50 mg of solid EDAC (1-ethyl-3-(dimethylaminopropyl)) carbodiimide HCl. The pH will rise to about 5.3 and remain constant.

15. Stir the mixture occasionally for an hour at room temperature, then let stand for 20 hours without stirring.
16. Wash the gel with 1 M NaCl.
17. Wash the gel with 1 M NaCl containing 0.1 M Tris-HCl buffer (pH 6.8).
18. Wash the gel with 1 M NaCl containing 0.05 M sodium acetate buffer (pH 3.0).
19. Repeat step 17.
20. Wash the gel with water.
21. Wash the gel with PBS.
22. The gel (about 6 mL) is ready for use. If it cannot be used now, store it in 2 M NaCl.

Part C - Separating the Lectin

23. Place 1.0 g of Sephadex G-25 in a 100 mL beaker and add 15 to 20 mL of PBS solution. Let this stand during the following steps to allow the beads to swell. Go to the next step immediately. NOTE: your instructor may have this gel prepared already.
24. Add enough of the affinity gel to a 7 cm x 1 cm glass or plastic column to make a bed 1 to 1.5 cm high and drain the excess liquid to the top of the gel layer.
25. Wash the gel with 5 mL of PBS and drain to the top of the gel layer. Use a flow rate of about 2 mL/min.
26. Add the soybean extract to the column and allow it to drain at 1.5 to 2.0 mL/min. If this is done properly, the lectin is retained and about 99% of the other materials are washed away.
27. Wash the column with 10 to 15 mL of PBS solution to remove the unwanted protein.
28. The completeness of washing can be determined if desired by either uv or visible spectroscopy. An absorbance of 0.02 or less at 280 nm indicates the protein has been removed from the column. See part F for the visible range method.

Part D - Elution of the Lectin

29. Elute the lectin with 10 mL of 0.5 M galactose in PBS. Use a flow rate of 1.0 mL/min. Collect the eluate in a small test tube.
30. If the column is to be used again, wash the gel with 5 mL of 1 M galactose in PBS followed by 10 mL of PBS + 0.02% sodium azide (preservative).

    At this stage of the separation, the excess galactose used to elute the lectin significantly reduces the lectin's agglutinating activity and must be removed. The large concentrations of galactose (if only by mass action alone) cover the active sites and prevent other compounds from reacting as well, even if they have a better affinity. The removal of the excess galactose can be done either by dialysis, which requires about 2 days, or by a size exclusion technique, which requires less than an hour.

31. Place a Pharmacia K 9/15 column (or equivalent) on a ring stand and adjust it so a 10 mL graduated cylinder can be placed under it. A long thin gel bed is preferred here rather than the short gel bed used previously.
32. Add the G-25 gel to the column, draining the excess liquid into a small beaker. Discard the liquid.
33. Add the eluted sample to the column and collect it in a graduated cylinder so the volume can be measured. The large protein molecules (lectin) should pass through as void volume, while the smaller galactose molecules are retained.
34. Rinse the column with 15 mL of PBS solution while you are doing the next steps. An absorbance of <0.02 at 280 nm indicates that the protein has been washed through the column.

    The lectin probably will still have some galactose associated with it, but the large excesses are removed, and much of the agglutinating activity should be restored.

Part E - Agglutination Assays

The success of a separation of this type is judged not so much by how much product was obtained, but by how much activity the compound has. The method used to measure activity in this case is the formation of a precipitate with rabbit red blood cells. *Agglutination activity* is expressed in µg/mL and is determined from serial dilutions of a 1 mg/mL solution in phosphate buffered saline, pH 6.8.

A portion of the purified eluate is diluted successively until no precipitate forms. The degree of agglutination is determined using the serological scale of ++++ (maximum agglutination) to 0 (no agglutination). One SBA unit of activity is the amount that will cause half maximal (++) agglutination after 1 hour incubation at 25 °C.

35. Dilute the standard the same as the sample. Obtain the standard (1mg/mL) from the instructor.

36. Obtain a test tube rack for 10 x 75 mm test tubes and arrange three rows of six. Record the positions in your notebook, leaving space to record your observations. One row is for the unpurified crude extract. The middle row is for a set of standards, and the other row is for the purified extract.

37. Use a pipet and place 1.8 mL of PBS solution in the first tube in each row and 1.0 mL in each of the other tubes.

38. Add 0.2 mL of crude extract to the first tube in its row. Add 0.2 mL of the standard and 0.2 mL of the purified extract to the first tubes in their rows.

39. Stopper with a small cork and gently invert the liquid to ensure mixing. Do not cover the tubes with your bare fingers when inverting the tubes; lectins can be dangerous if the skin has breaks..

40. Flush the pipet with PBS solution after each of the following dilutions. Add 1.0 mL of the liquid from tube 1 to tube 2, stopper, and mix.

41. Add 1.0 mL of the liquid from tube 2 to tube 3, stopper, and mix.

42. Repeat the above 1:1 dilutions for all six tubes.

43. Arrange two spot plates so you have three rows of seven depressions.

44. Add 0.1 mL from each dilution to a well in the spotting plate and then 0.1 mL of SBA solution to the last well as a blank.

45. With a pipet, add 0.2 mL of 1.5% rabbit red blood cells in PBS (resuspend the cells before pipetting) to each well.

46. Mix the samples by gently shaking the plate and let stand for 1 hour.

47. Determine the degree of agglutination. This is your best estimate, and you must decide what is ++++ (maximum on the standard) and then what is +++, ++, and +.

48. Based on the concentration of the standard, calculate the μg of material present after each dilution.

49. Determine the activity of the crude and purified products..

Part F - Determination of the Protein Content and Amount of Purification

Two ways are described to determine the protein content, a uv method and a visible spectroscopy method. The uv method is a direct measurement at 280 nm. For those that do not have a uv instrument, the protein is bound to a dye and measured with a visible range instrument. (see Figure E21-1)

UV Method

50. Place 0.2 mL of an SBA standard (1 mg/mL) and 1.8 mL of PBS solution in a 1 cm thick quartz cell. Thoroughly mix the contents.

51. Use PBS as a blank and determine the absorbance of the standard at 280 nm.

52. Calculate the absorptivity using concentration units of μg/mL.

53. Determine the absorbance (A) of the crude extract.

54. Determine the absorbance (A) of the purified product at the dilution that was ++.

55. Calculate the protein concentration in each sample.

56. Calculate a ratio of activity (μg/mL) vs protein (μg/mL) for both the crude extract and the purified product. Compare these to determine the purification enhancement.

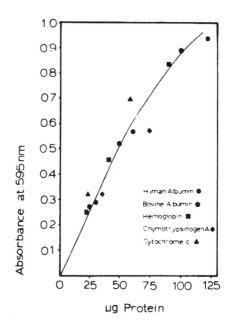

Figure E21-1. Protein dye binding response pattern for various proteins
(Courtesy - Bradford, M.M., *Anal. Biochem.*, **72**, 251, 1976)

57. There is a general rule that "One absorbance unit ~ 1 mg protein/mL at 280 nm". How do your results compare to this generalization?

58. *Turn off the instrument, replace all chemicals in the refrigerator that need to be kept cool, and clean up the area. Thank you.*

Visible region method (Bradford, M.M., *Anal. Biochem.*, **72**, 248, 1976).

Coomassie Brilliant Blue G-250 exists in two different color forms, red (595 nm) and blue (465 nm). The red form is converted to the blue form upon binding of the dye to the protein. The binding reaction is complete within 2 minutes, and the color is stable for about an hour.

59. Place 0.2 mL of an SBA standard (mg/mL) and 5.0 mL of protein reagent in a small test tube and shake to mix.

60. Use protein reagent for the blank and determine the absorbance of the standard at 595 nm.

61. Repeat steps 52 through 58. Refer to Figure E21-1, p. 571, to compare your results.

EXPERIMENT 22. SIZE EXCLUSION CHROMATOGRAPHY

Gel filtration chromatography - The separation of vitamin B_{12} from dextrans by means of Sephadex G-100.

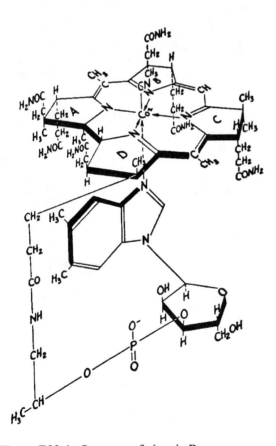

Figure E22-1. Structure of vitamin B_{12}
(Courtesy - *Merck Index*, Merck and Co., Rahway, NJ, 1989)

BACKGROUND

Vitamins compose 0.0002 to 0.005 % of the human diet. Vitamins are compounds that the body requires in such small amounts that it has no mechanism to synthesize them, but relies on outside sources for supply. Analogously, you need gasoline to run your car, but it is easier to buy the gasoline already prepared than to build your own oil refinery. Vitamin B_{12} is synthesized by bacteria in our intestinal tract. We need 0.000002 g/day to prevent pernicious anemia.

Minot and Murphy showed in 1927 that people ill with pernicious anemia could be cured by eating liver. In 1948, the substance now known as vitamin B_{12} was isolated in crystalline form from *Streptomyces griseus*. Alexander Todd's team and Dorothy Hodgkin's team in England and Folkers in the United States identified the structure in 1956 (Figure E22-1). It is the only vitamin to contain a metal ion and cyanide. Liver and kidneys are good sources of this vitamin. Vitamin B_{12} appears to be necessary to produce ribonucleoproteins involved in protein metabolism. It appears to function by enhancing the utilization of circulating amino acids for building body tissues.

The blood level of B_{12} varies widely in different species; from about 0.22 μg/mL in humans, dogs, and calves to 1.7 in alligators, 6.5 in chickens, and 40.6 in rabbits.

YOUR PROBLEM

Several grams of vitamin B_{12} were added inadvertently to a mixture of dextrans. It is desired to recover the vitamin and a gel filtration separation has been proposed to do this. You are given a small amount of the sample and told to determine if the gel filtration system will work.

EQUIPMENT

- 1 Beaker, 100 mL
- 1 Bottle, wash, 250 mL
- 1 Brush, long handle
- 1 Bulb, pipet, small
- 1 Bulb, pipet, large
- 1 Clamp, buret
- 1 Clamp, versatile
- 1 Column, Sephadex, K9/30
- 1 Cylinder, graduated, 25 mL
- 1 Dish, weighing, Al
- 1 Flask, Erlenmeyer, 25 mL
- 1 Flask, Erlenmeyer, 250 mL
- 1 Flask, volumetric, 3 mL
- 1 Holder, clamp
- 1 Level
- 1 Pipet
- 1 Plate, spot
- 1 Rod, stirring, 25 cm
- 1 Spatula
- 1 Stand, ring
- 1 Stopper, rubber, 2 hole, no. 6.5
- 1 Syringe, 1.0 mL
- 1 Towel, cloth
- 1 Tubing, glass, 6 mm x 20 cm
- 1 Tubing, rubber, small size

CHEMICALS   S = Student

Sodium chloride, NaCl, 0.1 M, 200 mL/S
Sephadex G-100, 0.75g/exp
Blue dextran   9 mg   } Dilute these three compounds to 3 mL and use 0.3 mL/exp.
Yellow dextran 9 mg   } (Order the Dye Kit from Pharmacia Biotech, Piscataway, NJ)
Vitamin B_{12} 2.7 mg

APPARATUS SETUP

The apparatus is set up as shown in Figure E22-2.

BIG PICTURE

Do not wash the column with organic solvents. Add 0.75 g of Sephadex G-100 to 15 mL of hot 0.1M NaCl. After the gel swells, the column is filled and rinsed with 0.1 M NaCl, and the sample is added. The sample is then eluted with 0.1 M NaCl. Blue dextran, m.w. 2,000,000, yellow dextran m.w. 20,000, and vitamin B_{12} m.w. 1,357 will be separated. Blue dextran is used to determine the bed void volume.

THIS EXPERIMENT TAKES ABOUT 4 HOURS. The last 2 hours require very little effort, and it is suggested that you plan on doing another experiment or work on something else.

DIRECTIONS

1. Turn on a hot plate to a high position.

2. Remove the column from the ring stand, if it is attached.

3. Start with the bottom end first. This is the part that has a small diameter plastic tube 7 to 10 cm long sticking out of the end. Unscrew the red section. Notice a small black O-ring that is on the column. This must be greased, or it may leak. Place a small amount of vacuum grease on the O-ring.

4. Look inside of the red plastic base and notice the fine mesh plastic sieve that supports the column packing.

5. Reassemble the base parts.

6. Go to the top of the column. There should be a silver screw about 1.5 cm long in the hole at the top of the column. If it is not there, locate it and place it in that hole. This is the filling port.

7. Unscrew the red top assembly and lubricate that O-ring. Do not reassemble this part.

8. Place the column vertically on a ring stand.

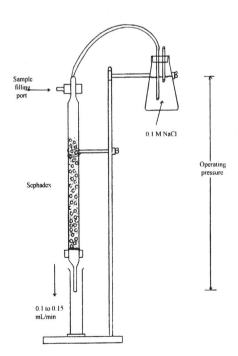

Figure E22-2. Apparatus for a gel filtration.

9. Place about 50 mL of 0.1 M NaCl into a 100 mL beaker, place it on a hot plate, and bring it to a boil Go to step 10 immediately.

10. The object of this step is to fill the column just below the plastic frit with liquid so air bubbles will not form there. They can cause the column to stop flowing. Fill a syringe with a few mL of cold 0.1 M NaCl solution, attach it to the end of the plastic tube on the outlet, and slowly inject NaCl into the base until it seeps through the frit. Remove the syringe and seal the end of the plastic tube with a small wire plug.

11. Place 0.8 g of Sephadex G-100 into a 25 mL graduated cylinder.

12. Place a cloth towel around the beaker containing the boiling NaCl solution to protect your hands and add 15 mL of the hot solution to the graduated cylinder.

13. Gently stir the mixture with a stirring rod for about 10 minutes. Add another 2 to 3 mL of the NaCl solution.

14. Set the gel aside for about 15 minutes to swell and to settle.

15. Make up a sample solution, if one is not already prepared. Do this by adding 3 mg of the sample to a 3 mL volumetric flask and dilute to the mark with 0.1 M NaCl.

16. Remove the excess water from above the Sephadex beads in the graduated cylinder with a syringe.

17. Carefully, pour the Sephadex into the column. Pour it down the sidewalls so you do not get any air bubbles in the column. (Normally the column would be filled with liquid before the Sephadex is added. However, it takes a long time to drain away the excess and that is not comparable with the time allotted for this experiment). (NOTE: if an air bubble forms in the column, pour out the contents and try again).

18. Let the contents settle for about 5 minutes.

19. Place the top back on the column and make sure the column is still vertical.

20. Place about 200 mL of cold 0.1 M NaCl solution in the reservoir flask and adjust the flask so its bottom is about 30 cm above the outlet of the plastic tube at the bottom of the column.

21. Connect the rubber tubing from the reservoir flask to the top of the column.

22. You need to remove the air from the connecting tubing. Remove the silver metal plug from the filling port.

23. Place a pipet bulb on the glass tube in the stopper covering the 250 mL reservoir flask.

24. Gently squeeze the rubber bulb and force liquid from the reservoir into the column. Do this slowly to minimize the entrapment of air bubbles. As .soon as liquid comes out of the filling port, place the metal plug in it, but do not release the pressure on the pipet bulb until the plug is tight. If you release the pressure before the filling port is closed, then you will get air bubbles in the column. Remove the pipet bulb.

25. Remove the plug from the drain line at the bottom of the column, place the line into a beaker, and let the column drain for about 25 to 30 minutes. This flushes out the column and settles the beads. (The bed volume will reduce to about half of its original volume).

26. Disconnect the tubing from the reservoir and stop the flow.

27. Remove the plastic fitting from the top of the column.

28. Use a pipet and carefully withdraw as much liquid from the top of the beads as you can without disturbing them.

29. Like me and most others, you probably stopped too soon. To correct this, unplug the bottom drain tube and let the liquid drain until it is just at the top of the beads. Close the tubing.

30. Place 0.3 mL of sample solution in a syringe and add the liquid dropwise to the top of the beads.

31. Remove the drain plug and allow the sample liquid to drain into the beads until the liquid is just at the top of the beads. Close the drain.

32. Add three 4-drop portions of salt solution to the top of the bed to wash the sample into the beads and level the top of the beads.

33. Drain off this excess liquid.

34. Carefully fill the column above the beads with salt solution as you did previously.

35. Place an empty 25 mL graduated cylinder under the drain tube and allow the sample to be eluted. The normal flow rate is about 2 to 3 drops/min.

36. Measure the volume of eluent collected to completely remove the blue dextran from the column. This is the *void volume* of the column, because the blue dextran molecules are too large to be retained.

37. Continue the separation only as far beyond this point as necessary to see that the yellow dextran is separated from the red vitamin B_{12}.

38. Clean up the column as follows. Remove the top, insert a syringe filled with water into the outlet tubing, tip the column upside down, and force the water slowly into the column. You may throw away the Sephadex, or you may take it home as a token of our friendship. *Do not use organic solvents to rinse out the column, or our friendship ends.*

39. Reassemble the column so the parts won't get lost. Leave the salt solution in the reservoir. Clean up the area. Thank you.

EXPERIMENT 23. FLASH CHROMATOGRAPHY
Separation of vanillin from vanillin extract.
(Courtesy - Buszek, K., Department of Chemistry, Kansas State University)

BACKGROUND
Vanillin comes from a water-ethanol extract of the dried and aged beans of the vanilla vine, a member of the orchard family (*Vanilla planifolia*) naturally occurring in Mexico. Vanillin also occurs in potato parings and in Siam benzoin. It is made synthetically from eugenol or guaiacol and from the waste lignin from the paper industry. It is a favorite flavoring agent.

The vine requires a hot and wet climate and grows in the shade. In the winter, it produces bean pods about 20 cm long. These have no aroma. The pods are collected and when sweated nightly and exposed daily to sunlight for 10 days, they become chocolate brown and attain their aroma. The pods are dried and cured for 4 to 5 months, ground, and extracted with ethanol. The quality of the beans is indicated by the number of small crystals of vanillin on the outside of the pod. Vanillin is not naturally present in the exterior of the pod, but is secreted through hair-like papillae and becomes a part of the oily coating. Cured pods contain about 2% vanilla. Vanilla is formed in the plant by an enzymatic action on coniferin as shown below.

YOUR PROBLEM
The local police have some powdered vegetable material that they suspect contains marijuana, and they want to use the Duquenois test to prove it. However, they are out of reagent, and the local chemistry laboratory is out of vanillin, a key part of the formula. It is the beginning of a holiday weekend, and it may be several days before they can get some vanillin the normal way. To use vanilla extract is unacceptable, because the dark brown color will obscure the Duquenois test. You believe that you can purify enough extract by a flash chromatography separation to permit the police to prepare the reagent, because only 20 to 25 mg would be necessary. You know enough chemistry to use the pure extract, not the imitation (FDA regulation designation) type, because it contains propylene glycol, which will ruin the chromatographic separation.

EQUIPMENT
1	Air line	1	Clamp, 3-way, small
1	Beaker, 10 mL	1	Cutter, glass
1	Beaker, 50 mL	1	Cylinder, graduated, 50 mL
1	Beaker, 100 mL	2	Flask, Erlenmeyer, 25 mL
1	Bottom, burner	1	Flask, Erlenmeyer, 50 mL
1	Box, fluorescence viewing	1	Flask, Erlenmeyer, 125 mL
1	Bulb, pipet, 2 mL	1	Funnel, powder, small
1	Burner, Bunsen	1	Funnel, separatory, 125 mL

1	Forceps	1	T, plastic or glass
1	Holder, clamp	6	Thin layer plates, 2 cm x 5 cm Silica Gel 60 F_{254}
1	Holder, culture tubes		
1	Hotplate	2	Towels, paper
1 bx	Kimwipes	18	Tubes, culture, 10 x 75 mm
1	Lamp, uv	3 ft	Tubing, burner
1 bx	Matches	8 ft	Tubing, plastic, 8 mm
1	Paper, filter, 11 cm	1	Valve, to regulate air flow
1	Pencil	3	Vials, 1 dram
4	Pipets, disposable, 5.5 inches	1	Watch glass, 5 cm
1	Ruler, metric		
1	Sponge		
1	Stand, ring		

CHEMICALS   S = Student

Acetone, $H_3C(C=O)CH_3$, 25 mL/S
Cotton, 0.01 g/S
Developing solutions: 50%, 30 %, 20 %, 10 % v/v ethyl acetate, hexanes, 30 mL/S
Dip reagents:
    Anisaldehyde: Store at room temperature.
        1,350 mL Absolute ethanol, H_3CCH_2OH
        50 mL Sulfuric acid, H_2SO_4
        15 mL Glacial acetic acid, H_3CCOOH
        37 mL Anisaldehyde, $OHC-C_6H_4-OCH_3$
    Universal dip reagent   (optional)
        100 mL Water
        5 g (2.84 mL) Sulfuric acid, H_2SO_4
        3 g Phosphomolybdic acid, $20MoO_3-P_2O_5-51H_2O$
        Pinch of ceric sulfate, $Ce(SO_4)_2$, (0.5%)
        Stir overnight.
Cotton, 0.01 g/S
Ethyl acetate, $H_3C-C(=O)-CH_2CH_3$, 75 mL/S
Hexanes (petroleum ether), 50 mL/S
Magnesium sulfate, anhydrous, $MgSO_4$, 10 g/S
Sea sand, 0.1g/S
Silica Gel 60, 230 to 400 mesh ASTM, 5 g/S
Thin layer plates, Silica Gel 60 with 254 nm fluorescent dye added, 250 μm thick.
Vanillin standard, 10 mg/S
Vanillin extract, pure, not imitation, 10 mL/S

APPARATUS SETUP

The apparatus is set up as shown in Figure 18-7, p. 182. A 5.5 inch disposable pipet, packed as described below, is held in place with a small 3-pronged clamp mounted on a ring stand. A 25 mL Erlenmeyer flask for eluting solvent is placed nearby. A series of 10 x 75 mm test tubes in a plastic rack is placed by the outlet, and one is shown in place. The air line is held in place as needed. A flask for rinse solvent is to the upper left.

THE BIG PICTURE

The sample and standards are spotted on a TLC plate and developed with each of the developing solvents, dried, and examined under uv radiation. The R_f's are determined, and the solvent providing an R_f between 0.2 and 0.35 is selected. A microscale column is prepared, and the sample added. 15 to 18 fractions are collected and spotted on two TLC plates, dried, examined with uv radiation and stained. The fractions containing the desired compound are combined, the solvent is evaporated and the compounds are collected.

DIRECTIONS

This experiment should be done in a hood, but the amounts of chemicals are so small that it can be done safely on a bench top as well - your choice.

1. Place 10 mL of pure vanilla extract in a 125 mL separatory funnel and add 15 mL of water.
2. Extract three times with ethyl acetate and combine the extracts in a 125 mL Erlenmeyer flask.
3. Add about 10 g of anhydrous magnesium sulfate to the combined extracts and let stand for 0.5 hr.
4. Transfer to a 100 mL beaker (filter through a plug of cotton in a powder funnel) and place on a hot plate at medium heat until the solvent has been removed, but not to dryness.
5. Prepare several TLC plates, if they are not already available. Place a 20 x 20 cm TLC plate upside down on a paper towel on a clean bench top. Place a straight edge (board or ruler) on it and score it about every 2 cm with a glass cutter. Turn the plate and score it about every 5 cm. Break the pieces apart and store them in a box. Try to touch the ends and edges only, or wear gloves.
6. Turn on a hot plate (hood preferred). Make sure that the dipping solvents are nearby.
7. Prepare the air pressure line. If this is not available, cut an 8 to 9 ft length of 8 mm plastic tubing into three nearly equal length pieces. Connect each piece to the same T. Connect one open end to the bottom of a burner. This is to be the pressure release valve. Connect one open end to the air line. The air line outlet should have a needle valve attached to it to adjust the air flow. The third open end is to be used to force the solvent through the column.
8. Test the air line. You want to have a gentle pressure, with any excess going out the needle valve of the burner bottom. Open the needle valve part way. Turn on the air line and hold your finger over the end to be placed on the column. Adjust the needle valve so you get a gentle pressure (experience helps), then leave it alone and remember how far you turned on the air line valve. You should not be able to hear the air flow, but you should feel pressure if you hold your finger over the tubing, and you should feel air flow out the needle valve.
9. Attach a small 3-prong clamp to a ring stand with a clamp holder.
10. Prepare the sample spotting tubes if they are not already available. NOTE: normal capillary m.p. tubes and the tips of disposable pipets are too large and unsatisfactory. Turn on a Bunsen burner and heat a 2-cm section of a disposable pipet about 5 to 6 cm from the tip end. Pull out this capillary for about 30 to 40 cm. Break it into 10 to 12 cm lengths and store them upright in a 50 mL Erlenmeyer flask.
11. Obtain three 1 dram vials, a 25 mL Erlenmeyer flask with 5-10 mL of acetone in it, the sample, and the standards. Use the top end of a disposable pipet, dip it into the sample, and obtain 2 to 3 mg. Tap this off into one of the vials, then rinse the pipet by dipping it into acetone. Repeat this for each of the standards, placing each into a separate vial.
12. Use a disposable pipet and add a *few drops* of acetone to each vial. Swirl the vial to dissolve the solid. The liquid level should be about 2 to 3 mm in the bottom of the vial. Keep it as concentrated as possible.
13. Place about 5 mL of the 50% ethyl acetate/hexanes solvent in the bottom of either a 50 mL beaker and cover it with a small watch glass or use a 60 mL weighing bottle. The level should not be more than 3 to 4 mm from the bottom.
14. Place a TLC plate on a piece of paper toweling. Handle it with forceps on what you will use as the top of the plate.
15. Use a sharp pencil and, with a straight edge, make a pencil mark (light pressure, just so you can see it) about 5 mm from the bottom across the bottom of the TLC plate. This is your spotting line. Place a small dot with the pencil at three places along this line with the outer ones being at least 2 mm from the edge of the plate to reduce edge effects.
16. Dip the end of your spotting pipet into one of the vials, and it will draw in sample by capillary action. Withdraw the pipet, wipe off the end with your fingers or a Kimwipe, and place a single drop of liquid on the marked spot. Do not let the pipet tip stay on the spot, just touch it and withdraw it. Blow on the spot to dry it (about 3 to 4 seconds) and spot it again. Do this three times to build up the concentration, yet keep the spot small.
17. Rinse the pipet tip in acetone and use it for the next sample.
18. Repeat step 10 for each sample.
19. Use a forceps and place the TLC plate in the developing jar. Be sure the solvent does not cover the spots at the start. A 120 mL weighing bottle or a 150 mL beaker are suitable developing tanks.
20. When the solvent has risen to about 1 cm from the top, remove the plate (forceps) and place a pencil mark where the solvent front is.

Experiment 23

21. Take the plate to a uv viewing box, if available, or to a dark place otherwise. Irradiate the plate with uv radiation and locate the spots. Circle each one with a pencil mark.
22. Place the plate on a hot plate for about 5 seconds. If it is on too long, it will crack during the next step.
23. Dip the plate into the anisaldehyde solution (usually in a small weighing bottle) twice for about 2 seconds each time.
24. Wipe off the bottom of the plate with a Kimwipe and place the plate on the hot plate again. After a few minutes, any spots should begin to develop. After about 2 to 3 minutes, remove the plate and wipe off the bottom with a Kimwipe. You often can see the development better from the bottom side.
25. Determine the R_f of each spot (measure from the front of the spot; remember it is R_f). If the spot you want has an R_f between 0.2 to 0.3, then you are through. If not, then repeat steps 9 to 21 with the other solvents until you find a best one.
26. Place 18 10 x 75 mm culture tubes in a rack and set them aside.
27. Place a *very small* piece of cotton in the end of a 5.5 inch long disposable pipet and tamp it down with a second pipet so it extends into the pipet about 3 to 4 mm, 5 mm at most.
28. Place the top end of the pipet into a jar of sea sand and pick up a few grains of sea sand. Tip up the pipet and allow the sand to settle on top of the cotton (2 mm at most). This keeps the silica gel from falling through the cotton.
29. Place the pipet in a 3-prong holder and adjust its height so the tip will just enter a 25 mL Erlenmeyer flask placed under it.
30. Obtain two 25 mL Erlenmeyer flasks and place 20 mL of the chosen solvent into one.
31. Roll a 11 cm piece of filter paper into a funnel and place it in the top of the second Erlenmeyer flask.
32. Add about 5 grams of silica gel to the flask.
33. Add about 5 mL of the chosen solvent to the silica gel and swirl to mix.
34. You will now slurry pack a micro flash chromatography column. Place a small pipet bulb on a disposable pipet and pipet solvent into the column to a height of about 2 cm. With the same pipet, pipet several drops of the silica gel slurry into the column. Allow this to settle by gravity, then continue the addition until the silica gel is just to the neck at the top of the column. If air pressure is used at this step it can pack the gel too tightly.
35. Let the excess solvent drain out, then add solvent to the very top of the column to rinse it. Air pressure can be used now to adjust the flow. You want to apply enough pressure to have about 2 drops/second come out the bottom.
36. Use the top end of a pipet as a spatula and place about 10 mg of sample in a vial. Then with a pipet, add the *minimum amount* of solvent to dissolve it. This minimizes tailing. If necessary, draw the mixture up into the pipet and flush it out several times to help dissolve the sample.
37. Pipet this onto the top of the column and let it settle by gravity. Rinse the pipet with acetone and store it upright in a 125 mL Erlenmeyer flask.
38. Place a 10 x 75 mm culture tube under the column and adjust the column so the tip enters the tube about 3-4 mm.
39. Place a small pipet bulb on a pipet and fill the column reservoir with eluting solvent.
40. Apply air pressure to force the solvent through the column at about 2 drops/second. You can adjust this by how hard you hold the tubing on the top of the column.
41. Remove the tube and place it in the rack. Replace it with another tube and repeat steps 35 to 37.
42. Continue with steps 35 to 37 until you have collected 18 tubes.
43. Prepare two TLC plates as before, except mark 9 places for spotting on each plate.
44. Spot as before, taking a sample from each tube and spot them in order from 1 to 18 on the two plates.
45. Develop the first plate as before and view it under uv radiation. Place a pencil mark around the developed spots
46. Heat the plate and then dip it into the anisaldehyde reagent as before. Record which tubes have sample in them.
47. Repeat steps 41 to 42 with the second plate.

48. Combine the contents of the tubes containing the desired compound in a 10 mL beaker, warm to evaporate the solvent, and collect the purified product.

49. Shut off the hot plate and the air supply. Rinse all of the pipets with acetone and store them in the 125 mL Erlenmeyer flask. Empty the culture tubes into a waste container, rinse them with acetone, and store them for future use. You may take the column home as a token of our friendship.

50. Take a look at the work area and make sure it is left cleaner and better organized than it was when you started. Thank you.

EXPERIMENT 24. HIGH PERFORMANCE LIQUID CHROMATOGRAPHY
Detecting explosive residues

BACKGROUND

As an example, in the United States there is an average of 1500 bombings yearly involving explosives of various kinds. The detonation of a bomb leaves in its wake a large number of fragments and a lot of debris, not only from the bomb but from the surrounding area. In order for the investigation to proceed, the basic components of the bomb must be reconstructed. One of those components is the type of explosive used.

As a general rule, the bomb investigator searches the debris for pieces of those materials that the bomb was constructed from and then examines those pieces for crystals of unburned explosive, so he can determine the nature of the bomb and its sophistication. The more sophisticated the bomb is the fewer the suspects.

Chemical spot tests often are used to determine which pieces of debris contain explosive residues. This then is followed either by a microscopic examination of the crystals for a positive identification or to use HPLC for identification.

The debris from a bomb blast site usually contains traces of the undetonated explosive. Therefore, the location of the explosive at the point of detonation must be located and the debris and closely surrounding soil should be removed, packaged, and labeled. The objects located near the explosion are good sources of explosive residue. Wood, insulation, rubber, or other soft materials which are readily penetrated often collect traces of the explosive.

Particles of the explosive are usually found adhering to the pipe cap or to the pipe threads in pipe bombs. This occurs either by handling during the construction or being impacted into the metal by the force of the explosion.

Types of Explosives

There are many ways to classify explosives, some of which are: military, civilian, high-order, or low-order. Few bombings involve military explosives (e.g., PETN, RDX, TNT), because their availability is severely restricted. The most commonly used explosives are dynamite and homemade mixtures containing black powder, smokeless powder, or chlorate and sugar. Table E24-1 lists several explosives and their common abbreviations. The structures of several of these are shown in Table E24-2, p. 580.

Table E24-1. Abbreviations of common explosives

2,4-DNT	2,4-Dinitrotoluene	NS	Nitrostarch
2,6-DNT	2,6-Dinitrotoluene	PETN	Pentaerythritol tetranitrate
EGDN	Ethyleneglycol	RDX	Cyclotrimethylene trinitramine
MMAN	Monomethylamine nitrate	TNT	2,4,6-Trinitrotoluene
NC	Nitrocellulose	TETRA	2,4,6-Trinitrophenyl
NG	Nitroglycerin		methylnitramine
		TNT	2,4,6-Trinitrotoluene

Dynamite

Dynamite is by far the most commonly encountered high-order explosive used in destructive devices associated with criminal acts in this country. There are a number of types of dynamites, including straight dynamite, ammonia dynamite, blasting gelatin, ammonia gelatin, and nitrostarch dynamite. The color of these explosive compounds ranges widely - from off-white to nearly black.

In spite of the variety of dynamite formulations, only a few components need to be identified to confirm the presence of dynamite residues in bomb debris. These components are sulfur, ammonium nitrate, sodium nitrate, and the

Table E24-2. Structures of several common explosives.

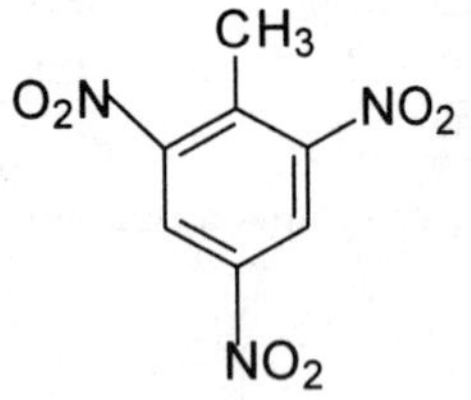

[Structures shown: TNT, Picric Acid, Styphnic Acid, Tetryl, RDX, HMX, EGDN, NG, PETN, Nitrocellulose (2,000 - 3,000)]

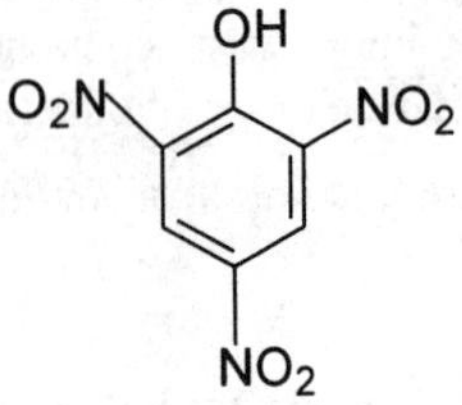

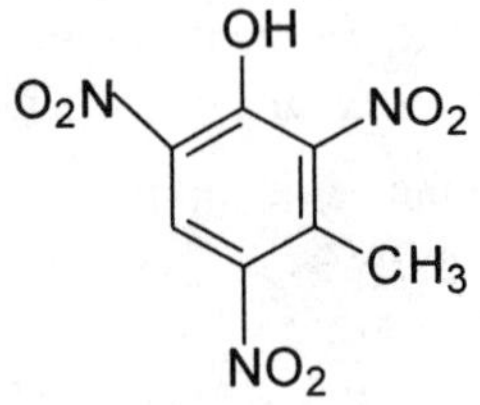

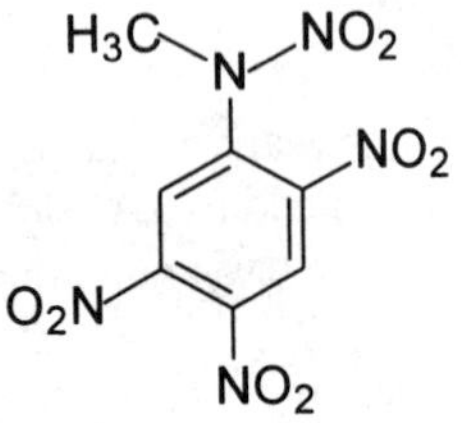

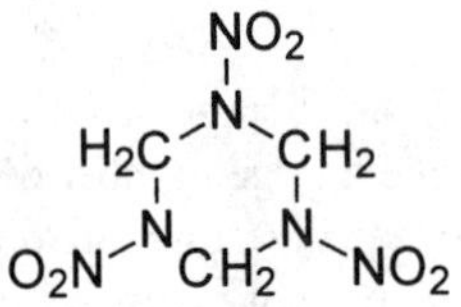

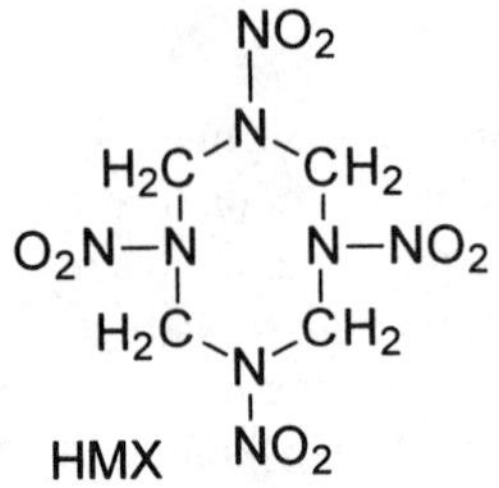

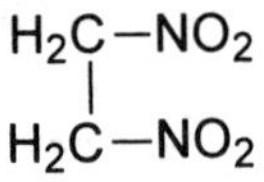

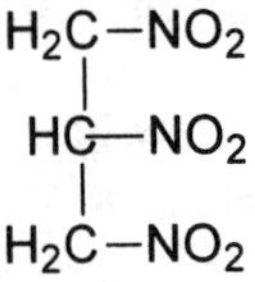

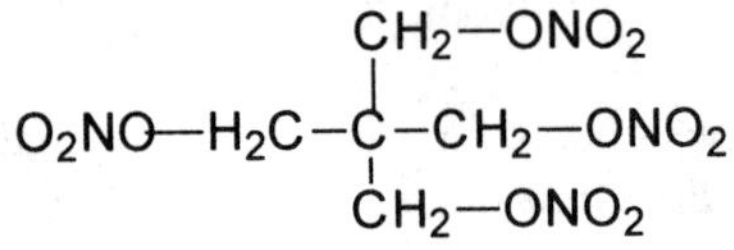

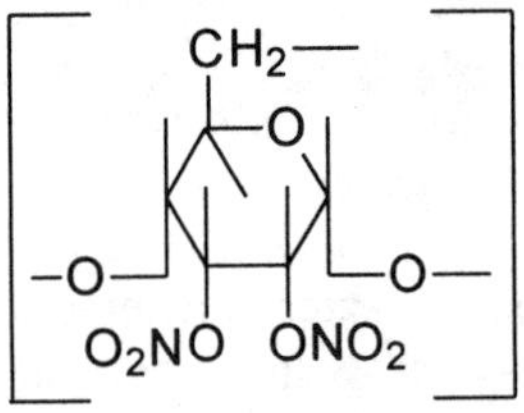

explosive oils [nitroglycerin (NG) and ethylene glycol dinitrate (EGDN)] absorbed on the binder, or "dope", which makes up the bulk of the dynamite compound.

Most dynamites contain sodium nitrate, but only ammonia dynamite contains ammonium nitrate. All commercial dynamites, with the exception of nitrostarch dynamite, contain explosive oils in their binders.

Improvised Mixtures

Low-order explosive mixtures made from commonly available chemicals have been termed *improvised mixtures*. The majority of the improvised mixtures used in criminal bombings in this country fall into two categories: homemade black powder and homemade flash powders. These mixtures usually are confined in some container, such as a section of capped pipe, and initiated with an external burning pyrotechnic or homemade fuse. Homemade black powder is a mixture of potassium nitrate, sulfur, and charcoal. It is usually poorly made, and unconsumed particles are abundant in the bomb debris. The components of this mixture can be identified chemically using techniques described previously.

Homemade flash powder mixtures have a variety of formulations. They usually contain finely divided aluminum or magnesium metal and an oxidizing agent, such as ammonium chlorate, potassium chlorate, or potassium perchlorate, with possible additions of sulfur and/or sawdust.

YOUR PROBLEM

A bomb exploded yesterday outside the marine recruiting station. A huge crowd gathered immediately, and they destroyed nearly as much as the bomb. A detective managed to gather up several pieces of debris and put them in a box.

They were examined by another chemist, and one piece that tested positive to aromatic nitrates has been given to you to determine which nitrate is present.

EQUIPMENT

1	Adapter, suction filter	1	Funnel, filter, 50 mL M sintered glass
1	Adapter, vacuum, 24/40		
1	Aspirator, water	1 pr	Gloves, plastic
1	Beaker, 30 mL	1 pr	Goggles, safety
3	Bottles, 500 mL	1 pk	Paper, filter, 4.25 cm
1	Bottle, trap, 200 mL	1	Stopper, rubber, 1 hole
1	Chromatograph, HPLC, Allteck 153	1	Stopper, rubber, no. 4, 1 hole
1	Cylinder, graduated, 500 mL	1	Syringe, 25 µL
2	Elbows, glass elbows, 7 mm	2	Syringes, 10 mL
1	Flask, volumetric, 100 mL	3 ft	Tubing, vacuum

CHEMICALS   S = Student

Methanol, H_3C-OH, HPLC grade, 100 mL/S
Water, HPLC grade, 400 mL/S
Standards:
    o-Nitrotoluene; O_2N-C_6H_4-CH_3, 8.6 µL = 10 mg, 10 mg/100 mL 70% MeOH-H_2O
    Diphenylamine; $(C_6H_5)_2NH$, same as above.
    Picric acid; $(NO_2)_3$-C_6H_2-OH, same as above.
    m-Dinitrobenzene; $(NO_2)_2$-C_6H_4, same as above.
Unknown: any combination of the above.
    NOTE: Any aromatic nitro group type explosive can be used. These are not always easy to obtain, but small amounts of almost any explosive can be dissolved in acetone and stored safely.

THE BIG PICTURE

You first will become familiar with a way to prepare HPLC grade solvents in the event you either cannot afford to purchase them or you run out in the middle of a project. You then will become familiar with the operation of a single pump HPLC and then separate and identify several explosive materials with an "isocratic solvent". You then will identify the components in an unknown mixture.

APPARATUS SETUP

Figure E24-1 shows an apparatus for degassing a solvent. Figure E24-2, p. 582, is a block diagram for the component parts of an Alltech Model 153 HPLC. This is probably the simplest HPLC available and is selected because it is modular and has minimum computer control, so the principles can be more easily understood. The wavelength is 254 nm.

DIRECTIONS

The following directions are based on an Alltech Model 153 HPLC. This is one of the simplest HPLC's made. Yours probably will be different and your instructor will tell you what changes to make. Do not fear handling solutions of these explosives. When present as dilute solutions in methanol or acetone, they are not explosive.

1. Locate the apparatus and identify each item by referring to Figure E24-2. *Have your instructor demonstrate how to inject a sample into the system.*

2. Turn on the equipment so it can warm up. This is done as follows:

    A. Detector power supply - wavelength fixed at 254 nm- push in the red button.

    B. Pump - power switch on. Make sure the flow rate is at zero - zero. You decrease the flow rate by pushing the button at the bottom just below the numbers.

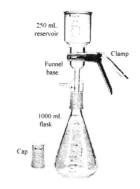

Figure E24-1. Apparatus for degassing a solvent.

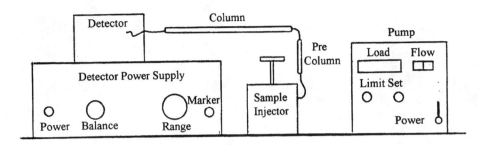

Figure E24-2. Block diagram of the HPLC apparatus.

    C. Recorder - Turn to "Standby," set the dampening control to "maximum," turn the "sensitivity" control completely counterclockwise, and set the "range" to 25 mv.

3. Make sure the waste reservoir jug (on the floor) is at least half empty. Discard any excess into the sink and flush with water. Reinsert the drain tube.

4. Take the degassing apparatus and a 500 mL bottle to a water aspirator.

5. Examine the apparatus (Figure E24-1, p. 581). Place a piece of fine filter paper on top of the fine-pore, sintered glass disc to remove the larger particles and keep the sintered glass from plugging.

6. Connect the apparatus to a water aspirator with a trap between the aspirator and the collection bottle. A distilled solvent is placed in the funnel and suction filtered directly into the solvent container. The air is removed during the evacuation process. The trap is used to prevent water contamination in the event the water pressure should decrease.

7. Obtain a 500 mL graduated cylinder and a bottle of HPLC grade methanol.

8. Filter 360 mL of methanol into the 500 mL bottle.

9. Obtain 90 mL of deionized-distilled water. Filter this as in step 8, adding it to the methanol already in the 500 mL bottle.

10. Keep the vacuum on and swirl the contents of the bottle to mix them.

11. Remove the vacuum tubing and then shut off the aspirator.

12. The solvent in the instrument will now be changed. Remove the solvent inlet filter, which is the metal piece about 4 cm long and 1 cm in diameter and place it in a small beaker.

13. The plastic tube connected to that metal filter goes to a black screw connection. Unscrew that piece of plastic tubing, remove the black part and blow out (mouth) all of the solvent that is in that tubing. You can tell when the solvent has been removed, because it will stop foaming at the metal frit.

14. Set the bottle of 70% methanol-30% water on top of the pump and place the filter into the bottle.

15. Prime the pump. This is done by attaching a 10 mL syringe with a screw-type tip to the sinker tubing to the screw end of the white plastic tubing just cleaned out.

16. Withdraw 7 to 8 mL of liquid into the syringe, so the plastic tubing is filled with this new solvent.

17. Disconnect the syringe and connect the plastic tube back onto the black plastic tip from which it was previously removed.

18. Eject the solvent in the syringe into a small beaker.

19. Turn on the pump to 1.0 mL/min. You can increase the flow rate by pushing the buttons on the top of the double zero-zero and you decrease the flow rate by pushing the buttons below the zero-zero.

20. Let the pump operate for 10 to 15 minutes to flush out the system and establish a steady baseline.

21. Flush out the sampling valve. Use a 5 to 10 mL syringe with a *flat-tipped* needle. Pour about 10 mL of solvent into a clean small beaker and fill the syringe with 3-4 mL of solvent.

22. Place the syringe needle into the syringe holder and gently tighten the holder. Do not tighten it too much or the inner seal can be broken.
23. Slowly add solvent into the sample valve until it runs out the bottom of the white plastic tube dipping into the 100 mL beaker, then remove the syringe.
24. Empty the excess solvent into a small beaker.
25. Turn the left recorder knob to the "servo" position.
24. Use the balance knob on the left side of the detector power supply and adjust the pen on the recorder to about 10% full scale. Do not put the recorder pen on the paper.
27. Change the flow rate to 2.0 mL/min.
28. Notice that the column pressure is slowly increasing. It should get in the range of 3200 to 3300 p.s.i. and stop.
29. The dials below the pressure gauge are the limit set and will shut off the pump if the pressure reaches the set limit. Set the right hand adjustment to 3400 to 3,500 p.s.i., and the left hand adjustment to the "reset" position.
30. Place the chart pen on the paper.
31. Turn the recorder knob from "servo" to "chart." The recorder speed is 1 inch/min. Be sure that the paper is moving and the pen is working.
32. Adjust the pen to the about 10% full scale.
33. It usually will require a few minutes for the system to reach equilibrium. If a stable baseline has not been obtained within 15 minutes, get help from the instructor.
34. Obtain a 25 µL syringe with a *flat tip* and the volumetric flasks containing the standards and the unknown.
35. Rinse the syringe as follows: Place a few mL of the solvent in a small beaker and flush the syringe 5 to 6 times.
36. Obtain 10 µL of any one of the standards as follows. Draw a sample into the syringe and push it out. Repeat this 3 to 4 times until all of the air bubbles are removed from the barrel. When the syringe is filled with liquid and no air bubbles are present, adjust the volume to 10 µL.
37. Add the sample as demonstrated by your instructor. As soon as the sample is added, mark on the chart paper the starting point. The marker for the chart paper is controlled by the white button on the right side of the detector. When the sample is injected, push this button, and the pen should deflect a small amount. If not, make a small mark on the paper with a pen.
38. Notice the peak height and the attenuator setting. If the peak height is not at least half the width of the chart paper, change the setting for the next sample.
39. Repeat the process for each standard and obtain a retention time for each.
40. Inject 10 µL of the unknown and identify the components in the unknown.
41. Estimate the amount of each component present. The standards contained 10 mg/100 mL. Determine the area obtained per µg of standard.
42. Remove the syringe from the sample loop and flush it thoroughly with solvent.
43. Shut off the recorder, remove the pen from the paper, and place the cover on the pen.
44. Shut off the pump; the red light will slowly go out - don't worry.
45. Shut off the detector power supply and *clean up the area.* Thank you.
46. Your instructor will tell you where to find more elaborate HPLC's. Examine at least one of them and identify the components.

EXPERIMENT 25. HIGH PERFORMANCE LIQUID CHROMATOGRAPHY
Separation of aspirin, acetaminophen, and caffeine in Excedrin

BACKGROUND
Extra-strength Excedrin contains aspirin, acetaminophen, caffeine, microcrystalline cellulose, and stearic acid.

"Aspirin", acetylsalicylic acid, was obtained by Charles Gerhardt, a German chemist, as a natural byproduct of coal tar in 1853. Its medical value was not recognized until 1899 by Heinrich Dreser, a German chemist. It is an analgesic, antipyretic, and antirheumatic.

"Acetaminophen", 4-hydroxyacetanilide, was first prepared in 1878 by Morse in Germany. It is an analgesic and antipyretic.

"Caffeine", 1,3,7-trimethylxanthine, has been discussed previously and is used as a CNS, respiratory, and cardiac stimulant.

YOUR PROBLEM
One of the jobs of the FDA is to monitor drug formulations to ensure that the amount of each drug listed on the label is actually present. The tolerances are from 95 to 105% of the declared amount. You are a new trainee, and to familiarize you with HPLC as well as some sample preparation techniques, you are given a sample of Excedrin and told to determine if the drugs listed on the label are in fact present and if they are within the accepted range.

EQUIPMENT

1	Balance, analytical		1	HPLC, like Alltech Model 153
1	Bath, steam		1	Mortar and pestle
4	Beakers, 100 mL		1	Paper, filter, 11 cm
1	Bottle, wash		1	Pipet, delivery, 10 mL
1	Bulb, pipet		1	Pipet, delivery, 25 mL
1	Column, 30 cm x 3.9 mm µ Bondapak C-18		1	Rack, funnel
			1	Recorder, Heathkit
1	Cylinder, graduated, 100 mL		1	Spatula
6	Flasks, volumetric, 100 mL		1	Stand, ring
1	Funnel, 60°, short stem		1	Syringe, 10 mL, w/ millipore filter
1 pr	Goggles		1	Towel, cloth

CHEMICALS   S = Student
Acetaminophen, 0.05 mg/mL in ethanol std.
Acetic acid, HOAc, conc., 1 mL/S
Acetonitrile, CH_3CH_2CN, HPLC grade, 100 mL/S
Aspirin, 0.5 mg/mL in ethanol std.
Caffeine, 0.1 mg/mL in ethanol std.
Ethanol, 95%, 100 mL/S

APPARATUS SETUP
The apparatus consists of a degassing system and a single pump, single wavelength detector HPLC the same as that described in Experiment 24, p. 579.

THE BIG PICTURE

A weighed tablet is powdered in a mortar and pestle. While it is being dissolved over a steam bath, the separating solvent is being filtered and degassed. The apparatus is brought to equilibrium, and a set of standards is run. The dissolved tablet then is examined, and the amount of each of the active components is determined.

DIRECTIONS

The directions below are based on FDA Laboratory Information Bulletin 2405.

1. Locate the apparatus and identify each item by referring to Figure E24-2, p. 582. *Have your instructor demonstrate how to inject a sample into the system.*

2. Turn on the equipment so it can warm up. This is done as follows:
    A. Detector power supply - wavelength fixed at 254 nm- push in the red button
    B. Pump - power switch on. Make sure the flow rate is at zero - zero. You decrease the flow rate by pushing the button at the bottom just below the numbers.
    C. Recorder - Turn to "Standby" set the dampening control to "maximu," turn the "sensitivity" control completely counterclockwise, and set the "range" to 25 mv.

3. Make sure the waste reservoir jug (on the floor) is at least half empty. Discard any excess into the sink and flush with water. Reinsert the drain tube.

4. Weigh a single table of Excedrin to 0.1 mg.

5. Powder the tablet with a mortar and pestle.

6. Transfer a weighed amount of the powdered sample to a 100 mL beaker and add 50 mL of ethanol.

7. Place the beaker on a steam bath (or a beaker of water covered with a metal plate and heated with a hot plate) to dissolve the caffeine.

8. Take the degassing apparatus and a 500 mL bottle to a water aspirator.

9. Examine the apparatus (Figure E24-1, p. 581). Place a piece of fine filter paper on top of the fine-pore, sintered glass disc to remove the larger particles and keep the sintered glass from plugging.

10. Connect the apparatus to a water aspirator with a trap between the aspirator and the collection bottle. A distilled solvent is placed in the funnel and suction filtered directly into the solvent container. The air is removed during the evacuation process. The trap is used to prevent water contamination in the event the water pressure should decrease.

11. Obtain a 100 mL graduated cylinder and a bottle of HPLC grade acetonitrile.

12. Filter 16 mL of acetonitrile into the 500 mL bottle.

13. Obtain 188 mL of deionized-distilled water. Filter this as in step 10, adding it to the acetonitrile already in the 500 mL bottle.

14. Filter 2 mL of concentrated acetic acid into the 500 mL bottle.

15. Keep the vacuum on and swirl the contents of the bottle to mix them.

16. Remove the vacuum tubing and then shut off the aspirator.

17. Remove the sample from the steam bath. It must be filtered to remove the microcrystalline cellulose, which eventually would plug any HPLC column.

18. Place a millipore filter on the end of a 10 mL syringe.

19. Transfer several mL of the sample to the syringe and filter it, collecting the sample in a 100 mL volumetric flask. Repeat the above as many times as necessary, until the entire sample has been filtered.

20. Dilute to volume with ethanol.

21. Refer to the amount of sample you dissolved and determine how much further dilution will be necessary to make the sample concentrations close to those of the standards. Read the label on the bottle to determine the amounts that should be present. For extra-strength Excedrin, each tablet contains 250 mg of aspirin and acetaminophen and 65 mg of caffeine.

    For example: 0.5 g of sample would contain about 185 mg of aspirin. This would be 1.85 mg/mL as presently diluted. 25 mL further diluted to 100 mL would be 0.46 mg/mL. Select a proper pipet and dilute as needed.

22. Follow steps 12 through 46 of experiment 25, substituting the solvent, standards and sample required for this experiment.

The conditions established by LIB 2405 were:

Column:	30 cm x 3.9 mm i.d. μ Bondapak C-18
Mobile phase:	8% acetonitrile in water containing 1% HOAc
Column oven:	35 °C
Flow rate:	2.3 mL/min
Sensitivity:	0.32 absorbance units full scale (AUFS) for acetaminophen
	0.16 AUFS for caffeine and aspirin
Chart speed:	5 min/inch
Pressure:	1300 p.s.i..
Detector:	uv @ 254 nm
Sample loop:	20 μL

The approximate retention times for these parameters are acetaminophen: 2.8 min (linear from 0.021 to 0.073 mg/mL); caffeine: 5.3 min (linear from 0.045 mg/mL to 0.18 mg/mL); and aspirin: 11 min (linear from 0.2 mg/mL to 0.8 mg/mL).

EXAMPLE CALCULATION

A tablet of extra-strength Excedrin weighed 0.674 g. It was treated as described above, and after it was powdered, 0.593 g was transferred to a 100 mL beaker to be dissolved. The dissolved sample first was diluted to 100 mL and then 20 mL again diluted to 100 mL. 20 μL of a standard containing 0.62 mg aspirin/mL had a peak height of 100 mm. 20 μL of the sample had a peak height of 70 mm. What is the concentration of aspirin in this sample?

ANSWER

Determine the sample concentration in mg/mL, correct it for dilution, and then determine the amount per tablet.

A. $$\frac{100 \text{ mm}}{0.62 \text{ mg/mL}} = \frac{70 \text{ mm}}{X \text{ mg/mL}}; \quad X = 0.43 \text{ mg/mL}$$

B. $$\frac{0.43 \text{ mg/mL}}{20 \text{ mL}} \times 100 \text{ mL} \times 100 \text{ mL} = 215 \text{ mg in } 0.593 \text{ g of sample}$$

C. $$\frac{215 \text{ mg}}{0.593 \text{ g}} \times \frac{0.674 \text{ g}}{\text{Tablet}} = 244 \text{ mg/tablet}$$

D. $$\frac{244 \text{ mg}}{250 \text{ mg}} \times 100 = 98 \% \text{ of declared}$$

EXPERIMENT 26. GAS-LIQUID CHROMATOGRAPHY- ARSON INVESTIGATION

Detection of accelerants in debris by head space analysis.

BACKGROUND

Arson is one of the highest dollar crimes committed in the United States. In 1996 1,975,000 fires caused $9.4 billion worth of damage, 25,550 people were injured, and there were 4990 deaths. Of these, 85,000 fires were considered arsons causing $1,405,000,000 in losses. Firemen usually can tell a suspicious fire almost immediately. Two questions must be answered; (1) was an accelerant used, and (2) what type is it? The answer to the first question is necessary to determine if legal proceedings will be required and for insurance payments. The answer to the second question is of importance in helping the detectives locate the arsonist.

Fires burn up, outward, and downwind, and accidental fires start in only one place. If an accelerant has been poured on wood, then the wood burns hotter and forms an *alligator scale* pattern. If an accelerant has been poured on the floor, carpeting burns more completely than the surrounding carpet, and floor tiles often rise where the accelerant was. This is known as the *pour pattern*, and if you stand back and look at a large area, you can usually see it.

Unless it was a very hot fire such as one fanned by a high wind, some accelerant almost always soak into the wood, under a sill, below the carpet, into a crack in the cement, under floor tile, into the stuffing of a car seat, or into the ground, and remains unburned or partially unburned. .

The fire marshal will determine the point of origin and look for starting devices such as partially burned candles, clocks, and unburned matches. He then will collect samples of debris at the suspect point. *Note*: GET A BLANK, or you probably will lose in court.

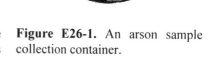

Figure E26-1. An arson sample collection container.

BLANKS: A blank is of prime importance. Many materials can produce combustion products that look like accelerant residues. For example, glues from carpets, hydrocarbons from floor tiles, or essential oils from grasses. I once had a house fire to investigate that supposedly was set by pouring gasoline on the side of the house. The sample was some grass and soil, and there were many peaks that looked liked possible gasoline peaks. However, the blank of grass and soil showed that those peaks were the oils from the grass, and an accelerant had not been used.

The blank should be taken from a place as close to where you suspect the fire started and be of the same material as the sample. It should be from a burned area, so the combustion products will be similar.

A criminalistics laboratory will have a file (that they prepare for their instrument) of chromatograms for several types of gasolines, kerosenes, lantern fluids, lighter fluids, and almost anything that can be easily purchased and used by an arsonist.

Samples are placed either in new quart paint cans or quart fruit jars (Figure E26-1). The containers containing the samples are heated in an oven to about 120 °C for 20 minutes to drive the traces of accelerant from the sample. A hole is punched in the lid (do this before heating and cover it with a septum), and 2 mL of head space gas is examined with a gas chromatograph fitted with an FID detector.

This is called headspace analysis and can be applied to a variety of samples. It is a common commercial technique.

YOUR PROBLEM

A house caught fire late one evening, but the fire was put out after burning part of one side, where a fireplace was located. It could have started from an overheated fireplace, or it could have been arson. Upon questioning by the police, the owner indicated that his daughter had had a quarrel with her boyfriend a few days earlier. The police obtained a warrant and found an empty can of gasoline in the trunk of the boyfriend's car. They were able to obtain a few mL of gasoline from the can, some partially burned boards from the side of the house, and some soil from along side of the house on the burned side. Appropriate blanks were obtained. The police give you these samples, and you are asked to determine if an accelerant is present.

EQUIPMENT

1 Chromatograph, gas, with FID detector	1 Recorder, 1 mv full scale
1 Column, 1.8 m x 3 mm, SS	3 Regulators, gas, H_2, air, He
3 Holder, gas cylinder	1 Syringe, 10 mL, gas tight
4 Jars, fruit, 1 qt, fitted with silicone septums on the lid	1 Syringe, 10 µL, liquid
	1 Oven, 130 °C

CHEMICALS   S = Student

1 Cylinder, H_2	1 Tube, Super Glue
1 Cylinder, Air	SE-30, 10% on Chromosorb-W, 60-80 mesh
1 Cylinder, He	Gasoline samples, a variety of gasolines, kerosenes, lighter fluids; store in small plastic vials.

APPARATUS SETUP

The apparatus is a gas chromatograph with an FID and temperature programming. The recorder is 1 mv full scale. Your instructor will tell you what modifications are needed. However, the directions given here show the sequence needed, and the sample preparation is the same regardless.

Chromatograph settings:
- Injection port temperature: 270 °C
- FID detector temperature: 270 °C
- Column temperatures: Initial 35 °C; 4 min initial hold; ramp at 12 °C/min to 250 °C; 1 min final hold
- Detector attenuation: 10
- Helium flow rate: 25 mL/min
- Hydrogen flow rate: 30 mL/min
- Air flow rate: 60 mL/min
- Recorder chart speed: 1 cm/min

THE BIG PICTURE

The containers of samples are fitted with a septum on the lid so a syringe needle can be inserted. The containers are placed (in turn) in an oven at 130 °C for 20 minutes to evaporate the accelerants from the debris.

A gas-tight syringe is placed in the septum covering, and a 2.0 mL sample of head space gas is obtained and injected into the chromatograph. A 1.8 m x 3 mm i.d. SS column packed with 10 % SE-30 on 60-80 mesh Chromosorb-W is used to separate the components. Be sure to do the blank in the same manner, and do as many standards as you feel necessary.

Figure E26-2 shows what one brand of pure gasoline looks like. This is 5 µL of liquid, input attenuation 10X, 25 mL He/min, 30 mL H$_2$/min, 60 mL air/min, FID 270°, Inj 270°, 4 min hold at 35°, program at 12 °C/min to 250°, 1 min final hold. Notice that the attenuation is changed several times during the run, from 2048 to 256 and to 8. An arson chromatograph usually will not look exactly like this figure. Normally, the low boilers have burned away, and you usually get few accelerant peaks until after about 6 minutes, then the peaks will be small. In addition, there will be several peaks from the burned material that must be subtracted. You need to subtract the blank, then match retention times and look at the overall pattern to decide if an accelerant was used.

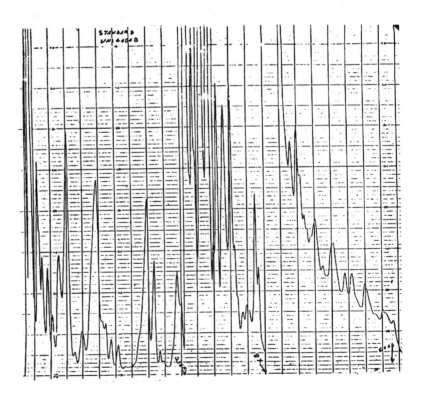

Figure E26-2. Chromatogram of Standard unleaded gasoline. 6', SS, 1/8", 10% SE-30 on 60 to 80 mesh Chromasorb W.

DIRECTIONS

These are generic directions as your chromatograph will be different from the authors. However, the sequence is likely to be the same and a strip chart recorder is used because it is less expensive and the author believes it is a better teaching tool than a software computer.

1. Turn on the main cylinder valve (counterclockwise) on the helium tank. Check the high pressure side of the regulators. If there is less than 400 p.s.i., have the TA change the cylinder. The pressure will drop 200 to 300 p.s.i. during this experiment.

2. Turn the handle on the regulator valve (clockwise) until a pressure of 18 to 22 p.s.i. is indicated on the low pressure gauge.

3. Repeat steps 1 and 2 for the hydrogen cylinder (red cylinder) and set the *low* side at 18 to 20 p.s.i..

4. Repeat steps 1 and 2 for the compressed air cylinder (yellow cylinder) and set the *low* side to about 20 p.s.i..

5. Repeat steps 1 and 2 for the nitrogen cylinder (orange cylinder) -Nitrogen is used to operate the cooling valve.

6. If there is a main control valve for the gases, turn it on. This controls the flow of gases to the chromatograph.

7. Readjust the low side gauges, if necessary, to the 18 to 22 p.s.i. region.

8. Turn on the main power switch.

9. Turn on the oven blower if has a separate switch.

10. Turn the recorder on.

11. You now will make fine adjustments of the flow rates of each gas. Adjust the He flow to 25 mL/min. This is a setting of 3.5 on the author's rotameter.

12. Adjust the H_2 to a flow rate of 30 mL/min. This is a setting of 30 on the author's rotameter.

13. Adjust the air to read about 0.6, (60 mL/min).

14. Ignite the flame of the FID by momentarily pushing the switch. There may be a "pop" when it lights. Remove the cover and test it to be sure it is on by holding a mirror or shiny spatula about 2 cm above the top of the cylinder. A fogging from condensed water should occur.

15. The chromatographic conditions will now be set on the microprocessor. NOTE: usually, every step must have a value, even if it is 0, or the program will not work. Use the values listed in the apparatus setup.

16. Set the "detector temperature" at 270 °C.

17. Set the "injection port temperature" at 270 °C.

18. Set the "initial oven temperature" at 35 °C.

19. Set the "final oven temperature" at 250 °C.

20. Set the "program rate" at 12 °C/min.

21. Set the "initial hold" for 4 minutes.

22. Set the "final hold" for 1 minute.

23. Examine the columns in the oven while the instrument is heating. Open the oven door just a few inches. Locate the fan and the columns and notice that they are tagged and how they are connected. Close the door.

24. Locate the detector attenuation controls. The numbers 1, 10, 100, 1000 are multipliers. If you push in 1000, then you DECREASE the signal from the detector 1000 times. This is called "attenuation" and is used to help keep the recorder pen on scale. It reduces large signals and increases weak signals. Push in on the "10" button.

25. Some instruments have "automatic attenuation." It automatically will alter the peak height to keep it on scale. The difficulty is that you do not know the attenuation and cannot do quantitative work. Turn this off. Obtain the most sensitive signal and watch the pen on the recorder.

26. Use the zero adjust, try to adjust the pen to "10" on the chart paper. Try for about 15 seconds, then go to step 27.

27. Tough wasn't it? Move the "zero range" switch to a less sensitive position and try again.

28. Get a 10 µL GC syringe. A GC syringe has a sharp needle tip; an HPLC syringe has a flat needle tip.

29. The sample is injected into a heated injection port. By now it is HOT; be careful.

30. Practice a few injections. Inject straight down.
31. Get a 10 mL gas-tight syringe. This has a control knob just before the needle. Practice working the control adjust. It costs about $85.00; handle accordingly.
32. Obtain a piece of paper and write down these conditions for later reference. The column is 1.8 m stainless steel, 3 mm o.d., packed with 10% SE-30 on 60 to 80 mesh Chromosorb-W. The helium flow rate is 25 mL/min., the gas flow rates for the detector are 30 mL/min for H_2) and 60 mL/min of air. The initial oven temperature is 35 °C with a 4 minute hold. The ramp rate is 12 °C/min to 250 °C, with a final hold of 1 min. Both the inlet port and detector block temperatures are 270 °C. The "input" attenuation is initially at 10, and the "output" attenuation is initially at 2048. After 10 minutes, the output attenuation will be changed to 256 for 5 minutes, then changed to 8 and left there until the programmer reaches the cool cycle.
33. Find a 20 mL beaker and pour 1 to 2 mL of gasoline into it. Obtain a second 20 mL beaker for waste.
34. Remove the cover from the recorder pen and place it in a small beaker taped to the side of the recorder.
35. Push the "pen lift" switch to "on", and the pen will lower onto the paper.
36. Turn the "chart" to "on" and set the "chart speed" to 1 cm/min.
37. When you inject a sample, there are two things you must do as quickly as possible. Immediately after the sample injection, you must start the temperature programing (the initial hold will take care of the timing) and you must make a mark on the chart paper when the sample was injected.
38. Readjust the recorder pen to 10 if it has drifted away.
39. Inject a 5 µL sample.
40. Write on the chart paper what the sample is, the size, the attenuation, the column, and the temperatures.
41. After 10 minutes, switch the attenuation to 256, and then after 5 minutes, switch it to 8.
42. When the instrument goes into the COOL cycle, open up the front of the oven and push "cool" on the programmer. This turns on an auxiliary fan and more quickly gets the oven back to room temperature.
43. Inject a 5 µL sample of any of the following: lighter fluid, lantern fluid, or kerosene. Flush the syringe several times with the new liquid before you inject it. IMMEDIATELY GO TO STEP 44 AFTER THE INJECTION.

    The arson sample. You must bake out the syringe and heat the sample and blank to 130 °C for at least 20 minutes. Obtain a sample and a blank from your instructor. These will be in 1 quart glass jars with a rubber septum sealed onto the lid with Super Glue. (Figure E26-1, p. 587).

44. Remove the plunger from a 10 mL gas-tight syringe and place both pieces on a watch glass in the oven. Place the jars containing the sample and the blank in the oven and note the time, being sure the lids are screwed on tightly.

    NOTE: Bake out the syringe between samples. It is easy to contaminate a gas syringe. Place it in the oven and bake it while the sample is heating. Handle it with gloves, so you won't get burned when you take it out.

45. Allow the chromatograph to cool down from the second liquid sample.
46. Obtain a cloth towel and remove the syringe from the oven. After a minute or so, replace the plunger and push it completely forward to remove all of the air. *Note*: to save time, you will not test the syringe to see if it is clean. Ordinarily, that would be done.
47. Remove the jars containing the sample.
48. Check the chromatograph and determine when it is back to 37 to 38 °C.
49. Push on the "green" button to open the syringe and obtain about 3.0 mL sample of head space gas from the sample jar. Push in the "red" button to seal in the sample.
50. The attenuation settings must be changed, because even a 2.0 mL gas sample won't contain as much sample as 5 µL of liquid. Set the "input" to "1" and the "output" to 8.
51. Readjust the recorder pen to 10, if necessary.

52. You will inject a 2.0 mL sample into the chromatograph. Push on the "green" button to open the needle, then carefully push in the plunger to a 2.0 mL volume. Inject the sample into the chromatograph, start the temperature program, and mark the chart.

53. Do not change the attenuations during this run.

54. Bake out the syringe while this sample is running.

55. Repeat steps 44 to 52 as appropriate for the blank.

56. Put the syringe needles back in their boxes and back in the drawer.

57. "Lift" the recorder pen and replace the cover over the tip.

58. Turn the "lift" off, the chart off, and the recorder "power" off.

59. Shut off the main "power" switch and the "blower" switch on the chromatograph.

60. Turn off the main cylinder valves on each gas cylinder. *Do not* turn off the toggle valves. This will allow the gas to bleed from the regulators, allowing them to last longer, and the hydrogen flame will *slowly* go out.

61. Clean up the area. Leave the sample heating oven on.

62. Compare your chromatograms. Tear them apart, place the blank under the sample, and hold them up to the light or place them on a window. Match the starting times and mark those peaks that are from the blank. Figure E26-2 shows a chromatogram of Standard unleaded gasoline.

63. Repeat step 62 with your reference chromatograms. What you look for are the overall patterns. Match peak ratios and the patterns of a group of peaks. The sample has been subjected to high temperatures but the standards have not, so they won't be exactly the same. However, if an accelerant was used, then you may see some sort of a pattern remaining.

EXPERIMENT 27. GAS-LIQUID CHROMATOGRAPHY- DETERMINATION OF ANTIOXIDANTS
The determination of BHA and BHT antioxidants in breakfast cereals
(Based on Takahashi, D.M., *J. Assoc. Off. Anal. Chem.*, **50** (4), 880, 1967)

BACKGROUND
Oxygen reacts with lipids in foods to form acids and carbonyl compounds that have unpleasant tastes and odors, and the food is said to be *rancid*. The reaction, called *autoxidation,* is a *free radical chain reaction* that involves the formation of peroxides. Once it begins, it proceeds rapidly.

Initiation:
$$RH + O_2 \rightarrow R\cdot + HO_2\cdot$$
$$n\,ROOH \rightarrow RO\cdot + RO_2\cdot + HO\cdot$$

Propagation:
$$R\cdot + O_2 \rightarrow RO_2\cdot$$
$$RO_2\cdot + RH \rightarrow ROOH + R\cdot$$

Termination:
$$2\,R\cdot$$
$$RO_2\cdot + R\cdot \rightarrow \text{Stable products}$$
$$2\,RO_2\cdot$$

The initiation reactions can be accelerated by traces of metal ions that can oxidize and reduce the hydroperoxides.
$$M^n + ROOH \rightarrow M^{n+1} + RO\cdot + OH^-$$
$$M^{n+1} + ROOH \rightarrow M^n + RO_2\cdot + H^+$$

Many foods contain compounds such as vitamins C and E to act as antioxidants, but their amounts are small, and their effects are short lived.

Spices and herbs have been used for centuries to mask unpleasant tastes and odors and to slow down spoilage. It has been found that sage, cloves, oregano, rosemary, and thyme prevent peroxide formation. Food chemists have found that they can synthesize compounds that work more effectively, cost less, are of known purity, and are free of insect contamination. Those compounds used most often are butylated hydroxy anisole (BHA), butylated hydroxy toluene (BHT), and propylgallate shown below.

Experiment 27

[Structures: Vitamin C (Ascorbic Acid); Vitamin E (Alpha-Tocopherol); BHA (5-10% and 90-95% isomers); BHT; Propyl Gallate]

These structures have excellent solubilities in food oils and fats. All are essentially nontoxic and leave the body within 24 hours after ingestion if not used. Tests with mice indicated that those fed these compounds as a part of their diet lived 40 % longer than the controls. This is understandable, because one mechanism of cell destruction is by oxidation, and these compounds are antioxidants. It is difficult for the author to understand how people can decry the addition of "preservatives" on the one hand and complain about wasted food on the other hand.

One mechanism of an antioxidant is

$$ROO\cdot + InH \longrightarrow ROOH + In\cdot$$

$$In\cdot + ROO\cdot \longrightarrow InOOR$$

For example:

[Reaction scheme: BHT + 2 ROO· → quinone-type product with OOR group + ROO·]

Citric, ascorbic, and phosphoric acids are added to chelate the trace metals, particularly copper and iron, to retard the initiation step. Mixtures of antioxidants and chelating agents act *synergistically,* that is, the combination is more effective than the sum of the individual compounds. A commercial mixture recommended by the American Meat Institute and sold as *Tenox* or *Sustane* is 20% BHA, 6% propyl gallate, and 4% citric acid in propylene glycol. The propylene glycol is added to solubilize the citric acid in the oil.

These compounds are used as antioxidants in food for humans, animal feeds, plastics, soaps, petroleum products, synthetic rubber, and animal and vegetable oils. The amounts produced annually (metric tons) for use in the United States are BHA (1400); BHT (450); PG (230), and the sum of others including ascorbic acid, sodium and calcium ascorbates, erythorbic acid, and gum guaic (< 230).

YOUR PROBLEM

Breakfast cereals are particularly susceptible to the formation of rancidity once the box is opened, because it may be weeks before the product is completely consumed. Antioxidants thus prevent considerable waste. The favorite antioxidant is BHT. You have been given two samples of breakfast food, one that is supposed to contain "preservatives" and one that is "natural". Your job is to determine if the labels are correct. You may use either column or, if desired, both columns.

EQUIPMENT

1	Balance, triple beam	1 pr	Goggles
2	Beakers, 100 mL	1	Hot plate
1	Bottle, wash, plastic, 250 mL	1 bx	Kimwipes
1	Chromatograph, gas, with FID detector	1	Mortar and pestle
		1	Pipet, 1 mL
2	Clamps, buret	1	Recorder, 1 mv full scale
2	Columns, glass, 25 x 200 mm	1	Rod, glass, 1 cm x 30 cm
1	Cylinder of air	1	Ruler, mm scale
1	Cylinder of hydrogen	1	Sieve, 20 mesh
1	Cylinder of nitrogen	1	Spatula
1	Cylinder, graduated, 100 mL	1	Stand, ring
1	Dish, weighing, Al	3	Supports, gas cylinder
2	Flasks, volumetric, 5 mL	2	Syringes, Hamilton, 10 µL
2	Flasks, volumetric, 50 mL, low actinic	1	Towel, cloth

CHEMICALS   S = Student

Either:

Column, glass, 1.2 m x 4 mm. Wash column and glass wool plugs with 5% (v/v) dichlorodimethyl silane in toluene (in a hood) before packing with 1 g Apiezon L/ 20 g 80 to 100 mesh Gas Chrom Q. Condition for 3 days at 200 °C at 10 mL N_2/min or until a steady baseline is obtained. Order of elution: CS_2, BHA, BHT, di-BHA. This column will separate the 2 and 3 isomers of BHA.

OR

Column, glass, 1.8 m x 4 mm. Wash as above and pack with 2 g QF-1 (FS1265)/20 g 80 to 100 mesh Gas Chrom Q. Condition as above. Order of elution: CS_2, BHT, BHA, di-BHA.

Internal standard solution:

5 mg 3,5-Di-*t*-butyl hydroxyanisole/ 50 mL CS_2 (0.1 µg/µL). Store in low actinic glassware.

Standard solution mixture: (0.02 µg/µL)

1.0 mg of each/50 mL of CS_2. Store in a low actinic flask.

Butylated hydroxy anisole
Butylated hydroxy toluene
Di-BHA

Carbon disulfide, CS_2, 150 mL/S.

Glass wool, 2 g/S.

Sample: any cereal that lists BHA or BHT as an added preservative and one that doesn't.

APPARATUS SETUP

The apparatus is a gas chromatograph with an FID and temperature programming. The recorder is 1 mv full scale. Your instructor will tell you what modifications are needed. However, the directions given here show the sequence needed, and the sample preparation is the same regardless.

Chromatograph settings:

Injection port temperature: 220 °C
FID detector temperature: 220 °C
Column temperatures: 160 °C
Detector attenuation: 10
Nitrogen flow rate: 57 mL/min
Hydrogen flow rate: 40 mL/min (Apiezon L) or 25 mL/min (QF-1).
Air flow rate: 515 mL/min
Recorder chart speed: 1 cm/min

BIG PICTURE

Twenty grams of each of the cereals are ground separately to a fine powder in a mortar and pestle and placed in 20 cm long glass columns. The BHA and BHT are extracted with CS_2, evaporated to 2 to 3 mL, 0.8 mL of an internal standard added, and diluted to 4.0 mL. Standards are injected to establish retention times and calibrate the peaks. The unknown is injected, and the results are calculated.

594 Experiment 27

DIRECTIONS

Many of the directions for this experiment related to the gas chromatograph are similar. Read over the steps 1 to 31 of Experiment 26, p. 589, remembering that nitrogen is the carrier gas now and that no temperature programming will be done.

1. Turn on the instrument and the recorder (standby) to let the electronics warm up.

2. Turn on the main gas cylinder valves for H_2, N_2, and air.

3. Adjust the nitrogen flow rate to 57 mL/min (5.7 on the rotameter).

4. Adjust the air flow rate to 515 mL/min (9.8 on the rotameter).

5. Adjust the hydrogen flow rate to either 40 mL/min (Apiezon L) or 25 mL/min (QF-1).

6. Ignite the flame.

7. Adjust the column temperature to 220 °C.

8. Adjust the injection port temperature to 220 °C.

9. Adjust the column temperature to 160 °C.

Do steps 10 to 18 and 20 to 21 simultaneously for both cereals.

10. Weigh 20 g of cereal (± 0.01 g), record the weight, and place the sample in a mortar.

11. Grind the sample to a fine powder with the pestle.

12. Attach a 20 x 200 mm glass columns to a ring stand with a buret clamp.

13. Place a small amount of glass wool in the bottom of the column. Use a glass rod to tamp it in place, if necessary.

14. Add the ground cereal to the column and place a small piece of glass wool on top of it. Tamp the sample firmly with a glass rod to reduce the possibility of channeling.

15. Place a 100 mL graduated beaker under the column and about 60 mL of CS_2 in a graduated cylinder. (*Caution: CS_2 is very flammable (b.p. = 46 °C) and can be ignited by hot steam lines*). The extraction and concentration steps should be done in a hood.

16. Add the CS_2 to the top of the column as follows: 5 mL and allow it to completely wet the sample. Repeat two more times, then add 10 mL at a time until 50 mL has eluted.

17. Place the beaker of eluent on a hot plate (low) and evaporate the CS_2 until only 1 to 2 mL remain. Be careful - *if you have collected more than 50 mL, then divide the sample.* The only time an accident occurred was when a student collected 75 mL, and evaporated it too fast, spilled over, and caught fire from the hot coils of the hot plate. Know where the fire extinguisher is located, but don't panic.

18. While the evaporation is taking place, check the recorder, particularly the pen to make sure it writes.

19. If everything is going well and you believe you have time, you can inject your standards into the chromatograph while the CS_2 finishes evaporating - but don't let it go to dryness.

20. Transfer the residue to a 5.0 mL volumetric flask with no more than 1-2 mL of CS_2 from a wash bottle.

21. Add 1.0 mL of the di-BHA internal standard and make to final volume with CS_2. Mix thoroughly.

22. Check the chromatograph and make sure everything is functioning and that the flame is still on. If the recorder went off scale, it was probably due to the flame going out.

23. The recorder should now be adjusted so the pen is within 1 cm of the 0 line and the chart speed is set at 1 cm/min.

24. Obtain a 10 μL syringe and flush it several times with the standard solution.

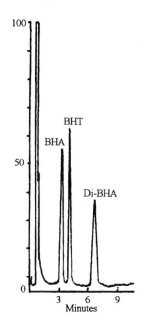

Figure E27-1. Chromatogram of antioxidant standards. (Courtesy - Takahashi, D.M., *J. Assoc. Off. Anal. Chem.*, **50**, 880, 1967).

25. Inject a 3.0 to 9.0 μL sample (record the amount) of the standard mixture and mark on the recorder paper when the injection was made. Figure E27-1, p. 594, shows a typical chromatogram from an Apiezon L column.

26. Inject a 3.0 to 9.0 μL sample of the cereal extract (record the amount).

27. Measure each peak height in mm.

28. Calculate the ppm antioxidant present, correcting for the internal standard, as follows:

ppm BHA or BHT = *(H/H') x (C'/C) x (H'$_i$/H$_i$) x (C$_i$/C'$_i$)* (E27-1)

where H = height in mm of the sample. H' = height in mm of the standard. H_i = height in mm of the internal standard in the sample. H'_i = height in mm of the internal standard in the standard. C = conc of sample (g/μL). C' = conc of standard (μg/μL). C_i = conc (μg/μl) of the internal standard in the sample. C'_i = conc (μg/μl) of the internal standard in the standard.

EXAMPLE CALCULATION

A sample of cereal, ground so fine as to be unrecognizable, was sent to an analyst as part of a quality assurance program to test his ability. He extracted a 20.7 g sample and followed the directions in Experiment 28. A 6.0 μL sample of the extract showed a BHT peak of 41 mm and a di-BHA peak of 34 mm. 5.0 μL of the standards produced a 43 mm peak height for BHA and a 28 mm height for di-BHA. How many ppm of BHT does this sample of cereal contain?

ANSWER

H = 123 mm; H' = 30 mm: C' = 0.02 μg/μL; C = 20.7 g/5.0 mL = 0.00414 g/μL; H'_i = 29 mm; H_i = 23 mm; C_i = 0.02 μg/mL (1 mL of 0.1 μg/μL diluted to 5.0 mL); C'_i = 0.02 μg/μL.

$$ppm\ BHT = \frac{123\ mm}{30\ mm} \times \frac{0.02\ \mu g/\mu L}{0.00414\ \mu g/\mu L} \times \frac{29\ mm}{23\ mm} \times \frac{0.02\ \mu g/\mu L}{0.02\ \mu g/mL} = 25\ ppm$$

EXPERIMENT 28.   PAPER CHROMATOGRAPHY

The determination of sulfamethazine, sulfametrazine, and sulfadiazine in Trisulfa tablets.
(*Pharmacopea of the United States*, 7th ed. pp. 735-37.)

Note: This experiment takes about 2 hours of actual laboratory time to complete, but there is a 16 to 20 hour (overnight) development time in between.

BACKGROUND

The following is taken in part from the World Book Encyclopedia. Sulfonamides are synthetic antimicrobial agents with a wide spectrum encompassing most gram + and gram - organisms. In 1935, the German bacteriologist, Gerhard Domagk, reported that Prontosil killed streptococcal bacteria. He was awarded the Nobel Prize in 1939, but the Nazie's would not let him accept it.

H_2N—⌬—COOH H_2N—⌬—SO_2—NH—⌬(N=/=N)

PABA   Sulfadiazine

Sulfa drugs do not kill bacteria, but stop them from reproducing. Bacteria need para amino benzoic acid (PABA) to reproduce. The sulfonamide drugs have a chemical structure similar to that of PABA, but they contain an S atom where PABA has a C atom. Bacteria cannot tell the difference and use the sulfa compound. The S atoms then stop one or more of the growth processes, and the bacteria cannot multiply. The body's defense kills the remaining bacteria.

Sulfa drugs reduced death rates from 50 out of 100 to 5 out of 100 for one form of meningitis and from 12 out of 100 to 5 out of 100 for pneumococcal pneumonia.

Sulfa drugs are used to treat pneumonia, dysentery, meningitis, blood poisoning, urinary tract infections, erysipelas, cellulitis, bubonic plague, cholera, rheumatic fever, and some venereal diseases.

Sulfa drugs must be present for long periods; therefore, the doses are high and continuous. Because some sulfa drugs have a tendency to crystallize and cause kidney damage, sufficient liquids must be taken to produce 1,200 to 1,500 mL of urine daily. Normal dosages are 1 gram, 4 to 6 times daily.

Multiple-ingredient tablets were quite popular for several years, because each drug could be designed to attack bacteria at various places throughout the body. This was effective when the source of the infection could not be located and time was of the essence. However, a few people became concerned that a person could develop a resistance to sulfa drugs and raised such a fuss that the FDA decided to permit only single drugs to be administered at one time. One trisulfa tablet, Neotrizine, was left in use primarily for infants, who cannot tell you where it hurts. It is short acting and provides therapeutic levels with a reduced risk of renal damage. Neotrizine contains sulfadiazine (shown above), sulfamerazine, and sulfamethazine.

$H_2N-\text{C}_6H_4-SO_2-NH-\text{(pyrimidine-}CH_3\text{)}$

Sulfamerazine

$H_2N-\text{C}_6H_4-SO_2-NH-\text{(pyrimidine-}(CH_3)_2\text{)}$

Sulfamethazine

YOUR PROBLEM

Counterfeit drug formulations, which often contain inexpensive mild tranquilizers to calm the patient, are sometimes substituted for real drugs. You are given a tablet that is reported to be a trisulfa containing sulfamethazine, sulfamerazine, and sulfadiazine. Your job is to determine if the tablet is valid.

EQUIPMENT

	1	Beaker, 50 mL	1	Paper, chromatographic, 2.5 x 70 cm
	1	Beaker, 100 mL	1 pk	Paper, filter, 11 cm.
	1	Bottle, wash	1	Pipet, lambda, 10
	1	Bulb, pipet, small	1	Rack, funnel
	1	Casserole, size 1	1	Rod, glass, stirring
	1	Chamber, chromatography, 10 cm x 75 cm, descending	1	Rod, glass, 30 cm, 8 to 10 mm in dia.
	1	Clamp, buret	1 pr	Scissors
	1	Clamp, versatile	1	Spatula
	1	Clothes pin	2	Sprayers, chromatographic
	1	Cylinder, graduated, 10 mL	1	Stand, ring
	1	Dryer, hair	1	Stopper, rubber, no. 5, 1 hole
	1	Funnel, 60°	2	Stoppers, rubber, no. 11, solid, 1 1 hole.
1 pr		Gloves, plastic	1	Towel or sponge
1 pr		Goggles, safety	1	Tube, glass, 75 cm x 10 cm
	1	Heat gun		
	1	Holder, clamp		
	1	Mortar and pestle, small		

CHEMICALS   S = Student

n-Butanol, $H_3C(CH_2)_3$-OH, saturated with an equal volume of $NH_3 + H_2O$ (8 + 93). Keep the top and bottom layers, 100 mL/S

Butyl nitrite, $H_3C(CH_2)_3$-ONO, 50 mL of n-butanol + 50 mL of butyl nitrite + 3 mL acetic acid. 100 mL/S

N-(1-Naphthyl) ethylenediamine dihydrochloride 100 mg in 93 mL of n-butanol. 10 mL/S

HCl, Conc. 5 mL/S

Trisulfa tablets - Neotrizine, 167 mg/tablet/drug. Requires a doctor's prescription. 1/S

BIG PICTURE

A tablet is ground in a mortar and pestle, and the sulfa drugs are dissolved in HCl and filtered from the residue. The descending technique is used. n-Butanol saturated with NH_3-H_2O (3 + 97) is the mobile phase. The development time is 16 to 18 hours. The sulfa drugs are located by first treating the paper with butyl nitrite and then adding

N-(1-naphthyl)ethylenediamine dihydro chloride. Purple spots appear in a few minutes.

Note: This experiment takes about 45 minutes to set up and then about 45 minutes at the end to develop the colors. The 18 hours in between can be scheduled rather liberally because they are not that critical. It is common to start at 4:00 p.m. one day and develop the strip the next morning before lunch. The spots are then separated about 1 cm from each other and are about midway down the paper.

APPARATUS SETUP

The apparatus consists of a 10 cm diameter glass tube about 75 cm long, held vertically on a ring stand by a chain clamp. The solvent chamber is made from 25 mm glass tubing with one end closed and flattened. The antsiphon bar and the paper holder bar are made from 8 mm glass rod.

DIRECTIONS

1. Refer to Figure E28-1 for the basic apparatus setup. Check the list of equipment and be sure all of the necessary items are present.

2. Carefully loosen the chain clamp that is around the equilibrium chamber, raise the chamber vertically, and remove the beaker resting on the bottom rubber stopper.

3. Fill the beaker 1/2 to 3/4 full with the 1-butanol:ammonia:water *lower layer*.

4. Place the beaker back in the apparatus and secure the equilibrium chamber.

5. Very carefully remove the stopper holding the mobile phase tank from the top of the apparatus. Fill that container to within 2 cm of the top with the 1-butanol:ammonia:water, *top layer*, and loosely set the tank back on top of the chamber.

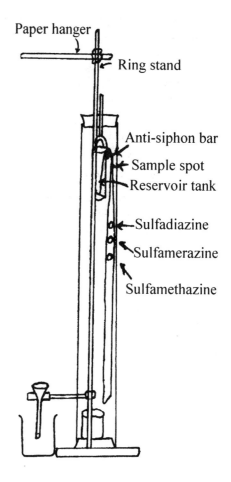

Figure E28-1. Apparatus for descending paper chromatography.

6. Obtain one of the test sulfa tablets, place it in a mortar, and grind it to a fine powder with the pestle.

7. Transfer the powdered sample to a small casserole and add 5 mL of water and 2 mL of concentrated HCl.

8. Stir the mixture with a glass rod for 5 to 10 minutes to dissolve the drug out of the excipient materials. Crush the larger particles with the end of the glass rod as you stir the mixture.

9. Use a fast-flow filter paper and filter the undissolved material away from the dissolved drugs. Collect the filtrate in a small beaker.

10. Use tweezers or finger cots or wear plastic gloves and obtain a 65 to 70 cm length of 2.5 cm wide paper. Return the unused paper to its container immediately to minimize contamination from room fumes and dust.

11. Take the curl out of the paper by pulling it through your fingers. The paper doesn't have to be perfectly straight, but it needs to be straight enough not to touch the sides of the equilibrium chamber.

12. Use a soft pencil and place a mark in the center of the paper about 15 cm from one end. This marks the place where you will spot the sample. Once the sample has been developed, it is sometimes hard to determine where it was spotted, and R_f values are difficult to calculate.

13. Obtain a 10 µL (or lambda) pipet. Fill it with the filtered sample and spot the sample on the pencil dot.

14. Use a hair dryer and warm the spot until the solvent evaporates.

15. Repeat steps 13 and 14. Make every effort to keep the total area of the spot as small as possible.

16. Place the spotted end of the paper in the solvent tank and add a small glass rod to keep the paper from falling out. Adjust the strip so the sample spot is 1 to 2 cm below the anti-siphon bar.

When the mobile phase washes over the sample spot and starts on down the paper, it will begin to siphon when its level gets below the solvent level in the solvent tank. This will cause the solvent to move too fast for the desired separation equilibrium to be attained, and streaking will occur. If a small-diameter rod is placed so the entire width of the paper makes contact with it, then a considerable adsorption force is developed between the solvent and the rod. This is sufficient to stop the siphoning action, and the solvent moves down the paper by capillary action alone.

17. Place the paper in the equilibrium chamber, adjust it so it does not touch the sides (except maybe the very end of the paper), and press the stopper on rather tightly.

18. It will take 16 to 20 hours to separate the components. Come back tomorrow and finish the experiment. It will take about 45 minutes then.

19. Welcome back. Very carefully remove the top rubber stopper, the solvent tank, and the paper from the equilibrium chamber.

20. Before the solvent front drys, mark its position with a soft lead pencil. If good paper and reagents have been used, then it will sometimes be impossible to determine where the solvent front was when the paper is dry.

21. Put a paper clip on the sample spot end and hang it from the support rod on top of the ring stand to dry. This will take about 10 minutes.

22. The detection of the sample components requires a two-spray development. The first step is to add nitrite and make a diazonium salt of the sulfa drugs.

$$C_4H_9ONO + H_2N-\underset{}{\bigcirc}-SO_2-NH-\underset{}{\bigcirc} \longrightarrow \overset{+}{N}\equiv N-\underset{}{\bigcirc}-SO_2-NH-\underset{}{\bigcirc}$$
$$\text{I}$$

The diazonium salt then is coupled with naphthylethylenediamine to produce the reddish purple color.

I + (naphthylethylenediamine) $\longrightarrow$ (coupled product with $N=N-\bigcirc-SO_2-NH-\bigcirc$)

23. It is best to spray the paper in a hood with a piece of cardboard behind the paper to collect the extra spray. Either move the apparatus to a hood or use a second apparatus and hang the paper from the rod.

24. Refer to p. 263 for the correct spraying technique and spray the paper with the butyl nitrite solution until it is moist, but does not drip.

25. Let the paper dry for a few minutes, then spray it with the naphthyl-ethylenediamine. Three spots should form at about the middle of the paper.

26. Measure the distance the solvent front moved and then the distance from the front of each spot to the pencil mark where they were spotted.

27. Calculate the R_f values for each compound.

28. Clean up the area. Pour the used solvents into the waste solvent container. If you are using Devilbis sprayers, remove the liquids from the reservoirs and flush out the sprayer tips with water.

29. Thank you.

EXPERIMENT 29. THIN LAYER CHROMATOGRAPHY

Ascending TLC separations requiring color development - vasodilator drugs- the aliphatic nitrates

BACKGROUND

Organic nitrates have been used for over 100 years to relax the smooth muscle cells of blood vessels and allow them to dilate, thus reducing blood pressure. This results in reducing the pain and relieving the symptoms of those people subject to heart problems. The first such drug used was amyl nitrate. Much work needs to be done in the area of the mechanism of the reactions involved, but apparently it is the *nitrite* ion that is effective in the cell. If nitrite ion is administered in large doses, then methemoglobin can form, which reduces the blood's ability to transport oxygen. Inorganic nitrates are more likely to be reduced to nitrite than organic nitrates. Therefore, organic nitrates are preferred, because they are less likely to be reduced before they reach the site of their reactivity and appear to penetrate the cell wall easier than nitrites. Nitrite then is apparently formed within the cell, but how it reacts to relax the muscle is not yet known. Aliphatic nitrates are preferred over aromatic nitrates, because they are absorbed readily in the mouth before they can be reduced in the intestinal tract, and it is easier to remove the nitrate groups from them at the pH of the cells. In fact, a patch of time-release capsules looking somewhat like a Band-Aid can be taped to a patient's arm, and the compound will slowly penetrate through the skin.

The compounds shown below currently are being used today to treat heart problem patients. Notice that all are designed to carry a maximum number of nitro groups, that can form nitrate or nitrite when removed. You also may recognize that each of these compounds is a potent explosive. This is not a problem, because the compounds are diluted 9 to 1 with lactose and only a few mg of the nitrate are present in a capsule.

YOUR PROBLEM

Drug counterfeiters have been known to substitute KNO_3 , a much cheaper and less reactive compound, for the organic nitrates. As a result, a fast analytical method was needed to determine if a substitution had been made and to simply check the drug to see if it contains the mixture reported on the label. You have been given a suspect sample, and you must determine if it contains an organic nitrate.

Nitroglycerine

Erythritol Tetranitrate

Mannitol Hexanitrate

Pentaerythritol Tetranitrate

Isosorbide Dinitrate

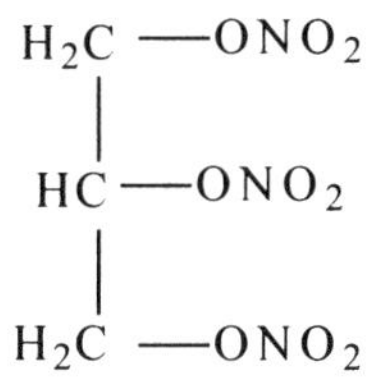

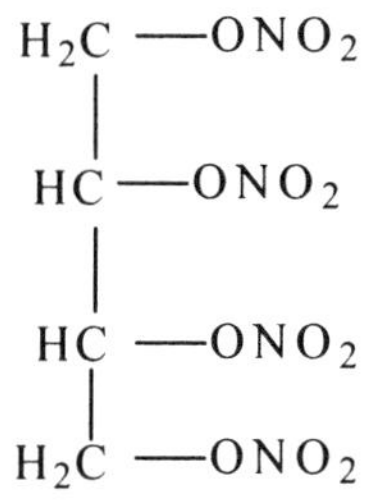

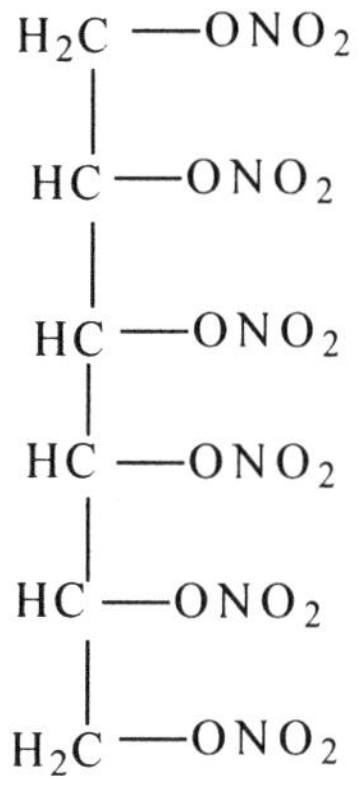

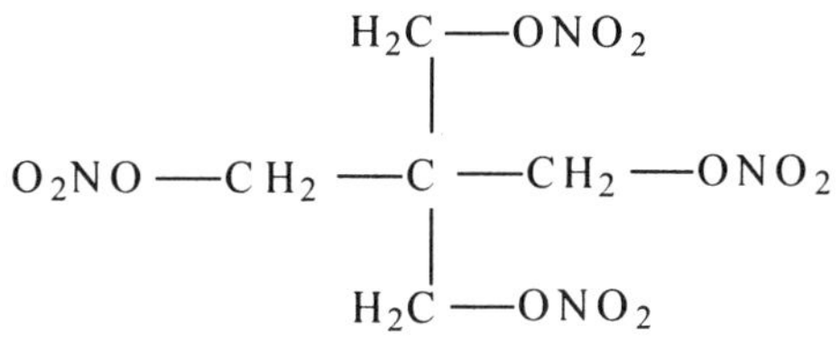

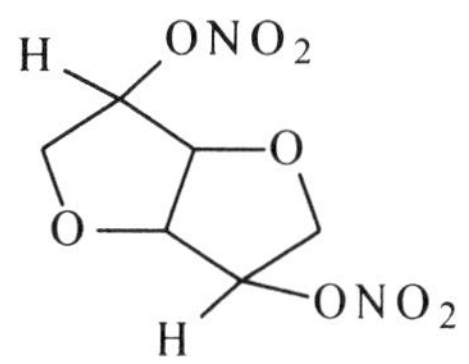

EQUIPMENT

1	Apparatus, fluorescence viewing box with short and long wavelength lamps
1	Bottle, wide mouth, screw cap, 8 oz
6	Bulbs, rubber for the long-stem disposable pipets
1	Desiccator, to store activated plates
1 pr	Goggles, safety
1	Guide, plastic, sample spotting
2	Holder, test tube
1	Mortar and pestle, small
1	Oven, drying, 110 °C, to activate plates
1	Pipet, 10 lambda
1	Rack, test tube
1	Razor blade, single edged, scalpel, or other sharp-edged instrument
1	Ruler, metric
1pr	Scissors
2	Sprayer, reagent, Devilbiss or pressure can
1	TLC plate, 5 x 10 cm, SilicaGel F
6	Tubes, capillary melting point tubes or long-stem disposable pipets
1	Tube, test tube, 100 x 10 mm

CHEMICALS   S = Student

Acetone, (CH_3COCH_3), 10 mL/S
Diphenylamine, $(C_6H_5)_2NH$ (2 % in 95 % ethanol) Spray reagent, 25 mL/S
Organic nitrate tablets or capsules containing one or more of the following
Nitroglycerine
    Mannitol hexanitrate
    Isosorbide dinitrate
    Erythritol tetranitrate
    Pentaerythritol tetranitrate
Standards of the above nitrates, 1 mg/mL
Toluene, C_6H_5-CH_3, 30 mL/S
p-Toluenesulfonic acid, HO_3S-C_6H_4-CH_3, (5% in 95% ethanol) Spray reagent, 25 mL/S

APPARATUS SETUP

The apparatus is quite simple. It consists of an 8 oz glass jar, that serves as the developing chamber, and a 5 cm x 10 cm silica gel plate coated with a fluorescent dye.

BIG PICTURE

During the past several months, a counterfeit drug has appeared that contains only KNO_3. The nitrate gives a positive test the same as the organic nitrates when the standard method is used. The capsules, inexpensive by comparison to quality drugs, passed inspection and were sold. A new TLC screening method has been developed to eliminate this problem. You have a capsule and are to test it for the presence of organic nitrates.

DIRECTIONS

This procedure was developed by D.J. Winters, FDA, Cincinnati District.

1. Obtain an amount of powdered sample such that you can obtain an extract of about 1 mg/mL.

2. Place the sample in a 100 mm x 10 mm test tube and add sufficient acetone to produce a final concentration of about 1 mg/mL.

3. Allow the solids to settle and spot 5 µL on the left side of a prepared 10 cm x 5 cm Silica Gel G or GF 0.25 mm plate (need not be activated). Spot the standards to the right of the sample and spaced equidistantly.

4. Place the plate in an 8 oz developing tank saturated with toluene and allow the solvent to reach nearly the top of the plate.

5. Let the plate air dry until the odor of toluene is no longer present, then spray the plate until moist with reagent no 1 (2% solution of diphenylamine in 95% ethanol).

6. Place the plate under long or short wavelength ultraviolet radiation until the spots appear light brown on a white background (ca 5 to 10 minutes).

7. Spray the plate with reagent no 2 (5% solution of p-toluenesulfonic acid in 95% ethanol). The spots appear as green on a white to yellow background.

8. The R_f values determined by this system were as follows:

Mannitol hexanitrate	2 spots obtained	0.03 and 0.27
Isosorbide dinitrate		0.33
Erythritol tetranitrate		0.42
Nitroglycerine		0.48
Pentaerythritol tetranitrate		0.52

EXPERIMENT 30. "INSTANT" THIN LAYER CHROMATOGRAPHY

The determination of antihistamines in drug preparations
(Private communication, Food and Drug Adm/LIB 5, No. 4171)

BACKGROUND

The body has three basic ways to defend itself against foreign agents. If the agent is large, such as a bacterium, then antibodies are produced to react with it. If the compounds are of much less molecular weight, then they are transported to the liver (85%), the kidneys (10%), or the body cells (5%) for detoxification. In the liver, a glucoside is formed with the toxic compound to make it water soluble. The third method, for very small molecular weight compounds, such as histamine, it to flush them away. This is why when people have hay fever, their eyes water and their noses drip liquid.

Histamine was discovered in 1910 and is released from cells when they are injured. An injured cell also stimulates production of histamine by the action of histadine carboxylase. The freed histamine causes capillaries to dilate and smooth muscles to contract. This results in a reddening of the area, itching or pain by a reaction with nerve endings, and a local edema from increased cell wall permeability.

The important group of histamine is believed to be the ethylamine. The first antihistamine that produced more therapeutic response than toxic side effects was pyrilamine, developed in 1944. The antihistamines apparently compete with histamine for the cell site.

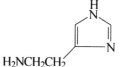

EQUIPMENT

	10	Beakers, 5 mL		
	1	Beaker, 10 mL		
	2	Chromatosprayers	1 pk	Paper, ITLCsheets, Gelman SAF 20 cm x 20 cm
		(Drummond Scientific Co., Bromall, PA 13008)	1	Paper, filter, Whatman 3 mm
			10	Pipets, 5 µL disposable
	1	Cylinder, graduated, 50 mL	4	Plates, glass plates, 8 x 8 inch
	1	Cylinder, graduated cylinder, 100 mL	1	Ring, iron, 3 inch
	1	Dryer, hair	1	Rod, glass, stirring
	9	Flasks, volumetric, 4 mL	1	Spatula
	1	Funnel, separatory, 250 mL	1	Spotting marker
1 pr		Goggles, safety	1	Stand, ring
	1	Holder, plate	2	Tanks, developing, 9 x 9 x 3.5 inch
	1	Mortar and pestle, small	1	Towel or sponge
	1	Oven, drying (100 °C)	1	Viewer, uv

CHEMICALS   S = Student

Acid developer
    Ethanol, H_3CCH_2OH,   These are mixed 10:5:1
    Acetic acid, H_3CCOOH,
    Sulfuric acid, H_2SO_4, 60%

Base developer
    Diethyl ether, H_3CCH_2-O-CH_2CH_3,   100 mL + 25 mL NH_4OH, the top layer is used.
    Ammonium hydroxide, NH_4OH, Conc.

Chloroform, CH_3Cl
Methylene chloride, CH_2Cl_2
Nitrogen, cylinder, optional

Experiment 30

Potassium iodide, KI, 6%

Platinum chloride, $H_2PtCl_6 \cdot H_2O$, 10%, Mix 3 mL of this with 97 mL H_2O and 100 mL of the KI solution.

Standards: All 1 mg/mL $CHCl_3$. The numbers refer to the scheme below.

    Chloropheniramine maleate (1)
    Chlorphentermine HCl (4)
    Chlorpromazine HCl (5)
    Dextromethorphan HBr (6)
    Diphenylhydramine HCl (2)
    Pheniramine maleate (7)
    Phenyltoloxamine dihydrogen citrate (8)
    Pyrilamine maleate (9)
    Tripelannamine HCl (3)

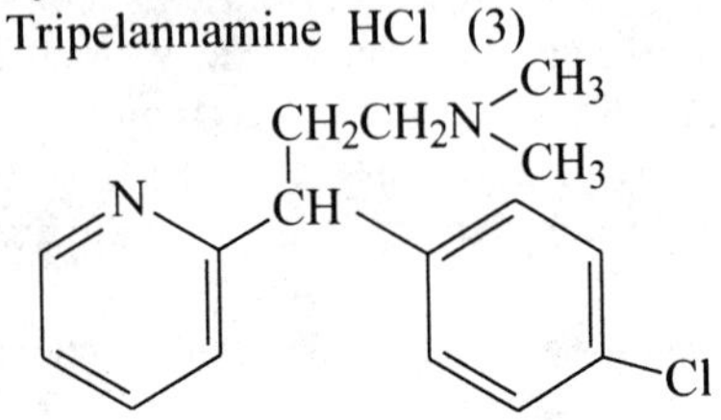

Chloropheniramine

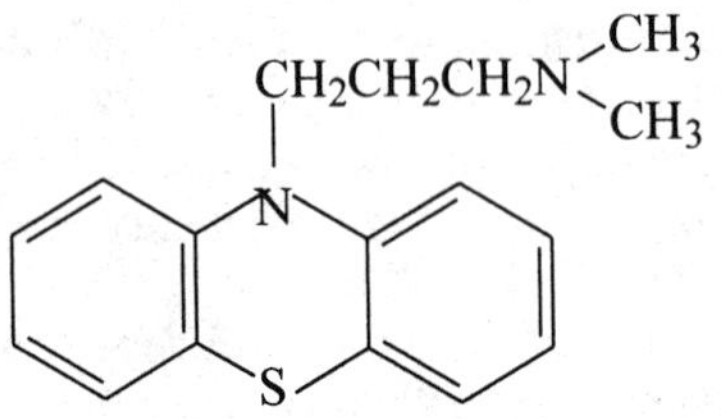

Chlorpromazine

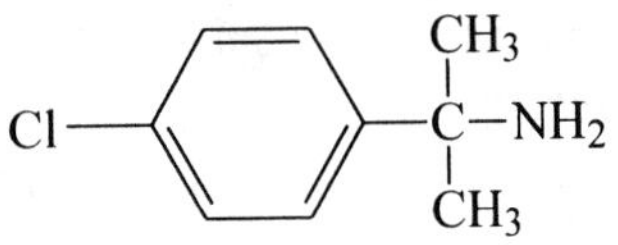

Chlorphenteramine

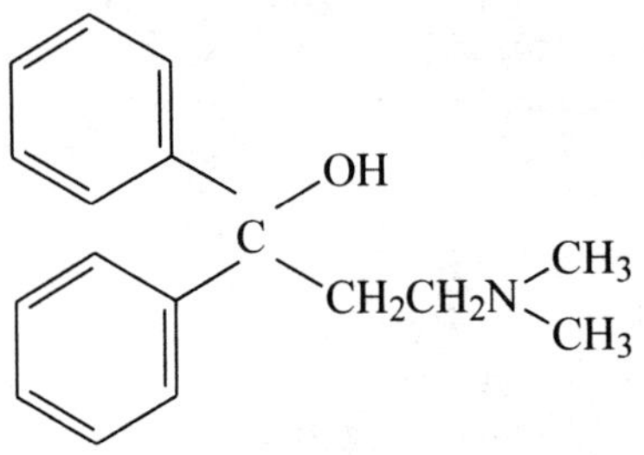

Diphenylhydramine

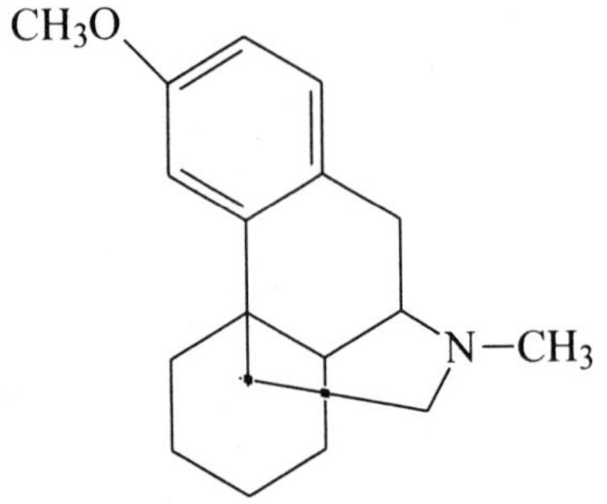

Dextromethorphan

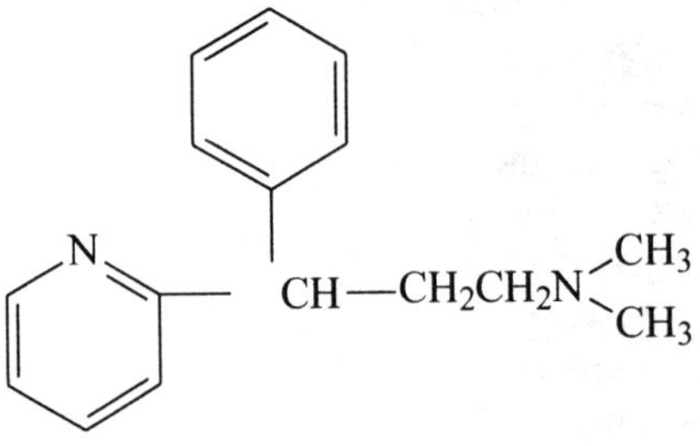

Pheniramine

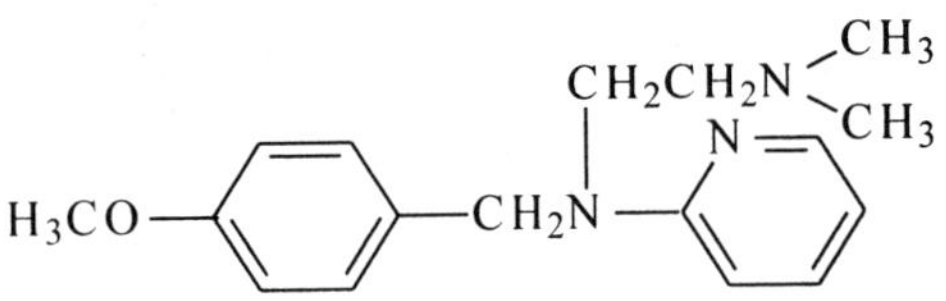

Pyrilamine

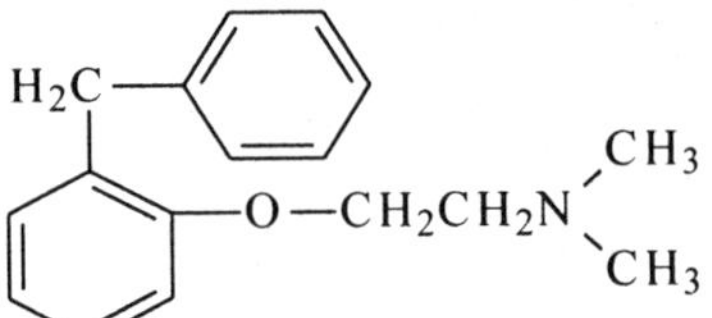

Phenyltoloxamine

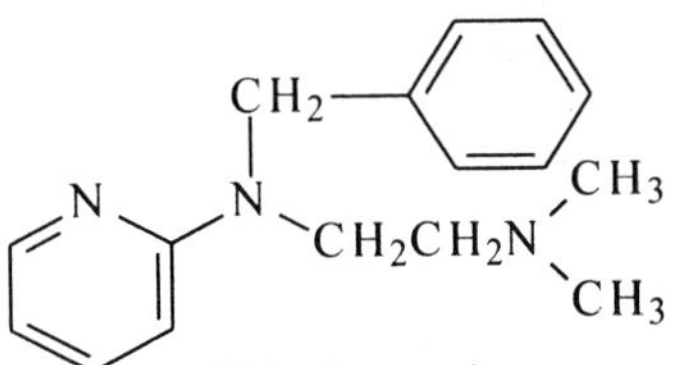

Tripelennamine

APPARATUS SETUP

The apparatus consists of two TLC developing tanks including racks and covers as described earlier, a uv viewing box, and two sprayers. A cylinder of nitrogen is arranged so a stream can be directed through the marking guide and keep the plate dry.

THE BIG PICTURE

Thin layer plates are activated at 100 °C for 30 minutes and spotted with the unknown and standards. The unknown tablet is ground to a fine powder, and the antihistamines are extracted with $CHCl_3$. It takes about 5 minutes to develop the plate.

With uv quenching, 1, 3, 5, 7, and 9 are detectable at 0.05 µg. The others at levels of 10 to 15 µg.

All amines are easily detectable at the 0.05 µg level with the platinate spray.

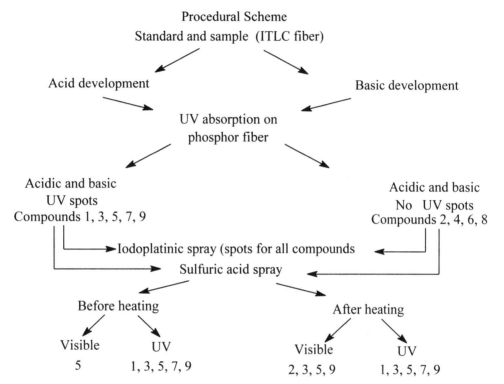

After the fiber is sprayed with the sulfuric acid spray reagent, 1 µg of chlorpromazine is immediately visible as a pink spot. Under uv radiation, 1, 3, 7, and 9 appear as bright blue fluorescent spots, and 5 remains a dark spot. After heating, 5 is pink, and 2, 3, and 9 are faint yellow, light brown, and light pink, respectively. Under uv radiation, 1 and 7 are dull blue, and 3 and 9 are bright blue.

DIRECTIONS

Instant thin layer chromatography (ITLC) plates are prepared from ground-up glass fibers impregnated with silica gel. This provides a uniform plate, readily wetted by the more polar solvents so capillary action is rapid. The *plates are fragile* and should be handled accordingly.

1. Locate a plate carrier and a metal box with several glass plates in it. Put two glass plates in it.

2. Place a 20 x 20 cm ITLC sheet on each plate.

3. Place the holder containing the plates in a oven set at about 110 °C - note the time.

4. Cover the unused ITLC plates and return the box to its storage place.

5. Go to the "acid" tank; gently remove the cover plate and lay it on the bench top.

6. Add 150 mL of the acid solvent. The tank should have a piece of chromatographic paper that covers three sides and a plate holder. If neither are there, put them in place.

7. Put the cover plate back on the tank and set the tank off to the side, so the inner chamber can come to equilibrium.
8. Go to the "base" tank and remove the cover from it.
9. The base tank solvent has to be made fresh daily. This is done as follows: Add 100 mL of diethyl ether to a 250 mL separatory funnel. Add 25 mL of NH_4OH to the funnel. Shake gently a few times, invert the funnel, and release the pressure. Continue the mixing for 1.5 to 2 minutes and set the funnel back in the holder to let the two layers separate.
10. Drain off the bottom layer and discard it into a recycle solvent container.
11. Let the ether layer drain into the base tank.
12. Repeat steps 9 to 11.
13. Select an antihistamine tablet from the choices provided. If you choose a capsule, take it apart and pour the powder into a 10 mL beaker. If you choose a tablet, powder it in a mortar with a pestle and then transfer the powder to a 10 mL beaker.
14. Add 2 to 3 mL of $CHCl_3$ to the beaker and stir the contents for a couple of minutes.
15. Locate two DeVilbiss sprayers or something similar. These look a lot like fly sprayers with gray rubber bulbs on them. If they are not clean, then clean them now. Pressurized spray cans are preferred, but the DeVilbiss sprayers are much less expensive.
16. Place the iodoplatinate liquid in one of them and the 60% H_2SO_4 solution in the other. Spray each one a few times (into a sink) to make sure they are working.
17. If at least a half an hour has passed since you put the thin layer sheets in the drying oven, then remove them carefully; use a towel to hold the metal frame.
18. If you haven't already done so, put the metal foil cover on the unused ITLC sheets to keep as much room air and dust away from them as possible.
19. Lay one of the glass plates in front of the acid tank with the paper sheet on top of it and with the holes away from you.
20. Place it in the marking guide and turn on the N_2 to a low flow rate. The stream of N_2 is to form a positive pressure and keep moist air from deactivating the plate.
21. Adjust the sheet so the sample spots will be about 4 cm from the bottom of the guide.
22. Place the other plate back into the oven.
23. Refer to the previous description, and you will see that the antihistamines are numbered 1 through 9. You may not have all nine of them, but several will be available. It will simplify matters to line up the flasks containing the knowns in the same order as they are listed above.
24. Use a 5 µL disposable pipet and spot the unknowns on the left end at the first black mark about 4 cm from the end of the sheet.
25. Spot each of the available standards in the designated order. It is suggested that you skip a space between each one. Spot the unknown again. Make a pencil dot to the side of the sample spot.
26. Hang the sheet in the plate holder in the acid developing tank *so it does not touch the solvent*.
27. Cover the chamber and let the sheets hang for about 15 minutes to come to equilibrium with the solvent.
28. Repeat steps 19 to 26 with the other sheet and place it in the base tank.
29. Lower each sheet into the solvent.
30. Let the plates develop until the solvent has risen about 6 to 7 cm. This usually will take less than 10 minutes.
31. Remove each sheet from the tank and lay it on the glass plates. Mark the solvent front with a light pencil line.
32. Put the plates in the plate carrier and place them in the oven for 4 to 5 minutes to dry.

    You will now use a variety of methods to detect where the sample spots are.
33. Remove the sheets from the oven.
34. Locate the uv viewing cabinet or the uv lamp.
35. If no viewing cabinet is available, turn off the room lights above your bench or go to a dark place.

36. The ITLC plate has been impregnated with a fluorescent dye, which will glow when irradiated with uv radiation. Some compounds react with the dye to produce a different color, and others quench the fluorescence and produce a dark spot. This is a very sensitive method for detecting many compounds. Irradiate each plate with uv radiation. Circle each visible spot with a pencil mark. Pretty aren't they? Refer to Table E30-1 to aid in the identification.

37. Use an R_f plate or a mm rule and determine the R_f values. Compare them with the literature values to determine how close you agree. Tentatively identify the compounds in the unknown.

38. Turn off the uv lamp.

Table E30-1. R_f values and direct uv absorption

Compound	$R_f \times 100$ Basic Solvent	$R_f \times 100$ Acidic Solvent	Color with Direct uv absorption
1. Chlorpheniramine maleate	51	19	Dull blue
2. Diphenhydramine HCl	75	43	
3. Tripelannamine HCl	73	29	Dull blue
4. Chlorphentermine HCl	39	52	
5. Chlorpromazine HCl	76	48	Dull blue
6. Dextromethorphan HBr	55	48	
7. Pheniramine maleate	49	2	Dull blue
8. Phenyltoloxamine dihydrogen citrate	72	43	
9. Pyrilamine maleate	65	27	Dull blue

39. Turn on the room lights.

    Color-developing reagents will now be used.

40. The iodoplatinate solution is a compound that will react with amines. (This is a very expensive reagent. Don't waste it and don't throw away what is left.) Place each sheet separately in a spraying chamber, if one is available, and spray them with proper technique until the sheet is wet, but not so the liquid runs.

41. Place the moist plates on glass plates and place them in the drying oven for a few minutes.

42. Every spot should show up. This reagent is sensitive to 0.05 µg of amine. Circle the spots with pencil marks. The spots tend to fade with time. If this was a sample involved in a court case, then the plate should be photographed immediately.

    Those of you that did the paper chromatography experiment will recall how large the spots were because an 18-hour development allowed time for lateral diffusion. Here, with only 5-10 minutes of developing time, the spots don't have time to diffuse so they are small. The more uniform spots make both quantitative and qualitative analysis easier.

43. Refer to Table E30-2, p. 606, for aid in identifying the unknowns.

44. Sulfuric acid is a good dehydrating agent and will char most organic compounds, producing a black spot. Spray the LOWER HALF of the plates, but *don't* dry them. Look at them with both uv radiation and no uv radiation. Refer to Table 30-3, p. 606. Record what you see.

45. Place the sheets on glass plates, place them in the oven, and heat them for 10 minutes. Look at them again with both normal and uv radiation. Record the results.

46. Clean up the area. Pour the developing solvents into the recycling containers (CAREFUL - those developing tanks cost nearly $100 each), but leave the wall sheets inside.

47. Remove the racks, so they will air dry.

48. Clean out the DeVilbiss sprayer containing the sulfuric acid. Rinse it with distilled water - CAREFUL - use plenty of water initially to reduce heat formation. Fill the sprayer with water and spray it a few times to clean out the tube. Pour out that water.

Table E30-2. Color patterns after spraying

	Iodoplatinate spray	Sulfuric Acid Spray			
		Before Heating		After Heating	
Compound		Visible color	UV color	Visible color	UV color
1. Chloropheniramine maleate	Blue gray	------	Br blue	-----	Dull blue
2. Diphenhydramine HCl	Purple	------	-----	Lt yellow	-----
3. Triplennamine HCl	Blue gray	------	Br blue	Lt brown	Br blue
4. Chlorphenteramine HCl	White	------	------	------	-------
5. Chlorpromazine HCl	Blue brown	Pink	Dr pink	Pink	Dr pink
6. Dextromethorphan Hbr	Blue gray	------	------	------	--------
7. Pheniramine maleate	Purple	------	Br blue	------	Dull blue
8. Phenytoloxamine	Blue	------	------	------	-----
9. Pyrilamine maleate	Blue gray	------	Br blue	Lt pink	Br blue

Table E30-3. Sulfuric acid spray

Before Heating		After Heating	
Visible	uv	Visible	uv
5	1,3,5,7,9	2,3,5,9	1,3,5,7,9

49. Pour the unused iodoplatinate reagent back into the reagent bottle and flush the sprayer with distilled water.

50. Clean the separatory funnel and be sure the ITLC box is closed.

51. *Be sure any spilled liquids are wiped off of the bench top.* Thank you.

EXPERIMENT 31. TWO-DIMENSIONAL CHROMATOGRAPHY
Separation of sulfamethazine or carbadox from medicated feeds

BACKGROUND
Medicated feeds are used in the production of all red meat animals and poultry to reduce the losses from disease, to produce a better quality final food product, and to reduce the amount of feed grains required to bring an animal to market weight. The end result is a higher quality product at less cost. *Withdrawal* feeds are those that contain no drugs and are fed to animals a few days before slaughter to ensure that the drugs in the medicated feeds have passed through the animal's system and will not be in the meat served to the public.

Drugs are added to animal feed for two main purposes: (1) to be absorbed into the system through the digestive tract and prevent disease, and (2) to kill intestinal bacteria that consume food intended for the animal. This latter can amount to as much as 20% in a steer.

Sulfamethazine is added at 100 g/ton to control bacteria.   *Carbadox* is added to control bacteria.

YOUR PROBLEM
You have been given a medicated feed that is supposed to contain sulfamethazine/carbadox at feeding levels. You are to determine if the farmer received a feed that contained the drug stated on the label.

NOTE: If you desire, a more involved procedure can be performed to test a withdrawal feed for sulfamethazine or carbadox contaminants. The equipment and chemicals needed are listed below and indicated as such.

EQUIPMENT

1	Aspirator, water	1	Oven, drying
1	Balance, top loading	1	Plate, 20 x 20 cm, Silica gel GF
1	Beaker, 250 mL	1	Tank, developing
1	Cylinder, graduated, 50 mL	4 ft	Tubing, vacuum, rubber (Carbadox only)
1	Dryer, hair		
1	Flask, filter, 125 mL	1	Pipet, spotting
1	Funnel, Buchner, 5 cm	1	Rod, glass, stirring, 15 cm
1 pr	Goggles, safety	1	Spatula
1	Hot plate	1	Sprayer
1	Lamp, uv long wavelength	1	Towel, cloth

For withdrawal feeds

1	Blender, Waring	1	Ring stand
1	Clamp, buret	1	Steambath
1	Column, glass, 2 cm *x* 15 cm	1	pH meter
1	Cylinder, compressed air or N_2	1	Test tube, conical, 5 mL
1	Funnel, separatory, 250 mL	3 ft	Tubing, 6 mm diameter

CHEMICALS   S = Student

Methanol, CH_3OH, 50 mL

Sulfamethazine

Solvent A - Chloroform, $CHCl_3$ (90): Methanol (10), 100 mL/S
Solvent B - Chloroform (20): Ethanol, CH_3CH_2OH (10):
       Heptane, C_7H_{16} (20), 100 mL/S
Sodium nitrite, $NaNO_2$, 5 g/S
Hydrochloric acid, HCl, conc, 2 mL/S
Marshall reagent, 0.4 % N-(1-naphthyl)-ethylene diammonium dichloride in methanol
Sulfamethazine, standards, 0.1, 0.5, 1.0 µg/5µL in methanol

Carbadox

Solvent A - Chloroform (90): Methanol (10), 100 mL/S
Solvent B - Chloroform (50): Acetone (50), 100 mL/S
Carbadox, standards, 0.1, 0.5, 1.0 µg/5µL in methanol

Withdrawal Feeds

NH_4OH, 4 %, in methanol, 200 mL/S
Acetone, $CH_3C(O)CH_3$, 200 mL/S
Chloroform, 200 mL/S
NaCl, 10 % in 0.1 N NaOH, 50 mL/S
Potassium dihydrogen phosphate, KH_2PO_4, 1 M, 15 mL/S
Sodium sulfate, Na_2SO_4, solid, 10 g/S
Sulfamethazine or Carbadox standards, 0.02, 0.05, 0.1 µg/µL in methanol

APPARATUS SETUP

The developing tank is the same as that in Experiment 29, and the plate is handled as shown in Figure 22-24, p 265.

THE BIG PICTURE

A 10 g sample of medicated feed is extracted with 50 mL of methanol, vacuum filtered, and concentrated by evaporation, and the drug separated by two-dimensional TLC.

DIRECTIONS

Three procedures are presented: (1) sulfamethazine in medicated feeds, (2) Carbadox in medicated feeds, and (3) the additional cleanup required if withdrawal feeds are to be checked for contamination. You select the one you want.

The directions below are copied exactly as they appear in the FDA manual.

Withdrawal feeds

Skip this part, if you are doing a medicated feed. The concentration is so high that interferences are of no consequence.

Blend 20 to 100 g of prepared sample with 200 mL of 4% ammonia in acetone for 2 min. Filter with vacuum. Repeat with 2/100 mL portions of acetone. Evaporate the extract on a steambath with a current of air, almost to dryness. Transfer to a 250 mL separatory funnel with ca 60 mL of chloroform, add 50 mL of 10% NaCl in 0.1 N NaOH and shake gently. Discard the chloroform layer. Wash the alkaline layer with several 20 mL portions of chloroform until the wash is colorless, discarding the chloroform washes. Adjust the pH of the alkaline soln to ca 6 with 1 M KH_2PO_4 (approx 12 mLs) and extract with 3/30 mL portions of chloroform. Filter the combined extracts through a 1 inch layer of sodium sulfate and evaporate almost to dryness. Transfer to a 5 mL conical test tube and make to a suitable volume for TLC or HPLC.

Two-dimensional TLC for sulfamethazine
    Plate - Analtech uniplate 250 μm Silica Gel GF (activate for 1/2 hr at 105 °C)
    Solvent   A - Chloroform (90): methanol (10) to 12 cm
               B - Chloroform (20): ethanol (10): heptane (20)
    Detection - Expose plate to nitrous acid fumes (add several drops of HCl to ca 5 grams of sodium nitrite in a convenient container) followed by spraying with 0.1% Marshall's reagent in methanol.

Spotting order - Spot and develop plate as follows:
    Spots 1, 2, 3 and 5, 6, 7 --- Standards 0.02 to 0.1 μg.
    Spot 4 -- Sample equivalent of 5 to 50 g.
    All spots ca 2 cm from edge of plate. After development and spraying, sulfas appear as bright red spots.

(*Note*: If the withdrawal feed cleanup procedure was not done, then penicillin (from procaine penicillin, at 50 g/ton) may produce a second spot.

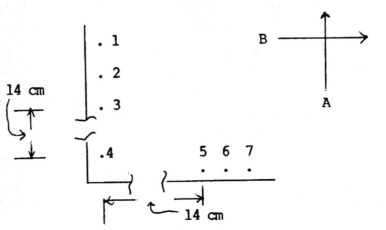

Two-dimensional TLC for Carbadox
    Plate - same as above.
    Solvent   A - 10% methanol in chloroform to 12 cm
               B - 50% acetone in chloroform
    Detection - Long wave uv transilluminator.
    Spotting order - Spot sample and standard solutions as described for sulfa determination. After development, examine plate on the transilluminator. Carbadox appears as a yellow-green fluorescent spot.

EXPERIMENT 32. ION EXCHANGE
Separations of several cations in mineral water, pond water, or low acid foods

BACKGROUND
Originally, the hardness of water was understood to be a measure of the capacity of the water for precipitating soap. Soap is precipitated chiefly by the calcium and magnesium ions commonly present in water but can be precipitated by other polyvalent ions such as aluminum, iron, manganese, strontium, and zinc.

$$CH_3(CH_2)_7CH=CH(CH_2)_7COOH + M^{+n} = [CH_3(CH_2)_7CH=CH(CH_2)_7COO]_nM$$

    (Oleic acid, "soap")

Because all but the first two compounds are usually present in insignificant concentrations in natural waters, hardness is defined as a characteristic of water that represents the total concentration of just the calcium and magnesium ions *expressed as calcium carbonate*. The main salts responsible for hardness in waters are the hydrogen carbonates (bicarbonates) and the sulfates of calcium and magnesium. If water is brought to a boil, the bicarbonates decompose, some of the CO_2 boils away, and the rest forms insoluble precipitates. Because the carbonates can be removed, this is known as "temporary hardness." The sulfates do not boil away or precipitate during boiling, so they are known as "permanent hardness."

Total hardness: The total of all of the cations present.
Temporary hardness: Bicarbonates and carbonates.
Permanent hardness: Sulfates.

What is good water? If you run a pressure boiler, then 10 ppm total hardness (as $CaCO_3$) is needed. If you bottle soft drinks, then 250 ppm constitutes good water. Generally speaking, water in the 100 to 200 ppm range is considered good drinking water. If less than that is present you have a hard time removing the soap when you take a shower (soft water), and if more than that is present (hard water) cooked fruits and vegetables get quite hard. In this country, about 10% of the population uses water with more than 1,000 ppm and, in one instance, a water had over 4000 ppm. A term used by engineers involved with water quality is *grain*.

$$1 \text{ grain/gal (U.S.)} = 17.12 \text{ ppm} \qquad (E32\text{-}1)$$

Cation exchange for softening water. The *American Water Works Association* (AWWA) lists the following equations as being primary in the water softening process. Notice that the insoluble carbonate formers, Ca^{+2} and Mg^{+2}, are replaced by Na^+, which form soluble carbonates.

$$Ca(HCO_3)_2 + Na_2R = CaR + 2\,NaHCO_3$$
$$CaSO_4 + Na_2R = CaR + Na_2SO_4$$
$$CaCl_2 + Na_2R = CaR + 2\,NaCl$$
$$Mg(HCO_3)_2 + Na_2R = MgR + 2\,NaHCO_3$$
$$MgSO_4 + Na_2R = MgR + Na_2SO_4$$
$$MgCl_2 + Na_2R = MgR + NaCl$$

For regeneration:

$$CaR + MgR + NaCl \text{ (whopping excess)} = CaCl_2 + MgCl_2 + Na_2R$$

Table E32-1 from the AWWA manual shows the soap savings from reducing the hardness to 85 ppm when using 100 gallons of water per day per person.

Table E32-1. Soap saved by a family of five per year when hardness is reduced to 85 ppm.

Raw Water Hardness	Pounds of Soap Saved
150	28
300	94
400	138
500	182
750	328
1000	400
1500	620

A city water plant can remove total hardness for about 1/4 the cost of the individual home owner. Usually, only the hot water line is connected to the water softener to hold the cost down and to keep the sodium content of drinking water as low as possible. This is important for people with high blood pressure or who are hypertensive.

$CaCl_2$ often is added to low acid foods such as corn, beans, and peas to prevent them from becoming too mushy when they are cooked. The calcium, being a dication, can associate with carboxyl groups between two cell walls and help hold the cells together, producing a firmness and a better mouth feel.

Years ago, the indicator used for the titration of calcium and magnesium was a soap solution. As long as calcium and magnesium were present, no soap suds would form. The solution was shaken continuously as the titrant was added, and the first appearance of soap suds was the endpoint. This was a difficult endpoint to reproduce. The discovery by Swartzenback that ethylenediaminetetraacetic acid (EDTA) would chelate with calcium and magnesium ended the soap titrations. In 1948, Walter Biedermann discovered Eriochrome Black T. It is also known as Erio-T, F-241, or C.I. 203. This was a tremendous improvement for the endpoint detection and allowed water analysis to become a routine operation by technicians. However, the reagent was stable only for about 1 week. In 1960, Lindstrom and Diehl at Iowa State University prepared Calmagite, which is water soluble, is stable indefinitely, and gives a sharper color change at the endpoint than does Erio-T.

Eriochrome Black T   Calmagite

Erio-T is a tribasic acid and forms a colored complex with magnesium ions. This is a beautiful example of how a knowledge of equilibrium constants can be used to solve a problem. The formation constants of Ca^{+2} and Mg^{+2} with EDTA are 3.0×10^{10} and 5.0×10^{8}. The magnesium complex with the indicator is more stable (1×10^{7}) than the complex with the calcium (2.5×10^{5}) but less stable than the Mg EDTA complex. Thus, during a titration, the EDTA reacts first with the free Ca^{+2} ions, then with the free Mg^{+2} ions, and finally with the Mg in the indicator complex

$$Mg^{+2} + \underset{\text{(Blue)}}{HIn} = \underset{\text{(Red)}}{MgIn^-} + H^+$$
$$+$$
$$HY^{-3}$$
$$\updownarrow$$
$$MgY^-$$
$$+$$
$$H^+$$

When the end point is reached, the solution turns from pink to blue.

INTERFERENCES

Any metal ion that has an EDTA formation constant higher than calcium or magnesium will interfere. Cyanide complexes strongly with copper, cobalt, nickel, zinc, and ferrous iron. Hydroxylamine or ascorbic acid is added to reduce iron to the ferrous state. If the solution is buffered to pH 10 before the indicator is added, then iron will not interfere because it precipitates as the hydroxide before it can react with the indicator or the EDTA.

The aluminum interference with the indicator is removed by adding triethanolamine: $N(CH_2CH_2OH)_3$. Manganese interferes, but it is usually present is such small amounts that no attempt is made to remove it.

YOUR PROBLEM

You are a chemist in a private analytical laboratory, and your specialty is determining ions in aqueous liquid samples such as mineral waters or pond water and the added ions in low acid foods. You are prepared to handle any of these samples. You have been given one of these by your boss as a quality control sample.

EQUIPMENT

1	Beaker, 250 mL		1	Holder, buret
2	Beakers, 400 mL		1	Lamp, uv
1	Bulb, pipet, 60 mL		1	Opener, can
1	Buret, 50 mL		1	Paper, filter, 11 cm, fast 1/S
1	Clamp, buret		1	Pipet, 10 mL
1	Column, ion exchange, 1 x 15 cm		1	Pipet, 25 mL
1	Darkroom or equivalent		1	Plate, glass, amber or yellow
3	Flasks, Erlenmeyer, 250 mL		1	Plate, titrating plate, 15 x 15 cm
1	Flask, volumetric, 250 mL		1	Rack, funnel
1	Funnel, long stem		2	Stands, ring
1 pr	Goggles, safety			

CHEMICALS   S = Student

Ascorbic acid, 50 mg/S

Buffer pH 10, 6.75 g NH_4Cl with 57.0 mL conc NH_4OH, dilute to 100 mL, 5 mL/S

Calcein indicator, 0.2 g in 25 mL of 1 N NaOH and dilute to 100 mL. Store in the dark, if possible.

Dowex 50-8X, 20-50 mesh, cation resin or its equivalent, 10 g/S

EDTA, 0.01 N. *Note*: this will be prepared and available for your use.
> 3.7 g of disodium EDTA·2 H_2O/L. To this add 125 mg of $MgCl_2·6H_2O$/L. Standardize against a standard calcium solution made by dissolving 1.000 g of $CaCO_3$ in a little HCl and diluting to 1,000 mL. 1 mL = 1 mg $CaCO_3$.

Either: Eriochrome black T, 0.05 % in ethanol. (prepare each week) or Calmagite, 0.05 % in water. (stable indefinitely).

Glass wool, 1 g/S

Hydrochloric acid, HCl, (1:1), 25 mL/S

Mixed indicator, Methylene blue : Methyl red, pH change 5.4
   (base form - green; acid form - lavender)

pH indicator paper, wide range, 10 cm/S

Potassium hydroxide, KOH, 1 N (5.6 g/100 mL) containing 1 g KCN/100 mL (DANGER), 15 mL/S

Sodium cyanide, NaCN, 5%, 20 mL/S (DANGER)

Sodium hydroxide, 0.01 N, (standardized is much preferred) 1,000 mL/S

Sodium thiosulfate, $Na_2S_2O_3$, 0.1 N, 0.5 mL/S

Triethanolamine, $N(CH_2CH_2OH)_3$, 20% in water, 20 mL/S

APPARATUS SETUP

The apparatus consists of an ion exchange column and a 50 mL buret held on a ring stand by a buret holder.

THE BIG PICTURE

A strong acid cation exchange resin is converted to the H^+ form. This then is used to remove the cations from a liquid sample such as mineral water, pond water, or the juices from low acid foods and exchange them with H^+. An aliquot of the sample containing H^+ will be titrated with standard base to determine the total amount of ions present. A second aliquot will be titrated with EDTA to determine the total calcium and magnesium ions present, and a third aliquot will be used to determine magnesium, again with an EDTA titration.

DIRECTIONS

Preparing an Ion Exchange Column

Ion exchange resins should be treated before they are used to be sure the exchangeable ion is the one you want. In this case, you want to be sure it is H^+. The procedure is to add an excess of acid, allow it time to exchange, then wash away the excess.

1. Place 10 g of resin in a 250 mL beaker, add 25 mL of HCl (1:1), swirl, and let stand for 5 minutes.

2. Decant, then rinse the excess HCl from the resin by washing five or six times with 25 mL portions of distilled water. Discard the washings.

3. After the fifth wash, test a few drops of the wash solution with wide-range pH indicator paper. If the wash solution is more acidic than the distilled water, continue washing the resin until the excess acid is removed.

4. Attach your ion exchange column (the bottom half of a broken 50 mL buret works well) on a ring stand with a buret holder and adjust it so it is vertical. Close the stopcock.

5. Add a small glass wool plug, push it to the bottom of the column, and then add your resin as a water slurry. Drain the water slowly so the beads pack uniformly, but do not let the water level ever fall below the top of the resin beads.

6. From the information on the label on the bottle of resin, determine the equivalents of exchange capacity of your column.

Determining the Total Hardness in the Sample

In this part, you will titrate the sample as it is to determine the original acidity. This will be considered as the blank. An aliquot of the sample then will be passed through the resin, and the cations allowed to exchange with H^+ ions, which will be eluted from the column. This solution then will be titrated, and the increase in acidity will be due to the cations exchanged. You then will know the total milliequivalents of cations in your sample, but you will not know what they are. Later steps will allow you to determine that portion that is calcium and magnesium.

Sample Preparation

A wide variety of samples can be used in this experiment, some of which will contain a lot of suspended matter that must be removed by filtration.

7. Set up a funnel rack, a 60° long stem funnel, and place a piece of fast filtering paper in it.

8. If you have a sample of canned fruits or vegetables, then open the can and drain the juice into a beaker. You will need 125 mL of canned goods juice or 225 mL of a water sample. Use as many cans as necessary.

9. Filter the juice or water sample and collect it in a 400 mL beaker.

10. Pipet 10.0 mL of your sample onto the resin column and allow it to drain at about 1 drop/2 seconds. Collect the effluent in a 250 mL Erlenmeyer flask.

11. Wash the column with five 25 mL portions of distilled-deionized water. While this is being done, go immediately to the next step.

12. Pipet 10.0 mL of your sample (*that has not been run through the resin*) into a clean, dry, 250 mL Erlenmeyer flask, add 50-75 mL of distilled-deionized water, 2-3 drops of mixed indicator, and titrate with 0.01 N NaOH.

    Note: Natural waters are slightly basic and only 1 drop of base will turn the indicator. Be careful.

    Note: About 10% of the students in a normal class will be colorblind. The endpoint is often easier to see by colorblind people if mixed indicators are used. If you cannot see the endpoint, don't be afraid to ask for help.

13. Repeat step 12 when time permits.

14. Add 2 to 3 drops of the mixed indicator to your collected sample and titrate it with 0.01 N NaOH.

15. Repeat steps 10, 11, and 14.

16. Calculate the number of milliequivalents of acid in the original sample (this is the blank) and then in the sample that has been passed through the resin. Subtract the two values to determine the number of milliequivalents of cations in the sample. See the example calculation.

    Caution: Does this value equal the capacity of the resin you determined in step 6? If it does, then probably your sample has more ions than the resin has capacity to exchange. If this is the case, then regenerate the resin and either (1) use more resin or (2) use a smaller sample.

Determination of Both Calcium and Magnesium

HANDLING CYANIDE: The polyvalent ions that might interfere, such as iron and copper, are removed by forming the cyanide complexes, which are more stable than the EDTA chelates. Cyanide is very toxic and should not be handled unless you wear gloves, if you have any cuts or scratches on your hands. *Do not place KCN or NaCN in any acid solution* because the gas HCN is formed (weak acid being displaced by a strong acid), which is highly toxic. Cyanides in basic solution are safe, if handled properly. Have the TA present when you handle cyanide salts. THE ANTIDOTE IS AMYL NITRITE.

Estimation of Sample Size

You should do this part in duplicate, which will require 50 mL of the juices of canned goods or 100 mL of a water sample. If you intend to do steps 26 to 32, those will require the same amount of sample.

17. If you are in doubt as to how much sample you have, measure it with a graduated cylinder. If it is less than 100 mL, then either get more sample or reduce the size of sample listed below accordingly.

Standardization of the EDTA

18. Pipet 50.0 mL of the 1.00 mg/mL $CaCl_2$ standard into a 250 mL Erlenmeyer flask.

19. Add 10 mL of the pH 10 buffer to precipitate any iron and then add 4 drops of either Erio-T or Calmagite indicator.

20. Titrate with the EDTA until the color changes from red to blue. In normal daylight, the blue color is seen easily, but under tungsten filament lamps, it can be almost invisible.

21. Calculate the normality of the EDTA:

$$[mL_{Ca}][N_{Ca}] = [mL_{EDTA}][N_{EDTA}]$$

Titration of the Sample

22. Pipet 25.0 mL of a canned goods juice sample or 50.0 mL of a water sample into a 250 mL Erlenmeyer flask.

    If you are titrating the juices from canned goods, they most likely contain iron and aluminum. If you are not sure if your sample contains interferences, then a simple way to test is to titrate a small amount of the sample and notice the endpoint color change. If it is sharp, then no interferences are present, but if it is not sharp or the solution becomes cloudy, then interferences are present and must be removed. We will assume that interferences are present.

23. Add in the following order: 10 mg of ascorbic acid, 5 mL of the 5 % cyanide solution (*Careful*), 5 mL of 20 % triethanolamine, 5 mL of the pH 10 buffer, and 4 drops of the indicator.

24. Titrate with the EDTA to a blue endpoint.

25. Calculate the hardness in ppm $CaCO_3$ and in grains/gallon. (If the EDTA is made exactly 0.01 N and a 50.0 mL sample was taken, then the ppm = 20 x the buret reading)

Determination of Calcium in the Presence of Magnesium

The titration of calcium only in the presence of magnesium is carried out at a pH of 12, conditions under which the magnesium is converted to magnesium hydroxide, precipitates, and is rendered unreactive toward EDTA. Calcein, the indicator used in this titration, itself forms a nondissociated ion with calcium, which is yellow-green. In the alkaline solution, the indicator itself is brown. At the endpoint in the titration, the calcium is extracted from the Calcein-calcium complex by the EDTA and the color changes from yellow-green to brown. Magnesium does not form a nondissociated ion with the indicator.

The Calcein-calcium chelate (Talanta **2**, 277)

Large amounts of Na salts do not affect the titration. Sr and Ba interfere and are titrated along with Ca. Cu and Fe interfere with the endpoint, but such interferences are removed easily by the addition of cyanide. The titration of Ca can be performed in the presence of chloride, nitrate, acetate, and sulfate.

26. Pipet 25.0 mL of the juices from a canned goods sample or 50.0 mL of a water sample into a 250 mL Erlenmeyer flask.

27. Add in this order with mixing after each addition: 10 mg of ascorbic acid, 5 mL of 20% ethanolamine, 5 mL of 1 N KOH containing KCN, and 1 drop of the Calcein indicator.

    There are two ways to do the titration: (1) the standard method and (2) the fluorescence method. You may do whichever you choose, although the latter is something new for you.

The Standard Method

28. Titrate the sample with 0.01N EDTA until the yellow-green to brown endpoint is reached.

29. Calculate the Mg present in ppm and in grains/gallon.

The Fluorescent Method

The Calcein-calcium complex fluoresces in ultraviolet radiation, but the uncomplexed Calcein does not. The sample to be titrated is placed in a dark room or in a box and exposed to uv radiation. The sample then is titrated until the fluorescence disappears. This is a very sharp endpoint and must be approached slowly near the end. The color of the fluorescence is enhanced by viewing though an amber glass or a piece of yellow cellophane, which removes the small amounts of visible blue light from the lamp.

30. Take your sample, ring stand, buret, and EDTA solution to a photographic darkroom or a closet.

31. Set up to do the titration, turn on the uv lamp, and then turn out the lights.

32. Titrate the sample and calculate the results as ppm Mg and as grains of Mg/gallon. Refer to the example calculation on p. 276.

EXPERIMENT 33.   ION CHROMATOGRAPHY
The separation of corrosion inhibitors and contaminants in engine coolants
(The directions are abstracted from an application bulletin from Dionex Corp., Sunnyvale, CA)

BACKGROUND
About one third of the heat produced by an engine is lost through the exhaust, another third to produce propulsion, and the last third is removed by the coolant. Water has the highest heat capacity, so less needs to be used. However, it freezes easily and corrodes engines quickly. Early antifreezes were salt brines. They reduced the freezing problem, but enhanced the corrosion problem. Methanol and ethanol then were used, but they boiled away too easily and were quite flammable. In 1927, Prestone was introduced, which was ethylene glycol ($HO-CH_2CH_2OH$) mixed with anti-corrosion additives.

Ethylene glycol provides the best freezing-point and heat-transfer characteristics. However, the quality of the water it is mixed with and the high temperatures involved cause many problems that chemists must control. Water usually contains Ca^{+2}, Mg^{+2}, Cl^-, SO_4^{-2}, and dissolved O_2. The engine and radiator are made of many metals, all of which can corrode. Corrosion is an oxidation process, so O_2 must be removed. Acids produced also tend to dissolve away the metal. This corrosion scale reduces the heat transfer and, if the scale falls away, then these particles can bombard other sections of the metal and either remove more scale or "sand blast" away any inhibitor film and expose more metal to corrosion. These particles eventually plug up the radiator or wear out the water pump.

Flowing particles cause *erosion corrosion* by breaking down the inhibitor layer. *Transport deposition* occurs when something is dissolved in one area (engine) and deposited in another (radiator). This deposition reduces heat transfer just where it shouldn't. *Cavitation corrosion* is caused by air bubbles in a foam breaking on a metal surface.

Additives are used to coat the metal surfaces and prevent O_2 and acids from reacting and to prevent foaming. These include borates, molybdates, nitrates, nitrites, phosphates, silicates, amines, triazoles, and thiazoles. Acids are formed by Cl^-, SO_4^{-2} and the oxidation products of ethylene glycol at high temperatures. Foaming is caused by entrapped air and is retarded by silicones, polyglycols, or oil. Most of the air comes from an improperly filled overflow container.

Silicates, phosphates, and molybdates coat iron and prevent corrosion. However, the latter two also form precipitates with Ca^{+2} and Mg^{+2}, which form a scale that reduces heat transfer. Silicates form gels that can clog the radiator.

Please be careful. Ethylene glycol is sweet tasting, and animals will lick up any that is eventually spilled. Ethylene glycol is converted into oxalic acid and will kill the animals. It takes about 95 mL of ethylene glycol orally ingested to kill a 150 lb person. Although the author has no knowledge of anyone consuming that much, the Swiss require propylene glycol to be used, which requires about 479 mL/ 150 lb person to be fatal.

Many antifreeze formulations are proprietary. This is to prevent lawsuits by consumers who claim that their engine was ruined by using product X. The consumer must prove that product X was used, and the company can tell by the additives present if such is the case.

YOUR PROBLEM
The market in your area is being targeted by a company that is selling recycled antifreeze at a very low price. You are being hired to check this antifreeze to see what it contains in the way of additives. You have never done this before, but you know that ion chromatography is the best way to scan for multiple anions. To check your technique and to provide quality assurance to your employer, you are going to examine a sample of nonrecycled antifreeze by ion chromatography.

EQUIPMENT (Dionex part numbers)
- 1 AutoRegen (39594)
- 1 Balance, analytical
- 2 Beakers, 100 mL
- 1 Bottle, wash
- 1 Bulb, pipet, 6 mL
- 1 Cartridge, anion, AutoRegen (39564)
- 1 Diffuser, gas, for helium
- 1 Dish, weighing
- 13 Flasks, volumetric, 100 mL
- 1 pr Goggles, safety
- 1 Kit, micro-membrane installation (38018)
- 1 Pipet, Mohr, 1.0 mL

1 Chromatograph, ion, like Dionex 4000i, 2000i or 2000i/SP.
1 Clamp, gas cylinder
1 Column, anion, HPIC-AS4A (37041)
1 Column, guard, HPIC-AG4A (37042)
1 Detector, conductivity (Dionex)
1 Pipet, Mohr, 5.0 mL
1 Recorder, strip chart
1 Sponge
1 Suppressor, anion micro-membrane, AMMS, (38019)
1 Syringe, 100 µL with 0.45 µm filter.

CHEMICALS   S = Student

Eluent:
  Sodium carbonate, Na_2CO_3, 1.8 mM when used. 19.1 g/L as a 100 x concentrate. 10 mL/S
  Sodium hydrogen carbonate, $NaHCO_3$, 1.7 mM when used. 14.3 g/L as a 100 x concentrate; 10 mL/S
Helium, for degassing the eluent.
Regenerant:
  Sulfuric acid, H_2SO_4, 25 mN when used. 0.66 mL/L 4 L/S
Standards: Stock solutions are all 1,000 ppm in the anion.
  A. Sodium benzoate, C_6H_5-COONa, 0.119 g/100 mL.
  B. Sodium chloride, NaCl, 0.165 g/100 mL.
  C. Glycolic acid, $HO-CH_2CH_2$-COOH, 0.100 g/100 mL.
  D. Sodium molybdate dihydrate, $Na_2MoO_4 \cdot 2\ H_2O$, 0.151 g/2100 mL.
  E. Sodium nitrate, $NaNO_3$, 0.137 g/100 mL
  F. Sodium nitrite, $NaNO_2$, 0.150 g/100 mL
  G. Sodium oxalate, NaOOC-COONa, 0.152 g/100 mL
  H. Potassium dihydrogen phosphate, KH_2PO_4, 0.143 g/100 mL
  I. Sodium sulfate, Na_2SO_4, 0.148 g/100 mL
Sample: A few mL of any unused and used commercial antifreeze. 1 g each/S
  The ingestion of 100 mL of ethylene glycol can result in acute poisoning which is characterized by abdominal distress, central nervous system depression, and respiratory or renal shutdown. The coolant is mildly irritating to the eyes and skin.

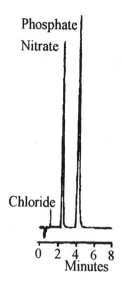

Figure E33-1. Fresh engine coolant (1:100).
Courtesy - Dionex Corp., Sunnyvale, CA)

THE BIG PICTURE

The anions shown in Figures E33-1 and E33-2 are separated isocratically by anion exchange chromatography and detected by conductivity with chemical eluent suppression. Dilute coolants 100-fold by weight and filter through a 0.45 µm filter before injection. Analysis of the entire list of anions in one sample may require two injections at different dilution volumes. Anions such as nitrite, phosphate, benzoate, and molybdate, added as corrosion inhibitors, require relatively higher sample dilution, whereas trace contaminants such as chloride, sulfate, and oxalate require a lower sample dilution. A set of standards will be run, and the instrument adjusted accordingly. A sample of unused engine coolant will be run, followed by a sample of used coolant.

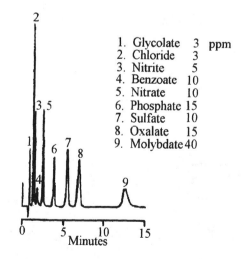

Figure E33-2. Anion standards for engine coolant analysis.
(Courtesy - Dionex Corp. Sunnyvale, CA)

APPARATUS SETUP

The directions are based on a Dionex instrument. However, the following descriptions are presented so any instrument can be adapted.

Sample loop: 50 μL
Guard column: HPIC-AG4A
Separator column: HPIC-AS4A
Eluent:
    1.8 mM Na_2CO_3
    1.7 mM $NaHCO_3$
Eluent flow rate: 2.0 mL/min.

Suppressor: Dionex AMMS
Regenerant: 25 mN H_2SO_4
    flow rate: 4 mL/min.
Detection: Conductivity, 20 μS full scale
Pressure drop: 1040 psi

DIRECTIONS

1. Prepare a standards mixture daily. To a 100 mL volumetric flask, add 0.30 mL of A and B (3 ppm); 0.5 mL of C (5 ppm); 1.0 mL of D, E, and G (10 ppm); 1.5 mL of E and H (15 ppm); and 4.0 mL of I (40 ppm). Dilute to volume with water.

2. Fill a 2 L eluent container with ~ 1.8 L of deionized water and degas with helium.

3. Add 20 mL of combined carbonate, hydrogencarbonate eluent concentrate and dilute to 2 L.

4. Add 2.8 mL of sulfuric acid to a 4 L container and dilute with deionized water.

5. Fill a 100 mL volumetric flask about half full with deionized water.

6. Weigh out 1.00 g of engine coolant (weighing dish) and transfer it to the flask with a stream of deionized water. Dilute to volume. Do the same with the used coolant in another flask.

7. Filter a few mL of each into separate 100 beakers and cover them with a weighing paper.

8. Ask your instructor for an explanation of how your instrument operates.

9. Turn on the instrument and be sure the pumps are filled with liquid. They may need priming if liquids have been changed.

10. Depending upon your instrument and its software package, set the following initially: Eluent flow rate - 2.0 mL/min. Regenerant flow rate - 4.0 mL/min. Detector - 20 μS full scale.

11. Inject 50 μL of the standards and adjust the recorder similar to that in Figure E33-2, p. 615.

12. Inject a second sample of standards, if necessary, to verify that every compound can be detected.

13. Inject 50 μL of unused coolant and record the chromatogram. Identify the components by comparing retention times.

14. Inject 50 μL of used coolant and record the chromatogram. Identify the components by comparing retention times.

15. Shut down the apparatus. Flush the sample engine coolant down the drain (it is dilute and biodegradable) and *clean up the area so it is better than when you found it.* Thank you.

16. Measure the peak heights in mm unless an electronic integrator is a part of your instrument, and determine the ppm of each component present.

EXAMPLE CALCULATION

The peak height for a 5 ppm nitrate standard was 60 mm. The peak height for 0.945 g/100 mL of a used coolant was 46 mm. What is the concentration of nitrate in the used coolant? Assume 1 mL = 1 g of the diluted coolant.

ANSWER

$$5\ ppm \times \frac{46\ mm}{60\ mm} \times \frac{100\ g}{0.945\ g} = 405\ ppm$$

EXPERIMENT 34. ION RETARDATION
Separating salt and urea from urine.

BACKGROUND
Investigations into metabolic disorders frequently require the screening of large numbers of urine samples for abnormalities in the excretion pattern of amino acids. In order to obtain good resolution without tailing or deformation when doing TLC separations of the amino acids, the dissolved salts must be removed. Ion retardation also removes certain peptides that interfere and are not removed by other desalting methods. Urea is usually present is such large amounts that it can be used to indicate when the organic portion has been separated from the salt.

An easy test for urea is to use a combination of bromthymol blue and the enzyme, urease, on a test paper. The urease converts urea into two molecules of NH_3 and one of CO_2. In water these form NH_4OH and H_2CO_3. The solution then is slightly basic. A small piece of test paper is placed on a microscope slide, a drop of sample solution is added, and a cover glass placed on top. Bromthymol blue is orange in acid solution and blue in basic solution (pH 6.0 to 7.6). By placing a cover glass on the test paper, the ammonia is trapped, and the solution becomes basic, the orange paper turning blue if urea is present.

YOUR PROBLEM
A new test for phenylketonuria in urine requires that the salt be removed. You are given a sample and asked to remove the salt by ion retardation.

EQUIPMENT
2	Beakers, 50 mL	1	Pencil, grease or marking
1	Brush, test tube, small	1	Pipet, delivery, 2 mL
1	Column, 1.5 x 30 cm	1	Rack, test tube
2	Cylinder, graduated, 10 mL	30	Slides, microscope
1	Cylinder, graduated, 100 mL	1	Sponge
1	Dropper, medicine	1	Stand, ring
1	Holder, buret	15	Tubes, test, 10 x 100 mm
1 pr	Goggles, safety		

CHEMICALS S = Student
Ammonium chloride, NH_4Cl, 1 M, 50 mL/S

Ammonium hydroxide, NH_4OH, 1N, + 0.5 N NH_4Cl, 50 mL/S

Bromothymol blue, 0.15 g + 2.4 mL of 0.1 N NaOH + 47.6 mL of water. The solution should be green. 15 test strips /S - see below.

Glass wool, ~ 2 g /S

Hydrochloric acid, HCl, 1 N, 50 mL/S

Resin, AG11A8, 30 g/S

Sample: urine, 2 mL/S

Silver nitrate, $AgNO_3$, 1 % in water in a dropping bottle, 10 mL/S

Test strips, 10 mL of bromothymol blue solution + 10 mL of urease solution. Dip paper strips into the reagent, let dry naturally. The strips should be orange. Store in a dark bottle. 15/S

Urease, 0.2 g urease powder to a few drops of water to make a paste then dilute to 10 mL with water.

THE BIG PICTURE
2 mL of urine is passed through an ion retardation column, eluting it with distilled water, discarding the first 10 mL and collecting the next 30 mL in 2 mL fractions. To test for urea, one drop of eluate is spotted on a urease-bromthymol blue test paper, covered with a microscope slide and allowed to set for 5 minutes. A blue color indicates the presence of urea. To test for chloride, one drop of $AgNO_3$ is carefully added to each fraction without shaking, and any white precipitate is an indication of chloride.

APPARATUS SETUP
A 1.5 x 30 cm column held onto a ring stand by a buret holder. A test tube rack holding up to fifteen 10 cm test tubes, each marked at the 4 mL height.

DIRECTIONS

1. Prepare the ion retardation column (1.5 x 30 cm), if it is not already prepared. (30 g soaked in 50 mL of water). Place the column on a ring stand and adjust its height so a 10 mL graduated cylinder will fit below the outlet. Place a small amount of glass wool in the bottom, then add wet AG11A8 resin to a height of about 12 to 13 cm of. Drain away the excess water to 1 to 2 mm from the top of the packing.

2. Test a few mL of your distilled deionized water with a drop or two of $AgNO_3$ to make sure it is chloride free. If a slight haze forms, set it aside as a blank.

3. Fresh AG11A8 resin will require about 4 mL of water/cm^3 of wet resin to remove residual chloride. Your instructor, or the previous student, may have already cleaned the resin. However, test it by adding 100 mL of deionized water to the column, let it drain about 12-15 mL/min and test the effluent with a few drops of $AgNO_3$. Continue until the chloride is removed or is the same as the blank. Leave about 1 mm of liquid on top of the resin.

   Note: If the resin persists in eluting chloride, then regenerate it as described in Table 25-1, p. 291, using about 25 mL of each solution.

5. Test your urease test strips as follows. Tear off about a 1 cm piece of the urease-bromthymol blue test paper and place it on a microscope slide. Pipet one drop of urine onto the test paper, and cover it with a second microscope slide, or cover plate. If urea is present, the paper will turn blue in a few minutes.

6. Place a test tube rack containing 15 to 20 test tubes (10 cm) under the column.

7. Carefully pipet 1 mL of urine onto the top of the resin, disturbing the resin as little as possible.

8. Drain the sample into the resin until the liquid is 1 mm or less above the top of the resin.

9. Elute the urea with deionized water at a rate of about 1 drop/second, collecting the eluent in 10 mL test tubes, 3 mL in each tube.

10. Tear off a 1 cm piece of the urease-bromthymol blue test paper and place it on a microscope slide. Pipet one drop of fraction 1 onto the test paper, and cover it with a second microscope slide or cover plate. If urea is present, the liquid will turn blue in a few minutes. Record the fraction number and the results. Do not throw away this fraction until it has been tested for chloride.

11. The fractions come off slow enough so that each fraction can be tested while another is being collected. Do step 10 with each fraction. Do not throw away any fraction until it has been tested for chloride.

12. Test each fraction for chloride with $AgNO_3$. One drop of $AgNO_3$ is carefully added to each fraction, without shaking, and any white precipitate is an indication of chloride.

13. The urea usually elutes rapidly, in the first 12 to 15 mL. There then is a clear period, and then the chloride elutes taking about 120 mL total. Collect 8 to 10 mL fractions and test each one for chloride. Continue until the column is clean.

14. If you are sure that the column is clean, then stop the flow. If after passing 150 mL of water through the column, and it is still not clean, then regenerate it as described in Table 25-1, p. 291. Test it for chloride with $AgNO_3$ after about 100 mL, and occasionally after that until it is Cl^- negative.

15. Plot tube number vs. urea and chloride, and draw a conclusion.

16. *Clean up the area and arrange the apparatus so it looks like a professional used it last.* Thank you.

EXPERIMENT 35 ION EXCLUSION

Determination of total vitamin C in applesauce.
(Courtesy - Kim, H., *J. Assoc. Off. Anal. Chem.*, **72** (4), 681, 1989)

BACKGROUND

The recommended daily allowance (RDA) for vitamin C (ascorbic and dehydroascorbic acids) is 60 mg. A deficiency of vitamin C causes a disease called scurvy. The cells of capillaries weaken, resulting in hemorrhages, especially in the skin in areas of stress. Bleeding gums are an early sign followed by loosening of teeth. Edema may occur, especially in the legs. Collagen is affected, resulting in structural weakness. Wounds are slow to heal. Eventually the inability of the mesenchymal (connecting) supporting tissues to form and maintain their intercellular substances causes death. These symptoms take four months to begin (why old-time sailors got the disease), but can start to be reversed within 24 hours if ascorbic acid is ingested. A well balanced diet including fresh fruits and vegetables easily provides an adequate supply. However, vitamin C is readily oxidized during storage and loses its activity. Rose hips are 40 times more potent than orange juice (30 to 56 mg/100 mL).

Ascorbic Acid   Dehydroascorbic Acid   1,4-Dithiothreitol: $HSCH_2(CHOH)_2CH_2SH$

Ascorbic acid: $K_{a1} = 7.94 \times 10^{-5}$, $K_{a2} = 1.62 \times 10^{-12}$

YOUR PROBLEM

A local argument is going on, with one person claiming that generic labeled applesauce is poorer quality than the name brands. It is decided that a measure of the vitamin C (ascorbic acid) in each would settle the argument and because you are a chemist you have been selected to make the measurements. Vitamin C is not enough. Because dehydroascorbic acid is reduced to ascorbic acid upon standing, it is desired to determine the total amount and the amount of each.

EQUIPMENT

1	Balance, analytical		5	Flasks, volumetric, 100 mL
2	Beakers, 250 mL		1 pr	Goggles, safety
1	Bulb, pipet		1	Integrator, computing (optional)
1	Centrifuge, 100 mL minimum capacity		1	Mixer, Polytron type preferred
1	Chromatograph, ion, like Wescan		1	Mixer, vortex type preferred
1	Column, guard, anion exclusion		1	Opener, can
1	Column, anion exclusion/HS 100 x 4.6 mm		1	Pipet, Mohr, 1 mL
1	Cylinder, graduated, 10 mL		1	Pipet, Mohr, 10 mL
1	Detector, electrochemical, like Wescan Model 271		4	Rods, stirring, glass, 15 cm
			1	Spatula
			1	Syringe, 25 mL
1	Filter, membrane, 0.45 μm, Nylon 66		4	Tubes, centrifuge, 100 mL total capacity

CHEMICALS   S = Student

Ascorbic acid, (AA) 100 ppm stock standards. 10 mg ascorbic acid in 100 mL of 0.1 M H_2SO_4, (0.55 mL/100 mL) 0.01% EDTA (34 mg disodium salt/100 mL) solution, 2 mL/S
Buffer, pH 7; 50 mM Na_3PO_4, (8.2 g/L) adjusted to pH 7, 2 mL/S
Dehydroascorbic acid, 10 mM; (17.4 mg/100 mL), 1 mL/S
Dithiothreitol, (DTT) 10 mM in pH 7 buffer (154 mg/100 mL), 2 mL/S
Extracting solution, 20 mM H_2SO_4 (1.11 mL/L) degassed under vacuum, 400 mL/S
Sample: 1 small can each of generic and brand name applesauce, 10 g/S

APPARATUS SETUP

The apparatus is a regular HPLC with an ion exclusion column and an electrochemical detector.

THE BIG PICTURE

The ascorbic acid is extracted from a sample of a generic applesauce and a brand name applesauce with 0.1 M H_2SO_4, 0.01% EDTA and filtered; then 20 μL are passed through an ion exclusion column with 20 mM H_2SO_4 as the eluant. A Pt electrode at +0.8V vs. Ag/AgCl is the detector.

A second set of samples is extracted with 20 mM H_2SO_4 in a Polytron homogenizer for 1 minute, filtered, and reduced with 10 mM DTT for 2 minutes. The ascorbic acid is separated and detected as before.

DIRECTIONS

1. Check out the ion chromatograph and clean and stabilize the electrochemical detector as follows:
    a. Turn on the ion chromatograph.
    b. Make sure the eluant bottle contains 0.20 mM H_2SO_4.
    c. Set the sensitivity range at 100 nA full scale.
    d. Determine what pressure setting will produce a flow rate of 0.6 mL/min.
    e. Clean the detector as follows: You can start the sample preparation while this is being done.
        Apply -1.0 V for several minutes followed by +1.8 V for several minutes before re-equilibrating at +0.8 V

2. Open a can of generic applesauce and a can of name brand applesauce.

3. Weigh a 10 g sample of each into separate 250 mL beakers.

4. Dilute each with 90 mL of 20 mM H_2SO_4. Stir to mix thoroughly.

5. Homogenize for 1 minute on a Polytron homogenizer at a setting of 6 (or something equivalent).

6. Centrifuge for 2 minutes.

7. Decant the supernatant liquid and filter through a disposable membrane filter (0.45 μm, nylon 66) attached to a syringe. Collect each in a 100 mL volumetric flask and dilute to volume with 20 mM H_2SO_4.

8. Pipet 0.5 mL of the ascorbic acid standard (1.0 mL Mohr pipet) into a 100 mL volumetric flask and dilute to the mark with 20 mM H_2SO_4 (now 0.5 ppm).

9. Inject 20 μL of this standard into the ion chromatograph and adjust the sensitivity so that the peak is about half scale. Figure E35-1 shows typical chromatograms. The detector output for the ascorbic acid (AA) on the left chromatogram was 15 nA.

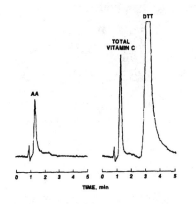

Figure E35-1. AA (left) and total vitamin C (right) in vegetable juice stored at 4 °C for 20 hr.

10. Inject 20 μL of one of the samples into the ion chromatograph as a trial run to see how high the peak is. If it is off scale, then dilute the sample further (measure how much) before the second sample is injected.

11. Repeat step 10 with a sample that will stay on scale and do it with both samples.

12. Pipet 0.2 mL of sample into a 10 x 100 mm test tube. Pipet in 0.6 mL of pH 7 buffer, 0.2 mL of 10 mM dithiothreitol and mix, preferably with a vortex mixer. Let stand for a minimum of 2 minutes at room temperature before injecting this reduced sample into the ion chromatograph.

13. Inject 20 μL of this reduced sample into the ion chromatograph.

14. Repeat for the other sample. Table E35-1 shows the results for several beverages and fruits.

15. *Clean up the area and leave it better than you received it.* Thank you.

Table E35-1. Concentration of vitamin C in selected foods

Sample	Vitamin C, mg/100g		
	Total	AA	DHAA
Applesauce, fresh	40.2 (± 1.0)	41.2 (± 0.8)	0.0
old	40.2 (± 2.2)	31.5 (± 0.5)	8.7
Vegetable juice, fresh	15.7 (± 0.9)	14.7 (± 0.5)	1.0
old	16.0 (± 0.8)	8.0 (± 0.1)	8.0
Orange juice	72.6	62.5	10.1
Grape juice	28.8	27.2	1.6

EXPERIMENT 36. LIGAND EXCHANGE

Determination of hydrazine, methylhydrazine, and dimethylhydrazine in liquid rocket fuels.
(Based on Shimomura, K., Dickson, L., and Walton, H.F. *Anal. Chim. Acta,*, **37**, 102, 1967)

BACKGROUND

The determination of the components in liquid rocket fuels is quite important, because they are not stable and the composition determines how much thrust a rocket can attain for a given volume of fuel. These materials can be treated either as weak bases and titrated or as an oxidation-reduction system. In either case, there is no way to differentiate them. The technique of ligand exchange is a way of separating these materials for further analysis.

YOUR PROBLEM

The fuel for the Aerobee rocket is a mixture of hydrazine (H_2N-NH_2), methylhydrazine ($H_3C-NH-NH_2$), and unsymmetrical dimethylhydrazine, $((CH_3)_2N-NH_2)$. You are to follow the ligand exchange method described in *Anal. Chim. Acta.* **37**, 102 (1967) and determine the hydrazines by the method of R.A. Penneman and L.F. Audrieth, *Anal. Chem.*, **20**, 1058 (1948).

A sulfonic acid resin in the Ni^{+2} form is used. Unsymmetrical dimethylhydrazine is not retained at all, methylhydrazine is eluted with 0.4 M NH_4OH, and hydrazine is eluted last with 5 M NH_4OH.

The titration step is (Penneman and Audrieth)

$$N_2H_4 + KIO_3 + 2\,HCl \longrightarrow KCl + ICl + N_2 + 3\,H_2O$$

In the presence of concentrated HCl (>4N), hydrazine can be titrated directly with standard iodate solution. The addition of iodate is continued until the iodine color is discharged. Actually, the initial reaction involves reduction of iodate to iodine chloride, resulting in the disappearance of the iodine color. A few milliliters of chloroform or carbon tetrachloride are added to dissolve the iodine, and the endpoint is reached when the solvent layer is decolorized.

In acid solution, iodate reacts with hydrazines in a sequence.

$$N_2H_2 \longrightarrow N_2 + 4H^+ + 4e^-$$

$$5e^- + IO_3^- + 6H^+ \longrightarrow 3H_2O + 1/2\,I_2$$

$$1/2\,I_2 + Cl^- \longrightarrow ICl + e^-$$

The acid formed reacts with the iodate to liberate iodine, and the solution turns brown. Then the iodine reacts with the chloride from the HCl to form ICl, which is soluble. The overall acidity of the solution is rather important and should be between 3 to 5 M. If it is less than 3 M, the endpoint is not sharp, and if it is greater than 7 M, the formation of ICl_2^- stops the original reaction, and the result is wrong.

EQUIPMENT

- 1 Beaker, 250 mL
- 1 Beaker, 100 mL
- 1 Bottle, aspirator, 500 mL
- 1 Bottle, wash
- 1 Brush, buret
- 1 Bulb, pipet
- 1 Buret, 25 mL, self-filling preferred
- 3 Clamps, versatile
- 1 Column, ion exchange, 15 cm
- 2 Cylinders, graduated, 10 mL
- 6 Flasks, Erlenmeyer, 125 mL
- 1 Flask, volumetric, 50 mL
- 1 Funnel, powder, small
- 1 pr Goggles, safety

3   Holders, clamp
1   Holder, buret
1   Pipet, 2 mL
1   Pipet, Mohr, 10 mL
1   Pipet, Mohr, 5 mL
1   Plate, titrating

1   Stand, ring
1   Towel, cloth
3 m Tubing, condenser
30 cm Wire, copper

CHEMICALS   S = Student

Ammonium hydroxide, NH_4OH, 0.4 M (50 mL/S) and 5.0 M),100 mL/S
Carbon tetrachloride, CCl_4, 75 mL/S
Glycerine, a few drops/S
Hydrochloric acid, HCl, conc, 150 mL/S
Ion exchange resin, sulfonic acid type
Nickel nitrate, $Ni(NO_3)_2$, saturated, 50 mL/S
Potassium iodate, KIO_3, 0.07 M, 100 mL/S
Stopcock grease
Sample: 2 mL/S
    Hydrazine   2 to 3 mM/mL
    Methylhydrazine   2 to 3 mM/mL   Store these 3 in a refrigerator
    Dimethylhydrazine   2 to 3 mM/mL   when not being used.

THE BIG PICTURE

This first part is done by the instructor, so the resin is ready for the student. The resin is soaked overnight in a Ni solution and then packed into a 1 cm x 15 cm column. It is washed with 5 M NH_4OH and again allowed to set overnight. The rocket fuel is diluted with water to make a solution approximately 2 to 3 millimolar per mL. A little HCl is added to stabilize the solution. Two mL of this solution is added to the column and eluted with 0.4 M NH_4OH. The dimethylhydrazine comes off first, followed by methyl hydrazine. Then 5.0 M NH_4OH is used to elute the hydrazine. 5 mL fractions of eluent are collected, and transferred to a 125 mL Erlenmeyer flask, and 5 mL of CCl_4 and 8 mL of conc HCl are added. The solution is titrated until the violet color is removed from the CCl_4 layer. You will not have to prepare the resin.

APPARATUS SETUP

Refer to Figure E36-1. It can all be set up on one ring stand. The apparatus often will become covered with a white solid. This is ammonium chloride and will wipe away easily with a wet towel and is not harmful.

DIRECTIONS

A 10 mL buret is preferred to improve the titration accuracy and self filling and automatic zeroing speed up the titrations.

1. Fill the buret reservoir bottle with the standard KIO_3 and fill the buret.

2. Place about 5 mL of the 0.4 M NH_4OH on top of the ion exchange bed.

3. Drain this into a 125 mL Erlenmeyer flask until the liquid is 2 to 3 mm above the bed. This removes the 5 M NH_4OH used to store the beads.

4. Pipet 10.0 mL of concentrated HCl and 5.0 mL of CCl_4 into each of six 125 mL Erlenmeyer flasks. Be sure to use a pipet bulb. Refill these whenever you can as the titrations proceed. CCl_4 is toxic. However, the narrow-necked flasks and being covered with HCl offer adequate protection.

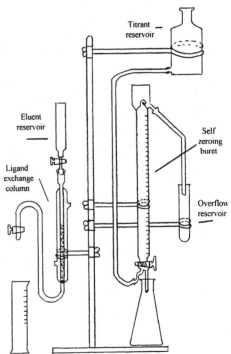

Figure E36-1. Diagram of the apparatus used for this experiment.

5. Prepare a page in your notebook by writing in one column; 5 mL, 10 mL, 15 mL, along the length of the page. This is mL of eluent. Record the mL of titrant used at each of these points.

6. Pipet 2.0 mL of the rocket fuel onto the top of the resin.

7. If its not already there, place the reservoir on top of the column and fill it to within 2 cm of the top with 0.4 M NH_4OH.

8. Obtain two 10.0 mL graduated cylinders; put one under the column outlet and have the other nearby.

A review of the next steps is helpful. You are going to collect a 5.0 mL fraction in a graduated cylinder. Then move that graduated cylinder out of the way and slip another one underneath the outlet. Transfer that 5.0 mL to one of the 125 mL flasks containing the acid and CCl_4. Don't bother to rinse out the graduated cylinder. It will be a reproducible error. Titrate that sample with the iodate solution. While you are titrating this sample, another sample will be filling. A drop every 6 to 10 seconds is about right.

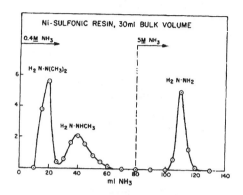

Figure E36-2. Elution sequence for the separation of hydrazine rocket fuels.
(Courtesy - Shimomura, K., Dickson, L. and Walton, H.F., *Anal. Chim. Acta.*, **37**, 102, 1967).

The CCl_4 in the first flask probably will just turn yellow when the iodate is added. The others will usually turn yellow with the first drop and then a brown precipitate will form. Keep titrating and the precipitate will dissolve and turns light yellow. The CCl_4 layer will turn purple. Shake the solutions vigorously and titrate until the purple color disappears.

9. The dimethylhydrazine will elute immediately and be followed by the methylhydrazine. Continue titrating until the methylhydrazine is eluted. Figure E36-2 shows a typical elution pattern.

10. Much stronger NH_4OH is required to remove the hydrazine. Shut off the ion exchange column and empty the reservoir.

11. Replace the reservoir and fill it to about 2 cm from the top with 5 M NH_4OH.

12. *Note*: 5 M NH_4OH fumes from the column plus HCl fumes from the flasks make for an interesting afternoon. The white smoke is NH_4Cl and is harmless, although you look like a mad scientist to all who pass by. Continue the titration as before, until all of the hydrazine has been eluted.

13. Remove the 5 M NH_4OH from the reservoir and be sure there is at least 1 cm of liquid above the bed.

14. Rinse out all of the flasks and pipets. You may collect the CCl_4 for recycling if you desire, otherwise dispose of it in a chlorinated solvents waste container.

15. With a damp cloth, wipe the NH_4Cl off of the column and anywhere else it is present.

16. *Clean up the area in general and leave it cleaner than when you arrived.* Thank you.

17. Plot the mL of IO_3^- vs. volume of eluent for each fraction to obtain an elution curve.

EXPERIMENT 37. ELECTRODEPOSITION
The determination of copper and nickel in a U.S. 5 cent piece.

BACKGROUND
If you have samples of alloys that contain high percentages of metals, then often the best way to separate and quantitate these metals is to selectively electroplate them. Copper and nickel are on opposite sides of hydrogen in the activity series and, hence, can be separated electrically by control of the pH.

Experiment 37

YOUR PROBLEM

You are hired as an outside consultant to check the purity of some of our coinage metals. Your specific task is to determine if the U.S. nickel contains 75% copper and 25% nickel as it is supposed to. This technique in the hands of an experienced person is capable of 1 part in 5000 accuracy.

THE BIG PICTURE

The gauze electrode and the coin are weighed. The coin (nickel) is dissolved in 10 mL of H_2SO_4 plus 20 mL of HNO_3 plus 40 mL of water. The NO_2 then is driven off, and the solution diluted to 100 mL. A 10 mL aliquot is diluted to 100 mL in an electrolytic beaker (tall and narrow with no pouring lip). 0.5 g of urea is added to remove any nitrite formed, and the copper plated out at 1.5 amperes. After 1 hour, the electrode is removed and weighed. The solution will be a light green. The solution remaining is evaporated to dryness, and 25 mL of water added as well as 15 mL of NH_4OH. The resulting blue solution is examined for an iron hydroxide precipitate. If a brown ppt is formed, this is filtered off. The nickel solution then is electroplated for 1 hour or for 10 minutes after the color is gone. The electrode is weighed, and the percent Ni determined.

EQUIPMENT

1	Anode, Pt wire coil		1	Electrode, Pt wire, anode
1	Apparatus, electrodeposition		1	Funnel, 60° long stem
1	Balance, analytical		1	Flask, volumetric, 100 mL
1	Beaker, electrolytic, 180 mL, Berzelius type		1	Oven, drying
1	Beaker, 150 mL		1 pk	Paper, filter 11 cm
1	Beaker, 1000 mL		1	Pipet, delivery, 10 mL
2	Bottles, wash		1	Rack, funnel
1	Brush, tooth brush type		1	Rod, stirring, 15 cm
1	Bulb, pipet		1	Stand, ring
1	Cylinder, graduated, 10 mL		1	Towel, cloth - or sponge
1	Cylinder, graduated, 100 mL		1	Watch glass, 7.5 cm dia
1	Dish, Al weighing		6 ft	Wire, insulated
1	Electrode, Pt gauze, cathode			

CHEMICALS   S = Student

Acetone, $H_3C(C=O)CH_3$, 50 mL/S
Ammonium hydroxide, NH_4OH, conc., 20 mL/S
Distilled water
Nitric acid, HNO_3, conc., 30 mL/S
Sulfuric acid, H_2SO_4, conc., 5 mL/S
Urea, $(H_2N)_2C=O$, 0.5 g/S

APPARATUS SETUP

Figure 26-2, p. 300, shows the Eberback electrolytic analyzer that will be used for this experiment. The Eberbach Corporation, Box 1024, Ann Arbor, MI 48106-1024, has a complete stock of electrodeposition apparatus, electrodes, chemicals and glassware.

DIRECTIONS

1. The apparatus described is an Eberbach electrolytic analyzer unit that is a duplicate system. The two meters at the top left (A) control the left electrode pair, and the two meters on the right (B), the right electrode pair. You will be using the right electrode system. The switch (C) just below the left set of meters is the power switch; when the power is on, the green light to its right (D) is lit. The switch further to the right (E) is to turn on the stirrers; when they are on, the yellow light (F) is lit.

    There is a polarity switch (G) for each set of electrodes. When the switch is in the middle position, the current is shut off to the electrodes. You want your metals to plate out on the negative electrode. The amount of current is adjusted by turning the dial below the polarity switch (H). Turning it clockwise increases the current. This apparatus has no control over the voltage. Set the polarity switch in the middle on both sets of electrodes and turn the current control full counterclockwise.

    Look below the overhang, and you will see two sets of electrode holders (I). Electrodes are inserted in these by pushing in on the front of the holder, inserting the electrode into the slot, and releasing the pressure. The stirrer (J) is inserted up into the holder and is held by a compression fit. The beaker holder is adjusted to any height desired by using either a coin or a screwdriver to loosen the collar (L) and then moving it to the height desired. You will use one like the one on the left. The one on the right is a magnetic stirrer.

2. Get a nickel and a small brush (toothbrush is fine).
3. Put a few drops of acetone on the nickel and scrub it to remove any dirt and grease.
4. Weigh that nickel to the nearest mg and record the weighing.
5. Place the nickel in a 150 mL beaker and add 15 mL of water, 5 mL of concentrated sulfuric acid, and 10 mL of concentrated nitric acid.
6. Place the beaker on a hot plate in a hood and turn it to the high position.
7. Place three glass hooks over the edge of the beaker and then set a watch glass on the hooks to prevent spattering later on.
8. While the nickel is dissolving, you must clean and weigh the platinum gauze electrode. It is very important that you clean it carefully, because it may have a nickel residue on it from a previous experiment that will interfere with your results. Also, fingerprints on the gauze will cause the copper to deposit erratically. Get two platinum electrodes. One is a cylinder of gauze and is to be the negative electrode, and the other is a coiled platinum wire and is to be the positive electrode.
9. Get an electrolytic beaker. They are about 12 cm tall and about 4 to 5 cm in diameter. They have this shape so that you can completely cover your electrode and yet not use a very large volume of liquid.
10. Place 10 to 15 mL of concentrated nitric acid into that beaker.
10. Set the platinum gauze in that beaker and swirl it around, then tip the beaker until it's almost horizontal, or until the nitric acid covers the upper edge of that gauze. Then slowly rotate the gauze in the beaker so that you get nitric acid over all of the gauze. Use distilled water to rinse the gauze, then rinse it with acetone. Let the gauze air dry and then weigh it.
11. Insert the gauze into one of the electrode holders by pushing in on the holder, inserting the electrode, and then release the pressure.
13. Insert the wire electrode into the other holder.
14. Adjust the polarity so that the gauze electrode is the negative electrode.
15. Heat the dissolved nickel solution until the brown fumes completely cease coming off.
16. Transfer that solution to a 100 mL volumetric flask, using a few mL of distilled water to help in the transfer.
17. Because this is a hot solution and will contract on cooling, dilute it to 3 mm above the mark on the 100 mL volumetric flask. Stopper the flask and thoroughly mix the contents.
18. Pipet 10 mL of this solution into an electrolytic beaker.
19. Add 0.5 gram of urea.

    Nitrites are quite harmful to the separation because they will redissolve the copper from the electrode. How these ions are formed and how they are removed are shown in the following equations:

    $3\ Cu + 8\ HNO_3 \rightarrow 3\ Cu(NO_3)_2 + 2\ NO + 4\ H_2O$

    $2\ NO + O_2 \rightarrow 2\ NO_2$

    $NO_2 + NO + H_2O \rightarrow 2\ HNO_2$ This dissolves the Cu

    $2\ HNO_2 + (NH_2)_2CO \rightarrow CO_2 + 2\ N_2 + 3\ H_2O$

    or

    $HNO_2 + HOSO_2NH_2 \rightarrow H_2SO_4 + N_2 + H_2O$
    Sulfamic acid

20. Place the beaker under the electrodes and adjust it so that the stirrer just misses the bottom of the beaker.

21. Dilute the solution with water until it covers *no more than half* of the gauze electrode.

22. Be sure the apparatus is plugged in.

23. Turn on the stirring motor and make sure that the solution is being stirred and that the electrodes are not touching each other.

24. Look at the bottom of the beaker and make sure the solution is getting stirred down there. If not, then raise the beaker a little bit. *Under no conditions cover that entire gauze.* Leave some uncovered gauze for later.

25. Adjust the current to about 0.7 amperes.

26. In a minute or two, you should notice some gas bubbles forming in the beaker, and after a few minutes, you should see a salmon-colored deposit forming on the gauze electrode. If the deposit is forming on the coiled wire electrode, then the polarity of the electrode is wrong. Change the polarity.

27. Look at the texture of the copper plating out. If you wanted to make this really smooth, you would use a current of 0.2 or 0.3 amps, but it would take a long time to finish the experiment. A current of 0.7 amps will produce a nice deposit on the electrode.

28. Turn the current control clockwise and raise the current to somewhere between 1.5 and 1.7 amperes.

    Look around the bottom edge of that gauze electrode and you will notice that suddenly it gets darker brown. What is happening is that the copper is plating out so fast that it is very finely divided and spongy. This changes the amount of radiation that it will reflect and it looks darker. You can stop this by lowering the current back to around 0.7 or 0.5 amps. You will see that, in just a few minutes, this will clear up and it will become a nice salmon pink color again.

29. When you have finished adjusting the current up and down, set it at about 1.5 amperes and let this solution plate for about half an hour.

30. The solution will be a light green because of the nickel that is in the solution. To determine if the copper is plated out, raise the beaker about 5 to 6 mm and look at the fresh gauze that is exposed to the solution.

    If the copper has all been plated out, that fresh gauze will stay shiny. If the gauze turns salmon pink, then continue the electrolysis for another 10 minutes. Repeat the process until you can no longer see copper plating out, then continue the electrolysis for another 5 minutes.

31. Do not shut off the voltage to the electrodes. If you shut off the voltage to this system too soon, the acid in the solution will redissolve the copper off of the electrodes before you can get them out of the solution. You have to keep the voltage on until you get all of the acid washed off. Move the support stand out of the way and slowly lower the beaker. As you do, spray a stream of water from a wash bottle over the electrodes to wash off the acid.

32. Turn the voltage off, set the polarity switch to the middle, and shut off the stirring motor.

33. Leave the electrode alone for the moment. Transfer the solution that's in your beaker to a 1 liter beaker. Use as few mL of water as possible to make a transfer.

34. Put the beaker on the hot plate, turn the hot plate to the high position, and evaporate the solution nearly to dryness. While the solution is evaporating, determine how much copper is in the coin.

35. Take that copper-coated electrode over to the sink and spray it with a stream of acetone.

36. Let it air dry a few minutes, weigh it to the nearest mg, and record the weight. You can dry the electrode in an oven, but don't leave it in there very long, because it will air oxidize and begin to turn black. This will increase the weight and cause an error.

37. You now have three options to plate the nickel: (1) deposit the nickel on top of the copper, (2) change the pH and plate it on the anode, or (3) remove the copper and plate it on the cathode. The third option will be done. The reason is that if you don't get the solution strength just right for the next step, some of that copper can dissolve and ruin your nickel analysis. Place the wire gauze with the copper on it in a small beaker, add 10 or 15 mL of nitric acid, and clean off the gauze like you did before.

38. It is not necessary to rinse the electrode with acetone. Be sure the polarity switch is in the center and that the current is at zero. Put the electrode back on the instrument just as you had it before.

39. Go to the solution that is evaporating. As it gets close to dryness, watch it very carefully, and turn the heat down to medium if it looks like it is going to spatter a great deal.

40. Turn off the hot plate.

41. Remove the beaker from the hot plate and very carefully add about 25 mL of water to it from a wash bottle. Be sure to wash down the sides.

42. You should have a very light green solution possibly with a few clear particles in the bottom. These particles are hard to dissolve in water. Add about 20 mL of concentrated ammonium hydroxide to this solution. Swirl the solution, and those particles should dissolve.

    One other problem sometimes occurs. One of these alloys sometimes contains a little bit of iron. Iron will plate out as nickel and cause an error. As soon as you added base to this solution, it should have turned blue. Iron will precipitate as a reddish-brown precipitate. Look at your solution, and if you have any reddish-brown particles, filter that solution in the hood. Collect the filtrate in your electrolytic beaker. Use 1 + 300 ammonium hydroxide as a wash solution.

43. If you have no iron precipitate, transfer the solution to an electrolytic beaker.

44. Place that solution on the electrolysis apparatus the same as you did for copper.

45. Again, adjust the polarity to make the gauze electrode the negative electrode, and turn on the stirrer.

46. Set the current for 1.5 amperes and plate this system until all of that blue color is gone.

47. It is very difficult to tell when all of the nickel is out. When all of the blue color is gone, then continue for about 10 more minutes.

48. Wash the electrode as before. Remember, leave the voltage on and slowly lower the beaker as you wash the electrode.

49. Turn the polarity switch to the middle position, turn the current control full counterclockwise, and turn off the stirrer.

50. Take the electrode off of the apparatus, rinse it with acetone, air dry it, and weigh it.

51. Dissolve the nickel off of the electrode the same as you did the copper.

52. *Clean up around the area.* Thank you.

53. Return the electrodes to the TA.

54. Calculate the percentages of copper and nickel in that coin. A U.S. nickel is about 75% copper and 25% nickel.

EXAMPLE CALCULATION

Assuming the nickel is 75% Cu and 25% Ni and that the current efficiency is 100%, calculate the time it should take to plate out each metal if the nickel weighs 5.000 g and the current is 1.5 amperes. Remember, you took a 1/10 aliquot.

ANSWER

$$Cu: 0.375 \text{ g} = \frac{1.5 \text{ amp} \times t \times 63.55 \text{ g}}{2\,e^- \times 96{,}485 \text{ coulombs}} = 759 \text{ sec}; \quad Ni: 0.125 \text{ g} = \frac{1.5 \text{ amp} \times t \times 58.70 \text{ g}}{2\,e^- \times 96{,}485 \text{ coulombs}} = 274 \text{ sec}$$

EXAMPLE CALCULATION

What are the % Cu and % Ni in a U.S. nickel if it weighed 5.000 g, the weight of Cu and the Pt gauze = 19.9832 g, the weight of the gauze = 19.6112 g, the weight of Ni and the Pt wire = 9.7334, and the weight of the anode wire = 9.6114 g.

ANSWER

Remember that a 1/10 aliquot was taken.

$$\%Cu = \frac{(19.9832 \text{ g} - 19.6112 \text{ g}) \times 100}{0.5000 \text{ g}} = 74.4\%; \quad \%Ni = \frac{(9.7334 \text{ g} - 9.6114 \text{ g}) \times 100}{0.5000 \text{ g}} = 24.4\%$$

EXPERIMENT 38. AGAR HORIZONTAL STRIP ELECTROPHORESIS

The separation of the isoenzymes of lactic dehydrogenase (LDH) in blood serum.
(Wright, E.J., Cawley, L.P., Eberhardt, L., *Am. J. Clin. Path.*, 737, 1966; Strandjord, P.E., Clayton, K.J., and Freir, E.F., *J. Am. Med. Assoc.*, **182**, 1099, 1962).

BACKGROUND

LDH is a tetramer made up of two monomers (called A and B here). The total m. w. is about 135,000, and the system is a complex polypeptide. There are five ways to combine two monomers into a tetramer.

LDH5 = A4B0; LDH4 = A3B1; LDH3 = A2B2; LDH2 = A1B3; LDH1 = A0B4

These are called *isoenzymes*. By definition, isoenzymes are electrophoretically distinguishable, but with similar substrate specifications.

A normal cell will produce these enzymes in a definite ratio. However, when the body is sick, the cells produce a different ratio. By examining the intensity ratios, some diseases can be detected. Figure E38-1 shows several patterns for human serum.

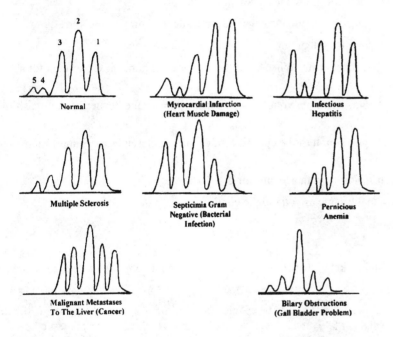

Figure E38-1. LDH isoenzyme patterns for several human diseases. (Courtesy - Gelman Instrument Co., Ann Arbor, MI)

Lactate   Pyruvate   Phenozine methosulfate

Visual localization of electrophoretically separated LDH isoenzymes is accomplished by the reduction of nitro-tetrazolium blue as the electron acceptor (terminal) in a medium containing phenazine methosulfate and NAD. The linked reaction is as follows:

Lactate + NAD $\longrightarrow$ Pyruvate + NADH$_2$

NADH$_2$ + PMS $\longrightarrow$ NAD + reduced PSM

Reduced PMS + NBT $\longrightarrow$ Formazan + PSM

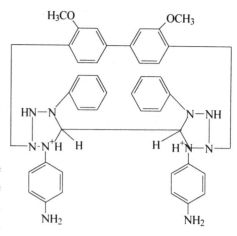

Nitro blue (its structure is close to the diagram to the right).

YOUR PROBLEM

You have been asked by a local veterinarian to help him diagnose a sick horse. He does not have an electrophoresis apparatus nor the expertise to use it, but he knows that it might provide the data he needs. You obtain about 10 mL of blood from his horse, take it back to the lab, and perform an LDH isoenzyme separation. (heparin not needed, wear gloves). If you worry about handling blood or don't have a readily available source, you can use a set of standards obtained from Sigma Scientific Co.

EQUIPMENT

1	Apparatus, electrophoresis	2	Ovens, drying, one at 56 °C, one at 100 °C
1	Balance, analytical		
1	Balance, triple beam	2	Pans, aluminum about 30 cm x 30 cm x 2.5 cm
1	Beaker, 50 mL		
1	Beaker, plastic, 250 mL	1	Pipet, Mohr, 10 mL
1	Brush, test tube, small	1	Pipet, delivery, 20 mL
1	Bulb, pipet	1	Razor blade, single edge
1	Centrifuge, 15 mL tubes	1	Refrigerator
1	Cylinder, graduated, 100 mL	1 pr	Scissors
2	Dishes, weighing, Al or weighing papers	1	Slide, microscope, 2.5 cm x 10 cm
		1	Spatula
5	Film, cellulose acetate, 35 mm Dupont P-40 B leader, 150 cm/S	2	Stoppers, rubber, no. 0, solid
		2	Stoppers, rubber, no. 4, solid
1	Flask, Erlenmeyer, 125 mL	1	Syringe, Hamilton, 10 µL
1	Flask, Erlenmeyer, low actinic, 125 mL	1	Syringe, 1 mL
1	Flask, volumetric, 4 mL	2	Towels, cloth
1 bx	Foil, aluminum	1 pk	Towels, paper
1 pr	Goggles, safety	2	Tubes, centrifuge, 15 mL
1	Hotplate	1	Vial, 1 dram

CHEMICALS   S = Student

Acetic acid, H$_3$C-COOH, 10%, 200 mL/S

Agarose, 250 mg/S

Bromophenol blue or crystal violet, (Marker dye) 0.5 mL/S

Buffer pH 7.5; 15.7 g KH$_2$PO$_4$ + 68.8 g Na$_2$HPO$_4$/4 L H$_2$O, 900 mL/S

β-Diphosphopyridine nucleotide (DPN), 40 mg/S

Nitro blue tetrazolium (NBT), 20 mg/S

Lithium lactate, H₃C-CHOH-COOLi,	3264 mg	Dissolve in 240 mL of buffer. It can be
Sodium cyanide, NaCN,	288 mg	used immediately, but works best if aged
Magnesium chloride, MgCl₂·6H₂O	203 mg	3 weeks. Store in a black bottle.

Phenazine methosulfate (PMS), 1 mg/mL

Add the DPN and the NBT to 20 mL of the buffer and then add 1 mL of the PMS solution. This is a light-sensitive reagent. Store in a low actinic bottle.

Sample: Do not get this sample until you are actually doing the experiment. 10 mL of blood will be required. (No heparin needed) Horse blood is preferred, but any type will do.

THE BIG PICTURE

You will first prepare five acetate strips, two will be spares. Weigh 250 mg of agarose into a 125 mL flask. Add 50 mL of buffer and heat until dissolved, let cool a few minutes. Place a thin layer of this solution on the acetate strips, and let cool for 5 minutes, place in a tray, and set in a refrigerator.

Get a 15 mL centrifuge tube fitted with a rubber stopper and go to a veterinary hospital. Ask them for 10 mL of blood, horse blood being preferred.

Centrifuge the blood and transfer 5 µL of the serum to the acetate strip. Mix 3 drops of marker dye with 0.25 mL of the serum and place a small spot on the edge of the acetate strip. Place the strip in the electrophoresis apparatus and set at 180 volts, but no more than 12 ma/strip. Adjust the voltage accordingly. Refer to Figure E38-2.

Run the separation until the marker dye has moved 2.5 cm. Remove and gently dab dry on a paper towel. Scrape away the excess agarose from each end. Cover the remaining agarose layer with the overlay solution, place the strips in a foil-covered pan, and incubate at 56°C for 45 minutes. Soak in acetic acid for 1 hour and dry at 100°C. The excess dye is removed with acetic acid. The relative intensity of the bands is estimated.

Serum - the watery portion of blood after coagulation.

Plasma - the fluid part of blood.

APPARATUS SETUP

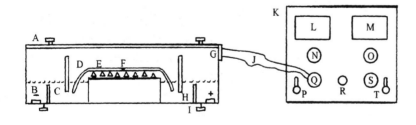

Figure E38-2. Apparatus for this experiment. (A) cover plate. (B) electrode. (C) baffles. (D) acetate strip. (E) agarose layer. (F) sample spot. (G) electrode plug-in. (H) buffer. (I) leveling screws. (J) electrical connections. (K) power supply. (L) amperes. (M) volts. (N) coarse adjust. (O) fine adjust. (P) main switch. (Q,S) output plugs. (R) pilot light. (T) polarity switch.

DIRECTIONS

1. Inspect the apparatus. If it has a multiple unit power supply, be sure the unused sections are shorted out so the remaining units will operate. Be sure the power is OFF, then remove the cover and locate the platinum electrodes, the support rack, the leveling screws, and the buffer level marker.

2. Turn on a hot plate to medium.

3. Put about 400 mL of buffer into each side of the buffer chamber or until it reaches the leveling mark.

4. Look carefully at the leveling marks and the buffer level on both sides and adjust the leveling screws if necessary to level the tank.

5. Put the cover back on the developing chamber.

6. Go to the power supply and locate the voltage side, the amperage side, and the polarity switch.

7. Note which electrode is positive, so you can determine the charge on your compounds when they are separated.

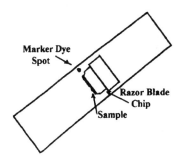

Figure E38-3. The use of a broken razor blade to apply a strip of sample.

You need to practice two techniques now for use later on. These are the use of a syringe and how to apply a streak of sample on an acetate strip.

8. Locate a 1 mL syringe and a small beaker. Add some water to the beaker. Never touch the inside ground glass part of a syringe. Fingerprints on it may make it stick after a period of time. *You do not oil or grease a syringe; it is lubricated with the liquid placed in it.*

9. Draw a small amount of water into the syringe, withdraw it, and tilt the needle upward. Do you notice a small air bubble? That has to be removed, if accurate dispensing is to be achieved.

10. Push the plunger to eject the air. With small-bore syringes, you may have to tap the barrel. Place the syringe needle back into the water and fill the syringe and eject any excess. If this is a liquid different from a previously used one, then fill and eject the liquid several times to flush out the barrel. Be sure no air is present and then adjust the plunger to the volume desired. Do this several times until you can do it correctly.

11. Locate a µL syringe and repeat the technique. A fingerprint on these is a disaster, and these are expensive syringes.

12. Locate the single edge razor blade used to apply the sample. Figure E38-3 illustrates the technique for applying a strip of sample to the buffer.

The technique is to use a µL syringe and apply a thin strip of sample to the cutting edge of the razor blade. The blade then is inserted vertically into the gelled buffer until it touches the acetate strip. The blade then is tilted away from the side where the sample was applied, allowing the sample to fall into the slit. The blade then is tilted in the opposite direction and slowly removed, scraping the sample off and leaving it behind as a strip in the buffer.

13. Cut from a roll or obtain 3 cellulose acetate strips about 20 cm long and 35 mm wide. Lay them on the table top with the curved ends upward.

14. Place 250 mg of agarose in a 125 mL Erlenmeyer flask.

14. Add 50 mL of buffer to the flask, place the flask on a hot plate, and heat the mixture until it becomes a clear liquid, but do not let it boil. When this liquid cools, it will form a gel to hold the buffer on the acetate strips.

16. Remove the hot liquid from the hot plate and let cool for 4 to 5 minutes.

17. Fill a 10 mL Mohr pipet with the hot buffer and spread a thin layer of it on each acetate strip. Be careful. Try not to let it run over the side, but if it does, let it cool before you try to remove it.

18. Find a metal pan about 3 cm x 20 x 30 cm, or equivalent. Place a cloth towel in the bottom and add enough distilled water to make it extra damp, but not soaking wet.

19. Lay the acetate strips on the cloth, cover the pan with Al foil, and place the pan in a refrigerator. They work better if cool because less buffer evaporates when the current is applied.

20. LOCAL OPTION. Get blood where your instructor directs. Take a 15 mL centrifuge tube with a rubber stopper and go to the veterinary hospital. Have them draw a few mL of blood, preferably from a sick horse. *Do not add heparin.*

    Note: Detergent residue ruins this experiment. Be sure you rinse this tube several times.

21. Locate a centrifuge and a second centrifuge tube.

22. Place as much water in this second tube as the amount of blood you have, stopper it, and place it in the centrifuge.

23. Remove the stopper from the tube containing the blood sample. Use a spatula and cut loose the film on the top of the sample, replace the stopper, and place the tube in the centrifuge *opposite* the other tube.

24. Centrifuge the sample for about 5 minutes.

25. Remove the acetate strips from the foil-covered pan and lay them on a paper towel.

26. Very gently dab across the top of them with a paper towel to remove the moisture droplets.

27. Remove your blood sample from the centrifuge. Use a spatula and break up any gel that has formed on the top.

28. Use a µL syringe and obtain 5 µL of the top serum of the blood sample.

29. Place this on the edge of the razor blade, then place it on the middle of the acetate strip. Do not streak across the entire strip, but leave a small space on each side for the marker dye to be added later. Do this to all three strips.

30. Use a 1.0 mL syringe and get 0.2 mL of serum. Place it in the 1 cm x 4 cm screw-cap bottle.

31. Add 3 drops of bromophenol blue into the bottle and mix it into the serum. It should turn the serum blue. This is the marker dye.

32. Use a syringe and place a small drop of that dye mixture to the side along one edge of the cellulose acetate strip at the same place that you spotted your sample. Poke it into the gel, but keep the area small. This dyes the albumin in the blood serum. The albumin is the fastest moving component, and the dye is used to tell you how far the separation has progressed. Usually, two spots are formed; use the darkest one.

33. Be sure the power supply to the electrophoresis apparatus is turned off.

34. Add the 3 strips to the separation chamber. Spread them apart and be sure the ends touch the buffer.

35. Put the cover on the apparatus and connect the power line.

36. Turn the power on and adjust the voltage to about 150 volts.

37. You need about 6 to 7 milliamperes per strip. Try to adjust the voltage so that you get the best combination of 6 to 7 ma/strip and 150 volts overall. Never go over 7 ma/strip. With too much current, the strips will get hot and burn through. If that happens, you are ruined.

38. Monitor the amperage carefully. If one strip burns out, then the others will follow quickly because of the increased current.

39. The separation will take from 1 to 1.5 hours. While you are waiting, go to step 40. After 1 hour, SHUT OFF THE POWER, open the cover and see how far it has moved. If it has not moved 2.5 cm, then continue the separation.

40. Go to the refrigerator and get the beaker that contains the chemicals that are used next.

41. Weigh 4 mg of phenazenemethosulfate on an analytical balance and place it into a 4 mL volumetric flask.

42. Dilute to volume with water and store in a closed drawer to keep it dark.

43. Obtain a low actinic (red) 125 mL flask or a flask painted black.

44. Add 40 mg of β-diphosphopyridine nucleotide and 20 mg of blue tetrazolium to the flask.

45. Add 20 mL of solution from the black bottle and 1 mL of phenazenemethosulfate, mix the contents thoroughly, and store in a refrigerator until needed.

46. When the separation is finished, turn off the power, remove the strips, and lay them on a paper towel.

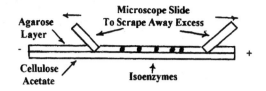

Figure 38-4. Use of a microscope slide to remove excess gel from an acetate strip.

47. Dab them gently with a damp towel to remove the excess moisture.

48. Refer to Figure E38-4 for the technique to remove the excess gel on each strip.

    Start about 1 cm away from where you applied the sample in the opposite direction from where the dye moved. Lay the sharp edge of the microscope slide on the gel and, with one swipe, scrape away the excess. Do the same to the other side of the strip about 1 cm beyond where the marker dye is.

49. With a pair of scissors, cut off the excess cellulose acetate on each end, so you have 3-4 cm to hold onto.

50. Adding the overlay. There are certainly more than 5 compounds on that strip. What is needed is to locate the 5 LDH isoenzymes. This is done by using a selective dye and the technique is called *hstochemistry*. The result is a *zymogram*. With a pipet, spread 1-2 mL of the solution in the low actinic flask over the top of the gel. This is called an *overlay*.

51. Carefully, so the overlay doesn't run off, transfer the strips to the Al pan, moistening the cloth if necessary, and cover the pan with the foil so the strips do not dry out.

52. Place the pan in an oven, set at 56 °C, for 5 minutes.

53. Remove the pan (use a towel) from the oven, remove the foil, and have a look.

54. Disappointing, isn't it? There are no bands. What you need to do is wash away the nonreacted dye. Locate another pan and add 10% acetic acid to a depth of 1 cm.

55. Place the strips in this solution for 1 hour at room temperature (overnight if you want to).

56. Remove the strips, pour out the acetic acid, wipe out the pan, place the strips back in, and place the pan in an oven at 100 °C for 10 - 15 minutes.

57. When the agarose layer has flattened so you can hardly tell that its there, remove the pan from the oven, remove the strips, and you have a *zymogram*.

58. With horse blood, the middle band is the strongest. The patterns for horses are different than those shown for humans. What you would now do is to measure the darkness of each band with an instrument called a *densitometer*. This is nothing more that a high quality spectrophotometer turned sideways so the light beam passes through the flat strip. You then would compare the pattern with known diseases to make your diagnosis. Figure E38-5 shows the order of appearance.

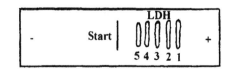

59. Clean out the syringes by pumping water through them.

Figure E38-5. The order of appearance of the LDH isoenzymes.

60. Clean out the pipet being sure it is not plugged.

61. The solution in the red flask and the 4 mL volumetric can be discarded.

62. Clean out the marker dye serum vial.

63. Discard the buffer into the sink.

64. Clean out the tube containing the blood. BE CAREFUL - If any detergent is not rinsed away it will ruin the next students experiment.

65. *Clean up everything and put things back in order, including the reagents in the refrigerator.*

EXPERIMENT 39 AGAROSE GEL HORIZONTAL STRIP ELECTROPHORESIS
DNA fingerprinting - A paternity case
(EDVOTEK Kit #109, West Bethesda, MD 1-800-EDVOTEK. THIS KIT DOES NOT CONTAIN HUMAN DNA.)

BACKGROUND

DNA typing (also called DNA profile analysis or DNA fingerprinting) is a recently developed method that allows for the unambiguous identification of the source of unknown DNA samples. The method has become very important in forensic biochemical laboratories, where it has been used to provide evidence in paternity and criminal cases. In contrast to the more conventional methodologies, such as blood typing, which can only exclude a suspect, DNA fingerprinting can provide positive identification with great accuracy.

DNA fingerprinting involves the electrophoretic analysis of DNA fragment sizes generated by restriction enzymes. Restriction enzymes are endonucleases that catalyze the cleavage of phosphodiester bonds within both strands of DNA. The points of cleavage occur in or near very specific sequences of bases called recognition sites, which are generally 4 to 8 base pairs (bp) in length. The two most commonly used restriction enzymes for DNA profile analysis are *Hae* III and Hinf 1, which are 4-base and 5-base cutting enzymes. The following examples show recognition sites for various restriction enzymes.

Bam HI	↓	*Hae* III	↓
	5'-GGATCC-3'		5'-GGCC-3'
	3'-CCTAGG-5'		3'-CCGG-5'
	↑		↑
Pst I	↓	*Hinf* I	↓
	5'-CTGCAG-3'		5'-GANTC-3'
	3'-GACGTC-5'		3'-CTNAG-5'
	↑		↑

The size of the DNA fragments generated depends on the distance between the recognition sites. In general, the longer the DNA molecule, the greater the probability that a given recognition site will occur. The DNA of an average human chromosome is very large, containing over 100 million base pairs. A restriction enzyme having a 6-base pair recognition site, such as *Eco* RI, would be expected to cut human DNA into approximately 750,000 different fragments.

No two individuals have exactly the same pattern of restriction enzyme recognition sites. There are several reasons for this fact. A large number of *alleles* exist in the population. Alleles are alternate forms of a gene. Alleles result in alternative expressions of genetic traits, which can be dominant or recessive. Chromosomes occur in matching pairs, one of maternal and the other of paternal origin. The two copies of a gene (which can be alleles) at a given chromosomal locus, which represent a composite of the parental genes, constitute an individual's unique genotype. It follows that alleles have differences in their base sequences, which consequently creates differences in the distribution and frequencies of restriction enzyme recognition sites. Other differences in base sequences between individuals can occur because of mutations and deletions. Such changes also can create or eliminate a recognition site. The example in Figure E39-1 shows how a silent mutation can eliminate a recognition site, but leave a protein product unchanged.

Individual variations in the distances between recognition sites in chromosomal DNA often are caused by intervening repetitive base sequences. Repetitious sequences constitute a large fraction of the mammalian genome and have no known genetic function. These sequences can occur between genes or are adjacent to them. They also are found within introns. Ten to 15% of mammalian DNA consists of sets of repeated, short sequences of bases that are tandemly arranged in arrays. The length of these arrays (the amount of repeated sets) aries between individuals at different chromosomal loci.

TGTTTA/TGTTTA/TGTTTA/......   variable number

When these arrays are flanked by recognition sites, the length of the repeat will determine the size of the restriction enzyme fragment generated. Several types of these short, repetitive sequences exist and they have been cloned and purified. Refer to Figure E39-2.

The variations in DNA sequences between individuals as determined by differences in restriction enzyme cleavage patterns are known as *restriction fragment length polymorphisms* (RFLPs). RFLPs are manifestations of the unique molecular genetic profile, or *fingerprint*, or an individual's DNA.

Agarose gel electrophoresis is a procedure used to analyze DNA fragments generated by restriction enzymes. The agarose gel consists of microscopic pores that act as a molecular sieve. Samples of DNA are loaded into wells made in the gel during casting. Because DNA has a strong negative charge at neutral pH, it migrates through the gel towards the positive electrode during electrophoresis. DNA fragments are separated by the gel according to their size. The smaller the fragment, the faster it migrates. After electrophoresis, the DNA can be visualized by staining the gel with dyes. Restriction enzyme cleavage of relatively small DNA molecules, such as plasmids and viral DNAs, usually results in discrete banding patterns of the DNA fragments after electrophoresis. However, cleavage of large and complex DNA, such as human chromosomal DNA, generates so many differently sized fragments that the resolving capacity of the gel is exceeded. Consequently, the cleaved DNA is visualized as a smear after staining and has no obvious banding patterns.

RFLP analysis of genomic DNA is facilitated by Southern blot analysis, p. 326. After electrophoresis, the DNA fragments in the gel are denatured by soaking in an alkali solution.

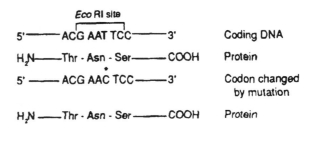

Figure E39-1. Elimination of a recognition site.
(Courtesy - EDVOTEK, West Bethesda, MD)

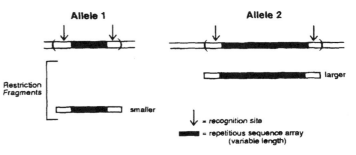

Figure E39-2. Enzyme Cleavage Patterns Based on Recognition Site Position.
(Courtesy - EDVOTEK, West Bethesda, MD)

This causes double-stranded fragments to be converted into single-stranded form (no longer base-paired in a double helix). A replica of the electrophoretic pattern of DNA fragments in the gel is made by transferring (blotting) them to a sheet of nitrocellulose or nylon membrane. This is done by placing the membrane on the gel after electrophoresis and transferring the fragments to the membrane by capillary action or suction by vacuum. The DNA, which is not visible, becomes permanently adsorbed to the membrane, which can be manipulated much more easily than gels.

Analysis of the blotted DNA is done by hybridization with a radioactive DNA probe. In forensic RFLP analysis, the probe is a DNA fragment containing base sequences that are complementary to the variable arrays of tandemly repeated sequences found in the human chromosomes. Probes can be labeled with isotopic or nonisotopic reporter molecules that are used for detection. A solution containing the single-stranded form of the probe is incubated with the membrane containing the blotted, single-stranded (denatured) DNA fragments. Under the proper conditions, the probe will only base pair (hybridize) to those fragments containing the complementary repeated sequences. The membrane then is washed to remove the excess probe. If the probe is isotopically labeled to the membrane, it then is placed on an x-ray film for several hours. This process is known as autoradiography. Only DNA fragments that have hybridized to the probe will reveal their positions on the film because the localized areas of radioactivity cause exposure. The hybridized fragments appear as discrete bands (fingerprint) on the film and are in the same relative positions as they were in the agarose gel after electrophoresis. Only specific DNA fragments, of the hundreds of thousands of fragments present, will hybridize with the probe because of the selective nature of the hybridization process. Because autoradiography is an extremely sensitive technique, only small amounts of DNA samples are required.

In forensic cases, DNA samples can be extracted and purified from small specimens of skin, blood, semen, or hair roots collected at the crime scene. DNA that is suitable for analysis even can be obtained from dried stains of semen and blood. The RFLP analysis performed on these samples then is compared to those performed on samples obtained from the suspect. If the RFLP patterns match, it is then beyond reasonable doubt that the suspect was at the crime scene. In practice, several different probes containing different types of repetitious sequences are used in the hybridizations in order to satisfy certain statistical criteria for absolute, positive identification. The use of different restriction enzymes allow for accuracies in positive identifications of greater than one in 100 million. In recent years, polymerase chain reaction (PCR), a method that amplifies DNA, has made it possible for very small amounts of DNA found at crime scenes to be amplified for DNA fingerprinting analysis. Using specific probes to prime DNA polymerase, many copies of the targeted areas of DNA can be synthesized in vitro and subsequently analyzed.

YOUR PROBLEM

A lady was raped, and in the struggle, she scratched the assailant. Based on a general description, the police have taken into custody two suspects, neither of whom can the victim positively identify. The police have obtained material from under the victim's fingernails, and a blood type match has been attempted. Both suspects and the victim have A+ blood. More specific information is needed, and DNA analysis is prescribed. In this case, DNA obtained from the two suspects and a sample from the crime scene are cleaved with two restriction enzymes in separate reactions. The objective is to analyze and match the DNA fragmentation patterns after agarose gel electrophoresis and determine if either Suspect 1, Suspect 2, or neither was at the crime scene.

EQUIPMENT

1	Apparatus, horizontal gel electrophoresis	1 pr	Gloves, asbestos or similar
1	Beaker, 600 mL	1 pr	Gloves, disposable
1	Cylinder, graduated	1 pr	Goggles, safety
1	Flask, Erlenmeyer, 250 mL	1	Graduated cylinder, 100 mL

1 Micropipet, automatic with 6 tips	1 Thermometer, 110 °C
1 Pipet, delivery, 1 mL	1 Visualization system, DNA (white light with Methylene Blue Plus ® staining)
1 Pipet pump	
1 Power supply, D.C., 0-250 v	
1 Sponge	

CHEMICALS

This kit contains ready-to-load DNA samples and reagents. Store the entire kit in the refrigerator.

Contents

- A DNA sample from crime scene cut with Enzyme 1
- B DNA sample from crime scene cut with Enzyme 2
- C DNA sample from Suspect 1 cut with Enzyme 1
- D DNA sample from Suspect 1 cut with Enzyme 2
- E DNA sample from Suspect 2 cut with Enzyme 1
- F DNA sample from Suspect 2 cut with Enzyme 2
- 1 Tube Practice Gel Loading Solution
- 1 Bottle of UltraSpec-Agarose powder (2 to 5 g)
- 1 Bottle of 50*x* concentrated electrophoresis buffer
- 1 Bottle of concentrated Methylene Blue Plus stain

   Distilled or deionized. water

THE BIG PICTURE

In this experiment, DNAs are cut by restriction enzymes and the fragmentation patterns serve as the individual fingerprint. The DNA fragmentation patterns are simple enough to analyze directly in the stained agarose gel, which eliminates the need for a Southern blot.

APPARATUS SETUP

Any standard horizontal flat-bed apparatus can be used.

DIRECTIONS (By EDVOTEK and based on their kit equipment and solutions.)

Preparing the Gel Bed

1. Close off the open ends of a clean and dry gel bed by using rubber dams or tape.

   A. Using rubber dams: Place a rubber dam in each end of the bed. Make sure the rubber dam sits firmly in contact with the sides and bottom of the bed.

   B. Taping with labeling or masking tape: Use 3/4 inch wide tape and extend the tape over the sides and bottom edge of the bed. Fold the extended edges of the tape back onto the sides and bottom. Press the contact points firmly to form a good seal.

2. Place the well-forming template (comb) in the appropriate set of notches as indicated in the experiment instructions. The comb should sit firmly and evenly across the bed. This experiment requires a gel with 6 sample wells. Place a well-former template (comb) in the first set of notches nearest the end of the gel bed. Make sure the comb sits firmly and evenly across the bed.

Casting the Gel

Prepare an 0.8% agarose gel for your size gel bed (See Table 39-1, p. 638).

3. Use a 250 mL flask to prepare the diluted gel buffer. With a 1 mL pipet, measure the buffer concentrate and add the distilled water as indicated in Table 39-1.

4. Add the required amount of agarose powder. Swirl to disperse large clumps.

Table 39-1. Guidelines for preparing individual 0.8% agarose gels
(Courtesy - EDVOTEK, West Bethesda, MD)

EDVOTEK Model #	Approximate Gel Bed Dimensions (W x L)	Amt of Agarose	Volume of Buffer (50x)	+	Distilled Water	=	Total Volume
M6 or M36	7 x 7 cm	0.24 gm	0.6 ml		29.4 ml		30 ml
M12	7 x 15 cm	0.48 gm	1.2 ml		58.8 ml		60 ml
M20	10.5 x 14 cm	0.8 gm	2.0 ml		98.0 ml		100 ml

5. With a marking pen, indicate the level of the solution volume on the outside of the flask.

6. Heat the mixture to dissolve the agarose powder. The final solution should be clear (like water) without any undissolved particles left.

   A. Microwave method:
   - Cover the flask with plastic wrap to minimize evaporation.
   - Heat the mixture on High for 1 minute.
   - Swirl the mixture and heat on High in bursts of 25 seconds, until all the agarose is completely dissolved.

   B. Hot plate or burner method:
   - Cover the flask with foil to prevent excess evaporation.
   - Heat the mixture to boiling over a burner with occasional swirling. Boil until all the agarose is completely dissolved.

Caution: DO NOT POUR BOILING HOT AGAROSE INTO THE GEL BED.
Hot agarose solution may irreversibly warp the bed.

7. Cool the agarose solution to 55 °C with swirling to promote an even dissipation of heat. If detectable evaporation has occurred, add distilled water to bring the solution up to the original volume as marked on the flask in step 3.

After the gel is cooled to 55 °C

If using rubber dams, go to step 9. If using tape, continue with step 8.

8. Seal the interface of the gel bed and tape it to prevent the agarose solution from leaking.

   Use a transfer pipet to deposit a small amount of cooled agarose to both inside ends of the bed. Wait approximately 1 minute for the agarose to solidify.

9. Pour the cooled agarose solution into the bed. Make sure the bed is on a level surface.

10. Allow the gel to completely solidify. It will become firm and cool to the touch after approximately 20 minutes.

Preparing the Solidified Gel for Electrophoresis

11. Carefully and slowly remove the rubber dams or tape.

    Be careful not to damage or tear the gel when removing the rubber dams. A thin plastic knife or spatula can be inserted between the gel and the dams to break possible surface tension.

Table E39-2. Electrophoresis (chamber) buffer
(Courtesy - EDVOTEK, West Bethesda, MD)

EDVOTEK Model #	Volume of Buffer (50x) +	Distilled Water =	Total Volume
M6	4 ml	196 ml	200 ml
M12	6 ml	294 ml	300 ml
M20	8 ml	392 ml	400 ml
M36	8 ml	392 ml	400 ml*

12. Remove the comb by slowly pulling it straight up. Do this carefully and evenly to prevent tearing the sample wells.

13. Fill the chamber of the electrophoresis apparatus with the required volume of diluted buffer as outlined in Table E39-2. The solidified gel should be completely submerged under buffer. Additional buffer may be required if running the model #M36 with lass than 6 gel beds.

14. Place the gel in the electrophoresis chamber, properly oriented, centered and level on the platform.

For DNA analysis, the same EDVOTEK 50x Electrophoresis Buffer is used for preparing both the agarose gel buffer and the chamber buffer. To dilute: add 1 volume of buffer concentrate to every 49 volumes of distilled or deionized water.

Hint

For optimal DNA fragment separation, voltages higher than 125 volts should be avoided. Higher voltages can overheat and melt the gels.

15. Load the samples in the wells and conduct the electrophoresis according to experiment instructions. See Table E39-3 for time and voltage guidelines.

Preparing the Electrophoresis Buffer

The electrophoresis (chamber) buffer recommended is Tris-acetate EDTA (20 mM tris, 6 mM sodium acetate, 1 mM disodium ethylene-diamine tetraacetic acid) pH 7.8. Prepare the buffer as required for your electrophoresis apparatus (see Table E39-2. To dilute EDVOTEK (50x) concentrated buffer, add 1 volume of buffer concentrate to every 49 volumes of distilled or deionized water.

Electrophoresis Time and Voltage

Your schedule will dictate the length of time samples will be separated by electrophoresis. General guidelines are presented in Table E39-3. Please refer to the experiment instructions for specific requirements.

The DNA samples contain tracking dye, which moves through the gel ahead of most DNA (except extremely small fragments). Tracking dye migration will become visible in the gel after approximately 15 minutes. Do not move the apparatus during electrophoresis.

Table E39-3. Guidelines for times and voltage.
(Courtesy - EDVOTEK, West Bethesda, MD)

Volts	Recommended Time	
	Minimum	Optimal
15	8.0 hrs	10.0 hrs
25	2.0 hrs	3.0 hrs
50	1.0 hr	2.0 hrs
75	40 min	1.5 hrs
125	30 min	1.0 hrs

Reminder:

During electrophoresis, the dye samples migrate through the agarose gel towards either the positive or negative electrode depending upon their charge. Load the dye samples in wells in the middle of the gel.

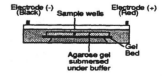

In cases where DNA fragments are similar in size, the fragments will migrate close to one another during electrophoresis. In general, longer electrophoretic runs will increase the separation between fragments of similar size. Experiments that involve measurement of fragment molecular size or weight should be run at the recommended optimal time to ensure adequate separation.

Terminate the electrophoresis when the tracking dye has moved several centimeters from the sample wells and before it moves off the gel. For convenience, the power source can be connected to a household light timer and set to stop at the appropriate time. The gel can remain overnight in the apparatus under buffer and stained the next day. For optimal results, the gel should be stained immediately after electrophoresis.

General Preparations

1. Place the gel on its bed into the apparatus. Make sure the gel is oriented properly as shown in the diagram at left. The sample wells should be closest to the negative electrode.

2. Fill the apparatus with the appropriate amount of electrophoresis buffer. The gel should be submerged completely under buffer.

3. Before loading samples, position the apparatus so that when the cover is placed on the unit, the leads reach the power source. This will avoid the necessity of moving the apparatus after samples are loaded, which could cause the samples to spill out of the wells.

Loading DNA Samples

4. Load each of the samples in tubes A-F into the wells in consecutive order. It is best to center the samples on the gel. Skip the first two wells beginning from the left side of the gel and place sample A in the third well from the left.

Running the Gel

5. After the samples are loaded, carefully snap the cover down onto the electrode terminals.

6. Make sure that the negative and positive indicators on the cover and apparatus chamber are oriented properly.

7. Insert the plug of the black wire into the black input of the power source (negative input). Insert the plug of the red wire into the red input of the power source (positive input).

8. Set the power source at the required voltage and run the electrophoresis for the length of time as determined by your instructor (or Table 39-3, p. 639). When current is flowing properly, you should see bubbles forming on the electrodes.

9. After the electrophoresis is completed, turn off the power, unplug the power source, disconnect the leads, and remove the cover.

10. Remove the gel on its bed. Place your hands on each end of the gel to prevent it from slipping off the bed.

11. Transfer the gel to the Gel Staining Tray to stain the gel for DNA Visualization with Methylene Blue Plus.

Visualization of Nucleic Acids

EDVOTEK Series 100 and 200 DNA electrophoresis experiments are designed for staining with EDVOTEK's special chemical formulation of Methylene Blue Plus ®. It is a biological stain that interacts with DNA and is formulated to yield optimal sensitivity when used in conjunction with EDVOTEK equipment and reagents. Methylene Blue Plus stain is considered safe and does not require ultraviolet light to view the gel. The separated DNA fragments will appear as dark blue bands against a light blue background after destaining.

As with any biological stain, care should be taken when handling solutions or gels containing Methylene Blue Plus stain. *Gloves and goggles always should be worn.*

Methylene Blue Plus Stain Preparation

Add 60 mL of the concentrated Methylene Blue Plus Staining Solution to 540 mL of distilled water. Gently mix.

Protocol Hints

To accelerate destaining of the gel, destain for 5 minutes with occasional stirring and replace the destain solution with fresh distilled water.

Methylene Blue Plus Staining Procedure

12. Remove the gel from its bed and totally submerse it in a tray containing diluted Methylene Blue Plus stain. *Do not stain gels in the electrophoresis apparatus.*

13. Stain for a minimum of 20 to 30 minutes, with occasional stirring.

14. To destain, completely submerse the gel in fresh distilled water. Bands of larger DNA molecules become visible within 30 minutes.

15. For optimal results, destain for several hours until the bands are clearly visible. If the gel is too dark, you can replace the destain solution with fresh distilled water. You also can destain overnight.

16. Carefully remove the gel from the destain solution and examine the gel on a Visible Light Gel Visualization System. To optimize visibility, use the amber filter provided with EDVOTEK equipment.

    Note: If the gel is too light and the bands are difficult to see, repeat the staining and destaining procedures outlined in steps 2 and 3.

17. Compare the DNA profiles of the scene and suspects. Can you match a suspect to the scene? Report your conclusions on the data sheet.

Storage and Disposal of Gel and Stain

A gel stained with Methylene Blue Plus can be stored in the refrigerator for several weeks. Place the gel in a Gel-Save Cassette (Cat. # 526) or sealable plastic bag with several drops of destaining liquid. DO NOT FREEZE AGAROSE GELS. Stained gels that are not kept can be discarded in the trash.

Avoiding Common Pitfalls

Potential pitfalls and/or problems can be avoided by following the suggestions and reminders listed below.

- To ensure that DNA bands are well resolved, make sure the gel formulation is correct (see Table 39-1) and that electrophoresis is conducted for the optimal recommended amount of time.

- Correctly dilute the concentrated buffer for preparation of both the gel and electrophoresis (chamber) buffer. Remember that without buffer in the gel, there will be no DNA mobility. Use only distilled water to prepare buffers. Do not use tap water.

- For optimal results, use fresh electrophoresis buffer and Methylene Blue Plus stain prepared according to instructions.

- Before performing the actual experiment, practice sample delivery techniques to avoid diluting the sample with buffer during gel loading.

- To avoid loss of DNA fragments into the buffer, make sure the gel is oriented properly so the samples are not electrophoresed in the wrong direction off the gel.

- To avoid gels that are missing small DNA fragments (small fragments move faster), remember that the tracking dye in the sample moves through the gel ahead of the smallest DNA fragments. Terminate the electrophoresis before the tracking dye moves off the end of the gel.

- If DNA bands appear faint after staining and destaining, repeat the procedure. Although 20 to 30 minutes is sufficient, staining for a longer period of time will not harm the gel.

EXPERIMENT 40. IMMUNOELECTROPHORESIS

Immunoelectrophoresis of human proteins in urine or serum

(Courtesy - Helena Laboratories, Box 752, Beaumont, TX, 77704-0752, Kit No. 3047)

BACKGROUND

Immunoelectrophoresis (IEP) is a reliable and accurate method for routine protein evaluations, detecting both structural abnormalities and concentration changes. The most common application of IEP is the diagnosis of *monoclonal gammopathies*. A monoclonal gammopathy is a condition in which a single clone of plasma cells produces elevated levels of a single class and type of immunoglobulin. The elevated immunoglobulin is referred to as a monoclonal protein, M-protein, or paraprotein. Monoclonal gammopathies may indicate a malignancy such as multiple myeloma or macroglobulinemia. The class (heavy chain) and type (light chain) must be established, because the patient's prognosis and treatment may differ depending on the immunoglobulin involved. Differentiation also must be made between monoclonal and polyclonal gammopathies.

A polyclonal gammopathy is a secondary disease state caused by disorders such as liver disease, collagen disorders, rheumatoid arthritis, and chronic infection. It is characterized by the elevation of two or more (often all) immunoglobulins by several clones of plasma cells. Polyclonal increases are usually twice normal levels.

YOUR PROBLEM

You suspect that something is wrong with you, but you are afraid to go to the doctor. You decide to do a little preliminary work first and collect a sample of your urine.

EQUIPMENT

Although any medium size flat-bed electrophoresis apparatus will work, the directions given are for the Helena Laboratories apparatus and supporting chemicals.

10 Titan Gel IEP plate No. 3047. Plates contain 1% agarose (w/v) in barbital-sodium barbital buffer with 0.1% sodium azide added as a preservative. The plates are ready for use as packaged and should be stored horizontally at 15 to 30 °C. Discard plates if they appear cloudy, show bacterial growth, or have been frozen or in excessive heat.

5	Blotter C (1 x 40)	1	Microdispenser and tubes, No. 6210
5	Blotter E (2 x 40)		1 to 10 µL, 1 to 25 µL, No. 6225
1	Chamber, gel, No. 4063	1	Power supply, No. 1520
1	Chamber, humidity, No. 9036	1	Rack, plate holder
1	Destaining tray	1	Rotator, serological
1	Development weight, No. 5014	1 pk	Sponge wicks
1	Incubator, Dryer, No. 5116	1	Ziptrol and tubes, No. 6009

CHEMICALS S = Student

The reagents can be prepared or ordered from Helena. The stock no. is provided. All contain 0.1 sodium azide as a preservative and should not be mixed with acidic solutions. All are negative for Hepatitus B antigen and HIV antibody.

Antiserum to Human IgG	No. 9232	
Antiserum to Human IgA	No. 9231	
Antiserum to Human IgM	No. 9234	
Antiserum to Human IgD	No. 9249	
Antiserum to Human IgE	No. 9250	
Antiserum to Human Kappa light chain	No. 9262	
Antiserum to Human Lambda light chain	No. 9257	
Trivalent antiserum to Human immunoglobulins, heavy chain, IgG, IgA, IgM	No. 9236	
Antiserum to Human Serum	No. 9233	
Pentavalent antiserum, heavy chain, IgG, IgA, IgM, IgD, IgE	No. 9251	

All are ready to use and should be stored at 2 to 6 °C.

Normal Human Serum Control	No. 9010
Add 2 drops of albumin marker before using.	
Buffer, pH 8.3-8.7	No. 5016
Barbital/sodium barbital	
Albumin marker, 0.5 % bromphenol blue	No. 9011
Stain, Coomassie Brilliant Blue	No. 3049
Dissolve in 450 mL of ethanol. Add 450 mL of water and 100 mL of acetic acid. Filter before use.	
Saline solution, 0.85 %	
Destaining solution, 675 mL 95% ethanol, 675 mL water, 150 mL acetic acid.	

THE BIG PICTURE

The patient's serum or urine sample and a Normal Human Serum Control are electrophoresed on the agarose plate. Antisera then are added to troughs in the plate and allowed to diffuse into the agarose support medium. When a favorable antigen-to-antibody ratio exists, a precipitin arc will form on the plate. Diffusion is halted by rinsing the plate in 0.85 % saline. Unbound protein is washed from the plate by the saline, and the antigen/antibody precipitin arcs are stained with a protein-sensitive stain. The precipitin arcs formed by the patient's sample and the control are compared for a semiquantitative protein analysis.

DIRECTIONS

Three wash periods are available; 24 hr, 6 hr, and quick. You will use the quick to conserve time, knowing that the background will not be as clear.

1. Dissolve one package of buffer in 1 L of deionized or distilled water.

2. Pour 100 mL of buffer into each outer section of the chamber. The buffer can

3. be reused one time by reversing the polarity of the chamber.

4. Then place the IEP plates with the wells on the left, anodic side.

5. Place a long IEP sponge wick in each buffer-filled compartment. Allow the sponges to become saturated with buffer. Place the sponges against the chamber

SUMMARY OF CONDITIONS

Plate	TITAN GEL IEPlate
Electra® B, Buffer	1 pkg. dissolved in 1.0 L deionized or distilled water
Buffer Volume	100 mL each side
Sample volume	2 µL
Migration distance	35 mm
Electrophoresis time	40-50 minutes
Voltage	100 volts
Antisera volume	25 µL
Incubation time	18 to 24 hrs.
Incubation temperature	15 to 30°C
Wash solution	0.85% saline
Six Hour Wash	
Wash time	6 hrs., change every 1 hr.
Drying temperature	60°C to 70°C
Drying time	3 to 5 minutes
Quick Wash	
Total wash/press time	35 minutes
Press	4x, 5 minutes each
Wash between presses	3x, 5 minutes each
Drying temperature	60°C to 70°C
Drying time	3 to 5 minutes
Stain	TITAN GEL IEP Stain
Staining time	4 minutes
Destaining time	4 to 6 minutes
Drying temperature	60°C to 70°C
Drying time	3 to 5 minutes

Figure E40-1. The chamber.

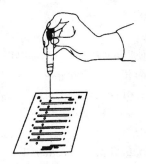

Figure E40-2. Cleaning the wells.

walls as shown in Figure E40-1.

5. Cover the chamber until ready to use.

6. Add two drops of albumin marker to a vial of Normal Human Serum control.

7. Remove the plate/s from the protective packaging and save the plastic holder for later use in the incubation chamber.

8. If the wells in the plate contain moisture or the agarose surface appears excessively wet, allow the plate to lie on the countertop for approximately 10 minutes before applying samples. Moisture remaining in the wells can be removed by placing a capillary tube in the well, taking care not to damage the agarose, and allowing the liquid to flow into the tube by capillary action. Refer to Figure E40-2.

9. Apply 2 µL of the control to wells labeled "C" (Figure E40-3) using a microdispenser or Ziptrol. Take care not to damage the wells during sample application.

10. Apply 2 µL of the patient's sample to wells labeled "P" in the same manner.

11. Quickly put the plate/s in the electrophoresis chamber, *agarose side down*, with the wells toward the cathode (-). Make sure that the agarose is in good contact with the sponge wicks. Two plates can be electro- phoresed in one chamber.

12. Put the cover on the electrophoresis chamber and wait 30 to 60 seconds before applying current. This allows the plate/s to equilibrate with the buffer.

13. Electrophorese the plate/s at 100 volts for a migration distance of 35 mm. This requires approximately 40 to 50 minutes. Migration distance can be verified by observing the position of the albumin marker.

14. Remove the plate/s from the chamber and put them on a flat surface, *agarose side up.*

15. Apply 25 µL of the appropriate antiserum to each trough in the plate. Fill the troughs by placing the tip of a microdispenser in the end of the trough furthest from the sample well. Holding the microdispenser in place, slowly depress the plunger and dispense the antiserum into the trough. The antiserum will flow down the trough by capillary action. The troughs easily hold 25 µL of antiserum without overflowing. Severe overflowing can cause antisera to contaminate other troughs.

16. Before moving the plates, allow the antisera to absorb for approximately 3 to 5 minutes.

17. Put each plate in the plastic holders saved from step 7.

18. Stack the plates (within the holders) in a humidity chamber containing a moist paper wick.

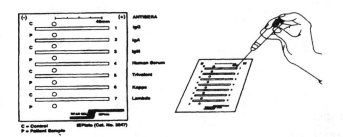

Figure E40-3. Antisera application.

19. Incubate the plates at room temperature (15 to 30 °C) for 18 to 24 hours. *The minimum incubation time is 18 hours.* Optimum precipitation will occur with 2 µL of sample and at least 25 µL of antiserum following 18 to 24 hours.

20. At the end of the incubation period, remove the plates from the plastic holders and put them in 0.85% saline. The saline stops the diffusion reaction and washes out the unbound protein.

21. Wash and press-dry the plate using the Quick Wash (Figure E40-4).

22. Press the plate dry for 5 minutes as follows: Lay the plate on a flat surface, *agarose side up*. Place one Blotter C directly on the plate, followed by two blotter E's. Then place a development weight on top of the plate and blotters for 5 minutes.

23. Remove the weight and discard the blotters.

24. Place the plate in a shallow dish containing sufficient 0.85 % saline to achieve a level of approximately 1/2 inch. Place the dish on a laboratory rotator at slow speed.

25. Wash the plate in 0.85% saline for 5-10 minutes.

26. Repeat the press and wash steps two more times. After the third wash, press again.

27. Upon completion of the washing/pressing steps, place the plate on a blotter, *agarose side up*, in the incubator/dryer or other drying oven at 60 to 70 °C for 3 to 5 minutes or until dry. The plate must be completely dry in order to stain and destain properly.

28. Fill a staining chamber with stain and fill the destaining chambers with destaining solution.

29. Place the dried plate/s in a stacking rack and lower it into the staining chamber. Stain the plates for 4 minutes.

30. Remove the staining rack containing the plate/s and put it on a paper towel to drain off surplus stain.

31. Raise and lower the rack 4 times in the first destaining chamber to further remove the excess stain.

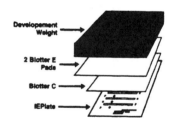

Figure E40-4. "Quick wash".

32. Repeat step 31 in the second destaining chamber. The background should be clear after this wash; if not, put the plates in the third destaining chamber until the background is clear, no more than 2 to 3 minutes. *Do not destain further or the stain in the IgM arc will be lost.* Plates can be restained if necessary.

33. Remove the plates from the rack and put each plate on a blotter, *agarose side up*.

34. Dry the plates in the incubator/dryer or other labor-atory drying oven at 60 to 70 °C for 3 to 5 minutes or until dry

35. Examine the plates and draw a conclusion. Compare your results with those shown in Figure E40-5.

36. *Clean up the area, leaving it better than it was when you started.* Thank you.

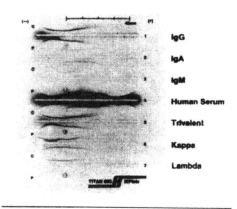

C = Control P = Patient Sample
IgG Monoclonal Gammopathy, Kappa Type

Figure E40-5. The "school solution".

EXPERIMENT 41. DISC ELECTROPHORESIS
Identification of fish species by their amino acid pattern

Ornstein, L., and Davis, B.J., Ann. *New York Acad. Sci.* **121**, 404 (1964); Chu, R., *J. A.O.A .C.*, **51**, 743 (1968); Payne, N.R., *J. A. O. A. C.,* **46** 1003, 1963)

(NOTE: this experiment takes about 6 hours if all goes well)

BACKGROUND

This technique has found application in all areas of medicinal and biological chemistry and even in the food industry. The experiment you will do involves the procedure to detect a fraud. Cheap fish often are substituted for the more expensive kinds. Substitution's of carp for red snapper and stingray for scallops have been detected by the FDA However, each species of fish has a characteristic amino acid pattern. The disc technique permits such high resolution of the amino acid components that it is very easy to detect one type of fish's pattern in another. This same technique has been used to detect kangaroo and deer in hamburger.

YOUR PROBLEM

You are to determine if a particular fish fillet is what it is supposed to be. You will run a sample and a standard and see if they are the same.

EQUIPMENT

1	Apparatus, disc electrophoresis		1 pk	Paper, filter, Whatman no. 1
6	Beakers, 10 mL		1	Pencil
1	Beaker, 100 mL		1 bx	Q-tips
1	Beaker, plastic, 400 mL		1	Rack, funnel
1	Blender, Waring		1	Refrigerator
1	Board, 5 cm x 20 cm *x* 6 mm		1	Saw, hack
9	Bottles, dropping, 8 oz		1	Spatula
2	Bottles, wash, 250 mL		1	Stand, ring
1	Brush, test tube, small		6	Stoppers, rubber, no. 1
1	Bucket, plastic, 1/2 gallon		6	Stoppers, rubber, no. 2
1	Bulb, pipet, small		6	Syringes, 1 or 2 mL
1	Centrifuge (50 mL tubes)		1	Syringe, plastic
3 dz	Corks, for 10 *x* 75 mm test tube		1	Towel, cloth
3	Cylinders, graduated, glass stoppered, 10 mL		1 pk	Towels, paper
1	Cylinder, graduated, 250 mL		6	Tube, bottoms, rubber
1	Dish, weighing, Al		4	Tubes, centrifuge, 50 mL
4	Funnels, plastic, 60°, short stem		6	Tubes, glass, 8 mm *x* 7.5 cm tapered bottoms
1	Funnel, powder, 110 mm		6	Tubes, glass, 9 mm *x* 7.5 cm
1 pr	Goggles, safety		3 dz	Tubes, test, 10 mm *x* 7.5 cm
1	Lamp fluorescent			

CHEMICALS   S = Student

Ice
Fish fillets, frozen, students choice
Rubber cement, 2 mL/S

Solutions: Store A -F in brown bottles in a refrigerator
A: HCl 1 N 48 mL
   TRIS (tris(hydroxymethyl)amino methane) 36.6 g
   TEMED (N,N,N,N,-tetramethylethylenediamine) 0.23 mL
   Water to 100 mL
B: HCl 1 N 48 mL
   TRIS   5.98 g
   TEMED   0.46 mL
   Water to 100 mL
C: Acrylamide 28.0 g
   BIS (N,N-methylene bis acrylamide) 0.735 g
   Water to 100 mL
D: Acrylamide 10.0 g
   BIS 2.5 g
   Water to 100 mL
E: Riboflavin 4.0 mg
   Water to 100 mL
F: Sucrose   40 mg
   Water to 100 mL

Destaining solution: Acetic acid, 7%
Fixative stain solution: Coomassie brilliant blue

WORKING SOLUTIONS
Solution 1, small pore gel
    1 part A
    2 parts C   pH 8.8 to 9.0
    1 part water
Solution 2, small pore gel
    Ammonium persulfate 0.14 g
    Water to 100 mL

Make fresh every week and store in a refrigerator. If polymerization does not take place, check this first.

Large pore gel solution
    1 part B
    2 parts D   pH 6.6 to 6.8
    1 part E
    4 parts F

Buffer solution: Lower tank
    TRIS   6.0 g
    Glycine 28.8 g   pH 8.3
    Water to 1 L

Buffer solution: Upper tank
Same as lower, but with 1 mL of 0.001% bromophenol blue

THE BIG PICTURE

Samples of fish, up to six kinds, are obtained. Frozen fillets are preferred. Cut off a piece about 2 cm square and grind it in a blender. Filter and store in a refrigerator until it is to be used.

Prepare the small pore gel solution and fill each tube to about 18 mm from the top. Photopolymerize under a fluorescent lamp until set. Prepare the large pore spacer gel and add about 6 to 8 mm to the top of the other gel. Photopolymerize this. Add 40 μL of sample to 100 μL of large pore gel, place this on top of the spacer gel, and photopolymerize. Place the tubes in the rubber holders in the bottom of the upper tank. Fill the tanks with the proper buffers. Connect the electrodes, making sure the bottom tank is +, and run at 3 mA/tube until the blue disc just gets to the small pore gel end, then increase the current to 5 mA/tube. Run until the blue marker disc moves at least 30 mm into the small pore gel. Remove the gel by placing the tubes in a bucket of ice water and reaming them out. Transfer to a small test tube and add 2 mL of staining dye. Let stand for about 15 minutes. Pour off the excess, add the destaining solution and destain until the discs are well developed. Remove and rinse with water. Place in small test tubes filled with water for storage. You may keep these as a token of our friendship.

Note: ground up fish develops a strong odor in a short time. Be sure to wash every bit of it down the drain and put plenty of rinse water after it. You can lose friends quickly otherwise.

APPARATUS SETUP

An apparatus of the type shown in Figure 29-2, p. 346, will be used.

DETAILED DIRECTIONS

1. Take a bucket and fill it about half full of ice.

2. Find the Waring blender and plug in the power cord to the wall outlet.

3. If you have frozen fish, find the hack saw and cut off about 2 cm square of your fish. If you have four or five samples, do them one at a time. Saw right through the paper and the cardboard. Remove the paper and put the chunk of fish in the Waring blender.

4. Add about 10 pieces of ice to the Waring blender. Put the cover on and turn it on. Let it go for 4 or 5 minutes or until all of the fish is completely cut up and the ice looks like it is melted.

5. Fold a filter paper and put it in a funnel.

6. Find the plastic centrifuge tubes that are about 2.5 cm in diameter and about 10 cm long and place the one that is marked #1 beneath this funnel.

7. Transfer your fish from the Waring blender to the filter paper and let the material filter into the centrifuge tube.

8. Get rid of the extra fish that is in the Waring blender as follows. Put a little water in the Waring blender and then carefully pour all those extra fish particles down the drain. Make sure that it goes down the center of the drain.

9. If you've got a second or third kind of fish, repeat the process until all of your fish samples are cut up and filtered.

10. Remove the centrifuge tubes. Put the lid on the centrifuge and place the centrifuge tube in the ice that is left in the ice bucket.

11. Take those funnels of fish, this is the fish that you have just filtered, to the restroom and flush them down the toilet.

12. Rinse out the plastic funnels.

13. Now you have to start preparing the gels for the separation. Locate six gray rubber bottoms. They are about 1.5 cm in diameter at the bottom and about 1 cm in diameter at the top and have little ridges in them.

14. Find the board that has the graph paper on the back side of it. Refer to Figure E41-1.

15. Get the bottle of rubber cement. Put about one drop of rubber cement on the wood in the center of those little circles and stick one of those rubber bases on the wood support by each one of those holes.

16. Find six pieces of glass tubing about 6 millimeters in diameter and about 6 to 7 cm long. Make sure that they are dry and clean.

17. Stick one of those tubes in each one of the rubber holders so that the tubes are all setting vertical.

18. Use all six tubes, even if you don't have six fish samples. You'll need at least one and maybe two tubes later on for practice. You are now going to make the small pore gel. Go to the refrigerator and get nine brown bottles that have the solutions in them for this experiment and bring them to the lab.

19. Make up the following solutions. Add 1 mL of solution A to a glass stoppered 10 mL graduated cylinder.

20. Add 2 mL of solution C to the graduated cylinder.

21. Add 1 mL of water.

22. Put the stopper on and mix it.

23. You have just finished making what we call small pore solution #1. Find small pore solution #2. The troubles that you will run into with this experiment occur with the ammonium persulfate. It has to be made up fresh every week. *If you do not get a polymerization within half an hour, then*

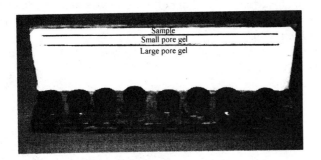

Figure E41-1. A marker gauge for the polymerization process.

make up a fresh ammonium persulfate solution. Add 4 mL of the ammonium persulfate solution, or the small pore solution #2, to the 4 mL of small pore solution #1 that you have just prepared in the graduated cylinder. Thoroughly mix them.

24. Take a syringe and put this solution into each one of the six tubes and fill them to that bottom mark where it says small pore gel.

25. Clean out that syringe immediately.

25. You do not want a meniscus to form on top of that gel or the separated bands will be disc shaped rather than a flat disc. By putting a drop of water on top of the gel, the meniscus is in the water layer and not in the gel, and you get a nice flat layer on top of the gel. Put one or two drops of water on top of the gel solution, so that the gel flattens out and the water has the meniscus.

Figure E41-2. A recommended size uv lamp.

26. Place this set of tubes underneath the uv lamp (Figure E41-2 or similar). Lower the lamp so that it is just to the side of the tubes and leave it there for at least 20 minutes.

27. Please empty the rest of the gel solution that is in the graduated cylinder. Getting it out is quite difficult once it has polymerized.

28. After about 20 minutes check and see if your gels are polymerized by tipping the wood block. Look at the interface between the water and the gel. If that interface doesn't move, but the water moves, then you know the gel is polymerized. If it hasn't polymerized, then let it set for another 5 or 10 minutes. If it hasn't polymerized within half an hour, make up a new ammonium persulfate solution, which is small pore solution #2, and redo the entire step.

29. Before you can put the spacer gel on, you've got to get rid of the water that is in there. Take a paper towel and lay it on the bench top.

30. Tip the wooden rack over so that the ends of the tubes will lay on the paper towel. The water should run down to the end and blot out onto the paper. If it doesn't, then tear a little piece off the end of a paper towel and stick it up the tubes as a wick. Be careful that you don't stick it into the gel and disturb the gel face.

31. The spacer gel is large pore gel, i.e., it contains less crosslinking and has bigger holes. You will make enough of this for both the spacer gel and the sample gel. To store it, place it in one of the drawers by the apparatus to keep light off of it. Take a clean 10 milliliter stoppered, glass cylinder and add 1 mL of solution B.

33. Add 2 mL of solution D.

34. Add 1 mL of solution E.

35. Add 4 mL of solution F. Mix the contents thoroughly.

36. Use a syringe, and being really careful, add between 3-4 mm of the spacer gel to the tube so that it does not disturb the layer of the small pore gel.

37. Put the unused solution in a drawer and close it to keep the light off of it.

38. Again put a water layer on the gel. Be careful so that you don't mix the two together.

39. Put the tubes back under the light and leave them there for 15 or 20 minutes, the same as you did previously.

40. While the spacer gel is polymerizing, you should centrifuge your fish samples. Take the tubes of fish samples out of the ice bucket, put them in the centrifuge, turn the centrifuge on, and let them spin for 20 to 25 minutes.

41. The sample gel is the same gel solution as the spacer gel, except that you mix your sample in with it. Take a 5 mL beaker and put about 1 mL of the spacer gel in it. That is the gel that you now have stored in a drawer.

42. Add one large drop or two small drops of your fish sample and mix these together thoroughly.

43. Fill the #1 tube with this sample gel.

44. You don't need a water layer on top of this, but generally a little of the upper buffer solution is placed on it.

45. Repeat this procedure for each one of your fish samples. Please, when you are done, wash out the syringe very carefully so that the solution doesn't polymerize in the needle and plug it up. Thank you.

46. Put this set of samples under the uv lamp and let them polymerize for 15 or 20 minutes.

47. Examine the electrophoresis equipment necessary for this experiment. It has two major components: the power supply, which is the blue box, and the electrophoresis chambers, which are the clear plastic cylinders about 20 cm tall. Pull up on the red and black electrodes, and the top cover will come loose. Remove the top cover.

48. Notice that you cannot put your hands into this conducting solution unless the power is off. This is a safety feature and a good one. Carefully remove the upper chamber and take a close look at it. This is called the upper buffer chamber. Notice that it has 18 rubber-stoppered ports in it, so 18 samples can be done at the same time. If you are only going to do two samples, then the rest of the ports are plugged with rubber stoppers. If any glass tubes have been left in the apparatus, carefully remove them.

49. Your sample tubes are placed where the rubber stoppers are and go down into the lower buffer chamber. Practice removing the plugs. Only the red rubber portion is removed. Pull it out from the bottom and then put it back by twisting it a little and then finally pushing it in with a glass stirring rod.

50. You want to maintain electric equivalence, so you do not use holes next to each other. Use holes opposite to each other as much as possible. Examine the lower buffer chamber. It has a water jacket around it so that the samples can be cooled if necessary. The upper buffer and the lower buffer are separated as stock solutions, because the upper buffer has a marker dye in it that the lower one does not have. These buffers can be used again. Notice on the buffer stock solutions that the label tells how many people have used them. You can use this solution four different times before you have to change it. Please initial the label, showing that you have used it. Fill the lower buffer chamber to within 1 cm of the rim.

51. Look at the power supply. On the front of it you will see the off-on switch; two output buttons, one for constant voltage and one for constant current; two buttons for the display, one for volts and one for milliamperes; an output adjust knob; and a output terminal block with two pairs of openings. Look at the connecting cords. The ends have what are called banana plugs on them. Plug one end of the red cord into the positive side of the terminal block and plug the black cord into the negative side of one pair of openings.

52. You now have to put your samples in the rubber rings in the bottom of the upper tank. The first problem you run into is the fact that the sample gel may not be very hard. The reason for this is that the sample material will keep the polymer from polymerizing very tightly. Be careful that you don't tip these tubes upside down, or your sample may fall off. When you pull these tubes out of the gray holders, you may find that the gel will slide down a little bit and stick out of the bottom maybe 1 or 2 mm. This is a normal thing to have happen. If it slides out 2 to 3 mm or so, then put the rubber bottom on the end of the tube to keep it from coming out. Pull a tube off of the board.

53. Take a wash bottle with water in it and wet the hole in the stopper, so that the tube will slip a little bit easier. Then put your sample tube into that rubber with the sample part up.

54. Put all of your samples in place. Take off any bottom gray stoppers and then set this chamber over the lower buffer tank. Make sure that all of the rubber stoppers are tight. Do this carefully, so that you don't break the chamber or spill any of your sample.

55. The upper buffer has dye in it. It's very dilute, one thousandth of a percent bromphenol blue. You can hardly see if there's anything there, but this will concentrate in the sample tube and should give you a beautiful little blue disc that will mark the progress of the electrophoretic separation. Put enough buffer in the upper chamber, so that it will cover the wire that is wrapped around the bottom part of the center chamber.

56. Check for leaks.

57. Look at where the electrodes are to be plugged in, and you will see that one is marked with a red dot. Be sure to connect the red cord to that electrode and then put the cover on.

58. Plug in both electrodes.

59. Turn the power switch on.

60. You need 3 milliamperes of current for each sample that you have in place. If you have four tubes, then you need 12 milliamps. Go to the black output adjust control knob and slowly turn it clockwise to adjust the current to the value you need.

61. Push in the constant current button, if it is not already in.

62. Now you can sit back and relax for a few minutes. In about 10 minutes you should see a little blue disc come down below the rubber stopper and go through the spacer gel. As soon as that blue disc gets to the small pour gel, raise the current to 5 milliamperes per tube. It will take about 10 or 15 minutes for that disc to move down that far. Notice how beautifully sharp that disc will be. For some unknown reason, some students don't get the blue disc. If that should happen, then continue for 1.5 hours. Continue the separation until the blue disc gets to within 1 cm of the bottom of the tube. While this is developing, you will practice on the next technique. Get one of the tubes that was filled with gel but did not have a sample in it.

63. The technique of removing the polymer from the tube is called rimming. This is done as follows. Go to the plastic bucket with the ice in it and add a little more water to it, so that you can completely submerge your hands and sample tube under the water. Hold the tube at about a 45° angle under the ice water, place the needle of a syringe very carefully between the gel and the glass to a depth of about 3 to 4 mm, and then roll the tube with your fingers. Then move the needle down about 2 to 3 mm, again roll it, and continue this until the gel is free. You may have to turn the tube over and go in the other end. If you do this rimming under cold water, the gel contracts and pulls away from the glass tube. Also, water lubricates the gel, so that it will slide out of the tube. Rim a tube for practice.

64. Get a 7.5 x 75 mm test tube.

65. The next step is to force the gel out of the tube into the test tube. This is done by first filling a very small pipet bulb with water. The pipet bulb then is attached to the bottom end of the glass tube holding the sample, and the top end is placed in the test tube. It will be upside down in the test tube because later on, after it is stained, you have to take it out of this test tube and put it into another tube with the sample down in the bottom. Try it a time or two, depending on how many practice tubes you've got.

66. Check to see where the marker disc is. If it is not yet within 1 cm of the bottom, then let it continue for a while. If you cannot see a disc, then let the system go for no more than 1.5 hours.

67. Generally, about the only thing you will see now is the little blue marker disc. Shut off the power.

68. Unplug the banana plugs from the cover.

69. Get several of the 7.5 x 75 mm test tubes and place a mark on them with a grease pencil to identify which sample is to be placed in them so that you won't get them mixed up.

70. Pour out the upper buffer and place it back into the upper buffer bottle.

71. Take out one of your sample tubes and using the same technique that you just practiced, rim out the gel and transfer it to a test tube with the sample end towards the bottom of the test tube. Do this for all of your samples and put the test tubes in the test tube rack.

72. To make the protein bands visible you are going to forst fix them and then stain them with Coomassie brilliant blue. Use a syringe and add 20% sulfo-salicylic acid to each gel test tube until it covers the small pore gel.

73. The sulfosalicylic acid fixes the proteins so that they can be stained. Leave this solution in the tubes for about 10 minutes and then carefully pour it off and down the drain.

74. Add the Coomassie blue solution to each tube.

75. How long it takes to stain the gels will depend upon the protein concentration. Leave the gels in the stain for about 30 minutes.

76. Pour the staining solution off one gel.

77. Rinse with water.

78. Examine the gel and decide if it is stained dark enough to see the protein bands clearly.

79. If not, return it to the staining solution and check at 10-minute intervals. It will be better to leave the gel in longer than necessary, so if you are in doubt, put it back in. When you are finished, return the Coomassie blue to the reagent bottle.

80. Rinse the gels with distilled water to remove any excess reagent.

81. Notice that each species of fish has a different protein band pattern. This is how you can identify mixtures of these fish. You can keep your gels as a souvenir. *Please clean up the mess.*

82. Pour the bottom buffer back into the bottle.

83. *Get the equipment back into shape for the next person. Please be sure that you wash down the drain any extra fish that you have in the centrifuge tubes.* Thank you.

EXPERIMENT 42. ION FOCUSING ELECTROPHORESIS

Ion focusing of ferric leghemoglobins from soybean root nodules

(Fuchsman, W. H. and Appleby, C. A. *Biochim. Biophys. Acta.*, **597**, 314, 1979 and directions by L. Davis, Kansas State University Department of Biochemistry)

Note: This experiment requires soybean nodules which can be obtained by growing a small pot of soybeans, obtaining them from a farmer, or from a college or university laboratory. If the soybeans are grown in soil that has already had soybeans grown in it, then nodules will form. If not, then the soil must be inoculated with nitrogen-fixing bacteria. The directions for all of this are presented below.

BACKGROUND

Leghemoglobin, first isolated in 1938 by J. Pietz, is a hemoglobin found in legume nitrogen-fixing nodules and is a mixture consisting of two major components, a and c and two minor components, b and d, with c and d having two components, c_1 and c_2, and d_1 and d_2. Later experiments have shown the presence of c_3 and d_3. It is responsible for maintaining the oxygen level in nitrogen-fixing nodules. Its structure is similar to that of myoglobins from animal cells and is therefore, judged to be homologous to animal globins. The structure sequence of the major component, according to Ellfolk and Sievers (*Acta Chemica Scand.* **25**, 3532, 1971) shows 142 amino acid residues and a mw of 15, 775.

The sequence is H₂N-Val-Ala-Phe-Thr-Glu-(5)-Lys-Gln-Asp-Ala-Leu-(10-Val-Ser-Ser-Ser-Phe-(15)-Glu-Ala-Phe-Lys-Ala-(20)-Asn-Ile-Pro-Gln-Tyr-(25)-Ser-Val-Val-Phe-Tyr-(30)-Thr-Ser-Ile-Leu-Glu-(35)-Lys-Ala-Pro-Ala-Ala-(40)-Lys-Asp-Leu-Phe-Ser-(45)-Phe-Leu-Ala-Asn-Pro-(50)-Thr-Asp-Gly-Val-Asn-(55)-Pro-Lys-Leu-Thr-Gly-(60)-His-Ala-Glu-Lys-Leu-(65)-Phe-Ala-Leu-Val-Arg-(70)-Asp-Ser-Ala-Gly-Gln-(75)-Leu-Lys-Ala-Ser-Gly-(80)-Thr-Val-Val-Ala-Asp-(85)-Ala-Ala-Leu-Gly-Ser-(90)-Val-His-Ala-Gln-Lys-(95)-Ala-Val-Thr-Asn-Pro-(100)-Glu-Phe-Val-Val-Lys-(105)-Glu-Ala-Leu-Leu-Lys(-110)-Thr-Ile-Lys-Ala-Ala-(115)-Val-Gly-Asp-Lys-Trp-(120)-Ser-Asp-Glu-Leu-Ser-(125)-Arg-Ala-Trp-Glu-Val-(130)-Ala-Tyr-Asp-Glu-Leu-(135)-Ala-Ala-Ala-Ile-Lys-(140)-Ala-Lys-COOH. Iron is believed to coordinate between the histidines at 61 and 92.

The compounds are red, so the focusing process can be followed visually as it proceeds. Figure E42-1 shows what a nodule looks like, and Figure E42-2 shows the results of a flat-bed prep scale separation of ferric leghemoglobins as the nicotinate complexes.

Figure E42-1. Nitrogen fixing nodules on soybean roots.

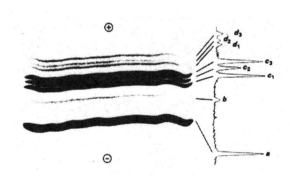

Figure E42-2. Prep scale separation of the ferric leghemoglobins in soybean nodules.
(Courtesy - Fushman, W.H. and Appleby, C.A., *Biochim. Biophys. Acta,* **579**, 314, 1979)

YOUR PROBLEM

A research study of the structure of ferric leghemoglobin is to be undertaken. Your job is to provide a sample of each leghemoglobin for this research, and it is to be done by isoelectric focusing of a soybean extract.

EQUIPMENT

The disc electrophoresis apparatus is the same as that described on p. 346.

- 1 Apparatus, disc electrophoresis
- 1 Aspirator, water
- 1 Beaker, 100 mL
- 1 Beaker, 600 mL
- 1 Board, 5 cm x 20 cm x 6 mm
- 1 Bottles, wash, 250 mL
- 1 Brush, test tube, small
- 1 Bucket, plastic, 1/2 gallon
- 1 Bulb, pipet, small
- 1 Centrifuge, micro preferred
- 1 Dish, weighing, Al
- 1 pr Electrodes, pH
- 1 Flask, filtering, 50 mL
- 1 pr Goggles, safety
- 1 Lamp, uv

1 Meter, pH	1 Syringe, plastic, 1-2 mL
1 Paper, graph	1 Towel, cloth
1 Pencil	6 Tube, bottoms, rubber
2 Pots, clay, flower, ~1 gallon	4 Tubes, centrifuge, 1.5 mL
1 Rod, glass, stirring, tapered end	6 Tubes, glass, 8 mm x 7.5 cm tapered bottoms
1 Spatula	4 Tubes, test, 7.5 x 75 mm
6 Stoppers, rubber, no. 1	3 ft Tubing, rubber, vacuum
6 Stoppers, rubber, no. 2	
1 Syringe, 100 μL	

CHEMICALS   S = Student

Acetic acid, H_3CCOOH, 0.5 M, 28.7 mL/L, 500 mL/S

Acrylamide, $H_2C=CHCONH_2$, (38% monomer), 1 mL/S

Ammonium persulfate, $(NH_4)_2S_2O_8$, 10%, 50 μL/S

Ampholines, (8 parts pH 4.2 to 4.9 plus 1 part 4.0 to 6.5), 0.5 mL/S

BIS (N,N-methylene bis acrylamide), 2% in final mixture with 38% acrylamide, 1 mL/S

Disodium phosphate, Na_2HPO_4, 0.05 M (pH 7), ordinary pH 7 buffer to calibrate a pH meter, 0.5 mL/S

Glycerol, $HO-CH_2CHOHCH_2-OH$, 1 mL/S

Glycine, H_2N-CH_2COOH, 0.05 M, 3.76 g/L, adjust to pH 6.5, 500 mL/S

Nicotinic acid, $C_5H_5N-COOH$, 0.1 M, 1.23 g/100 mL, 0.5 mL/S

Potassium ferricyanide, K_3FeCN_6, 0.1 M, 3.3 g/100 mL, 0.5 mL/S

Rubber cement, 2 mL/S

THE BIG PICTURE

Soybean plants are grown, and after at least 3 weeks growth, the nodules are removed from the roots. The nodules are crushed, and the leghemoglobins extracted with pH 7 phosphate buffer. The debris is centrifuged away, and 15 μL are placed on top of a polymerized acrylamide gel containing an 8:1 ratio of pH 4.2 to 4.6 and 4.0 to 6.5 ampholines. 0.5 M acetic acid is placed in the lower chamber and 0.05 M glycine in the upper chamber. The current is adjusted to 3 mA/tube, and the system operated overnight. The gels are removed by rimming, the spacing is plotted on graph paper, and the gels placed in small test tubes to save if desired.

APPARATUS SETUP

The apparatus is the same as that in Experiment 41, on p. 646.

DIRECTIONS

These directions assume you have nothing prepared and must grow your soybeans and extract them.

1. Prepare a mixture of about 50% soil, 25% Pearlite (an expanded clay product) and 25% peat moss sufficient to fill two large flower pots. If the soil has not been used previously for soybeans, then it is necessary to inoculate it with nitrogen-fixing bacteria, some variety of *Rhizobium japonicum*, which can be obtained at a Co-oP. Mix a small handful into the soil in each pot.

2. Obtain a small handful of soybean seeds from a Co-oP or a local farmer. Plant the soybean seeds about 2 cm deep and water as needed to keep the soil moist. They usually will take about 3-4 days to germinate, and nodules will begin to form about 10 days later. Plants 3 weeks old should contain sufficient nodules for this experiment.

3. Refer to Figure E42-1, p. 653, to identify what nodules look like. Carefully wash the plant roots and brush off the nodules. Place several nodules having an estimated volume of 0.5 mL into a disposable, conical-bottom, microcentrifuge tube and cover with 0.25 mL of pH 7 phosphate buffer. Do two tubes.

4. Crush the nodules thoroughly by using a tapered glass stirring rod that will press them down against the bottom of the conical centrifuge tube.

5. Centrifuge for 2 minutes, homogenize again, and centrifuge again for a few minutes.

6. Decant the supernatant liquid into two fresh tubes. To each tube, add 0.1 volume of 0.1 M potassium ferricyanide and 0.1 M nicotinic acid. The solution should be pink to red at this point. The purpose of adding ferricyanide is to convert the heme to a common state of charge and ligand binding. That way, any differences seen on the isoelectric focusing gel are caused by differences in amino acid sequence, not by the oxidation state of the iron in the heme or the ligand bound to it. The purpose of the nicotinic acid is to form a complex that moves faster and separates better during the electrofocusing step.

7. Centrifuge for a few minutes to remove any debris. You should have about 0.25 mL of clear solution.

8. Refer to Experiment 41, p. 646. Locate six gray rubber bottoms. They are about 1.5 cm in diameter at the bottom and about 1 cm in diameter at the top and have little ridges in them.

9. Find the board that has the graph paper on the back side of it.

10. Get the bottle of rubber cement. Put about one drop of rubber cement on the wood in the center of those little circles and stick one of those rubber bases on the wood support by each one of those holes.

11. Find six pieces of glass tubing about 6 millimeters in diameter and about 6 to 7 cm long. Make sure that they are dry and clean.

12. Stick one of those tubes in each one of the rubber holders, so that the tubes are all setting vertical.

13. Use all six tubes. You'll need at least one and maybe two tubes later on for practice.

14. To prepare polymer for six tubes, mix the following in a 25 or 50 mL filtering flask:

    10.8 mL of water; 1.5 mL (g) glycerol (weighed out for simplicity, rather than pipetted); 0.75 mL of ampholines (8 parts pH 4.2 to 4.9 and 1 part 4.0 to 6.5); 1.9 mL 40% acrylamide (38% acrylamide, 2% *bis* acrylamide).

15. Degas this mixture for 10 minutes with a water aspirator vacuum.

16. Add 75 µL of a 10% ammonium persulfate solution. Swirl gently, then add this mixture to the tubes to within 2 mm from the top. Polymerization can be hastened by irradiation from a uv lamp, if one is available. *If you do not get a polymerization within half an hour, then make up a fresh ammonium persulfate solution.*

17. Rinse out the filtering flask immediately before the mixture polymerizes in it.

18. Examine the electrophoresis equipment necessary for this experiment. There are two major components: the power supply, which is the blue box, and the electrophoresis chambers, which are the clear plastic cylinders about 20 cm tall. Pull up on the red and black electrodes and the top cover will come loose. Remove the top cover.

19. Notice that you cannot put your hands into this conducting solution unless the power is off. This is a safety feature and a good one. Carefully remove the upper chamber and take a close look at it. This is called the upper buffer chamber. Notice that it has 18 rubber-stoppered ports in it, so 18 samples can be done at the same time. If you are only going to do two samples, then the rest of the ports are plugged with rubber stoppers. If any glass tubes have been left in the apparatus, carefully remove them.

Your sample tubes are placed where the rubber stoppers are and go down into the lower buffer chamber. Practice removing the plugs. Only the red rubber portion is removed. Pull it out from the bottom and then put it back by twisting it a little and then finally pushing it in with a glass stirring rod.

20. You want to maintain electric equivalence, so you do not use holes next to each other. Use holes opposite to each other as much as possible and plug the other holes.

21. Examine the lower buffer chamber. It has a water jacket around it, so that the samples can be cooled if necessary. Fill the lower buffer chamber to within 1 cm of the rim with 0.05 M acetic acid. This aids in cooling.

22. Look at the power supply. On the front of it you will see the off-on switch; two output buttons, one for constant voltage and one for constant current; two buttons for the display, one for volts and one for milliamperes; an output adjust knob; and a output terminal block with two pairs of openings. Look at the connecting cords. The ends have what are called banana plugs on them. Plug one end of the red cord into the positive side of the terminal block and plug the black cord into the negative side of one pair of openings.

23. When you pull the tubes out of the gray holders, you may find that the gel will slide down a little bit and stick out of the bottom maybe 1 or 2 mm. This is a normal thing to have happen. If it slides out 2 to 3 mm or so, then put the rubber bottom on the end of the tube to keep it from coming out. Pull a tube off of the board.

24. Take a wash bottle with water in it and wet the hole in the stopper, so that the tube will slip a little bit easier. Then put your sample tube into that hole with the 2 mm space up.

25. Put all of your samples in place. Take off any bottom gray stoppers and then set this chamber over the lower buffer tank. Make sure that all of the rubber stoppers are tight. Do this carefully, so that you don't break the chamber or spill any of your sample.

26. With a µL syringe, place about 150 uL of sample onto the top of each tube. It contains sufficient protein and other chemicals that its density will keep it in place when the glycine is added.

27. Put enough 0.05 M glycine buffered to pH 6.5 in the upper chamber, so that it will cover the wire that is wrapped around the bottom part of the center chamber. If you add this carefully, you will not disturb the sample on top of the gel.

28. Check for leaks.

29. Look at where the electrodes are to be plugged in, and you will see that one is marked with a red dot. Be sure to connect the red cord to that electrode and then put the cover on.

30. Plug in both electrodes.

31. Turn the power switch on.

32. You need 3 milliamperes of current for each sample that you have in place. If you have four tubes, then you need 12 milliamps. Go to the black output adjust control knob and slowly turn it clockwise to adjust the current to the value you need. Review the section on heating and calculate the heat you are producing. The current will decrease as the separation progresses and, in fact, go to zero when the ampholines and sample compounds are all in place.

33. It takes several hours for the ampholines to adjust and the proteins to follow suit. You should be able to observe this amazing process take place as the hours pass. Continue the experiment until you are satisfied that you have a decent separation. Do not worry; it will stabilize if you do not get back to it for some time. Go to step 34 now.

34. The technique of removing the polymer from the tube is called rimming. This is done as follows. Fill at least a 0.5 gallon plastic bucket about 1/4 full of crushed ice and then fill it to 3/4 with water, so that you can completely

submerge your hands and sample tube under the water. Hold the tube at about a 45° angle under the ice water, place the needle of a syringe very carefully between the gel and the glass to a depth of about 3 to 4 mm, and then roll the tube with your fingers. Then move the needle down about 2 to 3 mm, again roll it, and continue this until the gel is free. You may have to turn the tube over and go in the other end. If you do this rimming under cold water, the gel contracts and pulls away from the glass tube. Also, water lubricates the gel, so that it will slide out of the tube. Rim a tube for practice.

35. Get a 7.5 x 75 mm test tube.

36. The next step is to force the gel out of the tube into the test tube. This is done by first filling a very small pipet bulb with water. The pipet bulb then is attached to the end of the glass tube holding the sample, and the other end is placed in the test tube. Apply gentle pressure on the pipet bulb, and the water should force the gel out of the tube into the test tube. Try it a time or two, depending on how many practice tubes you've got.

38. After you have a satisfactory separation, shut off the power.

39. Unplug the banana plugs from the cover.

40. Pour out the upper buffer and place it back into the upper buffer bottle.

41. Take out one of your sample tubes and using the same technique that you just practiced, rim out the gel and transfer it to a test tube. Do this for all of your samples and put the test tubes in the test tube rack.

42. Further staining of the bands will not be done, because the bands of interest in this case are highly colored already.

43. Use graph paper and make a sketch of the distribution of the bands in your gel system. Indicate relative intensities by darkness of the pencil line. Show where the electrodes are placed.

44. Pour the acetic acid solution back into the bottle to be reused.

45. *Get the equipment back into shape for the next person.* Thank you.

EXPERIMENT 43. CAPILLARY ZONE ELECTROPHORESIS
Micellular capillary electrophoresis separation of aspartame breakdown products in soft drinks
(Courtesy - Bello, A. C., Philadelphia District Food and Drug Laboratory, *Laboratory Information Bulletin* 3866)

BACKGROUND
Many years ago, two major non-nutritive artificial sweeteners were marketed, sodium cyclamate ($30x$ sweeter than sucrose) and sodium saccharin ($500x$ sweeter than sucrose). Cyclamate was erroneously thought to cause cancer and was banned. Saccharin is believed not to cause cancer, but to activate other compounds to be potential carcinogens. The result was a search for another compound. Nutrasweet ($160x$ sweeter than sucrose), a brand name of aspartame (N-L-a-aspartyl-L-phenylalanine-1-methyl ester), is a widely used artificial sweetener. This small peptide has two main routes of degradation, both leading to compounds with no sweetening properties. One, which has limited its application in baked products because increased high temperature breakdown, results in the formation of 5-benzyl-3,6-dioxo-2-piperazine acetic acid. The second pathway leads to the nonesterified form of the peptide, aspartyl phenylalanine, and, eventually, to its component amino acids, aspartic acid and phenylalanine. This latter route of degradation is accelerated greatly by extreme pH conditions and therefore, is of interest when analyzing diet soft drinks. A pH of about 4 offers the maximum stability, whereas that of a soft drink can be as low as 2.7. Temperature and time are also important factors influencing the rate of the hydrolytic process.

Experiment 43

Aspartame (Nutrasweet) → *5-Benzyl-3,6-Dioxo-2-Piperazine Acetic Acid*

Aspartylphenylalanine → *Aspartic Acid* + *Phenylalanine*

Capillary electrophoresis using surfactant-containing buffers is useful not only in resolving neutral analytes, but also in separating ionic compounds according to their charge and degree of hydrophobicity. This technique, which controls, but does not obliterate, the electro-osmotic flow, results in more reproducible migration times and peak areas than those obtained with straight buffers. The present procedure is based on methodology provided by Beckman Instruments and adapted to this purpose.

YOUR PROBLEM

A consumer has complained that the contents of one can of their diet soft drink had a peculiar taste. It has been given to you to determine what is causing this off-taste. The can has no expiration date so it could be just "old pop", or it could be something more serious. You decide to determine if the problem could be just sweetener breakdown by using a micellar capillary electrophoresis technique.

EQUIPMENT

1	Balance, analytical	2	Flasks, volumetric, 1 L
2	Beakers, 250 mL	1	Filter, syringe, 0.45 μm
1	Bulb, pipet	1	Funnel, long stem
1	Capillary, 50 cm x 75 μm	1 pr	Goggles, safety
1	Capillary electrophoresis system like Beckman Model 2050	1	Holder, funnel
1	Cylinder, graduated, 100 mL	1	Meter, pH
1	Data integration, like Spectra Physics SP4270	1	Paper, filter, 11 cm, medium
		1	Pipet, Mohr, 5 mL
7	Flasks, volumetric, 25 mL	1	Sponge or cloth towel
1	Flask, volumetric, 50 mL	1	Stand, ring
2	Flasks, volumetric, 100 mL	1	Stirrer, magnetic
2	Flasks, volumetric, 250 mL	1	Syringe, 10 mL

CHEMICALS   S = Student

Sodium borate, 200 mM; Weigh 19.07 g of Borax ($Na_2B_4O_7 \cdot 10H_2O$, m.w. 381.43) and dissolve in about 900 mL of water. Transfer to a 1 liter volumetric flask and bring to volume with water. A 0.05 M solution of Borax is 0.2 M in borate.

    To dilute to 80 mM: Using a graduated cylinder, pour 100 mL of the above stock solution into a 250 mL volumetric flask and bring to volume with water.

Boric acid, H_3BO_3, 200 mM: Weigh 12.37 g of boric acid and dissolve in about 900 mL of water. Transfer to 1 liter volumetric flask and bring to volume with water. To dilute to 80 mM: Using a graduated cylinder pour 100 mL of the above stock solution into a 250 mL volumetric flask and bring to volume with water.

Borate buffer pH 8.8 to 9, 80 mM: Place 200 mL of 80 mM Na borate in a beaker and add enough 80 mM boric acid to reach a pH of 8.8 to 9 (start with 10 to 12 mL). Monitor the addition with a pH meter while stirring the solution with a magnetic bar.

Sodium dodecyl sulfate (SDS), 100 mM in 80 mM Na borate pH 9: Weigh 3.42 g SDS and transfer to a 250 mL beaker. Add a magnetic bar and pour 150 mL of 80 mM Na borate pH 9. Stir very gently, so as not to cause excessive foaming. Some solids might remain. Filter the resulting solution first through paper and then through a 0.45-micron membrane filter disk before use. Keep at room temperature.

Standards:

Aspartame:

Stock solution: Weigh about 25 mg of Aspartame, transfer to a 50 mL volumetric flask, dissolve in and take to volume with water. (About 0.5 mg/mL).

Working standards (WS):

WS B: Measure 5.0 mL of Aspartame STOCK solution into a 25 mL volumetric flask and take to volume with water (about 0.1 mg/mL). Use this solution for reproducibility of areas and migration times based on 5 consecutive injections.

WS A: Measure 3.0 mL of Aspartame STOCK solution into a 25 mL volumetric flask and take to volume with water (about 0.06 mg/mL).

WS C: Measure 7.0 mL of Aspartame STOCK solution into a 25 mL volumetric flask and take to volume with water (about 0.14 mg/mL).

Caffeine:

Stock solution: Weigh about 50 mg of caffeine and transfer to a 100 mL volumetric flask. Dissolve in and take to volume with water (about 0.5 mg/mL).

C-1: Measure 5.0 mL STOCK solution into a 25 mL volumetric flask and take to volume with water. (About 0.1 mg/mL).

C-2: Measure 5.0 mL of C-1 into a 25 mL volumetric flask and take to volume with water (about 0.02 mg/mL).

Sodium or potassium borate:

Stock solution: Weigh about 50 mg of Na or K benzoate and transfer to a 100 mL volumetric flask. Dissolve in and take to volume with water (about 0.5 mg/mL)

B-1: Measure 5.0 mL of STOCK solution into a 25 mL volumetric flask and take to volume with water (about 0.1 mg/mL).

B-2: Measure 8.0 mL of B-1 solution into a 25 mL volumetric flask and take to volume with water. (About 0.03 mg/mL).

THE BIG PICTURE

The instructor will show you how to use the instrument. The linearity of the range is verified by injecting, in duplicate, WS A, B, and C and plotting the average areas *vs.* actual concentrations. One injection of C-2 is made to identify the caffeine peak by its migration time. One injection of B-2 is made to identify the benzoate peak by its migration time. One injection of the caffeine/aspartame/benzoate standard mix is made to show the resolution between the 3 peaks. A sample of fresh soda then is injected. A sample of old soda then is injected. The recovery of Aspartame from the matrix (soda) is verified. Two injections of the soda alone (diluted 1 to 5 as per instructions) and two injections of the recovery solution are made.

APPARATUS SETUP

Beckman Capillary Eletrophoresis System 2050.

Capillary: open tubular, 50 cm (to the detector), 75 mµ diameter

Data integration: Spectra Physics SP4270, at Attn. 16, PW = 2 and PT = 100

Detection: 214 nm Range: 0.020

Separation: 10 minutes, constant voltage: 20 kv

Temperature: 30° C

DIRECTIONS

The following software settings and those in Table E43-1, are based on the P/ACE 2000 series version 3.0 Beckman Instr. Inc.

    Vial contents:

        1, 11, 33, SDS buffer; 10, Empty; 21, Sample; 34, Water.

Display channel A with grid lines.

Time: 0.00 to 10.00 minutes

Channel A   -0.005 to 0.020 absorbance.

1. Have your instructor show you how to use your apparatus and to recommend sample size.

2. Verify the linearity of the range by injecting, in duplicate, WS A, B, and C. Plot the average areas vs. actual concentrations.

3. Make one injection of C-2 to identify the caffeine peak by its migration time.

4. Make one injection of B-2 to identify the benzoate peak by its migration time.

5. Caffeine/aspartame/benzoate standard mix. In a 25 mL volumetric flask, mix 4.0 mL of Aspartame STOCK solution, 5.0 mL of C-1, and 8.0 mL of B-1. Take to volume with water. Make one injection to show the resolution between the 3 peaks.

6. Inject a sample of fresh soda.

Table E43-1. Software settings (Courtesy - Bello, A. C., Philadelphia District Food and Drug Laboratory, *Laboratory Information Bulletin* 3866)

Step	Process	Duration	Inlet	Outlet	Control Summary
1	Set temp				Temp: 30 °C
2	Set detection				uv: 214 nm rate: 2 Hz rise: 1.0 sec
					Range: 0.020 - 10 zero: 1.0 min
3	Rinse	1.0 min	33	10	Forward: high pressure
4	Inject	4.0 sec	21	1	
5	Separate	10.0 min	11	1	Constant voltage: 20.00 kv
					Current limit: 250 uA.
					Integrator on: current data
6	Rinse	1.0 min	34	10	Forward: high pressure
7	Rinse	1.0 min	33	10	Forward: high pressure

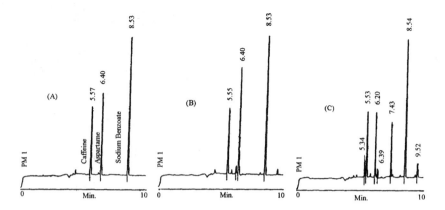

Figure E43-1. Electropherogram of (A) standards. (B) fresh soft drink. (C) degraded soft drink.

7. Inject a sample of old soda. (Note to the instructor: If you cannot find an old soda, a fresh one can be degraded rather rapidly at a pH below 2 or above 7).

8. Verify the recovery of Aspartame from the matrix (soda) by mixing 3.0 mL of fresh soda with 3.0 mL of Aspartame STOCK (about 1.5 mg) in a 25 mL volumetric flask and take to volume with water. Make two injections of the soda alone (diluted 1 to 5 as per instructions) and two injections of the recovery solution.

9. *Shut down the instrument,, and clean up the area.* Thank you.

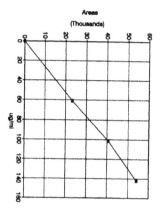

Figure E43-2. Calibration curve.

Figure E43-1 shows the electropherograms of a fresh soft drink and one that has been in the intact can from 1 year to 18 months under uncontrolled storage conditions. The can had no expiration date. The level of aspartame found in the fresh drink was 417 ppm, which approximates the 500 ppm claimed by the manufacturer. In the old product, the sweetener seems to have completely disappeared, giving rise to a series of unidentified small peaks. Neither aspartic nor phenylalanine was detectable at this wavelength and level of sensitivity. The other major peaks in the electropherogram of the product correspond to caffeine and sodium benzoate. Figure E43-2 shows the calibration curve.

EXAMPLE CALCULATION

Using the data in Table E43-2, prepare a calibration curve.

Table E43-2. Calibration curve data

Standard No., µg/mL	Concentration	Response Area
1	0.000	0.000
2	60.80	23,218
3	101.41	40,496
4	142.00	54,100

ANSWER
See Figure E43-2, p. 661.

EXAMPLE CALCULATION
A 3.0 mL sample of fresh soda, diluted 1:5, had a peak area of 32273, and a degraded sample had a peak area of 2327. A sample spiked with 3.0 mL of 0.507 mg/mL aspartame diluted to 25 mL had an area of 43882. What are the amounts of aspartame in the sodas, and what is the % recovery?

ANSWER
Interpolate from Table E43-2, p. 661, and check using Figure E43-2, p. 661. 101.4 ppm - 60.8 ppm = 40.6 ppm over 40496 - 23218 or 17278 units. This is 425.6 units/ppm. 32273 - 23218 = 9055 units or 21.3 ppm. 21.3 ppm + 60.8 ppm = 82.1 ppm for the fresh soda. The old soda [(23218 - 0) units/ 60.8 ppm= 381.9 units/ppm]). 2327 units/381.9 units/ppm = 6.1 ppm. The spike (54100 - 40496 = 13604 units. 13604 units/ (142.0 ppm - 101.4 ppm) = 335.1 units/ppm. 43882 - 40496 = 3386 units or 10.1 ppm. 101.4 + 10.1 = 111.5 ppm.

The soda contains 82.1 µg aspartame/mL, the degraded soda contains 6.1 µg/mL, and the spike contains 110.5 µg/mL

The % recovery : Found x 100/Theoretical
Found = 110.5 ppm.
Theoretical: from the spike = 3.0 mL x 0.507 mg/mL = 1.521 mg in 25 mL or 060.8 mg/mL or 60.8 µg/mL.
  from the soda = 82.1 x 5 = 410.5 µg/mL or 0.4105 mg/mL
    x 3.0 mL in 25 mL = 49.26 µg/mL. Theoretical then = 60.8 ppm + 49.3 ppm or 110.1 ppm
% recovery = (110.5 x 100)/110.1 = 100.4%

EXPERIMENT 44.   FIELD FLOW FRACTIONATION
Determination of the size/mass of latex spheres using sedimentation FFF.
(Courtesy - Caldwell, K., Biotechnology Department, University of Utah, Salt Lake City, UT)

BACKGROUND
Synthetic latex polymers got their first main push during WWI when the Allies blockaded Germany, cutting her off from the South American rubber plantations. It got its second big push in WWII when Japan occupied the South East Asian plantations. Today, manufacture of synthetic latex polymers is a $13 billion industry of which $5.5 billion is from the 604 million gallons of paint made annually. The paint manufacturing industry in 1993 employed 58,200 people in 800 to 900 plants. The average costs of a gallon of latex paint in 1993 were $9.15 wholesale and $13.25 retail.

Latex describes the type of resin or binder used. Exterior paint usually contains 100% acrylic latex emulsions, which usually are copolymers of methylmethacrylate with butylacrylate or 2-ethylhexylacrylate. This combination stands up well against uv radiation. Interior paint has no uv requirement, and the best combination for cost is a copolymer of vinylacetate and acrylates. Many other formulations are available depending on the use and quality desired.

The usual process for paint is a free radical emulsion polymerization, in which the monomer mixture is mixed in with a surfactant in a water solution, and the process started with an initiator, all under a blanket of nitrogen. The result is the preparation of beads 0.1 to 1 µm in diameter.

One surfactant is sodium lauryl sulfate, and the initiator is usually $(NH_4)_2S_2O_8$. This forms the sulfate radical anion, which starts the process. Table E44-1, p. 663, shows a typical charge.

$$\underset{\text{Methylmethacrylate}}{H_2C=C(CH_3)-C(=O)OCH_3} \qquad \underset{\text{Butylacrylate}}{H_2C=CH-C(=O)OCH_2CH_2CH_2CH_3}$$

$$H_2C-\underset{OCH_3}{\overset{CH_3}{C}}-C(=O)OCH_3 \;+\; H_2C=CH-C(=O)OCH_2CH_2CH_2CH_3 \xrightarrow[N_2\;\;85\;°C]{(NH_4)_2S_2O_8} H_3C-\underset{COOCH_3}{\overset{CH_3}{C}}-CH_2-\underset{COO(C_4H_9)}{CH}$$

The monomer charge is prepared by adding the reagents in the order listed and degassed with N_2. The reactor charge is heated to 85 °C under a blanket of N_2. The initiator is added, followed by the slow addition of the monomer charge over 2.5 hrs, maintaining 85 °C. Oxygen stops the polymerization.

The particle size distribution of a latex paint polymer determines its performance related to its film-forming properties, stability, rheology (flow), and morphology (form). The particles are usually spherical. Larger particles and a narrow distribution usually provide a more stabile system, because smaller particles with a wider distribution tend to interact and coagulate. SFFF is the most popular method to determine particle size distribution.

Table E44-1. A typical charge to produce an emulsion polymerization of a latex paint resin.
(Courtesy - Kirk, R.E., and Othmer, D.F., *Encyclopedia of Chemical Technology*, Wiley-Interscience, New York, NY, 1995)

Monomer Emulsion Charge	Parts
Distilled water	13.65
Sodium lauryl sulfate	0.11
Methylmethacrylate	22.50
Butylacrylate	22.05
Methacrylic acid	0.45
Initiator Charge	
Ammonium persulfate	0.23
Reactor Charge	
Distilled water	30.90
Sodium lauryl sulfate	0.11

YOUR PROBLEM

Some freezing tests are to be done on a sample of latex paint, and you are to determine the particle size distribution and surface characteristics before the tests.

EQUIPMENT

- 1 Apparatus, sedimentation FFF Channel: 90 cm long, 0.0254 cm thick, 2 cm wide.
- 3 Beakers, 250 mL
- 1 Control and data acquisition, ZEOS 386SX PC
- 1 Densitometer, Paar M602.
- 1 Detector: Linear106-uv (254 nm)
- 1 pr Goggles, safety
- 1 Meter, pH
- 1 Pump: Eldex AA-100S metering

CHEMICALS   S = Student

Carrier fluid: for bare PS latex, DI water with 0.1 % FL-70 (surfactant for coated particles, 20 mM imadazole)

β-Casein; 10 mg/mL in 20 mM imidazole, density 1.37 g/cm^3. Sigma, St. Louis, MO

Imidazole, $C_3H_4N_2$, f.w. = 68.1, 20 mM, 13.6 mg/100 mL

Standards: Polystyrene (PS) PS 252 nm particle mixture; 2% (w/v), density 1.053 g/cm^3. Duke Scientific, Palo Alto, CA or Seradyn Inc., Indianapolis, IN

THE BIG PICTURE

The unknown latex sample is injected and the centrifuge spun in the absence of flow to provide for the establishment of the particle equilibrium distribution (relaxation period). The density of the carrier solution. is determined. The flow is started and a fractogram is collected. The elution positions for the resolved components are determined and their sizes calculated corresponding to the observed retentions (no calibrations needed). The adsorption complex between a protein and one of the sample latices is formed. The above procedure is repeated using a buffer suitable for maintaining the coated particles in suspension and with an ionic strength such that particle repulsion is minimal during fractionation. The surface density of the adsorbed protein is calculated.

APPARATUS SETUP

See p. 375, Figure 32-7.

DIRECTIONS

Instrument parameters and settings: maximum field strength: 1600 x g (,000 rpm); r_o = 15.5 cm. The TA should have prepared all of the carrier fluids. If not, then

1. Prepare the carrier fluid for bare PS latex (2% w/v in 0.1%FL-70 in DI water).

2. Repeat the above for the coated particles in 20 mM imidazole (0.1 mL casein solution, (10 mg/mL); mix with PS 252 for 30 minutes. Check the pH.

3. Check the power switches on the control box, detector, and the computer of the SFFF system.

4. Run the control program SEDU; key in the run parameters. The operating conditions are:

    Relaxation time: 12 minutes
    Carrier flow rate: 2.6 mL/min
    rpm of the channel: 1200
    λ of the uv detector: 254 nm
    Temperature: 24 °C, or what ever it is

5. Both bare particle and coated particle samples are injected manually. With a low flow rate, 0.2 mL/min, add 5 μL of sample and maintain the flow for 30 seconds.

6. After the relaxation time, the time used to force all of the particles against the wall, begin a flow of 2.6 mL/min and 1200 rpm.

7. One minute before the end of the relaxation time, turn on the pump (while bypassing the channel). This will help to stabilize the detector.

8. Determine the density of the carrier solutions.

9. When finished, save the data.

10. A shift in carrier will now be performed. Use the new carrier of 20 mM imidazole solution.

11. Inject the coated sample and separate it at 1200 rpm.

12. Run the sample at 1200 rpm. Estimate the relaxation time (under selected flow conditions).

13. Shut off the instrument. *Have the TA assist you in cleaning out the coil. Clean up the area and arrange everything in an orderly manner.* Thank you

14. Determine the elution volume. Refer to the example calculations in Chapter 32, p. 376, related to sedimentation field flow fractionation, then calculate the total adsorption and layer thickness.

EXPERIMENT 45. PURGE AND TRAP
Separation of aroma and flavor compounds in fresh and aged beer.
(Abstracted from - Wampler, T.S., Washall, J.W and Matheson, M.J., *Anal. Lab.*, 18T, July 1996)

BACKGROUND
The human sense of taste can detect only four flavors - sweet, sour, bitter, and salty. The remaining flavor notes detected are, in reality, aromas. Therefore, many flavor-producing compounds are volatile, making them amenable to purge and trap GLC. This is particularly applicable to the study of alcoholic beverages, which are perfect examples of a food that relies heavily on both aroma and taste for product acceptance. There are dozens of compounds involved in flavor. Some of those are propanol, ethyl butyrate, isoamyl acetate, ethyl caproate, ethyl caprylate and ethyl caprate.

"In recent years, the importance of sulfur-containing compounds in beer flavor and aroma has been recognized. These are formed during the brewing process from the sulfur present in the proteins of hops. Beer that is exposed to strong sunlight, even for a few hours, often has an off flavor and is known as "sun struck". It is also known as "skunky beer". H_2S is found at about 5 ppb in fresh beer and decreases during storage, while thiols are formed. Light catalyzes a number of chemical reactions in beer, which is why most beer is stored in dark bottles. Miller beer, which is sold in clear glass bottles, utilizes a special strain of hops that prevents light-catalyzed reactions. Some imported beer, such as Corona, is sold in clear glass bottles, but without attempting to prevent these reactions. Dimethyl sulfide, dimethyl disulfide, dimethyl trisulfide, dimethyl tetrasulfide, methanethiol, ethanethiol, and 3-methylbutenethiol are some of the sulfur compounds that have been identified in beer. Because only a few ppb can cause a problem, it is necessary to have some method of concentrating these compounds as well as separating them from the product.

YOUR PROBLEM
You can take your choice of either determining several nonsulfur-containing compounds by using a FID detector or try to determine the sulfur-containing compounds by using a Hall detector. In either case, you will need to examine fresh beer and old beer, and if you are determining sulfur compounds, then set an open glass of beer in the sun for a few hours and measure it again.

EQUIPMENT

1	Apparatus, purge and trap GC, like Figure 33-6, p. 388.	1	Cylinder, graduated, 100 mL
		1	Detector, GLC, FID.
1	Bulb, pipet, 6 mL	1	Detector, Hall, S-mode
4	Clamps, gas cylinder	1	Generator, hydrogen
1	Column, 60 m, 0.53 mm RTX volatiles Restek Corp., Bellefonte, PA	1 pr	Goggles, safety
		1	Pipet, delivery, 2 mL
1	Column, 50 m 0.25 mm OV-225B, Quadrex Corp, New Haven, CT	1	Program, Peakmaster from CDS
		1	Recorder, similar to HP 3390A

1 Regulator, gas, helium	1 Regulator, gas, air
1 Regulator, gas, nitrogen	1 Sponge
1 Regulator, gas, hydrogen	1 Trap, Tenax
1 Regulator, gas, hydrogen, ultrapure.	

CHEMICALS   S = Student

Air, cylinder, 1/lab   Nitrogen, cylinder, 1/lab
Helium, cylinder, 1/lab   Liquid nitrogen, 1L/S
Hydrogen, cylinder, 1/lab
Hydrogen, cylinder, ultrapure or a hydrogen generator, 1/lab
Standards, as many as desired. <1 mL/S

For the FID portion;

Ethanol, H_3CCH_2OH Ethyl caprylate, $H_3C(CH_2)_6COOC_2H_5$
Ethyl butyrate, $H_3C(CH_2)_2COOC_2CH_5$ Isoamyl acetate, $H_3CCOOCH_2CH_2CH(CH_3)_2$
Ethyl caprate, $H_3C(CH_2)_8COOC_2H_5$ 1-Propanol, $H_3C(CH_2)_5$-OH
Ethyl caproate, $H_3C(CH_2)_4COOC_2H_5$

For the Hall portion;

Dimethyl sulfide, H_3C-S-CH_3 Dimethyl tetrasulfide, H_3C-SSSS-CH_3
Dimethyl disulfide, H_3C-SS-CH_3 Ethane thiol, H_3CCH_2SH
Dimethyl trisulfide, H_3C-SSS-CH_3 Methanethiol, H_3CSH
                                          3-Methylbutenethiol, $(CH_3)_2CHCH=CHSH$

THE BIG PICTURE

Depending upon both the time and equipment available, 2 mL samples of two beers, preferably one domestic and one foreign, will be purged and trapped, and the components separated and detected. These beers will then be set in strong sunlight for at least 1 hour, and the process repeated, particularly if sulfur compounds are of interest. Isoamyl acetate should be used to set full-scale deflection, if the FID is used, and dimethyl sulfide, if the Hall detector is used.

DIRECTIONS

1. Set the following parameters if the FID detector is used: purge with He for 10 minutes; flow rate, 40 mL/min; trap rest 35 °C; dry purge for 2 minutes; trap desorb at 250 °C for 2 minutes; and bake at 275 °C for 4 minutes.

2. Column: 60 m, 0.53 mm RTX volatiles. Set the program as: start at 35 °; hold for 2 minutes; ramp at 6 °C/minute to 240 °C.

3. Try to pipet 2 mL of beer into the purge vessel. After you fail, getting mostly foam, then do the following. Fill a 100 mL graduated cylinder with beer and slowly lower a 2 mL pipet to the bottom. Let it fill above the mark. Then raise it slowly until the beer is at the volume mark. Transfer it to the purge vessel.

4. Set the following parameters, if the Hall detector is used:

    Same as above for the purge except add a 4 minute cryofocusing step at the head of the GLC column.

5. Use a 50 m, 0.25 mm OV-225B column, but with the same parameters as above.

6. If available, a GLC/MS could be used to aid in identifying the components.

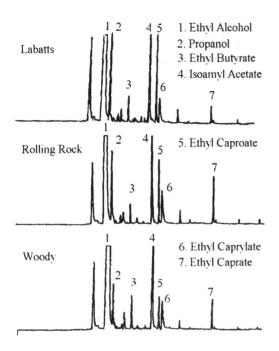

Figure E45-1. Chromatograms of beer obtained by purge and trap.
(Courtesy - Wampler, T.P., Washall, J.W., and Matheson, M., *Anal. Lab.*, 18T, July 1997)

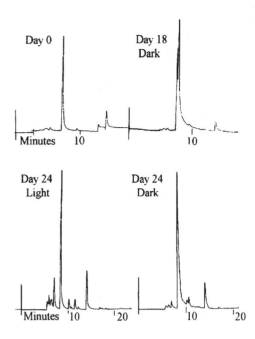

Figure E45-2. Hall detector chromatographs of Negro Modelo beer (*top*) and Genessee 12 Horse beer (*bottom*).
(Courtesy - Wampler, T.P., Washall, J.W., and Matheson, M., *Anal. Lab.*, 18T, July 1997)

EXAMPLES

Figure E45-1 shows the chromatograms of three beers using an FID detector. Figure E45-2 shows chromatographs of one beer under different storage time, light and dark exposure, and using a Hall detector.

EXPERIMENT 46. FOAM FRACTIONATION

Separation of detergents from waste water
(Courtesy - Skomoroski, R.M., *J. Chem. Ed.*, **40** (9), 471, 1963)

BACKGROUND

Synthetic detergents were developed to overcome the major disadvantage of soaps, that of forming insoluble residues when reacting with dissolved metal ions, particularly calcium and magnesium (the old ring around the bath tub). The -COOH group on soaps was replaced with a -SO_3H group on detergents. This group forms water-soluble complexes with the metal ions, and no deposit forms.

The first commercial detergent, Dreft, was produced in the mid 1930's by Procter and Gamble. However, it was not until 1947 at Mt. Penn, PA, that detergents were recognized as causing a pollution problem that we were aware of. On Friday, a door-to-door distribution of a new detergent was made. By Monday, foam 5 feet high greeted the workers of the sewage plant.

1 μg of detergent/mL (1 ppm) will cause frothing in rivers and streams so the tolerance is set at 0.5 ppm. The old type detergents, containing alkylbenzene sulfonate (ABS), were made from petroleum wastes so it was inexpensive and

used a waste product. However, the chain was highly branched, and bacteria could not biodegrade it very well because they degraded it two C's at a time and stopped at the branch. Chemists then made a linear chain detergent (LAS), which was biodegradable.

Alkylbenzene Sulfonate
Branched Chain   Hard To Degrade

Linear Chain   Easy To Degrade

The following statement by N. Tharapiwattananon (*Sepn. Sci. Tech.*, **31** (9), 1233, 1996) illustrates the current problem. "Surfactants are widely used in many industries. As environmental regulations tighten, there is increasing concern about reducing the surfactant concentration in effluent streams. One source of these streams is surfactant-based separations, which are being increasingly used to remove pollutants from wastewater and ground water. In these processes, surfactants are added to a contaminated stream to effect removal of pollutants. Sometimes a stream emitted from one of these processes contains a low surfactant concentration. In addition to satisfying environmental regulations, the value of the surfactant being emitted sometimes makes recovery operations economical."

There are 3,057 possible isomers of C_{12} and about 80,000 for the C_{10} to C_{15} range. Therefore, any method must be capable of determining many similar compounds. The standard method for determining detergents in water is to develop a color with methylene blue. This is a sensitive method (<0.1 ppm) and requires simple instrumentation. Methylene blue reacts with benzene-containing detergents to form a salt, which is soluble in chloroform and can be used over the range 0.1 to 4 ppm. Organic sulfates, sulfonates, carboxylates, phosphates, and phenols interfere and give high results, whereas amines cause low results because they compete with the methylene blue. Inorganic Cl^{-1}, NO_3^{-1}, SO_4^{-2}, PO_4^{-3}, SCN^{-1}, and S^{-2} do not interfere.

Methylene Blue + (LAS) → 1 : 1 Blue Complex

YOUR PROBLEM

It has come to the attention of the local city engineer that the city waste water may contain sufficient detergent to consider reclaiming it by a foam fractionation process. You are to obtain some preliminary data on the amount of detergent originally present and if foam fractionation causes a sufficient enrichment to make further consideration reasonable. You are to obtain a sample of waste water prior to sewage treatment and analyze it for detergents and then perform a laboratory-scale foam fractionation to determine if enrichment occurs and how much.

EQUIPMENT

The equipment list below is based on Figure 34-11, p. 401, plus the spectrophotometric measuring system. An alternate is shown in Figure E47-1, p. 672, with its equipment list and is the system described in the directions. The equipment listed below is in addition to that required for the apparatus shown in Figure E47-1.

1	Apparatus, foam fractionation 10 cm x 100 cm	1	Flask, volumetric, 1,000 mL
		5	Funnels, filtering, 60°, short stem
1	Balance, analytical	10	Funnel, separatory, 250 mL
5	Beakers, 10 mL	1 pr	Goggles, safety
1	Beaker, 2 L or similar container	1 bx	Kimwipes
1	Bottle, wash, 250 mL	6	Papers, filter, 11 cm
1	Brush, test tube, large	1	Pipet, Mohr, 10 mL,
1	Bulb, pipet, 6 mL	3	Racks, funnel
2	Cells, Pyrex, 1 cm	1	Spectrophotometer, 654 nm
5	Flasks, volumetric, 100 mL	1	Sponge

CHEMICALS   S = Student   Air, cylinder
Chloroform, $CHCl_3$, 1 L/S
Docecylbenzenesulfonate, $H_3C(CH_2)_{11}C_6H_5OSO_3H$, commercial grade, 10 µg/mL
Glass wool, extracted with $CHCl_3$ prior to use, 5 g/S
Grease, stopcock
Methylene blue; Dissolve 100 mg in 100 mL of water. Transfer 30 mL to a 1 L flask; add 500 mL H_2O, 41 mL of 6 N H_2SO_4, and 50 g $NaH_2PO_4 \cdot H_2O$, dilute to 1 L with H_2O, 150 mL/S
Monosodium dihydrogen phosphate monohydrate; $NaH_2PO_4 \cdot H_2O$, 100 g/S
Phenolphthalein indicator, 1 mL/S
Sample, any waste water sample, preferably just before it reaches the sewage treatment plant or dish washing water, 2 L/S
Sodium hydroxide, NaOH, 1 N (40 g/L), 10 mL/S
Sulfuric acid, H_2SO_4,, 1N and 6 N (2.8 and 16.6 mL/100 mL, respectively), 1 mL and 82 mL/S
Wash solution, Add 41 mL of 6 N H_2SO_4 to 500 mL of water in a 1000 mL flask. Add 50 g of $NaH_2PO_4 \cdot H_2O$ and shake until dissolved. Dilute to 1000 mL

THE BIG PICTURE

A small sample of the waste water is saved to be tested as the untreated blank. Two liters of waste water is poured into the fractionator, the air turned on and the flow adjusted to produce a foam. The foam is collected in a crystallizing dish and broken with a drop or two of octanol. Samples of the aereated water are taken every 10 to 15 minutes for an hour. The detergent remaining at each time period is determined by developing the methylene blue complex and measuring its absorbance at 652 nm.

APPARATUS SETUP

Three apparatus setups are required: (1) a visible range, 652 nm, spectrophotometer; (2) a ring stand with three funnel racks for filtration and extraction, and (3) a foam fractionation apparatus similar to that shown in Figure E47-1, p. 672. This apparatus is used in preference to that in Figure 34-11, p. 401, because it can be used for two types of experiments.

DIRECTIONS

1. Replace the buret with a stopper containing a stopcock to facilitate taking samples during the foaming process.

2. Obtain about two gallons of dish washing water or from a waste water treatment plant.

3. Check the apparatus to make sure it is arranged correctly. Shake the charcoal jar to remove any channels and fill the water jar to within 5 cm of the top. Secure each stopper.

4. Turn on the air pump to insure it works and to be sure the stoppers do not blow off of the jars.

5. Filter away any floating debris from the water sample. Collect in a 2 L beaker or suitable container.

6. Be sure the air is on and add 2 liters of water to the flotation tube. Place about 10 mL of this into a small beaker for later spectrophotometric measurement.

7. If water enters the air line, the tubing is disconnected from the water saturator bottle. Any water in the line between the water saturator and the dispersion tube is blown out and the line then reattached to the saturator. If this is not done then the air will usually not pass into the flotation tube and the increased air pressure can blow off the rubber stoppers on the charcoal and the water saturator jars.

8. Adjust the air flow with either a rotameter or a variable transformer (preferred) so a steady stream of bubbles are passing through the solution.

9. Observe the clarity and consistency of the water. After 10 minutes take another 10 to 15 mL sample (side stopcock) and place it in a beaker for later analysis.

10. Repeat step 9 every 10 to 15 minutes for an hour.

11. Shut off the air.

12. Loosen the chain clamp and raise the flotation tube about 1 foot.

13. Disconnect the tubing from the water saturator and drain the majority of the water from the flotation tube into any suitable container.

14. Remove the collection vessel and observe the material that was removed from your sample then clean the vessel and replace it on the stand.
15. Carefully - remove the flotation tube, rinse it out, and replace it.
16. The amount of detergent in the water is determined by forming the methylene blue complex, a standard method for linear detergents.

Preparation of a calibration curve.

17. Prepare a series of five 250-mL separatory funnels containing 0, 5, 10, 15, and 20 mL of the 1 ppm standard.
18. Add water to make 100 mL.
19. Adjust the pH. Add 3 to 4 drops of phenolphthalein. If colorless, make alkaline with a few drops of 1 N NaOH, then just colorless with a few drops of 1 N H_2SO_4.
20. Add 10 mL of $CHCl_3$ and 25 mL of the methylene blue reagent.
21. Invert the funnel repeatedly for 30 seconds and then let the phases separate. If an emulsion should form, it can usually be broken by swirling or adding a few drops of isopropyl alcohol. If it is added to one, it must be added to all
22. Draw off the $CHCl_3$ layer into a second separatory funnel, rinsing the delivery tube with a small stream of $CHCl_3$.
23. Repeat the extraction two more times. If the water layer looses all of its blue color, start over with a smaller sample.
24. Combine all of the extracts in the second separatory funnel. Add 50 mL of water and shake for 30 seconds. Emulsions do not usually occur now.
25. Drain the $CHCl_3$ layer through a funnel containing a plug of glass wool (cleaned by washing with $CHCl_3$) into 100 mL volumetric flasks.
26. Extract the wash solution twice more with 10 mL each of $CHCl_3$, adding the extract to the volumetric flasks.
27. Rinse the glass wool and funnel with $CHCl_3$. Dilute to the mark with $CHCl_3$, stopper and mix.
28. Determine the absorbance at 652 nm against a $CHCl_3$ blank. If the absorbance is greater than 1.5, then dilute the $CHCl_3$ extract sufficiently and record the dilution so the calculations can be corrected.
29. Plot A vs. ppm LAS. Draw the best straight line through the points.

Examination of the sample.

30. Clean all of the glassware used for the calibration curve preparation.
31. Pipet 2.0 mL of each sample into a separatory funnel.
32. Do steps 18 to 28 for each sample.
33. Plot A vs. ppm LAS in the sample as it is removed. Draw the best curved line through the points to indicate the rate of clean up.
34. *Clean up the area in general and leave it better than you received it.* Thank you.

EXAMPLE CALCULATION

From a calibration curve a 2 mL sample of untreated dish wash water had an absorbance of 0.78 and after 40 minutes of foaming it had a value of 0.16. These values corresponded to 90 ppm and 19 ppm, respectively. What is the concentration in ppm of LAS in each sample?

ANSWER

The equation for calculating the methylene blue active substances (MBAS) is:

$$mg\ MBAS/L = \frac{\mu g\ apparent\ LAS}{mL\ of\ original\ sample} \tag{E46-1}$$

$$mg\ MBAS/L = \frac{90\ \mu g}{2\ mL} = 45\ mg/L\ or\ 45\ ppm$$

The concentration of the foamed sample is then 9.5 ppm

EXPERIMENT 47. GAS-ASSISTED FLOTATION
Separating suspended solids from river water
(Based on an article by R.B. Grieves, *J. Am. Water Works Assoc.*, **59**, 859, 1967)

BACKGROUND
One factor in assessing the quality of surface waters is the amount of suspended solids or, "how muddy is it?" For example: the standard for many rivers and streams is 100 mg/L. If more than this occurs that segment of the stream is declared "impaired." Much of this solid material is colloidal and cannot be filtered under ordinary means. In water treatment plants, aluminum hydroxide is used to surface adsorb or simply trap the solids, and this floc then either settles out or is filtered by a gravel bed. It has been found that cationic surfactants will surface adsorb to most of the solids suspended in water, and this coating will attach to air bubbles passing through the system. Some surfactants proven effective are cetyldimethylethylammonium chloride, cetyldimethylbenzylammonium chloride, cetylpyridinium chloride, and ethylhexadecyldimethyl ammonium bromide. These are all antiseptics as well and are used in many mouthwashes. Only a few ppm are necessary (10 to 60) to provide clarification, and the LD_{50}'s are in the range of 500 mg/kg.

$$[(CH_3)_2-\overset{+}{N}-(CH_2)_{15}CH_3 \quad | \quad CH_2CH_3] \quad Cl^-$$

Cetyldimethylethyl ammonium chloride

$$[\text{pyridinium}-\overset{+}{N}-(CH_2)_{15}CH_3] \quad Cl^-$$

Cetylpyridinium ammonium chloride

YOUR PROBLEM
The army needs a system to clarify medium amounts of stream and river water for drinking, cooking, and washing. The use of aluminum and gravel filtration is not very portable, and an alternative is being sought. You are to demonstrate that gas-assisted flotation will work and will use water from around your area to prove it. In addition you will use a quaternary amine as the surfactant, knowing that any excess will act as a disinfectant in the water.

EQUIPMENT

1	Apparatus as shown in Figure E47-1, p. 672	1	Transformer, variable
2	Bottles, wide mouth, 32 oz	1	Tube, addition
2	Clamp, chain type	1	Tube, gas dispersion
1	Dish, crystallizing	1	Tube, glass, 8 mm, 4' long
1	Fractionator, glass, 10 cm x 105 cm	1	Valve
1	Pump, air	4 ft	Tubing, glass, 8 mm
1	Rotameter, air	2 ft	Tubing, Tygon, 8 mm
1	Stand, ring	5	Beaker, 100 mL
1	Stopcock, 2 mm bore	1	Beaker 1000 mL
2	Stoppers, rubber, 2 hole, no. 13	1	Bucket, 2.5 gallon
2	Stoppers, rubber, 1 hole, no. 4	1	Cylinder, graduated, 10 mL
1	Support for the crystallizing dish	1	Spectrophotometer, visible region

CHEMICALS   S = student
Cetyldimethylethylanmmonium chloride, 1 g/L, 60 mL/S
Charcoal, activated, 20 g/apparatus
Cotton, 10 g/ apparatus
Sample, 1.5-2.0 gallons of muddy river water, 2.0 L/S

APPARATUS SETUP
The apparatus is set up as shown in Figure E47-1, p. 672, with these modifications. The rotameter was found not to be necessary and in fact made air flow more difficult. Instead, a variable transformer was used to control the rate of the pump. The air enters the bottom of the tube through a gas dispersion tube and surfactant is added through a buret attached as a side tube. The foam is collected in a crystallizing dish. Samples of clarified water are taken by immersing a glass tube into the water from the top of the tube.

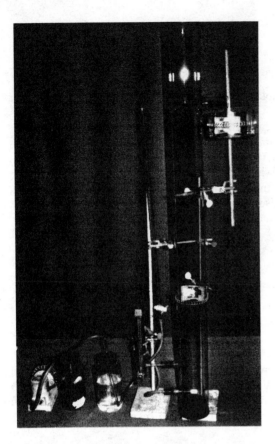

Figure E47-1. Gas assisted fractionation unit.

THE BIG PICTURE

Skim the floating debris from two liters of water, and add it to the flotation tube. The tubing is disconnected from the water saturator bottle and the air is turned on. Any water in the line between the water saturator and the dispersion tube is blown out and the line then reattached to the saturator. A sample of this water is taken to serve as the original. The quaternary amine solution is added to the addition tube, and 20 mL are added to the water. The system is allowed to run for a few minutes to see if foaming occurs. If not then further additions are made until foaming occurs. The foam is collected, and after about 15 minutes, or until the foaming has subsided, another sample is taken from the tube. More quaternary amine is added, and the process repeated two more times. The absorbance of the water is measured by a spectrophotometer set at 400 nm. The goal is to get water below 0.015 absorbance units.

DIRECTIONS

1. Go to any stream or lake in the area and obtain about 2 gallons of water. Do not stir up the water to make it muddier than it is.

2. Check the apparatus to make sure it is arranged correctly. Shake the charcoal jar to remove any channels and fill the water jar to within 5 cm of the top. Secure each stopper.

3. Turn on the air pump to ensure that it works and the stoppers do not blow off of the jars.

4. Fill the quaternary amine addition tube (buret) with the quaternary amine solution.

5. Filter away any floating debris from the water sample.

6. Add 2 liters of water to the flotation tube. Place about 10 mL of this into a small beaker for later spectrophotometric measurement.

7. The tubing is disconnected from the water saturator bottle. Any water in the line between the water saturator and the dispersion tube is blown out, and the line then reattached to the saturator. If this is not done, then the air will usually not pass into the flotation tube, and the increased air pressure can blow off the rubber stoppers on the charcoal and the water saturator jars.

8. Adjust the air flow with either a rotameter or a variable transformer (preferred) so a steady stream of bubbles are passing through the solution.

9. Observe the clarity and consistency of the water. After a few minutes, add 20 mL of the quaternary amine solution to the addition tube. It will go into the water, because of the pressure difference. This will be 10 ppm in the water. Check the calculations to verify it.

10. The system is allowed to run for a few minutes to see if foaming occurs. If not, then further additions are made until foaming occurs. The foam is collected, and after about 15 minutes or until the foaming has subsided, another sample is taken from the tube.

11. More quaternary amine is added, and the process repeated two more times.

12. The absorbance of the water is measured by a spectrophotometer set at 400 nm. Use tap water for the blank, and determine the absorbance of each sample. The goal is to get water below 0.015 A.

13. Decision time. Look at the water and see how clear it is. Will you drink it? If it is satisfactory to you then go to step 14. If not, and you have time, then repeat steps 9 and 10 as you desire.

14. Shut off the air.

15. Loosen the chain clamp and raise the flotation tube about 1 foot.

16. Disconnect the tubing from the water saturator and drain the majority of the water from the flotation tube into any suitable container.

17. Remove the collection vessel and observe the material that was removed from your sample then clean it and replace it.

18. *Carefully* - remove the flotation tube, rinse it out, and replace it.

19. Plot A vs. ppm of quaternary amine added. Most waters are clarified by 10 to 60 ppm of amine. Does your data verify this?

20. *Clean up the area in general and leave it better than you received it. Thank you.*

EXPERIMENT 48. LIQUID-ASSISTED FLOTATION - TOMATOES
Separating fly eggs and maggots from canned tomatoes

(*Association of Official Analytical Chemists, International*, Method 955.46); The insect photos, diagrams, and much of the descriptions were obtained from O.L. Kurtz and K.L. Harris, *Micro-Analytical Entomology for Food Sanitation Control,* Association of Official Analytical Chemists, Washington, D.C., 1960.

BACKGROUND

In this case, you are going to separate fly eggs and maggots (which are higher in protein molecules) from tomato pulp (which is higher in carbohydrate molecules). This is more of an entrapment than a surface coating. If you place a peeled tomato in water, it sinks. If you place a peeled tomato in petroleum ether, it sinks. If you pulverize a peeled tomato, place it in water, and shake it vigorously, the pieces are dispersed,, and eventually settle out. If you pulverize a peeled tomato in petroleum ether and shake it vigorously, the pieces will sink. However, if you add the pulverized tomato to a system of water and petroleum ether and shake it vigorously, the tomato pulp will all rise to the bottom of the petroleum ether layer. The petroleum ether (hexanes) should not wet the highly cellulosic cell walls of the tomato, so what does the hexane do? When the pulp is shaken with the hexane, some of the hexane (density 0.8) becomes trapped in the ruptured cells and causes the pulp to float. It collects at the interface because the surfaces of the tomato cells are hydrophilic and stay in the water layer. The eggs, maggots, and seeds are not ruptured, apparently do not trap hexane, and slowly sink. These then are drained off, examined under a microscope, and identified.

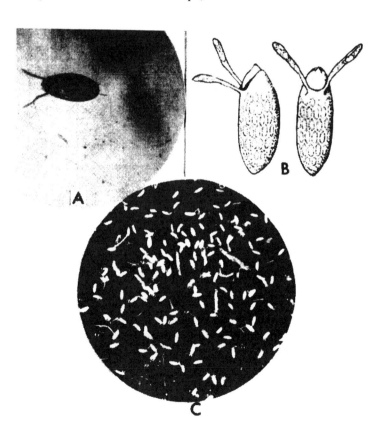

Figure E48-1. Photograph of fruit fly eggs and maggots.

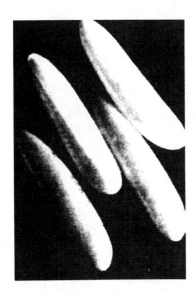

Figure E48-2. House fly eggs.

Figure E48-3. House fly maggots.

Fly Eggs and Maggots

The current guidelines are that no more than 10 fly eggs or maggot segments less than 2 mm can be present in 1 lb of canned tomatoes. This may seem like a large number, but if one fly can lay 200 to 400 eggs in a season, then a lot more eggs could be present. About half of those eggs will get to the maggot stage if left undisturbed.

Some people have demanded that zero fly eggs be present. If this were the law, then tomatoes would be so expensive that no one could afford them, and no factory would even try to meet that specification. The common sense approach is to determine what good manufacturing practice will provide and regulate on that basis.

A maggot is the larva stage of a fly. An egg takes from 2 to 5 days to develop into a maggot. The maggot will grow to full size, enter a pupa stage, and emerge as a fly, usually within 1 month. How do they get into tomatoes? If a tomato grows too fast, it will crack. If this happens out in the field, fruit flies may deposit their eggs in that soft tomato pulp. To protect itself, the tomato will heal the wound. The fly eggs are now trapped on the inside, where they can develop into maggots and emerge. If the canner processes those tomatoes, you run the risk of having eggs and maggots present. Fruit fly eggs are white and look like little footballs with a TV antenna attached at one end. These are 1/32 in. long. These are shown in Figure E48-1, p. 673. There are three maggots in the top right quarter of the lower photograph.

Now let's bring good whole tomatoes to the cannery. Tomatoes sometimes get bruised during shipping and split open. Flies are always around canneries because of the tomato juices, and those flies, called *scavenger or houseflies* can deposit their eggs in the tomatoes. These eggs are like little white cigars as shown in Figure E48-2. They have no antennae and are about 1/16 in. long Figure E48-3 shows the maggots. Higher magnification shows that the maggots are segmented, and they tend to break apart at the segment divisions. In general then, two different kinds of eggs can be present. Whenever this is seen, you know that either poor quality tomatoes are being canned or few precautions have been taken in the factory, such as putting up screens to keep the flies away. Generally, you are more likely to get this contamination with the less expensive brands of tomatoes. By knowing the kind of eggs present in the tomatoes, a food inspector will know where to begin to look for the problem.

You will look for maggots, fruit fly eggs, and scavenger fly eggs. Eggs and maggots fluoresce and are easier to separate from immature seeds this way.

YOUR PROBLEM

You have been informed that a reassessment of good manufacturing practices is to be done on canned tomatoes and that you will have to examine dozens of cans of tomatoes in the next few weeks. It has been awhile since you did this, so you have gone to a grocery store and obtained a 1 lb can of tomatoes to practice on. You also have visited a university laboratory and obtained samples of fruit and house fly eggs and their young maggots.

EQUIPMENT

- 1 Aspirator, water
- 2 Beakers, 1000 mL
- 1 Beaker, plastic, 100 mL
- 1 Bolting cloth 10 XX (from a local grain mill or use cotton batiste)
- Dye black, cut into 5.5-cm circles
- 2 Bottles, wash, 1 water, 1 ethanol
- 1 Brush, test tube
- 2 Clamps, versatile
- 1 Cylinder, graduated, 100 mL
- 1 Dish, Petrie, both top and bottom
- 1 Flask, filtering, 2000 mL
- 1 Funnel, 270-310 mm
- 1 Funnel, Buchner, 5.5 cm
- 1 Funnel, separatory, 4 liter preferred, but at least 2 L.
- 1 pr Goggles, safety
- 2 Holders, clamp
- 1 Mercury lamp, short wavelength

1	Microscope, low power 10 or 20x	1	Spatula
1	Microscope, high power 80 to 400x	2	Stands, ring
1	Opener, can	1	Stopper, rubber, no. 14
1	Probe, dissecting, wooden handled	4 ft	Tubing, vacuum
1	Ring, iron, 10 cm,	1 pr	Tweezers
1	Sieve, standard no. 8, 9, or 10		

CHEMICALS   S = Student

1 can of the LEAST EXPENSIVE canned tomatoes you can find, 1 lb needed.

Ethanol, $H_3C\text{-}CH_2\text{-}OH$, 95% 15 mL/S

Glycerine, $H_2C(OH)CH(OH)CH_2\text{-}OH$, 1:1 with water 5 mL/S

Petroleum ether, primarily $H_3C(CH_2)_4CH_3$, 145 mL/S

THE BIG PICTURE

One pound of canned tomatoes is forced through an 8- or 10-mesh sieve and collected in a 1 L beaker. This pulp is transferred to a large separatory funnel and 145 mL of petroleum ether are added. This mixture is shaken vigorously for several minutes; then the beaker placed in a ring on a ring stand, and the mixture is allowed to separate. The tomato pulp will float, and the fly eggs, maggots, and seeds will sink. This lower level is drained and filtered on a black cloth. Glycerine, water, and ethanol are added to keep the fragments intact and then examined under a microscope. The tomatoes are filtered as shown in Figure E48-4 and are separated as shown in Figure E48-5.

APPARATUS SETUP

This is shown in Figures E48-4 and E48-5.

Figure E48-4. Setup for processing tomatoes.

DIRECTIONS

1. Arrange a large funnel and a sieve on a ring stand as shown in Figure E48-4.

2. Open a can of tomatoes and push the sections through the sieve with a large rubber stopper.

3. Turn the sieve upside down and wash the pulp off with a stream of water from a wash bottle. Collect the washings in the beaker.

4. Put the large separatory funnel through the iron ring on the ring stand and transfer the pulp to it (Figure E48-5).

5. Add 145 mL of petroleum ether (hexanes).

6. Stopper the separatory funnel. *Caution*: big funnels are expensive, so hold it very tightly as you shake it. After shaking for a few seconds, release the pressure.

7. *The success of the analysis depends on how hard you shake.* The tomato seeds have fine hairs that hold the eggs in place, and you simply have to shake the eggs off. Shake the funnel until you're tired and then shake some more - for about 10 min.

8. Put the funnel back in the ring and let the layers separate. Go immediately to Step 9.

9. Go to the microscope and, with help from your instructor, learn to identify fly eggs and maggots. Look at fly eggs under a uv lamp and notice that they fluoresce, whereas tomato seeds do not.

Figure E48-5. Apparatus for floating the tomato pulp.

10. Attach a funnel (Hirsh preferred, but Buchner acceptable) to a 2 liter filtering flask and connect it to a water aspirator.

11. Place a piece of black bolting cloth in the funnel and wet it down with water.

12. We do not want the seeds that are in the bottom of the separatory funnel to come through onto the bolting cloth so place the sieve on top of the Hirsh funnel. Hold the separatory funnel over the Hirsh funnel (remove the stopper) and very carefully allow the lower layer to drain into the funnel. Turn on the water aspirator, but do not allow that black filter pad on the funnel to become dry, because the eggs tend to collapse.

13. Immediately transfer the black pad to a Petri dish. Put about 3 or 4 mL of water in the Petri dish to cover the black cloth and then add 1 or 2 mL of alcohol. The alcohol is to keep molds from growing so that you can keep the sample for a long period of time if you desire.

14. Examine your sample. *Stir it gently* with a dissecting needle to turn the eggs in different directions. Look for the TV antennae on the end of the fruit fly eggs. If you find anything be prepared to show the rest of the class.

15. Compare what you see with standards if they are available.

16. Visit other groups and see what they have found.

17. Place a cover on the Petri dish, tape it in place, and label it for others to view.

EXPERIMENT 49.   LIQUID-ASSISTED FLOTATION - CORNMEAL AND FLOUR
Separating insect fragments and rodent hairs from cornmeal and flour.
(Courtesy - *A.O.A.C. Intl.*, 15th Ed. 970.71, 971.32 and 980.27) The insect photos, diagrams, and much of the descriptions were obtained from O.L. Kurtz and K.L. Harris, *Micro-Analytical Entomology for Food Sanitation Control*, Association of Official Analytical Chemists, Washington, D.C., 1960.

BACKGROUND
In the tomato and fly eggs experiment the unwanted material was floated to the top. In this case, the waxy surface of the chitinous insect fragments or the oil-covered rodent hairs are wetted by mineral oil and floated away from the more polar starchy cornmeal and flour. The water and solid residue are drained away, and the oil layer filtered, leaving the insect fragments and some bran particles behind. These then are examined with a microscope.

Each insect has a covering on its exoskeleton of a variety of waxes and other lipids. These vary from insect to insect so much that an insect can be identified by this composition. A wax is an ester made from a long chain (C_{18} - C_{24}) alcohol and a long chain (C_{24} - C_{36}) acid. It is hydrophobic so insects are not wetted very well by water. However, many are slightly more dense than water, and they or their fragments will eventually sink.

Mineral oil is a mixture of long chain hydrocarbons obtained from petroleum and will easily wet the waxy surface of insects. Mineral oil has a density of 0.80 to 0.83, so it easily floats on water.

Rodent Hairs
Rodent hairs are bad to have in meal or food of any kind. Generally, they don't get there because the animal is shedding. They get there from rodent pellets. These pellets are contaminated with rodent hairs (Figure E49-1, p. 677) plus possible pathogenic bacteria. During processing of cornmeal, the feces dissolves or is spread throughout the cornmeal so it can't be detected. It is known that if rodent hairs are present, rodent filth must have been present. Obviously then, when a rodent hair is present, the food is suspected as being a carrier of disease and it is taken off the market. The rule is this. If you find 2 rodent hairs or 2 insect fragments in 45 g of cornmeal then that lot is taken off the market. Other guidelines are 2 fragments per 450 g of bread or 2 fragments per 100 g of peanut butter. Figure E49-1, p. 677, shows a rodent pellet with the hairs embedded. Look carefully at the top and lower side.

Rodent hairs can be distinguished from human hairs by their dark spotted appearance (Figure E49-2, p. 677) At least a 80x microscope is required to see the medulla (dark center section) of a hair. Mice hairs have the medulla sprinkled throughout whereas rat hairs have elongated clumps from one side to the other. Animal hairs usually have half or more of the hair width as medulla, whereas human hairs have less than half. Synthetic fibers have no medulla. Figure E49-2, p. 677, shows the medulla pattern for mice and rats.

Insect Fragments
Insect fragments are more difficult to assess than fly eggs. The most serious situation would be if all stages of life were found: egg, larva, pupa, and adult. This would indicate that the grain was infested heavily rather than

Figure E49-1. A rodent pellet contaminated with hair.

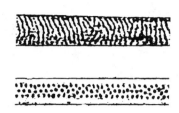

Figure E49-2. Rodent hairs 80x; upper - mouse hair, lower rat hair.

Figure E49-3. Rice weevil.

having an insect wandering through. *The best way to determine if a brown particle is an insect fragment or a particle of bran is to probe it with a dissection needle.* Insect fragments are generally brown and are hard if you touch them with a wire probe. A soft brown mass attached to a starch particle is not an insect fragment, it is bran.

The most common insects found are rice, granary, and maize weevils; red flour, confused, saw tooth, and flat grain beetles; and the lesser grain borer. Weevils have long snouts, like a short elephant trunk. Figure E49-3 shows the reddish-brown rice weevil. Figure E49-4 shows the chestnut brown granary weevil (note the snout). Figure E49-5 and Figures E49-6 and E49-7, p. 678.

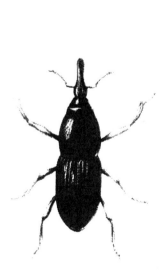

Figure E49-4. Granary weevil.

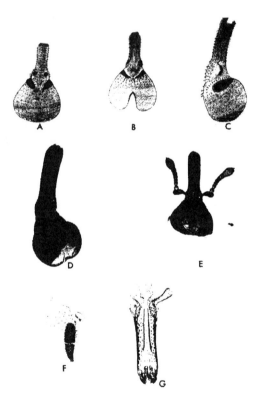

Figure E49-5. Rice and granary weevils. (A) Head (dorsal view, drawing). (B) Head (ventral view drawing). (C) Head (lateral view drawing). (D) Head (lateral view, actual). (E) Head and antennae (ventral view). (F) Compound eye. (G) Snout with mandibles at the tip.

Figure E49-6. Rice and granary weevils. (A) Prothoracic leg. (B) Mesothoracic Leg. (C) Mid Coxa. (D) Hind coxa. (E) Fore coxa with trochanter and femur attachment. (F) Fore coxa. (G) Proximal articulation of tibia.

Figure E49-7. Rice and granary weevils. (A and B) First tarsal segment. (C) Tibia-tarsus. (D) Tarsus (dorsal view). (E) Tibial fragment. (F) Hind tibia. (G) Fore or mid tibia.

The red flour beetle (*Tribolium castaneum*) is a common household insect. This is usually the one that infests the flour and flour products stored in your home. It is reddish-brown, 3.0 to 3.73 mm long, and 0.97 to 1.26 mm wide. You can tell immediately if you have an infestation, usually at the worst time. During the holidays when visitors are present and you are making gravy, these insects are wetted by the grease the same as the mineral oil in this experiment and float to the top in the clear grease, where you can see them. Just pour off the grease, and no one will ever know the difference. This insect is shown at the bottom right in Figure E49-8, p. 679. This figure shows some of the common fragments.

YOUR PROBLEM

A friend of yours has observed that a certain brand of cornmeal seems to have strange looking particles in it and wants you to examine it.

EQUIPMENT

1 Aspirator, water	1 Clamp, pinch, screw type
1 Balance, triple beam or top loader	1 Cylinder, graduated, 100 mL
1 Beaker, 100 mL	1 Dish, Petrie, both top and bottom
1 Beaker, plastic, 100 mL	1 Flask, filtering, 1 L
1 Beaker, 1000 mL	1 Flask, Kilborn, 1 liter
3 Bottles, wash, 1 water, 1 ethanol, 1 soap	1 Funnel, Buchner, 5.5 cm filter section
1 Brush, test tube	1 Funnel, powder
1 Clamp, chain	1 pr Goggles, safety
	1 Hot plate

1	Knife, chopping	1	Spatula
1	Microscope, low power 10 or 20x	1	Stand, ring
1	Microscope, high power 80 to 400x	1	Stirrer, magnetic
1 pk	Paper, filtering, 5.5-cm preferably ruled	1	Tubing, rubber, 2-in. piece
1	Probe, dissecting, wooden handled	4 ft	Tubing, vacuum
1	Rod, stirring, 6 "	1 pr	Tweezers

CHEMICALS   S = Student

Ethanol, 95 %, 245 mL/S
Glycerine, 1:1 with water 5 mL/S
Hydrochloric acid, HCl 7 + 93, 400 mL/S
Liquid household detergent (2 + 450 mL H_2O) 20-30 mL/S
Mineral oil, paraffin oil NF white light, 30 mL/S
1 box of the LEAST EXPENSIVE cornmeal or flour you can find, 45 g/S

THE BIG PICTURE

45 g of corn meal, flour or other similar product is placed in a 1 L beaker, 400 mL of 7% HCl and 30 mL of mineral oil are added, and the mixture brought to a rolling boil. The mixture is transferred to a Kilborn flask, the solids are drained away, and the oil is vacuum filtered onto ruled filter paper. The paper is transferred to a Petrie dish and examined with a microscope.

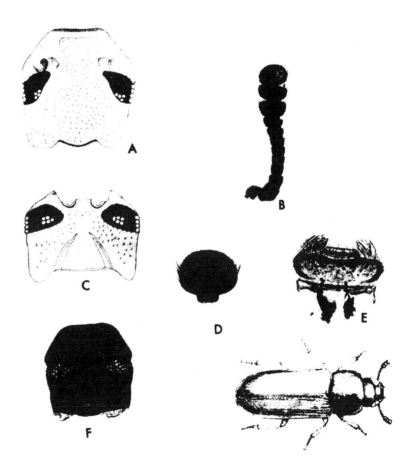

Figure E49-8. Red flour beetle. (A) Head (dorsal view, drawing). (B) Antenna 45x. (C) Head (ventral view, drawing). (D) Terminal antennal segment 100x. (E) Labrum 100x. (F) Head, actual. *Bottom right*: Adult beetle.

APPARATUS SETUP

The apparatus is set up as shown in either Figures 34-21, p. 408 or Figure 34-2 p. 408.

DIRECTIONS

1. Take a 100 mL plastic beaker, put it on a triple beam balance, and weigh into it 45 g of either your sample of ground cornmeal or flour or our sample.
2. Place this in a 1000 mL beaker and add 400 mL of 7% HCl. Stir this until all of the sample is wetted and you have a slurry. The purpose of the HCl is to denature the protein so it will settle faster. 7% HCl is used, because 100% would dissolve the chitin.
3. Add 30 mL of mineral oil to the beaker.
4. Put this on a hot plate, bring it to a boil and let it boil for 10 minutes..
5. Remove the beaker and let it cool.
6. Attach a Kilborn flask to a ring stand. Screw the pinch clamp on the rubber tubing tightly so that the flask will hold a solution. Stir the cornmeal and oil mixture and pour it into the Kilborn flask. Use a wash bottle and rinse out the beaker. If the mixture is up to within 1 cm of the top of the Kilborn flask, then forget about it. You can catch it next time.
7. Slowly stir the contents of the Kilborn flask for a minute or so and then let the contents settle for another couple of minutes.
8. Place a 1000 mL beaker under the Kilborn flask. Unscrew the pinchcock on the rubbing tubing. The cornmeal may plug the rubber tubing, and you will have to take your fingers and squeeze the tubing so the cornmeal will come down. Be prepared to pinch it off immediately, because once it unplugs, the mixture drains rapidly. Drain the water and the cornmeal mixture until about 3 to 5 cm of water are left in the Kilborn flask.
9. If all of your sample was not transferred from the original beaker, then clean out the beaker at this time and fill the flask within 1/2 inch from the top with water.
10. Stir the mixture gently, then let the mixture settle, and again drain off the water to about 3 to 5 cm from the bottom.
11. Add water and continue to repeat this process until the bottom wash water is very nearly clear.
12. Filter the oil layer as follows: Take a piece of 5.5 cm white filter paper (lined type preferable) and put it in a small Buchner funnel. Moisten it with a little water and gently turn on the suction.
13. Take the Kilborn flask off the ring stand and drain most of the water into the sink.
14. Put it over the funnel, turn the water to the aspirator on so that you get maximum suction, and filter both layers.
15. Rinse out the Kilborn flask, particularly the sides, with distilled water and then with a little ethanol. Filter these washings.
16. Wash out the Kilborn flask with a soap solution and filter these washings.
17. Rinse the whole apparatus again with distilled water and filter these washings. Alcohol was used to dissolve the mineral oil and help free the fragments and rodent hairs. Soap solution was used to dissolve the oil film.
18. Take the filter paper out of the funnel and put it into the bottom of a Petri dish. Add 10 drops of a 1 to 1 glycerin-water mixture to keep the starch moist.
19. Examine the contents on the filter paper under a low power microscope.
20. Go around the lab and see what other students are finding.

EXPERIMENT 50.   LIQUID-ASSISTED FLOTATION-LEAFY VEGETABLES
Separating mites, aphids, and thrips from leafy vegetables
(Courtesy - *A.O.A.C., Intl.* 16th Ed., 945.82, 974.33, 975.47). The insect photos, diagrams, and much of the descriptions were obtained from *Micro-Analytical Entomology or Food Sanitation Control*, by O.L. Kurtz and K.L. Harris, Association of Official Analytical Chemists, 1960, Washington, D.C.

BACKGROUND
Many types of insects feed on food items. Mites, aphids, and thrips are looked for because they are small and can be retained easily in the folds of leafy vegetables. One way to assess whether good manufacturing practices are being followed by an industry is to examine these foods. If the small insects are either gone or mostly gone, then the crop was either not infested or the industry used good washing techniques. You can examine canned items, but students have more success with fresh items; broccoli and brussels sprouts are the best, with spinach and collards next.

Mites
"Many consumers have had the experience of returning flour to the grocer because it was heavily infested with mites. These mites are small (less than 1 mm) and usually whitish and occur in a wide variety of foods. They have come to be known under such common names as 'cheese mites', 'flour mites', 'the bulb mite', and the 'dried fruit mite'. They also are found in smoked and dried meats and fish.

Dermatitis among handlers of mite-infested foodstuffs is not an uncommon occurrence. Some foods frequently involved are copra, cheese, figs, prunes, and wheat pollards. Such terms as copra itch or cheese itch typically are applied to the respective dermatitis."

According to Terbush, "Though mites comprise but one order, Acarina, of arthropods within the class Archnida, they are one of the largest groups of animals in terms of species and individuals. G.W. Wharton has estimated that 600,000-1,000,000 species exist, a figure comparable to that for the entire class Insecta. Mites occupy almost every type of ecological habitat imaginable from polar ice and snow to hot springs; from relatively arid soil to fresh and marine waters; from vegetation to the mantle cavities of clams and the lungs of vertebrates. Ecological studies demonstrate the existence of mites that are predaceous, parasitic, phytophagus, fungivorous, and saprophagus.

"We are probably all familiar with the spider mites and many of the other economically important plant feeders. Likewise, we are all familiar with ticks and chiggers, though not everyone is aware that ticks are essentially large mites, and that chiggers are simply the larvae of a particular mite family.

"Great numbers of mites live in the feathers of birds, the hair of mammals, and under the scales of reptiles. The follicle mite, occurs on a variety of mammalian hosts, including man. It can be found in the pores of the face of a large majority of humans (about 70 to 80%), but rarely is the source of reported irritation."

"The body is stout and white or off white. Their antenna are longer than thrips, their heads are more triangular compared to a more square head for thrips. In stored grains they cause great economic loss, due mainly to the growth

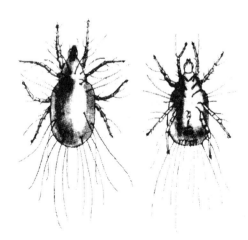

Figure E50-1. Grain or flour mite.

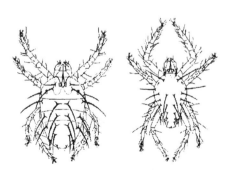

Figure E50-2. Spider mites.
Courtesy - Kurtz, O.L., and Harris, K.L., *Micro-Analytical Entomology*, A.O.A.C., Intl., Washington, D.C.,)

of molds caused by the change in moisture content. The grain mite or flour mite (*Acarus siro*) Figure E50-1, p. 681 is quite common. Spider mites, Figure E50-2, p. 681, vary in color from yellowish or greenish to redish, are pear shaped with the narrow portion posteriorly and have a pair of eyes located on each side of the head."

Aphids

Aphids are small, soft bodied, pear-shaped insects either with or without wings, with oval flat abdomens, and inconspicuous heads. They infest raw agricultural fruits and vegetables, particularly leafy vegetables. They are so small that they seldom break apart in unprocessed foods, but can fragment in comminuted products. Compared to mites and thrips, these are black and easily spotted. Aphids and some parts are shown in Figures E50-3 through E50-7.

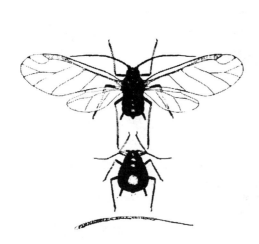

Figure E50-3. Aphids.

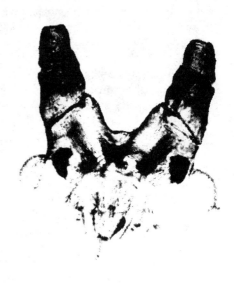

Figure E50-4. Aphid head, rostrum missing.

Figure E50-5. Aphid leg.

Figure E50-6. Aphid femur.

Figure E50-7. Aphid tarsus.

Thrips

Figures E50-8 through E50-12, p. 683, show thrips and various parts. Thrips are from 0.5 to 5 mm long, nearly transparent with dark, usually red, eyes. Their antenna are short, usually only 6 to 10 segments long. Many thrips are plant feeders, attacking flowers, leaves, fruit, twigs, or buds and destroying the plant cells by feeding on them. Some act as vectors for plant disease. They are easily carried into a food processing plant and incorporated into the finished product. When looking for thrips, look for the eyes first, then look for a nearly transparent body attached to them. Figure E50-10 is an excellent photograph. The arrow in Figure E50-9 points to the mouth cone of a thrip. This is usually found intact. The wings are long, slender, membranous, and fringed with close-set hairs and have are few veins. The antennae point forward, are close together and have from 6 to 9 segments. This is different from aphids and easily seen under a microscope.

Table E50-1 lists several defect action levels for some of the samples used in this experiment. If the product contains more than the level allowed, the product is taken from the market.

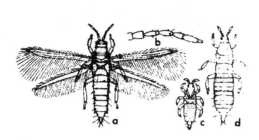

Figure E50-8. Thrips tabaci. a, Adult. b, Antenna. c, Nymph, ventral view. d, Nymph, dorsal view.

Figure E50-9. Thrip head.

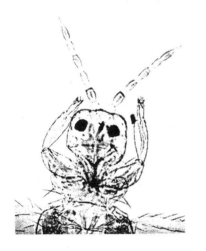

Figure E50-10. A thrip head.

Figure E50-11. Thrip fore wing.

Figure E50-12. Thrip antenna.

Table E50-1. Defect action levels

Product	Defect Action Level
Broccoli, frozen	Average of 60 aphids, thrips, and/or mites per 100 grams.
Brussels sprouts, frozen	Average is more than 30 aphids and/or thrips per 100 grams.
Spinach, canned or frozen	Average of 45 aphids, thrips and/or mites per 100 grams; or 2 or more 3 mm larvae and/or larval fragments of spinach worms (caterpillars) whose aggregate length exceeds 12 mm in 24 pounds; or an average of 8 per 100 grams leaf miners of any size or, an average of 4 per 100 grams of leaf miners 3 mm or longer; or average of 10% leaves by count or weight showing areas of at least 1/2" diameter affected by mildew or other type of decomposition.

EQUIPMENT

- 1 Aspirator, water
- 1 Balance, triple beam or top loader
- 1 Beaker, 100 mL
- 1 Beaker, plastic, 100 mL
- 3 Bottles, wash, 1 water, 1 ethanol, 1 soap
- 1 Brush, test tube
- 1 Cylinder, graduated, 100 mL

Experiment 50

1	Dish, Petrie, both top and bottom	1 pkg	Paper, filter, 5.5 cm, preferably ruled
1	Flask, Erlenmeyer, 1000 mL	1	Probe, dissecting, wooden handled
1	Flask, filtering 1000 mL	1	Ring, iron, 10 cm
1	Funnel, Buchner, 5.5 cm	1	Rod, stirring, 6 in.
1	Funnel, powder	1	Spatula
1 pr	Goggles, safety	1	Stand, ring
1	Hotplate	1	Stirrer, magnetic
1	Knives, chopping	1	Trap, Wildman
1	Microscope, low power 10 or 20x	4 ft	Tubing, vacuum
1	Microscope, high power 80 to 400x	1 pr	Tweezers
1	Opener, can		

CHEMICALS   S = Student

Ethanol, 95%, 245 mL/S
Glycerine, 1:1 with water, 5 mL/S
Hydrochloric acid, HCl, 2% (v/v), 1000 mL/S
Liquid household detergent (2 + 450 mL H_2O), 20 to 30 mL/S
Mineral oil, paraffin oil NF white light, 35 mL/S
Can or box, frozen or raw, of the least expensive spinach, broccoli, or brussels sprouts you can find, 100 g/S

YOUR PROBLEM

A local fruit and vegetable grower wants to have some idea of the insect contamination of their fresh produce. You have volunteered to check some of the spinach, broccoli, or brussels sprouts for mites, aphids, and thrips.

APPARATUS SETUP.

The apparatus is set up as shown in Figures 34-23 and 34-24, 409.

THE BIG PICTURE

100 g of a well-chopped vegetable is placed in a 1 L Erlenmeyer flask. 600 mL of 2% HCl and 35 mL of mineral oil are added. The entire mixture is boiled for 10 minutes. After cooling, a Wildman trap then is placed in the flask, the trap is set in place, and the oil layer vacuum filtered. The filter paper is examined under a microscope.

DIRECTIONS

1. Chop frozen or canned, drained leaves into pieces 3 to 5 cm long, and weigh 100 g of the well-mixed sample into a 1-liter Erlenmeyer flask.

2. Add about 600 mL of 2% HCl (v/v) and a magnetic stirring bar.

3. Mix about 10 minutes with a magnetic stirrer at a medium speed or until the frozen spinach leaves turn dark and are completely thawed.

4. Add about 35 mL of mineral oil. Place on a hot plate and heat to boiling for 10 minutes. While you are waiting, go immediately to step 5.

5. Go to the microscope and examine the examples of mites and aphids. Get help from your instructor if you need it. Adult aphids are black. Other stages of aphids and mites are almost colorless but have very pronounced red eyes which you look for first (Figure E50-10, p. 683). Because they are nearly transparent, you have to search carefully for them.

6. Remove the flask from the hot plate and let the contents cool for a few minutes.

7. Moisten the rubber end of a Wildman trap rod and insert it into the flask. *Caution*: Hold the flask firmly to avoid getting hot oil on you (Figure 34-24, p. 409).

8. Fill the flask to the narrow part of the neck with 2% HCl. Let stand for 10 minutes, stirring gently with the rod several times during the first 5 minutes of the settling period.

9. Secure the rod (Figure 34-24, p. 409) above the settled debris. Check with the instructor to make sure you have it correct before you tip the flask.

10. Place a piece of filter paper (lined type preferred) in a small Buchner funnel and moisten it with water. Attach a water aspirator and turn it on full force.

11. Pour off the oil onto the filter paper. Use the wash bottles present and wash the neck of the flask with ethanol.

12. Lower the trap rod. Add 25 mL of mineral oil, stir for a few minutes, and repeat the trapping and alcohol rinse.

13. Remove the filter paper, moisten it with glycerine-H_2O (1:1), and examine the residue with a microscope. Visit your neighbors and see what they found.

EXPERIMENT 51.   OSMOSIS
Determining the osmotic pressure of a potato
(Courtesy - Edvotek, West Bethesda, MD, Kit 281)

BACKGROUND
The following material is abstracted from Edvotek product literature by permission. "The water potential of a system, ψ, is the chemical potential of water at specified conditions of temperature, pressure, and volume. The chemical potential is the amount of free energy, or energy available to do work, per mole of water. The chemical potential is a measure of how the energy of a phase changes with the number of moles of a component while other components, such as temperature and pressure, are held constant. Water moves from a region of higher water potential to a region of lower water potential. At constant atmospheric pressure, the water potential, the water potential can be expressed as:

$$\psi = \psi_p \qquad [E51\text{-}1]$$

"ψ_p is the osmotic pressure. The osmotic pressure is the change in presure caused by the diffusion of water due to solute concentration, i.e. osmosis. The osmotic pressure for a dilute, non-ionizing solute is approximated by the equation of van't Hoff:

$$\psi_p = -cRT \qquad [E51\text{-}2]$$

where c = the molar concentration of the solute. R = the gas constant, 0.0821 L atm/K mol. T = K (°C + 273).

"The addition of solute to pure water lowers the concentration of water, which lowers the osmotic pressure (makes it more negative).

"If potato cells are placed in pure water, there will be a net influx of water into the cells, since ψ cell is lower due to the cytoplasmic solutes. The cell will swell and gain mass. The water flow stops when ψ cell equals ψ water. If the potato cells are placed in sucrose solutions where ψ cell is higher than ψ solution, there will be a flow of water out of the cells. The cells will shrink and lose mass. At some intermediate concentration of sucrose, ψ cell and ψ solution will be equivalent and no net change in cellular mass or volume will occur. At this concentration of sucrose, the water potential of the cell can be determined by calculating ψ_p."

YOUR PROBLEM
A discussion has occurred involving the osmotic pressure of cells. One person is adamant that all cells have the same osmotic pressure under the same conditions. Others disagree. It is decided to test this by measuring the osmotic pressure of several varieties of potato. Each person is to test one type of potato.

EQUIPMENT

1	Balance, top loading	1	Paper, weighing
1	Beaker, 1,000 mL	1	Paper, graph
8	Beakers, 250 mL	1	Rod, glass stirring, 20 cm
1	Cylinder, graduated, 100 mL	1	Sponge
1	Cylinder, graduated, 50 mL	1	Thermometer
1 pr	Goggles, safety	1 pr	Tongs, crucible
1	Knife, long blade	4	Towels, paper

CHEMICALS   S = Student
Sucrose, $C_{12}H_{22}O_{11}$, 342 g/S
Potatoes, several types

Experiment 51

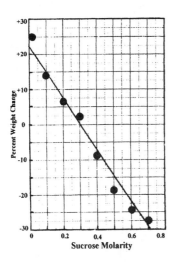

Figure E51-1. A plot of weight change vs. sucrose molarity.

THE BIG PICTURE

Eight solutions of various sugar concentrations are prepared. A 5-10 g section of a potato is weighed and placed in each solution and let stand for 24 hours. The sections then are damp dried and reweighed. The percent gain or loss in weight is plotted against the concentration, and the concentration at 0 % gain or loss is used to calculate the osmotic pressure.

DIRECTIONS

1. Add 342 g of sucrose to a 1,000 mL beaker, dissolve it in water, and dilute to 1,000 mL (1 M).

    You will do this in molarity. However, what would be the molality of this solution assuming 1 mL = 1 g?

2. Set out eight 250 mL beakers in a row.

3. Prepare each as follows:
    a. 100 mL distilled water (DW) (0.0 M)
    b. 90 mL distilled water + 10 mL 1 M sucrose (0.1 M)
    c. 80 mL distilled water + 20 mL 1 M sucrose (0.2 M)
    d. 70 mL distilled water + 30 mL 1 M sucrose (0.3 M)
    e. 60 mL distilled water + 40 mL 1 M sucrose (0.4 M)
    f. 50 mL distilled water + 50 mL 1 M sucrose (0.5 M)
    g. 40 mL distilled water + 60 mL 1 M sucrose (0.6 M)
    h. 30 mL distilled water + 70 mL 1 M sucrose (0.7 M)

4. Place a thermometer in one of the beakers.

5. With a long blade knife, slice eight sections, 2 to 3 mm thick, from your potato.

6. Place each section on a paper towel and dab away the surface moisture.

7. Weigh each piece, record its weight, and place it into one of the beakers. Let stand overnight.

8. Use a pair of tongs and remove each piece of potato from its solution, dab it dry on a paper towel, weigh it, and record the weight.

9. Measure and record the temperature of the solutions.

10. Calculate the percent gain or loss in weight.

11. Plot the sucrose concentration on the x-axis and the percent gain on the y-axis and draw the best straight line through the points.

12. Determine the sucrose concentration where the is 0% change and use equation E51-2, p. 685, to calculate the osmotic pressure.

13. Compare your results with those of others in the class and make a decision if there is a difference in potatoes.

14. *Clean out the beakers containing the sugar solutions. The solutions can be poured down a drain. Clean up the area so it is ready for the next person.* Thank you.

EXAMPLE CALCULATION

The data in Table E51-1 were obtained from eight slices of an Idaho Russet potato tested as above. What is the osmotic pressure of the cells in this variety of potato? $T = 22$ °C.

Table E51-1. Weight of potato slices in sucrose

Molarity	Initial weight	Final weight	Molarity	Initial weight	Final weight
0.0	6.51	8.14	0.4	8.01	7.31
0.1	7.09	8.07	0.5	5.36	4.35
0.2	7.05	7.49	0.6	7.01	5.35
0.3	6.50	6.65	0.7	8.91	6.52

ANSWER

A plot of the data is shown in Figure E51-1. This shows that c of 0 change is at 0.30 M.
$\psi_p = -0.30 \times 0.0821 \times (273 + 22) = 7.25$ atmospheres.

Because of the wide range of values, it is preferred to calculate the equation for the line rather than to estimate it. Table E51-2 is a summary of the data needed to do this for the above case. X = the sucrose molarity, Y = the % weight change.

$$a = \frac{\sum (x'_n y'_n)}{\sum (x'_n)^2}; \quad x'_n = x_n - \overline{x_n}; \quad y'_n = y_n - \overline{y_n}; \quad b = \overline{y} - a\overline{x}$$

The equation for a straight line is $Y = aX + b$, where a is the slope and b is the Y intercept.

Table 51-2. Data needed to determine a straight line

X	Y	x'	y'	(x')²	(y')²	x'y'
0.0	25.00	-0.35	28.84	0.1225	831.7	-10.09
0.1	13.80	-0.25	17.64	0.0625	311.2	-4.41
0.2	6.24	-0.15	10.08	0.0225	101.6	-1.51
0.3	2.30	-0.05	6.14	0.0025	37.70	-0.30
0.4	-8.70	+0.05	-4.86	0.0025	23.62	-0.24
0.5	-18.84	+0.15	-15.0	0.0225	225.0	-2.25
0.6	-23.70	+0.25	-19.86	0.0625	394.4	-4.96
0.7	-26.82	+0.35	-22.98	0.1225	528.0	-8.04
2.8	-30.72	0.00	0.00	0.00	2453.2	-31.80
x = 0.35	y = -3.84					

$a = -31.80/+0.420 = -75.7$
$b = -3.84 - (-75.7)(0.35) = 22.66$

$Y = aX + b;$ $Y = (-75.7)X + 22.66;$ If $Y = 0$, then $X = 0.299$ molar

EXPERIMENT 52. DIALYSIS

The determination of phosphatase in milk to determine proper pasteurization
(*J. Assoc. Off. Anal Chem.* **48**, 811, 1965)
Note: This experiment takes about 5 hours

BACKGROUND

Phosphatase is an enzyme that occurs naturally in milk. When the milk is pasteurized, this enzyme's activity is destroyed. If the milk is properly pasteurized, no phosphatase should be left. Therefore, a measure of residual phosphatase is a check on the quality of pasteurization.

The method described below is based on the following principles. The test milk sample is mixed with disodiumphenyl phosphate and a carbonate buffer. This solution is placed in a dialysis bag, warmed to 37 °C, and held there for 1 hour. Any phosphatase present will react with the phenylphosphate to liberate phenol, which will dialyze through the bag. The phenol then is reacted with copper sulfate and 2,6-dichloroquinonechloramine (CQC) to produce a blue solution, which is measured with a spectrophotometer at 650 nm. The dialysis step separates the phenol from the proteins, which interfere with the color development.

THE BIG PICTURE

One sample of pasteurized milk, one containing a trace of raw milk, and a blank are run. A total of 15 mL (buffer, sample, raw milk) is placed in a dialysis bag, rinsed. and placed in a test tube containing 10 mL of copper sulfate. It is incubated for 1 hour at 37 °C. CQC is added to the diaslyate, and the color developed compared to a set of standards at 650 nm.

YOUR PROBLEM

A batch of milk was being pasteurized and a power failure occurred. The workers assumed that it had been heated long enough to be safe, because there was a safety factor in the heating time. However, the supervisor wasn't so sure, and he has asked you to sample the milk and determine if it has been heated properly.

EQUIPMENT

1	Bath, water, 37 °C	1 bx	Kimwipes
1	Beaker, 250 mL	1	Pipet, delivery, 5 mL
1	Beaker, 1,000 mL	1	Pipet, delivery, 10 mL
1	Bottle, wash, 250 mL	1	Pipet, Mohr, 1 mL
1	Brush, test tube	1	Pipet, Mohr, 10 mL
1	Bulb, pipet	2	Racks, test tube
2	Cells, spectrophotometer, 1 cm, Pyrex	1 pr	Scissors
1	Cylinder, graduated, 50 mL	1	Spectrophotometer, visible range
1	Flask, Erlenmeyer, 125 mL	1	Sponge
1	Flask, volumetric, 1,000 mL	1 m	String
1	Funnel, separatory, 60 mL	1	Thermometer, 110 °C
1 pr	Gloves	12	Tubes, test, 25 mm x 150 mm
1 pr	Goggles, safety	1 m	Tubing, dialysis - Visking no. 27 DC 0.001 thickness, 21 mm diameter.
1	Hot plate		

CHEMICALS   S = Student

n-Butanol, H_3C-CH_2-CH_2-CH_2OH, 10 mL/S
Copper sulfate, $CuSO_4$, 200 mg/S
2,6-Dichloroquinonechloramine (Sigma D6511), 50 mg/S
Disodiumphenylphosphate - keep refrigerated (Sigma P7751), 1.1 g/S
Ethanol, H_3C-CH_2OH, absolute, 10 mL/S
Milk, pasteurized, 5 mL
Milk, raw, 2-3 mL is all. Get from a milk company or a Dairy Bar. This requires a letter from the instructor
  explaining what is to be done, because it is illegal to sell raw milk.
Phenol, C_6H_5-OH, 1 g/S

Sodium bicarbonate, NaHCO$_3$, 10.2 g/S
Sodium carbonate, Na$_2$CO$_3$, 11.5 g/S

Solutions to be Prepared

A. 2,6-Dichloroquinonechloramine (CQC):
  Dissolve 50 mg CQC in 10 mL absolute ethanol and store in a dark brown bottle in a freezing compartment.

B. Carbonate buffer substrate:
  Place 11.5 g Na$_2$CO$_3$ and 10.2 g NaHCO$_3$ in a 1 L beaker and dilute to about 950 mL with water. Dissolve 1.1 g of disodiumphenyl phosphate in 10 mL of this solution in a separatory funnel. Add 4 drops of CQC reagent, swirl, and let stand 30 minutes. Extract the color with two 5 mL portions of n-butanol, letting the layers separate. Add the colorless aqueous layer quantitatively to the remainder of the buffer. Heat this solution to 85 °C for 2 minutes, cool, and transfer to a 1000 mL volumetric flask. This is stable for 1 year, if it is withdrawn with minimum exposure to the air. Develop the color and re-extract before use, if necessary.

C. Carbonate buffer solution:
  Dissolve 11.5 g sodium carbonate, 10.2 g sodium bicarbonate, and 0.1 g CuSO$_4$ in water and dilute to 1 L.

D. Copper sulfate solution:
  Dissolve 100 mg of CuSO$_4$·5H$_2$O and dilute to 1 L.

E. Dilute phenol solution:
  Dilute 2 mL phenol stock solution (1.000 g/L) to 500 mL with carbonate buffer solution. 1.0 mL = 4 µg phenol.

F. Dialyzing tubing:
  Cut three 20 cm lengths of tubing and immerse in water for not < 30 seconds. Remove, and wrinkle one end with your fingers. Twist and tie this end tightly into a leak-proof knot and cut the excess cellulose beyond the knot with scissors. Immerse the bag in water until ready to use. During use, take precautions against phenol or enzyme contamination from contact with fingers or phenol-containing substances, for example, plastics.

DIRECTIONS

1. Check the temperature of the water bath. It should be 37 ± 2 °C. If not, make an adjustment. Check the water level. It should be within 2 cm from the top. Adjust it as necessary.

2. You will need some raw milk. Either get some from your instructor or go to a dairy and get a few mL. You will need a letter from the instructor. Take a 125 mL Erlenmeyer flask with you.

3. Perform a reagent blank determination on the buffer substrate. Add 2 drops of CQC to 10 mL of the buffer substrate and incubate for 5 minutes at room tempersture. If the substrate turns blue, the reagents must be repurified.

4. Find a roll of dialysis tubing and cut off three pieces 18 to 20 cm long.

5. Place them in the water bath and let them soak for at least 30 seconds.

6. Place one end of the tubing between your fingers and rub your fingers back and forth to open the tubing. Once it starts to come open, put it back in the bath and rub it a little more to open it up completely.

7. Flush water through the tubing by holding it on one end and pulling it back an forth through the water a few times.

8. Tie a knot in one end of the tubing close to the end. Pull it tight, but don't punch a hole in it with a sharp fingernail.

9. Do this to all three tubes.

10. Place a 250 mL beaker nearby.

11. Place three 25 x 150 test tubes in a rack and pipet 10 mL of the copper sulfate solution into each one.

12. Pipet 15 mL of the carbonate buffer substrate solution into the bag and set it upright in the beaker.

13. Use 18 to 20 cm of string and tie off the top of the bag. The top of the beaker is about the right height. Tie it tight so it won't leak and leave some free space in the bag. Cut off some of the excess string, but not all of it, because you will use it for a handle later on.

14. Rinse the outside of the bag with distilled water.

15. Place this bag (Blank) in a 25 x 150 mm test tube containing 10 mL of copper sulfate solution and set the tube in a test tube rack. The bag does not go in easily. Tap the bottom of the test tube in the palm of your hand to shake the bag down.

16. Pipet 10 mL of carbonate buffer solution and 5.0 mL of pasteurized milk into another dialysis bag prepared as before. Tie it off, rinse it off, and invert it several times to get thorough mixing, then place it in the test tube rack. *If, during filling, milk inadvertantly gets on the outside of the bag and is not rinsed away, turbidity appears in the copper sulfate solution and the determination must be repeated.*

17. To the third bag, add 10 mL of carbonate buffer, 5.0 mL of pasteurized milk, and 0.1 mL of raw milk. Tie it off, rinse it, mix it and place it in a test tube and put the tube in the rack.

18. Place the test tube rack containing the tubes in the water bath and note the time. It should dialyze for 1 hour.

19. Place nine 25 x 150 mm test tubes in a test tube rack and place a small label on the top of each one to indicate the concentration of phenol present.

20. Prepare standards as follows:

Carbonate buffer soln mL	Dilute Phenol soln mL	Phenol conc µg/10 mL	Carbonate buffer soln mL	Dilute Phenol soln mL	Phenol conc µg/10 mL
0.0	10.0	40	9.0	1.0	4
2.5	7.5	30	9.5	0.5	2
6.0	4.0	16	9.75	0.25	1
7.0	3.0	12	10.0 (blank)	0.0	0
8.0	2.0	8			

Place a total of 10 mL in 25 x 150 mm test tubes, add 2 drops CQC to each, and swirl to mix. Add 0.1 to 0.2 mL of the carbonate buffer to the 40 µg standard. This has been found to be necessary or no blue color will form at this concentration. Incubate for 5 minutes.

21. Take this set of test tubes, a small beaker, and a wash bottle to a spectrophotometer. Read each solution at 650 nm within 30 minutes.

22. The directions that follow are for a Bausch and Lomb Model 2000. Your spectrophotometer will likely be different, but read these directions to get a feel for the sequence of events, then go to step 39.

23. As you face the instrument, a key pad with an LED (light emitting diode) display is on the right side and the cell compartment is on the left.

24. Open the cell compartment door and notice two holders, one for a single reference cell and a multiple sample cell holder. Push the lever on the front of the chamber and watch how the cell holder turns.

25. Find at least two Pyrex cells. Place a cell in the holder and move it into the beam. Make sure the clear side is in the beam, not the frosted side. Remember how it is placed in the holder.

26. On top of the instrument is a roll of heat-sensitive paper to print out the results. Paper must be in there or the instrument will not work. Ask an assistant for help, if it is not there.

27. Turn on the power. The switch is on the back side at the upper right corner and is pushed up.

28. Fifteen seconds after the power is on, the instrument will beep five times if it is functioning properly. The LED display will give a value of approximately 695 nm in red numbers with an absorbance value of about 0.4. To change the wavelength, use the white buttons and enter the desired value, in this case 650.

29. Press the "Go to Lambda" button, a brown button on the left side. After a few seconds, the LED should indicate that the instrument is at 650 nm.

30. The instrument is fitted with two lamps, a tungsten lamp that goes from 800 nm to 355 nm and a deuterium lamp that will go from 355 to 200 nm. Both are on at the start. Turn off the deuterium lamp by pressing the button labeled "Deut".

31. If you want to take your data in percent transmittance, press the button labeled "%T", or if you want absorbance (A), press the button "ABS". Chemists prefer *A*, because it is linear with concentration. You will use absorbance, so press the ABS button.

32. Fill a Pyrex cell within 3 to 4 mm of the top with the blank solution. This is the carbonate buffer solution. Tilt the cell about 45° and fill it carefully, so that you do not get any air bubbles in it. If an air bubble is trapped, tap the cell until it floats to the top.

33. Wipe the sides dry and fingerprint-free with a Kimwipe.

34. Use a few mL of the *most dilute* phenol standard to rinse out another Pyrex cell. If you have a series of solutions to measure, it is good technique to measure the lowest concentration first, so that if a small drop remains in the cell, it is overcome by the addition of the more concentrated solution.

35. Fill this cell, wipe it clean, and insert it into position 1, so that the clear side will be in the beam.

36. Close the door, and the LED readout will give you the absorbance value for this standard. The instrument automatically subtracts the blank. Record this value.

37. Rinse out the cell with a few mL of the next higher concentration standard and repeat steps 35 and 36.

38. Repeat steps 35 and 36 for each of the standards.

39. Plot absorbance vs. μg of phenol/10 mL.

40. Remove the samples from the water bath after 1 hour and remove the bags from the test tubes. They can be trashed.

41. Add 2 drops of the CQC reagent to the clear solution in each test tube and swirl each one gently.

42. Place the test tubes back in the rack and the rack back in the water bath and let the color develop for about 5 minutes.

43. Measure the absorbance of each of these solutions. Use this blank solution for the blank.

44. Refer to your calibration curve and determine how many μg of phenol would be in your sample. You may have noticed that when you removed the bags from the test tubes, you did not need to rinse them nor did you dilute to a known volume. This is such a sensitive test that actual quantitative results are not needed in court. If you get a blue color, the sample is violative - period.

45. Be sure to turn off the spectrophotometer, rinse out the cells with distilled water, and place them back in their box.

46. Clean out all of the test tubes.

47. Put the CQC and the substrate back in the refrigerator.

48. The raw milk must be thrown away. You can either drink the extra pasteurized milk or throw it away.

49. *Do not* shut off the water bath.

50. *Look at the area - make it cleaner and neater than you found it.* Thank you.

EXPERIMENT 53. AIR FILTRATION
Separating pollen and molds from air.

BACKGROUND

An *allergy is an exaggerated or pathological reaction (as by sneezing, respiratory embarrassment, itching, or skin rashes) to substances, situations, or physical states that are without comparable effect on the average individual.* The substance causing this condition is an *allergen*. It is estimated that one of 10 people suffer an allergy of some type, and that and additional six of 10 will suffer some type of allergy during their lifetime. Many molds and plant pollens cause human allergies; 300 ragweed pollen grains/m^3/day will cause severe allergic reactions. The average person breathes about 10 m^3 of air/day.

A mold is a plant of the fungi group. It has no chlorophyll, so it cannot produce its own food and must live on decaying organic matter or on living organisms. Figure E53-1, p. 692, shows several as they look under a microscope. These usually cause more problems in the summer. They are on the surface of the ground and when it rains, the rain drops throw up the dust and molds, and the mold spores get into the air. The "fresh smell of rain" is actually the smell of molds. Two major molds are *Alternaria* and *Cladosporium,* and their numbers are reported. All the rest are usually called "other molds".

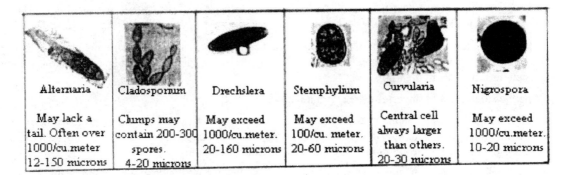

Figure E53-1. Photomicrographs of Several Molds.
(Courtesy - Smith, E.G., *Sampling and Identifying Allergenic Pollens and Molds*, Blewstone Press, San Antonio, TX)

Pollen is a mass of microspores in a seed plant appearing usually as a fine dust. Figure E53-2, p. 693, shows several trees grains of pollen plus ragweed. Tree pollen occurs in the spring, grass pollen in the late summer, and weed pollen in the summer and early fall. Those plants that are pollinated by insects do not produce as much pollen as those that depend upon the wind. Corn is wind pollinated, and one plant may produce as many as 50,000,000 pollen grains.

For a reason not yet understood, large protein and carbohydrate molecules in the pollen react with the cells they come in contact with and cause an antigen-antibody reaction, which then causes *histamine* to be released into the blood stream. Histamine in the cell is harmless, but in the blood stream, it causes tissues to swell. One ppm is harmful. The histamine molecule is so small that the body determines that it is not worth taking to the liver and detoxifying nor is an antigen-antibody reaction called for, but it can simply be washed away. This is why people's eyes water and their noses run. Thus, allergy medicines are called antihistamines.

YOUR PROBLEM

A local TV station has contacted you about doing pollen and mold counts as a service to their viewers. You know very little about it but would like to see what can be done. You do not have an air sampling apparatus, so you have two choices: one, you can assemble one using a sampler like that shown in , Figure 37-2 , 432, and add a pump to it (Figure 37-3, p. 432), which you can calibrate (Figure 37-5, p. 433), or you can buy a commercial apparatus. The TV station decides to buy you a commercial apparatus, and they buy the Allergenco Air Sampler shown in Figure E53-3, p. 693, and opened up in Figure E53-4, p. 693.

EQUIPMENT

1 Sampler, air
1 Slide, microscope
1 Glass, cover
1 Q-tip
1 Microscope

CHEMICALS   S = Student

Calberla's stain, modified; 1 mL/S
 5 mL glycerine, 10 mL 95 % ethanol plus 2 drops of saturated aqueous basic fuchsin. The normal stain is too runny, so glycerine is added.
Silicone grease; Hexsilicon is preferred because it has no trapped air bubbles, which may look like pollen grains under a microscope.

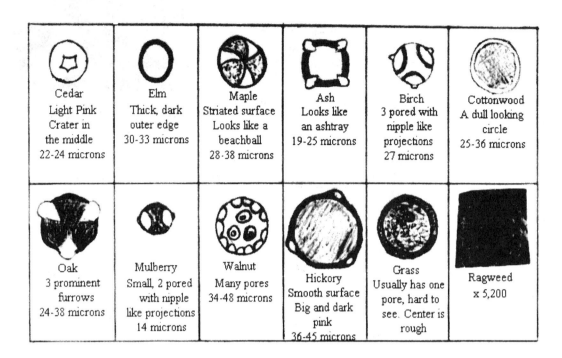

Figure E53-2. Drawings of common pollen grains.
(Courtesy - Topeka Allergy Center, Topeka, KS)

THE BIG PICTURE

A microscope slide is covered with a very thin film of grease and placed into the apparatus. The timer is set. After collection, the slide is removed, stained with 2 drops of Calberla's stain, and allowed to set for 1 hour. It then is examined with a microscope.

APPARATUS SETUP

The apparatus is shown in Figure E53-3 and with the top open in Figure E53-4.

Figure E53-3. Air Sampler model MK-3.
(Courtesy - Allergen LLC, San Antonio, TX)

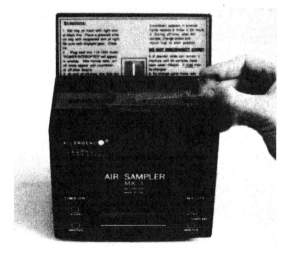

Figure E53-4. MK-3 Air Sampler opened.
(Courtesy - AllergenLLC, San Antonio, TX)

DIRECTIONS

Either assemble your own collector or use a commercial self-contained unit. Place it 6 to 8 feet above ground and as clear of overhanging trees or structures as possible. However, all outside air samplers must be protected from rain, so it should be placed under some type of small cover.

Figure E53-5. Eight collections on a slide. (Courtesy - Topeka Allergy Center, Topeka, KS)

1. Obtain a microscope slide, a Q-Tips, and a jar of silicone grease.
2. Cover about 3/4 of the slide with a very thin coat of grease, applied with a Q-Tips.
3. Look at the sampler. The small slot in the top is where the air sample enters. The slide is placed below it. The pollen will stick along a narrow strip. For 8 collections in 24 hours, the strips are 0.040 inches wide by 14,400 microns long and 2 mm from center to center. Figure E53-5 shows a photograph of a slide with 8 collections.
4. Lift open the top cover, and you will see a brass slide holder. This moves on a track and is removable.
5. Lift up the holder and move it all of the way to the right and adjust it until you feel the gears mesh.
6. Carefully place the slide in the holder, slipping the right edge below the lips on the holder. Close the lid firmly.
7. Plug in the sampler.
8. The words "POWER INTERRUPTED" will immediately appear in the LCD display window, and 1 minute later, 00.10 will appear in the left and 00.19 at the right side of the display window. You have 19 minutes to make any setting changes.
9. The Sampler is set to sample for 10 minutes at the factory. Leave this alone.
10. To obtain 8 samples/day, set the time-off for 2 hrs 50 min. If you start sampling at 9:00 A.M., the first sample will be at 11:50 and finished at 12:00. You can change slides any time in between without affecting the cycle in the least.
11. After 1 day, sometime between cycles, remove the slide. If it does not have a light end as the one shown so that end can be identified, make a mark on the right end with a marker. *The first collection is the one farthest from the right.*
12. Place 3 drops of Calberla's stain across the center and two smaller ones above and below. Carefully place a cover slip over the liquid, one edge first, to squeeze the air out and prevent air bubbles from being trapped.
13. Go to lunch. The stain needs at least 1 hour to do an adequate job. Pollens are readily stained; molds only slightly.
14. Obtain the book by E. Grant Smith, *Sampling and Identifying Allergenic Pollens and Molds*: An Illustrated Manual for Air Samplers (1990), Blewstone Press, Box 8571, San Antonio, TX 78209. Until then, refer to Figures E53-1, p. 692, and E53-2, p. 693, to help you identify pollens and molds.
15. Refer to the example calculation in the air sampling section, p. 433, and calculate the pollen grains/m^3/day. Use the velocity data supplied with your apparatus.

EXPERIMENT 54. DENSITY GRADIENTS FOR SOILS
Comparing soils at a hit-and-run accident

BACKGROUND

Soil is a common form of physical evidence found at the scene of such crimes as hit-and-run accidents, automobile collisions, rapes, and burglaries. Soil from the crime scene may be picked up by an automobile, thus providing a valuable link between the car and the crime. Similarly, soil or mud found adhering to clothing or shoes may provide the clue that can link a suspect to a particular crime site. Figure 38-1, p. 452, shows a density gradient column for determining a soil profile, made by mixing various proportions of xylene ($d = 0.88$) with bromoform ($d = 2.89$).

For our experiment, a soil sample is dried at 100 °C for 1 hour and then sieved; a small amount of the 30 to 45 mesh fraction is placed on the top of the column. Heavy particles will settle to their level in a few minutes, but light particles may take a few hours to stop moving. (*Note:* The size of a particle does not change its density, only how fast it will settle. Why?) The soil profile at the scene is then compared with the soil profile from the suspect.

YOUR PROBLEM

A woman has been beaten and sexually assaulted in a park while walking back to her apartment from work. The crime took place at approximately 11:00 p.m. The investigating officers have obtained a description of the assailant and placed into custody three young men answering the description. The assault took place in an area where the ground was damp, and the officers investigating the scene noticed that the soil was compressed in one area, as if the assailant's knee had gouged into a soft place in the earth. The victim has picked one of the three men out of the lineup, stating that he was her attacker. The officers assigned to the case have searched the accused man's apartment and found a pair of trousers with soil particles lodged in the fabric of the knee area. The laboratory analyst's task is to attempt to determine whether the soil in the trouser fabric and the soil from the compressed area at the scene do indeed have a common origin. You are the analyst. Remember, one can get dirt on the knees of a pair of trousers in many ways. You also may determine that the accused is innocent. This is only one piece of evidence and not conclusive in itself, even if positive.

EQUIPMENT

1	Beaker (graduated), 100 mL	2	Pipets, 5 mL
20	Bottles, sample, 2 oz, for known and unknown soil samples	1	Rack, test tube
		1	Ruler, metric
1	Brush, test tube, small diameter	1	Stand, ring
4	Corks, small	1	Sieves, set, ranging in size from 25 to 80 mesh
7	Droppers, medicine		
1	Funnel, glass, 35 mm	1	Spatula
1 pr	Goggles, safety	3	Tubes, glass, 30 cm x 10 mm (closed at one end)
2	Holders, buret		
1	Mortar and pestle	7	Tubes, test, 10 cm
1	Oven, drying	7	Vials, 4-dram
1	Pencil, grease		

CHEMICALS   S = Student

Soil samples of differing composition (gravel, peat, loam, fine sand)
Bromoform, $CHBr_3$, 15 mL/S
Xylene, C_8H_{10}, 15 mL/S

DIRECTIONS

The first step in soil analysis is to obtain a sample composed of uniformly sized particles. This is done by sieving the gross sample obtained from the crime scene or from the possessions of the suspect.

1. Dry the samples, if they are not already dry.

2. Powder the samples in a mortar and pestle. If you have a small sample, then sieving is not necessary or practical. If the sample is large enough to sieve, then sieve it as follows.

3. Arrange the set of sieves in numerical order with the smallest number at the top and the largest at the bottom (the smaller the number, the larger the mesh of the wire screen in the sieve).

4. Place the sieves on the shaker (if available), and pour your sample into the top sieve.

5. Place the cover on the upper sieve, fasten the binding straps, and turn on the power switch of the shaker. (If a power shaker is not available, the shaking of the sieves can be done manually.) Allow the shaker to operate for 5 minutes. The time is not crucial; only the proper separation of particles is important.

6. Turn off the shaker and release the binding straps.

7. Remove the sieve cover, and separate the sieves. The only portion of sample to be used in the analysis is one taken from a sieve with a screen mesh in the middle range (45 mesh works well). The other portions should be recombined and saved for possible future needs. Be certain to label them.

This analysis depends upon the principle that an object will be suspended in a liquid whose density matches its own; it will sink in a liquid that is less dense, and it will float in a liquid that is more dense.

8. Prepare the tube in the following manner. Obtain bromoform and xylene and make mixtures of the two liquids in the proportions listed below (3 mL of each solution minimum):

   a. Pure bromoform (density 2.89)
   b. 0.5 mL xylene, 2.5 mL bromoform
   c. 1.0 mL xylene, 2.0 mL bromoform
   d. 1.5 mL xylene, 1.5 mL bromoform
   e. 2.0 mL xylene, 1.0 mL bromoform
   f. 2.5 mL xylene, 0.5 mL bromoform
   g. Pure xylene (density (0.88)

9. Mark off seven equal increments on three 30 cm glass tubes. A tube that has a solid glass seal on one end is preferred. However, the end can be closed with a cork, but the *cork must be either taped or waxed in so it won't fall out*.

10. Carefully pour each of the solutions into each tube in the order listed in step 8. Begin with the bromoform the most dense liquid.

11. Close the top of the column with a small cork to prevent evaporation and place the tube in a buret holder.

12. Repeat the above, filling a second and third tube, and place them in a holder also.

13. Place a small amount (0.1 g) of the suspect sample soil into two of the tubes and an equal amount of soil found at the scene into the third tube. Be certain to label each tube. Most particles will reach their depth within a few minutes, but allow the tubes to stand for 24 hours for minor adjustments, and then compare the levels at which the soil particles have become suspended. If they are the same, the scene and suspect soils have the same density distribution. If they are found at different levels, the soils probably are not the same.

Note: if you get a definite layer or layers of particles after 24 hours at the boundaries where you added the different solutions, then the column has not reached density equilibrium and should be allowed to set for a while longer.

EXPERIMENT 55. CENTRIFUGATION
The separation of fat from milk and ice cream (Based on *A.O.A.C.* method 989.04)

BACKGROUND
Traditionally, milk fat has been the most valued and most variable component of milk. The market value of milk depends upon its fat content. Prior to 1890, no rapid method was available to estimate the volume of fat in milk. Thus, milk was sold by volume only, and adulteration was a problem. It was easy to add water to the milk to increase the volume, and if heavily adulterated, to add calcium carbonate to increase the whiteness.

Milk Fat
Triacylglycerols are dominant and constitute about 98% of milk fat, together with small amounts of di- and monoacylglycerols and free fatty acids. Small quantities of phospholipids, cholesterol, and cholesterol esters are also present as well as the fat-soluble vitamins A, D, and E. Lipid molecules in milk associate to form large spherical globules, which are surrounded by a phospholipid layer, the globule membrane, from the proteins in the milk. This membrane stabilizes the hydrophobic lipid in the aqueous phase of the milk. The emulsion must be broken and the protein film removed before the fat can be separated and determined volumetrically.

Ice Cream
Standards of identity of ice cream can be found in the Code of Federal Regulations (21 CFR 135). Ice cream is a frozen mixture of the components of milk, sweeteners, stabilizers, emulsifiers, and flavoring. In the U.S., ice cream is defined by federal standards as containing not less than 10% milk fat and 20% total milk solids, weighing not less than 4.5 lb/gal, containing not less than 1.6 lb of food solids/gal, and containing not more than 0.5% stabilizer.

Milk fat helps to give body to ice cream, produces a smooth texture, and increases the richness of flavor. *Texture* is the attribute of a substance relating to its finer structure - the size, shape, and arrangement of small particles. *Body* is its consistency and firmness, and in the case of ice cream, its melting resistance. Higher amounts of milk fat in ice cream enhances both the texture as well as the body. Insufficient amounts result in a coarse or icy texture.

Water ices were first used by the Chinese who made them from snow and the juices of lemons, oranges, and pomegranates. Marco Polo brought back the recipe to Italy, and either he changed it by adding milk to make sherberts or someone else did.

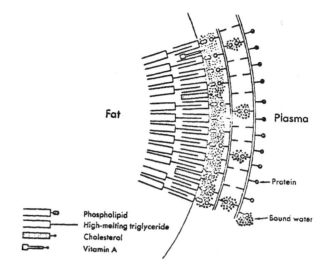

Figure E55-1. Diagram of a fat-water interface in milk.
(Courtesy - Charley, H., *Food Science*, Ronald Press, New York, NY, 1970)

The hand-cranked freezer with a dasher (paddle) was invented by Nancy Johnson of the U.S. in 1846. This made it possible for everyone to make ice cream. The dasher beat air into it which gave it a better texture and then acted as an insulator, so the ice cream did not appear to be as cold as it was.

The first ice cream factory in the U.S. was established in 1851 in Baltimore, MD, by Jacob Fussel. The ice cream soda was introduced in 1879 at the Centennial Exposition in Philadelphia. The ice cream cone was introduced at the Louisiana Purchase Exposition in St. Louis in 1904. A quart of ice cream provides about 1170 calories, and a quart of ice milk about 820 calories. The record amount of ice cream eaten by one person is 7 lb, 13 oz in 16 minutes by Archie Leggett, 22 yrs old, of Hamilton, Scotland, on February 9, 1973.

Classifications Based on Fat Content

Figure E55-1 is a diagram of the fat-water interface in milk.

Reduced fat: Ice cream made with 25% less fat than the reference ice cream (10% fat).
Light or Lite: Ice cream made with 50% less fat or 1/3 fewer calories than the reference ice cream, provided that in the case of the caloric reduction, less than 50% of the calories are derived from fat.
Lowfat   Ice cream containing not more than 3 g of milk fat per serving.
Nonfat   Ice cream containing less than 0.5 g of fat per serving.

The Babcock Method for Determining Fat in Milk and Milk Products

Dr. S.M. Babcock (1843-1931), an agricultural chemist at the University of Wisconsin, developed a method to determine fat in milk in 1890. When sulfuric acid is added, nearly all components except the fat are destroyed, and the heat produced melts the fat, which then is separated by centrifugation. If a measured volume of milk and acid is placed in a special bottle, the fat can be measured directly in the calibrated portion of the neck. Sulfuric acid is much heavier than milk (specific gravity difference 1.84 *vs.* 0.93) and will combine with the serum (nonfat phase of the milk) to have an overall specific gravity of about 1.43. This great difference in specific gravity will cause the fat to physically separate. Centrifugation is used to enhance the separation phase. Water is added to raise the fat into the neck of the bottle, where it can be measured volumetrically. Fat then can be calculated as a percent.

In ice cream, the presence of stabilizers require prolonged acid digestion, yet the concentration of the acid must be reduced to avoid excessive charring of the sugar-rich food.

Figure E55-2. Dr. Stephen Moulton Babcock.
(Courtesy - Department of Biochemistry, University of Wisconsin)

YOUR PROBLEM

Workers in the local dairy are noticing a problem with some batches of milk and the ice cream made from them. The fat content appears to be low, indicating an adulteration. You are given a sample of milk, chocolate milk, and two different brands of ice cream and are to determine the fat content in each.

EQUIPMENT

1	Balance, top loading		1	Cylinder, graduated, 25 mL
1	Bath, water (60 °C)		1	Dropper, medicine
2	Beakers, 100 mL preferred		1 pr	Gloves
1	Bottle, 17.5 mL marked (for acid)		1 pr	Goggles, safety
2	Bottles, Babcock, ice cream		1	Heater, hot water
2	Bottles, Babcock, milk		1	Pipet, milk, 17.6 mL
1	Brush, for Babcock bottles		1	Pipet, Mohr, 10 mL
1	Bulb, pipet		1	Sampling tube or dipper
1 pr	Calipers		1	Sponge
1	Centrifuge, 50 mL bucket rotor		1	Thermometer
1	Cloth, cheese, ~25 x 25 cm		1	Timer

CHEMICALS   S = Student

Ammonium hydroxide, NH_4OH, Conc., 6 mL/S
1-Butanol, 11 mL/S
Glymol (mineral oil), 1 mL/S
Sulfuric acid, H_2SO_4, conc., 20 mL/S (milk)
Sulfuric acid, Sp. Gr. 1.78 - 1.80. Add 1000 mL H_2SO_4 to 160 mL of water, 20 mL/S (chocolate milk)
Sulfuric acid, (3.5 : 1 v/v) 40 mL/S (ice cream)
Samples: milk, chocolate milk, and any two kinds of ice cream.
Soap, antimicrobial preferred, but not required.

THE BIG PICTURE

The milk components are destroyed with H_2SO_4 and the fat is separated by centrifugation. Chocolate milk and ice cream must be treated first with NH_4OH and 1-butanol to destroy any mucilage added as a thickener. A constant temperature water bath is used to bring all fat readings to a reproducible reference state.

APPARATUS SETUP

A centrifuge with 50 mL buckets capable of holding Babcock bottles is required. A hot water bath is needed and set at 60 °C as is some sort of hot water supply. This may be a bottle of water setting in the water bath or a separately heated supply.

DIRECTIONS

White Homogenized Milk

This is a volumetric method. A Babcock pipet will deliver 17.6 mL of milk to the Babcock bottle. The specific gravity of milk is 1.032. Thus, 1.032 x 17.6 will generate 18.16 g of milk in the bottle.

1. Ensure that a water bath with water at 55 to 60 °C (130 to 140 °F) is available and that it has a support rack for Babcock bottles to sit at the right height, which is just above the fat level to keep it melted.

2. Pipet 17.6 mL of milk (59 to 68 °F) into a Babcock bottle.

3. Fill the acid measuring bottle to 17.5 mL with concentrated sulfuric acid or place 17.5 mL of acid in a 25 mL graduated cylinder

4. Add about 4 mL of H_2SO_4 down the side of the neck of the bottle to wash any traces of milk into the bottle and shake well. Mix the milk and acid thoroughly. The solution will start to turn brown.

5. Repeat three more times for a total of 17.5 mL. Use vigorous shaking as needed to ensure that all traces of curd have disappeared. The solution will be dark brown. The bottle gets quite hot, so you may want to wear gloves.

6. Immediately transfer the bottle to a centrifuge, counterbalance it, and after the proper speed has been obtained, (Table E55-1, p. 699) centrifuge for 5 minutes. Do not let the contents cool, or the fat will solidify and not separate well.

Table E55-1. Recommended RPM's /rotor diameter
(Courtesy - M.C. Matt, A Manual for Dairy Testing, Kimble Glass Co., Vineland, NJ)

Rotor Diameter, inches	RPM's	Rotor Diameter, inches	RPM's
14	909	20	759
16	848	22	724
18	800	24	693

The distance is measured from the bottoms of the extended bottles.

7. Add hot distilled water (140 °F or above) until the bottle is filled to the neck. A bottle filled with water and placed in the water bath will suffice.

8. Immediately centrifuge for 2 minutes.

9. Add hot water until the fat is brought well within the scale on the bottle neck.

10. Centrifuge for 1 minute.

11. Transfer the bottle to a water bath (130 to 140 °F). Immerse it to the level of the top of the fat column for 5 minutes.

12. Remove the bottle from the bath, wipe it, and with the aid of a pair of calipers, measure the fat column. Measure from D to A, not C to B. See Figures E57-3 and E57-4.

13. Discard the contents to a waste container, wash out the bottle with soap and water, and rinse it well before the next test.

When the test is complete, five things must occur to ensure that the test is done properly.
1. The milk fat should be a yellow color.
2. The ends of the yellow color should be distinctive.
3. The fat needs to be free from flecks and sediment.
4. The water below the fat column should be clear.
5. All fat needs to be in the column of the bottle.
Duplicates should agree to within 0.1% of one another.

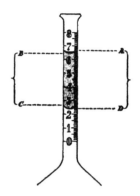

Figure E57-3. Reading the results of the test on milk.
(Courtesy - Matt, M.C., *A Manual for Dairy Testing,* Kimble Glass Co., Vineland, NJ)

The most common problems with a Babcock test are the first two. If the fat is not yellow or has flecks in it, the fat was not extracted properly. This occurs when the acid is too strong, the acid is poured directly into the milk, the acid and milk are not mixed properly, or the acid or milk is too warm. If the fat appears to be curdy in the bottle, then the acid was too weak, an insufficient amount of acid was added, the milk or acid was too cold, or the curd was not dissolved before centrifuging.

Chocolate Milk

The addition of mucilaginous gums as thickening and suspending agents has made it necessary to modify the existing methods. This is a modified Pennsylvania Method for fat in chocolate milk and chocolate drinks. The results include both milk fat and cocoa butter fat.

1. Stir the product carefully to make it uniform.

2. Pipet 17.6 mL into a Babcock bottle.

3. Add 1.5 mL of ammonium hydroxide (a buret is preferred, but a Mohr pipet is acceptable.). *Careful*: Concentrated NH_4OH is hazardous. Do in a hood if possible, or else very quickly.

4. Mix for about 30 seconds.

5. Add 5 mL of *n*-butanol and mix for about 1 minute.

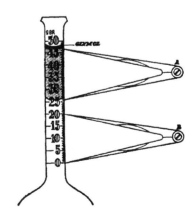

Figure E57-4. Reading the results of the test of fat in chocolate milk.
(Courtesy - Matt, M.C., *A Manual for Dairy Testing,* Kimble Glass Co., Vineland, NJ)

6. Add 17.5 mL of diluted H_2SO_4 in three divided portions, mixing the acid and chocolate milk by a rotary motion after each dilution until digestion is completed.

7. Immediately centrifuge for 5 minutes.

8. Add hot water (130 to 140 °F) with a pipet to bring the contents to 1/4th inch of the base of the neck of the bottle.

9. Immediately centrifuge for 2 minutes.

10. Add enough hot water to keep the fat within the graduated portion of the bottle until read.

11. Centrifuge for 1 minute.

12. Place the bottle in a water bath (130 to 140 °F) for 5 minutes.

13. Tilt the Babcock bottle slightly and add a few drops of Glymol to the top of the graduated portion so it slowly slides down the neck to the top of the fat column. Glymol is a diluted mineral oil with a red dye in it.

14. Measure the length of the fat column from the bottom of the lower meniscus to the sharp line between the glymol and the fat. See Figure E57-4 "A", p. 699. Then lower the calipers to "B" and determine the % fat.

15. Discard the contents to a waste container, wash out the bottle with soap and water, and rinse it well before the next test.

Ice Cream

This is the Pennsylvania modification of the Babcock method for measuring fat in ice cream. The Paley type Babcock test bottle is used. The bottle is so constructed that the graduated reading tube is placed on one side in order to provide an opening on the other side of the bottle that is sufficiently large to permit lumpy, thick, or viscous materials to be deposited without difficulty. The required amount of material is added directly into the bottle and the opening is closed with a rubber stopper. The calibrated reading tube is left open for the addition of the reagents and water.

1. Place several spoonfuls of ice cream into a 100 mL beaker and let it melt to room temperature. If necessary, warm it to eliminate the foam.

2. If nuts or fruit are present, filter the product through cheesecloth into a 100 mL beaker.

3. Use a 20 % bottle. Add 9 g of melted ice cream and 5 mL of water to the bottle.

4. Add 2 mL of NH_4OH and 3 mL of 1-butanol. Swirl to mix..

5. Add 17.6 mL of diluted sulfuric acid (3.5 : 1 v/v) in four parts to the bottle. Gently shake after each addition. The solution will turn black and get quite hot. Wear gloves if you want to.

6. Immediately centrifuge for 5 minutes.

7. Add hot water (130 to 140 °F) to bring the contents to the bottom of the neck.

8. Immediately centrifuge for 2 minutes.

9. Add hot water (130 to 140 °F) with a pipet to bring the fat well within the graduated scale.

10. Centrifuge for 1 minute.

11. Place the bottle in a water bath (130 to140 °F) for 5 minutes.

12. Add a few drops of Glymol. Read as shown in Figure E57-4, p. 699.

13. Discard the contents to a waste container, wash out the bottle with soap and water, and rinse it well before the next test.

EXPERIMENT 56. MASKING AND SEQUESTERING AGENTS

The determination of Zn accelerators in rubber products
(Courtesy - Hunter, T.L., *Anal. Chem.*, **37**, 1436, 1965)

BACKGROUND

When rubber products are made many inorganic ions get into the system. Natural rubber latex, which is a secretion from a tree, will contain traces of metals. Much larger concentrations are added during the curing process as accelerator or plasticizers and to keep the rubber from sticking to the molds. Many steps in the manufacture of a tire or other rubber products involve inorganic ions.

ZnO commonly is added to rubber stock to aid in the curing process. How fast a tire will cure depends upon how much zinc accelerator is present. A bicycle tire, for example, which is pretty thin rubber, needs a different amount of Zn than an earth mover tire. A fast method of analysis is necessary for quality control. The American Society for Testing Materials (ASTM) standard method (D 95-61T sec 40 F, 1964) involves the titration of Zn with potassium ferrocyanide using uranyl acetate as the external indicator. Roberts method (*Trans Inst. Rubber Ind.*, **33**, 97, 1957) used EDTA at pH = 10 and eriochrome black T as an indicator to get Ca, Mg, and Zn. Then another titration was used to determine Ca and Mg after masking the Zn with CN.

Both require a hydroxide precipitation of Fe, Al, and Ti, which is time consuming. The former method gives high results, and the second takes two titrations. Here, use is made of the fact that F^- will react to form stronger bonds than EDTA with all of the metals except Zn and large amounts of Fe.

YOUR PROBLEM

Pick any rubber product that interests you and determine the total inorganic content and the amount of zinc accelerator in it.

EQUIPMENT

1	Asbestos square, 10 x 10 cm or equivalent		1 pr	Gloves, temperature resistant
1	Balance, analytical		1	Holder, buret
3	Beakers, 250 mL		1	Hot plate
1	Bottle, wash		1	Pipet, delivery, 2 mL
1	Brush, buret		2	Pipets, delivery, 3 mL
1	Bulb, pipet		1	Pipet, delivery, 5 mL
1	Buret, 10 mL preferred		3	Pipets, delivery, 10 mL
1	Casserole, porcelain, size 1		1	Plate, titration
1	Cylinder, graduated, 50 mL		1	Policeman
1	Dish, weighing, Al		1	Razor blade, single edge
3	Flasks, Erlenmeyer, 125 mL		1	Rod, stirring
1	Flask, volumetric, 50 mL		1	Stand, ring
1	Furnace, muffle, capable of 550 °C		1	Tongs, long handle
1 pr	Glasses, safety		1	Towel, cloth

CHEMICALS   S = Student

Acetic acid, H_3CCOOH, 1 M
Ammonium acetate, $H_3CCOONH_4$, 1 M   A 1:1 mixture of these two serve as the buffer solution, 20 mL/S
Aluminum chloride, $AlCl_3$, 0.1 M, 2 mL/S
Ammonium fluoride, NH_4F, 3 M, 6 mL/S
Ammonium hydroxide, NH_4OH, conc
Dithizone, $C_{13}H_{12}N_4S$ 0.1% in acetone, 1 mL/S
Ethanol, H_3CCH_2OH, 95%, 130 mL/S
Ethylenediaminetetraacetic acid (EDTA) 0.01 M, 50 mL/S
Hydrochloric acid, HCl, conc., 10 mL/S
Methyl orange indicator, 0.01% in water, 1 mL/S
Nitric acid, HNO_3, conc., 3 mL/S
2,4-pentanedione, $H_3C-CO-CH_2-CO-CH_3$, 10% in ethanol v/v, 10 mL/S
Sulfuric acid, H_2SO_4, conc., 3 mL/S

THE BIG PICTURE

A 0.5 g sample of rubber (bring your own, any type) is placed in a casserole and decomposed with HNO_3 and H_2SO_4 and then ashed at 550 °C. The Zn is taken up in HCl and diluted to 50 mL. Three 10 mL aliquots are placed into three flasks. 50 mL of ethanol, 10 mL of buffer, and 6 drops of dithizone are added to the first flask. 5 mL of 10 % 2,4-pentanedione is added to the second and third flask. Add one drop of methyl orange and neutralize with conc NH_4OH. Add 10 mL of buffer, 40 mL of ethanol and 6 drops of dithizone.

All three of these solutions are titrated with 0.01 M EDTA. The first flask will show the total Zn, Al, Fe, Ti, Sb, Si, Ca, and Mg, which are generally found in rubber products. (Surprised that it isn't all rubber?) The second flask - using the F^- ion as an inorganic masking agent - will remove all but Zn and extra large quantities of iron. The third flask uses an organic chelating agent to mask the extra large quantities of iron, if they are present.

APPARATUS SETUP

A 10 mL buret on a ring stand, a hot plate in a hood, and a muffle furnace set at 550 °C are needed.

DIRECTIONS

1. Check the muffle furnace to be sure it is 550 °C *and no hotter*. If not, adjust it. Turn on a hot plate in the hood
2. With a razor blade, cut a piece of rubber of about 1 to 2 g from your sample.
3. Weigh a 0.5 g sample, but not more than 0.6 g.
4. Transfer this sample to a casserole and add 3 mL of concentrated H_2SO_4 and 3 mL of concentrated HNO_3.

    Note: Many types of rubber contain Cl^-, and $ZnCl_2$ is volatile at the muffle furnace temperature. Nitric acid is added to partially decompose the rubber so the Cl^- leaches out. The weaker HCl and HNO_3 acids are displaced by the strong acid H_2SO_4.

5. Place the casserole containing the sample on a hot plate in a hood and leave it there until it goes to dryness.
6. Using a pair of long-handled tongs and wearing high temperature gloves, place the casserole in a muffle furnace at 550 °C. Leave it there for 20 minutes.
7. If the rubber has not all burned away, leave it for an additional 10 minutes, but no longer.
8. Carefully remove the casserole and place it on an asbestos mat or similar type material.
9. Add about 5 to 10 mL of concentrated HCl in 2 to 3 mL portions to the casserole to convert the $ZnSO_4$ to $ZnCl_2$, which is more soluble, and transfer the contents to a 250 mL beaker. Use a policeman.
10. Wipe down the sides of the casserole to make a quantitative transfer to the beaker.
11. Evaporate this solution down to 4 to 5 mL on the hot plate.

    Note: All of the metals present are now chlorides. Any silicon was dehydrated by the HCl to SiO_2 and will settle to the bottom of the beaker. The solution may have some color, depending upon the type of rubber. It usually will have a few white or gray particles floating on it, which are sulfur particles and possibly a few black specks, which are carbon particles that didn't get oxidized. It is not necessary to filter this.

12. Remove the beaker from the hot plate (*Careful*) and cool the solution with 5 to 10 mL of water from a wash bottle. Turn off the hot plate.
13. Transfer the contents to a 50 mL volumetric flask. Use a wash bottle to wash down the sides of the beaker to make a quantitative transfer, then dilute to volume with distilled-deionized water.
14. Place a *clean* 10 mL buret in a buret holder, fill it and zero it with 0.01 N EDTA solution.
15. Pipet 10.0 mL of the sample into a 125 mL Erlenmeyer flask. (This is a 1/5 aliquot for your calculations). No masking or sequestering agents are to be added. This is to determine the total metals present.
16. Add 40 to 50 mL of 95% ethanol using a graduated cylinder. Add 10 mL of buffer and 6 drops of dithizone indicator. When you shake the solution, it should turn pink to red.
17. Place a titrating plate or something light colored under the flask and titrate to a green end point. Only 3 or 4 mL will be needed, so be careful.
18. Pipet 10.0 mL of sample into a second and third 125 mL Erlenmeyer flask.
19. Pipet 10 mL of the 3 M NH_4F solution into each flask. CARE: Wash your hands immediately after you use this solution. In the presence of strong acids, HF can form and can cause painful burns. This should not happen here, but be careful.
20. Add 2 mL of the $AlCl_3$ solution.
21. To flask three only, add 5 mL of 2,4-pentanedione.
22. Add one drop of methyl orange indicator to each flask, and the solutions should turn orange.
23. With a medicine dropper, add concentrated NH_4OH (hood) until the solution turns red.
24. Add 10 mL of buffer and 40 to 50 mL of 95% ethanol to each flask.

25. Add 6 drops of dithizone indicator to flask two and titrate it to a green end point. The fluoride acts as a masking agent to complex iron, titanium, silicon, calcium, and magnesium.

26. Repeat step 25 with flask three. This is done in case extra iron is present.

27. Clean the buret and place all of the reagents in an orderly fashion near the ring stand. Clean up the area. Thank you.

28. Do the calculations at home.

EXAMPLE CALCULATION
A 0.5908 g sample of a black rubber stopper took 2.27 mL of 0.01 N EDTA to reach a green end point. When NH_4F was added it required, 1.10 mL of EDTA and when the 2,4-pentanedione was added, it required 1.04 mL of EDTA. What are the % of total metal and the amount of other metals? Was excess iron present? What is the best value for Zn? Remember - you took a 1/5 aliquot.

ANSWER
The a.w. of Zn is 65.4. 1 mL of 0.01 M EDTA = 0.000654 g of Zn.

$$\% \text{ Metals} = \frac{(2.27 \text{ mL} \times 0.000654 \text{ g/mL})(100)(5)}{0.5908 \text{ g}} = 1.25\%$$

$$\% \text{ Zn} + \text{Fe} = \frac{(1.10 \text{ mL} \times 0.000654 \text{ g/mL})(100)(5)}{0.5908 \text{ g}} = 0.608\%$$

$$\% \text{ Zn} = \frac{(1.04 \text{ mL} \times 0.000654 \text{ g/mL})(100)(5)}{0.5908 \text{ g}} = 0.58\%$$

EXPERIMENT 57. SOLUBILITY OF PLASTIC MATERIALS
Separation and identification of thermoplastic polymers.
(Based on an article by Cloutier, H., and Prud'homme, R.E., *J. Chem. Ed.*, **62** (9), 815, 1985)

BACKGROUND
Recycling plastic is now a popular issue. In 1989, 58×10^9 lb of resin were produced in the United States. Of this, it is estimated that 14×10^9 lb were used for packaging and 29×10^9 lb became solid waste. At that time, only 340 to 400×10^6 lb were recovered.

A system of code numbers has been recommended by the Institute of Plastic Manufacturers to help identify the various common plastics. These are shown in Table 41-13, p. 486.

YOUR PROBLEM
Many plastics are not marked as to type, either because the manufacturer chooses not to or because the piece is too small or the marking would be unsightly. You are given such a piece and are to tentatively identify it by a series of solubility tests.

EQUIPMENT
1	Balance, ± 0.01g		1	Marker, pencil type
1	Blade, razor, single edge		1 bx	Matches
1	Brush, test tube		1	Rack, test tube
1	Burner, Bunsen		1	Sponge
1	Clamp, buret		1	Stand, ring
1	Cylinder, graduated, 10 mL		1	Towel, cloth
1 pr	Goggles, safety		20	Tubes, test, 18 x 150 mm
1	Holder, test tube		3 ft	Tubing, burner

CHEMICALS   S = Student

The amounts listed assume every test as shown in Figure 41-6, p. 487, is done on a single plastic.

Acetic acid, $H_3C-COOH$, 10 mL/S
Amyl acetate, $H_3C-COOC_5H_{12}$, 10 mL/S
Carbon tetrachloride, CCl_4, 30 mL/S
Chloroform, $CHCl_3$, 10 mL/S
Cresol, $HO-C_6H_4-CH_3$, 10 mL/S
Cyclohexanone, $C_6H_5=O$, 10 mL/S
Dimethylformamide, $HCOON(CH_3)_2$ 20 mL/S
Ethyl acetate, $H_3C-COOC_2H_5$, 10 mL/S
Methanol, H_3COH, 10 mL/S
Methylene chloride, CH_2Cl_2, 10 mL/S
Toluene, $C_6H_5-CH_3$, 10 mL/S
Xylene, $C_6H_4-(CH_3)_2$, 10 mL/S

THE BIG PICTURE

0.5 g sample is heated with 10 mL of solvent in a test tube for 5 minutes. If it does not produce a homogeneous and transparent solution, it is considered insoluble. The chart in Figure 41-6, p. 487, is followed.

APPARATUS SETUP

A test tube rack with 20 test tubes, a Bunsen burner, and a ring stand with a clamp to hold test tubes are needed.

DIRECTIONS

1. Sketch a chart like that in Figure 41-6 and label each solvent in sequence either 1, 2, 3, etc., or A, B, C, etc.

2. Arrange a test tube in a holder on a ring stand so its bottom will be above the flame of a burner.

3. Use a marker to label a test tube for each solvent, and place them in a test tube rack in a reasonable order.

4. If needed, cut your sample plastic into several small pieces, each of which will fit into an 18 x 150 mm test tube.

5. Weigh approximately a 0.5 g sample and place it in the first test tube.

6. Place a cloth towel where you can grab it immediately if necessary. This is your fire extinguisher.

7. Preferably in a hood, add 10 mL of toluene to the sample and heat it gently until the solvent boils. Adjust the height of the test tube so it gently boils. Continue the heating for about 5 minutes. If the plastic does not form a homogeneous and transparent solution, it is considered insoluble. *Careful*: if the solvent should catch on fire, do not panic. Simply cover the end of the test tube with a cloth towel to smother the flame, then continue on. Record your results. Make a list of either what it cannot be or what it may be.

8. If the sample was soluble, then weigh a fresh sample and repeat step 6 with the next solvent in the scheme. Continue as necessary, recording your observations at each step.

9. If the sample was insoluble, it is usually easier to use a fresh sample rather than to rinse away the previous solvent. Repeat step 6 with the next solvent.

10. Make a final list of what you eliminated and what you believe this plastic could be.

11. Place all chlorinated solvents in the chlorinated solvent waste container and the other solvents in the non-chlorinated waste container.

12. *Clean up the area leaving it better than it was when you began.* Thank you.

EXPERIMENT 58.   SOLUBILITY OF FIBERS

Identification of textile fibers by solubility measurements
Note: This is a long experiment if all of the solutions are used.

BACKGROUND

The need for the identification of textile fibers in forensic science is a common one. Many types of crimes, when investigated, yield fibers as physical evidence. Crimes in which fibers may be deposited at the scene include robbery, breaking and entering, homicide, rape, assault, and hit-and-run accidents involving a victim.

In many cases, matching fibers from the scene with those from the clothing of a suspect or a victim is the required analysis. In others, the identification of the particular type of fiber is important. In the first instance, microscopy is the usual technique employed; in the second, chemical testing is necessary. Recorded police files contain numerous cases on in which the conclusion as to guilt or innocence is based primarily on fiber analysis. This exercise is limited to the identification of textile fibers as to general type by solubility measurements.

There is no set procedure for fiber identification. The following is a suggested approach.

1. Use the burning test on fibers, yarns, and fabrics.
2. Use color tests on boiled off-white or striped samples, including the xanthroproteic test.
3. Try fiber microscopy: longitudinal or cross-sectional, or both.
4. Perform the applicable solubility tests. Use acetone first. Time, temperature, and reagent concentration are important, as well as the size of the fiber bunch, yarn, or fabric piece. Use the smallest sample possible in order to speed up the test results.

YOUR PROBLEM

A robbery occurred in which the thief or thieves entered through a skylight. One of the thieves caught his shirt on a wood splinter and several fibers were torn loose. The police have a suspect, but they have no evidence except the fibers. Other scientists are comparing the fibers by microscopy, infrared, x-ray diffraction, dying, and burning tests. Your part is to do solubility tests to see if fibers from the suspect match those at the crime scene.

EQUIPMENT

7	Beakers, 30 mL		1 pr	Goggles, safety
1	Blade, razor, single edge		1	Hot plate
18	Bulbs, pipet, small		18	Pipets, disposable
1	Cylinder, graduated, 10 mL		1	Thermometer, to 110 °C
1 pr	Forceps		1	Towel, cloth
11	Glasses, watch, 10 cm			

CHEMICALS   S = Student

Acetic acid, H_3C-COOH, 100%, 10 mL/S
Acetone, H_3C-C=O-CH_3, 100%, 5 mL/S
Calcium thiocyanate, $Ca(SCN)_2$, 10 mL/S
Chloroform, $CHCl_3$, 5 mL/S
m-Cresol, HO-C_6H_4-CH_3, 10 mL/S
Dimethylformamide, HC=ON$(CH_3)_2$
    52 to 60%, 5 mL/S
Formic acid, HCOOH, 90%, 5 mL/S
Hydrochloric acid, HCl, conc., 5 mL/S

Hydrochloric acid, 1:1, 5 mL/S
Monochlorobenzene, Cl-C_6H_5, 10 mL/S
Phenol, HO-C_6H_5, 90%, 5 mL/S
Sodium hydroxide, NaOH, 5%, 10 mL/S
Sodium hydroxide, 45%, 10 mL/S
Sodium hypochlorite, NaOCl, 5.25%, 5 mL/S
Sulfuric acid, H_2SO_4, 60%, 5 mL/S
Sulfuric acid, 70%, 5 mL/S
Zinc chloride, $ZnCl_2$, 10 mL/S, 67%

THE BIG PICTURE

Suspect and scene fibers are tested with 18 solvents, seven of which are heated on a small hot plate.

APPARATUS SETUP

A hot plate with eight 30 mL beakers and 11 watch glasses in a row off to the side, all in a hood.

DIRECTIONS

Note: Do in a hood.

1. Turn on a hot plate of 20 cm x 20 cm or equivalent area to medium.
2. Arrange 11 watch glasses in a row off to one side and place a solvent bottle behind each one.
3. Arrange the solvents in the following order:
   1. 100% acetone (room temperature): applied to mixed fibers.
      Soluble: acetate, Arnel, Dynel, vinyon, and Verel at 104 °C.

   2. 90% phenol (room temperature).
      Soluble: acetate, Fortisan slowly, Arnel, Dynel, nylon; Dacron in warm phenol; vinyon and polyvinyl chloride (Rhovyl) in boiling phenol; viscose.

   3. 5.25% sodium hypochlorite (room temperature).
      Soluble: protein fibers, hair, silk, wool (Vicra about 50% discontinued).

4. 90% formic acid (room temperature).
   Soluble: acetate, Arnel, nylon at 100 °F, Fortisan.

5. 1:1 hydrochloric acid (room temperature).
   Soluble: nylon.

6. Concentrated hydrochloric acid (room temperature): Fortisan.
   Soluble: pure silk, acetate, Arnel, nylon; cupra and viscose slowly.

7. Concentrated nitric acid (room temperature).
   Soluble: acetate, Acrilan, nylon, Arnel, vinyon, Zefran, Darvan, Creslan, Orlon.

8. 60% sulfuric acid: (room temperature) used mainly to separate viscose (soluble) from cotton.
   Soluble: cupra and viscose rayons, acetate, Arnel, silk and nylon.

9. 70% sulfuric acid (room temperature): Creslan.
   Soluble: cotton, linen, and all those fibers dissolving in 60% H_2SO_4.

10. 52 to 60% dimethylformamide (room temperature).
    Soluble: acetate, Arnel, Dacron, Darvan, nylon, vinyon, Saran, Acrilan, Orlon, Dynel, and Verel; breaks up.

11. Chloroform (room temperature).
    Soluble: Arnel, vinyon, acetate.

4. Place seven 30 mL beakers on the hot plate and set a solvent bottle off to the side by each one in the following order:

    12. 100% acetic acid (boiling): acetate dissolves at room temperature
        Soluble: acetate, Arnel, nylon (slowly).

    13. *m*-Cresol (80 to 99 °C): Arnel, vicose, silk, Orlon, nylon
        Soluble: acetate, Fortisan, vinyon, Dynel lumps, Saran decomposes.

    14. Calcium thiocyanate (hot)
        Soluble: Acrilan, Creslan, Orlon, Darvan.

    15. Monochlorobenzene (boiling)
        Soluble: vinyon, Saran.

    16. 5% sodium hydroxide (boiling)
        Soluble: cultivated silk, hairs, wool, Darvan
        Partly soluble: tussah silk (wild), regenerated proteins, acetate, Arnel.

    17. 45% sodium hydroxide (boiling)
        Soluble: cultivated silk, tussah silk, hairs, wool, Darvan, Arnel, acetate; Dacron in 40 minutes
        Partly soluble: Orlon (Dynel and Saran melt).

    18. 67% zinc chloride (40 to 46 °C)
        Soluble: cotton (not mercerized), viscose, acetate, Arnel, silk, Acrilan, Orlon, Creslan, and cupra.

5. With a single-edge razor blade, cut 2 to 3 mm-long sections of the suspect and scene fibers and place them on a clean sheet of paper out of a draft. Handle them with forceps.

6. With a disposable pipet, place 1 to 2 mL of solvent on the respective watch glasses, but not in the beakers on the hot plates.

7. Place a section of the fiber in each solvent. Observe it carefully, because it may be a mixed fiber and only part will dissolve in a particular solvent. Record your observations.

8. With a disposable pipet, place 8 to 10 mL of solvent in the respective beakers on the hot plate.

9. Drop a fiber section into each beaker and observe what happens. You can check the temperature with a thermometer, if necessary. Generally, these do not dissolve immediately, so watch them for several minutes. Record your observations.

10. Based on solubility alone, reach a conclusion about the nature of the fibers.

11. Turn off the hot plate.

12. Rinse off the glassware and dispose of the pipets. *Leave the area cleaner than when you arrived. Thank you.*

EXPERIMENT 59.   GAS ANALYSIS
Separation of CO_2, O_2, CO, and N_2 from car exhaust gas fumes
This experiment should take less than 1 hour after you have the sample.

BACKGROUND
Read the Techniques section on the Portable Orsat apparatus, p. 494.

YOUR PROBLEM
You are having a discussion with a group of friends, and the topic of air pollution from car exhaust comes up. One member of the group has an eight cylinder car, and he says that if one cylinder is not functioning well, it will not be noticed. You dispute this and say it will be noticed even with a six cylinder engine. To get evidence for your assertion, you go to his car and collect a sample of exhaust gas and determine the CO_2, CO, and O_2 contents. You then remove one spark plug, repeat the analysis, and compare the results.

EQUIPMENT
    1 Apparatus, portable Orsat
    1 Bottle, gas sampling, 200 mL
    1 Cylinder, graduated, 2,000 mL
    1 Engine, gasoline, any kind, (car, motorcycle, etc.)
1 pr Goggles, safety

CHEMICALS   S = Student
Acid cuprous chloride, Cu_2Cl_2, 150 g Cu_2Cl_2/L of 2:1 HCl-H_2O, 200 mL/apparatus
Alkaline pyrogallol, (1,2,3-trihydroxybenzene) 15%; mix 1:1 with 30% KOH, 200 mL/apparatus
Grease, joint, 1 tube/apparatus
Glycerine, small bottle/lab
Indicator, methyl red, 1 mL/apparatus
Potassium hydroxide, KOH, 30% (180 g/420 mL water), 200 mL/apparatus
Sulfuric acid, H_2SO_4, 10 mL/L/ 2 L/apparatus

APPARATUS SETUP
See Figure 42-3, p. 495.

THE BIG PICTURE
A 200 mL sample of car exhaust gas is obtained using a gas sampling bottle - *Be careful*: take the sample fast and don't breathe any more of the exhaust than you have to. One cup of CO inside of a car can eventually produce death, and exhaust fumes are sometimes as high as 8%. Place 100 mL of sample in the gas buret. Remove the CO_2 first by absorbing it in 30% KOH. Remove the O_2 next by absorbing it in alkaline pyrogallol. Remove the CO by absorbing it in acid cuprous chloride. Assume that the rest is N_2 and determine it by difference.

DIRECTIONS
1. Look at the apparatus. Remove the front and back covers. Lift them straight up and set them along side of the lab bench and out of the way. The apparatus should be placed so that the red solution is on your right.

2. Look at the scales on the gas buret and notice that on one side, the scale starts at 0 and goes down to 100, and on the other side the scale starts at 100 and goes up to 0. From the bottom of the gas buret, is a hose leads to a small bottle setting in a metal holder. This is known as a leveling bulb.

3. Go to the absorption bulbs to the left of the gas buret. You may see several different types there. One likely will be filled with glass tubes, which provide a large surface area. When the solution drains away from those tubes, it leaves a thin film on them, so any gas in there will be absorbed very quickly. When you bubble gas into that absorption bulb, it forces the liquid out of the front bulb up into the back bulb. The excess air is forced into the little gas bag on the back. You don't want these solutions to come in contact with the air, because they will absorb the gases from the air and be spent before you get a chance to make a measurement.

4. There's generally a mark right up by the stopcock or just below the connecting tube, which indicates a measuring place. Look at those absorption tubes and see if you can find a little reference mark. If not, then pick some reference point that you can use for adjusting the solution level.

5. Go to the glass capillary across the top with six stopcocks on it. This is called a manifold. On the upper left is a stopcock that goes to the outside of the apparatus. It's a 3-way stopcock. When that stopcock is turned to a horizontal position, you can add a sample. When it is turned vertically you open up the system to the outside air.

6. The first absorption bulb to the left of the gas buret contains alkaline pyrogallol. This is used to remove oxygen. Alkaline pyrogallol can absorb eight times its volume of oxygen. Hang a tag on the pyrogallol bulb, and every time oxygen is absorbed, write down the volume absorbed. That bulb will hold 200 mL of pyrogallol, so when you have absorbed 1600 mL, it's time to replace the pyrogallol.

7. Most of the absorbers for CO involve reduced copper salts, and cuprous chloride generally is used. This can be in acid, which is the case here, or in base. The only difference is in the order that you take out the gases. Cuprous sulfate can be used as an absorber for CO, but it absorbs slowly unless beta naphthol is added to it to act as a catalyst. The capacity of acid cuprous chloride for CO is generally pretty high. One mL will absorb 18 mL of CO gas.

8. For CO_2, you use KOH.

9. Find a gas sampling bulb. This is a bulb about 18 to 20 cm long and about 4 cm in diameter with a stopcock and a short piece of rubber tubing on each end.

10. Get a 2000 mL graduated cylinder and fill it to within 4 cm of the top with the dilute sulfuric acid solution.

    How are you going to get a gas sample into a bulb that's already full of air without mixing it with the air? One way to do this is to put something like water in the bulb that will force the air out. If it's ordinary water, it will absorb the acid gases, such as CO_2. However, if the water is slightly acidic, then the acid gases are not absorbed when the sample is obtained.

11. Turn the stopcocks on the sampling bulb to open both stopcocks.

12. Very carefully stick one end of the sampling bulb into the graduated cylinder and let the acid solution in the graduated cylinder fill up that sampling bulb, then close the top stopcock.

13. Withdraw the sampling bulb and close the bottom stopcock.

14. Go to the exhaust pipe of any type of gasoline or diesel engine. Place the gas sampling bulb into the exhaust pipe as far as you can and then open both stopcocks and let the water drain out. As soon as all of the water has drained out, close the stopcocks on the sample bulb. Take the sample back to the lab.

15. If this sample were to be used for legal purposes, one additional step would be necessary. If you're going to get a sample to be used in a court case, then you've got to take a witness along, someone that can testify on the witness stand how you got this sample. In addition, you have to put a tag on it that tells the date, the time, the circumstances for getting the sample, and how much you got. Then, every time that the sample is opened, a record must be kept on the tag and it should be witnessed. A lot of times, gas samples, particularly flue gases, such as carbon monoxide poisoning in houses are used in court cases. Other cases can involve mine poisoning, or a workman in a sewer getting a little too much H_2S and being overcome. These things do occur. Keep a record on that sample of everything that you do to it.

16. Go to the slow combustion chamber; that's the one with the platinum coil. Close the stopcock.

    To ensure that stopcocks are turned correctly, the ends can be painted. This has been done on this apparatus.

17. At the very left edge of the manifold, there is a stopcock with a yellow dot on it, and then on the stopcock itself, there is a yellow dot at about the 2 o'clock position and a white dot at about the 12 o'clock position. Turn this stopcock to about a 45° angle, so that the yellow dots line up.

18. Check all of the absorption tubes and make sure that all stopcocks are turned off. They will be horizontal.

19. Go to the leveling bulb on the gas buret. This is the bulb that has the red solution in it. Raise that bulb up to the top of the gas apparatus and open stopcock B and fill the gas buret to about 2 cm from the top. Then close the stopcock.

20. Place the leveling bulb down on the bench top. If the red solution in the gas buret lowers, that means that you have a leak in the system. If you have a leak, test each stopcock until you find it.

21. You must set all of the absorbing solutions at a 0 position. Start with stopcock G. These next steps need to be done with a bit of finesse. If you open stopcock G, the water level in the gas buret will lower, which will reduce the pressure in the absorption tube. That will cause the solution in the front bulb to rise. Look on the capillary about 2 cm below the rubber tubing and you will see a hairline mark. You have to adjust the solution to come up to that mark. It's done in this manner. Take the leveling bulb in your right hand and hold it up by the gas buret, at about the same level as the solution in tube L and then very slowly open the stopcock and watch the level of the solution in that front tube. By raising and lowering the level of the leveling bulb, get the solution in that front bulb to come up to that hairline mark. If the gas bag in the back is empty, the solution will not rise. Simply remove the stopper and let some air in the bag and then restopper it. Adjust the solution level to the hairline mark and close the stopcock.

22. Using the same technique, do this with stopcocks F and E. Bring the solution up to where the capillary starts below the stopcock.

23. The next step is to fill the gas buret with the red solution. Take the leveling bulb in your right hand, and using your left hand to control stopcock B, bring the red solution right up underneath the black stopper where the capillary just enters the gas buret; then shut off the system.

24. Put the leveling bulb back on the table top.

25. You are now ready to add a sample. Read this step entirely before you add the sample. On the top left of the apparatus by stopcock B, notice that the manifold extends out through a hole. That's where you are going to connect the sample tube. Take the short piece of rubber tubing that is on the sampling bulb and put a drop or two of water in the tubing, so that it will go over the glass easily and make a tight fit. Put the sampling bulb on the manifold at A, hold it there with one hand and, with your other hand, turn stopcock B to open it to the manifold. Open the stopcock on the sample bottle closest to stopcock B. *Very carefully* open the other stopcock on the sampling bottle. Because the leveling bulb is sitting on the table, the solution in the gas buret is going to lower and draw your sample into the gas buret. Do this rather slowly. If you do this too fast, the air rushing into your sample tube will mix with the sample and dilute it. Watch the red solution in the gas buret. Allow the solution to completely drain out of the gas buret. Take a bit more sample than 100 mL.

26. Shut off the sample bottle stopcocks.

27. Take off the gas sampling bottle.

28. Open stopcock B again. Look at the bottom of the glass buret at the graduations. There will be 0 mL on one side and 100 mL on the other. *Very carefully* raise the leveling bulb until the red solution just comes level with that bottom graduation. If you raise the leveling bulb too high beyond the 100 mL mark, you force sample out into the air; then, when the solution comes back down, it will draw air into the system instead of your exhaust gas. Good technique is very important here. When you get it level, shut off stopcock B.

29. You have now taken a 100 mL sample. It is saturated with water vapor, and it is at atmospheric pressure. This is the standard way to make all volumetric measurements. Go to stopcock E, this is the KOH tube, and open that stopcock.

30. The solution will drop a little. Raise the leveling bulb until the red solution in the gas buret goes up to about 1 cm from the top. Then lower the leveling bulb until the solution comes all the way back down to about 0. Then raise and lower it two more times.

31. You have scrubbed your sample through the KOH solution, and all of the CO should be removed. Now you must get the solution in that KOH scrubber back to where it was originally. Lower the leveling bulb until the KOH solution comes back up to the same place that it was when you first started this experiment. You may have to put the bulb below the level of the table top. As soon as you adjust it, then shut off the stopcock E.

32. You now have to measure how much CO_2 was in the system. Place the leveling bulb right up next to the gas buret, so you can just read the numbers. Then raise or lower the leveling bulb until the solution in the leveling bulb is just exactly level with the solution in the gas buret. The pressure inside the gas buret is back to atmospheric pressure, the same as it was when you started. Read the gas buret and see how many mL of gas remain. The difference between your first reading of 100 and what you have now is the volume % of CO_2.

33. The next gas that is removed will be O_2. This is stopcock G. Use the same procedure. Wash your gas sample back and forth in your pyrogallol solution 3 times.

34. By lowering the leveling bulb below the level of the table top, bring that pyrogallol solution back up to the original mark and shut off stopcock G.

35. Measure how much O_2 was absorbed. Put the leveling bulb next to the gas buret and make sure that they are level and then read the gas buret. Subtract this value from the one that you had previously and determine the amount of O_2 in your sample.

36. Now to get CO. This is bulb K and stopcock F. Some engines don't put out very much CO, and measuring it is very hard with this kind of apparatus. Other engines will put out as much at 10% CO, depending on how new and how efficient the engine is. Whether you get something here or not will depend a lot on the car that you used to get your sample from. Determine the CO.

37. You have removed three gases; CO_2, O_2, and CO. The rest of the gas is mainly nitrogen. There are very few hydrocarbons that are not combusted.

38. *Clean up the apparatus and put the covers back on. Clean up the area around the system and leave it in good shape for the next person.* Thank you.

IF IT EXISTS - CHEMISTRY IS INVOLVED
IF YOU CAN BUY IT - A CHEMIST WAS INVOLVED SOMEWHERE

APPENDIX B
ANSWERS TO THE REVIEW QUESTIONS

CHAPTER 1 - VOLATILIZATION
1. (1) Direct heating, (2) Displacement, (3) Oxidation , (4) Reduction.
2. By displacement by a stronger base such as NaOH.
3. Combust the compound in O_2, convert the H to H_2O, and absorb it on Anhydrone.
4. The sample is digested in acid, then the Hg^{+2} is reduced to Hg by $SnCl_2$. The Hg vapor is separated from the liquid by passing a stream of air or He through it, and the Hg passes into the gas bubbles.
5. No flame is used to treat the sample. The evolved gas is passed through a tube placed in the beam of the AA instrument.
6. The anhydrous cobalt salt is blue, and the hydrated form is pink.
7. $NaHCO_3 + HCl \rightarrow NaCl + H_2O + CO_2$
8. It is probably not simply $2\,HCl + CuSO_4 \rightarrow CuCl_2 + H_2SO_4$. Another possibility is $CuSO_4 \cdot 5H_2O$ (blue) $+ HCl + O_2 \rightarrow CuCl_2 \cdot 3CuO \cdot 4H_2O$ (Brunswick green).
9. Fortunately, very little S occurs in limestone. If it is oxidized to SO_2 it then can react with water to form H_2SO_3 and condense in the Z tube or the first two traps. If it forms H_2S, a weakly acid gas, it is said to be adsorbed on the $CuSO_4$, but if not, then it will be weighed the same as the CO_2.
10. These are believed to condense in the first two traps.
11. A representative sample is one that contains a proportional amount of limestone from each layer. If one layer is three times thicker than the one below it, you should have three times the amount of sample from that layer. Unless you took along a pointed hammer and a rough scale, you probably will not get it as representative as you would like when you collect only a bucket full. However, you will do the best you can.
12. $CO_2 + 2\,NaOH \rightarrow Na_2CO_3 + H_2O$
13. For each molecule of CO_2 (m.w. = 44), you are losing one molecule of H_2O (m.w. = 18), so the loss is $18/44 \times 100$ or 40.9%.
14. No. Fingerprints vary considerably, but most weigh less than 1 mg and require a microbalance to weigh accurately. The U tubes weigh much more than a fingerprint, as does the CO_2 gas. However, it is just good technique to remove any fingerprints, because several fingerprints may be present and some may be oily and pick up dust.
15. If you connect the inlet water at the top, it can run out of the condenser faster than it enters, so keeping the condenser filled is hard. In addition, the cold water is now cooling the hottest vapors, and this large temperature difference may cause the condenser to crack if it is old.
16. Bernoulli's principle. High velocity creates low pressure. A stream of water is forced rapidly past a small opening at a right angle. This high velocity water draws out any water or gas from the side arm creating a vacuum.
17. No, if both tubes are weighed. However, by following the directions, you have a primary system and a backup system, which should be more reliable than just a single system.
18. Any time they are handled, particularly if they are wiped with a towel.
19. It would give an erratic weighing, in that it would go up and down several mg.
20. Ground the tube by holding it against a water pipe or just let it sit for about 20 minutes.

21. The limestone contains Mg. Its m.w. is only 24 compared to 40 for Ca. If the sample is calculated as if it were all $CaCO_3$, the results will be high. Some Kansas limestones can calculate as 108% as $CaCO_3$.

22. $$\%CaCO_3 = \frac{\text{Weight of } CO_2 \times 100/44 \times 100}{\text{Sample weight}} = \frac{0.3365 \text{ g} \times 100/44 \times 100}{0.9565 \text{ g}} = 79.95\%$$

23. "As received" calculations:
$$\% CO_2 = \frac{\text{Weight of } CO_2 \times 100}{\text{Sample weight}} = \frac{0.1577 \text{g} \times 100}{1.1593 \text{ g}} = 13.60\%$$

"Oven dried" calculations:
% moisture = (0.0096/1.1593) x 100 = 0.83 %

$$13.60\% \frac{\% CO_2}{} = \frac{\%\text{"oven dried"} - \%\text{"oven dried"} (0.83\%)}{100}$$

$13.60\% = 1X - 0.0083X$ $X = 13.71\%$

24. 3.5×10^{-52} $[M/L]^2$. This is a very strong bond.
25. The C can reduce the Hg^{+2} to Hg, and it can be lost.
26. Use a soap film flowmeter.
27. A Mohr pipet has graduations down the length of the tube, whereas a delivery pipet has only one graduation at the top.
28. Disconnect the "out" side of the pump and open the vent on the filtering flask *while the pump is still on*.
29. $$\text{ppm} = \frac{0.75 \text{ µg}}{1.4420 \text{ g}} = 0.52$$

30. $\frac{1.4420 \text{ µg}}{1.6732 \text{ µg}} = \frac{0.85 \text{ µg}}{X \text{ µg}}$; $X = 0.99$ µg   1.55 µg - 0.99 µg = 0.56 µg recovered.

$$\frac{0.56 \text{ µg} \times 100}{0.60 \text{ µg}} = 93\% \text{ recovery}$$

CHAPTER 2 - ZONE MELTING

1. To produce very high purity metals or organic compounds or to regulate the level of a desired impurity.
2. In an effort to reach minimum energy, the molecules or ions at points of stress will tend to dissolve faster than those not under stress.
3. Use equation 2-1
   $C = 0.07 \times 0.040 (1 - 0.05)^{0.07-1} = 0.0032$ M;   Try $0.95^{-0.93} = 1/0.95^{+0.93}$.
4. Normal freezing involves one surface, whereas zone melting involves two surfaces.
5. It works because the concentration of impurities in the solid phase is different than that in the liquid phase.
6. (1) Where the concentration is decreasing, (2) where it is constant, and (3) where it is increasing.
7. $$\frac{C}{0.25} = 1 - (1\ 0.15)\ 2.71^{-0.15 \times 15/1} = 0.23\ M$$

8. $$\frac{C_f}{12.0\ ppm} = \frac{1}{1 + \frac{0.9\ cm}{40.0\ cm}\left(\frac{1}{0.12} - 1\right)} = 1.60\ ppm$$

9. Use more than one zone at a time.

Chapter 3   713

CHAPTER 3 - DISTILLATION-GENERAL INFORMATION

1. The collisions of moving atoms and molecules against a surface.
2. 1 Newton/m².
3. A molecule that has a partial charge separation that is unsymmetrical within the molecule.
4. Water under normal conditions is a liquid, so the gaseous phase is called a vapor.
5. The container in which the mixture to be separated is placed.
6. A column with sufficient surface area to permit more than one equilibrium step to take place.
7. The number of degrees of freedom you have to alter the system as you desire.
8. $C = FeCl_4^-$, HCl, water, and diethyl ether = 4. P = water and diethyl ether. T and P are constant so F = 4 - 2 + 0 = 2. These are the concentrations of Fe and HCl.
9. Ethanol: $(29.6 \times 0.789) \div 46.0 = 0.508$ moles. Water: $(4 \times 29.6 \times 1.00) \div 18 = 6.58$ mol.
   Ethanol: $0.508/(0.508 + 6.58) = 0.072$ and 0.072×78.4 torr = 5.6 torr.
   Water: $6.58/(0.508 + 6.58) = 0.928$ and 0.928×100 torr = 92.8 torr. $P_t = 5.6 + 92.8 = 98.4$ or 98 torr.
10. Height equivalent to a theoretical plate.
11. It is the number of equilibrium steps taking place in a distillation column. It is used to determine the efficiency of the column and to compare columns.
12. 135 °C - 111 °C = 24 °C difference. Referring to Table 3-2, p.26, probably 8 to 9 plates.
13. The ratio of the mole fraction of a compound in the vapor phase to the mole fraction in the liquid phase.
14. Use Trouton's equation. $\Delta H = (54 + 273) \times 21$ cal deg^{-1} mol^{-1} or 6867 cal deg^{-1} mol^{-1}; about 6.8 kcal or 28.7 kJ.
15. Apply equation 3-9.

$$\log \frac{37 \; torr}{P_2 \; torr} = \frac{-43.9 \; kJ/mol \; (278-311)}{2.3 \times 0.00831 \; kJ/mol/K \; (278 \times 311)} = 0.8766$$

$\log 37 - \log P_2 = 0.8766$; $1.5682 - 0.8766 = \log P_2$; $P_2 = 4.9$ torr, a 7.5 reduction in vapor pressure.

16. α is the relative volatility of two compounds. The larger the number, the easier it should be to separate them. Usually α should be greater than 1.02, which would be a difficult separation by distillation.
17.

$$\log \alpha = \frac{8.9 \; (273 + 98.5) - (273 + 85)}{(273 + 98.5) + (273 + 85)}$$

$$\log \alpha = \frac{120.1}{729.5} = 0.1646 \qquad \alpha = 1.46$$

18. Refluxing is returning the vapors back down the column to be redistilled. This improves the separation, and in another situation where a chemical reaction is taking place in a solvent, the solvent is returned to the reaction flask and not lost.
19. Use equation 3-12. $Y_n = 0.990$; $X_s = 0.50$ $\alpha = 760$ torr/508 torr = 1.496

$$\frac{0.990}{1 - 0.990} = 1.496^{(n+1)} \frac{0.50}{1 - 0.50} \qquad (n+1) \log 1.496 = \log 99$$

$1.995/0.175 = 11.4$; $11.4 = (n+1)$ $n = 10.4$ or 11 plates.

20. $V = L + D = 10 = 9 + 1$ $Y_{n-1} = 0.9X_n + 0.1X_D$
    $Y_n = X_D$ because it is nearly pure at the top of the column. 0.990 is nearly 1.0. If $X_D = Y_n = 0.990$

$$\frac{Y_n}{1 - Y_n} = \alpha \frac{X_n}{1 - X_n} \qquad\qquad \frac{0.990}{1 - 0.990} = 1.496 \frac{X_n}{1 - X_n}$$

$X_n = 0.9851$; $Y_{n-1} = 0.9(0.9851) + 0.1(0.99)$
$= 0.9856$

Follow the same pattern as in the example:

X_{n-1}	= 0.9786	X_{n-2}	= 0.9683
Y_{n-2}	= 0.9(0.9786) + 0.1(0.990)	Y_{n-3}	= 0.9(0.9683) + 0.1(0.990)
	= 0.9797		= 0.9705
X_{n-3}	= 0.9565	X_{n-4}	= 0.9410
Y_{n-4}	= 0.9(0.9565) + 0.1(0.990)	Y_{n-5}	= 0.9(0.9410) + 0.1(0.990)
	= 0.9598		= 0.9459
X_{n-5}	= 0.9213	X_{n-6}	= 0.8962
Y_{n-6}	= 0.9(0.9213) + 0.1(0.990)	Y_{n-7}	= 0.9(0.8962) + 0.1(0.990)
	= 0.9281		= 0.9056
X_{n-7}	= 0.8652	X_{n-8}	= 0.8274
Y_{n-8}	= 0.9(0.8652) + 0.1(0.990)	Y_{n-9}	= 0.9(0.8274) + 0.1(0.990)
	= 0.8776		= 0.8437
X_{n-9}	= 0.7848	X_{n-10}	= 0.7343
Y_{n-10}	= 0.9(0.7848) + 0.1(0.990)	Y_{n-11}	= 0.9(0.7343) + 0.1(0.990)
	= 0.8053		= 0.7599
X_{n-11}	= 0.6790	X_{n-12}	= 0.6209
Y_{n-12}	= 0.9(0.6790) + 0.1(0.990)	Y_{n-13}	= 0.9(0.6209) + 0.1(0.990)
	= 0.7101		= 0.6578
X_{n-13}	= 0.5624	X_{n-14}	= 0.5063
Y_{n-14}	= 0.9(0.5624) + 0.1(0.990)	Y_{n-15}	= 0.9(0.5063) + 0.1(0.990)
	= 0.6051		= 0.5547

X_{n-15} = 0.4546
Y_{n-16} = 0.9(0.4546) + 0.1(0.990)
= 0.5081

The next step will go below 0.500, so n-17 or 18 plates will be required. Notice the large effect when changing α from 1.89 in the example to 1.49 here.

21. You do not have to do a calculation like that in problem 20.
22. It provides more even heating, especially for round-bottom flasks.
23. It is connected to a thermocouple inside of the mantle and can be used to measure the temperature.
24. To provide a stream of air bubbles to prevent bumping.
25. A rapid evolution of gas bubbles will occur that may blow the hot solvent out of the container.
26. They are nontoxic.
27. The amount of liquid in the column during the distillation.
28. Berl saddles.
29. A spiral of wire mesh.
30. Low holdup and a high throughput.
31. So the reflux line entirely bathes the bulb, and the bulb is where the condensate is being taken off.
32. To keep the condenser full and so the coldest water will not meet the hottest vapors, thus reducing the chances of cracking.
33. A wet floor, embarrassment, and maybe even death as soon as the instructor catches you.
34. To provide a controlled reflux ratio.

CHAPTER 4 - AZEOTROPIC AND EXTRACTIVE DISTILLATION

1. A liquid mixture that is characterized by a constant maximum or minimum boiling point that is lower or higher than that of any of the components and that distills without change in composition.
2. An *azeotropic distillation* involves the formation of an azeotrope with at least one of the components of a liquid mixture, which then can be separated more readily because of the resulting increase in the difference between the volatilities of the components of the mixture.
3. (1) the added component reduces the partial pressure of one of the original components.
   (2) The formed azeotrope is removed more easily than anything else in the pot.
4. 88%.
5. The third component added often is called an *entrainer*.
6. (1) Form a new azeotrope. (2) Add a salt. (3) Changing pressure.
7. 2 x 12.5% = 25 proof.
8. The ratio of the speed of light in a vacuum to that in another medium, usually a gas or liquid.
9. D shows that the sodium D line was used to determine this value, and the 20 indicates the temperature in °C at which the measurement was made.
10. Both will slowly dissolve the glue holding the prisms in place, and both evaporate so fast that they cool the prisms, making an accurate measurement difficult.
11. The refractive index of water at 20 °C is 1.3330. Measure it just before you measure your unknown liquid. Suppose it is 1.3336 or 0.0006 higher than normal. You should then subtract 0.0006 from the unknown to reduce it to 20 °C.
12. Refer to the example problem in Experiment 4.
13. Refer to the example problem in Experiment 4.
14. Assume a 100 g sample. 35.8 g water; 64.2 g m-xylene. 35.8 g /18 g/mol = 1.99 mol water; 64.2 g/106 g/mol = 0.608 mol m-xylene. X = 0.766 water; X = 0.234 m-xylene. 1.3330 x 0.766 = 1.021 for water; 0.234 x 1.4973 = 0.350 for m-xylene; total R.I. = 1.3710
15. 1.4190 - 1.3330 = 0.0860 R.I. difference. 1.4190 - 1.4185 = 0.0005 R.I. difference due to water. (0.0005/0.0860) x 100 = 0.58 mol % water impurity.
16. A third component, called a *solvent* in extractive distillation or an *entrainer* in azeotropic distillation, is added to increase the difference in volatility between the key components.
17. Extractive distillation is based on the *attraction* between the solvent and one or more of the components in the mixture; azeotropic distillation is based on the *repulsion* between the entrainer and one or more components in the mixture.
18. Extractive distillation offers a wider range of possibilities than does azeotropic distillation because (1) the phase relationships are not as critical, so more solvents are available and (2) less heat input is required because the solvents are not taken overhead. Complete separations with extractive distillations are difficult to obtain and, as a result, extractive distillation is used when large quantities of reasonably pure materials are needed. Azeotropic distillations are used when higher purity is required.
19. During the refluxing, the benzene and cyclohexane can remix, and if no aniline is present, an azeotrope will form.

CHAPTER 5 - STEAM AND IMMISCIBLE SOLVENTS DISTILLATION

1. It is limited to those compounds that are immiscible with water.
2. It has a high vapor pressure and a low molecular weight and is inexpensive.
3. The melting point.
4.
$$\frac{100 \text{ g water}}{X \text{ g diethylaniline}} = \frac{744 \times 18}{16 \times 149} \quad X = 17.8 \text{ g}$$

5.
   (a) ~99 °C. (v.p. ~ 20 torr).
   (b)
$$\frac{X \text{ g } H_2O}{10.0 \text{ g nitrobenzene}} = \frac{740 \times 18}{20 \times 123} \quad X = 54 \text{ g}$$

6. The apparatus eventually would blow apart somewhere.
7. There would be more bumping. Little room is left to accommodate the bumping, and the stopper usually comes loose. It takes too long to k\heat that much water, so the separation takes longer and it takes more heat to keep the water boiling.
8. (a) The water would boil at about 72 °C.
   (b) At 72 °C, the v.p. of eugenol is about 1 torr.

$$\frac{g\ H_2O}{g\ eugenol} = \frac{759 \times 18}{1 \times 164} = 83.3:1$$

9. To reduce bumping directly into the condenser.
10. The condensed water in the sample flask will be drawn back into the trap.
11. Turn off the condenser water.
12. The positive hydronium ions neutralize the negative ions on the emulsion droplets, so they will combine.
13. Silicone materials are hydrophobic (repel water), so water molecules do not penetrate the paper, but organic molecules will.
14. Na_2SO_4 (Glauber's salt) forms $Na_2SO_4 \cdot 10H_2O$ in water and does so very quickly. In fact, water is removed from solvents by passing them through a column of anhydrous Na_2SO_4.
15. The Barrett trap has a stopcock so the water can be drained and weighed if needed.
16. Raw potatoes contain an average of 79.8% water. Therefore to get 10.0 of potatoes:

$$10\ g \times \frac{100\%}{100\% - 79.8\%} = 49.5\ g\ of\ potatoes$$

17. The final results depend upon the water analysis, which is no better than 3 significant figures because of the evaporation rate of the water.
18. The toluene will dissolve it. If the toluene is to be evaporated to determine the toluene-soluble volatiles, this is then a contamination.
19. If the total volatiles are not to be determined.
20. The outer jacket on a West condenser is closer to the center tube, so the water flow rate is faster, and faster cooling is possible.
21. The density of toluene is 0.866 compared to 1.0 for water. This is actually a large difference, so the droplets are most likely to separate.
22.
```
    66.217 g              34.755 g
     2.186 g              20.660 g
    ----------            ----------
    64.031 g of sample    14.095 g of water
```

$$\frac{14.095 \times 100}{64.031} = 22.0\%\ water$$

23. At 71 °C, the v.p. of water = ~240 torr and benzene = ~520 torr.
24. Using equation 5-1:

$$\frac{14.095\ g\ water}{X\ g\ benzene} = \frac{240 \times 18}{520 \times 78} \qquad X = 132.3\ g\ benzene$$

$$132.3\ g\ benzene \times \frac{1\ mL}{0.879\ g\ benzene} = 151\ mL\ of\ benzene$$

25. Small amounts of other volatile materials also were distilled away from the orange peel.

Chapter 6

CHAPTER 6 - VACUUM DISTILLATION

1. Refer to Figure 6-3, p. 59. 22 torr.
2. Refer to Figure 6-3, p. 59. 5 mm
3. Refer to Figure 6-3, p. 59. Estimated at ~45 °C.
4. 460 °C.
5. 142 °C.
6. It prevent decomposition and polymerization and may break an azeotrope.
7. Reduce friction, seal the joints, and provide a smaller dead space for the exhaust cycle.
8. It vaporizes at the reduced pressure, creating a back pressure that limits the vacuum that can be produced.
9. It is a bottle within a bottle with the space inside of them evacuated. This produces an excellent insulation
10. The residual oxygen can cause an explosion in rare instances.
11. Yes. The dry ice should be placed in first, so the evaporated CO_2 can escape with less spattering.
12. $CoCl_2 \cdot H_2O$(blue) + 4 $H_2O \longrightarrow CoCl_2 \cdot 6H_2O$ (red).
13. To reduce the number of glass fragments that might fly away from the flask, if it should break when under a vacuum.
14. During an implosion, the fragments move to the center of the system, whereas in an explosion, they move away.
15. 0.01 torr.
16. See Figure 6-9, 63.
17. See Figure 6-12, p. 65. Gas at the desired pressure is trapped in the bulb on the right. If the pressure in the bulb at the left is less, the Hg rises in the left bulb, closing off the glass frit and stopping the pressure reduction. If the pressure is greater in the left bulb than the right bulb, the reverse is true.
18. Tilt it slightly and have just enough Hg present so only a small movement of Hg will close the frit.
19. A piece of ordinary filter paper is folded for filtering and a small hole, a pin hole (the end of a paper clip works better), is punched in the bottom. When the dirty Hg is added, the heavy Hg drains out the bottom and the lighter oxides float to the top and stick to the paper.
20. This is to reduce air being trapped in the joint when the two pieces of the joint are pressed together.
21. Place an O-ring between the pieces of the ball joint and squeeze the joint together with a ball joint clamp.
22. He had several hundred patents on alternating current electricity and is the one responsible for developing the A.C. electric motor.
23. It contains highly purified hydrocarbons and can be dissolved with $CHCl_3$.
24. Too much gas is evolved too soon, and then the chips lose their effectiveness.
25. Desiccant. Don't feel bad - most people miss spell it.
26.

$$P = \frac{\pi\, r^2\, L^2}{V}: \quad 8 = \frac{3.14\, (1\,)^2 \times (10\,)^2}{V}: \quad V = 39.25\ mL$$

CHAPTER 7 - MOLECULAR DISTILLATION AND SUBLIMATION

1. No refluxing occurs in a molecular distillation.
2. It reduces heat transfer.
3. (1) Fraction reaching the condenser, (2) fraction colliding and the fraction of that that reaches the condenser, (3) the fraction that reaches the condenser by normal collisions.
4. 1×10^{-6} g/cm^2/sec.
5. $C_8H_7NO_4 = 181$; $C_6H_2Cl_4O = 232$; $C_{15}H_{20}O = 222$

$$\frac{1}{\sqrt{232}} = 0.066\ :\ \frac{1}{\sqrt{222}} = 0.067\ :\ \frac{1}{\sqrt{181}} = 0.074 \quad Divide\ by\ 0.066;\quad 1 : 1.02 : 1.12$$

6.

$$\frac{40}{\sqrt{200}} : \frac{40}{\sqrt{162}} : \frac{40}{\sqrt{X}};\quad 1.28 : 1.15 : 1.00 \qquad \frac{\frac{40}{\sqrt{200}}}{1.00} = \frac{\frac{40}{\sqrt{X}}}{1.28};\quad X = 122;\quad C_8H_{10}$$

7. $P_1 = 760$ torr; $P_2 = 10^{-3}$ torr; $T_1 = 273$ K; $T_2 = (220 + 273)$ K

$$n = \frac{6.02 \times 10^{23} \text{ atoms}}{22,400 \text{ cm}^3 \times \frac{220 + 273}{273} \times \frac{760}{1 \times 10^{-3}}} = 1.96 \times 10^{13} \text{ atoms/cm}^3$$

8.
$$L = \frac{1}{\sqrt{2} \times 3.14 \times (22 \times 10^{-8} \text{ cm})^2 \times 1.96 \times 10^{13} \text{atoms/cm}^3} = 0.40 \text{ cm}$$

9. Provide a large evaporating surface and a continuous porocess.
10. 0.1 to 2.0 mm
11. To increase the evaporating surface and thin the film for better heat transfer.
12. To thin the film and reduce the contact time, which reduces decomposition.
13. 400 to 500 rpm and 250 gal/hr.
14. See Figure 7-7, p. 77.
15. Thermocouple gauge - it operates on a Wheatstone bridge circuit.
16. Sublimation is a process in which a solid becomes a gas without first becoming a liquid. A molecular distillation involves going from a liquid to a gas.
17. The ice melts. A thin film of water forms and acts as "grease" to allow you to slide along.
18. Tilt it nearly horizontal with one opening up. Add Hg to this opening, allowing time for air to escape. Then tip it completely upside down to force all of the air out. When you tilt it right side up, a vacuum should be sealed off in the top portion.
19. Not if it was filled correctly. However, some gases usually dissolve in the Hg, so it is wise to tilt it upside down again and let any gasses escape.
20. This is usually a glass wool tape with copper wire inside that can be heated electrically. Its main advantage is that it can be coiled around curved surfaces.
21. Yes.
22. The material sublimed is the sublimate, and the material it is sublimed from is the sublimand.
23. To reduce the partial pressure of the sublimate below the triple point pressure.

CHAPTER 8 - LYOPHILIZATION (FREEZE DRYING)

1. Acetone (Dry ice = -78.5 °C)   Isopropanol
    (a) -78.5 °C (freezes at -94 °C).   (a) -78.5 °C (freezes at -85 °C).
    (b) Flash point = -20 °C.   (b) Flash point +11.7 °C.
    (c) Dissolves acetate, Arnel, Dynel Vinyon, and Rayon.   (c) Does not dissolve polymers.
    (d) Dissolves it instantly.   (d) Does not dissolve polystyrene.
2. Reduces the reaction rate of reactions requiring water and reduces weight for shipping.
3. In case of a fast evolution of water, the main condenser may not be 100% efficient, and the coolant may disappear from the main condenser during an overnight run.
4. It will sublime faster, because less external heat is needed.
5. It is less expensive, because the refrigerant is recycled, and extremely cold temperatures are not needed at this stage.
6. Large ice crystals will form and can break cell walls and cause texture changes.
7. The heat loss by the subliming water usually will keep the sample frozen.
8. 90%.
9. The tube containing the liquid sample is tilted as far as possible without spilling the liquid and then rotated to freeze a layer of sample around the inside of the tube. This provides a large surface area.
10. Measure the electrical resistance between two probes in the sample as the sample is cooled. A large change in resistance occurs when a phase change occurs.

11. Sp. ht. of ice = 0.52 °C/g °C; of water, 1.0 cal/g °C. Heat of fusion = 80 cal/g; Heat of vaporization = 540 cal/g.

    0.52 °C/g °C x 1.0 g x 10 °C = 5.2 cal
    1.0 cal/g °C x 1.0 g x 100 °C = 100 cal
    80 cal/g x 1.0 g = 80 cal
    540 cal/g x 1.0 g = 540 cal
    ----
    725 cal

12. When the coffee solution freezes, the water turning to ice forces the coffee compounds into the liquid layer, thus concentrating it. This is the principle involved in zone refining explained in a later chapter.
13. The alcohol concentrated in the remaining water and became "antifreeze".
14. Two sample chambers are involved, and a thermocouple is placed in the sample in each chamber. A pure material (or a reference material) is placed in one chamber, and the impure sample in the other. Both are heated at the same rate, and the difference in the temperature of any phase change is recorded.
15. The sugar becomes like honey when the water is removed, and slight carmelization can occur in rare cases.
16. This is to prevent "freezer burn". Any food left uncovered and placed in a freezer will slowly freeze dry. The water in the bag is to prevent this.
17. 10% sucrose in the solution.

CHAPTER 9 - EXTRACTION
1. An alcohol solvent extract.
2. The number of components.
3.
$$D = \frac{Total\ g/mL\ solute\ in\ the\ organic\ (upper) phase}{Total\ g/mL\ solute\ in\ the\ water\ (lower)\ phase}$$

4.
$$\frac{83}{17} = 4.88$$

5. It must be neutral.
6. If the ligands form ring compounds, the material is a chelate.
7. Usually no.
8. The effect of two components in a mixture is greater than the sum of the two separate effects.

9.
$$\% = \frac{100\ x\ 3}{3 + \frac{25}{10}} = 55\%$$

10.
$$\% = \frac{100\ x\ 4.9}{4.9 + \frac{5}{X}} \qquad X = 50\ mL$$

11.

$$0.01 = 0.20\left(\frac{20}{(0.3 \times 50) + 20}\right)^n$$

Take the log of both sides.
$\log 0.01/0.2 = n \log 0.571$
$n = \sim 6$ extractions

12.

(A) $\dfrac{100 \times 4 \times V_o}{4 V_o + 10}$ $V_o = 2.5$ mL   (B) $\dfrac{100 \times 4 \times 5}{4 \times 5 + V_w}$ $V_w = 0.040$ mL!

13.

$$0.001 = 0.3\left(\frac{150}{(110 \times V_o) + 150}\right)^1 \quad V_o = 411 \text{ mL}$$

14.

$$0.001 = 0.20\left(\frac{50}{(D \times 25) + 50}\right)^3 \qquad 0.005 = \frac{8}{D^3 + 6D^2 + 12D + 8}$$

Use successive approximations.
If $D = 1$ $0.135 \neq 8$; If $D = 2$ $0.32 \neq 8$; If $D = 5$ $1.71 \neq 8$; If $D = 8$ $5 \neq 8$; If $D = 10$ $8.64 \neq 8$;
Therefore, somewhere just under 10.

15.

$$0.0001 = 0.24\left(\frac{50}{(1.4 \times 20) + 50}\right)^n; \quad 0.000416 = (0.64)^n; \quad \log 4.16 \times 10^{-4} = n \log 0.64; \quad n = 18$$

16.

$$X = 0.5\left(\frac{10}{(2 \times 20) + 10}\right)^1 \quad X = 0.10 \text{ g left} \qquad X = 0.5\left(\frac{10}{(2 \times 5) + 10}\right)^4 \quad X = 0.031 \text{ g left}$$

17. Stripping is the reverse of extraction in that the material to be extracted is from the organic layer. D's less than 1 are desired
18. A rotation of the wrist first clockwise then counterclockwise. Four to five inversions will bring nearly 100% equilibrium. You do not have to shake the layers for hours to get a partition.
19. If a ligand forms a complex with an ion, it is a masking agent. If the ligand forms a chelate with the ion, it is a sequestering agent.
20. To reduce large volumes of solvent and concentrate the residue in a very small volume without losing any of the desired constituents.

CHAPTER 10 - CONTINUOUS EXTRACTION

1. The Scheibel column (A): Alternate compartments are agitated with impellers, whereas the others are packed with an open woven wire mesh. The rotating disc contactor (B) uses the shearing action of a rapidly rotating disc to interdisperse the phases. The Oldshue-Rushton column (C) consists essentially of a number of compartments separated by horizontal stator-ring baffles, each fitted with vertical baffles and a turbine-type impeller mounted in a central shaft. The reciprocating plate extractor (D) is a vibrating type. The open perforated reciprocating plate

2. a. A large surface area of contact between the two phases. b. A long contact time. c. A high relative volume of extractant to solvent. d. A high D.

3.

$$\frac{2\ drops}{sec} \times \frac{0.05\ mL}{drop} \times \frac{60\ sec}{min} \times 50\ min = 300\ mL$$

4.

$$k = \frac{0.693\ W}{V}; \quad \frac{0.693 \times 50}{300} = 0.115$$

5. 1 ppm – 0.0001%; 2.21% = 22,100 ppm.
6. Any halogenated solvent; CH_2Cl_2, CCl_4, H_3C-$CHCl_2$.
7. 1 g/46 mL of water. (*Merck Index*).
8. 1 g/5.5 mL of $CHCl_3$ (*Merck Index*).
9.

$$\frac{0.225}{5.0\ ppm)} = \frac{0.532}{X\ ppm}; \quad X = 11.8\ ppm$$

10.

$$10.0\ g\ caffeine \times \frac{0.11\ g}{1\ g} = 0.110\ g\ caffeine = 110\ mg. \quad \frac{110\ mg}{5\ mg} = 22\ dilution\ factor. \quad \frac{50\ mL}{22} = 2.27\ mL$$

11. The center tube will overflow without forcing the solvent out of the bottom, and no extraction will take place.
12. To force the ascending solvent droplets to take a longer path and have more time for extraction.
13. They must be immiscible with water; benzene, toluene, xylenes.
14. See page 131.
15. An aqueous mixture is placed in a 2-liter flask on the lower side arm. A lighter than water solvent is placed in a 1-liter flask attached to the upper arm. Both liquids are heated to reflux, and vapors are condensed and passed through the capillary tube. The solute is transferred to the solvent, and both liquids are returned to their original flasks. If a heavier than water solvent is used, you need only reverse the position of the flasks.
16. It sometimes exploded if a spark occurred.
17. The diethyl ether floats on the water in the trap and slowly evaporates. Its vapors are dense ans displace air in the bottom of the sink. Any spark then will set if off.
18. 1 mL of acid = 6.0 mL of base.

$$\frac{31.60 - 2.20}{6.0} = 4.90\ mL\ of\ acid.$$

5.0 mL - 4.9 mL = 0.10 mL.
0.10 mL x 2.24 mg alkaloid/mL = 2.44 mg alkaloid.

The product is mislabeled and would be recalled by the FDA.

$$\frac{2.44\ mg\ actual \times 100}{14.0\ expected} = 17.4\%\ of\ declared.$$

19. Partially dissolving a solid material to remove a desired compound.
20. To spread out the extracting solvent and provide for a more uniform extraction.
21. The air pressure on top of the extraction liquid in the chamber is greater than the air pressure of the solvent at the outlet going back into the receiver.

22. Higher temperatures caused by an increased pressure.
23. 2450 MHz.
24. 4×10^9/second.
25. To provide a dipole, so the microwave energy can be absorbed.
26. The ability of a molecule to produce a dipole when placed in an electric field.
27. A proportionality constant in Coulomb's equation relating the force of repulsion to the medium the repulsion is in.
28. The temperature is controlled, which controls the pressure, and then a rupture membrane is present if all else fails.

CHAPTER 11 - COUNTERCURRENT (EXTRACTION) CHROMATOGRAPHY

1. No compartments as such exist and there is no stopping to transfer a segment of mobile phase to another section.
2. Revolutionary and planetary.
3. To provide a centrifugal gradient that further enhances the percent retention.
4. By coupling a pulley on the helix holder to a stationary pulley of equal diameter on the axis of the centrifuge drive.
5. In the HSES system, a stationary coil is placed horizontally. In the HDES system, the coil rotates around its own axis.
6. The mobile phase lacks wall surface affinity.
7. It has no inefficient free space that is completely occupied by the mobile phase.
8.

$$k_1 = \frac{(25\ mL - 15\ mL)}{(75\ mL - 15\ mL)} = 0.166;\quad k_2 = \frac{(40\ mL - 15\ mL)}{(75\ mL - 15\ mL)} = 0.41 \quad \alpha = \frac{0.41}{0.166} = 2.51$$

$$N_1 = 16\left(\frac{25\ mL}{3\ mL}\right)^2 = 1111\ plates;\quad N_2 = 16\left(\frac{40\ mL}{4\ mL}\right)^2 = 1600\ plates;\quad Avg. = 1356;\ \sqrt{1{,}356} = 36.8$$

$$R_s = 0.25\ (2.51 - 1)(36.8) \times \frac{0.166}{0.5 \times 0.166 \times 3.51 + \frac{1 - 0.80}{0.80}} = 4.27$$

9. N/t_o.
10. That resolution depends upon the volume of stationary phase retained in the column.
11.

$$Cpd\ 1:\ K_p = \frac{13.8\ mL - 2.8\ mL}{12.2\ mL - 2.8\ mL} = 1.17 \qquad Cpd\ 2:\ K_p = \frac{16.5\ mL - 2.8\ mL}{12.2\ mL - 2.8\ mL} = 1.46$$

12. The rotation and revolution of the holder are synchronized to give a rotation-revolution ratio of 1:1.
13. Analytical scale separations.
14. Efficiency rises in inverse proportion to the internal diameter.
15. For a given length of tube, the partition efficiency decreases with the increase of the helix diameter.
16. Up to 5% of the total volume of the medium.

CHAPTER 12 - SOLID PHASE EXTRACTION

1. It is a liquid-solid extraction with the sample being in the liquid and the extracting agent in the solid.
2. The SPE particles are much larger.
3. It is faster, requires less solvent, reduces the need for large concentration steps, and is automated more easily.
4. 10 to 50.
5. It alters the selectivity of the surface.

6. It measures how two components behave on a given column surface.
7. How long the compound is held on the column.
8. It is not retained very long.
9. PH and ionic strength.
10. Octyl-bonded silica.
11. LC-SAX
12. $10 \times 0.3 = 3.0$ mL.
13. The packing usually is not easily wetted by water.
14. Use equation 12-2, p. 132. $(2.0 - 0.2)/0.2 = 9$
15. To remove the organic solvent from the cartridge.
16. To remove as many interfering compounds as possible.
17. The smaller the particle size, the greater the number of plates, and the faster the flow rate, the lower the number of plates.
18. The flow rate varies as the square of the diameter so it will now be 1/4th as fast.
19. 3-A, 2-B, or 1-H (Supelco Bulletin)
20. A piece of fused silica rod coated with an adsorbent.
21. In a plastic bag and in a desiccator.

CHAPTER 13 - SUPERCRITICAL FLUID EXTRACTION

1. It has low viscosity, and near zero surface tension, and the solvent evaporates at the end.
2. It is at a T, that no matter how much pressure is applied, it will not condense to a gas.
3.

$$100 \, ft \times \frac{30.5 \, cm}{1.00 \, ft} \times \frac{13.6 \, g}{1.00 \, cm^3} \times \frac{1.00 \, lb}{454 \, g} \times \frac{6.45 \, cm^2}{1.00 \, in^2} = 589 \, p.s.i.$$

4. The solvent forms clusters around the solute molecule and the solute molecules cluster as well.
5. Methanol can hydrogen bond with solutes, and it is not suspected of destroying the ozone layer.
6. Solubility usually increases substantially with an increase in pressure.
7. The method is much faster, the recoveries are about the same, and the % RSD is lower, making the method more reproducible. Also, no solvents are used.
8. Hag AG Corp. In Germany to decaffeinate coffee.
9. Remove nicotine from tobacco.
10. Chelate them with fluorinated ligands, then extract them.
11. To increase the polarity of the system.
12. Pelletized diatomaceous earth.
13. Less than 70% water, therefore, a 1:1 ratio. 4 g.
14. So it is placed in the cartridge in the same direction each time to prevent fine particles from coming out if the flow rate is reversed.
15. Usually 5 to 10 void volumes of solvent are required to completely extract the solutes, so the smaller the volume, the quicker the extraction is.
16. Hydromatrix, Celite, glass wool, clean sand, and tiny glass beads.
17. The sample by itself leaves too much void volume, and if a filler is added, it should be mixed in with the sample to minimize channeling, thus getting a better extraction.
18. By measuring the distance the pump piston travels.
19. Five to 10 void volumes and 10 to 20 minutes.
20. No. You increase the length of the restrictor.
21. 1.5 mL/min.
22. It is at the end of the system, and the emerging gas is rapidly expanding and cooling. The frozen solute particles will plug the restrictor.
23. About 500 mL

CHAPTER 14 - CHROMATOGRAPHY - GENERAL THEORY

1. Adsorption and partition.

2. Powder size dry particles.
3. On corners and edges.
4. Its molecules are more strongly adsorbed to the inert phase than the sample molecules and many more of them are present.
5. A portion of the sample gets ahead of the main body by not reacting with the column particles. It is caused by loose or irregular packing.
6. The lower the concentration, the tighter the molecules will adsorb.
7. It protonates the nonbonded electrons on the extra oxygen atoms.
8. It does not dissolve in the mobile phase, and it adsorbs tightly to the inert phase.
9. The second law of thermodynamics - systems if left to themselves tend to maximum disorder.
10. Less time is available for diffusion which causes band broadening and makes separations more difficult.
11. Ten column volumes are needed to elute that component from the column.
12. You would feel like doing nothing to teach him a lesson, but solvents are expensive and time is valuable, so you should pour out about 40 cm of the packing, assuming that no smaller column is available, and let them begin. The column is way overdesigned. Compounds with P's of 8 and 2 should be separated easily.

CHAPTER 15 - DISPLACEMENT AND MULTIPLE COLUMN PARTITION CHROMATOGRAPHY
1. To minimize band broadening and curvature of the solute band as it develops.
2. Water is quite polar and usually will strip off most of the sample without separating it.
3. Add 2 to -3 g of anhydrous sodium sulfate to the top of the column packing.
4. To minimize band broadening.
5. A loosely packed column.
6. The packing usually is too loosely packed, so the liquid can move faster along the wall than inside of the column. Also the column wall is smooth and wets differently than the packing, so the mobile phase moves faster along the wall. This causes skewed bands.
7. Gradient elution is changing the polarity of the solvent from the beginning of the elution to the end. It can be established by slowly mixing a more polar solvent into a less polar solvent.
8. The liquids go around the air bubble rather than pushing it ahead of the mobile phase front. This causes band broadening and low recoveries.
9. Add mobile phase to cover the packing, then stopper the top of the column.
10. To allow you to push the packing out of the column.
11. Fill the bottom of a volumetric flask with the desired amount of mobile phase, then invert the combination into the top of the column. See Figure 15-5, p. 160.
12. $A = \pi r^2 = 3.14 \times (1.0 \text{ cm})^2 = 3.14 \text{ cm}^2$. $A_m + A_i + A_s$ should equal this or be close.
Volume of alumina gel = 83.0 g / 3.6 g/mL = 23.05 mL
    Butanol A_m = 61.0 cm³/30 cm = 2.03 cm²
    Alumina A_i = 23.05 cm³ / 30.0 cm = 0.768 cm²
    Water A_s = 9.6 cm³ / 30.0 cm = 0.320 cm²
                                                             3.12 cm²

$$\text{Phenol } R_f = \frac{2.03}{2.03 + (0.194 \times 0.320)} = 0.97; \text{ Salicylic acid } R_f = \frac{2.03}{2.03 + (3.42 \times 0.320)} = 0.65$$

13. Gallic acid - left; protocatechuic acid - right.

$$0.43 = \frac{2.03}{2.03 + P \times 0.32}; \quad P = 8.4 \qquad 0.74 = \frac{2.03}{2.03 + P \times 0.32}; \quad P = 2.2$$

14. $A = \pi r^2 = 3.14 \times (0.7 \text{ cm})^2 = 1.54 \text{ cm}^2$.

$$\frac{13.5 \text{ g}}{0.61 \text{ g/cm}^3} = 22.13 \text{ cm}^3; \quad A_i = \frac{22.13 \text{ cm}^3}{25 \text{ cm}} = 0.88 \text{ cm}^3$$

$A = A_m + A_i + A_s;$
$1.54 = A_m + 0.88 + 0.06;$
$A_m = 0.60$

$$A_s = \frac{1.5 \text{ g/cm}^3}{25 \text{ cm}} = 0.06 \text{ cm}^2$$

$$\text{Pyrogallol: } 0.51 = \frac{0.60}{0.60 + P \times 0.06}; \quad P = 9.6 \quad \text{Phenol: } 0.97 = \frac{0.60}{0.60 + P \times 0.06}; \quad P = 0.31$$

15.

$$\text{Citric acid: } R_f = \frac{0.60}{0.60 + 8.3 \times 0.06}; \quad R_f = 0.55; \quad \text{Lactic acid: } \frac{0.60}{0.60 + 0.31 \times 0.06}; \quad R_f = 0.72$$

16. (a)

$$R_f = \frac{0.20}{0.20 + 1.50 \times 0.05} = 0.73; \quad R_f = \frac{0.20}{0.20 + 1.55 \times 0.05} = 0.72$$

(b) $V_h = 0.005 (0.20 + (1.50 \times 0.05)) = 1.375 \times 10^{-3} \text{ cm}^3;$ $1.375 \times 10^{-3} \text{ cm}^3/\text{plate} \times 10.0 \text{ cm}/0.005 \text{ cm/plate} = 2.75 \text{ cm}^3$
$V_h = 0.005 (0.20 + (1.55 \times 0.05)) = 1.387 \times 10^{-3} \text{ cm}^3;$ $1.3875 \times 10^{-3} \text{ cm}^3/\text{plate} \times 10.0 \text{ cm}/0.005 \text{ cm/plate} = 2.775 \text{ cm}^3$

(c) $138.7/137.5 = 1.0091$ ratio. For a 30 cm column, $30.0 \text{ cm} \times 1.0091 = 30.27 \text{ cm}$. Therefore the maximum concentration of each peak occurs at 29.86 cm and 30.14 cm.
% = 0.26 (2 wings); $100.00 - 0.26 = 99.74$ or $t = 3.0$.

$$3 = \frac{V}{0.00138 \sqrt{r}} - \sqrt{r}; \quad r = 30.00/0.0005 = 6{,}000; \quad 3 = \frac{V}{0.1074} - 77.46; \quad V = 8.64 \text{ mL}$$

(d)

$$t = \frac{8.64}{0.001375 \sqrt{6000}} - \sqrt{6000}; \quad t = 3.66 \text{ or } 0.036\% \text{ or } 0.018\%/\text{wing.}$$

17. The chloroform must be displaced by air, and the air will enter only when the chloroform drains below the opening. As soon as the column fills up to the flask opening, the air is shut off, and the flow of chloroform stops.
18. Wiping the beaker with a small planchet of glass wool.
19. The cough syrup was not broken into small enough particles, and the solvent went around them without dissolving the sample.
20. 1 g of sample to 3 g of Celite.
21. Only one component is to be saved on each column, so no fine separation is taking place.
22. Air in a column always should be avoided. However, in this case, the column packing volume is so large compared to a small air bubble that the loss in recovery is small.
23. The sample components are salts and need water to be soluble.
24. I turn them upside down over a waste basket and blow them out, then rinse them out in the sink. However, a safer way is to connect the outlet end to a water line with a piece of tubing and slowly turn on the water.
25. If it is too small, it tends to stick in the small tube and is hard to remove.

26. Step 1, 81.7 µg/mL; Step 2, 2,042 µg/2 mL; Step 3, 5.1 mg or 102 % of declared.

CHAPTER 16 - AFFINITY CHROMATOGRAPHY
1. To move the active group away from the bead surface and reduce stearic hindrance.
2. The use of a dye immobilized on cellulose to selectively retain a compound.
3. Greater mechanical strength, insoluble in most solvents, does not swell and dries easy. It is heated until the silicon-rich and boron-rich phases separate and then cooled, and the boron-rich phase is dissolved away.
4. Treat the surface with a triethoxy silane.
5. It is a much faster and more complete reaction.
6. They cause coagulation of many biological systems.
7. Lectins are plant and animal proteins capable of binding selectively to certain carbohydrate groups on the cell surface. Lecithins are lipids containing phosphorous and are found in the tissues of many plants and animals.
8. Excess galactose may have been present.
9. It makes the next steps easier.
10. 9000 times the force of gravity. This depends upon the speed of the centrifuge. Check yours to be sure it turns fast enough.
11. It is a preservative.
12. It reduces the activity of the lectin.
13. The mixture is centrifuged, and the liquid decanted. More wash solution is added, and the solid stirred into it again to wash away more impurities. The solid is removed by centrifugation, and the wash solution containing impurities is decanted.
14. Size separation is not as specific as affinity separation, so a longer column is preferred to provide more steps to take place.
15. concentration of a sample is measured. This is then diluted (usually 10-fold) and examined again. It is diluted again and measured, and the process continued until nothing is detected.
16. Activity is used to express how potent a biological compound is. It is more informative than concentration, because concentration is not always easy to measure, and even if the compounds are present, they may have lost their activity during preparation.
17. A measure of platelet adhesion. Hematocrit is a tube calibrated to facilitate determination of the volume of red cells in centrifuged blood expressed as corpuscular volume percent.
18.

CHAPTER 17 - SIZE EXCLUSION CHROMATOGRAPHY
1. Glass has too many sharp edges and corners that can puncture these fragile beads causing the column to become plugged.
2. These beads contain large amounts of water, and the bead structure is not very rigid.
3. Sephadex type materials swell considerably when placed in water solutions, and this changes the column volume greatly and can ruin a separation, if it occurs after the sample has been added.
4. Plungers are used to control the bed volume. These particles are very light and tend to float or drift when eluents are added. The plungers confine the beads, so reproducible results can be obtained.
5. Particles are separated primarily based on their size; those that are the largest, are excluded more.
6. Place a small amount of Sephadex G-100 into a 25 mL graduated cylinder. Bring a 0.1 M solution of NaCl to boiling and add 15 mL of the hot solution to the graduated cylinder. Gently stir the mixture with a stirring rod for about 10 minutes. Add another 2-3 mL of the NaCl solution. Set the gel aside for about 15 minutes to swell and to settle.

7. Add HNO_2 to Enzacryl polythiol or AH to form the azide, then add the enzyme, which will bond to the polymer by the NH_2 group of lysine on the enzyme.

CHAPTER 18 - FLASH CHROMATOGRAPHY
1. Regular column chromatography with a low pressure applied.
2. Too much dead volume occurs below them.
3. The column packing is likely to fragment.
4. Balloons, fish aquarium pumps, and laboratory air lines (filter the air with cotton or glass wool to remove any oil).
5. To reduce the pressure of the incoming gas to that required for proper solvent flow.
6. 0.35 and impurities ± 0.15 away.
7. Ethyl acetate, 30 to 60 °C petroleum ether, or their mixtures.
8. 900 mg.

CHAPTER 19 - HIGH PERFORMANCE LIQUID CHROMATOGRAPHY
1. It is fast, flexible, and reproducible.
2. The sample, dissolved in a mobile phase, is passed over a solid surface. The solid surface is usually polar, the mobile phase is usually nonpolar, and the polar sample components are separated.
3. Either a negative ion is paired with a positive ion or the reverse.
4. Tetrabutyl ammonium phosphate, and tetramethyl ammonium hydrogen sulfate.
5. To change the polarity of the surface of the solid phase so nonpolar compounds can be separated.
6. Dimethyldichlorosilane is added to the solid phase to form a reactive chlorosilane on former -OH groups. Then water is added followed by a long chain dichlorosilane, and finally endcapped.
7. It must be free of dissolved air and fine particles.
8. He is passed through the solvent to remove the air based upon the Second Law of Thermodynamics. The "high concentration" air in the solvent passes into the "zero" concentration air in the He and is removed. The He is easy to compress and does not form bubbles in the chromatograph.
9. It is a syringe connected to a motor by a shaft. When the motor is turned on, the shaft pushes the piston of the syringe forward, forcing out the mobile phase. By gear reduction, very high pressures can be obtained.
10. Check valves are used to prevent liquids from flowing in the wrong direction. They are designed so that flow in one direction forces them to close, whereas flow in the opposite direction allows them to open.
11. To make compounds detectable by uv spectroscopy. See Table 19-5, p. 203.
12.

Fluorescamine   Dansyl Chloride

13. The surface area of half of a sphere.
14. It measures the difference in refractive index between the mobile phase and the mobile phase plus sample.
15. It measures the change in current between two electrodes when an ion passes between them.
16. To ensure that no fine particles pass into the main column and plug it up. The precolumn is inexpensive and easily can be cleaned and replaced, whereas the main column is expensive.
17. No. Use equation 19-14, p. 194. A 20,000 p.s.i. pump would be required.

$$\frac{800}{\frac{30 \, x \, \eta \, x \, v}{\theta \, (50)^2}} = \frac{X}{\frac{30 \, x \, \eta \, x \, v}{\theta \, (10)^2}}: \quad X = 20,000 \, p.s.i.$$

18. No. Use equation 19-14, p. 194, as in question 17. An 8880 p.s.i. pump would be required.

19. Viscosity varies as the mole fraction in mixtures. A 30 % methanol-70 % water solution would have a η of 0.0.92

centipoises, whereas 70% methanol-30% water would have a η of 0.77 centipoises. Refer to equation 19-14, p. 194. Therefore, the pressure would decrease.

20. The pressure is related directly to the viscosity from equation 19-14, p. 194.

$$\frac{5700 \; p.s.i.}{1.20 \; centipoises} = \frac{X \; p.s.i.}{0.32 \; centipoises}; \quad X = 1520 \; p.s.i.$$

21. Polar compounds
22. The nonpolar would not be retained, the medium polar retained somewhat, and the polar compound retained the most.
23. The normal polar surface has been changed to a nonpolar surface.
24. Octyldecyl silane, which is an 18-carbon-long compound.
25. The beads collapse under the pressure and plug up the column.
26. 1.5 pH units above the highest pK_a.
27. It is used to make a polar compound nonpolar so it can be separated on a reversed-phase column.
28. Sulfonic acids.
29. Water, aqueous buffers, and nonaqueous low viscosity solvents.
30. It must be passed through at least a 0.45 μm filter and have the air removed.
31. It forms bubbles during the decompression stroke of the pump and, if not redissolved on the compression stroke, causes baseline noise, interferes with the separation in the column, and gives an error in the detector.
32. Helium removes the air and is very soluble in all HPLC solvents, so any bubbles that form will redissolve.
33. To remove any particles, the same as was done with the solvent.
34. Refer to equation 19-14, p. 194:

$$700 = \frac{300 \times 0.50 \times 0.60}{\theta \times (10)^2} \qquad \theta = 0.00128$$

35. Use the θ calculated above.

$$P = \frac{300 \times 0.50 \times 0.60}{0.00128 \times (3)^2}$$

$P = 7812$ p.s.i.g. - not practical with the normal 6000 p.s.i.g. pumps. Either buy a 10000 p.s.i.g. pump or use a shorter column.

36. An advantage is that it can pump unlimited quantities of solvent. A disadvantage is that a pressure drop occurs when the pump recycles, which causes a fluctuation in the baseline.
37. An advantage is that a constant pressure can be obtained. A disadvantage is that it has a limited storage capacity.
38. When a new pump is started, when a seal is replaced and sometimes when solvents are changed.
39. If the solvent composition remains the same throughout the analysis, it is isocratic.
40. If the solvent composition changes from the beginning to the end of the analysis, it is gradient.
41. 5 to 50 μL.
42. The pressure difference it too great. With HPLC, you are injecting a sample at atmospheric pressure into a column at several hundred, maybe even thousand pounds of pressure. If done like a GLC, the syringe would blow back at you like a bullet.
43. To protect the main column from becoming plugged with particles.
44. Only a few microns of its surface is porous.
45. The mixing volume between the beads becomes smaller with smaller diameter beads.
46. To recondition the column. This washes away material that may have collected on the surface during the day's work.
47. A Hg vapor lamp.
48. -C=C-, -C=O, -N=O and -N=N-.
49. 3,5-Dinitrobenzoyl chloride.
50. 4-Bromo-7-methoxy coumarin.

51. A pulsed xenon lamp.
52. A femtogram. One femtogram = 1×10^{-15} g.
53. It is an apparatus to add a reagents to compounds emerging from an HPLC column. It adds NaOH to carbamates to convert one part to a methyl amine, then it adds o-phthalaldehyde and 2-mercaptoethanol to form a very intense fluorescing compound.
54. It can detect all compounds. Not all pesticides are at trace levels.
55. Any ion or any compound that has a pK_a or pK_b greater than 7.

CHAPTER 20 - GAS-LIQUID CHROMATOGRAPHY

1. If it vaporizes without decomposition below 350 °C.
2. The time, usually in minutes, from the time the sample was injected until half of the compound being investigated passes the detector (the top of the peak).
3. Refer to equation 20-2, p. 213.

    (A)
    $$N = \frac{16 \, (50)^2}{(4)} = 2500 \; plates$$

    Refer to equation 20-3.

    (B) 3 m = 3000 mm;   3000/2500 = 1.2 mm/plate.

4. $P_o = 1.0$; $P_i = 1.5$
    $$V_r^o = \frac{3}{2} \frac{(1.5/1.0)^2 - 1}{(1.5/1.0)^3 - 1} = 0.79$$

5.
    $$4.2 \; min \times \frac{10 \; mL}{20 \; sec} \times \frac{60 \; sec}{min} = 126 \; mL$$

6.
    $$I = 100 \times \frac{[0.72 - 0.60]}{[0.90 - 0.60]} + 100 \times 7 = 740$$

7. A, Eddy diffusion; B, molecular diffusion; C, resistance to mass transfer.
8. That an optimum flow rate for any column occurs at the minimum of the curve.
9. (1) smaller, (2) higher, (3) smaller.
10. Nitrogen and helium.
11. Impurities in the gas (oil, water, oxygen, hydrocarbons) cause baseline problems.
12. Open the main gas cylinder valve for a few seconds before you attach the gauge to blow away any dirt and oil.
13. You get a noisy baseline, the early eluters have a lower response than normal, and the baseline may start to rise if the temperature is programmed.
14. It permits you to obtain a sample in the field and return to the laboratory without losing it.
15. If you have someone with rough technique injecting samples who keeps bending plungers, or if you do not have an auto-injector and need to get reproducible injections.
16. Additional sample may evaporate from the needle tip, the results will be erratic, and tailing may result.
17. It is used all of the time regardless of the sample type. It was designed to be used if the sample contained a wide variety of boiling point compounds and to eliminate the problem of a slow injection.
18. Fused Silica Open Tubular.
19. To eliminate the "race track" effect.
20. Hot metal tends to dehydrohalogenate many pesticides.
21. They reduce the possibility of having the tubing collapse at the bend.
22. A ferrule is a tapered fitting placed over the end of the column that, when clamped in place, will make a gas-tight seal.

23. Vespel-graphite.
24. Dip the end in pentane to ensure that the connector is all the way in then heat it to 150 °C.
25. To protect the column (from pieces of septum) and to serve as a quick connect that is not necessarily gas tight.
26. AW = acid washed; S = silanized.
27. Dissolve the silylating agent in methylene chloride, hexane, or toluene to make a 5 to 10% solution.
28. DB-17, DB-225
29. Polysiloxane, 50% trifluoropropyl, 50% methyl.
30. The rotation causes many of the particles to break, and these "fines" cause peak broadening.
31. Connect one end to a water aspirator and apply a slight vacuum. Place the packing on a piece of folded paper and place this at the other end. Gently tap the paper when the vacuum is applied, and the packing will enter the column. Tap the column gently, but continuously until it is packed.
32. Column length, solid support, liquid phase and %, mesh range, and the date packed.
33. To remove any traces of moisture, organics, and adsorbed gases in the original packing.
34. To ensure that the impurities from the column do not get into the detector.
35. To keep it coiled and manageable. It wants to be straight.
36. To allow for the expansion of a large injection volume and to provide a surface on which the co-extractives can condense on.
37. It provides a square cut with no burrs that would disturb the gas flow.
38. They contain phosphates, which can contaminate the system and be detected by P type detectors.
39. The detector will become contaminated by compounds condensing on the surfaces.
40. Inject a nonretained gas into the column. If the peak is symmetrical, the column is OK.
41. They are conditioned with the detector connected and for only an hour or so.
42. No. The ball rises based on the pressure around it, and the gas viscosities are different.
43. Thermionic (TSD or N/P), electron capture (ECD), flame photometric (FPD), and the Hall (ELCD).
44. Measure the distance from the tops of the baseline signal to the bottom in scale divisions. This is the random fluctuation of the instrument and is the "noise".
45. Limit of detection. It is three times the noise level.
46. Limit of quantitation. It is 10 times the noise level.
47. When the value is above the LOD and below the LOQ.
48. Yes - CH_4, $C_4H_9O_2$, $C_8Cl_4N_2$. No - H_2O, NH_3 - inorganic. No - CO - highly oxidized.
49. Place a mirror over the top of the chamber or hold a piece of paper over the burner tip.
50. The N/P detector has a bead of KCl placed by the end of the burner.
51. Acetonitrile, cyano stationary phases, and tobacco smoke (nicotine). All contain N.
52. Halogenated compounds.
53. An electron emitted from the nucleus of an atom by radioactive decay.
54. ^{63}Ni electroplated onto a metal plate.
55. Remove the ECD and box it; mark the box radioactive and what it contains, then contact a safety officer to have it disposed of. Let him/her prepare all of the paperwork so it is correct. If you throw this away along with the chromatograph, you will think the world is coming to an end when the NRC finds out about it. Everything will be shut down until it can be located, and they will try to make you feel like the world's dumbest person.
56. The emission from the 526 nm line of HPO.
57. The lower flame decomposes the sample, and the upper flame produces the light emission.
58. Decrease the carrier flow and then check the detector gas flow rates.
59. Molecules are irradiated with high energy radiation. They dissociate into ions that then are collected, and their intensity can be measured.
60. He used a.c. rather than d.c., so space charges wouldn't build up and no electroplating would occur on the electrode's surfaces.
61. The sample is combusted in a stream of hydrogen. These products are dissolved in n-propanol and water to form ions, which change the electrical conductivity between two plates placed in the stream.
62. They are not ionized in the 50% n-propanol solution.
63. To remove the acid gases, yet let NH_3 pass through.
64. 99.9999% H_2 and 99.999% N_2.
65. A molecular jet separator connected to a high vacuum pump.
66. $$\frac{\text{mg sample}}{\mu\text{L injected}} = \frac{(80)(100)}{(200 + 88 - 10)(6)} = 4.8 \text{ mg}/\mu\text{L}$$

67. $$\frac{\text{mg sample}}{\mu\text{L injected}} = \frac{(80)(25)}{(350)(5)} = 1.14 \text{ mg}/\mu\text{L}$$

68. $$\frac{\text{mg fat}}{\mu\text{L injected}} = \frac{3.0}{6.0} = 0.50 \text{ mg}/\mu\text{L}$$

69. $$\frac{\text{mg sample}}{\mu\text{L injected}} = \frac{\mu g}{(6.0)(0.50)} = 1.0 \text{ mg}/\mu\text{L}$$

70. $$\frac{56 \text{ mm sample}}{45 \text{ mm standard}} \; x \; \frac{(3.0 \;\mu\text{L standard})(0.25 \text{ ng}/\mu\text{L standard})}{(3.0 \;\mu\text{L sample})(2.5 \text{ mg}/\mu\text{L sample})} = 0.12 \text{ ppm}$$

71. If the peak is small and/or its baseline width is < 10 mm.
72. Adjust the amount of standard injected, so its peak height is as close to the unknown's as possible.
73. At 0.45 h.
74. Construct a trapezoid.
75. 50 g x 0.35 = 17.5 mL of sample moisture.

$$\text{grams recovered} = \frac{(80)(50)}{(317.5)} = 12.6 \text{ g}$$

76. 50 g x 0.095 = 4.75 mL of sample moisture.

$$\text{grams recovered} = \frac{(80)(50)}{(355)} = 11.3 \text{ g}$$

77. The LOD It is the concentration of a compound that would produce a peak height of 3 x 4 or 12 mm and the LOQ It is 10 x 4 or 40 mm.
78. Use equation 20-24, p. 241, to calculate the total ppm in the spiked (or fortification) sample.

$$\frac{93}{74} \; x \; \frac{(4.5)(0.2)}{(2.0)(3.41)} = 0.166 \text{ ppm}$$

Then use equation 20-25, p. 242, to calculate the % recovery:

$$\frac{(0.166 - 0.075)(100)}{0.10} = 91\%$$

79. V = 0.98 x 200 + 0.92 x 0.225 x 100 + 0.62 x 64.2
    V = 257 mL

CHAPTER 21 - PAPER CHROMATOGRAPHY

1. The direction the paper is pulled through the rollers.
2. Sulfa drugs are basic and form salts with HCl.
3. More difficult separations can be made, because the mobile phase can travel further.
4. The polar surface of the paper is replaced with a nonpolar phase.
5. Dry the paper in an oven, then dip it into the reverse phase liquid and dab it dry with paper towels.
6. The oil causes the mobile phase to change its flow rate, which then skews the separation.
7. To ensure that the sample is run under the same conditions as the sample.
8. Adsorption on the inert phase.
9. The sequence is; spot, dry, spot over the same spot, dry again, and repeat as necessary.
10. Determining if good technique has been used is difficult later on.
11. The two layers were at equilibrium. If you establish equilibrium in the chamber with one layer and use the other layer for development, then the entire chamber is at equilibrium, and the concentrations on the paper will not change with time.
12. It permits for quicker equilibrium.
13. When you want to separate larger quantities of material for further work.

14. It is an antisiphon bar. It prevents the solvent from siphoning out of the tank. If siphoning occurs, it is usually so fast that equilibrium cannot be established.
15. Find a solvent in which the solutes are more soluble.
16. The faster the separation, the better the resolution is, providing it is not so fast that equilibrium cannot be established.
17. The ratio of the distance the solute moves to the distance the solvent moves based on the front edge of the solute spot.
18. Paper not at equilibrium, wrong solvent, or too fast of a development.

CHAPTER 22 - THIN LAYER CHROMATOGRAPHY
1. It has no machine direction, and it can withstand strong acids and bases.
2. By adding 7 to 8% $CaSO_4$ to the inert phase.
3. So the space inside will become saturated with vapor, providing for equilibrium conditions.
4. The system is not at equilibrium.
5. To reduce edge effects by the solvent moving up the plate.
6. To remove as much water as possible, because water is polar and covers up the active sites.
7. It ensures that the spots are placed uniformly on the plate. The circles are used to provide a semi-quantitative estimation of the sample.
8. The solvent front will be retarded at the spot.
9. The spot will form an elongated area.
10. Its composition can change by one portion being more soluble in the coating than others.
11. The more polar the solvent the greater the eluting power on polar plates is.
12. The pressurized can, because it provides a finer mist and is more reproducible.
13. So you do not double the concentration of the developing reagent on the sides of the plate.
14. After the plate is dried and replaced in the solvent, the solvent mobilizes the bottom edge of the spot, but not the top edge. This permits the bottom molecules to catch up to the top molecules.
15. If the R_f's are close, and if you are going to court.
16. It destroys the compounds.
17. It permits you to have a straight solvent front at the end of the development, and R_f's are easier to determine.
18. 1 g of platinum iodide was $80 in 1997.
19. Manganese-activated zinc silicate, zinc/cadmium sulfide, lead/manganese-activated calcium silicate.
20. 0.1 to 1 µg.

CHAPTER 23 - ION EXCHANGE
1. It will allow liquid from the column to flow until the liquid in the column is across from the outlet on the tip.
2. Ion exchange is the exchange of ions without regard to separating each type. It can be done in a bucket. Ion chromatography is done in a column, so each type of ion that has been exchanged can be separated from the others.
3. They have low capacity and are sensitive to acids and bases as well as temperature.
4. The formation of spherical particles of polymer.
5. The polymer exchange group is a carboxylic acid, so it is a weak acid. The resin cation is H^+; the beads will pass through a 30 mesh sieve, but not a 60 mesh. 4% divinyl benzene was added at the beginning of the polymerization. The exchange capacity is 3.6 meq/g when the resin is dry.
6. The anions will pass through as a group with the elution front. The remaining ions will be retained as $Co^{+2} > Ba^{+2} > Ag^+ > NH_4^+ > Na^+$ based on the higher charge being retained longer then the higher m.w. for those of the same charge.
7. The passing across the surface of the resin bead.
8. Less mixing volume exists between the beads, and the ions have less distance to move through the bead, so the band broadening is less.

CHAPTER 24 - ION CHROMATOGRAPHY
1. The addition of a suppressor column to eliminate the conductivity of the eluent.
2. All anions that have pK_a's less than 7 and all cations with pK_a's greater than 8.
3. 25,000 plates/m and ppb.
4. It is universal for all ionic species and is linear at low concentrations.

5. The anion of the acid exchanges with the OH⁻ which then reacts with the H⁺ of the acid to form non-ionic and nonconducting HOH.
6. It has a -2 charge and can be used at low concentrations.
7. It can form oxidation products that tend to poison the column and shorten its life.
8. Choose an eluent that does not dissociate strongly.
9. Benzoic acid.
10. Too much sample.
11. Their capacity is too large, requiring large amounts of eluent.
12. 1 to 2 meq/g to 0.04 meq/g.
13. They tend to break up at *pH*'s greater than 8, and they adsorb F⁻.
14. They are temperature sensitive.
15. By the fixed negative sulfonic acid functional groups in the membrane.
16. Tetrabutyl ammonium hydroxide.
17. A constant current power supply.
18. $2 H_2O \longrightarrow 4 H^+ + O_2 + 4 e^-$
19. $2 H_2O + 2e^- \longrightarrow 2 OH^- + H_2$
20. The conductance between two metal plates 1 cm apart with 1 L of 1 *N* solution between them.
21. The Siemen (S), $10^{-4} m^2 mol^{-1}$.
22. H⁺ (~350) and OH⁻ (198).
23. A blackened Pt indicator electrode with a Ag/AgCl reference electrode.
24. It removes the oxide film buildup on the electrode surface.
25. The system has a constant and measurable signal until the sample comes through, which creates a vacancy.
26. The sample loop is replaced with a small section of ion exchange resin, and the ions are concentrated. These then are eluted later in a much more concentrated manner.

CHAPTER 25 - ION RETARDATION, ION EXCLUSION, AND LIGAND EXCHANGE

1. It has both cation and anion groups adjacent to each other.
2. To separate ionic compounds from nonionic compounds.
3. Strands of either cation or anion resin.
4. The opposite charges and the mobility of the strands.
5. The cation and anion strands are cross-linked to each other.
6. Divinyl benzene, sorbic acid, and muconic acid.
7. 0.7 meq/mL.
8. Snake-cage is limited to 6 to 8 times whereas net-cage can go over 50 times without degradation.
9. The differences in solubility of ionic and nonionic compounds between the resin beads and the aqueous surroundings.
10. Either the cation on the resin must be the same as the major cation in the sample, or the anion on the resin must be the same as the sample.
11. The differences in solubility between the ionic compounds and the nonionic compounds in the resin.
12. In the aqueous phase outside of the resin beads.
13. Any nondissociated compound.
14. By adding a small amount of strong acid to the eluent.
15. It is less expensive, because there is no cost of regeneration of the resin.
16. 8% for ionics and 40% for nonionics.
17. The ionics are separated immediately, and the nonionics then can be separated.
18. It permits you to use an ordinary HPLC column, yet modify it temporarily for ion exclusion.
19. The size of the molecule, the chain length, and solution concentration.
20. Pellicular, 5 to 10 μm diameter beads with either strong acid or base functional groups.
21. Nonionic compounds can be exchanged.
22. Complexing metal ions.
23. Complex formation.
24. Ligands of higher valence are retained more.
25. Assign oxidation numbers to N. Assume CH_3 is neutral. $H_2NN(CH_3)_2$ (N = -1), H_2NNHCH_3 (N = -1.5) and H_2NNH_2 (N = -2).

CHAPTER 26 - ELECTRODEPOSITION

1. How fast a metal ion is deposited.
2.

$$5.0 \; g \times 0.75 = \frac{1.5 \; amp \times t \; sec \times 63.5 \; a.w.}{2 \; electrons \times 96{,}485 \; coulombs}; \quad t = 7597 \; seconds \qquad \text{Increase the current.}$$

3. The nucleus of each metal ion has a different affinity for their outer electrons.
4. Sir Thomas Browne, an English physician.
5. Benjamin Franklin, an American scientist, publisher, and politician.
6. H.L.F. von Helmholtz, a German physicist.
7. Joseph J. Thompson, an English physicist.
8. Write as reduction potentials with most + on top. The reaction will go spontaneously in a clockwise direction.
   $Ce^{+4} + e^- = Ce^{+3} \qquad E^o = +1.61 \; v$
   $Sn^{+4} + 2e^- = Sn^{+2} \qquad = +0.14 \; v$
   Ce^{+4} will be reduced and Sn^{+2} will be oxidized.
9. $+1.61 \; v + (-0.14 \; v) = +1.47 \; v$. Remember - the $+0.14$ now becomes -0.14 in the other direction.
10. The most + will plate out first. $Ag^+ = +0.80 \; v$, $Fe^{+3} = -0.036 \; v$, $Pb^{+2} = -0.126 \; v$, $Zn^{+2} = -0.76 \; v$.
11. Oxidation occurs at the anode therefore Cl_2 gas would form.
12. The single electrode at each electrode, the polarization potential at each electrode, and the IR drop across the cell. They are converted easily to S which coats the electrodes.
13. $E_{back} = E^o_{anode} - E^o_{cathode}$; $E_{back} = +1.23 \; v - (-0.71 \; v) = +1.94 \; v$.
14. To minimize concentration over voltage.
15. This ensures that each electron has time to do what you want it to, rather than a competing reaction.
16. The amount of charge buildup needed for an electron to transfer from the electrode to the ion.
17. It can form before the desired metal begins to plate out.
18. It tends to make the metal quite brittle.
19. Metals that are + to H_2 can be deposited in acid solution.
20. It reduces the concentration of the metal at the electrode surface.
21. Perchlorate first and nitrate second.
22. As the solution becomes more acidic the more + the E^o value becomes.
23.

$$E = +1.52 - \frac{0.059}{5} \log \frac{[10^{-5}][1]^4}{[0.1][0.01]^8}; \quad E = 1.52 \; v - 0.14 \; v = +1.38 \; v$$

24. A potential buffer is a mixture of a reduced form and an oxidized form of the metal, both as ions. One ion is reduced at one electrode, and the other is oxidized at the other, keeping the potential the same.
25. 0.2 v.
26. The working electrode where the deposition occurs, an auxiliary electrode to complete the circuit, and a reference electrode.
27. A device that maintains a set potential in a solution.
28. 0.020 amperes.
29. To keep from being electrocuted or getting a shock.
30. 1 ampere/second or a quantity of charge, which in one second crosses a section of a conductor that has a constant current of 1 ampere.
31. The anode.
32. Make the article the anode for a few minutes.
33. Make the article the anode for a few minutes in a polishing solution.
34. Cover the base metal with a more active metal or one that will adhere better than the final plating metal.

CHAPTER 27 - ELECTROPHORESIS - GENERAL

1. The ion with the smallest diameter.
2. The rate of movement of an ion in a liquid in an electrical field.
3. Reactions with acids and bases, dissociation by polar solvents, H-bonding, chemical reactions, polarization, and ion pair formation.
4. The viscosity of the solvent retards the ion.
5.

$$U = \frac{2 \times 4.8 \times 10^{-10}}{6 \times 3.14 \times 2.78 \times 10^{-8} \times 9.3 \times 10^{-3} \times 300} = 6.57 \times 10^{-4} \ cm^2 volt^{-1} sec^{-1}$$

6.

$$u = \frac{1}{2}\sum (0.0200 \times 2^2) + 2 \times 0.0200 \times 1^2) = 0.006$$

$$\frac{1}{1 + 2.78 \times 10^{-8} \times 0.233 \times 10^8 \times \sqrt{0.006}} = 0.95$$

7. V is in volts, E is in volts/cm.
8.

$$k = \frac{17.5 \, cm^{-1}}{8150 \, ohms} = 0.00215 \ ohm^{-1} cm^{-1}$$

9.

$$E = \frac{0.018}{(3.14 \times 0.5^2)(0.00215)} = 10.65 \ volts/cm$$

10.

$$U = \frac{5.20}{(196 \times 60)(10.65)} = 4.15 \times 10^{-5} \ cm^2 volt^{-1} sec^{-1}$$

11.

$$H = \frac{(0.018)^2}{(3.14 \times 0.5^2)^2 \, (0.00215)} = 0.192 \ W/cm^3$$

12. The movement of a liquid or liquid solution in an electric field.
13. Water moves to the negative electrode, indicating that water is net positive.
14. It has such a low mobility that it just drifts with the liquid.
15.

$$U = \frac{5.20 \ cm + 0.3 \, cm}{10.00 \ volts/cm \times (190 \, minutes \times 60 \, seconds/minute)} = 4.8 \times 10^{-5} \ cm^2 volt^{-1} sec^{-1}$$

16. To reduce the possibility of trapping air bubbles.
17. Agarocytes in algae found in the Pacific and Indian oceans and the Sea of Japan.
18. To make depressions for sample addition.
19. They are flexible, stronger and easier to handle than other gels, as well as being transparent to uv/vis radiation and having negligible electro-osmosis and diffusion.
20. To ensure that the net charge on the compounds stay the same from the beginning to the end of the separation.
21. Coomassie blue R-250 is 100 x less sensitive than silver, which is 150 times less sensitive than auto-radioography.
22. The compound is 2-methoxy-2,4-diphenyl-3(2H)-furanone and it can detect 1 ng of protein.
23. ^3H, ^{32}P, ^{35}S, ^{14}C and ^{125}I.
24. If you wanted to blot it.
25. To eliminate diffusion.
26. The compounds diffuse upward by capillary action.
27. To transfer DNA fragments from an agarose gel to a nitrocellulose membrane.
28. To cover all of the sites on the membrane that do not react with the compound you are interested in, so only your compounds will adhere.
29. To indicate the progression of the separation.
30. To reduce evaporation and as a safety precaution against electrocution.
31. To prevent siphoning.
32. It allows the molecules to become "unstuck" and move in another direction.
33. The free end of the molecule moves forward, freeing the other end, and then that end moves forward.
34. This design permits a uniform electric field across the gel to be obtained and the associated electronic switching can be very fast.
35. Refer to equation 27-5, p. 316.

$$\eta = \frac{-11.4 \times 4.8 \times 10^{-10}}{-2.0 \times 10^{-4} \times 6 \times 3.14 \times 2.78 \times 10^{-7} \times 300} = 1.74 \times 10^{-2} \; poises$$

36.

$$r = \frac{1 \times -4.8 \times 10^{-10}}{-2.49 \times 10^{-4} \times 6 \times 3.14 \times 9.6 \times 10^{-3} \times 300} = 3.55 \times 10^{-8} \; cm$$

37. Refer to equation 27-5, p. 316.

$$r = \frac{4.8 \times 10^{-10}}{6.09 \times 10^{-4} \times 6 \times 3.14 \times 8.28 \times 10^{-3} \times 300} = 1.68 \times 10^{-8} \; cm$$

38. Refer to equation 27-5, p. 316.

$$7.3 \times 10^{-5} = \frac{n \times 4.8 \times 10^{-10}}{6 \times 3.14 \times 1.44 \times 10^{-7} \times 8.05 \times 10^{-3} \times 300}; \quad n = +1$$

$$-1.46 \times 10^{-4} = \frac{n \times 4.8 \times 10^{-10}}{6 \times 3.14 \times 1.44 \times 10^{-7} \times 8.05 \times 10^{-3} \times 300}; \quad n = -2 \quad A\;3\;e^- \;change$$

39. Refer to equations 27-8 to 27-11, pp 317 to 318.
    $k = C/R = 24.7 \; cm^{-1}/10,500 \; ohms = 0.00235 \; ohm^{-1} \; cm^{-1}$.

$q = (0.55 \text{ cm})^2 \times 3.14 = 0.95 \text{ cm}^2; q^2 = 0.90 \text{ cm}^4$

$$H = \frac{(0.010 \text{ amp})^2}{0.90 \text{ cm}^4 \times 0.00235 \text{ ohm}^{-1} \text{ cm}^{-1}} = 0.047 \text{ W/cm}^3$$

$$E = \frac{0.01 \text{ amp}}{0.95 \text{ cm}^2 \times 0.00235 \text{ ohm}^{-1} \text{ cm}^{-1}} = 4.48 \text{ V/cm}$$

40. Refer to equations 27-8 to 27-11, pp 317 and 318.
    $k = C/R = 12.6 \text{ cm}^{-1}/4150 \text{ ohms} = 0.0030 \text{ ohm}^{-1} \text{ cm}^{-1}$.
    $q = (0.50 \text{ cm})^2 \times 3.14 = 0.785 \text{ cm}^2; q^2 = 0.616 \text{ cm}^4$

$$H = \frac{(0.010 \text{ amp})^2}{0.616 \text{ cm}^4 \times 0.0030 \text{ ohm}^{-1} \text{ cm}^{-1}} = 0.054 \text{ W/cm}^3$$

$$E = \frac{0.01 \text{ amp}}{0.785 \text{ cm}^2 \times 0.0030 \text{ ohm}^{-1} \text{ cm}^{-1}} = 4.25 \text{ V/cm}$$

$$U = \frac{+6.2 \text{ cm}}{(75 \text{ min} \times 60 \text{ sec/min}) \times 4.25 \text{ V/cm}} = +3.24 \times 10^{-4} \text{ cm}^2 \times \text{volt}^{-1} \text{ sec}^{-1}$$

41. Refer to equations 27-8 to 27-11, pp 317 and 318.
    $k = C/R = 15.2 \text{ cm}^{-1}/2500 \text{ ohms} = 0.0061 \text{ ohm}^{-1} \text{ cm}^{-1}$.
    $q = (0.40 \text{ cm})^2 \times 3.14 = 0.50 \text{ cm}^2; q^2 = 0.25 \text{ cm}^4$

$$H = \frac{(0.009 \text{ amp})^2}{0.25 \text{ cm}^4 \times 0.0061 \text{ ohm}^{-1} \text{ cm}^{-1}} = 0.053 \text{ W/cm}^3$$

$$E = \frac{0.009 \text{ amp}}{0.25 \text{ cm}^2 \times 0.0061 \text{ ohm}^{-1} \text{ cm}^{-1}} = 5.90 \text{ V/cm}$$

$$U = \frac{+5.8 \text{ cm}}{(13 \text{ hr} \times 60 \text{ min/hr} + 36 \text{ min}) \times 60 \text{ sec/min} \times 5.90 \text{ V/cm}} = +2.0 \times 10^{-5} \text{ cm}^2 \times \text{volt}^{-1} \text{ sec}^{-1}$$

42. Refer to equations 27-8 to 27-11, pp 317 and 318.
    $k = C/R = 18.9 \text{ cm}^{-1}/8140 \text{ ohms} = 0.0023 \text{ ohm}^{-1} \text{ cm}^{-1}$.
    $q = (0.50 \text{ cm})^2 \times 3.14 = 0.785 \text{ cm}^2; q^2 = 0.616 \text{ cm}^4$

$$H = \frac{(0.012 \text{ amp})^2}{0.616 \text{ cm}^4 \times 0.0023 \text{ ohm}^{-1} \text{ cm}^{-1}} = 0.10 \text{ W/cm}^3$$

$$E = \frac{0.012 \text{ amp}}{0.785 \text{ cm}^2 \times 0.0023 \text{ ohm}^{-1} \text{ cm}^{-1}} = 6.65 \text{ V/cm}$$

$$U = \frac{+3.8 \text{ cm}}{86.5 \text{ min} \times 60 \text{ sec/min} \times 6.65 \text{ V/cm}} = +1.10 \times 10^{-4} \text{ cm}^2 \times \text{volt}^{-1} \text{ sec}^{-1}$$

CHAPTER 28 - IMMUNOELECTROPHORESIS
1. Any compound or group of compounds that reacts with antibodies.
2. A material produced in the body to react with invading materials.
3. To isolate a specific compound.
4. Protein molecules that function as antibodies.
5. The bloodstream.
6. 6 to 7 days.
7. It is not fully emptied, preventing air being blown into the filled wells.
8. As the antigen moves forward it meets more antibody. But because it is now less concentrated, it will diffuse slower, so the line of precipitate is not as far from the center of migration.
9. To determine which collected fractions contain the compounds of interest.
10. It is a two-dimensional separation.
11. It permits adding an antibody specific for the compound of interest, making its identification in a complex mixture much easier.

CHAPTER 29- POLYACRYLAMIDE GEL DISC ELECTROPHORESIS
1. One on top of the other, rather than side by side.
2. The small pore separation gel on the bottom and the large pore spacer gel and a large pore sample gel on the top.
3. If two solutions of ions are layered one over the other with the solution of higher mobility on the bottom, then the boundary between the two will be sharp as they migrate in an oppositely charged electrical field.
4. The lower solution must be more dense than the top solution, and the polarity must such that the ions are pulled downward.
5. The glycine molecules are very slow moving and the chloride ions are fast moving compared to the protein, so the protein molecules pass through the glycine and become layered between the glycine and the chloride ions.
6. This acts as a marker dye and will form a blue disc in the tube; the disc moves faster than the protein fragments to indicate when to stop the separation.
7. Place a drop of water on top of the unpolymerized polymer before polymerization. Be sure to remove it afterwards.
8. Ultraviolet radiation and potassium persulfate.
9. The persulfate solution has decomposed and a fresh solution should be used.

CHAPTER 30 - ION FOCUSING, SDS, AND TWO-DIMENSIONAL ELECTROPHORESIS
1. The net charge on a molecule can change depending upon the pH of its surroundings, and such a molecule will travel in an electric field in a medium of varying pH until it is neutral.
2. The pH at which the net + charge equals the net - charge.
3. A molecule with both + and - charges, equally ionized.
4. By adding compounds called ampholytes to the mixture. These migrate to their respective pH's when current is applied.
5. 24 to 72 hours for nongel systems, and 1 to 3 hours for gels.
6. Mixtures of aliphatic polyamino-polycarboxylic acids with molecular weights between 300 to 700.
7. The sample molecules are usually much larger than the ampholytes and move slower, allowing the ampholytes time to establish the pH gradient.
8. These are derivatives of acrylamide that are immobilized by polymerizing them into the polyacrylamide gel. This permits stable and reproducible linear gradients.
9. Sb wire for the indicator electrode and a micro-calomel for the reference electrode.
10. 0.45 watts/40 cm^2 = 11.2 mW/cm^2 or 11.2×10^{-3} J/s/cm^2. Therefore, $11.2 \times 10^{-3} \times 900 \times 2.5 = 25.3$ °C.
11. They conduct electricity and would be a safety hazard.
12. As the ampholytes and the sample molecules become neutral, they do not conduct current.
13. A collection of polar molecules arranged such that, when placed in water, the nonpolar ends tend to congregate so the polar ends face outward towards the water.
14. It is a reducing agent and breaks the -S-S- bonds holding the protein together.
15. It is best used to separate either neutral or high-molecular-weight soluble compounds.
16. The pore size of the gel retards the larger molecules.
17. You can always straighten a curved line by plotting one of the axes as a log. In this case, the pore diameters vary as the square of their diameter, whereas the molecular weight is proportion to the cube of the diameter.

18. It produces narrow bands, so the second dimension can start with sharp divisions.
19. Proteins differing by only a single charge can be separated, and amounts from 10^{-4} to 10^{-5} of the total protein can be detected.

CHAPTER 31 - CAPILLARY ZONE ELECTROPHORESIS
1. 10 to 100 µm and 30 to 100 cm long.
2. Electrophoresis and electro-osmosis.
3. As the square of the radius.
4. The H^+ ions dissociate, and the surface becomes negatively charged.
5. To the cathode.
6. Buffer ions, surrounded by water, act as if they have a positive surface and migrate in bulk to the cathode when a voltage is applied.
7. Anions, neutrals, and cations.
8. The voltage.
9. High voltage and short columns.
10. The best resolution will be obtained when the electro-osmotic flow just balances the electrophoretic migration.
11. The side wall of the capillary is coated with trimethylchlorosilane to neutralize the surface.
12. 5 to 50 nL; by either a hydrostatic or electro-migrative method.
13. The preconcentration effect permits much lower detection limits, down to sub-ppb in some cases.
14. Use cyclodextrin. Optical isomers associate differently and one form will be retained more than the other.
15. Mobility increases about 2%/°C, so the warmer center of the liquid column is in the center and flows faster. With a capillary, this difference is small or nonexistant, so a flat front is formed.
16. To carry most of the current.
17. Reverse the polarity and use cationic electrolyte additives, such as hydroxybutyric acid.
18. Modifiers are compounds that alter the direction and rate of the electro-osmotic flow. *s*-benzyylthiouronium chloride.
19. A relative apparent mobility is a way of helping determine the type of buffer to use.
20. pH.
21. The entire column is filled with an readily detectable compound that is displaced by the sample, so a vacancy occurs when the sample passes the detector.
22. 10- to 1,000-fold.
23. Lengthwise fluorescence. The excitation beam is passed lengthwise through the column, and the sum of the fluorescence of all of the compounds is recorded. As a compound emerges and drops out, the fluorescence decreases.
24. Use a membrane system fraction collector like that shown in Figure 31-11, p. 369.
25. Too large of a sample or adsorption to the column surface.
26. Strongly acidic anions.

CHAPTER 32 - FIELD FLOW FRACTIONATION
1. Molecules are forced to one wall of a channel and then a parabolic front flow of mobile phase is passed over them. The molecules near the center of the channel are pushed by a faster flow than those near the edge, and this results in a separation.
2. A random walk.
3. The diffusion coefficient.
4. About 20 °C between the top and bottom plates.
5. Cooling the system over time rather than warming it.
6. Centrifugal force.
7. 0.05 to 100 µm.
8. It decreases the HETP, because it is in the numerator of equation 32-10, p. 375.
9. A semipermeable membrane, permitting small mobile phase molecules to pass.
10. Flow FFF is more selective.
11. The larger particles provide more resistance to the mobile phase flow, so are pushed more.

CHAPTER 33 - PURGE AND TRAP AND DYNAMIC HEADSPACE ANALYSIS

1. The removal of a gas, liquid, or volatile solid from a solution by the diffusion of these materials into a stream of gas.
2. By passing a stream of small gas bubbles of nitrogen through the solution for a few minutes.
3. The second law of thermodynamics.
4. Any organic compound that boils below 180 °C.
5. Vapor pressure, solubility, temperature, sample size, method of purge, and purge volume.
6. They are not very soluble in water.
7. 30 to 50 mL/minute.
8. Purge, trap, desorb, liq N_2 trap, warm up and inject into the GLC, separate, detect.
9. Slurries, soils, solids, foaming liquids, and viscous samples.
10. Tenax, silica gel, coconut charcoal, 3% OV-1 on Chromosorb WHP, molecular sieves, and carbo traps.
11. Weakly polar compounds and no water.
12. About 5 mm in diameter and 25 cm long.
13. The sample has been diluted 100-fold in the ethanol, so do the calculation based on 2,000 ppm rather than 20 ppm. Use the ethanol density, because it overwhelms that of the dimethyldisulfide.

$$\frac{1.23\ g}{L} \times 2.00\ L \times \frac{2.00\ g}{1,000\ g} \times \frac{1.00\ mL}{0.789\ g} \times \frac{1,000\ \mu L}{1.00\ mL} = 6.2\ \mu L$$

14. Originally, 128 industrial compounds that the EPA considered as their first priority to monitor in the nations streams, lakes, and surface water.

CHAPTER 34 - FOAM FRACTIONATION, GAS- AND LIQUID-ASSISTED FLOTATION

1. It includes all adsorptive bubble separation methods.
2. The ratio of the concentration in the collapsed foam to the concentration in the feed.
3. To concentrate materials in low concentration in liquids where other processes are not economical.
4. An irregular dispersion of gas in a liquid, with the largest volume usually being the gas.
5. It makes the surface of the gas bubble polar, so it can collect molecules of opposite polarity.
6. The gas pressure inside of a small bubble is greater than that inside of a large bubble. The gas dissolves first into the lamellae, then diffuses through it to a region of lower pressure.
7. A dodecahedron, a 12-sided bubble with pentagonal faces.
8. The micelles collect colligend on their surface, but they stay in solution rather than becoming a part of the foam, and this reduces the effectiveness of the separation.
9. Part 1: Applying equation 34-9, p. 398, to the surfactant,

$$\Gamma = \frac{[(4.7 \times 10^{-7}) - (1.8 \times 10^{-7})] \times 0.22 \times 0.12}{6 \times 4} = 3.2 \times 10^{-10}\ g\text{-}mol/cm^2$$

Part 2: Applying equation 34-9, p. 398, to the colligend,

$$\Gamma = \frac{[(4.7 \times 10^{-12}) - (1.8 \times 10^{-13})] \times 0.22 \times 0.12}{6 \times 4} = 5.0 \times 10^{-15}\ g\text{-}mol/cm^2$$

10. Step 1: For this system, $E = 1$. $K = \Gamma_1/C_1 = (5 \times 10^{-15})/(1.8 \times 10^{-13}) = 0.028$ cm.
    Step 2: Substituting into equation 34-6, p. 398:

$$\Gamma_2 = \frac{5 \times 10^{-15}}{1 + \frac{[(5 \times 10^{-7}) - (3.5 \times 10^{-7})] \times 0.028}{3.2 \times 10^{-10}}} = 3.5 \times 10^{-16}\ g\text{-}mol/cm^2$$

11. About 10^{-4} M.
12. Polyoxyethylene glycol fatty alcohol ethers.
13. Quaternary amines, silicones, and benzene.
14. The efficiency will increase until micelle formation occurs, then slowly decrease.
15. The selective separation of solid particles from a heterogeneous mixture of particles dispersed in a liquid.
16. the particle surface must be hydrophobic.
17. To convert the surface of a mineral particle to being hydrophobic.
18. A compound that promotes collection.
19. Pine oil, certain aliphatic alcohols, polypropylene glycol ethers, and cresylic acids.
20. A large contact angle and sharp particle edges.
21. It requires energy to alter the contact angle, and this usually does not occur.
22. The plate is immersed in the liquid and then tilted until there is no deformity where the liquid contacts the plate.
23. 0.02 to 2.0 lb.
24. 45 to 65 mesh.
25. By washing the particles with acid or base.
26. Sulfuric acid.
27. To prevent premature flocculation.
28. About equal amounts of oleic and linoleic acids.
29. To make it water soluble.
30. Heteropoly organic compounds having one or more water-repellent, nonpolar hydrocarbon groups attached to a hydrophilic polar group.
31. So that large particles of ground grain can be removed from the beneath the floating mineral oil layer.
32. A rubber disc about 5 cm in diameter attached to the end of a 3-mm-diameter brass rod.
33. To seal off the oil layer from the sample so it can be poured off without contamination from the sample.

CHAPTER 35 - OSMOSIS AND REVERSE OSMOSIS

1. To achieve maximum disorder for the entire system.
2. The swelling actually might increase, because the $MgSO_4$ solution would be high concentration water compared to the water content of the cells in the swollen ankle.
3. $E = q - w$.
4. A 1 M sucrose solution = 342 g dissolved and diluted to 1 L of water. A 1 m sucrose solution = 342 g + 1000 g of water.
5.
$$\pi = 0.0821 \times (22 + 273) \times \frac{3.5}{8400 \times 0.250} = 0.040 \ atm.$$

6. Applying sufficient back pressure to force the normal osmosis process to go in reverse.
7.
$$\pi = \frac{0.200}{(100 - 0.200)} \times \frac{82.1 \times (25 + 273)}{58.4} = 0.840 \ atm.$$

$E = 0.840$ atm x 50,000 cm^3 x 0.10133 J/atm cm^3 = 4254 J

8.
$$\pi = \frac{1000}{1000 - 50} \times \frac{0.200}{(100 - 0.200)} \times \frac{82.1 \times (25 + 273)}{58.4} = 0.884 \ atm.$$

$E = 0.884$ atm x 50,000 cm^3 x 0.10133 J/atm cm^3 = 4480 J

9. A film that will let small molecules through but not larger ones.

10. The symmetric membrane has a uniform composition throughout its thickness whereas the asymmetric membrane has a dense, but thin, top layer, and a strong, more porous sublayer beneath it.
11. 0.1 to 0.15 gallons/ft^2/day at 600 to 1,000 psi.
12.

$$85,000 \; ft^2 \; \times \; \frac{0.12 \; gallon}{ft^2 \; day} \; \times \; \frac{1 \; day}{24 \; hrs} \; \times \; 8 \; hrs \; = \; 3400 \; gallons$$

13. A pure water layer of about 10 Angstroms thick develops at the surface, which tends to repel salts as ion pairs.
14. A sieving action related to the size and shape of the molecules.
15. Formaldehyde, because it is a bactericide and is small enough to penetrate most membranes.
16. The diffusion coefficient, the permeability coefficient, the solubility constant, the filtration coefficient, the solute permeability coefficient, and the reflection coefficient.
17. c at 0% is estimated at 0.51 M. $\psi_p = 0.51 \times 0.0821 \times 295 = 12.5$ atm.

CHAPTER 36 - DIALYSIS AND ELECTRODIALYSIS

1. These were the compounds that Thomas Graham found would not pass through a membrane and were gummy when the solvent was evaporated.
2. The concentration differences between the opposite sides of the membrane.
3. Glycerol as a humectant, sulfides as plasticizers, and heavy metals.
4. So the bag will float above the stirring bar.
5. It is much faster than regular tubing dialysis.
6. It tries to hasten the separation of charges particles through the membrane by applying a voltage.
7. ACACACAC.
8. The Na$^+$ ions cannot pass through the anion membrane and remain in the compartment. The Cl$^-$ ions cannot pass through the cation membrane and remain in the compartment.
9. ACCACCACCA.
10. To replace one of the ions (cations) in the sample with a different ion (cation) or an anion with a different anion.
11. The polarity is reversed periodically to clean the membranes and prevent scaling and fouling.
12. Years ago, water was poured through sand, gravel, and vegetation fibers to remove the mud and much of the bad taste so that it was good enough to put in a pot and take home to drink. It was "potable".
13. 1297 ppm/61,470 ppm/N = 0.0211 N; 214 ppm/61,470 ppm/N = 0.0035 N. $\Delta N = 0.0176$ N.
    998.1 gal/min $\times$ 3.7 L/gal $\times$ 1.0 min/60 sec = 61.58 L/sec.
14.

$$I \; = \; \frac{96,500 \; \times \; 61.5 \; \times \; 0.0176}{93.1 \; \times \; 500} \; = \; 2.24 \; Amperes$$

15. 476 = 15.32 x ohms. Ohms = 31
16. $\Delta N = (0.052 - 0.0081) = 0.044$. Therefore:

$$e \; = \; \frac{26.8 \; \times \; 9235 \; \times \; 0.044 \; \times \; 100}{24.6 \; \times \; 450} \; = \; 98.4\%$$

17. To prevent the membranes from warping when pressure is applied.
18. The current is carried by the ions passing through the membrane.
19. It permits higher electric current flow per unit area and encourages mixing, so more ions can get to the membrane faster and the membrane then is more efficient.

CHAPTER 37 - FILTERING AND SIEVING

1. A barrier of interwoven wire or cloth in a regular pattern with uniform openings.
2. From about 1 µm to 0.01 µm.
3. High efficiency particulate air.
4. They can filter particles down to 0.3 µm with 99.96 % efficiency.
5. The number of particles per cubic meter per either 8 hours or 24 hours of exposure.
6.

$$11.7\ (avg\ of\ 12)\ x\ \frac{1000\ L}{10\ x\ 13.6\ L}\ x\ \frac{1440\ min}{12\ x\ 10\ min}\ =\ 1032\ pollens$$

7. It retains particles throughout the membrane.
8. So the filter paper does not have to be offset when it is folded.
9. Flutes provide a larger surface area for the solution to drain and increase the flow rate. They are about 17 % faster on a timed test.
10. It is treated with HF to volatilize the silicates.
11. The paper is treated with a silicone, which then makes it impermeable to water but not organic liquids.
12. No. 41 = 12 sec; no. 40 = 75 sec; no. 42 = 240 sec.
13. About 4 times faster.
14. The weight of the column of water hastens the filtering and draws out more solution from the precipitate.
15. The solution is forced to go around the bubble and its resistance against the glass wall is considerable.
16. First to prevent splashing and secondly, breaking the surface tension of the drops emerging from the tip probably will hasten the filtering.
17. A thin sheet of rubber used by dentists to keep a tooth dry and by chemists during a vacuum filtration to squeeze the liquid out of the cake without having cracks form.
18. There are 50 holes per linear inch or 50 x 50 = 2,500 holes per in^2.
19. Never, never, never. Nothing can ruin a screen faster except an acid bath.
20. A no. 100.
21. Use an ultrasonic cleaner.
22. Standard Reference Material.
23. Glass beads and silicon nitride particles.
24. Wet sieving is used to overcome particle agglomerates, caking and electrostatic attraction.
25. A dry sieving will permit about 1% of the particles to pass, whereas a Wet-Vac will pass 10 %.

CHAPTER 38 - DENSITY GRADIENTS

1. g/cm^3 for liquids and g/L for gases.
2. Specific gravity is the ratio of the density of a solid or liquid to the density of water. No units.
3.

$$\frac{1.032\ g/cm^3}{0.997\ g/cm^3}\ =\ 1.035$$

4.

$$Volume\ ft^3\ =\ \frac{226\ lb\ -\ 9.0\ lb}{62.4\ lb/ft^3}\ =\ \approx 3.5\ ft^3$$

5. The density of a column, of liquid in this case, varies in a regular manner from the top to the bottom.
6. It is because the gradient is not yet fully established. As time goes by, this will change. It is not good, because the pattern is not as unique as it can be. Have patience.
7. No. Each particle will still fall to its density. Because more particles are present, the pattern will be more reliable.

8.

$$\text{Density} = \frac{0.660 \ g/cm^3 \times 43 \ mL = 1.589 \ g/cm^3 \ (57 \ mL)}{100 \ mL} = 1.19 \ g/cm^3$$

CHAPTER 39 - CENTRIFUGATION
1. 0 to 15,000 $x \ g$.
2. Solvent friction and diffusion.
3. To keep the spinning sample cool by reducing friction.
4. Up and sideways.
5. A concentration gradient as a function of the distance r to the rotor axis.
6. The moving diaphragm was replaced by a cylindrical lens.
7. The use of an interferometer arrangement.
8. An interference pattern developed by Rayleigh optics.
9. The results are available immediately.
10. RCF = $1.118 \times 10^{-5} \times 5.0 \times 2.5 \times (3,000)^2$ = 1250 x g.
11. 9500 x g.
12.

$$V_s = \frac{(5.56 \ g/cm^3 - 1.062 \ g/cm^3)(2.5 \times 10^{-4} cm)^2 \left(\frac{2 \times 3.14 \times 1,500}{60 \sec}\right)^2 \times 3.5 \ in \times 2.54 \ cm/in}{18 \times 0.00948 \ g/cm-\sec} = 0.36 \ cm/\sec$$

13. A measure of the rate of settling and is equal to 1×10^{-13} seconds.
14. A container of known volume used to determine densities.
15. Log 12.840 = 1.1085; Log of 12.993 = 1.1137; $d \log x$ = 0.0052.

$$s = \frac{2.303 \times 0.0052}{60 \ (0.1047 \times 50,750)^2 \ (32.3 - 24.3)} = 8.94 \times 10^{-13} \text{ seconds or } 8.94 \ s$$

16.

$$D = \frac{7.86 \times 10^{-3} \ cm^2}{2350 \ \sec. \times 3.64} = 9.18 \times 10^{-7} \ cm^2/\sec.$$

17. To confine the parts of any tube that breaks.
18. A centrifuge tube designed to measure the fat in milk directly. The % fat is read on the stem.
19. Use the same density throughout, but increase the speed periodically.
20. The density is established prior to adding the sample.
21. Cesium chloride, Ficoll, and Percoll.
22. By adding color-coded calibration beads.

CHAPTER 40 - MASKING AND SEQUESTERING AGENTS
1. A masking agent forms a complex ion with a metal ion, and no rings are formed when the ligands are added. When ligands are added to a metal and rings are formed between the ligand and the metal, it is a sequestering agent.
2. By observing the color formed with uramildiacetic acid when washing glassware with hard water.
3. The molecule has six possible coordination positions.
4. It is water soluble when chelated with a metal.
5. Lead forms a more stable chelate and releases the calcium, which is then free to replace the lead in bones.
6. A muffle furnace.

7. They melt.
8. The high temperature can cause you to get burned before you can remove the items from the furnace.
9. To protect your eyes. UV radiation will cause your eyes to burn, and it is painful for some time.

CHAPTER 41 - SOLUBILITY
1. No. You will get most of it, but it will not be quantitative.
2. H_2SO_4; HCl; H_3PO_4 or cold $HClO_4$ < 50%.
3. The alkali metals, silver, and thallium (I).
4. Ions at the edges and corners contact more solvent, and more of these occur per unit area in small particles.
5.
$$\frac{1,000\ mL}{1.00\ L} \times \frac{1.42\ g}{1.00\ mL} \times \frac{72\ g}{100\ g} \times \frac{1\ mol}{63.0\ g} = 16.2\ M$$

6. Removing the excess is difficult.
7. Syrupy H_3PO_4, but it takes forever.
8. 130 °C and 50%.
9. A perchloric acid hood has a water wash down feature, because $HClO_4$ is not an oxidizing agent if it is dilute and cold.
10. Esters.
11. NH_4VO_3. This dissolves very slowly, often taking several hours.
12. Use a Bethge apparatus. Start the digestion and drain away the condensed water until the desired composition is present as determined by the temperature, then change the stopcock so the evaporated water is returned to the flask.
13. HNO : $HClO_4$ (3:1). The HNO_3 attacks the easily oxidizable compounds first, particularly those compounds that might form esters with the perchlorate ion.
14. Perfluorocarbons.
15. Diethylene glycol.
16. Polar solvents dissolve polar solutes, and nonpolar solvents dissolve nonpolar solutes.
17. The color of pyridinium-N-phenoxide changes in proportion to the polarity of the solvent it is in.
18. It is the energy necessary to break the solvent bonds and provide space for the solute molecules.
19. ΔH_{vap} = -2900 + 23.7 (273 + 285) + 0.020 $(273 + 285)^2$ = 16,552 cal/mol.
    V = 120.16 g/mol / 1.2606 g/mL = 95.32 mL/mol.
    $\delta = (16,552/95.32)^{1/2} = 13.2$.
20. $\delta = 1.2606\ [(16,552 - \{1.98 \times 558\})/120.12]^{1/2} = 14.3$
21. From Table 41-9, p. 481; E = 148 + 115 + 131.5 + 86 + 148 + 226 = 854.5. δ = 1.04 (854.5/90) = 9.8
22. Low molecular weight alcohols, aldehydes, cyclanones, ketones, and esters that may have more than nine carbon atoms.
23. Class A_2.
24. It is polystyrene and can be separated for recycling.
25. Cellulose acetate.
26. The crystalline phase is the hardest, so the bonds are broken more easily by heating the polymer.
27. Add LiCl or $ZnCl_2$ in methanol to separate the H-bonds.
28. The container is made out of polyethylene terephthalate and can be sorted for recycling.

CHAPTER 42 - GAS ANALYSIS BY PORTABLE ORSAT
1. The only measurements you need to obtain a quantitative result are volumes.
2. The gas sample is passed through a solution of an absorbent selective for one gas in a mixture, and the volume decrease is determined.
3. The gas is allowed to react with a solution selective for one of the components, and the excess solution is titrated.
4. Pyrogallol's capacity is low, and the amount used is recorded on the tag.
5. Acidic chromous chloride.
6. A Jones reductor uses amalgamated Zn to reduce ions.
7. Alkaline KI and sodium arsenite.
8. More than one answer is possible depending upon how you cancel the equations.

a. $H_2 = C - O_2$; $H_2 = 2/3(C - 2 O_2)$; $H_2 = 2 (O_2 - 2 CO_2)$; $CH_4 = 3/4(O_2 - 1/3 C)$; $CH_4 = CO_2$.
b. $CH_4 = 4/3 C - 5/3 CO_2$; $C_2H_6 = 2/3 (2 CO_2 - C)$.
c. $CO = CO_2$; $H_2 = C - O_2$.
d. $H_2 = -1/2 O_2 + 5/6 C - 1/3 CO_2$; $CO = 3/4 CO_2 - O_2 + 1/3 C$; $C_2H_6 = 1/2 O_2 - 7/12 C - 1/2 CO_2$.
e. $CH_4 = 7 O_2 - 11 CO_2 - C$; $C_2H_6 = 12 CO_2 - 6 O_2 - 2 C$; $C_3H_8 = 3 O2 - 4 CO_2 - C$.
f. $C_2H_4 = 5/4 CO_2 - C$; $C_2H_6 = C - 3/4 CO_2$.
g. $CO = O_2 + 4/3 CO_2 + 1/3 C$; $CH_4 = 5/3 C - 1/3 CO_2 - O2$; $C_2H_6 = O_2 - C$.

9. 64.7 mg.
10.

$$673 \, ppm = \frac{0.229 \, mg \, H_2 S}{mL} \times \frac{1 \, ppm}{1.23 \times 10^{-4} \, mg} \times mL; \quad mL = 0.36$$

11. One scale reads from the top down, and the other from the bottom up.
12. To prevent it from absorbing the acid gases.
13. It is placed alongside of the gas buret. This ensures that the pressure inside of the buret equals the pressure outside.
14. Fill the sampling bottle with water to force out the air and, then as it drains, it can draw in sample essentially free of air.
15. To provide a place for the gas above the liquid in the back portion to escape to without allowing air to enter.

SEE - I KNEW YOU COULD DO IT.
I HAD GREAT CONFIDENCE IN YOUR ABILITY TO GET THE JOB DONE.

Dialysis and electrodialysis, 412, 423, 687
  determining phosphatase in milk, 687
  electrodialysis, 424
    calculations, 426
    concentrating-diluting, 425
    electrodialysis reversal, 426
    electrolytic dialysis, 425
    ion substitution dialysis, 425
  principles, 423
  techniques, 423
    cell pairs and stacks, 428
    hollow-fibers, 424
    tubing and membranes, 423
Differential thermal analysis, 86, 87
Diffusion pumps, 78, 529
  mercury, 79
  oil, three-stage, 79
Disc electrophoresis, 298, 345, 646
  apparatus, 346
  identifying fish species, 646
  principles, 345
  techniques, 350
    marker dye, 350
    polymerization, 350
    rimming, 350
  theory, 345
Displacement chromatography, 148, 149, 155, 561
  acid washing, 151
  adsorbents, 155
  calculations, 156
  gradient elution, 160
  leading and trailing edges, 149
  mobile phases, 155
  principles, 155
  separating *cis-trans* azobenzene, 561
  silinizing, 151
Distillation, 4, 21
Distillation column, 36
  column packings, 36
  holdup, 36
  throughput, 36
Distilling Heads, 40
  reflux ratio, 40
  refluxing, 25, 40
  stopcock control type, 40
  swing funnel type, 40

Electrodeposition, 298, 299, 623
  commercial techniques, 311
  definitions, 310
  electrodes, 311
  principles, 299
    apparatus, 300
    complexation and chelation, 306
    constant current, 308
    constant voltage, 308
    early history, 300
    electrochemical cells, 301

electrolytic cells, 303
  Faraday's law, 299
  overpotentials, 304
    activation, 305
    back emf, 304
    concentration, 304
    hydrogen, 306
  reduction potentials table, 302
  safety precautions, 309
  separating Cu and Ni in a coin, 623
Electrophoresis, 298, 315, 628, 634
  apparatus, 327
    horizontal strip, 327, 634
      DNA fingerprinting, 634
      separating LDH isoenzymes, 628
    ridgepole, 328
    sandwich, 328
    submarine, 328
    vertical slab, 328
  continuous electrophoresis, 328
  contour clamped, 335
  crossed field pulsed, 333
  densitometers, 330
  preparative electrophoresis, 329
  power supplies, 330
  pulsed field electrophoresis, 330
    gradient, 331
  principles, 315
    electro-osmosis, 319
    heat production, 318
    smiling, 319
  techniques, 320
    autoradiography, 324
    blocking, 326
    blotting, 325
      electrical, 326
      Northern, 326
      Southern, 326
      Southwestern, 326
      vacuum, 327
      western, 326
    buffers, 322
    drying, 324
    film removal, 325
    fluorescence, 324
    sample addition, 322
    staining, 322
    supports, 320
      agar and agarose, 320
      cellulose acetate, 320
      paper strips, 320
      polyacrylamide gels, 321
      starch gels, 321
  theory, 315
  racheting, 334
  transverse alternating field, 332
Emulsion, 51
  techniques for breaking, 51 Extractive distillation, 4, 47

Extraction apparatus
  coil-planet centrifuge, 92
  Soxhlet extractor, 92
Extraction-general principles, 93
  adducts, 97
  batch extraction theory, 99
  concentrators, 102
    Kuderna-Danish, 102
    rotoevaporator, 102, 103
  distribution law, 94
  distribution ratio, 92, 94
    chemical bonds, 97
    electrostatic bonds, 97
    h-bonds, 97
    synergism, 98
  gibbs phase rule, 93
  multiple extraction theory, 99
  partition coefficient, 94
  p-values, 103
  percent equation, 98
Requirements, 96
  chelates, 96
  ion association, 96
  micelles, 96
  techniques, 100
    backwashing, 100
    choice of solvent, 100
    emulsions, 100
    masking agents, 100
    stripping, 100
    salting out agents, 100
Extractive distillation, 516
  cyclohexane from benzene, 509

Field flow fractionation, 298, 371, 662
  determining latex spheres, 662
  principles, 371
  techniques, 373
    electrical, 379
    flow, 378
    magnetic, 380
    sedimentation, 374
      calculations, 376
      HETP, 375
    steric, 380
    thermal, 373
  theory, 371
Filtration and sieving, 412, 431
  air, 431, 691
    barriers, 434
    calculations, 433
    filters, 432
      HEPA, 433
    pumps, 432
    separating pollen from air, 691
  liquids, 434
    filtering technique, 436
    funnels, 434
    gravity filtering, 434

INDEX

Abbe refractometer, 512
Affinity chromatography, 148, 567
  activating methods, 167
  inert matrix, 165
  principles, 165
  separating lectin from soybeans, 567
  spacer arms, 166
Antifoam, 36
Azeotrope, 43
  ways to break, 46
Azeotropic distillation, 4, 43, 509
  absolute ethanol from, 95% ethanol, 509
  azeotropes, 4
  binary azeotropic systems, 45
  boiling chip, 35
  maximum b.p., 44
  minimum b.p., 44
  temp.-comp. diagram, 45
  ternary azeotropic systems, 46

Bumping, 35
  boiling chip, 35

Capillary zone electrophoresis, 298, 359, 657
  aspartame breakdown products, 657
  current, 367
  detection, 368
    amperometry, 369
    conductivity, 369
    fluorescence, 368
    mass spectrometry, 369
    radioisotopes, 369
    uv, 368
  potential, 367
  principles, 359
    electro-osmosis, 359, 361
    electrophoresis, 360
  relative apparent mobilities, 367
  techniques, 365

    buffers, 367
    collecting fractions, 369
    columns, 365
    modifiers, 367
    neutral molecules, 366
    optical isomers, 366
    sample, 365
  temperature, 367
  theory, 362
Centrifugation, 448, 455, 696
  apparatus, 464
    adapters, 466
    rotors, 465
    safety shields, 465
    tubes, 466
  density gradient centrifugation, 467
  molecular weight determinations, 463
  nomograph, 459
  principles, 455
    diffusion coefficients, 462
    sedimentation coefficients, 460
  pycnometers, 461
  separating fat from milk, 696
  standards, 468
  theory, 458
  ultracentrifuge, 455
    optical systems, 456
      Philpot-Svensson, 457
      Rayleigh, 457
      Schlieren, 456
      spectrophotometric, 458
  viscometers, 461
Chromatography, general theory, 149
Clapyron equation, 28
Column packings, 36
  Berl saddles, 36
  glass beads, 36
  protruded metal clips, 36
  Raschig rings, 36
Columns
  Hemple, 37

  perforated Plate, 37
  Snyder, 37
  spinning band column, 37
  Vigreux, 37
Condensers, 38
  Allihn, 39
  coiled, 39
  connecting tubing, 39
  Friedrichs, 39
  Graham, 39
  inlet water, 38
  Liebig, 39
  West, 39
Continuous extractors, 92, 107, 548
  commercial extractors, 108, 109
  efficiencies, 107
  half extraction value, 108
  laboratory-size extractors, 109
    solvent heavier than water, 109, 540
      caffeine from cola drinks, 540
    solvent lighter than water, 110, 544
      alkaloids from Ipecac, 544
  Soxhlet extractor, 111, 113, 548
    oil from nutmeats, 548
      Phillips beaker, 549
      Thimples, 112
Countercurrent extraction, 92, 550
  flavors from oregano, 550

Dalton's law, 23, 27, 49
  partial pressure, 23
Density gradients, 448, 449, 694
  comparing soils at a hit and run, 694
  principles, 449
    specific gravity, 450
  techniques, 451
    density calibration, 452
    density gradients, 451
      forming apparatus, 453
      mixtures, 453
        calculations, 453

papers, 435
pressure filtering, 439
vessel, 441
vacuum filtering, 438
dental dam, 438
principles, 431
solids, 441
sieve standards, 442, 443
sonic sieving, 442
wet sieving, 443
flash chromatography, 148, 575
collection and detection, 182
column packing, 179
eluting solvents, 181
flow control, 181
pressurizing gases, 181
principles, 179
pumps, 180
sample addition, 181
separating vanilla from extracts, 575
small-scale systems, 182
foam fractionation, 384, 395, 667
separating detergents from water, 667
principles, 395
foams, 396
techniques, 399
defoamers, 400
foamers, 399
theory, 396
Fractional distillation, 22
apparatus, 25
condenser, 22
distilling head, 25
fractionating column, 23
reflux line, 25
Freeze drying, 4, 85, 538
connecting valves, 87
freezing rate effects, 86
ice crystal formation, 87
multiport condenser, 88
process, 85
sample holders, 87
water from beverages, 538

Gas analysis, 448, 489, 707
absorbents, 490
apparatus, 494
absorption bulbs, 496
gas buret, 495
gas sampling bottle, 496
manifold, 496
portable Orsat, 494
determination of car exhaust fumes, 707
methods, 489
fractional combustion, 492
mixtures, 491
slow combustion, 492
principles, 489
Tutweiler apparatus, 493

Gas–liquid chromatography, 148, 211, 586, 591
antioxidants in cereals, 591
arson investigation, 586
calculations, 239
correction for sugar, 240
fatty samples, 241
low-fat samples, 240
peak areas, 242
symmetrical peaks, 243
unsymmetrical peaks, 244
ppm residue, 241
recoveries, 242
strength injected, 239
detectors, 230
argon ionization, 234
Beilstein, 235
electron capture, 234
Ni-63, 234
flame photometric, 235
gas density balance, 232
Hall electrolytic, 236
hydrogen flame ionization, 232
mass spectrometer, 238
photoionization, 235
terms, 230
thermal conductivity, 230
thermionic specific, 233
principles, 211
theoretical plates, 213
techniques, 218
backflushing, 229
carrier gases, 218
columns, 221
capillary, 227
coating technique, 226
conditioning, 227
marking, 226
packing, 226
flow meter, 229
injection ports, 219
pressure regulators, 218
solid supports, 224
stationary phases, 225
syringes, 219
filling technique, 220
temperature programming, 229
tubing benders, 223
tubing connectors, 223
tubing cutters, 222
tubing liners, 224
theory, 214
retention volume, 215
Kovats retention index, 216
Van Deemter equation, 217
Gas-solid flotation, 384, 400, 671
collectors, 406
anionic, 406
cationic, 406
frothers, 407
principles, 400
separating solids from water, 671
theory, 402
contact angle measurement, 405
Gel filtration chromatography, 148, 171, 572
columns, 173
lipophilic Sephadex, 173
other products, 174
reference compounds, 174
separating B_{12}, from dextrans, 572
Sephacryl, 173
Sephadex, 171
Gel permeation chromatography, 148, 171, 175, 572
Bio-gel-P, 175
Enxacryl, 175
activating and binding, 175
Gibbs phase rule, 23

Heating devices, 34
steam bath, 34, 35
heating mantles, 34
heating tape, 38
hot plates, 34
insulated nichrome wire, 38
Height Equivalent to a Theoretical Plate, 26
HETP, 26
High performance liquid chromatography, 148, 183, 579
basic components, 192
column packings, 200
column packing recommendations, 188
gradients, 197
packing columns, 201
Pickering post column, 203, 204
precolumns, 198
pumps, 194
sample injection, 198
solvents, 192
detectors, 202
charged coupled devices, 205
electrochemical, 206
evaporative light scattering, 207
fluorescence, 203
derivatives, 203
mass spectrometer, 207, 209
photodiode arrays, 204
refractive index, 206
uv, 202
derivatives, 203
explosives separation, 579
Excedrin examination, 584
principles, 183
ion-exchange, 186
groups, 186

High performance liquid chromatography
(continued)
   liquid-solid, 183
   paired-ion, 186
   reversed phase, 184
     surfaces, 185
   theory, 187
     flow irregularities, 189
     mass transfer, 189, 190
     molecular diffusion, 189
     resolution, 190
       capacity factor, 190
       selectivity, 191
       theoretical plates, 191
Hydride generation
   As and Se in foods, 9

Immiscible solvents distillations, 4, 52, 521
   apparatus, 53
   separate water from fruits, 521
   traps, 52
     Barrett trap, 52, 53
     Bidwell-Stirling, 52
     Dean-Stark, 52
     solvent heavier than water, 52
   Water content found in several foods, 53
Immunoelectrophoresis, 298, 339, 642
   human proteins in urine, 642
   principles, 339
     antibodies, 339
     characteristics, 341
     immunoglobulins, 339
   techniques, 340
     crossed, 342
       spotting template, 343
     fused rocket, 342
       spotting template, 342
     rocket, 340
       spotting template, 342
     tandem-crossed, 343
       spotting template, 344
Ion chromatography, 268, 614
   corrosion inhibitors in coolants, 614
   detectors, 282
     conductivity, 282
     d.c. amperometry, 283
     pulsed amperometry, 285
     UV/vis, 285
   principles, 277
     single column, 279
     suppressed systems, 277
   techniques, 279
     columns, 279
       resin-based particles, 281
       silica-based particles, 279
     electrochemical regeneration, 281
     gradient elution, 287
     sample concentrator, 287
Ion Exchange, 268, 269, 608
   principles, 269
     cations in water, 608
     chelate resins, 274
     complex resins, 274
     donnen membrane theory, 273
     functional groups, 272
     general rules, 2272
     kinetics, 273
     structures, 271
   techniques, 274
     apparatus, 275
     membranes, 275
     porous particles, 275
Ion exclusion, 268, 291, 618
   applications, 292
   principles, 291
   techniques, 293
     column packing, 293
   theory, 293
   vitamin C in applesauce, 618
Ion focusing electrophoresis, 298, 351, 652
   leghemoglobins in soybeans, 652
   principles, 351
   techniques, 352
     ampholytes, 353
       immobilized, 353
       natural, 353
     determining pH, 354
     gradient development, 352
     heat production, 354
Ion retardation, 268, 289, 617
   principles, 289
   separating salt from urea, 617
   techniques, 290
     resin regeneration, 290
Ito coil planet centrifuge, 117, 550
   factors effecting separation, 124
   hydrostatic system, 118, 119
   hydrodynamic system, 118, 120
   oils from oregano leaves, 550
   seal free systems, 124, 125
   theory, 121

Ligand exchange, 268, 293, 621
   liquid rocket fuels, 621
   principles, 293
Liquid–liquid extraction, 92
Liquid–solid flotation, 384, 408, 673
   principles, 408
   separating aphids from spinach, 681
   separating fly eggs from tomatoes, 673
   separating insect fragments from flour, 676
   techniques, 408
     Kilborn flask, 408
     Wildman trap flask, 408
Lyophilization, 85 (see freeze drying)

Magnetic stirrer, 36
   air- or water-driven, 36
McCabe-Thiele Equation, 33
   operating line equation, 30
Manostats, 64
   Cartesian diver, 64
   Lewis-Nester, 64, 65
Masking and sequestering agents, 448, 469, 700
   complexometric titrations, 469
     EDTA, 469
     formation constants, 470
   determining Zn in rubber, 700
   muffle furnaces, 471
     glasses, 471
     gloves, 471
     tongs, 471
   principles, 469
Mean free path, 74
Mercury, 65
   cleaning metallic Hg, 66
   cleaning up Hg spills, 66
   diffusion pumps, 79
   mercury poisoning, 65
   pinholing, 66
Microwave assisted extraction, 112
   molecular polarizability, 113
   instrument, 114, 115
Molecular distillation, 4, 71, 527
   Langmuir equation, 71
   Luchak-Langstroth equation, 74
   nomograph, 73
   number of plates required
     graphical method, 27
     partial reflux, 27
     total retlux, 27
   partial reflux, 30
   pressure gauges, 79
     McLeod gauge, 79
     Pirani, 79, 80
     thermocouple, 79, 80
   Raoult's Law, 4, 23, 25, 27
     Mole fraction, 23
   receivers, 40
   Rose's equation, 28
   separation vitamin E from margarine, 527
Multiple column partition chromatography, 161, 564
   principles, 161
   separating drugs in cough syrup, 564
   techniques, 162

Osmosis and reverse osmosis, 412, 413, 685
   osmotic pressure of a potato, 685

principles, 413
reverse osmosis, 415
techniques, 418
membrane parameters, 420
cell, 419
membranes, 418
organic rejection, 419
salt rejection, 419
theory, 413

Paper chromatography, 148, 249, 595
components, 251
Chambers, 252
mobile phases, 253
paper, 251
R_f, 253
sample detection, 253
sample preparation, 252
spotting, 252
determining sulfa drugs, 595
principles, 249
techniques, 250
ascending, 250
descending, 250
horizontal, 250
reversed phase, 251
two-dimensional, 251
Weisz ring oven, 250
Partition chromatography, 148, 152
partition coefficient, 153
Purge and trap, 384, 665
determining aroma changes in aged beer, 665
principles, 385
purge efficiency, 386
priority pollutants, 390
techniques, 387
apparatus, 388
spargers, 387
split interface, 389
splitless interface, 390
traps, 387

Simple distillation, 22
Size exclusion chromatography, 148, 171, 172, 572, see *Gel permeation chromatography*
Sodium dodecylsulfonate electrophoresis, 354
principles, 354
techniques, 356
Solid phase extraction, 92, 129, 554
coatings, 129
effect of cartridge size, 133
effect of flow rate, 132
effect of k, 131
process, 130
separating herbicides from water, 554
Solid phase microextraction, 134

Solubility, 448, 473, 703
inorganic compounds, 473
HCl, 474
$HClO_4$, 475
Bethge apparatus, 476
liquid-fire reaction, 477
organic compounds, 475
HF, 474
HI, 475
HNO_3, 474
H_3PO_4, 474
H_2SO_4, 474
water, 473
organic compounds, 477
Hildebrand solubility parameter, 479
molar attraction constants, 481
solubility classes, 484
solubility values, nonpolar, 482
polymers, 485
code numbers, 486
separation of fibers, 704
thermoplastic polymers, 703
principles, 473
safety, 473
solubility values, moderately polar, 483
solubility values, H-bonds, 484
solvents, 478
solvent predictors, 479
Stills, 75
apparatus, 24
centrifugal still, 77, 78
falling film still, 75
flowing film still, 74
wiped-film still, 76, 77
Steam distillation, 4, 49, 50, 519
apparatus, 50
oil of clove separation, 519
standard boiling point, 49
uncorrected boiling point, 49
vapor pressure-temperature diagram, 50
Sublimation, 4, 81, 532
benzoic acid from saccharin, 532
entrainer sublimation, 4, 81
caffeine from coffee, 535
entrainer sublimators, 82
Supercritical fluid extraction, 92, 137, 556
cartridges, holders, 143
diffusivity, 138
drying agents, 143
fluid flow diagram, 142
modifiers, 142
pesticide recoveries, 140
phase diagram, 138
phase diagram for CO2
pumps, 144
recent applications, 141
restrictors, 144
separating oil from pecans, 556

solubility behavior, 139
viscosity behavior, 138

Tesla coil, 67
Thermometer placement, 38
Thin layer chromatography, 148, 255, 599
principles, 255
separating vasodilator drugs, 599
separating antihistamines, 601
techniques, 255
chambers, 256, 257, 260
coatings, 257, 258
detection, 262
charring, 263
spray reagents, 262
fluorescence, 263
multiple development, 263
overloading, 260
plates, 256
programmed multiple development, 264
quantitation, 263
solvents, 261
spotting, 259
two-dimensional, 264, 606
Medicated feeds, 606
Total reflux
quantitation, 263
refluxing, 28
Trouton's rule, 28
Two-dimensional electrophoresis, 356

Vacuum distillation, 57, 523
air bubblers, 67
ballast or surge tank, 62
screen wire cage, 62
taped bulb, 62
bleeding valves, 63
Dewar flasks, 61
filling technique, 62
handling dry ice, 62
polystyrene containers, 62
drying towers, 62
fraction collectors, 66
cow, 66
udder type, 66
general rules, 59
pressure measuring devices, 63
digital gauges, 63
McLeod gauges, 63
nomograph, 58
U-tube manometer, 63
Zimmerli, 63
pressure nomenclature, 57
Pascal, Pa, 57
purify DMF, 523
traps, 61
vacuum line joints, 66

Vacuum distillation *(continued)*
   ball and socket, 66
  vacuum pumps, 60
    diffusion pumps, 60
    mechanical pumps, 60
     duoseal, 60
     hyvac, 60
Vapor pressure, 21
  atmosphere, 21, 22
  pascal, 21
  pressure conversion factors, 21

Variable transformer, 35
  Powerstat, 35
  Variac, 35
Volatility, 27
Volatilization, 4, 5, 500, 504
  CO_2 in limestone, 500
  determination of mercury, 9
  inorganic compounds, 5
  Knorr alkalimeter, 8, 502
  organic compounds, 6
  mercury in hair by flameless AA, 504

  moisture determination, 6
   "as received" calculations, 7
   "oven-dried" calculations, 7

Zone leveling, 13, 16
Zone melting, 4, 13
Zone refining, 13, 14, 507
  flat tray zone refining, 19
   use of a solvent, 19
  methyl red from naphthalene, 507